HIGHWAY ENGINEERING

HIGHWAY ENGINEERING

Clarkson H. Oglesby

Silas Palmor Professor of Civil Engineering Emeritus, Stanford University

R. Gary Hicks

Professor of Civil Engineering, Oregon State University

FOURTH EDITION

JOHN WILEY & SONS
New York
Chichester
Brisbane
Toronto
Singapore

Library of Congress Cataloging in Publication Data:

Oglesby, Clarkson Hill, 1908–
Highway engineering.

Includes bibliographical references.
1. Highway engineering. I. Hicks, Russell G.
II. Title.
TE145.0675 1981 625.7 81-12949
ISBN 0-471-02936-X AACR2

Printed in the United States of America

10 9 8 7 6 5 4 3 2 1

PREFACE

This book is designed primarily as a text for junior, senior, and fifth-year college courses in highway and transportation engineering. It is also a starting point for advanced courses and individual study in these areas. Furthermore, it summarizes the state of the art and important new developments in transportation planning and highway technology to assist practicing engineers, administrators, and other decision makers.

The major focus of this work is on transporting people and goods by highway. The planning, financing, legal requirements, design, construction, operation, and maintenance of facilities and their economic, social, and environmental consequences are thoroughly discussed. Since, in the United States, urban public transportation in all but the largest cities is predominately by motor vehicle, many public transportation problems are also considered. Nonhighway urban public transportation also is addressed briefly.

Because highway and public transportation problems are many faceted and involve disciplines other than engineering, there are seldom single "right" answers. This makes the subject particularly troublesome to undergraduate engineering students. We have made every effort here to minimize this confusion. Wherever possible, the governing principles underlying a topic are developed first and then current practice is explained in terms of those principles. Finally, research findings that point up the shortcomings of present-day knowledge and practices are outlined to indicate probable changes to come.

This book can, in the main, be used by those who have very little background in technical engineering. However, some knowledge of route surveying, dynamics, and fluid mechanics is necessary for a complete understanding of certain topics. These topics are included because they are important to more advanced students and to practicing engineers.

For those who desire more detailed knowledge on specific topics, a substantial number of pertinent references have been cited. They include classics from the past and others giving recent findings. With a few exceptions, these references are from publications usually available in engineering libraries at colleges and universities in the United States and Canada. Sources include primarily the publications of the Transportation Research Board, the American Association of State Highway and Transportation Officials, the American Society of Civil Engineers, and the Institute of Transportation Engineers. We deeply regret that space limitations often made it necessary to omit the names of some or all the authors of individual articles.

In preparing this edition we have taken into account the changes to be expected in highway and public transportation in the years ahead while retaining and updating the form and substance of the previous edition. This has meant added attention to the following:

- Recognizing the shift from building entirely new highways to rehabilitating existing facilities. This calls for attention to maintenance management and advance planning as well as to new procedures, design methods, and techniques in pavement strengthening and reconstruction, and possibly to reclaiming scarce materials.
- Making better use of existing highway facilities through more efficient management and by employing more advanced means for traffic control.
- Devoting increased attention to the various forms of public transportation.
- Exploring the problems that energy shortages and cost increases have brought or will bring in highway financing, travel, vehicle design, and construction and maintenance practices.
- Expanding the discussion of highway safety.
- Including information on new materials such as fabrics and on expected or desirable changes in existing ones like aggregates, bitumen, cement, and concrete.
- Devoting more attention to the highway problems of developing nations.

We have compromised on the matter of English Engineering or U.S. Customary (FSS) versus S. I. Units. On graphs and tables, wherever possible, both have been given; at the least, multipliers to convert to S. I. units have been provided. This uneasy approach reflects current practice in the United States and still recognizes the needs of users in countries employing the S. I. system and of instructors who favor it.

The problems for this edition are placed at the ends of the individual chapters. To the greatest extent possible, they are based on real situations and actual data. We do not anticipate that instructors will assign all or even a large fraction of the total. However, with this number to choose from, examples should be available to illustrate most of the topics selected for special emphasis. Most of the problems apply to subjects covered in Chapters 4, 8 to 12, 14, 19, and 20, since the basic principles and design techniques discussed there may be illustrated to good advantage by carefully selected problem assignments. Problems for the remaining chapters fall largely into two classes: (1) those that require investigation and reporting of local situations, in contrast to the national viewpoint presented in the book, and (2) those that force the student to apply to a particular situation the principles or ideas developed in the text, by class discussion, or by supplementary reading. Where appropriate, parallel problems calling for full solutions or answers in S. I. units have been introduced. These are numbered with a "B" designation: for example, Problem 4-1B is to be solved in S. I. and local currency units.

We express our special thanks to Dr. L. I. Hewes, coauthor for the first two editions of this book, and to E. L. Grant, Professor of Economics of Engineering, Emeritus, Stanford University, for his continual encouragement during the years when all four editions were in preparation. Our wholehearted appreciation is also due to the host of engineering teachers, researchers, and practicing engineers who through their writings or personal contacts have contributed greatly. We are particularly indebted to Professor R. D. Layton of Oregon State University who reviewed the entire manuscript and proposed new problems, and to F. N. Finn, consultant, J. F. Shook of the Asphalt Institute, Grant J. Allen of the Arizona Department of Transportation, and G. W. Beecroft of the Oregon Department of Transportation for reviewing the sections on pavement design and rehabilitation. Thanks also are extended to Christine Fagin, Christine Krygier, and Elizabeth Clausen for their assistance in typing parts of the manuscript and reviewing galleys and page proofs. Finally, our wives Ardis and Joan helped immeasurably not only by participating actively in preparation of the manuscript and by reviewing galley and page proofs, but by their patience and forbearance during the long gestation and production periods that accompanies writing a book.

Stanford, California
Corvallis, Oregon

Clarkson H. Oglesby
R. Gary Hicks

CONTENTS

1 INTRODUCTION

THE IMPORTANCE OF MOTOR VEHICLE TRANSPORTATION[1]

Today in the United States, Canada, Europe, and other developed areas of the world, vehicles rolling over highways and streets are the principal means for transporting persons and goods. In the United States some 154 million motor vehicles,[2] including 33 million trucks and buses, travel 1.5 trillion miles annually on some 3.9 million miles of roads and streets. There is a motor vehicle for each 1.5 persons, and a passenger car for every two: enough to transport the entire population at once. Eighty-one percent of U.S. households own at least one automobile. Licensed drivers total 143 million, or 64% of the population. Some 97% of local trips (less than 30 miles) and 86% of longer trips are made by automobile.

Private motor transportation consumes 14% of the expenditures of individuals. To keep highway transport moving takes 24% of the nation's steel, 65% of its rubber, 55% of its petroleum, and vast amounts of many other products. In 1979, taxed gasoline consumption of about 108 billion gal, or 95% of all uses, was 500 gal for each man, woman, and child. Trucks and buses took 14 billion gal of diesel fuel, or 22% of total consumption. Assuring a continuing supply of fuel is a critical issue for the years ahead.

Table 1–1 dramatically illustrates in more detail the overall importance of motor vehicles to intercity passenger and freight movement in the United States. And the comparisons of overall ton miles for freight movements do not tell the whole story, since truck transportation predominates where goods have high value, where quick delivery is important, or where haul distances are short. For example, of the total tonnage of manufactured goods transported less than 50 mi, about 69% goes by highway, 17% by rail, 13% by water, and 1% by postal or parcel delivery services.

Urban areas likewise rely heavily on motor vehicle transportation. Countrywide, 95% of all urban passenger miles are by user-operated automobile, 3% by motor bus, and 2% on subways or elevated or surface railways. Not only do these figures indicate the predominant role of the private automobile in urban movements, but they also show that rubber-tire vehicles traveling on pavement

[1]Unless otherwise noted, the statistics cited here are for 1979. They come from a variety of sources.
[2]This total does not include 5 million registered and possibly an equal number of unregistered motorcycles and mopeds.

TABLE 1–1. Intercity Passenger and Freight Movements in the United States, 1977

Carrier	Passengers		Freight	
	Billions of Passenger Miles	*Percent of Total*	*Billions of Ton-Miles*	*Percent of Total*
Railroad	11	0.8	799	36.5
Motor vehicles				
Automobiles	1236	85.8	—	—
Buses*	25	1.7	—	—
Trucks	—	—	510	23.3
Pipelines	—	—	523	23.9
Inland waterways	4	0.3	352	16.1
Airways	165	11.4	4	0.2
Totals	1441	100.0	2188	100.0

*Excludes school buses.

carry a major share of public transit. Even so, such averages can be deceptive. On the one hand, some smaller urban areas are without effective public transportation even by bus, and rely almost entirely on the private automobile. At the other extreme, New York City, which generates about four-fifths of all rail and one-fifth of all bus traffic for the nation, would be crippled if the 62% of work trips made by transit were thrown onto the private automobile. Other very large cities with high population densities likewise depend heavily on public transportation, particularly for trips to and from work. However, in all but the largest cities, the private automobile is dominant; in the central areas of cities with populations over 100,000, 72% of the work trips are by private automobile. How much and how rapidly transit use will increase because of gasoline shortages and price increases is still unknown.

Motor vehicles also dominate goods movements in urban areas. First of all, they take the urban-bound intercity motor freight directly to its destination or to terminals for distribution by smaller trucks. Then they distribute all sorts of products within individual urban areas and consolidate loads for intercity transport.[3] In addition, the whole gamut of public services ranging from garbage and trash collection to maintenance of streets and utilities is based on truck transport.

Motor transportation has also changed the faces of our cities. Until some 50 yr ago, urban populations were concentrated in limited, tightly knit areas, largely because of the restrictions in movement imposed by rail-mounted or horse-drawn vehicles. The combined effects of the freedom of movement offered by motor transportation and the population shift from rural to urban areas have caused urban areas veritably to "explode." This trend began after World

[3]In the New York City urban complex, 73% of the tonnage is moved by truck, 25% by water, and 1% by rail, with water transport made up almost entirely of heavy or bulk products such as sand, gravel, or fuel. Few other areas have the water alternative available, so, for them, the truck percentage would be far higher.

War I and has rapidly accelerated since World War II. Today our urban areas have, for better or worse, assumed a new, dispersed form, geared to motor vehicle transportation. As discussed in Chapter 3, efforts are under way to slow or reverse this trend but the results will be seen slowly, if at all.

Highway transportation has brought great changes to rural areas. Practically all farm products are moved initially by motor vehicle. Many, such as milk, perishable food, and livestock, for which quick delivery is important, travel all the way to market in that manner. With the school bus, the consolidated school has replaced the one-room schoolhouse. Medical attention and similar services are almost as close at hand in the country as in town. There are increased opportunities for recreation, social contacts, and education. In fact, the rural mode of living has become much like that of town and city.

A massive industrial complex has grown to serve motor transportation. In each recent year, 9 to 12 (or more) million new vehicles, of which one-fourth are trucks and buses, have rolled off the assembly lines. To produce and sell these vehicles and to maintain, service, and supply them and the remaining fleet occupies some 4 million people. Added to these are the 9 million involved in trucking, the 300,000 in passenger transport, and the 740,000 in highway construction and maintenance. In all, highway transport uses 14 million people or 22.5% of the total U.S. work force.

Supplying highway transportation in the United States has its price. As indicated above, it consumes many economic resources. In addition, each year it takes more than 50,000 lives, brings injuries to almost 2 million people, and damages 30 million vehicles. It plays a major role in air and noise pollution. Some attribute much of our urban blight, crime, and other ills to it.

Highways, and transportation operating on them, also play an important role in the other "developed" countries of the world. The Canadian situation closely parallels that in the United States.[4] The countries of Western Europe and Japan generally have much greater population densities and more highly developed rail and bus systems for the transportation of people. For example, in London and Paris, transit rides per unit of population are over 250 per year; in the United States, with the exception of New York, the range is 50 to 80.

Automobile ownership is lower in other countries than the one car for every two persons in the United States. Ratios in other developed countries include Sweden, 1 to 2.8; West Germany, 1 to 3.0; France, 1 to 3.1; and the United Kingdom, 1 to 3.9. For comparison, the ratios for Japan are 1 to 5.7; Spain, 1 to 6.1; Greece, 1 to 15; Russia, 1 to 46; and China, 1 to 24,000.

Many of the developed countries have policies that discourage motor vehicle use, such as high taxes on them and on fuel. Furthermore, national policies may favor rail facilities over buses and trucks. Even so, motor vehicle manufacture is an important industry in the developed nations of the world. In 1978, production in millions of vehicles was Canada 1.8, Europe 16.2, and Asia 9.3.

In the underdeveloped nations of the world, improving transportation, primarily through providing motor vehicles and highways for them, is a major

[4]The United States and Canada, with 5% of the world's population, have 43% of the motor vehicles.

goal. Efficient movement of agricultural products, access to medical attention, and the ability to transport raw materials and finished products are all essential if they are to raise living standards above the subsistence level. None of these can be accomplished when transportation relies on what people or animals can carry on their backs or pull in carts or wagons. In a day, they can transport something like 60 or 300 lb, respectively, for 15 mi. Possibly a horse, mule, or elephant can draw half a ton an equal distance. But on all-weather roads, one person driving a diesel truck can move about 16 tons 200 mi daily.

Some developing nations have made substantial beginnings in developing roads; others have plans under way, often with support from the World Bank, the Agency for International Development, the United Nations, and other agencies. However, they lack motor vehicles. An extreme case is China where, with 40,000 automobiles and 700,000 trucks and buses, the ratios of automobiles and all vehicles to population are 1 to 24,000 and 1 to 1300, respectively. For Africa, excluding the Republic of South Africa, which has 44% of the continent's total vehicles, these ratios are 1 per 100 and 1 per 150, respectively. A major effort will be required to provide motor transport and the roads to accommodate it in many developing countries in the years ahead.

HOW MOTOR VEHICLE TRANSPORTATION IS PROVIDED

Motor vehicle transportation differs from rail, air, pipeline, and some other forms of transportation because it is not under unified control. Rather, in most of the world, almost all motor vehicles are privately owned and operated. With the exception of certain licensed passenger or freight carriers, there is a free selection among vehicles and the timing, route, and speed of travel, subject only to the restrictions brought by congestion and regulations imposed for the safety and welfare of others. Governments at various levels, as one of their primary functions, have provided and operated the roads over which these motor vehicles travel and have established agencies to finance, plan, construct, operate, and maintain them, to license motor vehicles and drivers, and to police their operation.

In the United States, roughly $0.08 of the motor-transport dollar is spent for the roads and streets on which vehicles move. This amounted to $36 billion in 1979. During the years 1921 through 1979, over $330 billion was invested. A goodly portion of this has come from taxes on motor vehicles and the fuels they consume.

HISTORY OF HIGHWAYS

Early Roads

Traces of early roads have been found which antedate recorded history. The first hard surfaces appeared in Mesopotamia soon after discovery of the wheel about 3500 B.C. On the island of Crete in the Mediterranean Sea a stone-surfaced road constructed before 1500 B.C. was found. The direction in the Bible (Isaiah 40:3–

5) "make straight in the desert a high road" refers to a road constructed soon after 539 B.C. between Babylon and Egypt. In the Western Hemisphere evidence exists of extensive road systems constructed by the Mayan, Aztec, and Incan people of Central and South America. Away from seas or rivers, facilities such as these were essential, not only to move armies for conquest or to defend against it, but also to move food and trade goods between and into cities.

The Romans bound their empire together with an extensive system of roads radiating in many directions from Rome. Some of these early roads were of elaborate construction. For example, the Appian Way, built southward about 312 B.C., illustrates one of the procedures used by the Romans. First a trench was excavated to such a depth that the finished surface would be at ground level. The pavement was placed in three courses: a layer of small broken stones, a layer of small stones mixed with mortar and firmly tamped into place, and a wearing course of massive stone blocks, set and bedded in mortar. Many of these roads are still in existence after 2000 years.

With the fall of the Roman Empire, road building became a lost art. It was not until the eighteenth century that Tresaguet (1716–1796) in France developed improved construction methods that at a later time, under Napoleon, made possible a great system of French roads. Highway development in England followed soon after. MacAdam (1756–1836) in particular was outstanding. A road surface that bears his name is still used.

Although little significant road building, as such, was done in England before the eighteenth century, the foundations of English and thus American highway law were being laid. Early Saxon laws imposed an obligation on all lands to perform three necessary duties: repair roads and bridges; maintain castles and garrisons; and aid in repelling invasion. Soon after the Norman conquest it was written that the king's highway was "a sacred thing, and he who has occupied any part thereof by exceeding the boundaries and limits of his land is said to have made encroachment on the King himself." Very early, applications of this law made clear that ownership of the roads actually was vested in all persons who wished to use them. Other statutes, dating as far back as the thirteenth century, required abutting property owners to drain the road and clip any bordering hedges, and to refrain from fencing, plowing, or from planting trees, bushes, or shrubs closer than specified distances from the center of carriageways. In these and other early statutes can be seen the rudiments of such present-day concepts as the government's responsibility for highways, the rights of the public to use them without interference, and the obligations of and restrictions on the owners of abutting property.

Early American Roads[5]

Few roads were built during the early history of the United States since most of the early settlements were located along bays or rivers and transportation was largely by water. Inland settlements were connected with the nearest wharf, but

[5]For details on these and later American roads see *American Highways—1776–1976*, published by FHWA.

the connecting road usually was just a clearing through the forest. Before the Revolutionary War, travel was mainly on foot or horseback, and roads were merely trails cleared to greater width. Development was extremely slow for a time after the war's end in 1783. For example, poor roads were the real cause of the Whisky Rebellion in Pennsylvania in 1794. The farmers objected to a tax on the whisky that they were making from grain. One historian has recorded that "a pack horse could carry only four bushels of grain over the mountains but in the form of whisky he could carry the product of twenty-four bushels." Construction of the Philadelphia-Lancaster Turnpike resulted from this incident. It was a toll road 62 mi long, 50 ft between fences, and surfaced to a width of 21 ft with hand-broken stone and gravel.

Between 1795 and 1830 numerous other turnpikes, particularly in the northeastern states, were built by companies organized to gain profits through toll collections. Few of them were financially successful. During this period many stagecoach lines and freight-hauling companies were organized.

The "Old National Pike" or "Cumberland Road" from Cumberland, Md., to Wheeling, W. Va., on the Ohio River was one of the few roads financed by the federal government. It was originally toll-free. The Cumberland-Wheeling section was authorized by Congress in 1806 and was completed 10 yr later. It was 20 ft in width, and consisted of a 12-in. bottom and a 6-in. top course of hand-broken stone. Some 20 more years elapsed before the road was completed to St. Louis. During this same period numerous canals were constructed, particularly along the Atlantic Seaboard; but they offered little competition to turnpike development since the terrain of most of the country was unsuited to canal construction.

The Railroad Era

The extension of turnpikes in the United States was abruptly halted by the development of the railroads. In 1830 Peter Cooper constructed America's first steam locomotive, the *Tom Thumb*, which at once demonstrated its superiority over horse-drawn vehicles. Rapid growth of the railroad for transportation over long distance followed. Cross-country turnpike construction practically ceased, and many already completed fell into disuse. Rural roads served mainly as feeders for the railroads; improvements primarily led to the nearest railroad station and were made largely by local authorities and were to low standards. However, the improvement of city streets progressed at a somewhat faster pace. Also, the development of the electric trolley in 1885 launched the trend toward public transportation.

Regarding highway development before 1900, federal road officials stated that[6]

> At the end of the century, approximately 300 years after first settlement, the United States could claim little distinction because of the character of its roads. As in most

[6]*Highway Practice in the United States*, Public Roads Administration (1949).

parts of the world, the roads were largely plain earth surfaces that were almost impassable in wet weather. Neither the Federal nor state governments had undertaken to provide funds on a scale that would permit general road improvement. Those seeking knowledge on road-building methods and administration turned to the countries of Europe for information.

The first two decades of the twentieth century saw the improvement of the motor vehicle from a "rich man's toy" to a fairly dependable method for transporting persons and goods. There were strong demands not only from farmers but from bicyclists through the League of American Wheelmen for rural road improvement, largely for roads a few miles in length connecting outlying farms with towns and railroad stations. This development has been aptly described as "getting the farmer out of the mud." Great improvements also were made on city streets.

In this period it was recognized that road improvement was a matter of federal and state concern rather than of purely local interest to be dealt with by county and city governing bodies. Federal and state highway organizations were established and small amounts of money appropriated by Congress and the state governments to deal with road problems.

Modern Highway Development

The period from 1920 at least into the late 1970s might well be called the "automobile age," for during this period highway transportation assumed a dominant role in America and the rest of the developed world. These countries can well be described as "nations on wheels." Figure 1–1 presents this development for the United States in graphical form. It shows a 15–fold increase in several measures of highway activity from 1920 to 1979 and a tripling between 1950 and 1979.

Because of inflation, the plot of highway expenditures on Fig. 1–1 is not realistic. In the years from 1950 to 1979, the dollar cost of appropriate units of highway construction and maintenance quadrupled. This is a geometric increase of about 5% per year. Unfortunately, inflation continues and at a considerably higher rate.

To give added perspective on the effects of inflation, expenditures converted to 1977 dollars have been plotted on Fig. 1–1. This illustrates among other things that real expenditures for highways were substantially lower in the late 1970s than in the 1960s, in spite of substantial increases in highway use. Values taken from this cost curve and that for motor vehicle registration show that the real annual expenditure for highways per motor vehicle, stated in 1977 dollars, had fallen from about $270 to $160 between 1950 and 1979.

Figure 1–1 also shows that during the 1920 to 1979 period, road and street mileage increased relatively little, possibly 20%. This growth resulted mainly because new roads and streets were built to serve areas where land use became more intensive, plus the addition of a relatively small mileage of major arteries, including freeways, on new alignments.

From 1920 to 1935, highway development was focused primarily on the completion of a network of all-weather rural roads comparable to the street systems undertaken by local governments. By 1935 cross-country travel by automobile in almost any direction was possible. Since 1935 highway activities in rural areas have been devoted mainly to an attempt to provide facilities of higher standards and with greater capacity and load-carrying ability. During the same period, increasing attention has been focused on urban areas, which have been struck simultaneously by rapidly increasing population, lower population densities resulting from a "flight to the suburbs," and a shift from mass transportation to the private automobile. Indications are that only minor additions to road mileage will be made in the future.

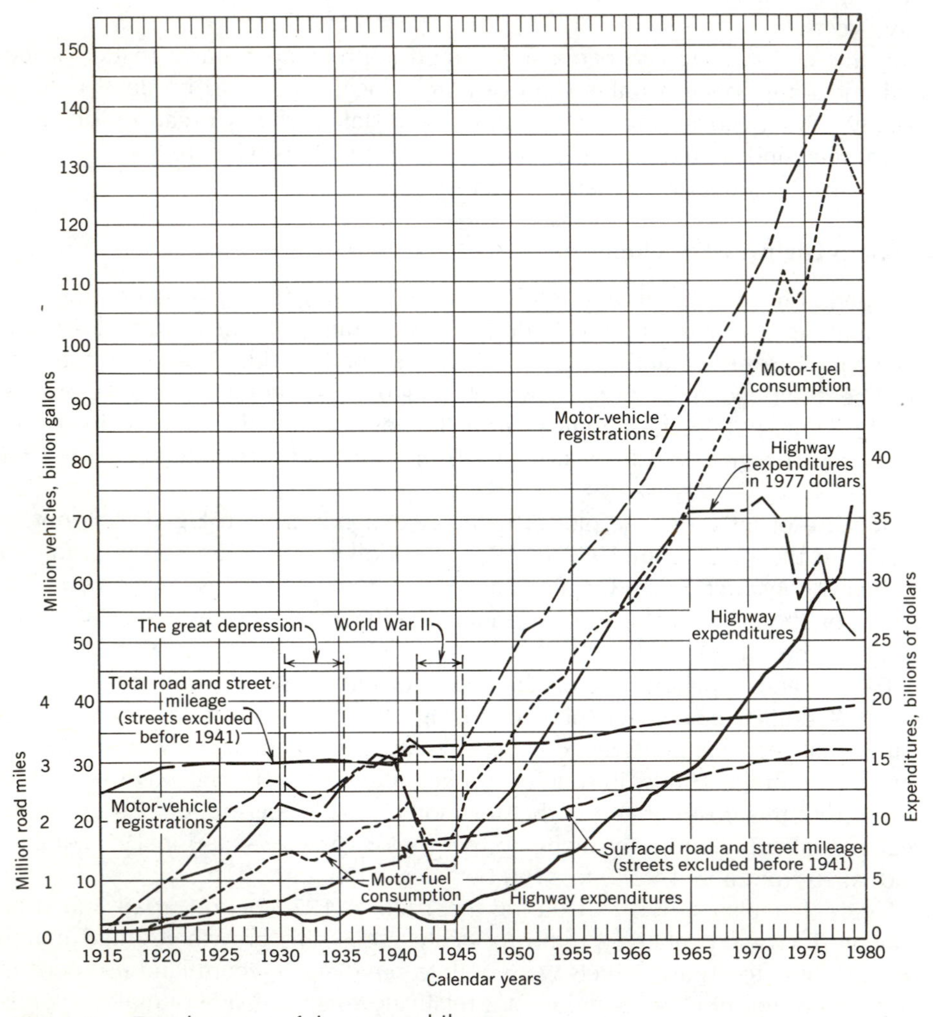

Fig. 1–1. Development of the automobile age.

FUTURE DEVELOPMENTS IN HIGHWAY TRANSPORTATION

Technological advance has been great during the age of modern highways and continues today. Knowledge has been extended in the fields of soils and other highway materials and the designs using them so that they are now more economical and reliable. Developments in machinery and management techniques have revolutionized construction and maintenance methods. The highway engineer has become increasingly conscious that a highway can be attractive and safe as well as useful and has learned much about roadside improvement, erosion control, and noise abatement. Entirely new approaches have been developed in the fields of highway and urban transportation planning, geometric and structural design, and traffic control. In all of these areas the computer has become an essential tool. Many challenges lie ahead for those interested in research, design, and administration as present practices are refined and new approaches are developed.

Possibly the most difficult problem now facing highway and transportation planners, engineers, and administrators is to define the role of the automobile, highway-based public transit, and other ways of moving people and goods in urban areas. Currently, critics are blaming the automobile for such problems as urban area expansion and wasteful land use, congestion and slum conditions in the central areas, and air and noise pollution. These problems are aggravated by a crisis in petroleum-based energy supply. It follows that those who will plan, construct, maintain, and manage our transportation facilities face a changed world; the problems in the 1980s are not those of the 1950s, 1960s, and early 1970s where the target was to build a system of freeways and other major arteries to accommodate expanding demands for better mobility by automobile. Rather, efforts will be directed toward making minor additions and adjustments to that system, rehabilitating it so that it does not completely fall apart under the ravages of time and heavy traffic, and operating it for maximum efficiency and safety. To search out, demonstrate, and implement viable approaches that will help solve these problems will challenge the ingenuity, abilities to deal with people, and staying power of all in the years ahead.

As indicated above, the decades from 1920 to the late 1970s have been called the "automobile age"; and the transition to primary dependence on other currently known systems of transportation or to new vehicles or means of propulsion will be evolutionary. But will motor vehicle use, as well as other demands for transportation in present or modified form, continue to increase as in the past? One viewpoint, that of the prestigious National Transportation Policy Commission, is that it will. A few of its projections for low-, medium-, and high-growth scenarios to the year 2000 are shown in Table 1–2. But these do not fully recognize the possible scarcity and already soaring costs of motor fuel. Furthermore, others claim that housing, health care, amenities, and control of the environment will consume a greater share of our resources, including energy, forcing a curtailment in nonessential travel. And it may be that some "essential" trips will become unnecessary as improved means of communication decrease the need to travel to the work place. Only one things seems clear; it is that the need for transportation and the highways and other facilities to serve them will be with us in the near future, but less certain as time passes.

TABLE 1–2. Projections of Highway-Based Service Demands and Capital Costs in or to the Year 2000*

	Intercity		*Urban*		*Total Financing Requirements Years 1976–2000 ($ in Billions)§*	
Year of Estimate	*Annual Passenger Miles (Trillions)*	*Annual Ton Miles† (Trillions)*	*Annual Passenger Miles‡ (Trillions)*	*Annual Freight Ton Miles (Trillions)*	*Highways‖*	*Local Public Transit*
1975	1.15	0.5	2.6	2.4	—	—
2000						
Low growth	1.53	0.9	4.2	4.0	289	58
Medium growth	1.86	1.3	4.6	6.3	294	58
High growth	2.18	1.6	5.0	7.7	297	58

*Source: *National Transportation Policies Through the Year 2000.*
†Based on highway freight movements being 23% of total movements.
‡Includes rail transit.
§Estimates are in 1977 dollars.
‖Highway costs only. Capital costs of vehicles are almost three times these amounts.

PROBLEMS

1–1. Secure data on national motor-vehicle registrations, motor-fuel consumption, and highway expenditures for the last year of record. Also compute expenditures adjusted for inflation by the FHWA construction cost index (suggested source is *Highway Statistics* published annually by the Federal Highway Administration). Plot this information on Fig. 1–1 of the textbook. Briefly outline the cause or causes of the significant changes that have occurred.

1–2. Secure the data called for in Problem 1–1 for the state in which your college is located.

2 HIGHWAY AND URBAN TRANSPORTATION SYSTEMS, ORGANIZATIONS, AND ASSOCIATIONS

LEGAL FOUNDATION FOR HIGHWAYS AND PUBLIC TRANSIT

In the United States, government has since early times assumed the responsibility for providing and regulating roads and streets for public use. This concept, and the principles of law that support it, developed in Great Britain and, even earlier, with the Romans.[1] Fundamental authority for and control over roads, excepting the 6% on federal lands, rest at the state rather than the federal level. Thus, the constitution of each state, and the acts of each state legislature in carrying out the provisions of that constitution, provide the foundations for highway policy. Within the limits of its constitutional powers, the legislature may delegate its authority for roads to a state highway or transportation commission or director, and to county, township, district, and city authorities. However, control over all highway matters in a given state rests primarily in the state constitution and the legislature. It follows that existing plans for highway administration, finance, and other affairs may be modified by suitable state legislative action. Likewise, with exceptions related to the regulation of interstate commerce, the individual states control highway use.

Public transportation also is controlled at the state or local and not the national level. Originally it was provided by private enterprises operating under franchises granted by state or local jurisdictions. However, today most public transportation is provided by public agencies under authority granted by the respective state legislatures.

The role of the federal government in most highway and transportation matters is almost completely different from that of the states. Congress, however,

[1]Toll roads and toll bridges may appear to be exceptions to the principle of governmental responsibility for roads. Actually, however, they operate either as agencies of government created by legislative action or have a franchise granted by government.

does exercise authority parallel to that of the state legislatures over the mileage on federal lands. On the other hand, it does not have jurisdiction over state and local roads in the several states. Its sole but very considerable power comes through control of the substantial sums of money granted to the individual states or local transit operations under the provisions of a continuing series of federal-aid highway and mass transportation acts, and, in 1978, a combined Surface Transportation Act.

A few among the many federal levers include: (a) restricting the use of federal-aid funds to designated groups of roads designed and constructed to approved standards; (b) requiring that the states provide matching funds; (c) possibly withholding allotted federal aid from a state that has given insufficient maintenance to a road constructed earlier with federal-aid funds or which violates some other federal provison, and (d) stipulating the conditions under which block grants for transit will be made. Thus, through curbs on the use of its money, the federal government has consistently given direction to the highway and transit policies of the individual states.

In contrast to the United States, where primary responsibility for highways and public transportation rests with the individual state governments, the central government of many countries retains direct control of at least the major highways and sometimes the transit systems.

HIGHWAY SYSTEMS[2]

Introduction

By legislative acts in the several states, roads and streets have been separated into "systems." Authority over each of these systems rests with an appropriate legislative or administrative body. It, in turn, makes provision for the planning, design, construction, maintenance, and operation of its particular group of highways. Table 2–1 classifies the mileage of roads and streets in the United States by states and by systems.

In the United States, 85% of the highway mileage, some 3,190,000 mi, lies in rural areas. Its distribution over the country varies with population and development; it ranges, excluding Alaska, from 0.40 mi of road per square mile for Arizona to 4.2 mi of road per square mile for New Jersey. Before 1890 this rural mileage was without system or classification. Responsibility for its establishment and upkeep was in the hands of local government; countries, townships, and towns took care of the roads. In general their condition was poor.

New Jersey, in 1891, first initiated state aid for rural roads, and by 1910 about half the states had set up state highway departments with varying degrees of authority. Finally, the Federal-Aid Act of 1916, which made participation in federal aid contingent on having a state highway organization, caused the remain-

[2]Mileage, travel, and financial data for the various highway systems are given in the annual and summary reports titled *Highway Statistics*, published by FHWA.

ing states to establish departments; in similar manner, the Federal-Aid Highway Act of 1921 (to be discussed subsequently) brought the concept of highway systems for rural roads to all the states.

The remaining 18% of the country's roads (694,000 mi) is in urban areas. Here the ratio of mileage to area is large; for example, Washington, D.C. has 16.5 mi of streets per square mile. As with rural roads, early responsibility for streets rested solely with local governments. However, in contrast to the situation with rural roads, state support for city streets did not begin in substantial amount until 1924. Not until 1934 were any city streets included in a state highway system.

The distinctions between highway systems, if merely "on paper," would be unimportant. However, they reach far deeper, particularly in the area of finance. Funds for highways are appropriated from designated sources to specific systems. Thus, funds for improvements to one system may be quite readily available while those for another are extremely scarce.

Figure 2–1 offers comparisons for mileages and traffic volumes among different highway systems. The contrast is particularly marked between the Interstate System and principal arterials with low mileage and heavy use on the one hand, and local roads and streets for which the reverse is true.

Federal-Aid Highway Systems

In 1912 Congress made the first in a continuing series of appropriations to the states for road construction. This appropriation, followed by annual grants for the years 1916 to 1920, was for "post roads," over which the mail was carried. It soon developed that this form of appropriation resulted in scattered improvements without any assurance of continuity or an ultimate system of improved highways. To remedy the situation, federal appropriations since 1920 have been made to limited mileages of roads of specified characteristics; thus the various federal-aid systems described in subsequent paragraphs are the result of instructions from Congress as to the use of federal funds. About 22% of all rural roads and 24% of the urban road networks, generally the more important ones, are on federal-aid systems.

Congress has, over time, redefined the various Federal-aid systems and changed federal-aid allocations with the aim of putting the money where it is most needed. It seems clear that in the years ahead, the systems described below, the pattern of federal-aid financing, and the permitted uses of funds will be substantially altered as conditions change.

THE FEDERAL-AID PRIMARY OR "A" SYSTEM. The Federal-Aid Highway Act of 1921 established the federal-aid primary system. It required the states to make an initial selection of 7% or less of their total rural mileage as a system of primary and Interstate highways. Selections were subject to the approval of the Secretary of Agriculture operating through an Office of Public Roads. Federal aid was restricted to these roads. After the 1921 act, the main trunk roads of the nation, totaling 180,000 mi in length, were quickly selected. This primary fed-

TABLE 2–1. Total Road and Street Mileage in the United States 1978.*

	Rural Mileage							*Municipal Mileage*			
	Under State Control										
State	*State Primary System*	*State Secondary Roads*	*Other State Roads*	*Total*	*Under Local Control*	*Under Federal Control*	*Total Rural Roads*	*Under State Control*	*Under Local Control. Local Municipal Streets*	*Total Municipal Mileage*	*Total Rural and Municipal Mileage*
Alabama	8,372	10,287	875	19,534	47,909	297	67,740	2,389	15,887	19,275	87,015
Alaska	3,722	1,018	—	4,740	1,766	1,767	8,273	508	1,149	1,657	9,930
Arizona	5,660	—	—	5,660	21,837	19,236	46,733	403	10,377	10,780	57,513
Arkansas	14,125	—	1	14,126	47,871	1,622	63,619	1,871	8,718	10,589	74,208
California	12,733	—	2,452	15,195	60,231	38,634	114,060	2,504	59,748	62,252	176,312
Colorado	8,464	—	—	8,464	68,533	1,204	78,201	691	8,406	9,087	87,288
Connecticut	327	1,004	119	1,450	4,022	—	5,472	2,372	11,260	13,632	19,104
Delaware	390	3,717	—	4,107	114	—	4,221	451	572	1,023	5,244
Dist. of Col.	—	—	—	—	—	—	—	—	1,101	1,101	1,101
Florida	9,784	—	—	9,784	59,568	1,184	70,536	2,225	24,359	26,584	97,120
Georgia	15,798	—	1	15,799	70,791	—	86,590	2,609	13,839	16,448	103,038
Hawaii	421	406	39	866	1,800	100	2,766	128	1,000	1,128	3,894
Idaho	4,635	—	158	4,793	25,044	30,575	60,412	359	3,220	3,579	63,991
Illinois	13,135	—	171	13,306	88,957	292	102,555	3,908	26,777	30,685	133,240
Indiana	9,761	—	211	9,972	64,562	1	74,535	1,474	14,849	16,323	90,858
Iowa	8,866	219	—	9,085	89,563	114	98,762	1,375	12,011	13,386	112,148
Kansas	9,671	—	405	10,076	112,869	70	123,015	783	11,057	11,840	134,855
Kentucky	4,217	19,437	174	23,828	39,374	312	62,514	1,171	5,096	6,267	68,781
Louisiana	4,168	10,540	—	14,708	29,768	544	45,020	1,618	8,757	10,375	55,395
Maine	3,504	7,163	234	10,901	7,748	166	18,815	401	2,581	2,982	21,797
Maryland	995	4,194	220	5,410	16,703	417	22,530	392	4,039	4,431	26,961
Massachusetts	615	—	239	854	5,557	14	6,425	2,227	24,932	27,159	33,584
Michigan	8,132	—	213	8,345	88,364	2,460	99,169	1,336	18,854	20,190	119,359
Minnesota	10,032	—	1,153	11,185	95,986	1,507	108,678	2,141	17,099	19,240	127,918

Mississippi	9,291	—	—	9,291	50,487	911	60,689	1,101	6,696	7,797	68,486
Missouri	6,898	22,940	—	29,838	68,806	700	99,344	2,333	15,873	18,206	117,550
Montana	6,500	—	54	6,554	59,162	9,456	75,172	180	2,500	2,680	77,852
Nebraska	9,312	—	522	9,834	79,120	406	89,360	547	6,752	7,299	96,659
Nevada	2,240	2,635	—	4,875	43,145	31	48,051	207	1,641	1,848	49,899
New Hampshire	1,231	1,708	32	2,971	7,056	121	10,148	1,459	3,962	5,421	15,569
New Jersey	774	—	867	1,641	11,860	13	13,514	1,443	18,116	19,559	33,073
New Mexico	4,943	6,806	39	11,788	47,454	7,285	66,527	1,123	4,378	5,501	72,028
New York	10,671	—	468	11,139	53,859	—	64,998	4,501	39,454	43,955	108,953
North Carolina	12,027	59,413	473	71,913	—	3,945	75,858	4,237	11,854	16,091	91,949
North Dakota	6,816	—	29	6,845	95,054	1,264	103,163	310	3,668	3,978	107,141
Ohio	15,999	—	1,225	17,224	69,427	29	86,680	3,220	21,284	24,504	111,184
Oklahoma	10,918	—	797	11,715	81,783	34	93,532	1,310	14,981	16,191	109,723
Oregon	4,242	2,447	2,811	9,500	34,446	59,833	103,779	879	7,571	8,450	112,229
Pennsylvania	12,681	23,669	3,914	40,264	43,366	851	84,481	8,654	25,359	34,013	118,494
Rhode Island	329	—	259	588	1,594	—	2,182	556	3,055	3,611	6,793
South Carolina	8,915	24,556	175	33,646	19,988	598	54,232	5,413	2,004	7,417	61,649
South Dakota	8,744	—	57	8,801	68,849	1,646	79,296	298	2,924	3,222	82,518
Tennessee	7,949	—	351	8,300	59,890	1,135	69,325	1,820	10,911	12,731	82,056
Texas	62,433	—	18	62,451	136,690	1,009	199,150	7,881	55,028	62,909	262,059
Utah	4,894	—	21	4,915	21,588	17,301	43,804	674	4,504	5,178	48,982
Vermont	2,383	—	200	2,583	9,606	238	12,427	152	1,400	1,552	13,979
Virginia	8,311	42,166	54	50,531	948	2,767	54,246	2,014	8,522	10,536	64,782
Washington	6,232	—	10,453	16,685	40,767	15,896	73,348	688	9,888	10,576	83,924
West Virginia	5,433	26,997	221	32,651	—	1,130	33,781	867	2,879	3,746	37,527
Wisconsin	9,672	—	—	9,672	79,990	374	90,036	1,738	14,464	16,202	106,238
Wyoming	6,012	—	6	6,018	23,654	3,361	33,033	178	1,291	1,469	34,502
Total	403,378	271,322	29,721	704,421	2,255,526	230,850	3,190,797	87,108	607,547	694,655	3,885,452

*Source: *Highway Statistics*, 1978 Edition, FHWA.

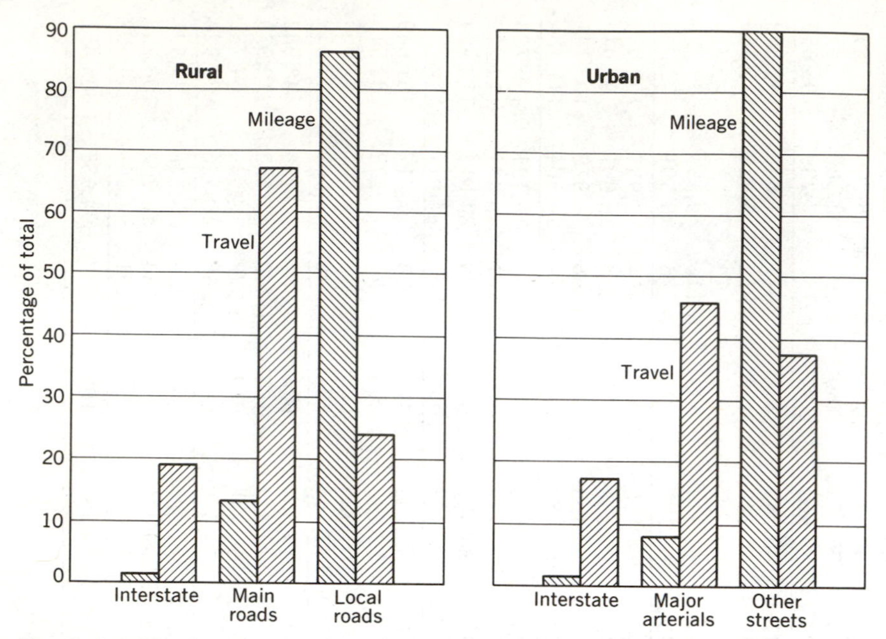

Fig. 2–1. Distribution of mileage and travel among general classes of highways and streets.

eral-aid system, often referred to as the "7% system," now totals 265,000 rural and 37,000 urban mi. In combination with the Interstate System, these roads form the core of our highway network.

THE FEDERAL-AID SECONDARY OR "B" SYSTEM. Congress established the federal-aid secondary system in 1944 to supplement the federal-aid primary system. Originally it had both rural and urban facilities, made up primarily of important county and secondary state highways. Selection of the routes has been the joint responsibility of local authorities, the state highway departments, and the Federal Highway Administration. Today's system totals 396,000 mi, all rural.

FEDERAL-AID URBAN SYSTEM.[3] Until World War II, construction and maintenance of streets and alleys were considered to be of local concern, and federal, state, and county spending was devoted almost exclusively to rural highways. As traffic in urban areas increased, more and more attention was focused on the problems it created; and federal, state, and sometimes county funds are now used to improve major routes in them. Federal participation began when Congress in 1944 authorized a special fund to aid in extending the rural federal-aid systems into urban areas. In 1970 it created a separate urban system consisting of important arterials that are not urban extensions of the rural system.

[3]An urban area is legally defined as "an area including or adjacent to a municipality or other urban place having a population of 5000 or more . . ."

Routes are selected by local officials subject to state and federal approval. Beginning in 1975, some of these urban funds could be used for mass transit. This system now totals 126,000 mi.

THE NATIONAL SYSTEM OF INTERSTATE AND DEFENSE HIGHWAYS. The Interstate System, when completed, will consist of 43,000 mi of the most important highways in the country. It connects and extends into most of our larger cities (See Fig. 2–2). Joint selection was first authorized by the Federal-Aid Highway Act of 1944, was approved by the state highway departments and the Commissioner of Public Roads, and was finally adopted[4] in August 1947. Included by 1980, were about 33,000 rural and 9400 urban radial and circumferential miles. Although 40,000 mi were in service, only 8200 mi were complete in every detail. Some 2300 mi are toll roads. Although it constitutes only 1.1% of all road mileage it carries roughly 20% of all motor-vehicle traffic. ADT (average daily traffic) in 1978 was about 16,000 vehicles on the rural and 46,000 on the urban sections. Initially, the total cost of the system was set at $26 billion. The 1979 estimate to complete a slightly extended mileage is $113 billion of which $46 billion is yet to be spent.

The Interstate System has been and is being designed to the highest standards appropriate for the terrain traversed and the traffic served. Rights of way are wide and access is controlled; some 85% of its length is on new locations, since existing developments, poor vertical and horizontal alignments, and other features of the present highways along the same routes could not be reconciled to the extremely high standards.

Congress has given construction of the Interstate System tremendous impetus by providing special funding. (See Chapter 5.) The original aim was that the entire system be finished in a 12 to 15 yr period, or by about 1972. However, that target has not been met. The 1978 Surface Transportation Act prohibits construction starts after September 30, 1986, and authorizes $3,625 million annually through 1990 for construction and reconstruction.

SPECIAL FEDERAL HIGHWAY SYSTEMS. The federal government has full responsibility for about 221,000 mi of forest highways, forest developmental roads, highways in national parks and Indian reservations, and in certain other federally administered areas. But, as indicated earlier, it has left the principal highway and mass transportation roles to the states.

State Highway Systems

In each state a system of roads has been designated by the legislature as a state highway system. As can be seen from Table 2–1, rural proportions of the total

[4]The Interstate System is an outgrowth of studies of nationwide routes begun in 1939 by the Public Roads Administration under a directive from Congress. The work was continued by a committee appointed by the president in April 1941. This committee, composed in part of outstanding highway engineers and assisted by a staff from the Public Roads Administration, completed its work in January 1944. Its report, "Interregional Highways," *House Document 379*, 78th Congress, 2d session, is outstanding.

Fig. 2–2. National System of Interstate and Defense Highways. Urban connections and some route designations are not shown. (Courtesy Federal Highway Administration.)

state road mileage under state jurisdiction vary from about 6 to 100%, with an average of 22%. At one extreme, state governments have kept control of only the most important arteries; at the other they have assumed responsibility for almost all rural roads, including some with gravel or soil surfaces. It is common for rural state systems to incorporate the Interstate System, the federal-aid primary system and some routes from the federal-aid secondary system, and possibly other important highways as well. The combined length of the 50 state highway systems is 704,000 mi rural and 87,000 mi urban. ADT, including Interstate System traffic, was roughly 2300 rural and 22,000 urban in 1978.

As shown in Table 2–1, only 12% of the urban mileage is on state highway systems. This is a continuation of the viewpoint from early days that streets are primarily a local responsibility.

County and Township Roads

In the 3000 counties of the United States there are 1.7 million mi of rural roads not in the state highway systems. A relatively small portion of this mileage is in the federal-aid secondary system. These are commonly classified as local rural roads, although in large urban areas, county roads may include major traffic arteries. In addition, there are some 17,000 townships and other jurisdictions that have distinct and separate rural road systems. Their mileage totals 519,000.

As indicated by Fig. 2–1, although the various secondary, federal, and local rural road systems constitute 86% of the nation's road mileage, the vehicle-miles accumulated on them are less than one-fourth the total. Their function is largely that of land service, and traffic averages about 110 vehicles per day or about one vehicle every three minutes in the peak hour. Many carry far lower volumes. In less populated areas, improvements are often of a low order; nationwide, some 651,000 mi are primitive or are only graded and drained. Land-use studies have revealed that many rural agencies have mileage in excess of that needed for proper land service and many are making efforts to abandon some of it.[5]

Urban Streets

As noted, some important urban streets have been incorporated into the federal-aid or state highway systems. There remains 607,000 mi of streets and alleys in 18,000 urban communities that are under local control. Some serve primarily as arteries for local traffic and others mainly provide access to property. Traffic on them varies widely, but averages about 800 vehicles per day. Volume on purely residential streets would be far lower.

Systems for Route Designation

Practically all the major highways in the United States are marked with route-designation signs which are shown on maps prepared for motorists. This route-

[5]See, for example, W. C. Hartwig, *TRB Record 716*.

designation system is usually separate and distinct from that used for highway-management purposes, and the two should not be confused. In sum, the route-designation scheme for the United States is as follows:

INTERSTATE ROUTES (see Fig. 2–2). A numbering system, consistent nationwide, developed by AASHTO. North–south trending routes have odd numbers, with numbers increasing from west to east. East–west trending routes have even numbers, with numbers increasing from south to north. Signs are reflectorized and in full colors—red, white, and blue.

UNITED STATES HIGHWAY ROUTES. A numbering system, reasonably consistent nationwide, that overlaps portions of the federal-aid and state highway systems. Odd and even numbers are for north–south and east–west tending routes, respectively, as with the Interstate System. However, the numbering begins in the east and north rather than west and south. The sign shape resembles a shield.

STATE HIGHWAY ROUTES. A numbering system for routes of some continuity within individual states. Each state has adopted a distinctive shape or pattern for its sign; for example, the Pennsylvania route marker resembles a keystone. Similar numbering schemes also have been developed by some local agencies.

SPECIAL DESIGNATIONS. At times, freeways or toll roads are given special designations to identify them more clearly in the minds of motorists. Examples are the New York Thruway, the Will Rogers Turnpike in Oklahoma, and the San Bernardino Freeway near Los Angeles, California. Such designations are an aid to those driving straight through but may confuse others who are looking for specific destinations along the route. A related problem is that of the motorist who, unfamiliar with the road, has trouble identifying an intended turnoff. One plan is to indicate each turnoff with the name of a nearby community; another is to number them consecutively, or in relation to the miles from some beginning point.

Research is under way to develop far more sophisticated motorist advisory systems. These are discussed in Chapter 10.

HIGHWAY AND TRANSPORTATION ORGANIZATIONS

Department of Transportation (DOT)

The Department of Transportation was established April 1, 1967 with the aim of incorporating all federal transportation activities in a single agency. Its seven separate operating agencies are the Federal Highway Administration (FHWA), the National Highway Traffic Safety Administration (NHTSA), the Urban Mass Transportation Administration (UMTA), the Federal Aviation Administration (FAA), the Federal Railroad Administration (FRA), the U.S. Coast Guard, and the Saint Lawrence Seaway Development Corporation. It also has its separate Research and Special Programs Administration and departments carrying out stipulated staff functions. The first three agencies listed above have as their particular concerns the highway and public transportation matters discussed in detail in this book.[6]

Proposals have been advanced to combine FHWA and UMTA and possibly NHTSA into a single organization with a multimodal focus. Passage by Congress in 1978 of a unified Surface Transportation Act has been seen by some as a harbinger of such a change. This merger had not happened by 1981. Some of the difficulties with such a reorganization can be envisioned when the difference in function and operating procedures outlined below are examined.

Federal Highway Administration (FHWA)

The Federal Highway Administration is the agency within the Department of Transportation designated by Congress to administer most of the highway programs of the federal government. It was created in 1893 as the Office of Road Inquiry of the Department of Agriculture. For 20 years, under several names, its function was to gather available knowledge and to teach others how to build roads. It was the Post Office Appropriations Act of 1912 that gave the first assignment of actual road building; functions were further expanded by the Federal-Aid Highway Act of 1916. In 1918 it became the Bureau of Public Roads of the Department of Agriculture. Under a federal reorganization effective July 1, 1939, it was transferred to the Federal Works Agency and became the Public Roads Administration. In 1949, under another reorganization of the government, it was transferred to the Department of Commerce and again named the Bureau of Public Roads. In 1967 it was transferred to the newly formed Department of Transportation. In 1970 the agency's name was again changed to Federal Highway Administration.

Activities of FHWA differ markedly from those of many other federal public-works agencies. As indicated earlier, most of the federal funds for highways are spent by or channeled through the state highway or transportation departments, with FHWA serving as adviser and monitor. In contrast, UMTA makes block grants to regional or local public transportation agencies. The Bureau of Reclamation and the Civil Works Division of the Army Engineers and the General Services Administration carry their projects through planning, design, and construction, and then operate and maintain them.

Functions of FHWA include the following:

1. Allocate federal-aid funds to the various states in accordance with laws enacted by Congress.
2. Supervise the manner in which allocated funds are spent. This involves, among other things, reviewing project selections, plans, and specifications, and monitoring con-

[6]Brief descriptions of these agencies and their assignments are given below. Financial aspects of their operations are summarized in Chapter 5. Their assignments are spelled out in detail in the continuing series of public laws enacted by Congress.

struction for all federal-aid projects. Division offices in each state capitol and regional offices located strategically over the country carry out this function.
3. Conduct research in the highway field. FHWA has its own research staff which conducts investigations in all phases of highway engineering, including traffic safety. In addition, it sponsors research by universities, institutes, and consultants.
4. Gather and disseminate information. *Highway Statistics* and the magazine *Public Roads*, both cited as references in this book, are illustrative of this function. FHWA also publishes the results of its in-house and sponsored research, and a variety of implementation reports. Some of the FHWA publications are also available from the National Technical Information Service.
5. Provide an educational program through its National Highway Institute.
6. Design and construct highways in national parks, on certain other federal lands, and for other agencies of the federal government.

National Highway Traffic Safety Administration (NHTSA)

NHTSA is a division of DOT whose entire effort is devoted to highway safety. Its activities take many forms, among them:

1. Administering federal grant programs for highway safety.
2. Developing and promulgating safe design standards for automobiles, buses, and trucks.
3. Working through state governments to develop state and local educational and informational programs to promote highway safety.
4. Carrying out research to improve many aspects of highway safety.

Urban Mass Transit Adminstration (UMTA)

UMTA is the division of DOT concerned with developing, improving, and promoting the use of public transportation. It carries out its mission primarily through a program of grants and loans to metropolitan or other local transit agencies. Among them are:

1. Discretionary grants in large sums for heavy or light rail or other facilities for large urban areas.
2. Financial aid, distributed by formula, for the purchase of equipment such as buses and other transit needs.
3. Grants for demonstration projects, from large projects such as people movers to experiments in paratransit (see Chapter 3).
4. Sponsored research and educational efforts.

Other Federal Highway and Transportation Activities

As shown in Table 2–1, a substantial mileage of highways is administered by federal agencies. U.S. Forest Service roads total more than 200,000 mi; the National Park Service, the Indian Service, the military, and other governmental units control road activities in their jurisdictions.

Many agencies of the federal government are involved in transportation as an adjunct to their primary functions. An example would be activities associated with the health and welfare of the disadvantaged. Again, administrative agencies such as the Office of Management and Budget are involved in appropriations. The Congress has its own advisory and investigative groups, such as the Advisory Commission on Intergovernmental Relations and the General Accounting Office. Also, independent regulatory agencies, for example the Interstate Commerce Commission, set and monitor rates and standards for vehicular safety. This list is only the beginning; its intent is to show how various facets of transportation, particularly over the highways, play a part in much governmental activity.

State Transportation and Highway Departments

As of 1979, some 40 states had followed the pattern of the federal government and organized Departments of Transportation. In them are centered state-level activities in highways, mass transit, airways, railways, and sometimes marine. They are also the center of overall transportation planning. In some instances, other functions such as driver licensing, vehicle licensing and inspection, and policing are included. A chart showing the organization for the state of New York is shown in Fig. 2–3. It illustrates the diversity of activities that are involved.

In state departments of transportation, the level of attention devoted to the various modes varies markedly. Almost always, the highway division will incorporate fully developed design, construction, maintenance, and operation activities and the laboratories and other facilities to support them. In some departments, planning activities are under the highway division; in others there is a common planning unit for highways and public transportation. Activity and staffing levels of the other divisions may be lower than for highways. For example, the public transportation, railroad, and aviation branches may function as advisors to state government and local agencies and as a mechanism for passing through or distributing funds from federal or state sources. These diversities at state level are to be expected, since situations among the 50 states are so different.

Ten states retain state highway departments as separate entities. Their activities parallel those of the highway divisions of state departments of transportation.

State organizations for carrying out the highway function all deal with similar technical problems, but on far different scales. Mileages administered range from less than 1100 to more than 76,000; annual expenditures from less than $70 million to more than $1.4 billion. Also, as can be seen in Table 2–1, some have responsibility only for a very limited mileage of principal arteries; on the other hand, a few administer all rural roads and many city streets.

State departments typically are headed either by a director, or by a commission of three to seven members appointed by the governor, which in turn selects a director or chief engineer. With an appointed director, responsibility is clear

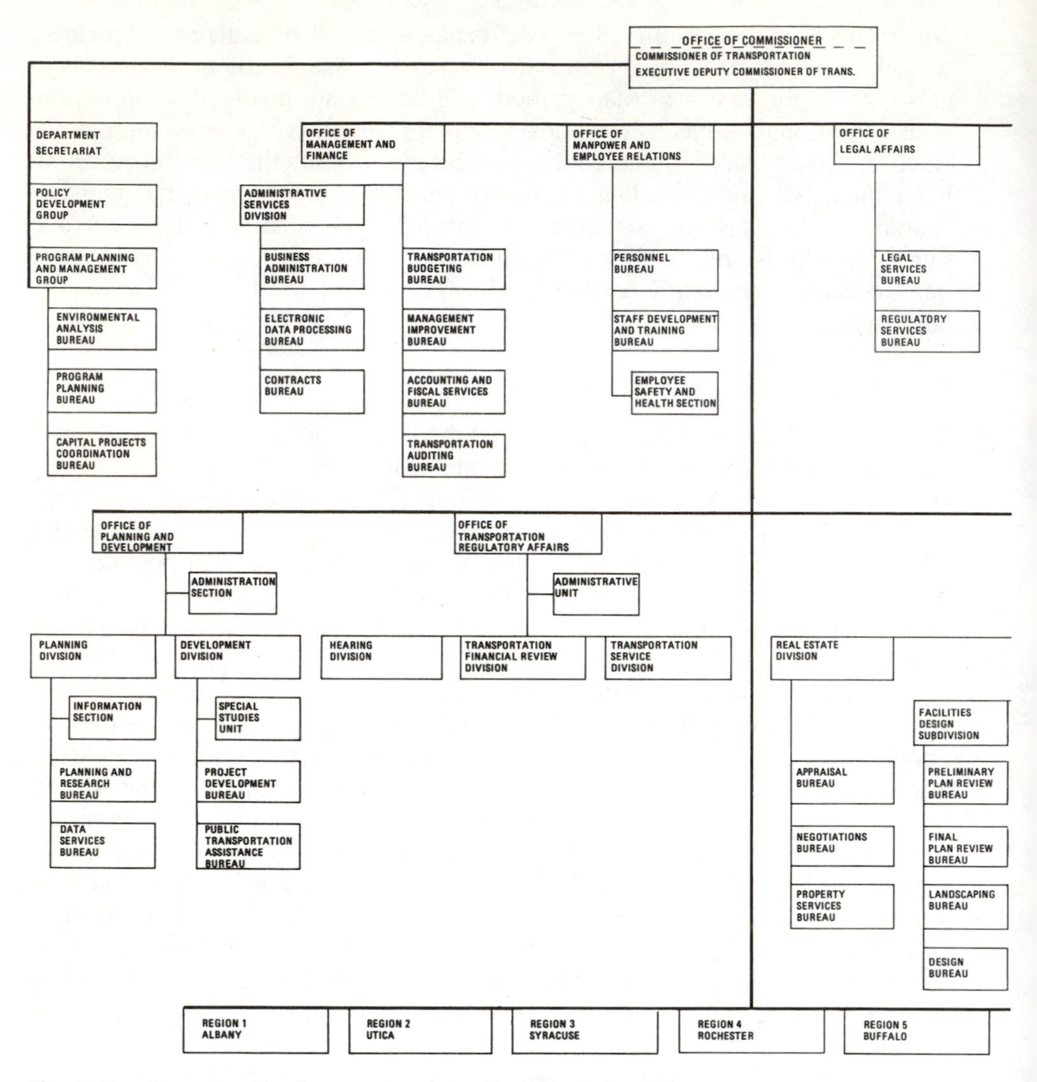

Fig. 2–3. Organizational structure of the New York State Department of Transportation.

and centralized, but political forces may be direct and strong. Under the commission form of organization, the chief officer is less vulnerable. Furthermore, if commissioners are appointed at large rather than by districts, and with staggered terms, there is a further cushion against partisan control and conflicts over where and how funds are to be spent. Today, all but a few state transportation or highway agencies operate on a highly professional basis and under civil service rather than the spoils system.

Many state agencies carry out research in all areas affecting their operations. All have well-staffed in-house or affiliated laboratories for research and testing.

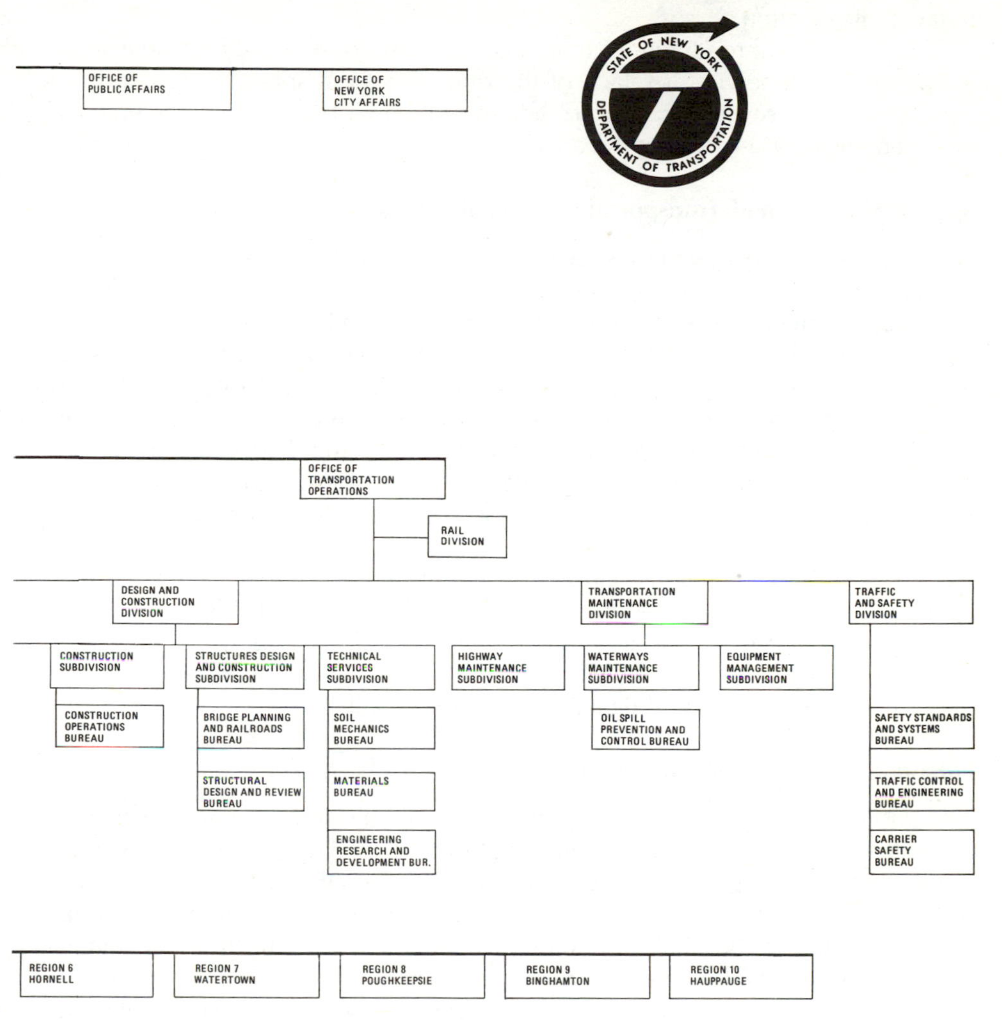

Fig. 2–3. *Continued.*

Local Road Organizations

Almost all the most heavily traveled roads in a state are administered by the state highway or transportation department. In marked contrast, some 38,000 county, town and township, urban, and special agencies administer local roads and streets. Sizes and organization of these agencies differ widely; some are larger than the smaller state departments while others administer only a few miles of lightly traveled roads or streets. Many of them have good engineering and administrative supervision; but the smaller ones may lack these abilities.

Again, in some instances, almost the entire personnel may be changed after an unfavorable election.

County and local road administration is being improved steadily. Widespread adoption of civil service in place of the once prevalent spoils system and the consolidation of several small road administrations into a single unit under a qualified engineer are among the advances.

Special Highway and Transportation Organizations

Numerous special organizations have been created by the legislatures of the individual states to carry out special transportation functions. These include the planning, financing, construction, and operation of toll roads, toll bridges and tunnels, and transit facilities. In addition, cooperative agreements across state lines have created several multistate regional organizations to coordinate and possibly operate some or all of the public transportation activities in large urban complexes. Some of these special organizations are empowered only to gather data and make recommendations and can neither levy taxes to implement their findings nor control the activities or other organizations. Others have more power. For example, the Port Authority of New York and New Jersey has jurisdiction over bridges, tunnels, airports, bus and rail lines, truck terminals, port facilities, and buildings in the New York metropolitan area. It seems clear that the agencies created to date are the forerunners of many more to come.

In 1977, highway-related tolls collected by these special agencies exceeded $800 million; outstanding obligations exceeded $7 billion.

Personnel for Transportation Agencies.

Administrative and technical positions in transportation and highway agencies are still largely in the hands of and controlled by civil engineers. In certain states, neither an engineering education nor professional registration is required of the chief executive officer; but the great majority of other key persons must be so qualified. In the more important positions, engineers have heavy administrative functions and are in constant contact with elected officials and the public. Recognition by educators that all engineers in key positions have such duties has brought added attention to nontechnical subjects such as written and spoken English and management skills.

Highway and public transportation agencies today need many skills other than engineering. A typical roster could well include technical specialists in fields such as planning, geology, computer science, economics, political science, anthropology, and possibly archaeology and others from the social sciences, as well as specialists in administration and business. To meet these needs, many agencies have developed in-service professional and preprofessional educational programs. The aim of many of the programs for professionals is to provide rapid development and rounding out of young engineers to fit them for planning, design, or administrative positions of responsibility. Such an approach is necessary since it is seldom possible today for young engineers to have and maintain technical and management competence on their own. Neither is it possible for them to serve a long apprenticeship under experienced supervisors. Several approaches are used. One is to encourage professionals to take nighttime extension courses offered by nearby colleges or to attend full-time short courses and conferences. Some agencies send career employees to graduate school for a year or more with salary and expenses paid. FHWA has a National Highway Institute which provides courses and training materials in a variety of forms.

Some highway agencies have instituted in-house management training programs for engineers and others in administrative positions. An alternative is to send them to management schools such as those sponsored by AASHTO or given by management consultants or universities.[7]

A degree-level education commonly is not required for preprofessional employees such as surveyors, draftsmen, inspectors, and computer programmers. Many of them begin employment after two years in a community college or through a training program. In many agencies, attractive civil-service progression leading to positions of reasonable responsibility and prestige have been developed for them. Procedures for certifying such personnel are available through FHWA.

Developments in such fields as photogrammetry and computer-based planning, design, computation, drafting, scheduling, and record keeping, coupled with a widespread substitution of technicians, draftsmen, and clerks for engineers on routine assignments, have drastically altered the role of engineers in transportation and highway agencies.[8] Further changes can be anticipated as new approaches to transportation develop and the emphasis shifts from design and construction toward rehabilitation, maintenance, and operation.

Consultants in Transportation and Highway Planning and Design

Many transportation and highway agencies employ consultants or consulting firms to make special studies, to design specific projects, and at times to supervise construction. Seldom, however, do consultants actually administer operating agencies. The degree to which consultants are employed varies widely among agencies. Furthermore, there are strong differences in opinion about the advisability of using consultants instead of doing this work in-house.

With highway planning, location, and design becoming more complex, difficult, and controversial, consultants from such fields as architecture, landscape architecture, environmental protection, and the social sciences often are engaged. At times a design team that includes these and possibly other specialists may be used (see Chapter 3). Accepted procedures for hiring consultants have been developed[9] but there is controversy over how much they should be used and over the selection procedures, particularly for price as a factor.

[7]See *HRB Record 266* and 387, NCHRP Synthesis 11, and *TRB Special Report 150* for added discussion. *Compendium 14* of Transportation Technology Support for Developing Countries, TRB, deals with training for those situations.
[8]See, for example, *HRB Special Report 128*.
[9]See *ASCE Manual and Reports on Engineering Practice No. 45*, and *ARTBA Bulletin 253*.

TRANSPORTATION AND HIGHWAY ASSOCIATIONS

American Association of State Highway and Transportation Officials (AASHTO)

The American Association of State Highway Officials was established in 1914 as an association of state, territorial, and District of Columbia highway departments and the Federal Highway Administration. In 1973 its name was expanded to bring in state departments of transportation. Officials of these agencies govern its operations. Engineering activities are carried on through standing committees which, among their other duties, prepare specifications, manuals, and standards representing current practice.

Publications of AASHTO include, among others, *Transportation Materials (Specifications and Tests); Specifications for Highway Bridges; Geometric Design Standards;* and numerous policy statements and guides. All these works are authoritative and frequent reference will be made to them in this book.

The association also publishes the *AASHTO Quarterly*. It reports on current highway and transportation subjects and reflects trends in thinking and legislation.

Transportation Research Board (TRB)

The Transportation Research Board (TRB), organized in 1920 as the Highway Research Board (HRB), is a private, nonprofit organization. It operates within the Commission on Sociotechnical Systems of the National Research Council, which, in turn, is a part of the National Academy of Sciences–National Academy of Engineering. TRB is supported financially by all the state transportation and highway departments, FHWA, UMTA, FAA, and FRA, numerous transportation and trade organizations, and many individuals.

The board's primary function is to encourage research in transportation and to provide a forum for the presentation, discussion, and publication of the results. This is done primarily through about 150 committees made up of roughly 1800 administrators, engineers, social scientists, and educators. Its annual meeting is by far the largest single gathering of specialists in transportation.

Another area of board activity is arranging workshops and conferences on special subjects of short-term or long-term importance. These provide a neutral forum where various viewpoints can be presented. The board has also, over the years, carried out a number of sponsored research projects, the most notable of which probably was the $27 million AASHO road test.

Board publications total some 6000 pages per year. Included are the *Record* series (which in 1962 succeeded the *HRB Bulletin* series and the annual *Proceedings*), the *Special Report* series, a magazine *Transportation Research News,* which reports six times yearly on current happenings, and a series of *Circulars* on specific topics. There are also publications on railroad, water, and air transportation topics.

In 1962, the board was assigned direction of the National Cooperative Highway Research Program under which the NCHRP staff administers contracts for research on specific topics selected by AASHTO. Results are published in a series of *NCHRP Reports, NCHRP Syntheses,* and *NCHRP Research Results Digests.* A parallel effort for transit, called the National Cooperative Transit Research and Development Program (NCTRP), was undertaken by TRB in 1980. Beginning in 1978, with financial support from the Agency for International Development (AID) it produced a series of *Compendiums* and *Syntheses* under the title of *Transportation Technology Support for Developing Countries.*

TRB also operates a computer-based Highway Research Information Services (HRIS) and cooperates in an on-line, computer-based Transportation Research Information Service Network (TRISNET). Through them, abstracts of past research and of research in progress can be obtained quickly. Among other outputs from this system are summaries of reports of completed projects and research in progress.[10] Indexes of TRB publications, compiled at possibly four-year intervals, also come from this data base. These are reported by subject, author, and title of each paper.

Without question, the publications of TRB are the most fruitful single source of advanced knowledge concerning highway and public transportation. References in this book are predominantly from its publications.

Associations of Local Governments

Local governments also have associations that sponsor activities in the highway and public transportation fields. An example is the National Association of Counties and its affiliated National Association of County Engineers (NACE). Among its other activities has been the publication of a series of technical and management manuals. The National League of Cities and the American Public Works Association operate in a parallel manner.

Foreign Research Organizations

The United States is not alone in transportation and highway research. An outstanding example of another nation's activities is the Transport and Road Research Laboratory, located near Crowthorne, Berkshire, England. Its studies have covered a wide range of subjects dealing not only with the problems of Great Britain, but also with those of the developing nations. It also offers a variety of education programs including lectures, seminars, and short courses. Lists of publications and individual research reports are available to officials, researchers, and educators on request. There are similar organizations, too numerous to mention, in Canada, continental Europe, and other countries around the world.

[10]See *Transportation Research News,* TRB May–June 1980 for details.

Institute of Transportation Engineers (ITE)

The Institute of Transportation Engineers is a society whose members have professional interests in all aspects of traffic and public transportation management. The institute publishes its *ITE Journal*[11] every month and sponsors the *Transportation and Traffic Engineering Handbook*. ITE also has excellent technical committees which, along with other studies, develop standards, particularly for traffic control devices.

American Road and Transportation Builders Association (ARTBA)

This association is a nonprofit, noncommercial organization whose memberships includes highway and urban transportation officials, engineers, teachers, equipment manufacturers and distributors, materials producers, and contractors. The association staff and special committees observe and report happenings of interest to its various constituencies. It publishes a newsletter which reports current legislative, administrative, and other events and, occasionally, technical or other reports.

Other Transportation and Highway Associations

Numerous trade associations interested in promoting the use of the products of their members are active in the transportation and highway fields. Two in this large group are the Asphalt Institute and the Portland Cement Association. Many of these associations publish technical bulletins and release other data. Some have field engineers located strategically over the country. Much reliable and useful information can be gained from these sources. Certain individual manufacturers also are active in a similar manner.

Another group of associations includes those having special areas of interest in transportation. One is the American Public Transit Association. The National Safety Council, concerned with accidents, has among its functions the collection, classification, and distribution of accident data. The Highway Users Federal for Safety and Mobility, supported by the automotive, oil, and trucking industries, has fostered research and education toward safe and efficient highway transportation. The federation's interests also include highway administration and planning, and it has made significant studies in both fields. The Insurance Institute for Highway Safety has particularly emphasized vehicle design and crash resistance and occupant protection. The Eno Foundation for Transportation, Westport, Connecticut, among other activities, publishes the excellent periodical *Traffic Quarterly*. The International Road Federation promotes education, information interchange, and understanding throughout the world.

[11]This publication was titled *Traffic Engineering* through May 1977, and *Transportation Engineering* for the year ending May 1978.

There are also associations whose concern is highway transportation as it affects their members. Representative of these are the American Automobile Association and the American Trucking Association. Most of these also publish magazines or bulletins.[12]

University and College Activities

Most engineering colleges have specialists in highway and public transportation on their teaching staffs and offer undergraduate courses in the subject. Some also offer graduate programs and provide extension courses, in-service training, and special conferences. A few have affiliated research institutes. In addition, many faculty members conduct research on transportation problems, often with the cooperation and financial support of governmental agencies or other interested sponsors.[13]

Research and Information Retrieval for Highways and Public Transportation

FHWA, UMTA, and many other federal agencies, the individual state highway departments, a number of universities, private or university-related research groups and individuals, and many of the other organizations and associations mentioned here conduct research on highway and transportation problems or provide financing for it. Many other agencies also carry out projects with strong transportation implications. The output from these efforts is large; one estimate places it at 30,000 titles per year. It follows that to find what research has been done or is under way is a difficult task.

Among the ways to find research results are through TRB publications and other services mentioned earlier or in such summaries as the annual *Transactions* of ASCE. Many other agencies, for example FHWA's Highway Institute, publish bibliographies from time to time. The subject indexes in good university and public libraries can be helpful. Unfortunately, many valuable research reports, particularly those issued by federal agencies, are not available in libraries unless they are depositories and even their collections may be incomplete.

Even after the titles and abstracts of research reports have been located, gaining access to the report itself may be difficult. Copies of some of the most significant can be purchased from the National Technical Information Service, Springfield, Va. 22151. Often, however, it may be necessary to contact the author or sponsoring agency. Because information retrieval from the vast number of publications is so difficult, those who wish to have their findings known should report them through TRB, ASCE, or other commonly available sources.[14]

[12]Because of space limitations only a few of the many associations and other organizations can be listed here.
[13]For discussions of university activities see *TRB Special Report 150* and *TRB Record 748* and *793*.
[14]See *TRB Record 738* for discussions of several aspects of technology transfer.

PROBLEMS

2–1. For the state where your college is located:

a. Compute the percentage of the total road and street mileage in each of the systems listed in Table 2–1.

b. Compute these percentages for the United States as a whole.

c. On the basis of your knowledge of your state and the country as a whole, briefly explain any substantial differences between the answers to parts *a* and *b*.

2–2. For the state or country where your college is located:

a. Determine the total mileages and vehicle-miles of travel in (1) rural and (2) urban areas. What percentage is each of the total?

b. Plot the data obtained on part *a* in a form similar to Fig. 2–1 of the textbook.

2–3. Locate and copy the organization chart of the state-highway or transportation department of the state in which your college is located. (Suggested sources are the reports of each state highway or transportation agency, prepared annually or biennially.)

2–4. Investigate the local road or street organization in the area where your college is located. In particular, what agency administers the roads or streets, and how is it organized; if an engineer is in charge, what limitations are there on his or her authority to establish standards and priorities for improvement; are the employees under civil service; which, if any, groups of employees belong to labor unions; what percentage of the construction work is done by contract?

2–5. Through the Transportation Research Board representative or librarian at your school, determine the availability of the TRB publications given as references in this book.

3 HIGHWAY AND URBAN TRANSPORTATION PLANNING

THE CHANGING ROLE OF HIGHWAY AND TRANSPORTATION PLANNING

In the United States before 1930 (and in many developing nations today) the primary attention of highway agencies was on rural areas; it focused on "getting out of the mud" by establishing a system of all-weather roads from rural areas to the nearest town. With this objective there seemed to be little need for "planning"; the problem was to get these roads built. Long-distance transport of people and goods was left to the railroads. Cities were responsible for the construction of their streets; public transportation was carried out by private enterprise, primarily with electric street cars.

About 1930 the attitude of highway agencies toward planning began to change. City streets were in relative distress, and many rural highways were inadequate. The practice of using all federal aid and the bulk of state highway funds for the improvement of main rural highways and letting urban areas fend for themselves needed examination. To get facts on which to base decisions, the so-called "highway planning surveys" were undertaken. Beginning with the Federal-Aid Act of 1934, Congress authorized expenditures, without matching state funds, not to exceed 1½% of federal-aid funds apportioned to each state for making a complete road inventory and for planning, surveys, and engineering investigations of projects for construction in the future. By 1940, all the state highway departments were assembling the facts necessary to develop long-range highway-improvement programs. But this effort and the accompanying federal and state financing continued, until after World War II, to have rural and highway focuses. Intercity transport was still left to the railroads. Cities were responsible for their streets. Public transportation was still largely in private hands, but was shifting from street car to bus. Integrated transportation planning was unknown.

The period from the late 1940s until about 1970 was one of great change in transportation facilities and their use. Motor-vehicle and heavy-truck travel measured in vehicle miles tripled in both rural and urban areas. Public consensus was that more and better highways were needed, including the Interstate Freeway System. At the same time, highway agencies expanded their programs to deal with urban congestion. Among the consequences of these practices were the veritable explosion of cities into the suburbs accompanied by decay of near-downtown areas, a two-third's decrease in public transit ridership, and financial distress and public takeover of many privately owned transit operations. Also, railroads lost most of their intercity passengers to the private automobile, buses, and airlines, and trucks assumed increasing importance in long-haul freight transportation. Even so, planning was still primarily directed at accommodating the demands of motor vehicles, with little attention to the interrelationships among modes or the deterioration of public transportation.

Since the early 1970s, the continued growth of motor vehicle use, the planning premises and approaches of highway or transportation agencies, and the proposals for highway improvement stemming from them have been challenged on many fronts. No longer is there a consensus that the private automobile should provide almost all transportation. In fact, some governmental agencies have been created whose functions are to make its use unattractive. In addition, relatively lower funding for highways and the increasing cost of maintaining and rehabilitating existing facilities are claiming larger and larger sums of the available money, leaving less for construction. Concerns over the environment and energy are reducing some forms of travel by automobile[1] and many governmental policies have been, at least until 1981 when President Reagan's fiscal policies began to be implemented, aimed at shifting some of the remainder to a variety of viable and acceptable forms of public transportation. Increasingly, planning is dealing with transportation and its implications as an integrated whole rather than separately by mode. Furthermore, where motor vehicles are concerned, less attention is being given to planning new facilities and more to short-term programs and system management, commonly called TSM, to make better or different uses of existing facilities through modifications and more effective management.

All in all, highway and public transportation planners have been forced to move away from the earlier engineering orientation. Instead they must concentrate on decision-making made in the political rather than the professional arena. They are still groping for approaches to this new set of problems.

This chapter is merely an introduction to these complex subjects. It approaches them by (1) looking at the planning dilemma, (2) outlining a possible approach to planning, (3) examining the urban problem and public transportation for it, (4) outlining data gathering procedures, and (5) treating several planning situations and the means for handling them. It draws on many sources, and gives references to those that, to the authors, seem most pertinent.

[1]For example, the gasoline shortage brought a decrease in visitors of about 25% to the more remote national parks in the summer of 1979.

THE PLANNING DILEMMA

As indicated in Chapter 1, the United States is, with a few exceptions, committed to highway travel for some years to come. Until the late 1960s, the attitudes of the public and their elected representatives toward the private automobile and highways, particularly freeways, was highly favorable.[2] Probably such attitudes generally still prevail toward highways serving rural areas, although some critics blame the automobile for such problems as the removal of land from productive use, the failure of it or alternative transportation schemes to provide for the movement of rural residents, particularly the disadvantaged, and the overcrowding and despoiling of recreational and scenic areas. The urban scene is different. A segment of the population of unknown size and many planners, social scientists, and politicians charge the motor vehicle and the freeways and streets that serve it with primary responsibility for such problems as air and noise pollution, urban sprawl, displacement of the poor and minorities from their homes, and the deterioration of the central city and close-in residential areas. More recently, a major share of the blame for the shortage and high cost of motor fuel is attributed to them. Thus, where in the 1960s agencies proceeded with their urban highway programs with little interference and with a feeling of certainty, now major freeway and highway programs are largely a thing of the past.

Many reasons can be offered for these possible changes in attitude toward the motor vehicle and for the uncertainties that result. A few of them are:

1. Urban problems have been greatly intensified by such predicaments as[3]:
   a. The movement of rural populations to urban areas, with many of the minorities and disadvantaged trapped in the deteriorated areas surrounding central business districts.
   b. The flight of the more affluent, and services for them, to new communities in the suburbs, leaving the cities with acute problems and lower revenues with which to handle them.
   c. Economic forces, including land prices, transportation costs, and many governmental policies that have favored dispersal.

[2]*NCHRP Reports 49 and 82,* published in 1969, presented the results of a national survey of transportation attitudes and behavior. Two of its many findings which reflect the overall results were: First, to the deliberately negative question: "The automobile pollutes the air and creates traffic congestion. Highway development demolishes homes and often destroys previously attractive landscapes. The increasing number of automobiles along with inadequate highways kill over 50,000 people every year. In your opinion, 'Is the contribution the automobile makes to our way of life worth this?' " The responses were "Yes, 84.5%, No, 15.5%." The second question related to the amount to be spent by public agencies for services to the public. For roads and highways, responses were: Spend much less 1%, less 2%, the same 44%, more 34%, much more 19%. Roughly equal importance was attached to expenditures for education, police and fire, public transportation, and health and hospital services. But less importance was given to air and water quality and substantially less to welfare and urban renewal. Reports from a survey made ten years later (see P. G. Koltnow, *Traffic Quarterly,* Apr. 1979) were in the same vein.

[3]*Civil Engineering,* Nov. 1978, offers an in-depth analysis of the problems of New York City. It provides an excellent if untypical case study.

d. Governments have been relatively ineffective in dealing with these and other complex urban problems. Authority has been divided among many agencies at several levels. Furthermore, inept approaches, long delays, or inadequate financing of some projects has tended to discredit all such efforts.

2. Arguments about urban development, including use of motor vehicles and other environmental and conservation issues, have offered a highly visible means for expressing public frustrations with the many urban problems.[4] Under such circumstances, effective political action is difficult.
3. Proposed actions are delayed or prohibited because laws with conflicting purposes and the procedures set up to carry them out, along with challenges in the courts, result in long lead times or in no action at all.
4. As discussed in more detail below, transportation is but one of many problems for which solutions are being sought, and suboptimizing to produce better transportation services, as transportation planners have often done in the past, is not feasible today.
5. Rational planning is out! No longer is planning an objective process based on complex computer-based analyses free of the emotional concerns of citizens. Rather, in a democratic society, decisions finally are made in the political arena.[5]
6. Unquestionably, the public has lost confidence not only in government's ability to solve problems but also in the technical professional. Neither do professionals have confidence in themselves nor in the solutions they offer.

In sum, it must be concluded that planners today must operate in a world of rapid change where there are few guideposts and in the political arena where "decisions by oracles" may carry equal weight with those of professionals.

A POSSIBLE APPROACH TO TRANSPORTATION PLANNING[6]

An Overview

Given the dilemmas of highway and urban transportation planning just outlined, what steps can be taken to provide meaningful and workable recommendations for action? The paragraphs that follow summarize the inputs for a possible approach. Then the remainder of this chapter and portions of Chapters 4 through 13 present current knowledge, primarily for highway-related transportation. These areas are (1) organizing, (2) setting out the variables, (3) financing and programming, (4) evaluating the alternatives, (5) the planner's role, (6) the mul-

[4]It can be argued that the automobile and freeways are the wrong villains. For example, it is unrealistic to make comparisons between present conditions and an imaginary world that never existed rather than with the situations of 20 or 40 years ago. It also can be argued that urban problems are far less tractable than the technical ones involved in putting man on the moon. Finally there is this prevalent notion that building new highways or increasing the capacity of existing ones inevitably brings large increases in traffic and adverse environmental impacts. This claim is not supported by the facts (see *NCHRP Research Results Digest 177*, Dec. 1980). But, in the present world where the "noisy" get attention, rebuttals such as these seem to carry little weight.

[5]See, for example, R. A. Burco, *TRB Record 563*.

[6]To simplify the discussion, the following presentation is framed largely in terms of highways and the transportation vehicles that use them. However, the same concerns apply equally well to other transportation modes and to their interrelationships.

tidisciplinary approach, (7) citizen and community participation, and (8) institutional constraints.

Organizing for Transportation Planning

Successful transportation planning requires, at the outset, clear definition of specific objectives and of the roles of those involved. Because several agencies, often at different levels of government, are usually concerned, the place of each in terms of inputs of effort, data provision, and financing must be spelled out. Clarification of which agencies will set up the project and participate in other ways is crucial. For example, will it be primarily a professional effort or will politicians and citizens be involved, in what way, and at what points in time? Expectations must be agreed upon to see that the study addresses the problems to be faced by the decision makers and has sufficient financing and proper direction to provide the information that they feel is necessary. Arrangements for feedback throughout the process must be made to keep all participants informed and interested. Finally, the responsible party or parties for implementing the findings, whoever they may be, should be agreed upon. This short list only indicates the level of detail required when a planning effort is launched. It must, of course, be fleshed out for each situation.[7]

Setting out the Variables

As indicated above, decision making for transportation or even highways alone, whether for rural or urban situations, is complex in itself. In addition, the transportation system, of whatever nature, plays an important part in the many interactions of modern society; these complex relationships are illustrated by Figs. 3–1*a* and 3–1*b*.

Figure 3–1*a* is a simple flow diagram of a single-system decision process—for example, it could be applied to the question of whether or not to build a given highway and to what standards. It involves defining goals, determining alternative courses of action that will fulfill or at least partially fulfill those goals, predicting (simulating) the performance of each proposed facility, and evaluating the economic, financial, and other definable consequences of its adoption. Finally, the most attractive alternative is found by comparing the consequences of the various courses of action. As mentioned, however, seldom can a final decision be based solely from the outcomes of such analyses. Rather it is reached in the often-irrational political arena on the basis of what is possible.

As has been indicated earlier, even the simplest decisions about either rural or urban transportation have ramifications beyond those on users or others who are directly affected. Some of the other functions that are impacted are listed around the periphery of the circle of Fig. 3–1*b*. Notice that the figure indicates that the single-system decision process of Fig. 3–1*a* goes on in each of the areas indicated. It would also be employed when evaluating the combined impacts.

[7]For further detail and references, see *NCHRP Report 167*.

Figure 3–1*b* serves as a useful guide to thinking in several regards. First of all, the maze of lines indicates the many interrelationships that must be explored if all implications of all public decisions are to be investigated. Secondly, it suggests the two-way nature of the decision-making process. For example, transportation decisions such as the level of funding determine the accessibility and cost of developing potential resources such as energy. On the other hand, a decision to develop an energy source, for example coal, affects not only plans for transportation, but also economic and social well-being, environment and health, housing, and other concerns. On the urban scene, decisions about transportation affect almost every other activity shown on the chart.

To date, we have little knowledge and almost no quantifiable measures of most of these ripple effects or of one on the others. A further complication is that the systems represented in Fig. 3–1 are dynamic and not static over time. Some of the responses to change come quickly, but others stretch over a decade or more. Thus, although Fig. 3–1*b* might suggest that by this form of planning it is possible to determine a best action or set of actions, the true situation is that even single-system decisions are seldom the optimum ones, and means for overall system optimization are far in the future.

Tables 3–1 and 3–2 refer more specifically to the economic and other impacts of highways, and provide a useful checklist of the factors that must be considered. Although the study from which they were taken was for urban freeways, with a few additions and some change in emphasis they could be applied to almost any highway or transit system, urban or rural, in the United States or to

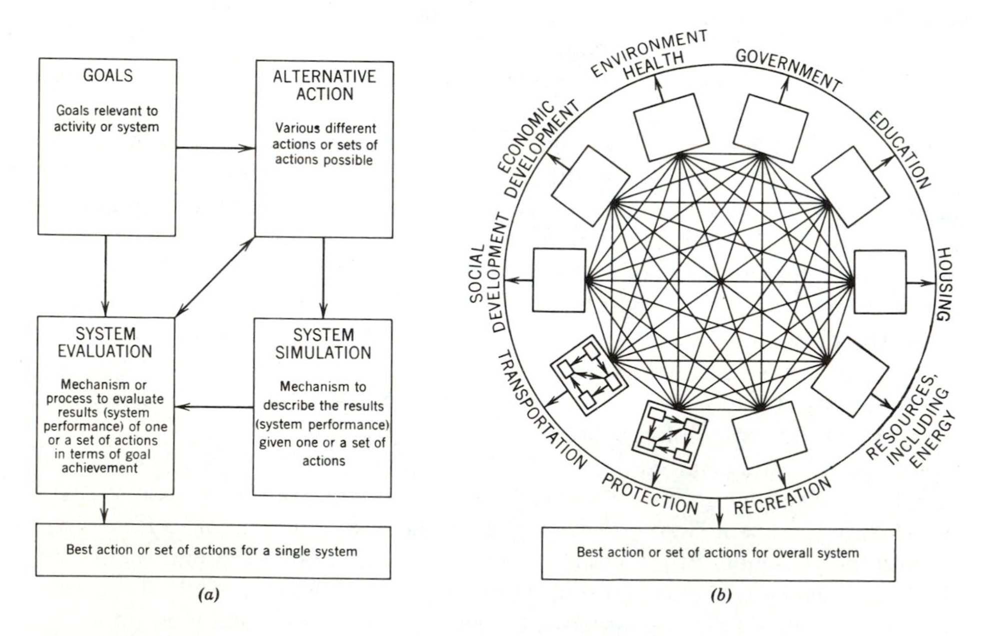

Fig. 3–1. Diagram of the planning process for single and multiple systems. (After J. R. Hamburg et al., *HRB Special Report 97.*)

TABLE 3–1. Direct Effects of Freeway Construction and Use*

Factor	Description	Units	Time Period (yr)
Quantifiable market values			
1. Cost of highway:			
a. Planning	Capital cost and annual cost of planning, constructing, maintaining, and operating the freeway		N.A.
b. Rights of way			20–40
c. Construction			20
d. Maintenance			Annual
e. Operation		Dollars	Annual
2. Costs (benefits) to highway user:			
a. Vehicle operating cost (including congestion costs)	Net increase (decrease) in costs of vehicle operation per year	Dollars	Annual
b. Travel time savings (commercial)	Net increase (decrease) in travel time, times dollar value of commercial travel time	Dollars	Annual
c. Motorist safety (economic cost of accidents)	Net change in expected number of accidents times average cost per accident	Dollars	Annual
Quantifiable nonmarket values			
3. Costs (benefits) to highway user:			
Travel time savings (noncommercial)	Minutes saved per vehicle trip	Minutes or hours	Annual
Nonquantifiable nonmarket values			
4. Costs (benefits) to highway user:			
a. Motorist safety	Accident costs of pain, suffering, and deprivation	?	Annual
b. Comfort and convenience	Discomfort, inconvenience, and strain of driving	?	Annual
c. Aesthetics from driver viewpoint	Benefit of pleasing views and scenery from the road	?	Annual

*From C. H. Oglesby et al., *HRB Record 305*.

any other nation where the broader problems of the relationships between overall economic development and land access must be considered. Tables 3–1 and 3–2 also demonstrate the "state of the art" of appraising the effects of highways. Although, with certain exceptions, units in which the various effects can be expressed are suggested, there is serious question that many of the community factors listed there can be measured accurately or that the proposed units are appropriate. This problem is discussed further in Chapters 4 and 13.[8]

[8]See *NCHRP Report 193* for details of a parallel study for freeways. See also *NCHRP Report 156, 216,* and *217*.

TABLE 3–2. Community Effects of Freeway Location and Use*

Factor	Measures and Suggested Measures: Description	Units	Time Period: Long Run	Time Period: Short Run
Local Transportation Effects Traffic service to community by freeway-highway capacity, O-D of trips, major traffic generators	1. Percent reduction of through traffic on city streets [(vehicles before—vehicles after)/ vehicles before]	Percent	x	
	2. Distance of freeway access from major traffic generators (e.g., academic, business, cultural, administrative centers) or as measured by road user or transportation costs	Miles	x	
	3. Corridor miles compatible with present or future public transportation development	Miles	x	
Effect on local transportation: city street circulation and public transit	1. Costs (savings) for improvement to city streets to provide for projected traffic volumes if freeway is not built	Dollars	x	
	2. Net change in parking space available as result of freeway	No. spaces	x	
	3. Number of interchanges with the community less streets closed	Number	x	
Access to regional facilities: recreation, education, culture, business, and employment	1. Travel time savings to regional activity centers [(minutes per vehicle) × (vehicles per day)] for each facility	Minutes per day	x	x
	2. Number of trips to community generated from outside	Vehicles per day	x	
Highway design standards: grades, alignment, and interchange location	1. Miles less than x percent grade	Miles	x	
	2. Miles of curvature less than y radius	Miles	x	
	3. Average distance between interchanges	Miles	x	
Community Planning and Environment Land use: land development, changes in use, multiple use, separation of uses	1. Land for potential development to which access is created	Acres	x	
	2. Miles of freeway separating incompatible land use minus miles driving compatible uses	Miles	x	
	3. Miles adjacent to or through land undergoing change in use	Miles per acre	x	
Aesthetic impact of freeway on community: depressed or elevated, land-scaping, structures	1. Miles depressed in residential areas plus miles elevated in commercial areas less miles at grade	Miles	x	
	2. Additional costs of aesthetic improvement in structures and landscaping	Dollars	x	
Noise	1. Increase in dB level weighted by miles residential, and numbers of schools, churches, and similar buildings adjacent to freeway	dB (weighted)	x	
	2. Additional cost of noise barriers in noise problem areas	Dollars	x	
Air pollution	1. Net change in noxious exhaust emissions for projected traffic with and without freeway	Percent	x	
Neighborhood and Social Structure Property values: changes in resale values	1. Increase or decrease (net) over normal trend in property value classified by type of use and distance from freeway	Dollars	x	
Neighborhood impacts: displacement and relocation of people, environmental qualities, neighborhood cohesivness and stability	1. Number of housing units displaced (or) number displaced as percent of community's total stock	Number Percent	x x	x x
	2. Number of people displaced (or) number displaced as percent of community's population	Number Percent	x x	x x
	3. Net loss of housing—units taken less vacant replacement housing in same price range with comparable financing less new construction planned on vacant land with financing	No. units	x	
	4. Cohesive neighborhoods severed by freeway (as determined by mapping neighborhood boundaries and social characteristics)	No. people	x	x
	5. Neighborhood stability (S. Hill and B. Frankland, *HRB Record 187*, pp. 32–42$	Index No.	x	x
Parks and recreational facilities	1. Acres of parks lost (gained) as percent of total available acres	Percent	x	
	2. Cost of park replacement less compensation	Dollars	x	
	3. Number of parks affected	Number	x	x
Cultural and religious institutions	1. Number of churches taken (or) total attendance affected	No. churches No. people		x x

TABLE 3–2. Community Effects of Freeway Location and Use*

Factor	Measures and Suggested Measures: Description	Units	Time Period: Long Run	Short Run
	2. Additional cost of relocation, excess over taking price	Dollars	x	
	3. Improved access or location for new church facilities	Minutes	x	
Historical sites and unique areas	1. Number of historical areas lost (total affected less those relocated)	Number	x	
	2. Value of monument measured by annual visits per year	Visits per year	x	
School system: attendance boundaries, school environment	1. Net loss (gain) in tax base for school system	Dollars	x	x
	2. Number of schools totally or partially taken (or affected)	Number		x
	3. Number of school attendance areas with access to school serious impaired where boundaries cannot be adjusted	No. pupils	x	x
	4. Increase (decrease) in cost of providing school services because of changes in busing	Dollars	x	x
	5. Net additional cost to the community of relocating schools affected by freeway (plus) cost of noise reduction in schools adjacent to freeway	Dollars	x	
Community Economic and Fiscal Structure				
Effect on tax base: Net change in assessed value of property on tax rolls	1. Loss of assessed valuation in right of way as percent of community total	Percent		x
	2. Loss of assessed valuation in right of way less increase of land values (assessed) caused by freeway impact	Dollars	x	x
	3. Net loss (gain) in tax revenue caused by freeway impact	Dollars	x	x
Community services: police and fire protection, utility services, water and garbage services	1. Net increase (decrease) in costs of providing fire and police protection and water, sewerage, and garbage service	Dollars	x	
Commercial activity: wholesale, retail	1. Net increase (decrease) over normal trend in gross wholesale and retail sales	Dollars	x	
	2. Net number of businesses located (displaced) by freeway	Number		x
Employment: creation of jobs, displacement of jobs	1. Net number of jobs located (displaced) as a result of freeway	Number		x
	2. Net gain (loss) in gross earnings from jobs located or displaced by the freeway	Dollars	x	x
	3. Net increase (decrease) in job opportunities caused by expanded commuting area less jobs available to outside commuting	Number	x	x

*From C. H. Oglesby et al., *HRB Record 305*.

Still another consideration too often neglected in evaluating the effects of highways and other transportation facilities is suggested in Tables 3–1 and 3–2. This is that some effects are long run and others are short run. For example, some of the strongest complaints against highways are related to noise and air pollution from today's motor vehicles. Some of the worst effects are already being controlled by strict enforcement of noise and air pollution regulations and by installing noise barriers. Furthermore, research may well produce engines and tire-pavement combinations that are far less noisy; likewise, modifications to today's engines or new means of propulsion are already substantially reducing air pollution. Thus it could be argued that sensible long-term investments to develop effective transport based on rubber rolling on pavement could be blocked by short-run effects that may largely disappear in the near future.

When choices among investments in alternative transportation modes are at issue, especially in large urban areas, a variety of factors omitted from or not particularly emphasized by Tables 3–1 and 3–2 may carry heavy weight. One relates to capital-intensive, long-term alternatives such as heavy or light rail, people movers, or freeways versus short-term ones including buses or other conveyances traveling primarily on existing highways. Another is the relative availability of funds for the various modes, particularly in the form of federal or other appropriations or grants. Furthermore, certain alternatives may provide greater short- or longer-term employment. For example, buses are short-lived and might be manufactured locally and their operation is labor-intensive. On the other hand, rail, once constructed, has long-lived trains which might be made elsewhere and which need relatively few operators.[9]

There are also land use concerns that influence the choice between transportation modes. One is a combination of policies designed to rehabilitate rundown areas and fill in vacant spaces in cities, preserve open space and agricultural land adjacent to them, and reduce energy consumption. This question is discussed in more detail later.

Establishing Economic and Environmental Viability

The foregoing discussion has focused on setting out in nonquantified form certain economic, community, and other consequences of (1) individual projects and alternatives within them and (2) alternative overall schemes. But procedures for selecting which projects to carry out should involve studies to determine economic viability or, in other words, studies to establish which among the proposals make the best use of available, limited resources. Detailed procedures for doing so in highway situations are outlined in Chapter 4.

The findings of economic analyses, if made at all, often have been largely disregarded for public undertakings in transportation. There are, of course, exceptions; for example, lending agencies such as the World Bank often make economic justification a precondition to financing. And they also are a part of the "alternatives analysis" process of UMTA. But all too often, only lip service is given to the results.[10]

Sometimes projects or alternatives that appear to be viable economically are ruled out because of environmental impacts such as noise, air, or water pollution. To mitigate these impacts may be too costly. Some of these impacts are discussed in Chapter 13.

Financing and Programming

Financing involves the determination of sources and amounts of money involved to construct, operate, and maintain candidate projects. Chapter 5 outlines the more important sources of income for highways and for federal contri-

[9]Economic impacts such as these are often measured by input-output models. Explanation of them is beyond the scope of this book.

[10]*National Transportation Policies Through the Year 2000* strongly urges greater use of the results of economic analysis in transportation decision making.

butions to transit. It also gives the basis for allocating them among the states and urban areas and provisions for matching when they exist.

In the past, highway agencies have had incomes sufficient to undertake a number of construction projects each year as well as to carry out continuing functions such as maintenance and operation. For them, planning often has been referred to as *programming*. Programming might be in two phases, a tentative, long-range view and a short detailed one which sets in motion the activities necessary to undertake and complete selected individual projects. Often the list of projects is developed by a highway needs study (see below).

A well-thought-out programming approach provides insights on many elements, including the following:

1. Assessment of available financial resources, both short- and long-term, by agency function such as construction, operation, and maintenance.
2. Recognition of legislative and administrative desires and constraints.
3. Establishing tentative priorities based on economic analysis, critical situations, present and expected future levels of traffic, and claims based on political subdivisions, population levels, and community needs and demands.
4. Providing for route and system continuity and for coordination with other transportation modes.
5. Project selection to balance construction capabilities, project durations, available materials and labor, and climatic constraints.
6. Scheduling for implementation of projects, recognizing lead times for such purposes as coordination with other agencies, acquiring rights of way, and producing designs, plans, and specifications.
7. Providing resources to cover emergencies such as floods or other natural disasters.

Based on these and other considerations, a tentative program and budget can be set forth for approval or modification by the appropriate administrator or administrative body.[11] The programming of operating transit systems would follow this same pattern.

For programming to be understood, and to avoid clouding or confusing important issues, it should be kept in mind that three separable sets of inputs are involved. These are (1) economic, which is the use of resources, (2) financial, which is who pays and who spends how much and when, and (3) political and administrative. In setting priorities, it usually makes sense first to select those projects that are most viable economically and then to check to see if they fit the financial and political criteria from the list given above. Only if a selected project fails either the financial or political test would it be modified or abandoned and a substitute selected.

In contrast to the continuing programs just outlined, major or new mass transit projects and improvements to existing ones often are financed by grants from the federal or state governments or by special local arrangements. In these instances, preliminary studies covering many of the points outlined above are

[11]Highway programming is far more exacting and complex than indicated here. Among many useful references are R. Winfrey, *Economic Analysis for Highways,* International Textbook Co.; 1969; *NCHRP Report 122* and *199; HRB Special Report 62; TRB Special Report 157; NCHRP Synthesis 48;* articles in *TRB Record 491, 585, 599, 680, 698,* and *742;* S. Rao et al., *Transportation Engineering Journal of ASCE,* May, 1977; L A. Neumann, ibid, Sept. 1980; and *TRB Circular 213.*

made and submitted to the granting agency. These, sometimes referred to as "alternative analyses," are expected to consider more than one approach and evaluate the benefits, costs, and environment impacts of each.[12] Also involved may be many activities necessary to assure matching funds. Setting the details of design and construction scheduling listed above under programming would follow completion of financing arrangements for part or all of the system.

Evaluating Alternatives

The foregoing discussion has dealt primarily with the difficulties in appraising the effects of proposed systems, routes, or individual links. At present, neither an approach nor data are available to evaluate all the impacts. Furthermore, so many factors are involved that rational judgments become difficult.[13] And yet, decisions as to whether to construct the proposed facility must be made.

One approach to making complex decisions is illustrated by Fig. 3–2 and Fig. 3–3. Figure 3–2 is a flow diagram of a decision process under which the quantifiable economic effects of each proposal are compared. Methods for making the economic analysis are outlined in Chapter 4. This paired-comparison approach has been criticized because it assumes that if alternative A is better than B and B is better than C, then A is better than C. But this dilemma occurs only if decision makers change their weighting of the various factors.[14] It should be emphasized that one alternative that should always be considered is the "no-build" one.[15]

Figure 3–3 suggests a procedure that can aid a decision maker or a decision-making group in weighing the comparative community effects of the various alternatives. The aim of such a diagram is to portray the many variables in such a form that rational comparisons can be made. It also provides a mechanism for defining areas of agreement and disagreement among contending parties. In this simple illustration it can be seen that, as far as the listed community factors are concerned, alternative 4 clearly dominates both alternatives 1 and 3. On the other hand, alternative 2 is somewhat better than alternative 4 in opening developable land and in its effect on the tax base. From the diagram, then, it might be possible first to secure agreement that alternatives 1 and 3 can be discarded and then center discussion on the pertinent differences between alternatives 2 and 4.[16]

[12]See P. Taylor and L. Howell, *Transportation Engineering Journal of ASCE,* Nov. 1977, for a report of an alternatives analysis for public transportation in Los Angeles.

[13]G. A. Miller, in his classic article "The Magical Number Seven, Plus or Minus Two," *The Psychological Review,* Mar. 1956, advanced the concept that short-term memory can accommodate only seven chunks or concepts at once. When more variables than this are present, some are ignored by decision makers or judgments become irrational. "Computer graphics" is one of the approaches being developed to overcome this human limitation (see Chapter 6).

[14]M. L. Manheim, *HRB Record 293* and *NCHRP Report 156* describe dynamic models for decision making and give extensive bibliographies.

[15]See *NCHRP Report 216* and *217.*

[16]T. E. Mulinazzi and G. T. Satterly, *TRB Record 518,* apply this "factor profile" approach to decisions about interchange conformations.

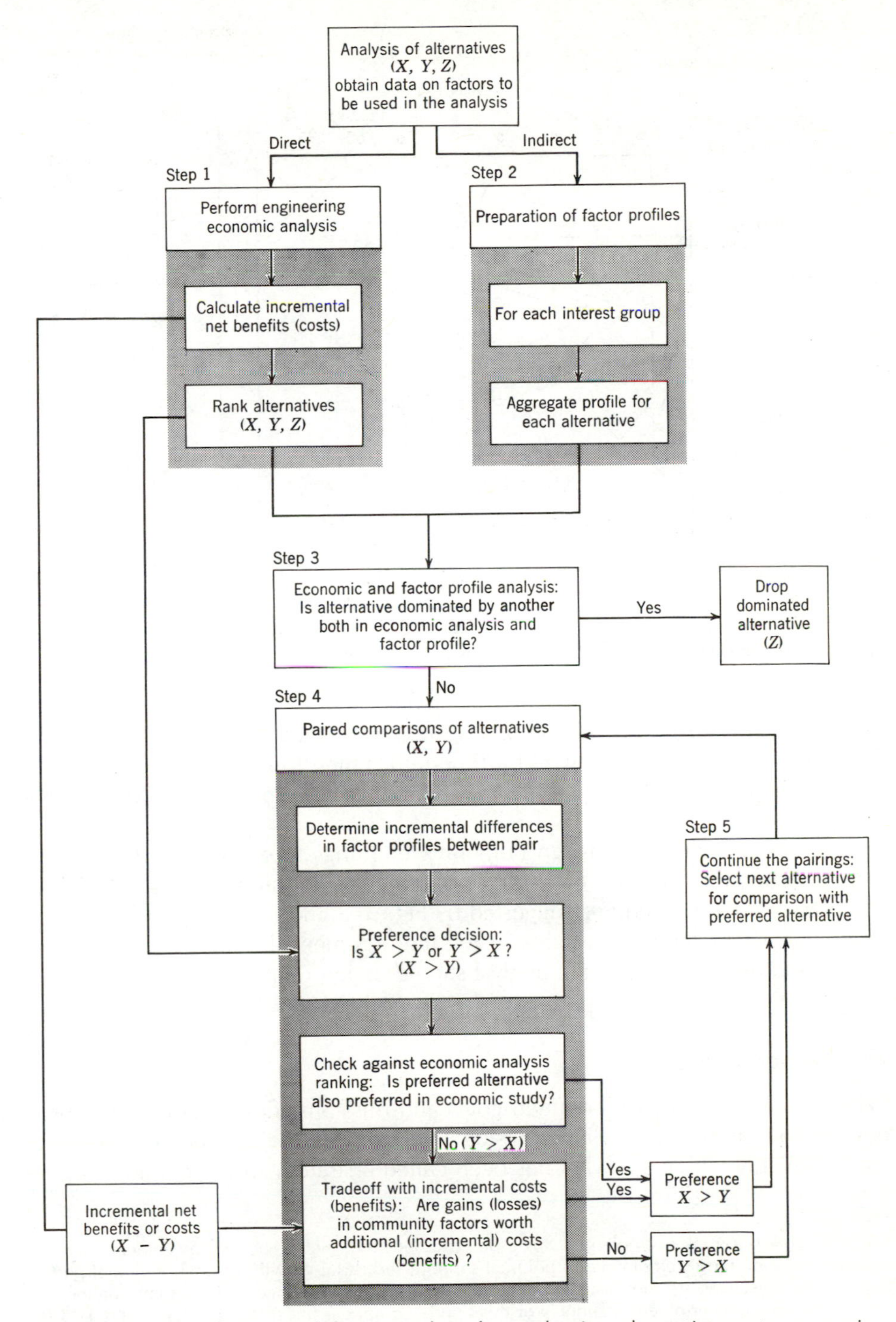

Fig. 3–2. Flow diagram of the procedure for evaluating alternative routes or projects. (After C. H. Oglesby et al., *HRB Record 305.*)

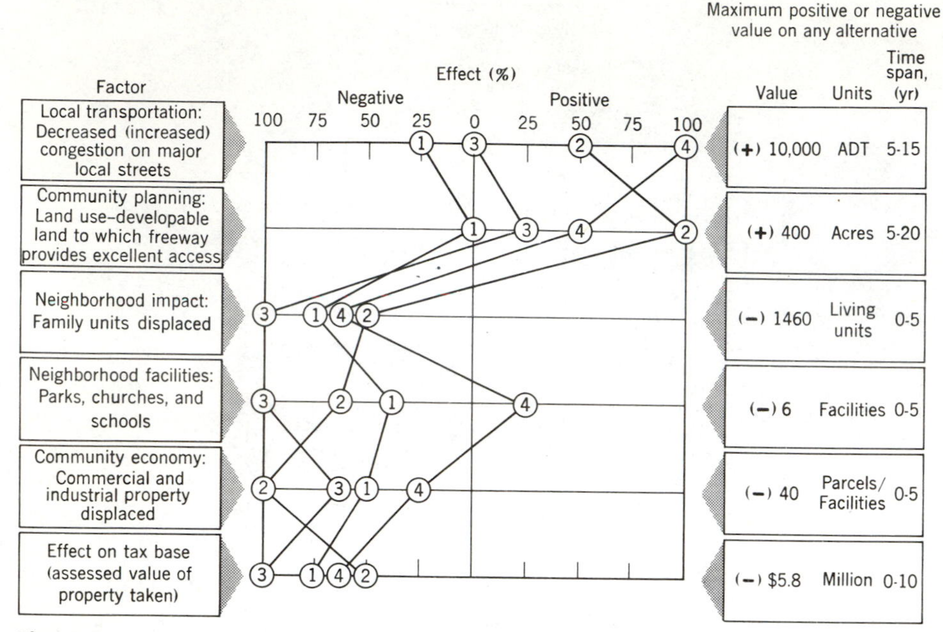

Fig. 3–3. Community factor profile—a method for comparing alternative transportation improvements. (After C. H. Oglesby et al., *HRB Record 305.*)

In the real world situations where a greater number of variables than those shown in Fig. 3–3 are to be compared, transparent overlays can be employed to compare the alternatives two at a time rather than four at once as the example shows. Or interactive graphics (see Chapter 6) can be used to facilitate comparisons.

Some analysts prefer to assign numeral weightings to each of the factors, including economic ones, as a way of simplifying the decision process.[17] It should be noted that the procedures suggested by Fig. 3–2 and Fig. 3–3 carefully avoid assigning overall numeral weights, using the argument that the decision maker or decision makers and not the analyst should assign relative weights to the various factors.

The Planner's Role

Where an agency has well-defined goals and missions, financing to carry them out, public support, and little outside interference, the planner or technical team probably will adopt what has been called a strategy of information or pos-

[17]Another way is to rank each plan by single-purpose objective such as access to centers of activity, maximum use of transit, employment, political support, industrial growth, or environmental protection. Other objectives are then listed in descending order of importance and, in some manner, an overall ranking is determined. Examples of this approach appear in *HRB Record 467* and *TRB Record 491, 574, 654, 680,* and 686. See also the section on Sufficiency and Deficiency Ratings and Priority Indexes in Chapter 4.

sibly a strategy of information with feedback as diagramed in Fig. 3–4*a*.[18] With such a strategy, decisions are made within the agency and any provision of information and the acceptance of feedback might be considered to be courtesies.

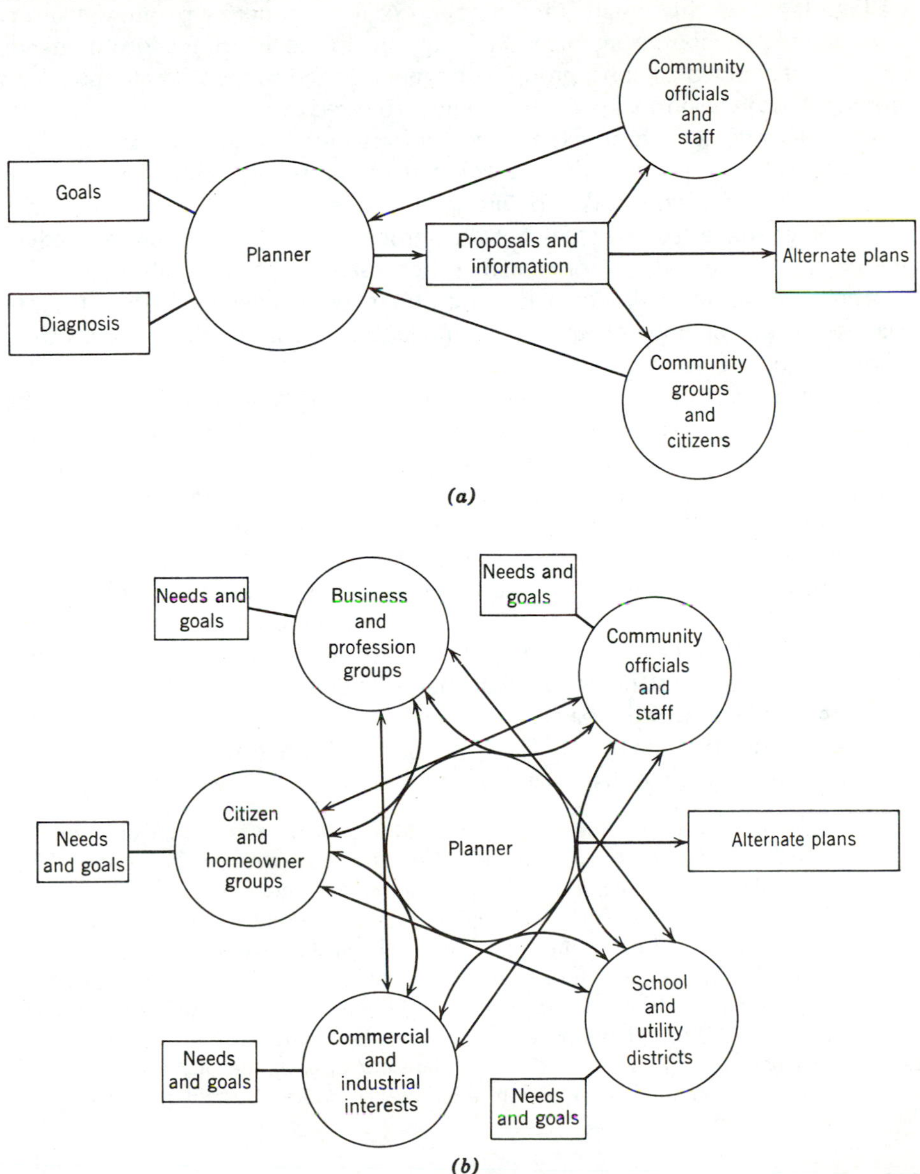

Fig. 3–4. Diagrams of two planning strategies. (After B. Bishop et al., *HRB Record 305*.) (a) Information with feedback. (*b*) Coordinator-catalyst.

[18]The effort to put a man on the moon is a classic example of this approach. This strategy also fitted highway planning until the late 1960s.

On the other hand, in instances where proposals may be challenged or blocked by other agencies or in the political arena or the courts, other strategies may be more appropriate. Among these are those of coordinator or coordinator-catalyst (see Fig. 3–4*b*). In this instance the function of the planner or planning team is to provide technical and organizational support, to receive inputs on the needs and goals of affected persons, groups, or agencies, and to incorporate them into planning and decision making. The catalyst role requires that arrangements be worked out under which an exchange of information on needs, goals, and alternative solutions of those who are affected are presented, and then, if possible, to aid the decision maker or group in reaching an acceptable solution. Among other suggested strategies not diagramed here are (1) community advocacy whereby an individual (ombudsman) represents all community interests or (2) arbitrative, which calls for a hearing officer who, after evaluating the approaches of the planner on the one hand and the community on the other, makes the final decision.[19]

The stances of planners or technical groups will be quite different under the various strategies. It would be logical where the strategy of information is used to take an authoritative, "I am the expert," approach. On the other hand, where planners are to be coordinators or catalysts or to work as ombudsmen or community advocates, they can be neither partisan nor advocate a particular solution. Their roles must be those of clarifiers, expeditors, conciliators, and impartial negotiators. They must avoid what is called the "Myth of rationality," which is to believe or cause others to believe that they think that, as professionals, they are uniquely qualified to adjudge what is best for society in their fields of competence. And they must recognize that their personal value systems and goals are not those of some of the participants.[20]

In the role of coordinator-catalyst, the suggested function of the planner or technical team during the planning process might be as follows[21]:

Step 1. *Study design*. Define the scope of the study and the initial work program. Establish the data base and acquire a basic understanding of the needs, plans, and objectives of potentially affected communities and citizens.

Step 2. *Exploration of the alternatives*. Continue data gathering. Establish contact with representatives of other affected agencies; begin to develop alternative project concepts and designs. There is extensive interaction with the public. It is hoped that, at the end of this stage, there will be heightened community understanding without affected groups taking hardened positions.

Step 3. *Detailed analysis*. Prepare detailed plans for appropriate alternatives through extensive community interaction; attempt to produce substantial agreement on alternatives.

[19]See B. Bishop et al., *HRB Record 305*.

[20]To assist their planners in maintaining such attitudes and to operate successfully, many agencies have developed manuals of instruction which suggest suitable procedures and conduct. One strategy involves the Delphi approach, under which confidential individual opinions are solicited and made known several times to participants as a means of arriving at a consensus. See, for example, D. W. Cravens et al., and C. M. Badger et al. in *TRB Record 563* and *617*, respectively.

[21]This framework is based on *NCHRP Report 156*.

Step 4. *Choice*. Secure formal ratification by the community and community officials and document the results. If all has gone well, this step is a formality.

From the viewpoint of professional ethics, planners must recognize that their role is to provide knowledge and an unbiased viewpoint. There is always the temptation to become partisan and emotionally involved. To do so can jeopardize the planners' credibility and, even more important, cause them to feel personal defeat and disillusionment with themselves and their profession if the solution they favor is not adopted.[22]

The Multidisciplinary Approach

The broad base for planning suggested here requires backgrounds and approaches broader than those that highway and transportation engineers have from their formal educations or on-the-job activities. Many transportation and urban planning agencies have added people from other engineering disciplines, archaeologists, architects, landscape architects, economists, political scientists, sociologists, and other specialists to their planning teams to expand their expertise and be sure that all facets of the various proposals are covered. Another aim is to represent the views of the several involved publics with their different sets of values, needs, and desires. This, done properly, generates public confidence in the agency's procedures and decisions.[23]

Citizen and Community Participation in Planning and Decision Making

To provide an opportunity for the public to be heard and provide its inputs to public works planning, it has been common to hold public hearings before major projects are approved. Such hearings on major decisions sometimes are required by law or by agency regulations. The criticism is often leveled that hearings are held after all major decisions have been made and are considered by officials to be a formality rather than an opportunity for community expression. However, particularly in the recent past, the objections raised by a coalition of opposition forces to proposed projects have been so strong that officials have found it necessary to reject the proposed plan or to make major changes in it. Thus, although public hearings may be required, their effectiveness in gaining community agreement and support has been seriously questioned.[24]

An alternative or complementary approach to public hearings is to involve the public from the time that planning begins. This approach, briefly summarized, is as follows:

1. Seeking out and soliciting the cooperation of public officials, influential individuals, and business, residential, or conservation groups or organizations that can speak for

[22]For further discussion, see J. H. Banks, *TRB Record 707*.
[23]The operations of in-house multidisciplinary staffs in four states are described in *TRB Circular 196*. See also J. R. Gordon, *TRB Record 551*, J. J. Meersman and L. Ortolano, *Transportation Engineering Journal of ASCE*, Jul. 1980, and *NCHRP Synthesis 40*.
[24]Two public hearings, one on the location and another on the design for Interstate routes have been required by federal regulation since 1969.

the community. Often a special staff is created to coordinate or carry out this function. Some agencies maintain continuous liaison with local community leadership rather than attempting to establish it when a project seems imminent. Again, surveys to determine community attitudes may be structured to identify informal as well as formal leaders. This approach is particularly valuable where the degree of influence of certain vocal individuals is unknown.

2. Creating the opportunities for this community leadership to participate continuously in planning, beginning at the earliest possible time.
3. Developing skills to organize and use constructive and continued participation of and with the public in group meetings, workshops, hearings, and many other activities.[25]

A classic example of a situation that led to an ongoing organization to promote citizen and community participation has been in Boston. There, in 1969, a combination of neighborhood and environmental groups and advocacy planners challenged the need for a section of the I–95 Interstate through downtown Boston. Based on the recommendation of a blue-ribbon task force created to make a re-study, the governor placed a freeze on highway construction inside Route 128, which circles Boston to the west on roughly a 10-mile radius. This study, which was set up to be participatory but decisive, multivalued, equitable, and to involve public participation by formal groups and in workshops, was established in a state office responsible to the governor. Ten percent of the study funds were allocated for community liaison, public participation, and technical assistance for these functions. The final recommendation and result was not to build the Interstate route and certain other highways but to expand transit. At the same time $1.1 billion of authorized Interstate funds were transferred to the transit program, as permitted by federal law.

Although the organization that reached this particular decision was disbanded, federal, state, and regional officials, wishing to maintain the highly successful participatory approach, institutionalized it into a Central Transportation Planning Staff under a state and regional Metropolitan Planning Organization. This group of professionals works under a steering group and provides services in systems analysis, design and environmental planning, policy and programming, community liaison, and area coordination to a Joint Regional Transportation Committee and its subarea forums and project committees. The aim of the program is to have centralized decision making based on sound technical approaches but, at the same time, to encourage inputs from well-informed citizens and local communities.[26]

[25]For a detailed listing and evaluation of techniques that have been valuable in developing citizen participation and a bibliography, see W. R. Torrey and F. W. Mills, *TRB Record 654*. See also articles in *TRB Special Report 142; TRB Record 528, 553, 555, 561, 617, 618, 654, 670, 677, 710, 731* and *747;* FHWA publications titled *Effective Citizen Participation* and *A Manual of Techniques for Citizens' Participation;* and *Guidelines on Citizen Participation,* AASHTO. Another source is the publications of the American Planning Association. For a review of a recent Federal study and regulations coming from it see *The Federal Register* Oct. 30, 1980 (Vol. 45, No. 212, pp. 71938ff). This states DOT policy on citizen participation and proposed guidelines which stipulate how any level of government that spends federal funds should provide for active and effective citizen participation in transportation planning.

[26]For details and a bibliography see J. S. Lane and K. E. Stein-Hudson, *TRB Record 654*.

Citizen and community participation is not a cure-all. All too often, administrators or planners have attempted to generate support for their plans by paying lip service to citizen participation while retaining the decision-making power. Almost without exception, this approach has had disastrous results, because such groups have rebelled against being used. The evidence seems clear that those who embark on the "citizens' participation" route must be wholeheartedly committed to it and be willing to accept not only inputs but even major changes in their proposals.

There are numerous examples of successes and failures of citizen and community participation in the literature.[27] Many of them indicate that it is difficult to gain attention unless the impacts are large and negative, can be directly identified, are concentrated rather than diffuse, are short-range rather than long-range, and create threats or a feeling of being victimized.

Institutional Constraints

In the United States, facilities and vehicles for transportation are provided in a variety of ways. Considering only highway-based situations, private owners provide almost all of the automobiles and trucks while government furnishes the bulk of the public transit vehicles. In sum, the provision and operation of these vehicles consume about 92 cents of the highway dollar. Highways, which take most of the remainder, are constructed, operated, and maintained by state and local governments. As indicated in Chapter 5, there are substantial grants of federal money. These federal funds and state subventions to local agencies are largely allocated by formula. In contrast, federal and state funds for public transportation go directly to urban transit agencies, often in the form of individual grants for specific purposes. Under these conditions it is to be expected that all granting agencies, politicians, and the public will demand a voice in how the funds are to be expended.

Superimposed on the wishes of and controls imposed by those concerned with transportation per se are the missions of and powers granted to other governmental units. For example, land-use controls primarily rest with local planning agencies which, as illustrated in Fig. 3–1, must deal with many considerations other than transportation. Again, transportation plans come under the scrutiny of agencies carrying out legislation and other directives about air, water, noise, visual, and other impacts of transportation on the environment. Although agencies such as these usually do not have control over financing, they can veto or force changes in plans through regulation. Yet another set of controlling influences comes through court or other actions by individuals, groups, or organizations. Common legal approaches are to challenge enabling legislation or regulations, to demand that an environmental impact statement be made, or to protest its adequacy if one has been done.

[27]See, for example, H. M. Steiner, *Conflict in Urban Transportation,* Lexington Books, Lexington, Mass., and *TRB Circular 205*. Much also can be learned from literature covering other areas of controversy in journals such as *Public Interest* and *Policy Analysis*.

Planning, to be effective, must recognize and deal with institutional constraints such as these and others not mentioned. As indicated earlier, this has been possible where a common purpose exists, as with construction of the rural Interstate System in the late 1950s and 1960s. There were, of course, differences in opinion among federal, state, and local interests over financing, location, and other details, but these were resolved and the system largely completed. However, when consensus is lacking as, for example, over the desirability of completing certain segments of the urban Interstate System, institutional constraints often block implementation of proposed plans or delay them for long periods of time. Issues such as funding, land use, regulation imposed on one governmental agency by another, and court actions, with these at more than one level of government, can all come into play. Then it becomes very difficult to carry major projects through, even with the most carefully developed approaches. Rather, today's focus is largely on small scale, short-range solutions, typified by those discussed later in this chapter under the headings of Shorter-Range Planning and Transportation System Management.[28]

TRANSPORTATION PLANNING AND THE URBAN PROBLEM[29]

The Past and Present Situation in Urban Transportation

Congestion has been cited as evidence that all is not well with our urban areas. But congestion is not a new phenomenon. For example, some historians have attributed the decline of Rome to it. In 1635 the high cost of living in London was attributed to extreme congestion and the costly delays in bringing in hay and provender; and even the London fire of 1666 did not improve the situation since reconstruction kept the old antiquated street layouts. Successive improvements in transport means, such as the electric street car followed by the automobile and motor bus, and now freeways and sophisticated traffic control schemes, have done little to alleviate the problem. Today, all large cities in both the developed and developing nations still have congestion problems, at least at morning and evening peak hours. But there may be at least one important difference from the past in the developed nations. In earlier years there was "poverty and congestion"; today there is, to a degree, "affluence and congestion," complicated by energy considerations. It follows that resources probably can be made available to apply corrective measures, but effective stratagems are yet to be proved and the political means to carry them out are only now evolving. And it may be that eliminating all congestion is impossible, since it seems to accompany healthy economic activity.

Today many planners, politicians, and others attribute urban congestion and

[28]For a detailed report and 18 case studies on strategies for dealing with institutional problems in cooperative efforts to reduce congestion, see *NCHRP Report 205*. See also *TRB Record 731*.

[29]An excellent reference is *Public Transportation* by G. E. Gray and L. A. Hoel, Prentice-Hall Englewood Cliffs, N.J., 1979. Other references, only a few of which are cited here, appear in, among other places, publications of TRB, ASCE, UMTA, FHWA, and ITE.

other ills of our cities to the private automobile and the construction of highways to serve them. Underlying this has been prosperity. After World War II, greater affluence made automobile and home ownership feasible for most of the population and brought the flight away from deteriorating close-in areas to the suburbs with their dispersed living and low-density land use. Then, to serve these suburbs, private investors brought shopping and other services and public agencies provided highways which made motor vehicle travel easy both downtown and around the circumference. In addition, newly developing light industry located in the suburbs, and many existing firms found it advantageous to leave the downtown areas.[30] The incentives in each case were cheaper land with easy, quick access and space for parking, a favorable tax situation, and attractive suburban living which made a supply of skilled labor available. For such reasons as these, many urban areas are almost completely oriented to the automobile. This is dramatically demonstrated by the statistic that, nationwide in 1976, private expenditures for transportation went 97% to the private auto, 2% to transit (*not* including its roughly equal public subsidy), and 1% to taxicabs.

Generalizations based on average statistics such as those given above, although they portray the overall situation, must be viewed with caution in evaluating individual situations. For example, in the United States as a whole, motor vehicle ownership, averaging about 1 for every 1.8 persons, ranges from about 1 for every 1.3 in predominately rural states to 1 for every 2.8 in the District of Columbia, which is almost wholly urban. In major European cities these ratios range from 1 to 3.2 in Munich to 1 in 5.2 for Copenhagen. London is 1 to 4.7.

Certainly availability of an automobile affects transit use. Again, 70% of the nationwide work trips are by auto; but for the larger, older, and denser urban areas, transit work trips are far more numerous. The 1970 census shows work trips by transit as follows: New York, 61%; Boston, 38%; Philadelphia, 37%; and Chicago, 36%. These cities all have rail as part of their systems. At the other extreme, work trips by transit were 18% in Detroit, 9% in Los Angeles, and 8% in Houston. Data for the San Francisco Bay area taken soon before the 1979 gasoline shortages and price increases show rail rapid transit (BART) carrying 5% of the to-and-from-work movements, buses and street cars 18%, and motor vehicles 77%. Transit use in the 14 largest metropolitan areas accounts for 70% of the nation's transit passengers, and New York alone accounts for 38%. These percentages are much lower for the more than 400 all-bus systems operating in the urban areas which have populations over 500,000 and the more than 500 systems in smaller communities.

As indicated, the shift to the private automobile began after World War II. This shift was not very great in the established, large, older cities, but was particularly strong in urban areas with populations under 500,000. Once started, the shift became a vicious circle. Comparing 1950 to 1972, the low in transit ridership, patronage dropped 62% in spite of a 37% population increase. There

[30]For a discussion of freeway locations to fit this pattern, see Chapter 6.

has been a 14% increase from 1972 through 1978 in comparison to a 7% population increase, but transit patronage is still far down from early levels. Over the last three decades, then, reduced patronage has led transit agencies to cut services. A second difficulty has been that transit operations suffered financially from tripled labor and equipment costs but recouped only 55% of them through fare increases, which were held down by both public and market pressures.[31] These cost increases resulted in part from inflation, in part because capacity had to be provided to accommodate the costly peak hour home-to-work movements while off-peak use fell substantially, and in part because of demands that service be provided at a loss to a dispersed suburbia and to the disadvantaged. Also, patronage fell because of the five-day and even four-day work week, suburban rather than downtown shopping, and the loss of night business attributable, at least in part, to television. Finally, community officials and the public expected transit to operate profitably or at least to break even and, at times, to pay taxes. In sum: in the United States, as of 1981, although transit use has increased, the industry is still financially distressed. Both privately and publicly operated companies are usually heavily subsidized by a reluctant public.[32] It is not yet certain to what degree the gasoline shortages and price increases that began in 1979, along with environmental concerns, will affect transit use and financing. However, there seems to be little chance that it will again be self-supporting.

The refusal of many Americans to use public transportation when it is available may not seem entirely rational. Often it is cheaper, nearly as or equally convenient, and does not carry the attendant problems such as parking and need for a second car for family use. And yet Americans, "en masse," still seem willing to pay quite dearly for the independence and the increments of convenience, freedom of movement, and time saving (often not realized) offered by the private automobile. From a public standpoint, also, use of the private automobile is hardly rational. It is extravagant of valuable street space, using three to six times as much as mass-transit vehicles[33] and it adds to difficult congestion and parking problems. Yet, in a democracy such as the United States, public officials have been almost powerless to check this apparently wasteful process. Proposals to apply "transportation pricing" to make driving more costly have been made (see Chapter 5) but to date have been applied in only a few instances.

Present and Future Urban Form[34]

Looking to the future, many urban planners approach transportation schemes as one of several mechanisms that can create new and satisfying urban settlements

[31]These and many other transit statistics are from the 1978–1979 *Transit Fact Book,* published by the American Transit Association. See also *TRB Record 759*.
[32]Of the $4.7 billion in revenues of all transit operations in 1978, 52% were from fares and other income sources, and 21%, 12%, and 15% from local, state, and federal sources, respectively.
[33]See. W. P. Walker and R. A. Flynt, *HRB Bulletin 167*.
[34]For a review of the development of cities over time and a discussion of many of the planning issues see A. B. Gallion and S. Eisner, *The Urban Pattern,* Van Nostrand, 1975. See also the publications of the American Planning Association.

with an improved "quality of life" and also as an aid in restructuring existing urban areas. They accuse transportation planners of thinking only of moving people and goods. In their view the primary problem becomes one of defining the urban form or forms that will function to fulfill human aspirations. The second step is to plan the facilities, including transportation, that will permit these urban forms to function effectively. Finally, governmental mechanisms will be created to coordinate and control public and private investments so that these urban forms and transportation to serve them can come into being. This approach will, of course, call for area-wide community planning and the development of either incentives for or restrictions against private investors and local government to a far greater degree than has existed in the past. There is evidence that such moves are being undertaken. For example, beginning in 1972, the Department of Transportation, using federal funds as the incentive, called for unified comprehensive plans for urban transportation before federal grants would be made for highway, airport, or mass-transit projects. Many other governmental actions at all levels have similar objectives.

There is no agreement among planners, politicians, or the public in general as to what new or modified urban forms would be most satisfactory. Concepts for them have been classed as strong-core, satellite, lineal or radial, and multiple center.[35] Strong-core schemes might even concentrate large populations in single structures in which they would live, work, and find educational, recreational, and cultural activities. One such scheme[36] envisions "arcologies," each of which could accommodate a population of several million. A first stage is to accommodate 5000 people on 15 acres under a single glass roof. For the ultimate plan, buildings would be several times higher than any in the world today with transportation by elevators, escalators, and moving walkways. Automobiles would be stored and used only for travel in the open space between arcologies or for vacations or similar purposes. Another proposal[37] employs the systems approach to urban design based on more effectively employing the third dimension (up and down) and the time dimension (around the clock) use of facilities. Each settlement would provide building space on eight circular platforms 8840 ft in diameter, spaced 30 ft vertically. Homes or apartments, including gardens, and facilities for industrial and commercial enterprises and for recreation could be built on the platforms. Vertical movement would be by ramps or elevators; horizontal travel could be on foot, bicycle, or in a small vehicle. All vehicle and pedestrian movements would be separated. Proponents argue for such proposals on economic grounds; they insist that savings in combined land, building, and transportation costs would more than offset the cost of the structure.

Any advanced scheme for large self-contained urban complexes must include

[35]See, for example, E. W. Walbridge, *TRB Record 658*, and R. G. Rice, *TRB Record 677*.
[36]Paolo Soleri, *Arcology, the City in the Image of Man*, M.I.T. Press, Cambridge, Mass., 1969, and *Engineering News-Record*, Aug. 17, 1978
[37]See G. B. Dantzig and T. L. Saaty, *Compact City*, Freeman, San Francisco, 1973.

provision for the movement of goods, which today account for roughly one-fourth of all weekday traffic movements. Proposals for goods handling, in addition to trucks, have included freight vehicles using passenger transport rails or guideways, special traffic tunnels, conveyers, and even large, air-actuated pipelines.[38]

Wilfred Owen, who refers to urban areas throughout the world as "accidental cities" or "unmanageable metropolises," proposes "regional cities" which are in the middle ground. They would be made up of interconnected high-density clusters surrounded by low-density land uses such as new towns (see below). Transportation in high-density areas would combine walking with people movers, elevators, and escalators; in low-density areas the automobile and bus would predominate.[39]

The *new town* concept which was implemented in Great Britain before World War II has urban satellite communities near to and associated with large urban centers. They are usually planned to be largely self-contained, with employment and all community services provided, but with easy access to the central city by rail or expressway bus. A number of such communities have been constructed with limited acceptance in the United States; successful ones include Reston, Va., and Columbia, Md., near Washington, D.C.[40] In Europe much of the population growth for such cities as Stockholm, Rome, and Belgrade among others is being cared for in such satellite communities. Canberra and Brasilia, capitals of Australia and Brazil, respectively, are other examples. Egypt has undertaken Sadat City, with a population of more than one million.[41]

In England, 33 new towns had been authorized by 1973. An example is Milton Keynes, a community some 40 miles northwest of London. It is designed for an ultimate population of 200,000 on 22,000 acres. Stations for a work force of 120,000 will be partially near the town center and partially on the perimeter; 90% of the families will own at least one car; total passenger journeys will be 80% by automobile and 20% by public carriers, but many short trips will be by walking on fully separated pedestrian ways. Provisions for transportation will use 11% of the total land area, as contrasted to 25 to 30% with the traditional grid street pattern. Housing will be in clusters of row cottages, 10 to 15 per acre, with common playgrounds; the areas between clusters will be open space.

To plan new communities effectively, there should be a reasonable expectation that the proposed plan will attract residents of the intended income and racial groups and mixes. To date, too little is known to predict such outcomes. And predicting them from past experience in the United States or Europe is dangerous. In the United States, it has already been demonstrated that some

[38]See, for example, J. J. Fruin, *Transportation Engineering Journal,* ASCE, Aug. 1972.

[39]See W. Owen, *The Accessible City,* Brookings Institution, 1972.

[40]As of 1978, 6 of the 13 financed in part by federally guaranteed bonds were judged by the government to be capable of continued development. Bonds on the remainder were paid off and no new federal support was proposed.

[41]See, for example, W. Owen, op. cit.; *Transportation Engineering,* Sept. 1977; I. W. Morison and W. G. Hansen, *HRB Record 229;* R. L. Morris, *HRB Record 293;* and *TRB Circular 199* for detailed information on new towns.

schemes for the disadvantaged, whether in new towns or by redevelopment, were not successful because they did not match the resident's expectations and modes of living. Successes that have been claimed, particularly those involving very small living units, may reflect the critical need for living space; they may become less acceptable as supply approaches demand.

Any such plans for new cities will have little impact on overall urban problems. Only a few new cities will spring full-blown in the near future. Rather, much of the activity in planning urban areas in the years ahead will be directed toward redeveloping and reconstructing deteriorated sections of our cities and carefully planning and possibly controlling urban expansion where this is desirable. This planning will be difficult because of the many and often conflicting goals of a variety of interests. First of all, although there are exceptions, most Americans seem to prefer owning a single-family home on its own plot of land in the suburbs as compared with occupying high-rise dwellings downtown. Next, those in the private sector feel free to profit from land development. This often involves expansion of housing, industry, shopping, and accompanying activities into areas on the urban borders.[42] Again, it may be more profitable to owners to let close-in property lie vacant or in a run-down condition than to develop it for business, industrial, or residential use. On the other hand, goals of the federal, certain state, and some local governments and numerous conservation and environmental interests have attempted to make near-in urban areas more compact and to limit expansion. Arguments for this approach include economic efficiency, cheaper public services, energy and land conservation, and reduced air pollution. Governmental strategies include inputs of highway and other money to abate air and noise pollution, rehabilitate run-down areas, and for joint development (see Chapter 6).[43] Even so, projections are for a continued decrease in urban population density in the years ahead.[44] It seems clear that no unanimity on policy for urban land development has emerged to date, and that, as discussed earlier under the heading of Institutional Constraints, any change of direction will be difficult and slow.[45]

Public Transportation and Urban Form

Added to the confusion over the relative desirability of different urban forms is the role of transportation in influencing them. Many planners and public officials believe that urban congestion can be reduced and other desirable public purposes achieved by a combination of transportation-related strategies. One is

[42]It has been estimated that 80 to 90% of private investments are outside city centers.

[43]For a recent study, see A. Politano and F. Mills, *TRB Record 747*.

[44]See, for example, *National Transportation—Trends and Choices,* U.S. DOT, Jan. 1977.

[45]Examples of the conflicting effects of governmental policies on land development abound. For example, the federal government, by guaranteeing loans and granting income tax deductions on interest payments, has encouraged individual home ownership, particularly in the suburbs. This negates efforts of other programs to improve central city housing. Again, in many areas, local property taxes on vacant land or dilapidated structures have been so low that owners find it advantageous not to undertake improvement of redevelopment, even with federal incentives. Among other instances are the taxing of fringe-area farm lands as though they were developed, thereby making farming unprofitable and subdivision the only feasible alternative.

to make cities more compact, thereby reducing the need to travel by a conveyance of any kind; another is to make transit more effective; a third is to restrict the use of the private automobile. The thrusts here would be first to improve public transit facilities and service; and second, to favor transit by using such strategies as (1) giving priority in traffic, (2) restructuring subsidies and using other pricing to make transit use financially attractive, (3) setting high parking and other automobile fees, (4) establishing automobile-free zones accessible only by transit or walking,[46] and (5) aggressively marketing transit.

There are those who are skeptical that approaches like these can reduce urban congestion or make our cities denser. They point out that efforts to date have been instituted primarily for political or prestige reasons, have taken a long time to carry out,[47] and may be ineffective if not counterproductive, because their effects are not understood. For example, a recent study by the Transportation and Road Research Laboratory found that travel times to work in cities in Europe had remained almost constant for 200 yr. This might be interpreted to mean that, because of the way people respond, better transportation by whatever means will result in dispersion. Does this mean that congestion must be maintained if cities are to remain compact? Furthermore, the notion that consequences will be the same in different cities has been challenged. For example, it was assumed that the effects of improved rail transit in Toronto would also apply in the San Francisco Bay Area. In Toronto, heavy business activities and denser residential development occurred in the zones of influence of the transit stations. Because of many differences, including growth patterns, climate, governmental structure, and greater competition from the private automobile on freeways, these changes, when they occurred at all in San Francisco, were less marked. Furthermore, in the San Francisco case, proximity to stations had an adverse effect on rehabilitation in and improvements to older neighborhoods. And, although proximity to the system stimulated a net increase in new housing, much of this was for single families in areas that earlier had been beyond reasonable commuting distance to the two principal central business districts. The net effect at these locations was dispersal and lower density land use. Also, transit use has been disappointing; to date it accommodates only 5% of the peak hour trips.[48] This is not to say that transportation systems cannot be employed to affect urban form, but that to date knowledge of the many influences and their effects are lacking so that outcomes cannot be predicted.[49]

Substantially reducing automobile use in favor of transit will be a slow proc-

[46]See, for example, C. Heaton and J. Goodman, *TRB Record 747*.

[47]To initiate major transit projects in the United States can be very time-consuming. For example, in the San Francisco Bay area a citizens' group was formed in 1952; the California legislature created the Bay Area Rapid Transit District (BART) in 1957; the citizens voted a bond issue guaranteed by the local property tax in 1962; and the first trains ran in 1972; 20 yr in all. The time required in Atlanta was almost this long.

[48]For reports on the impacts of BART on land development, travel, and planning see respectively M. V. Dyett and E. Escudero, *Transportation Engineering Journal of ASCE,* May 1978, and A. Sherret and J. Markowitz and L. S. Graebner and T. Higgins, ibid., Jul. 1978.

[49]For recent discussions see, among other sources, *TRB Special Report 181* and *183, TRB Record 634* and *677*, and *TRB Circular 199*.

ess, although it may be hastened by energy and environmental concerns. Among useful steps that can be taken is to preserve land corridors for projected transportation facilities.[50] This and similar actions calls for a high degree of cooperation among governmental agencies or their restructuring to get a unified approach to decision making. To date this cooperation or restructuring is happening very slowly. Furthermore, the planning scope must reach travel problems beyond the close-in urban area. For example, the land-side movements to and from airports have extremely high volumes and yet today are often poorly served by local transportation.[51] Finally, the influence of telecommunications, which can drastically reduce the need for central-city travel, must be carefully considered.[52]

Urban Goods Movement and Urban Form

Until World War II, raw materials entering urban areas were usually carried by railroad to near-in factories. Finished products not consumed locally then left by rail. Goods for local consumption also came by rail to warehouses near the downtown area. Distribution was at first by horse-drawn vehicles and later by small truck to nearby retail outlets. Now the outward explosion of cities and the motor truck have changed all this. Far fewer rail deliveries go to close-in factories or warehouses. In fact, in some cities rail lines and their downtown terminals have been abandoned. Instead, railroads and, increasingly, heavy trucks deliver goods originating elsewhere to factories or terminals located in outlying areas. Such products as now enter the close-in areas are first transferred to smaller, more easily maneuverable vans and trucks for delivery. But this is often difficult (see the discussion under parking in Chapter 10). At the same time, outlying shopping centers have become more competitive as costs of serving them with goods from other areas or outlying warehouses are lowered. In sum, change in the costs of moving freight and other goods in urban areas has been a strong force in suburbanizing our cities. And this may be a major obstacle to efforts to make the cities more compact.

PRESENT AND PROPOSED USES OF AND FACILITIES FOR URBAN PUBLIC TRANSPORTATION[53]

The Role of Public Transportation

Although, as noted earlier, public transportation presently is in financial distress, it serves functions that must be maintained. Much urban movement, par-

[50]See, for example, J. E. Leisch, *HRB Record 293.*

[51]For further discussion see *TRB Special Report 159* and *TRB Record 732.*

[52]See, for example, P. Gray et al., *Transportation Engineering,* Nov. 1977.

[53]For a far more detailed treatment of this subject see G. E. Gray and L. A. Hoel, *Public Transportation,* op. cit. TRB publications, particularly *Circular 212,* and those of the American Public Transit Assn., including its periodical *Transit Journal.* A detailed description of the transit system of Toronto appears in *ITE Journal,* Aug. 1979.

ticularly home to work, depends on it. Furthermore, some 50 to 70 million persons, primarily among the young, the infirm, the aged, and the financially disadvantaged, are totally dependent on it. There are also governmental policies directed at desirable social ends which are transportation-based. Among them are congestion reduction, urban consolidation and renewal, and energy conservation. All aim at getting some of the 95% of travelers who choose to go by auto to shift to transit. Because public transportation serves these many purposes, there is at least recognition that, if necessary, it should be subsidized; seldom can it operate as a profit-making venture and at the same time carry out so many money-losing public services.

Recently, governments at all levels have become involved in public transportation. The federal government has supplied and reluctantly may continue to provide large sums of money to local transportation agencies to help fund entirely new and sometimes grand schemes, to augment local financing so that existing operations can be salvaged or expanded, to buy equipment, and to carry out research and demonstration projects. State, regional, and local governments and a variety of special transit agencies also provide or distribute funds, carry out or participate in planning activities, or operate transit systems.

As of today, public transit is predominately a governmental function. Nine-tenths of the services are supplied by governmentally owned agencies. Buses and paratransit vehicles such as vans and taxis all travel on government-supplied highways and streets. In 20 of the largest urban areas, governmental agencies operate heavy, commuter, or light-rail facilities, augmented by buses.

The paragraphs that follow briefly describe the various transportation modes and their comparative advantages and disadvantages and discuss how they fit into the overall present and future transportation framework.

Buses

GENERAL ADVANTAGES AND DISADVANTAGES. The motor bus has both advantages and disadvantages when compared with heavy- or light-rail transit. It can share highways and local streets with the automobile and truck so that the cost of a separate traveled way can be avoided. Again, buses can travel on almost all highways and streets and so can be routed near enough to the riders' origins and destinations that they can walk the short remaining distances. Routes can be added, dropped, or modified easily to meet changing or special demands. In contrast, the routes of rail systems or trolley coaches are fixed and changes are costly and time-consuming. With heavy rail, where access is only at widely spaced stations, particularly in the suburbs, an automobile, bus, or bicycle often must be used to get to and from the station. Also, the passenger must leave one conveyance and wait for another. In some instances, buses may be more crowded and less comfortable. Furthermore, buses commonly must compete with the private automobile for space on congested arteries, while heavy and sometimes light rail travels free of other traffic.

Buses also have advantages and disadvantages when they are compared with an automobile that is driven alone or in a car or van pool. Advantages can include lower costs and relief from the frustrations and strains of driving or riding in heavy traffic. But the controlled timetable and fixed routes of buses, or waits of uncertain length with demand-responsive systems, do not offer the freedoms provided by the automobile or van. Commonly, the origin to destination time by bus is longer. Also, as noted, buses may be crowded and uncomfortable and offer no privacy, although with air conditioning and in some cases subscription service (reserved seats for a premium fare), some of these difficulties have been overcome.

Bus operations are not capital intensive; rather, seven-eighths of the annual outlays can be classed as operating expenses. This is in marked contrast to rail where capital investments are high but, once in, require less labor. Of course automobiles, being privately owned, require no public investment except for highways.

BUS SELECTION. Almost all buses, excepting some minibuses, are powered by fuel-efficient, time-proved diesel engines. Features vary: no one size or body conformation is best adapted to all applications, so that many styles exist in the more than 50,000 in use in public transit in the United States.

"Standard" bus lengths are 35 and 40 ft, and widths are 96 or 102 in. The wider ones, where permitted, give more comfortable seats and wider aisles. Seated capacities are, respectively, 41 to 45 and 49 to 53 passengers. Minibuses, which commonly seat 10 to 25 persons, are not a standard product and a wide variety, commonly based on truck or mobile home undercarriages, are in use.[54] As pointed out in Chapter 8, European high-capacity vehicles, for example the London double decker and various articulated designs, with up to 130 spaces, are gaining increasing attention in the United States.

Commonly heard arguments favoring minibuses over standard ones, particularly for demand-responsive uses, are maneuverability and low cost. Where street space is limited, maneuverability may be a problem with bigger units. However, the overriding concern may be the image of wastefulness with a large-capacity vehicle carrying only a few passengers. Actually, with drivers' wages at two-thirds or more of operating costs and other operating costs rising at a slower rate than capacity, cost savings with minibuses are not as substantial as they might appear.

Factors far too numerous to discuss here enter decisions on bus selection. One is the time required for passengers to load and unload. This is extremely important in downtown short-ride operations. Often doors at the rear or at mid-length are added to speed the process.[55] Arrangements for fare collection and controls over locations for entry and exit also can markedly affect operating efficiency.

Attention to the needs of the elderly and handicapped has led to the devel-

[54]See articles by B. D. Revis and G. Grimes, *TRB Record 696* and *TRB Circular 212*.

[55]For studies of bus loading and unloading times see *NCHRP Report 155*, W. H. Kraft and T. F. Bergen, *TRB Record 505*, and *TRB Circular 212*.

opment of a low-floor design "Transbus", which can be entered and left more easily. Early estimates were that it would cost possibly 50% more than conventional vehicles. In 1979 UMTA stipulated that buses purchased with federal funds be of this type. However, none of the manufacturers submitted bids for supplying them.[56] One of the issues in the years ahead, then, will be the extent and kind of arrangements made to provide services for such people. Some have suggested that it can be done better and more cheaply through special arrangements such as demand-responsive vans or taxis.

LOADING AND UNLOADING. The cheapest arrangements for loading and unloading bus passengers are at roadside or curbside locations. These may be midblock or at the near or far side of intersections, as determined by operational plans or street capacity considerations. Off-street loading or transfer lots or buildings are often provided. Favorable locations are in or adjacent to the downtown or shopping centers, or other "park and ride" or "kiss and ride" sites (see Chapter 10). In some cities, malls exclusively or almost exclusively for buses are being constructed. Shelters or other waiting areas at these off-street or mall locations provide a convenient place to give information on routes and schedules. Similar arrangements often are provided at transfer points from car or bus to rail transit.[57]

IMPROVING FACILITIES AND STRATEGIES FOR BUS OPERATION. Buses operating as a part of normal traffic streams suffer the same delays as other motor vehicles. However, many strategies are being employed to favor them. A few are described briefly here.

On freeways the most effective way to speed bus and possibly carpool traffic is to provide separate lanes called busways. Usually they are two lanes in the median area or possibly alongside the freeway. They operate only in the peak direction. Barriers prevent encroachment of other vehicles. Access to and egress from the lanes are by special ramps. A principal drawback is the very high first cost; another is the objection of motorists who observe the largely unoccupied lanes while they drive stop and go in the abutting freeway lanes.

In some cases, automobiles and vans carrying two three, four, or more occupants are allowed on busways. Some surveillance to control unauthorized entry is necessary.

Busway installations include, among others, the ones on the Shirley Highway in Virginia south of Washington D.C. and those on Interstate 10 east of Los Angeles. On the Shirley Highway in the peak hour 450 buses and 3500 carpools carry 30,000 people at speeds two to three times as fast as the other lanes.[58]

An alternative to permanently reserved lanes is possible where peak hour freeway traffic flows are very imbalanced. Then it has been possible to use the inside lane in the low-flow direction for buses traveling with the heavy flow.

[56]For an analysis of the effects of Transbus on operations see R. Casey, *TRB Record 746* and S. E. Polzin and J. L. Schofer, *Transportation Engineering Journal of ASCE*, Sept. 1979.
[57]See Gray and Hoel, op. cit., for added detail and references.
[58]See J. L. Glazer and J. Crain *TRB Record 718*, for a report on the Los Angeles busway.

Where four lanes are available in each direction, the second lane has served as a cleared area. Temporary markers, hand placed and removed morning and evening, serve to mark the reserved lanes.

In several experiments, the inside lane of an existing freeway or a wide median shoulder has been assigned to buses and carpools at peak hours. Results have been mixed. There is, of course, the high accident potential because of the large speed differential with the adjacent lane and as vehicles enter or leave the reserved lane. Also, users must weave through the other traffic when entering and leaving. Public acceptance has been mixed; in some cases violent protests of the other travelers or their representatives have led to abandonment or court action. In others, there has been little protest. In Los Angeles, one plan was given up following court action which cited the lack of an environmental impact statement. Experience to date indicates that public acceptance of reserved lanes probably can best be gained by adding to existing facilities, not by restricting already established usages.[59]

Another way to favor buses and possibly carpools traveling freeways is with priority entry ramps. For example, an existing or new ramp can be reserved for them. Again, an approach ramp may be striped two lanes, one solely for them. At the head of the ramp they are cleared onto the freeway before the other vehicles. Sometimes otherwise closed ramps are opened to favored vehicles or, where ramps are metered (see Chapter 10), priority vehicles go to the head of the waiting queue. Priority lanes have also been set up at the approaches and through toll gates for bridges and tunnels. Some take only buses, others permit carpools of predetermined size.

With either reserved lanes or ramp control, detecting and apprehending violators is a difficult and largely unsolved problem. Police are few in number and have many other concerns. And often the courts are lenient with offenders.

On conventional streets also, a variety of strategies are employed to favor buses and, at times, carpools and taxis. One is to reserve a curb lane exclusively for them. At times, in order not to disturb existing one-way traffic patterns and bus routes, this flow may be opposite to movement in other lanes. Also, as noted in Chapter 10, ways by which bus drivers can alter traffic signal timing are becoming common.[60]

Laying out bus routes and setting schedules and fares for them involves a variety of technical problems associated with the physical layout and capacity of highways, streets, and intersections, and setting realistic travel and dwell times. Arrangements for transfer between routes also must be developed. Then there are the tasks of selecting, purchasing, and maintaining the equipment fleet. Providing drivers, route supervisors, and maintenance personnel leads to many dif-

[59]For reports on reserved lanes see J. W. Billheimer, *TRB Record 663,* K. G. Courage et al., and M. P. Cilliers et al., *TRB Record 682* and *722,* respectively; and J. W. Erdman and E. J. Panuska, Jr., *Traffic Engineering,* Jul. 1976.

[60]Some of the many reports on approaches for expediting bus operation include D. A. Morin in Gray and Hoel, op. cit.; *NCHRP Synthesis 69; NCHRP Report 155; TRB Record 546, 606, 626, 630, 663, 746* and *761; TRB Special Report 153; Transportation Engineering Journal of ASCE,* Nov. 1976 and May and Jul. 1979; and *Traffic Engineering,* Nov. 1975 and Mar. 1976.

ficulties.[61] It should be clear that transit management, whether for bus or other facilities, requires a high level of technical and human skills.

Many other innovations to improve bus service have been developed. Among them are imaginative ways of route planning. For example, routes have traditionally been radial to serve the downtown movement. But for communities with heavy circumferential travel and a grid system of streets, routes along grid lines with easy and quick transfers may be advantageous. Often, "timing modules" of say 20 to 30 min with buses arriving and departing at these intervals are established to permit transfers without delay. Increasingly, pickup, transfer, and terminal points are weather-protected and in off-street facilities. Some of these are adjacent to shopping or other activity centers so that riders can combine trip purposes. These may also provide for park and ride. These examples merely suggest the wide range of approaches that can be employed under the diverse and unique situations that exist. In addition, by combining bus operations with demand-responsive and other special arrangements, many other innovations are possible.[62]

OTHER BUS OPERATIONS. Attention here has been focused on the bus as an urban transit vehicle. In addition, some 20,000 intercity buses provide the only public transportation between 15,000 communities. Sightseeing buses get tourists to and from points of interest in both rural and urban areas and chartered vehicles take groups to many events and exhibits. Most airports have bus service to city centers. In addition, the nation's schools operate some 300,000 vehicles, primarily standard-size buses. And, increasingly, private enterprises are providing bus service for employees. All these activities must be accommodated on the highways and be well managed if they are to operate efficiently. With the possible exception of intercity facilities, which are in financial distress, they have received comparatively little attention from highway and transit officials. Their peculiar problems are beyond the scope of this book.

Heavy-Rail and Commuter-Rail Systems

Heavy-rail systems usually consist of trains of several cars which travel between established stations at relatively high speeds on a fully protected right of way. In downtown urban areas they are commonly underground or elevated. If at grade, they are either separated from other traffic or have priority over it. In most systems, the trains run on steel rails similar to those of traditional railroads,

[61]Labor relations is among the most difficult problems facing transit managers. In the first place, labor problems and labor negotiations are highly politicized. Then, federal transit legislation imposes many conditions, among them reserving earlier worker rights and other stipulations which encourage unionization and add to union strength. Agreements barring or restricting split-shift work and part-time operators are common and drastically increase costs. Space limitations do not permit a detailed discussion of labor problems here. Starting points for further study are *TRB Special Report 181* and *TRB Record 573*.

[62]See, for example, several sections of Gray and Hoel, op. cit.; *TRB Special Report 184;* and *TRB Record 663*.

although in Montreal and a segment of the Paris system, rubber tires roll in a concrete guideway.

Commuter systems, which operate on the tracks of intercity railroads, have connected the suburbs with New York, Philadelphia, Chicago, San Francisco, and Toronto for many years. Trains for these systems commonly are coaches pulled by diesel or electric locomotives. In most instances operators have suffered heavy financial losses. Today, most of them are subsidized, or, in some instances, operated by transit authorities.

Subways, elevated systems, or both in Boston, Chicago, Cleveland, New York, and Philadelphia began operation before or near the turn of the century. However, with the coming of automobiles and buses, expansion stopped. Only recently have they been extended or upgraded.

Modern rapid transit did not come into being in the United States until 1962 with the creation of the San Francisco Bay Area Transit District (BART). Since that time, systems have been inaugurated on the Lindenwold line east of Philadelphia, and in Washington, D.C., and Atlanta. Others are under construction in Baltimore and Miami. All are rail systems, electrically propelled, and combine subway, elevated, and at-grade sections. Controversies have surrounded them. Among the issues are the need for them in the first place, the lack of standardization in facilities and trains between systems which could reduce costs, administrative snarls, underestimated construction costs, labor difficulties, and service problems including equipment failures and high operating costs. How many other systems will be undertaken is unsettled because the capital and operating costs of the existing rapid transit installations are high and continue to escalate. Probably those proposed for other large cities will be undertaken only with large subsidies from the federal government. Similarly, improvements or extensions of existing systems can very likely go forward only with federal assistance.[63]

In addition to objections to high capital costs and operating deficits, critics of rapid transit systems argue that they (and express bus systems also), were conceived by middle class people concerned with getting relatively well-to-do suburbanites between their homes and downtown work locations. They further contend that such systems are not designed to serve the needs of the urban poor, many of whom need transportation from the inner city to work sites in the sub-

[63]Capital costs for the BART system averaged about $20 million/mi. For planned completion of the 101-mi Washington, D.C. system, the 1980 estimate of total cost was $8 billion, or $80 million/mi, of which $3.2 billion will be federal grants.

For detailed descriptions of the physical features of several heavy and commuter rail facilities see Gray and Hoel, op. cit. For the Lindenwold line, see R. Schumacher, *Transportation Engineering Journal of ASCE,* Nov. 1970; for BART see W. A. Bugge, ibid., May 1974; and for Washington and Atlanta, respectively, see *Civil Engineering,* June and Jul. 1979. D. R. Bergmann, *HRB Record 449,* and V. R. Vuchic et al., *TRB Record 552,* give specific details of several systems. See also *TRB Record 627, 662,* and *760*. Bergmann, *Transportation Engineering Journal of ASCE,* Jan. 1977, offers a discussion of how freight and rapid transit services can operate on the same facility. J. L. Lammie and D. P. Shah, *Transportation Engineering Journal of ASCE,* Jul. 1980 describe the management structure employed in Atlanta.

urbs. Three reasons are given: (1) because the lines are radial, a trip to the downtown center and another back to the suburbs is required; (2) this trip will be very slow because of transfers and waits, since few trains are scheduled outbound morning and inbound evening; and (3) because transit stations in the suburbs must be widely spaced to maintain reasonable overall speeds,[64] a long walk or bus trip (if a bus is available) is required to reach the work destination.[65] Finally, it is contended that if bus services paralleling rail transit routes are abandoned, bus operations lose the profitable long-haul customers and have only the low-volume, high-cost feeder routes.[66] Also, the walk-bus-train combination takes so long that automobile travel becomes far more attractive. In sum, it can be said that the effectiveness of fixed-rail transportation in solving urban congestion and growth problems is still unsettled.[67]

Even for the suburb-downtown movement, fixed rail or guideway transit is feasible only where volumes are extremely high. Otherwise, buses traveling on existing arteries, reserved lanes on freeways, or on separate rights of way may be a better choice. Numerous studies comparing capacities, capital costs, and operating conditions and expenses have been made, but they are too complicated to be discussed here.[68]

Light-Rail Transit

Light-rail transit, once called street railways or street cars, is a principal form of urban transport in 300 cities over the world. This mode was also heavily used in the United States until supplanted by buses, beginning in the 1930s. It is again assuming increased importance.

In most instances light-rail units consist of one or at the most three or four electrically driven cars traveling on rails. They may share street space with automobiles, occupy a center mall or right of way alongside a street, or have an exclusive right of way at grade, in tunnels, or on an elevated structure.

Compared with heavy rail, light-rail stops generally are at more frequent intervals and are not limited to stations. Speeds are lower and vehicles are designed to negotiate sharper curves and steeper grades. Rolling stock costs are comparable with heavy rail, but other capital costs are much lower. Light rail lacks the flexible routing possible with buses. However, it is quieter and pro-

[64]For example, BART trains may have a top speed of 80 mph, but overall speed will be 50 mph or less. If stations were more closely spaced, the average speed would be reduced even further. For a far more detailed analysis of the station-spacing problem, see H. Permut, *Transportation Engineering Journal of ASCE,* May 1978.

[65]The home-to-station movement is not as difficult for suburban residents because if they drive to the station, they will find parking provided. Alternatively, an automobile will be available so that they can be taken to the station by a family member (called "kiss and ride"). These options usually are not available at the work end of the trip.

[66]For a discussion of this problem, see R. H. Pratt, *Transportation Engineering Journal of ASCE,* Feb. 1971.

[67]For the critic's point of view, see M. Wohl, *Civil Engineering,* Dec. 1971.

[68]Some of these comparisons appear in Gray and Hoel, op. cit., *HRB Record 293* and 459, and *TRB Record 559.*

duces far less air pollution because it is electrically propelled, as contrasted with diesel propulsion for buses.

Only recently have new light-rail vehicles been developed in the United States, and to date they have been produced in limited number. Most systems still use one of the PCC cars which were standardized in 1936; some equipment is even older.

Light-rail systems may fill some of the transportation needs of urban areas, large or small. Large cities in the United States that have them include Boston, San Francisco, Cleveland, San Diego, and Pittsburgh. Each is undertaking modernization. A recent installation has been made in Edmonton, Canada, and several other areas have studies or well-developed plans under way. However, the level of future use is still uncertain.[69]

Trolley Buses (Trolley Coaches)

Trolley buses combine certain features of street cars with others of buses. They run on the streets, so no tracks are necessary. Propulsion is by electric motors which receive power through overhead trolleys. They are quiet, operate smoothly, produce no air pollution, and do not use scarce diesel fuel. However, they cost more and lack the route flexibility of buses, since they are tied to the overhead power supply. Currently few of them are in use; whether or not they will be increasingly favored in the future is uncertain.[70]

Automated Guideways (People Movers)

More than 100 schemes to supplement or replace private automobiles, buses, or rapid transit in high-density areas have been put forward. All of them are intended to quickly bring passengers within easy walking distance of their destinations. They could be coordinated with elevators or escalators to provide a total transportation package. Because distances are relatively short, high speeds are not needed; rather, it is essential that points of entry, exit, or transfer be closely spaced.

Automated guideways are commonly small rubber-tired passenger or freight capsules rolling on a confining concrete track. The system at Morgantown, West Virginia, originally financed as a demonstration project by the Urban Mass Transit Administration, is typical. Its first stage serves five stations in Morgantown and on the University of West Virginia campus. Seventy-three rubber-tired, electric-powered cars travel a 3.3-mi route on elevated guideways at speeds up to 30 mph. Car capacity is 13 passengers, with 7 seated. Provision is made for both manual and automatic control. Total cost was $130 million. An installation similar to Morgantown provides for internal passenger movements at the Dallas–Fort Worth, Seattle, and Atlanta airports. Other systems, some of them cable supported, are being employed. As of 1979, federal grants had been

[69]For added detail see Gray and Hoel, op. cit.; *TRB Special Report 161* and *182;* M. Lenow, *Transportation Engineering Journal of ASCE,* May 1976; and G. D. Fox, *TRB Record 662.*
[70]For more detail, see J. W. Schumann and B. J. Hansen, *Traffic Quarterly,* Oct. 1979.

made for guideways at several other locations.[71] Means to reduce headways to as low as three seconds are being studied.

A variety of travel patterns can be developed for automated guideways. Shuttles back and forth on parallel alignments are simplest. But loops and bypasses coupled with automatic switching have also worked well.

Conveyors

Passenger conveyors are quite widely used in locations such as airports and shopping centers where walking distances are excessive. However, conventional single-belt designs are not suitable for longer distances, because their speed, limited by entering and leaving conditions, cannot be greater than about 3 ft/s and are generally less.

Accelerating walkways have speeds four, five, or six times as great. One design involves several parallel belts. At the entering location, each successive belt has higher speed until the main belt is reached. At the leaving location, deceleration is accomplished in a reverse manner. Several other approaches for accelerating beltways have been developed. The degree to which such conveyors will challenge other forms of mid-distance passenger movement is still uncertain.[72]

Paratransit

The term "paratransit" is defined in part as "Forms of public transportation services that are more flexible and personalized than conventional, fixed-schedule services. . . ."[73] It does not include conventional taxis nor chartered buses but does encompass other demand-responsive approaches, as discussed below.

Paratransit is not a new concept. For example, volunteers or paid drivers have for years gotten the disabled or disadvantaged to shopping, appointments, or events. Carpooling, arrangements under which several people ride to work, school, or special events in one vehicle, is another form of paratransit. What is new are the efforts of employers, government, and other interests to promote paratransit. These new efforts take a variety of forms. For example, in industry and government, employees are provided information on potential drivers and riders to assist in forming carpools. In some cases employers offer other incentives, among them favorable parking rates and close-in positions. Some governmental agencies promote carpooling by setting up and advertising a centralized

[71]For a detailed but early description of the Morgantown installation see F. E. Lo Presti, *Civil Engineering,* Nov. 1972. An update appears in *Engineering News-Record,* Jul. 19, 1979. See also C. Henderson in Gray and Hoel, op. cit.; D. R. Bergmann et al., *Transportation Planning and Technology* (Great Britain), 1978, Vol. 4; and L. C. Lavery and D. G. Stuart, *TRB Record 634*. Also, UMTA has sponsored a series of reports evaluating individual installations.

[72]For detailed descriptions of conveyor concepts see, for example, E. W. Walbridge, *Transportation Engineering Journal of ASCE,* Aug. 1975; J. K. Todd, *TRB Record 522;* and D. W. Bergmann, op. cit.

[73]This definition is from "Glossary of Urban Public Transportation Terms," *TRB Special Report 179*.

information source which can bring people from closely positioned work and home origins and destinations together. Financial aid for some of these schemes is provided from federal or state funds.

Studies of factors influencing the use of carpools show them to be complex. Perceptions of convenience, reliability, and time are weighed heavily while cost and public-interest issues such as congestion, energy, and air quality are of less concern. Demographic and travel characteristics have been poor indicators.[74]

Vanpools substitute larger vehicles for automobiles. To make participation attractive, some employers and governmental agencies encourage vanpools by purchasing the vans and making them available on favorable terms. Riders then can travel more cheaply than by driving alone or in a carpool. Often the driver has weekend and holiday use of the van free or at substantially reduced rates or is given ownership after a period of time.

It should be clear that many other paratransit options are available or will be developed. A few are mentioned later under the topic of demand-responsive transportation.[75]

Taxis

Taxis are a principal form of transportation in urban areas. They carry 75% as many passengers as buses and employ 600,000 people. Traditionally, they have operated as private enterprises and have served individual patrons or groups who engaged them at designated locations, by telephone, or by hailing them on the street. Until recently law or custom prevented drivers, once engaged, from picking up other customers along their routes.

Strong efforts are now under way to integrate taxis into overall transportation schemes. In some urban areas, the prohibition against picking up added patrons and other regulations have been relaxed. Taxis may be favored by permitting them to use lanes reserved for buses and carpools. Also, taxi service may be employed to feed the public transportation system or be subsidized so that the disadvantaged can use it at an affordable fare, either by engaging it directly or as a part of a demand-responsive system.[76]

Locally Generated Mass Transportation

In many cities of the world, mass transportation modes have developed to fit their peculiar needs and economic conditions. At one extreme is the rickshaw, propelled by a man on foot or bicycle, or sometimes motorized. Pickups or jeeps with the bodies occasionally modified to hold more passengers appear in

[74]See A. D. Horowitz and J. N. Sheth and discussants, *TRB Record 637*.

[75]Among the many recent references are *TRB Special Report 147, 154, 184,* and *186,* and *TRB Record 619, 650, 724 and 778;* FHWA has published a *Matching Guide* as an aid to those planning such arrangements. For data on how individual states regulate prearranged ridesharing, including provisions for insurance, see J. P. Womack, *TRB Special Report 181*.

[76]See *TRB Special Report 181, TRB Record 559, 618,* and *650* and the section on transportation for the disadvantaged for a few of the many references.

some cities. In other instances, automobiles or vans operate on favorable routes, often in competition with the regular transit system.

In most instances, these locally developed systems are operated privately. Whether or not they flourish or even survive depends on public policy and its enforcement. On the one hand it can be argued that they should be encouraged as they fill an important need at no cost to government. The counter-arguments are that they may be unsafe and that they take the most lucrative part of the transit market, making it more difficult for public transit to be self-supporting.[77]

Demand-Responsive Transportation Systems

Demand-responsive systems, sometimes referred to as "dial-a-ride" systems, operate in urban, suburban, or even rural situations. The basic concept is that potential riders notify dispatchers or dispatching devices of their desire to make a trip; then a vehicle, commonly a minibus, van, or taxi, is routed to the homes or pickup points and takes riders to their destinations. This scheme differs from the conventional taxicab in that it may operate on either a fixed or variable routing and schedule and may carry several individuals or groups at once rather than handling only one call at a time. Such a service has sometimes replaced the private automobile or fixed-route system for home to work and other movements. It also may be less costly than other means for transporting the disadvantaged.[78]

A variety of means for receiving messages and assigning vehicles are employed. The simplest approach is for a dispatcher to receive the requests for service and assign vehicles to them, probably by two-way radio. At the other extreme, requests are fed to a computer which can be programmed to match vehicles to passengers, determine sequence of pickup and delivery, or optimize some other aspect of the operation.[79]

Attempts to initiate and maintain demand-responsive transportation plans have often proved to be difficult. For example, all the buses used on a system of fixed routes in Santa Clara County, California, were transferred at one time to a computer-controlled demand-responsive scheme, since neither equipment nor finances permitted retention of both. Strong protests followed from citizens who had used the fixed routes and were dissatisfied with the service provided by the new system. As a result, the elected Board of Supervisors, which also functioned as the Transit Board, abandoned the new scheme before there was time to test it.[80] Five years later buses were still operating on fixed routes and schedules with no change in sight.

[77]For a few examples see S. Grava, *TRB Special Report 181.*

[78]A concern is that such systems, which will almost certainly be subsidized, may displace many effective organized or individual volunteer arrangements.

[79]Controlling and optimizing demand-responsive transportation has been studied intensively. Recent reports appear in *TRB Record 522, 563, 606, 608, 618,* and *650* and *TRB Special Report 147, 154, 164, 181,* and 184.

[80]See J. T. Pott, *TRB Record 608.* For other examples see *TRB Special Report 184.*

Innovations in Urban Public Transportation

Numerous proposals have been put forward for new approaches to public transportation. These range from entirely new systems to large or small changes in existing ones.

Possibly the most imaginative proposal is for an underground system which combines gravity and vacuums for propulsion.[81] Another all-new system is called "personalized rapid transit" (PRT) or "rental personalized automobile service" (PAS). It would employ small cars for individual movements in urban areas. These might travel on guideways and provide station-to-station service. In a dense downtown, several adjacent loops could cover the entire area. Another extension of the concept would be dual-mode: cars would travel on a guideway on the portion of their routes common to many other vehicles but run on the streets under their own power on the low-volume portion of their travel, such as from terminal to home. Even yet another approach is to store vehicles at closely spaced convenient locations ready for authorized drivers to use. They would drive over the streets and leave the vehicles at stations near their destinations. An arrangement to shuttle unused vehicles to points where high demand is predicted would be necessary. As with all entirely new schemes, these would require a substantial initial investment and a long phase-in period.[82] None have been implemented.

Innovations are also possible in other dual-mode schemes. For example, pallet approaches, such as are provided for automobiles on certain railroads, would have special heavy-rail cars traveling the main arteries carrying the small cars used at the origin and destination ends. A modest variant sometimes now employed has storage for bicycles on trains or buses, so that they are available at both ends of trips.[83]

A less elaborate and cheaper system involves slow-moving passenger vehicles that follow a buried cable along a fixed route. They would share walkways with pedestrians or streets with slow-moving vehicles. Numerous similar installations for transporting materials in factories have already been made.

INTERCITY AND RURAL PUBLIC TRANSPORTATION

As indicated earlier, some 20,000 intercity buses serve about 15,000 communities; only 700 of these have air passenger service; 500 are on Amtrak rail lines. These buses also serve rural residents along their routes. Although there are a few large companies such as Greyhound and Trailways, in all 1000 private companies are involved. Their market (340 million passengers in 1976) is predominantly the less affluent, the elderly, handicapped, and young.

As with most other transportation services, many of these bus companies are financially distressed. Over eight years, 1800 communities lost these services.

[81]See L. K. Edwards, *Transportation Engineering Journal of ASCE,* Feb. 1969.
[82]See C. Henderson in Gray and Hoel, op. cit.
[83]For more on this topic see *TRB Special Report 170.*

In 1978, the Congress authorized $30 million per year for three years to subsidize these companies, and some state and local governments are also involved.

Other efforts to improve rural public transportation include expanding school bus use to serve other purposes and special programs for the disadvantaged. A large number of experimental and demonstration programs have been undertaken with the aim of developing better approaches and overcoming financial, legal, and other barriers.[84]

TRANSPORTATION FOR THE DISADVANTAGED

In today's automobile-dependent society, the "disadvantaged" might be defined as those who suffer transportation disadvantage because they do not have the use of an automobile. These would include children and youth too young to drive, those who can not afford an automobile, the nondriving, the physically handicapped and disabled, and some but not all of the elderly. Another definition might classify the disadvantaged as those who make fewer trips than seem desirable for their well-being.

With only a few exceptions, transportation for the disadvantaged has not until recently been considered as a responsibility of either government or the public transportation system. Rather, these persons did not travel as much or were transported in some other manner. Today, however, transportation is one facet of a variety of governmental programs to provide special services of many sorts to the disadvantaged.

A classic example of the complexities in programs to help the disadvantaged occurred in Los Angeles. There was a large number of unemployed in the Watts area southeast of the city center. Job opportunities for them centered at the airport, which was located to the southwest. Bus service was radial, so getting to and from the airport involved two bus trips and waiting between them. To stimulate employment, a direct bus route from Watts to the airport was set up. It flourished for a short while. However, many of the bus users soon purchased automobiles and bus patronage fell drastically. In the broader sense of providing employment this transit experiment was a success; but it could be classified as a failure in developing and maintaining transit use.

Federal legislation granting financial assistance for public transportation requires special attention for the handicapped. For example, an UMTA rule promulgated in 1976 permits the grantee to choose among (1) providing at least ten round trips per week, (2) making, in time, half of the bus fleet wheelchair-accessible, or (3) devoting at least 5% of the funds to these efforts. Many other approaches also favor the disadvantaged. Included are reduced fares, at least at off-peak hours, special buses, demand-responsive service, subsidized taxis, and special transportation arrangements for those who cannot use these means. Even so, to date, programs for serving the handicapped are largely experimental.

[84]For added detail see R. J. Popper and J. W. Dickey, Gray and Hoel, op. cit., *TRB Record 696* and *718,* and the references given under the heading Transportation of the Disadvantaged.

Among the some 1000 urban and an equal number of rural public transportation operations involved, most of which are in financial difficulty, the levels of service provided the handicapped will vary from almost none to fairly good. Much is yet to be done and much to be learned before suitable, workable, and affordable service can be provided for the disadvantaged.[85]

SAFETY AND SECURITY IN PUBLIC TRANSPORTATION

On a passenger-mile basis, deaths of passengers on public transportation are about one-tenth those of travelers in private automobiles. Many reasons can be given, among them larger, heavier vehicles that protect the occupants, skilled operators, and governmentally prescribed vehicle design features and inspections. A detailed discussion of transit safety is beyond the scope of this book, although many common aspects of the problem are treated in Chapter 12.

Security for public transit passengers and operators and protection against vandalism are serious concerns. An estimate for 1971 placed crimes at about 36,000 and the cost of vandalism at $8.4 million. And, without question, fear of attack is a primary deterrent to transit use, particularly at night.

It has been reported that, with rail systems, most of the attacks occur on station platforms. Waiting bus passengers and drivers also are assaulted, but less frequently. Unfortunately, providing personnel to patrol all dangerous locations continuously appears to be far too costly. The most effective measure is to provide high visibility augmented by an alert system and surveillance with closed-circuit television. As with other criminal activities that plague our cities, there appears to be no easy solution.[86]

RURAL AND INTERCITY HIGHWAY FREIGHT MOVEMENT

As indicated earlier, trucks transport much of the goods that move through rural areas or from them to urban locations. Crops, dairy and meat products, and many other items go to market or processing plants over local and main roads. In addition, trucks monopolize transportation for perishable or lighter-weight, high-value products not only on short hauls but on long intercity ones as well, although "piggyback" (trailers on railroad cars) is increasing.

This vast goods movement is almost entirely in privately owned vehicles traveling on publicly constructed and maintained highways. Some trucks are owned and operated by licensed carriers including railroads, others by companies transporting their own products, still others by companies or individuals who provide transportation on a for-hire basis. Some truckers are regulated for rates and safety by federal and state agencies; others are largely unregulated.

[85]There is a large literature on transportation for the disadvantaged. See, for example, A. Saltzman in Gray and Hoel, op. cit.; articles in *TRB Record 516, 559, 563, 578, 608, 618, 637, 660, 661, 688, 696,* and *784*; NCHRP Synthesis 39, and *NCHRP Report 209*.

[86]For added discussion, see L. A. Hoel in Gray and Hoel, op cit. and L. G. Richards and L. A. Hoel, *Traffic Quarterly*, Jul. 1980.

Concerns of the highway engineer for this vast movement of goods center on the need to provide capacity and proper geometric and pavement designs for its efficient and safe movement. This topic is treated at appropriate places in this book. Management, operation, and regulation of the vehicle fleet is beyond its scope.

DATA GATHERING FOR PLANNING

Introduction

Certain data about past or present situations are fundamental for intelligent highway or urban transportation planning, and these are gathered by practically all agencies that plan, construct, operate, and maintain facilities. Inventories are kept of the physical state of existing facilities, accident experience, motor vehicle ownership and use, sources and distribution of funds, and traffic flows. Other data may be assembled when a special occasion arises, such as for urban transportation or corridor- or route-location studies. Some of these topics are discussed briefly below; information on road life is presented in Chapter 4; accident data collection is treated in Chapter 12; and vehicle ownership and use and finance are summarized in Chapter 5.

Physical Inventory

Many early rural roads were built without plans; or, if plans were available, they lacked detail. To provide data on these roads, a country-wide inventory was made in the 1930s. Observers recorded width, type, curvature, grades, and condition of roads and structures and located farms, dwellings, schools, churches, and other cultural features which were potential sources of traffic. This original road inventory cost somewhat less than $1/mi. When, about 1940, the data from the several states were assembled there was, for the first time, a detailed record of the country's rural highways and their condition. In general, cities had construction plans and other records and with them had or could assemble an inventory.

By building on these earlier records, coupled with plans for later construction (see Chapter 6), most highway and street agencies have fairly up-to-date inventories showing the major details of all their facilities. Many have expanded their inventories to include samplings of traffic volumes and speeds (see below), accident experience, and maintenance costs. Also, surveys to record the condition, remaining service lives, and surface-friction characteristics of pavements often have been made (see Chapter 21). Today much of this information may be in a computer-based data bank from which it can be retrieved or summarized.

Inventories such as those just discussed may lack certain information that had not been recorded. Even if available, recovering it might be costly. Furthermore, certain details such as restrictions to overhead clearances, signs, developments along the right of way, roadside encumbrances such as utility poles, trees, or shrubs, and particulars of pavement condition may be lacking. In the past, a special on-site survey would be required if, for example, high-accident locations were to be investigated or pavement-condition appraisals made. Today, this sort of information is often available for an entire highway system or as needed through a procedure known as photologging. For it, 35-mm photographs are taken in each direction from a moving vehicle at intervals such as 1⁄100 mi. Costs are low, possibly $15 per one-way mile. From them a quick viewing may provide an overall assessment of each road. Again, by analyzing individual photographs carefully, pertinent details can be obtained cheaply.[87]

Surveys of Goods Movement

Goods movements are very complex, with trucks often the only conveyer used at the beginning or end of the process. Their presence on rural and urban highways is recorded by traffic counts or other surveys. But there are many other concerns about goods movement which need to be recognized in overall transportation planning. One among them deals with the impacts from constructing new or abandoning existing transportation facilities, be they highways, railroads, waterways, or pipelines. For example, as the railroads abandon branch lines through agricultural areas or into the coal fields, the loads they have carried are shifted to trucks and result in many heavy axle loads. Already there are many cases of extensive damage to roads, particularly local ones not designed for such traffic. Only by having data on the extent of this traffic are highway officials in a position to plan necessary strengthening of the roads and the financing to pay for them.

Estimating the demand for trucking is difficult because it varies so greatly with, among other factors, location and land use. For example, commercial areas generate from 14 to 40 trips per acre per day, manufacturing from 15 to 40, and residential from 1 to 2.[88] Statistics such as these demonstrate only one aspect of a complex problem that cannot be discussed here.[89]

Rural Traffic Surveys

Early rural traffic surveys were done manually; that is, each individual vehicle passage was recorded by an observer. This task was commonly performed once or twice a year by maintenance personnel, sometimes supplemented by others hired solely to assist in the count. It is now common, however, to use a statistical approach to traffic counting. Detectors register traffic continuously and record the flow at a group of strategically located *control stations*. In addition,

[87]*Compendium 15* of Transportation Technology Support for Developing Countries, TRB, is an excellent reference on road and traffic inventories, particularly directed at low-volume roads. References on photologging include T. K. Datta and R. J. Labadie, *TRB Record 674* and several articles in *Transportation Engineering*, Apr. 1978.

[88]See J. D. Brogan, *TRB Record 716*.

[89]For added detail and extensive bibliographies see, for statewide situations, *NCHRP Reports 177* and *178*, and for urban ones, *Urban Transportation for Goods and Services*, FHWA and *TRB Record 747*.

portable counters, commonly actuated by pneumatic detectors laid across the roadway, note vehicle passings for periods of about 24 to 48 hr at a large number of *coverage stations*. These short counts are then projected statistically into close approximations of average-daily or peak-hour traffic. Thus, by combining control- and coverage-station data, traffic volume and its time characteristics can be developed for the entire highway system. In some states, data collected by the counters are transmitted to a central computer over telephone lines. In turn, the computer processes the raw data and calculates predicted volumes.[90]

Data from automatic counters tell nothing of the character of the traffic. This information is gathered through *loadometer studies*, which tell type of vehicle; rated capacity; gross weight; payload, width, height, and length of vehicle; commodity hauled; and origin and destination of trip. Loadometer studies provide the basis for many road-design and traffic-regulation decisions.

Quite naturally truck owners object to the delays that result when trucks are stopped or slowed to, say, 3 mph for weighing and measurement. Electronic devices that measure axle weights and spacings and speeds of vehicles in motion have been developed to minimize this problem.[91]

A traffic survey such as that just described may be sufficient for planning purposes in fully developed rural areas. However, it is of little value in, for example, a developing country where economic activity will follow after new roads are built or existing tracks are made into roads.

Inventory for Urban Transportation Planning

In rural areas, counts on individual or relatively parallel roads generally assess the present traffic volumes and classes quite accurately. In contrast, in the environs of large urban areas, movements of people and goods are far more complex. Alternate paths are often available and individual drivers or bus routes will follow the most favorable ones. Improvements on particular routes to make them more attractive will draw traffic from other courses. In effect, freeway and street capacity and condition as well as directness of route affect traffic volumes and traffic counts may give a distorted or partial measure of needs or desires. In addition, travel by public transportation or ride sharing either through choice or necessity is common and traffic counts would not show these details. Thus, in urban areas and on the roads approaching them, origins and destinations rather than volumes on specific routes are of primary importance.

The overall process for a comprehensive urban transportation plan is diagramed in Fig. 3–5. Across the top are listed the general topics on which inventory data are to be gathered. It is, in effect, an assessment of many aspects of the community. It can be seen that travel information is gathered not only to provide information as to today's conditions, but also to give a "year zero" accuracy check for trip generation models that will be employed to forecast the

[90]See *Compendium 15*, op. cit., for added detail on traffic surveys and the equipment for them.
[91]See for example R. B. Machemehl et al., *Transportation Engineering Journal of ASCE*, Nov. 1975; L. E. Welsh, *TRB Record 615*; and A. T. Bergan et al., *TRB Record 667*.

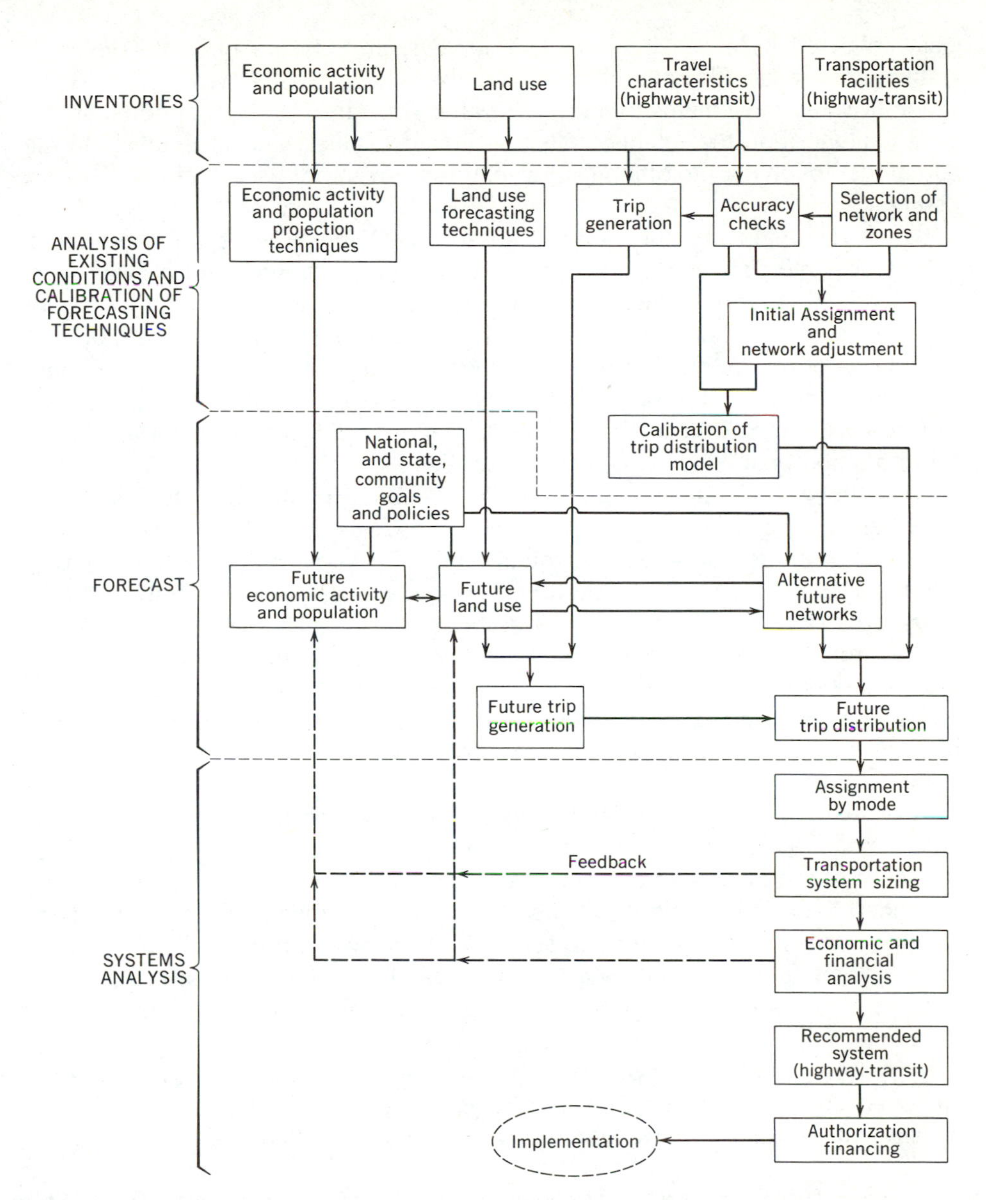

Fig. 3–5. Flow diagram of a comprehensive urban transportation planning process.

effects on travel of alternative plans for short- and long- range future investments in transportation facilities.

Techniques for assessing present and future economic activity and land use are not discussed here. However, travel-characteristic inventories are usually done by engineers; and are referred to as origin-destination (O-D) studies. A comprehensive study for a large urban area often combines home interviews or data collected by a census, truck-taxi surveys, and cordon line interviews or traffic counts for traffic entering or leaving the area. Special surveys also may be made to show movements to and from traffic generators such as airports, shopping centers, industrial plants, colleges and universities, and hospitals.[92] In all situations involving automobiles, a parking inventory (see below) also is essential.

Before data gathering for a comprehensive O-D survey is begun, the area is subdivided into significant geographical zones and subzones, and all trip ends are identified in terms of these units. At times the zones are selected to conform to topographic or other barriers that control travel routes; again, a system of rectangular coordinates may be used and the boundaries of individual zones and subzones stipulated in these terms. Today, zones often are related to those employed by the U.S. Bureau of the Census, since it collects much data useful for transportation planning.[93]

For cities with populations greater than possibly 50,000, O-D studies of the door-bell ringing type have been favored. Interviews are held in a statistically determined number of dwellings, roughly from 20% for cities of 50,000 population to 4% for those over 1,000,000. The interviewers determine the methods of transportation and the origins and destinations of each trip by each member of the family on the previous day. They may also gather information on family size, economic status, and vehicle ownership. Home interviews are supplemented by interviews or counts at cordon stations on routes into the area and by facts about bus, truck, and taxi movement gained from company records. Check counts of movement within the cities are made at *control points*, such as well-known bridges or other landmarks, or along *screen lines*, which intercept all arteries of travel.

Results of the interviews and surveys are coded into the computer. These are expanded into "trip tables" of zone to zone movements. Results are then checked against control-point and screen line counts.[94] In this way, the model for projecting future traffic flows (see below) can be verified. Also any cross correlations that are desired can be produced. For example, travel between zones by modes and by hours of the day or relationships between family income and trips by travel mode and trip purpose can be determined. These travel patterns are often complex, and depend in varying degrees on the layout, population density, affluence, and transportation facilities in each urban area.[95]

[92]For the use of aerial photographs for traffic surveys see *The Practicability of Aerial Traffic Data Collection* published by ITE.

[93]For the census, encoding files provide *x-y* coordinates for points such as street or road intersections. These are called DIME/GBF, standing for Dual Independent Map Encoding/Geographic Based File. See *HRB Special Report 145* and articles by J. M. Lutin, C. D. Gehner, P. S. Cornelio, and J. M. Manning, *TRB Record 677* and R. M. Somers and M. A. Jaro, *TRB Record 771*.

For a discussion of transportation planning data needed in or to be provided by the 1980 census see *ITE Journal,* Mar. 1979, and R. D. Dunfy, *TRB Record 701*.

[94]Trip-table volumes commonly fall within 10% of the checkpoint counts.

[95]Comprehensive O-D studies have been made in many communities and the techniques for them are well documented in publications of the Federal Highway and Urban Mass Transit Administrations. See also *NCHRP Report 120*. However, because of their high costs and today's emphasis on shorter-run solutions, comprehensive surveys are now made less frequently.

For urban areas with populations of less than roughly 5000, an external-cordon type survey may be appropriate. A cordon line is drawn encircling the area. Then, at stations on roads crossing this cordon line, a statistically determined percentage of vehicles is stopped and their drivers interviewed to obtain the desired information.[96] For urban areas in the 5000 to 50,000 population range, an internal cordon surrounding the central business district in addition to the external cordon is recommended. Check counts with automatic recorders permit reconciliation of data procured on different days.

At times, an external-cordon type of study may be made for a single purpose. For example, the residents or business interests of a small town may object if a main highway is to bypass their community. In this case, drivers are asked only if they plan to stop in the community and, if so, for what reason.

Return postcards, mailed to registered owners of all vehicles in the survey area, offer another O-D method. Spaces are provided on the return card for information about the driving done on a designated day. Although the postcard method provides less information than do other methods, it can be done more cheaply. Replies may be received from only 25% of the vehicle owners, some of them may be unusable, and it may be difficult to verify the results. However, by good publicity, returns of 50 to 60% have been generated.[97]

In some instances return postcards for O-D studies may be distributed to motorists at bridges or other points where traffic is confined to a few routes. Sometimes this method is quite reliable. In other cases, as when drivers from a particular area are extremely concerned by the implications of the survey, results may be badly distorted.

Urban Parking Surveys[98]

In the past, parking surveys were intended primarily to assess present and future supply and demand. Today, however, they provide data on which public policies can be set; for example, parking control may be employed as one approach to limiting automobile use in urban areas, particularly in central business districts.

A parking survey is not an easy task, because of the many complexities that should be recorded. Provisions for parking involve spaces at curbside and in lots and garages. Street parking is, of course, controlled by local government and competes for space with moving traffic. Off-street parking is in lots and garages, some provided by government and some by private enterprise. Some is subsidized and offered free, some is expected to pay its own way, and some is to make a profit. Sometimes use is open to all, sometimes it is restricted. Government enters the picture in a variety of ways. For example, through its control of zoning, it can dictate the amount of parking space to be provided for new or modified buildings or other developments. It also can set charges and turnover rates for public and some privately owned facilities. Only with knowledge of

[96]See, for example, D. Rhodes and T. Hillegass, *TRB Record 677*.
[97]See E. W. Waltz and W. L. Grecco, *HRB Record 472*.
[98]The legal, political, management, and physical aspects of parking are discussed in Chapter 10.

existing conditions can reasonable projections be made of the effects of parking controls and rates.

A traditional comprehensive parking survey for a downtown area commonly combines an inventory of available spaces and the rate structures for them combined with personal interviews giving vehicle data, time of arrival and departure, origin of trip, and downtown destination. Also, cordon counts are made on all streets entering the survey area. Where appropriate, parking outside the survey area is recorded. If the parking survey is to be in conjunction with an O-D survey, an alternative is to gather parking data as part of the home and cordon interviews. Studies such as these are costly and made infrequently today.

Because individual urban areas differ markedly in size, population density, physical layout, and automobile and transit use, comparisons with other apparently similar communities may not be reliable. Some generalizations from many studies are that the percentage of central-city curb parking decreases and that of garage parking increases with larger populations. Again, the percentage of central-city parking for work increases and that for shopping decreases with city size. Also, in most instances, curb spaces turn over two or more times as often as spaces in lots or garages. But in-depth knowledge of parking can come only by assessing it in each community.[99]

Among other less costly procedures for assessing parking use is with color aerial photographs taken at frequent (possibly 15 min) intervals from which essential data on street and lot parking can be taken. It would have to be augmented by information gathered from garages and areas covered with shelters or trees.[100]

Urban Traffic Surveys

As indicated, traffic counts on individual highways and streets may have only limited value for long-range planning, since traffic will shift with improvements or changes in conditions. On the other hand, they are essential as a basis for traffic control and operation and other short-range objectives. As pointed out in Chapter 10, decisions regarding the need for and type and timing of traffic signals and the desirability of turning lanes, parking controls, and stop signs must rest on traffic-flow data.

Techniques for urban traffic counts parallel those used for rural roads. Continuous counts at a limited number of control stations and short counts at coverage stations can be combined statistically to give the necessary data. At intersections, manual or recorded counts of individual movements and of delay times or an analysis of time-lapse photographs are the first steps in determining intersection capacity and setting signal cycle times and splits (see Chapter 10).

[99]Among references on conducting parking surveys and their results are the FHWA publication *Conducting a Comprehensive Parking Survey*; *HRB Special Report 125*; J. G. Schoon and H. S. Levinson, *Transportation Engineering Journal of ASCE*, Aug. 1974; and C. Okechuku and T. A. Lambe, *TRB Record 514*.

[100]See *The Practicability of Aerial Traffic Data Collection*, op. cit., and T. A. Syrakis and J. R. Platt, *HRB Record 267*.

Highway and Transit Impact Studies

As shown in Table 3–2, the construction of new highways or changes in highway locations or types, such as to freeway from conventional highway or street, have impacts other than on highway users. Transit systems have similar effects. These, commonly referred to as community or socioeconomic effects, have been studied carefully for a variety of situations. Some of the earliest analyses dealt with the impact of bypasses or freeways on businesses. Since then many investigations dealing with the effects of highways on land use and land values, on business and industry, on residential development, on rural economics and culture, on recreation, on social change, and on other public services have been made. Currently, topics, such as the effects of noise and air pollution on the urban environment and the impact of transportation on employment and living conditions, particularly for the urban and rural poor, referred to as "social accounting," are being given particular attention. Knowledge in these areas has now been accumulated that is far too detailed to be presented in this book. Some special findings are cited in appropriate places. This knowledge will help highway and transportation planners avoid mistakes made in the past so that designs for new facilities will have positive "nonuser" consequences.[101]

USES OF PLANNING DATA

Unless information gathered for planning is applied to decisions and actions, collecting it is a sterile activity. And yet, as indicated earlier in this chapter, decision making on complex public issues is difficult and often unrewarding. This is particularly true when long-term investments that will shape the future for 20 to 50 yr are at issue, but where a consensus on direction is lacking and the decision-making "climate" is one of crisis over short-run concerns. It has always been easy, looking back from the future, to decry past actions such as spending to get rural areas out of the mud at the expense of cities, letting transit decay after World War II, stretching highway dollars by building what later appeared to be substandard or capacity-deficient highways or freeways, displacing residents to build urban freeways, or even building urban freeways or costly rail-transit systems at all. Some of today's decisions likewise may, in A.D. 1995 or 2025, appear to have been foolish ones. Yet decisions must be made, using all the data and foresight provided by planning studies and perspectives such as those described earlier in this chapter. Eight uses of such findings are discussed briefly in the sections that follow.

Statewide and Nationwide Transportation Planning

In the decade 1930 to 1940 expenditures for highways failed to keep pace with road use as measured either by the number of motor vehicles or the miles dri-

[101]For a recent review and extensive bibliography on the environmental, social, and economic effects of freeways, see *NCHRP Report 193*. Also, the results of numerous studies have been published by FHWA.

ven. Then during World War II, roads were subject to intense use by large numbers of heavy vehicles while, at the same time, maintenance operations were curtailed. By 1945, much of the highway system lacked capacity and was in poor condition. After the war there was an immediate increase in motor traffic (see Fig. 1–1) and a public and legislative consensus that something must be done to better the highways. Since basic responsibility for these highways and streets rested with the states, many of them first made highway classification, needs, and financial studies. In essence, these attempted to (1) classify all roads and streets into systems (state, county, city), and assign them to appropriate jurisdictions; (2) determine the cost for bringing each system to acceptable standards; and (3) develop a financial plan to pay for these improvements. The approach taken in California was typical. In 1945, a legislative Joint Fact-Finding Committee on Highways, Streets, and Bridges was appointed. The committee secured excellent outside technical assistance and the cooperation of local, state, and federal government specialists. Based on facts developed from the highway-planning surveys and a number of special studies, an up-to-date inventory of all the road systems in the state was developed, a reclassification by function of all segments of the road system was proposed, current-and long-range needs were listed, a 10 yr construction program was outlined, and revisions to the highway tax structure were recommended to secure the needed funds. After a series of public hearings, the state legislature in 1947 enacted many but not all of the recommendations into law. Motor-vehicle taxes were increased to provide added funds for state and county roads and for major city streets; county highways were classed as primary and secondary with the use of some funds restricted to primary roads. Provision also was made for a more equitable division of statewide user revenues among the individual counties and cities. A most significant section provided special funds with which the counties could engage professional engineers. A series of subsequent studies to reassess the earlier findings provided a base for later legislative consideration as situations changed. For example, a statewide system of freeways was created.

More recent analyses in various states have expanded the focus of the earlier studies giving special attention to transportation system management, maintenance, public transit, public involvement, energy, and the environment.[102]

At the federal level, functional classification and needs studies have played an important role. For example, the so-called "Clay Report" (House Document No. 120, Eighty-fourth Congress, 1955), which preceded by 1 yr creation of the Interstate System, was a highway needs study for the entire country for the 30 yr 1955 to 1984. FHWA has made similar analyses every 2 yr beginning in 1966. In 1979, the National Transportation Study Commission, composed primarily of members of Congress but with representatives from federal agencies and industry, released its monumental report titled *National Transportation Policies Through the Year 2000*. Among its many findings are that about 40% of

[102]See H. Heckeroth, *TRB Record 654*; W. S. Weber, *TRB Record 677*; and C. Fleet et al., *TRB Record 710*.

urban arterials and freeways show some level of congestion and that the recent investment in transit has only arrested rather than reversed its decline. In terms of governmental capital expenditures for highways between 1976 and 2000, it proposes between $857 and $930 billion and for local public transit $167 billion, all in 1975 dollars. These amounts substantially exceed the present levels of spending detailed in Chapter 5.

From these examples, a notion of transportation planning at the state and national levels can be gained. It can be seen that such periodic studies on all phases of our transportation problems are essential if the public and its legislators are to make reasonably intelligent choices.[103]

Planning for the Interstate System

Planning for the Interstate System, now largely completed, has been an excellent example of federal, state, and local governmental cooperation. Activities begun before World War II and in the early years after it were devoted largely to selecting routes and developing design standards.[104] Studies to develop preliminary cost estimates and plans for financing the system culminated in the 1956 Federal-Aid Act, when Congress made very substantial funds available and stipulated their distribution for the first several years. Subsequent federal appropriations have seen the rural system and all but a few controversial urban links completed. Continuing concerns will include upgrading or rehabilitating the already completed sections and financing for segments which are currently toll roads and bridges.

It is doubtful that any new surge of massive freeway building will occur soon. But the Interstate program is an excellent example of what can be done when government sets out to accomplish a popular public purpose.

Planning Urban Transportation Systems

The discussion earlier in this chapter indicates the complexity of movements of people and goods in urban areas and some of the present and proposed methods for providing for them in the future. It also outlines the data-gathering techniques often employed to record these movements. But to plan transportation for the years ahead requires projections into the future, with all the uncertainties inherent in such forecasts. These include predictions of the forms that the individual urban areas will take, either through governmental control or under economic or other forces, the part that current or advanced types of transportation and communication systems will play, and the choices that transportation users will make under their financial circumstances and personal preferences, not only now but in the future. Such a task is not easy.

[103]For a detailed review of statewide planning efforts and the problems to be addressed see *NCHRP Synthesis 72, NCHRP Reports 179* and *199, TRB Special Report 189*, and *TRB Record 603*. For a discussion of long-range regional planning see D. F. Schulz, *TRB Record 707*.

[104]The classic report, *Interregional Highways*, House Document 379, Seventy-eighth Congress, second session (1944) led 12 yr later to the establishment of the Interstate System.

Until recently, most urban transportation planning could be classed as comprehensive or strategic, looking 20 or more years into the future to a completed master plan. Today, however, the emphasis is on short-range programs or "sketch planning" for the near future.

Comprehensive Urban Transportation Planning[105]

Comprehensive computer-based urban transportation planning was launched in 1962 when the Federal-Aid Highway Act specified that it be undertaken in urban areas with populations greater than 50,000 as a prerequisite to the use of federal highway funds. The steps involved for a larger urban area are diagrammed in Fig. 3–5.[106] As indicated earlier, an inventory is developed through origin-destination (O-D), parking, and other surveys, and data showing the present physical facilities. Given this inventory, the next steps are to develop a computer-based model for estimating future travel and to check its accuracy against existing inventory data. Then assessments are made of goals and policies, land uses, economic activity and population, and preliminary alternative transportation schemes and networks devised to serve the area.

Projections of future travel between zones in or adjacent to urban areas are then made for the proposed schemes in four steps, trip generation, trip distribution, modal split, and traffic assignment, to determine the use of each system link by mode. For some studies, the modal split analysis precedes trip distribution. For example, trips by members of low-income families, those who do not own cars, or those with easy access to public transportation may be assigned to it before trips between zones are distributed among routes or modes.

TRIP GENERATION. Trip generation estimates generally are based on projected land use and economic activity; for example, existing housing or open land that will be converted to it and other land uses will develop a certain number of trips during certain hours. These are determined from O-D survey or other data such as that compiled in studies of similar situations. In addition, estimates are made of trip generation by activities such as work, shopping, education, and recreation. These may then be expressed as trip rates or in equations.

Data giving average trip generation by residential type are given in Table 3–3. These are only to provide scale and cannot be taken as representative of a particular community. Some other factors that would affect the anticipated number of trips and mode of travel are the number of automobiles owned by the households,[107] population of and its concentration in the study area, the availability of alternatives to the private automobile, and income.

[105]For a review of the development of urban transportation planning see E. Weiner in Gray and Hoel, op. cit. See Chapter 6 for a report on planning now under way for Houston, Texas.

[106]Details for conducting such computer-based studies are given in PLANPAC of FHWA and UTPS, developed jointly by FHWA and UMTA. See also R. S. Dial and L. E. Quillian, *TRB Record 771*.

[107]In large, core-concentrated cities, increasing auto ownership from 0.8 to 1.3 vehicles per dwelling unit increases the mean average daily trips from 5 to 7. For large, multinucleated cities, the increase is from 6 to 9, and for medium-sized cities from roughly 6 to 10. For greater detail see Y. Chan and J. Perl, *Transportation Engineering Journal of ASCE*, Sept. 1979.

TABLE 3–3. Approximate Number of Trips per Day from Typical Dwelling Units, by Residence Type and Mode of Travel*†

Type of Residence	*Density (Units/ Acre)*	*Vehicle Trips per Unit per Day, to and from*	*Percent of Trips in Peak Hour* A.M.	P.M.	Typical Auto Occupancy	Percentage Transit of Total Person Trips
Single family	1–5	9.3–10.2	8.0	10.8	1.62	3.2
Medium density	5–15	7.0	8.0	10.8	1.57	5.6
Apartments	15–60	6.0	7.9	10.8	1.56	12.4
Retirement community	10–20	3.5	12.1	12.1	1.48	6.0
Condominiums	10–30	5.9	7.1	7.1	1.56	9.0

*Source: *NCHRP Report 187*.
†This table is given merely to suggest ranges, since they vary tremendously.

TRIP DISTRIBUTION. Given the number of "trip-ends" originating or ending in a given area at appropriate times, a "trip distribution" model, calibrated using present-day records, is employed to determine future travel. The most common of these for analyzing work trips is the "gravity" model. It is not necessarily recommended for distributing trips for other purposes.

Gravity model formulations are based on the hypothesis that trips produced at an origin and attracted to a destination are directly proportional to the total trip production at the origin, the total trip attraction at the destination, a calibrating "spatial impedance" factor, and possibly a socioeconomic adjustment factor. This is expressed by the formula

$$T_{ij} = \frac{P_i A_j F_{ij} K_{ij}}{\sum_{j=1}^{n} [A_j F_{ij} K_{ij}]} \quad (3\text{–}1)$$

where
T_{ij} = trips produced at i and attracted at j
P_i = total trip production at i
A_j = total trip attraction at j
F_{ij} = spatial impedance callibration factor for interchange ij
K_{ij} = socioeconomic adjustment factor for interchange ij
i = an original zone number; $i = 1, 2, \ldots, n$
n = number of zones

Analyses with the gravity model can be made either for person trips or vehicle trips. Commonly, trips having home as an origin or destination are segregated from others.[108] Minimum path travel time between zones, including terminal

[108]In some models, trips are broken down into internal, external, and through and by trip purpose such as work, other home-based, and nonhome based.

times, is the simplest measure of spatial impedance, with shorter times indicating less resistance to travel and thus greater attractiveness. In some instances, this impedance factor is weighted with a travel cost or travel distance factor (see below). The socioeconomic factor in the formula reflects special conditions unique to a particular zone. It is almost never used with manual analysis and may be disregarded for computer-based solutions.

As indicated, the gravity model is calibrated by finding the numerical values of the various multipliers that approximate the findings of the travel survey and other base-year data. An iterative process is employed until an acceptable level of agreement between model and survey is reached, as determined by regression analyses. Then a matrix showing trips between all pairs of zones called a "trip table" can be prepared.

The gravity model can be adjusted to reflect congestion which increases travel time. If, for example, the traffic assigned to a link exceeds or approaches its capacity, originally assigned travel times are adjusted and a new trip-distribution analysis made.

MODAL CHOICE. Modal choice or split involves procedures for allocating trips between the various proposed travel modes such as automobile versus bus, automobile versus rail or other fixed guideway transit, bus versus rail or other fixed guideway, and possibly vans, taxis, or jitneys, some of which might operate on demand. As indicated earlier, there are situations where certain travelers have little if any choice among modes; the trips of this group are appropriately assigned as a part of the trip-generation procedure.

The earliest models for modal split used the socioeconomic characteristics of the areas of origin and destination as predictors. A second approach incorporated the characteristics of the competing transportation systems as well. Today, however, strong efforts are toward behavioral choice and policy-sensitive models.

Behavioral choice models, called "disaggregated" or "trip-based," attempt to appraise the "driving forces" that affect choices about travel. They recognize that individuals, in the social and spatial ordering of their lives, subjectively weigh differently such factors as their occupation and educational levels; income; driving and parking versus transit charges; congestion; time consumed including walking and waiting and the values attached to the time spent on each; trip purpose which would reflect such needs as meeting schedules or carrying parcels, comfort and cleanliness of vehicles; privacy; accidents or other personal safety considerations; dependability; and the opportunity to carry out such activities as conversing, listening to the car radio, or working or reading. Energy or environmental concerns such as the difficulty of obtaining gasoline or an unwillingness to contribute to the overall national fuel shortage or to air pollution may also influence some choices.

For behavioral models to become workable requires the combined efforts and interactions between behavioral scientists, planners, and engineers. A notion of the complexities to be faced can be gained from the following list of areas that must be considered[109]:

1. Define travel choices and their environments.
2. Define the stimuli that will affect choices.
3. Relate the perceived values of these stimuli to their measured values. As an example, how long does a motorist or transit user perceive a 10 min wait to be?
4. How are choices based on these stimuli made? Are all the factors affecting a trip merged simultaneously in the person's mind, are several factors weighed separately, or is the choice of mode unrelated to other trip considerations such as purpose?
5. Given these stimuli, how does the learning process operate to change present or future choices?
6. What other influences, for example those within households, affect choice?

The complexities pointed out here should clarify why estimates of travel behavior based on simple, single measures often have gone awry.

The wide variation in the subjective values that automobile drivers assign to time illustrates some of the complexities of behavioral evaluation. These are shown in Table 3–4. The large differences, particularly with income, trip purpose, and length, and time and conditions of the delay should be noted. Another variable that has been measured is lack of dependability; one study found time priced at five times the average rate when buses did not arrive. It also attached an added nuisance value of 3 to 4 min to waiting times.

Although all the factors listed above affect modal choice, presently used models do not attempt to incorporate them. *NCHRP Report 186* recommends an exponential modal split formula for what it calls "impedance." It sums, for each mode, vehicle time, excess time at 2.5 times vehicle time, and a time equivalent of trip cost divided by income. Other modal split analyses permit comparisons directly in money terms. In such an approach, time is priced using figures such as those in Table 3–4.

"Policy sensitive" models attempt to measure the impact of public policies or policy shifts on such factors as travel demand or modal choice. A typical analysis would attempt to relate the level of fares or subsidies to modal choice. Another would appraise the effect of fuel prices on the behavior of automobile users. Still another, which would use subjective measures, would attempt to establish a relationship between the level of policing and transit ridership.

TRAFFIC ASSIGNMENT. Traffic assignment is the final step in forecasting future transportation needs or use and in determining the capacity that should be provided in various links.[110] First of all, decisions must be made (1) as to the specific network or networks that will be analyzed, and (2) whether sizing will be for peak hour or for average conditions and at what level of service. Also,

[109] This list is based on D. Brand, in *TRB Record 569*. His presentation offers a detailed analysis and an extensive bibliography.
[110] The discussion here is concerned with assignments to alternative routes or links, as contrasted with choices among modes, as discussed above.

TABLE 3–4. Suggested Dollar Values for Time Saving per Traveler Hour by Trip Purpose and Income Level*†‡

		Trip Purpose§‖	
Annual Family Income	*Time Saving (min)*	*Average Trips*	*Work Trips*
$5,000	0–5	$0.07	$0.15
	5–15	0.58	0.77
	Over 15	1.26	1.26
10,000	0–5	0.13	0.31
	5–15	1.16	1.55
	Over 15	2.52	2.52
15,000	0–5	0.21	0.48
(average)	5–15	1.80	2.40
	Over 15	3.90	3.90
20,000	0–5	0.27	0.62
	5–15	2.32	3.10
	Over 15	5.03	5.03
30,000	0–5	0.41	0.92
	5–15	3.48	4.65
	Over 15	7.55	7.55

*Source: *Red Book*, published by AASHTO.
†1975 values.
‡To convert to dollars per vehicle hour for passenger cars, the following multipliers in adults per car are suggested: Work, 1.22; social-recreational, 1.98; personal business, 1.64; average, 1.56.
§The *Red Book* proposes, for automobiles, an average value of $3 per vehicle-hour.
‖For transit passengers, the *Red Book* proposes 1.5 times the values given for waiting or walking times longer than 15 min. For initial waiting time or for walking or waiting where comfort or safety conditions are below average, the suggested multiplier is 2.0.

preliminary estimates of the facilities required on each link (for example, number of highway lanes or number of transit cars or trains) must be made. Travel times between origins and destinations next must be calculated.[111] Then the comparative travel times between origin and destination by each feasible route on each proposed system are calculated. For highways, some analysts would recommend increasing this travel time by introducing a "penalty" wherever there are traffic signals or left turns in the routes. On transit schemes, waiting-time estimates would be included. Finally, traffic is assigned to various routes. It is often assumed that travelers take the "least time" course on an "all or none" basis. Assumptions other than least time include least perceived or observed cost, least distance, or most safe, or some weighting of these and other factors. To better replicate trip-making behavior, a probabilistic approach that makes assignments to more than one route is sometimes employed. However, this further complicates an already complex analysis.

Finally, the zone to zone assignments must be examined to see that the volumes assigned to individual links, connectors, or intersections do not differ too greatly from those used for determining travel times. If they do, the modeling

[111]Data comparable to that in Figs. 8–6, 8–8, 8–9, and 8–10 are used for highway analyses. See also *TRB Record 682*.

process should be repeated with altered model parameters or with the travel network changed to correct its deficiencies. Again, if parking has not been an element of the overall O-D study, checks must be made to see if existing or planned parking facilities downtown or at outlying exchange points will handle the assigned vehicles; if not, modifications to modal split assignments must be made or the parking segment of the plan revised. For, when all is said and done, all models represent attempts to approximate and simulate reality in an intelligible manner, and checks to be sure the simulations are complete and balanced are a must.[112]

By referring back to Fig. 3–5, it can be seen that, after the proposed systems are outlined and financial and economic analyses are made, there should be feedback and a reanalysis of the transportation system's effects on economic activity, population, and land use. If the original projections are not verified, the study sequence should be repeated. Many earlier planning studies have been criticized because this reanalysis was not made, thus neglecting the dynamic effects of transportation. However, as indicated earlier, there is a paucity of knowledge from which the effects of transportation on land use and economic activity can be projected.

Given that all findings of a comprehensive urban transportation study have been positive and a scheme recommended, the concluding steps to implementation would be to establish the governmental mechanisms for financing, building, operating, and maintaining the system.

The planning scheme illustrated by Fig. 3–5 has, except by implication, ignored the place of citizen and political inputs to the planning process. A lesson learned repeatedly is that, without it, planning is largely a useless exercise.

Looking back over time, it seems safe to generalize that, in part because of federal requirements, urban freeway planning through the 1960s followed the complex scheme outlined above, except for the feedback loop. Similar planning approaches brought recommendations for rail rapid transit schemes. The first of these was BART. From today's perspective, that study could be considered deficient in at least two ways. The first is that it examined only two alternatives, rail transit or the private automobile on an expanded freeway system. The second was that estimates of costs and revenues for BART were far too optimistic. Instead of being self-sustaining as expected, the system requires an operating subsidy. In addition, property taxes must repay much of the capital cost. From the BART and subsequent experience, it seems clear that implementation of any new heavy-rail schemes will require substantial capital grants and subsidized operation.

[112]There is an extensive literature on the subject of modeling and forecasting urban travel. Findings are reported in, among other sources, P. R. Stopher and A. H. Meyburg, *Urban Transportation Modeling and Planning*, Lexington Books, Lexington, Mass., 1975; *TRB Record 526, 527, 534, 563, 569, 583, 587, 592, 599, 610, 637, 673, 707, 708, 723, 728, 750, 751, 765, 767, 771, and 775, TRB Special Report 149*; R. W. Lyles in *Transportation Engineering Journal of ASCE*, Mar. 1979; P. R. Stopher, and C. G. Wilmot, ibid. Nov. 1979; J. E. Bennett, U. Landau, and Y. Fedorowiez, and R. L. Smith Jr. and T. S. Brennan, ibid. Jan. 1980; and P. R. Stopher, ibid. Jul. 1980.

Shorter-Range Urban Transportation Planning

Comprehensive urban transportation planning, as outlined above, was carried out in many urban areas in the 1960s. It was intended to develop an "ultimate system." Only after the master plan was produced would more specific problems be tackled. This approach has been criticized for failing to recognize that details of the project, including the timing of its elements, also had to be considered from the beginning.[113] Furthermore, dissatisfaction with the single-focus "automobile" solution arose in the late 1960s, and with it came confusion in purpose and dissatisfaction with and de-emphasis of comprehensive planning.[114] Today there has been a shift in emphasis from planning for entire urban areas to planning for subareas and corridors, and to investigation of improvements to existing highways and transit operations. This means simpler and less time-consuming data gathering and analyses and short-range rather than long-range objectives. Little new O-D data are being gathered; rather, the old is being updated and behavior is being predicted from attitudes rather than from earlier findings. There is much more attention to social, environmental, and energy concerns and to participation by citizens and political decision makers. All these changes have focused attention on noncomputer techniques of analysis or at least on far simpler computer approaches coupled with simulation and graphics for quickly presenting the consequences of various alternatives.[115] Many of the findings are directed at Transportation Systems Management (TSM) as described in more detail below.

These simpler analyses have generally been referred to as "sketch" or "operational" planning. They look ahead possibly 5 yr and draw on the appropriate portions of the four steps of trip generation, trip distribution, modal choice, and traffic assignment discussed above. However, because specific problems of limited scope are being addressed, investigators have developed many approaches. These are far too numerous to outline here.[116]

Transportation Systems Management (TSM)

Since 1975, the Transportation Systems Management (TSM) approach has been required for small, short-term transportation investments involving federal funds. TSM now describes many similar activities carried out by numerous agencies of government. It starts with the premise that, in the main, our highway system is complete. It follows that, more and more, planning must focus on the question of how to use or modify this system to better accomplish the wide variety of

[113]See, for example, M. L. Manheim in Gray and Hoel, op. cit., and *NCHRP Report 156*.
[114]Some critics claim that comprehensive planning is based on concepts from the 1950s and desires and behaviors of the 1960s, and has no place in the 1980s.
[115]See, for example, *HRB Record 455* and *TRB Record 553, 559, 606, 619,* and *677*.
[116]Basic references include NCHRP Reports 208, 209, 210, 211, and *212*. Two others, in which 40 of these "sketch planning" procedures are outlined, are *NCHRP Reports 186* and *187*. Approaches for smaller communities are discussed in *NCHRP Report 167* and *TRB Special Report 187*. See also *TRB Record 569, 582, 634, 638,* and *730*, and the references given earlier in this section. J. W. Billheimer et al. *TRB Record 724*, assess models for paratransit.

public purposes that it serves or affects. Stated differently, the TSM focus is on operation, regulation, and management rather than on large investments. Some among the public purposes of TSM and the approaches to accomplish them are as follows.

1. Improve personal mobility on all modes of private and public transportation. Objectives would include reducing travel time, enhancing reliability and quality of service, and better fitting consumer needs. Strategies would include (a) bringing a shift from the private automobile to other forms of transportation to reduce congestion; (b) diminishing peaks by encouraging staggered work shifts and flextime; and (c) removing traffic bottlenecks and improving traffic control.
2. Conserve resources, particularly energy in all its aspects, by reducing both overall travel and private motor vehicle use. This will, in turn, decrease energy consumption and motor vehicle wear and tear. Approaches could include (a) improving traffic flow and reducing delays and congestion for automobiles and public transit vehicles, (b) encouraging carpooling, vanpooling, and greater use of public transportation as an alternative to the automobile,[117] and (c) making bicycle use and walking more feasible and attractive.
3. Improve public safety and health. Efforts can focus on reducing traffic accidents and mitigating the effects of those that do occur, lowering stress for both travelers and others, and reducing air pollution by decreasing vehicle use.
4. Provide suitable transportation for the disadvantaged. Strategies include (a) arranging for volunteers, (b) modifying equipment on transit, and (c) subsidizing service by vans or taxis.
5. Enhance the environment and community quality. Reducing noise and vibration, minimizing community disruption, improving area aesthetics, and pointing toward better land use are among the focuses.
6. Improve economic efficiency. This can be done by enhancing the capacity, efficiency, and safety of existing highways for moving both people and goods, reducing costs and saving time in urban travel, and minimizing adverse economic impacts on the surroundings. One largely overlooked approach is to reduce peak transit use, since it costs two to three times the average to provide for it.

The purposes and strategies listed above are not new. They represent sound professional objectives and practices that are being carried out today. What TSM may do is to force technical and management people from the wide spectrum of professions and agencies that deal with transportation to work more closely together. By so doing, many of the administrative, political, and human barriers that make change difficult may be overcome through common understanding and effort.

Certain TSM efforts may accomplish several public purposes. For example, improving personal mobility by traffic engineering measures may also conserve resources, improve public safety and health, enhance the environment, and improve economic efficiency. On the other hand, attempts to advance one of these purposes may mean that others will lose ground. For example, smog con-

[117]Increasing motor-vehicle fuel taxes, tolls, and parking charges, creating automobile-free areas, and granting carpools and buses priority are among the possible strategies for reducing private automobile use.

trol devices placed on automobiles to reduce air pollution increase energy use and make transportation more expensive. Again, providing suitable transportation for the disadvantaged by modifying bus design or providing special services will increase the operating costs and slow the operation of already financially distressed transit systems. And extending bus service into the suburbs or offering more frequent and extended-hour schedules to encourage transit ridership as an alternative to the automobile will further increase transit's financial difficulty and may not save energy.

The traditional comprehensive urban transportation planning approaches outlined above usually do not apply to TSM problems. First of all, comprehensive planning has primarily been concerned with capital-intensive, long-term investments. TSM is not. Furthermore, comprehensive planning has not dealt with many of the small, local, and often nonquantifiable issues which TSM raises and which, in most cases, are thrashed out in the political arena. On the other hand, many of the findings developed for comprehensive transportation planning can be applied to TSM problems.

Introducing TSM even within a single operating agency often is difficult because of conflicting interests. For example, attempts by a public transportation agency to employ volunteers or taxis to accommodate the disadvantaged, to eliminate or reduce nonprofitable services, or to encourage vanpools or carpools threatens the jobs of transit workers. Again, if a highway agency gives priority to buses and possibly carpools, motorists bring strong pressures against it.

Developing TSM programs that affect more than one governmental body or interest group raises even more difficult problems. For example, if a highway agency improves traffic flow, this may keep travelers in their automobiles, thereby reducing transit ridership and increasing air pollution to the distress of officials concerned with those matters. Again, prohibiting truck use of downtown streets at certain hours to improve traffic flow will create serious problems and increase costs for truckers and merchants.

As the few examples cited here should make clear, techniques are now available for instituting many TSM efforts. The difficulties lie in defining which among the multiple objectives listed above are to be given priority, getting agreement and cooperation from the often-competing interests, and overcoming the inertia and foot-dragging in the bureaucracies that control or affect transportation.[118]

[118]Many of the strategies that might be employed for TSM projects are outlined under specific topics in this book. Other sources of information include A. D. May, and R. Westland, *Supplement to Traffic Engineering and Control,* Feb. 1979; *TRB Special Report 153, 172, 184 and 190; NCHRP Report 205;* S. C. Lockwood in Gray and Hoel, op. cit.; *TRB Record 519, 719, 722, 746, 751, 761, 765, 767, and 772;* D. W. Jones Jr. and E. C. Sullivan, *Transportation Engineering Journal of ASCE,* Nov. 1978 (an assessment); M. A. Kennedy and W. Kudlick, ibid., Sept. 1979 (application to activity centers); *NCHRP Synthesis 73,* and A. A. Tannir and D. T. Hartgen, *TRB Record 677* (changed work schedules); R. Safavian and K. G. McLean, *Traffic Engineering,* Mar. 1975 (flextime); and S. R. Stokey et al. ibid, Jan. 1977 (ride sharing). The role of and strategies for applying TSM to parking are discussed by M. J. Demetsky and M. R. Parker in *TRB Record 682.* For a many-faceted TSM study for midtown Manhattan, see W. H. Kraft and S. I. Schwartz, *Transportation Engineering Journal of ASCF,* Mar. 1981.

"Marketing" of transportation can well be considered a TSM strategy. Its aim is both to "sell" transit and attract patrons to and keep them traveling by public transportation. It requires a continual focus on the public and the consumer rather than on the "product." This means promoting transit rather than reasoning that transit is not at fault but that travelers are irrational.

Marketing goals have sometimes been divided into long-run and short-run. Long-run goals are associated with community objectives such as having efficient and effective public transportation as a means of saving energy and improving the environment, providing mobility for the disadvantaged, and promoting proper community development. Efforts are directed toward convincing the public and its spokespeople that transit is desirable and merits political and financial support.

Short-run marketing objectives are, as mentioned, to provide services that will attract and keep patrons. First of all, they must be encouraged to try transit. Media publicity is one means. This is backed up by disseminating information on such details as routes, schedules, pickup points, and fares and fare-collection arrangements. Mechanisms include telephone- or mail-inquiry services and posted notices or displays at terminals and transfer stations, and along malls if they exist. Keeping patrons, once drawn, requires efficient, dependable, attractive, and safe vehicles and access and egress. Among the most difficult images to project is that transit patrons are secure and will not be victimized by hoodlums or gangs.

In the past there has been reluctance on the part of public transit agencies to engage aggressively in marketing. It seems clear, however, that this is a necessary ingredient of any such operation.[119]

Transportation brokerage is another marketing concept; with it a public agency is created to facilitate and coordinate arrangements for the effective use and sharing of all transportation facilities. These could include carpools, vanpools, employer-provided buses and vans, school and private buses, commuter subscription services, and taxis. It could also arrange and coordinate transportation for public and welfare services. The value of such an agency is that it cuts across many of the functional organizations such as highway and traffic engineering, transit planning, transit operations, schools, regulatory agencies, and social services. With its single focus on coordinating transportation, the brokerage agency may be able to get around many institutional, financial, and regulatory constraints imposed by individual interests and concerns.

A successful brokerage arrangement has been demonstrated in Knoxville, Tennessee, and others of more limited scope are being implemented. The promise is that, through such means, effective and less expensive transportation can be provided by marshaling, coordinating, and using to better advantage all the resources in each community.[120]

[119]This discussion is intended only to introduce this complex subject. References for further study include the publications of the American Public Transportation Association, several sections of Gray and Hoel, op. cit., *TRB Special Report 181* and *184*, *NCHRP Reports 208, 209, 210, 211,* and *212*, and *TRB Record 625* and *735*.

[120]See, for example, F. W. Davis, Jr., *TRB Special Report 181* and, with R. P. Aex, *TRB Special Report 184*. Also, see Gray and Hoel, op. cit. and *TRB Record 719*.

Planning Individual Routes and Segments

Planning-survey and parallel research data underlie many decisions of importance regarding individual highway routes or segments. For example, the road inventory of condition, accident experience, maintenance cost, and other characteristics of the existing road, coupled with traffic counts, provide the basis on which projections into the future can be made. From these, economy studies, programming, and other procedures for setting priorities for and levels of improvement can be established. Also they provide data on which designs of new facilities or modifications of existing ones can be made. These include the number of lanes, standards for alignment and grade, the geometry of intersections or interchanges, and, based on the expected number of trucks classified in terms of axle numbers and loads, the choice among pavement types and thicknesses. Only with this approach are reasonable designs possible.

As indicated earlier, forecasting future traffic volumes and types for the years ahead is done differently for rural and for urban areas and for small (50,000 or less) and larger urban areas.[121] Usually this estimate is subdivided into several elements, and separate projections are made for each. Commonly used classifications under a variety of titles include some or all of the following:

1. *Existing traffic*. This is the traffic currently using a facility that is being reconstructed or upgraded. In those instances where a new facility such as a freeway, expressway, or bypass roughly parallels an existing highway that will be left in service, existing traffic on the new facility will be zero.
2. *Area traffic*. This is traffic in the area which will not use the facility when it is completed. It may be affected favorably or unfavorably, however, through changes in traffic patterns and volumes.
3. *Normal growth traffic*. This is traffic that will come into being in the future as the result of expected trends in regional, state, and local growth, including population changes as well as area-wide changes in land use.
4. *Development traffic*. This traffic results because of shifts in land use in the general area served by the facility. The substantial increases in traffic brought by the shift of businesses from traditional locations to lands adjacent to freeways is a dramatic illustration of development traffic.
5. *Diverted or transferred traffic*. This is redistributed traffic, diverted to the new facility from other traffic arteries or from other forms of transportation.
6. *Generated (induced) traffic*. This traffic comes into being because the new facility is available. In some but not all instances, generated traffic on new arterials, ranging from 5 to 30% of existing traffic, has developed within 2 to 3 yr after the new facility was opened to traffic. For the Interstate System, generated traffic has been 30 and 60% for rural and urban segments, respectively.

Where a new facility is to generally parallel an existing one, the projected traffic traveling between zones of origin and destination must be assigned between the two routes. It must be recognized that driver choices are not made simply; they weigh such factors as time, distance, and congestion in making the decision. The difficulty in making predictions is illustrated by the earlier discus-

[121]See A. D. Jones and W. L. Grecco, *HRB Record 472*, for a discussion of planning for arterials in small urban areas. Estimates for larger areas come from some form of O-D survey.

sion of the variables in choices among modes. They are further complicated by a lack of knowledge of future changes in, for example, traffic control measures. A variety of assignment schemes are used by various agencies. The simplest is to base it "all or none" on travel time.

Transportation Planning for Resource and Recreational Development

When the main purpose of a highway or other transportation facility is to develop resources, the first consideration is that the economic gains from the development exceed the losses. Methods for making such comparisons are outlined in Chapter 4. The next is whether or not it can be financed and, if so, who will pay. Sources of funds are discussed in Chapter 5. Other impacts, such as those on the environment, may or may not be important depending on such factors as the impacts on other uses of the area being developed or traversed. Fewer objections to such facilities will be encountered in less affluent countries where economic development carries high priority. In the United States today, however, conservationists have challenged many economic developments on environmental grounds and have forced delays or changes in plans or have blocked them completely. Decisions about recreation also bring clashes from competing interests. For example, there are the conflicts between those promoting or wishing to use ski resorts, camp grounds, and other developments versus those wishing to leave the areas undisturbed. Heated arguments and political and judicial actions are a common result. It follows that planning for resource and recreational development is not an easy task and requires not only technical knowledge but also an understanding of the broader issues and skill at resolving conflicts.[122]

The U. S. Forest Service, which administers some 8% of the land area of the country, has been making intensive studies of its transportation and related problems. Researchers have developed a number of transportation models and approaches. One set, which involves network analysis techniques, includes procedures for estimating recreational demand, for allocating it among forest destinations, and for minimizing the cost of timber transportation. Other studies have simulated the relative economy of two-lane roads and one-lane roads with turnouts and have provided means for comparing vehicle running costs and times on different horizontal and vertical alignments.[123]

PROBLEMS

3–1. By visiting an appropriate official, determine the step-by-step procedure that is followed in carrying a proposed major highway or street improvement through the planning stage. Prepare a brief report on your findings including:

[122]For a discussion of the tradeoffs in environmental issues, see R. E. Rechel and R. Witherspoon, *HRB Record 408*. For other recreational and forest studies see *NCHRP Report 44, HRB Record 472, TRB Record 569, 582, 702,* and *710, TRB Special Report 160,* and *ITE Journal*, Jul. 1979.
[123]See E. C. Sullivan, *TRB Special Report 160*.

a. An appraisal of the strategy that was employed (use Fig. 3–4, as a basis for comparison).

b. A flow diagram, with time as the abscissa, showing the agencies or groups that are involved and when this involvement occurs.

3–2. By visiting an appropriate local official (engineer, planner, etc.) determine how his or her agency is meeting the federal or state requirements for:

a. A master plan for land use.

b. A comprehensive plan for transportation.

3–3. Investigate the financial solvency and the level of service provided by the public transit company or agency serving the community where your college is located. In particular, find out what innovations such as express buses or "demand-actuated" services have been or are being planned or employed.

3–4. Investigate the procedures and computer hardware and software employed by an appropriate local or state highway or transportation agency in predicting future movements of people and vehicles.

3–5. Investigate the procedures being used in the community where your college is located to encourage transit ridership, carpooling, and vanpooling.

3–6. What, if any, special provisions are made by the public transportation agency of your community to provide transportation for the handicapped?

3–7. Determine the status of plans for public transportation facility improvement or expansion in the community where your college is located. As directed by the instructor, this can be an individual or group project leading to an oral or written report dealing with appropriate items from the following list:

a. What agency leads in planning? Does this agency have in-house capabilities or does it rely on consultants?

b. What other agencies or groups are involved, when do they enter the planning process, and how much influence do they seem to have?

c. Are TSM approaches being employed and if so, which and how?

d. In which of these activities and to what degree do officials and/or faculty of your university or college participate?

3–8. Assume that a 20-acre parcel of land, now vacant, is being considered for residential development. Alternative acceptable uses are (1) medium-density housing, (2) a retirement community, or (3) condominiums. For each alternative, and employing data from Table 3–3, estimate:

a. The total number of vehicular and person trips per day for developments of minimum and maximum density.

b. The total number of such trips in the peak AM and PM hours for developments of minimum and maxium density.

c. At 45 seated passengers per vehicle, how many buses would be required during the morning peak hour to accommodate these trips based on the transit ridership percentages given in Table 3–3?

d. Answer part *c*, assuming that 50% of the travelers use transit.

4 ECONOMY AND RESOURCE USE

Governments have, of necessity, provided certain facilities that the private sector could not furnish. Among them are the highways on which private automobiles, trucks, and buses travel, and a large portion of public transportation. The intents of expenditures for highways and public transportation are to raise the level of the entire economy by providing for easy access to work places and ready transportation of goods[1]; to assist in problems of national defense; to make easier the provision of community services such as police and fire protection, medical care, schooling, and delivery of the mails; and to open added opportunities for recreation and travel. Transportation benefits the landowner because access makes property more valuable. Certain improvements to highways benefit motor-vehicle users through reduced cost of vehicle operation, savings in time, reduction in accidents, and increased comfort and ease of travel. On the other hand, such improvements consume resources, including land, that might be used for other productive purposes by individuals or by government. The vehicles traveling the highways produce air pollution and noise. From the point of view only of resource use, then, transportation improvements can be justified only if, in net sum, the consequences are favorable, that is, if cost reductions to users and other beneficiaries of the improvement exceed the costs, including some allowance for return on the money invested. There are, as has been indicated before, numerous other factors to be considered.

Highway economy was under discussion over a century ago. W. M. Gillespie, Professor of Civil Engineering at Union College, in his *Manual of the Principles and Practice of Road Making,* stated that "A minimum of expense is, of course, highly desirable; but the road which is truly the cheapest is not the one which has cost the least money, but the one which makes the most profitable returns in proportion to the amount expended upon it."

The first detailed attention to highway economy developed almost 50 yr ago at Iowa State College. It focused largely on the relative economy of various road surfacings and, later, on the costs of motor-vehicle operation. The advent of the

[1]Among numerous estimates of the cost per ton-mile for transporting goods by various means, stated at appropriate local wage rates and other costs, are: on human heads, 88 cents; by oxcart, 68 cents; by trucks on passable roads, 15 cents; by motor carrier in the United States, 5.5 cents.

statewide planning surveys with the masses of data developed by them brought attention to many other factors of importance to the overall problem. Even so, attention to highway economy as a topic for detailed research and as a factor in decision making has been relatively limited and often incompletely or incorrectly done.[2] An exception was that economic comparisons of alternative routes on the Interstate System were required by federal regulations. Many of these were based on the *Red Book,* developed by the AASHO Committee on Highway Design.[3] Economic analysis on federal-aid projects was one factor listed in the Federal-Aid Highway Act of 1970 (Sect. 136) which required that the Federal Highway Administration:

> promulgate guidelines designed to assure that adverse economic, social, and environmental effects . . . have been fully considered . . . and that final decisions on the project are made in the best overall public interest taking into consideration the need for fast, safe, and efficient transportation, public services, and the costs of eliminating such adverse effects as (1) air, noise, and water pollution; (2) destruction or disruption of man-made and natural resources, aesthetic values, community cohesion, and the availability of public facilities and services; (3) adverse employment effects, and tax and property value losses; (4) injurious displacement of people, businesses and farms; and (5) disruption of desirable community and regional growth.

In contrast to the relative inactivity in highway economy, federal agencies in the water field have highly developed procedures for economy studies. These began when Congress, in the Flood Control Act of 1936, specified that the benefits from proposed flood-control projects, to whomsoever they accrue, should be in excess of estimated costs. A large body of literature has developed which is a fruitful source for students of economy. Also, agencies such as the World Bank have insisted that projects to which they make grants or loans be justified, primarily on an economic basis.

Planners, political scientists, and environmentalists, in particular, have often been outspoken against economy studies because they misunderstand the purpose for which they are made. The strongest advocates of economic analysis recognize that where public policy is involved, economy studies are not intended to supply the final answer. Rather they aid decision making by determining the economic resources gained or lost if a given alternative, including that of doing nothing, is chosen. These can then be weighed against financial and political considerations and the irreducibles in reaching a decision. To illustrate, suppose that one of two feasible locations for a highway infringes on a park, but reduces the length of the highway by one-fourth of a mile and thereby offers lower construction and user costs. Knowing the cost differences, and other pos-

[2]For the results of a survey of state highway activities, see M. Roddin and D. Andersen, *TRB Record 550*.

[3]This document is titled *Road User Benefit Analysis for Highway Improvement*. The original version was issued in 1952 and an update in 1960. A new edition which provides much of the factual input for this chapter was issued in 1977 under the title *A Manual on User Benefit Analysis of Highway and Bus Transit Improvements*. Hereafter it will be referred to as the *Red Book*.

sible benefits such as energy saving and reduced air pollution, the agency that must make the decision, possibly with citizen involvement, can weigh these benefits to the public on the one hand against the damage to the park and neighborhood on the other. The decision might be to leave the park undisturbed. It could be to acquire another park site. Possibly the savings would outweigh the disruption and the park route would be adopted. In any event, the decision would have a rational or partially rational basis rather than being based solely on opinions and emotions. On the other hand, there are many design and administrative decisions that do not involve public policy. These should be made by selecting the alternative that is cheapest in the long run. In other words, the findings of an economy study, reasonably interpreted, should govern.[4]

This chapter focuses primarily on the economic aspects of highways, on which motor vehicles—automobiles, trucks, and buses—operate. However, with appropriate data, the approaches and techniques given here can be applied to other transportation modes or to comparisons among modes. Also of concern in today's world is the increasing cost and impending shortage of energy, particularly petroleum products needed to construct and maintain transportation facilities and to propel the vehicles on them. The last section of this chapter deals briefly with this topic.

A FRAMEWORK FOR ECONOMY STUDIES

Possibly the most difficult and error-prone phase of economy studies lies in placing the study in the proper framework or perspective. If this phase is done incorrectly, the most reliable data and flawless procedures for analysis will give erroneous results. Some of the guidelines to be followed in developing this framework are:

1. *Economy studies are concerned with forecasting the future consequences of possible investments of resources. Past happenings, unless they affect the future, are not considered.* This "forward" look is distinctly different from the "backward" look of accounting practice. This difference is illustrated by the discussion of incremental and sunk costs later in this chapter.
2. *Each alternative among which choices are to be made must be fully and clearly spelled out.* As an example, if a freeway is proposed to parallel a busy street, there will be vehicle operating cost savings not only to those diverted to the freeway but, possibly, also to the remaining travelers on the street. On the other hand, traffic using this same freeway could increase congestion and vehicle operating costs on other traffic arteries. This likewise should be recognized. Thus, the first step in analysis is to make a complete list of consequences, both economic and other. Such listings, for an urban freeway, are given in Tables 3–1 and 3–2.

[4]The presentation in this chapter is, of necessity, brief. For a far more comprehensive presentation, including extensive bibliographies, the reader is referred to R. Winfrey, *Economic Analysis for HIghways,* International Textbook, Scranton, Pa., 1969, and R. Winfrey and C. Zellner, *NCHRP Report 122*. See also *Synthesis 5* of Transportation Technology Support for Developing Countries, TRB.

3. *The viewpoint taken in the analysis must be defined and observed.* In the case of private individuals or businesses, for example, the viewpoint is narrow; the aim of the study is to determine the consequences of alternative courses of action primarily as they affect the individual or business. In the public-works field, however, when considered at the national level, the approach must be broad and all-inclusive; it must weigh the consequences to all in the nation who will be affected by the proposed improvement. The Flood Control Act of 1936, in which Congress stipulated that *the benefits, to whomsoever they accrue, shall exceed the costs,* expresses this viewpoint. Analysts for lower levels of government making economy studies for public projects that involve financing from higher levels often find themselves in a dilemma. From the viewpoint of local politicians, citizens, and officials, grants of money from higher levels often are viewed as costing nothing; in economic terms, they are a "free good." Taking this local viewpoint, the only consequences are the costs and benefits of the local community. At the same time, reviewers of such an analysis who represent federal or state financing agencies may insist that the study take their broader viewpoint. A possible resolution of the dilemma is, of course, to make two analyses, one from the local and the other from the higher viewpoint.

   A particularly troublesome aspect of the viewpoint dilemma is the *transfer* problem. To illustrate, suppose that one among several proposed freeway or transit plans for a large urban area permits the development of a large shopping center in a small city in the urban complex. This will increase land values and tax revenues in that community; but other nearby communities may suffer decreases in property values and tax revenues so that the gains to one community are largely offset by losses to the others. In this instance, from the viewpoints of the local communities, some will have economic and other gains; others will have losses. But from a regional, statewide, or national viewpoint, neither a gain nor loss will occur; rather there will be a transfer of resources between communities.[5]

4. *A clear distinction must be made between economic analysis (the use of resources) and financial considerations (the use of money).* It has already been indicated in Chapter 3 that decision making involves dealing with three elements in sequence. These are *(a)* economic, which is the use of resources; *(b)* financial, which deals with getting and expending money, and *(c)* political and administrative, a catchall phrase for the nonquantifiable forces that bear on the decision. It was also indicated that rational decisions were more likely to be reached if the best alternative from an economic point of view were tested in sequence for its financial and political and administrative viability. If this alternative failed either of these two tests the next most viable alternative would then be examined and so on.

   In the past, analysts sometimes erroneously have included financial considerations in economy studies. A first illustration is the practice of including interest as a cost *only* if money is to be borrowed to finance a project. But it can be seen that, regardless of the source of funds, the same resources are consumed in constructing, maintaining, and operating the proposed highway whether the project is financed with borrowed funds or with current revenues. Two more among the common situations where financial thinking can lead to errors in economy studies involve *allocated* and *sunk* costs. These are discussed in greater detail later.

[5]Some readers may prefer the view that freeways and other highway investments are damaging to the communities through which they may pass. Regardless of which assumption is made in this example, there will be gains or losses from the local viewpoint which will be viewed as "transfers" by higher levels.

Yet another source of error when economics and finance are mixed concerns situations where there is unemployment, or supply and demand are otherwise out of balance. Here, market prices may not be true measures of resource use; rather a *shadow price* should be employed. Consider, for example, the charge to be made against a project for an hour's work by a laborer. From a financial point of view, the appropriate charge would be the wage, plus all fringe benefits and all other costs of having the laborer on the job. But suppose this person would otherwise be unemployed. In this instance, there is no change in resource use because the person is put to useful work; it follows that the economic cost of an hour's work is zero. There might, of course, be costs involved in getting the person to work or in providing different or better food or living arrangements and it would be proper to include these costs in the economic analysis.

The shadow price concept has particular application when evaluating projects intended to provide work for the unemployed. Again, it can be very important in developing nations where there is a large idle work force coupled with a shortage of foreign exchange. In such instances, it may appear financially attractive to import labor-saving equipment; but from an economic (resource use) point of view, an analysis may show that it is advantageous to use labor-intensive methods.[6]

5. *Double counting of costs or benefits must be avoided.* In evaluating systems or projects, there is the hazard that costs or benefits will be included more than once in the analysis. For example, if a substantial improvement is made in the highway serving an area, the costs of vehicle operation to users will decrease; at the same time, land values in the affected area may increase because of the improved access. In this instance, only one of these two benefits should be included in the analysis; to include both is to double count. In other cases, there is the hazard that costs will be counted twice, as when both the total construction costs for all individual projects and the total construction expenditures of an agency would be included in comparing different transportation schemes.
6. *Taxes should not be included as costs in economy studies for public agencies.* Since taxes on vehicles, tires, fuel, and other items represent a substantial portion of the cost of owning and operating motor vehicles, their inclusion or exclusion in the cost of vehicle operation can make a substantial difference in the results of a study. But, from the public viewpoint, taxes of themselves do not consume resources; rather they represent transfers from private parties to the government. Even if the taxes all go to pay for roads, their inclusion in an analysis would be classified as double counting if the cost figures also include expenditures for construction, maintenance, and other agency costs. It is, of course, proper to include taxes in economic analyses for individuals or nonpublic firms or agencies. From their points of view, taxes do consume resources.
7. *Findings regarding resource consumption (market values) should be reported in the analysis separately from nonresource elements (nonmarket values).* Table 3–1 is titled "Direct Effects of Freeway Construction and Use." In this table these effects are segregated into market values, meaning resources whose value can be measured in the market place, and nonmarket values, which do not represent resource consumption so that the market cannot provide suitable measures. To illustrate, savings in travel time to commercial vehicles such as trucks and automobiles when traveling on busi-

[6]For a more detailed discussion of shadow pricing and several other aspects of developmental projects, see H. M. Steiner, *Public* and *Private Investments,* Wiley, New York, 1980.

ness missions represent (or can represent) a saving in resources. On the other hand, a reduction in the time of the journey to or from work does not save an economic resource. Rather the saved time goes to leisure or recreation. This value cannot be measured in the market; it must be imputed by observing people's behavior. Again, a human life saved by accident reduction measures has economic value which can be measured by the saved productive capacity of the individual. On the other hand, the social costs of pain, suffering, and deprivation cannot be measured in the market. Suggested measures such as legal settlements are, in an economic sense, transfers within society.

The viewpoint taken in this book is that market benefits and costs on the one hand and nonmarket ones on the other should be reported separately to the decision maker since one represents resource consumption while the other does not. This is not to argue that market consequences are more important than nonmarket ones. These could well vary with, among other factors, the degree of affluence of each society. Rather, by reporting the consequences separately, the decision maker rather than the analyst does the weighting of one against the other. There are some who argue that the nonmarket consequences, which are a measure of the public's desires, have equal validity to market ones and that the distinction proposed here is meaningless.

COSTS OF TRANSPORTATION FACILITIES

Determining Relevant Costs

The total first cost for improvements to a transportation system or segment includes engineering and design, expenditures for planning, the outlay for acquiring rights of way, and the costs of constructing roadway, structures, and pavements. Selection of the cost items to be included in and excluded from specific economy studies requires straight and careful thinking. A detailed discussion is beyond the scope of this book. However, four of the most important considerations are as follows:

1. In general, *allocated costs,* used for accounting purposes, should be omitted from economy studies. To illustrate, a given percentage may be added to estimated project costs for administration, planning, and engineering overhead. These costs probably will be incurred whether or not a specific project is undertaken; if so, they are not relevant in comparisons between possible courses of action. Stated differently, only the added or incremental costs are relevant.
2. Expenditures made before the time of the economy study should not be considered. These are called *sunk costs,* in that they cannot be recovered by any present or future action. For example, the roadway and pavement of an existing road may be in good condition and have a substantial "book value" in the records of the agency. Nevertheless, if one alternative proposal abandons the road, it would be an error to charge a value for it against any alternatives in the economy study. Again, it would be improper to include costs incurred earlier for preliminary planning and design.
3. All relevant costs must be included and all irrelevant changes excluded. In this regard, as mentioned earlier, *transferred costs* may be particularly troublesome. Assume, for example, that one of several plans for a proposed improvement requires a private utility company to move its facilities at its own expense. From a budgetary

standpoint this cost is not chargeable against the project. From a public-works economy-study standpoint, however, it is a proper charge. Economic resources are consumed, even though paid from private rather than public funds.

4. In certain types of economy studies, it is proper to make an allowance for the *salvage value* of a machine or structure at the end of its estimated useful life. As a general rule, salvage value should be neglected in economy studies for public investments. It is conjectural at best to assume that an investment in a facility will have great worth 20, 30, or 40 yr in the future. The *Red Book* proposes that for long-lived items such as embankments or structures, salvage value at the end of the study period be set using the ratio of remaining life to estimated life. One exception might be to assign salvage value to the land occupied by the road. Even in this situation, only the raw value of the land in its predicted future use, after deducting the cost of converting it to that use, would be included. Other costs associated with acquiring the land in the first place, such as legal expenses and the cost of clearing it of buildings, cannot be recovered and would not be a part of salvage value. For short-run projects such as the first step in stage construction, salvage value might be set at the difference in cost of building the next stage without and with the project under consideration.

Proposed improvements often will bring changes in annual maintenance and operating costs. For present conditions, data for these should be available from the cost records of the agency. Estimates of these costs for the proposed improvements must be projected. Here again, only the relevant costs are to be included; in particular allocated and sunk costs must be examined critically to be sure that only true cost differences are reflected.[7]

Service Lives of Transportation Elements

To develop information on service lives or life expectancies for all kinds of transportation facilities and their individual elements in various environments would be difficult. For example, there are many variables such as soil, climate, topography, and traffic volume that will affect differently the life of essentially the same type of highway in different places. In flat country the alignment may remain unchanged for many years. On the other hand, a road in rolling or mountain areas originally built on cheap crooked alignment often becomes obsolete because of restricted speeds and is relocated long before the life of the pavement is reached. Also, the art of highway building changes so that the date of construction will influence the probable life of a new highway of given type. Finally, it is common to incorporate portions of an existing facility into any reconstruction, as when pavements are resurfaced or lanes, new roadways, or grade separations are added to increase capacity. In these situations, the practice of classifying resurfacing or reconstruction as retirement, as shown in Table 4–1, might be questioned. In sum, forecasting service lives and time to retirement from historical data is difficult and the results questionable.

[7]For a more detailed presentation on this and other general aspects of economy studies, refer to E. L. Grant, W. G. Ireson, and R. S. Leavenworth, *Principles of Engineering Economy*, Wiley, New York, 1982, H. M. Steiner, op. cit. or one of the other textbooks in engineering economy. See also Winfrey, op. cit. for detailed discussions aimed particularly at highway situations.

TABLE 4–1. Average Service Lives and Retirement Causes for Highway Surfaces

Surface	Service Life (yr)	Method of Retirement (% of mileage) Resurfaced	Reconstructed	Abandoned	Retired
Low type					
Soil surface*	4.0	37.5	58.1	1.2	3.2
Gravel or stone*	7.5	58.0	30.4	2.5	9.1
Intermediate type					
Bituminous surface-treated†	14.0	58.5	32.6	2.5	6.4
Mixed bituminous†‡	12.0	59.8	30.3	2.0	7.9
Bituminous penetration†	17.0	46.0	34.1	6.1	13.8
High type					
Bituminous concrete†§	17.0	57.4	27.7	2.2	12.7
Portland cement concrete†	25.0	66.0	22.8	1.8	9.4

*Source: *Public Roads,* June, 1956.
†Source: R. Winfrey and P. D. Howell, *HRB Record 252.*
‡Thickness of surface and base less than 7 in.
§Mixed bituminous or bituminous penetration with thickness of surface and base in 7 or more.

Studies of the service lines and life expectancies of pavements were begun about 1935 and have been updated from time to time. Some of the earlier findings on past service lives and reasons for retirement of several pavement types are listed in Table 4–1. Several procedures for forecasting future retirements from such historical data have been developed. One of these, the annual rate method, employs data on retirements in a single year or band of years to develop "survivor" curves. In turn, "type survivor curves" selected from a family of such curves may be fitted to the plot. This approach is illustrated in Fig. 4–1. Another procedure is called the *"turnover method."* It compares the accumulated units in service with the accumulated retirements.[8]

Methods for forecasting service lives and life expectancies of pavements until reconstruction or resurfacing have also been developed.[9] Also, remaining service lives are being predicted by measuring the strength of existing pavements (see Chapter 19).

Where road-life studies have been made, as for road surfaces, the results (Table 4–1) provide a means for estimating economic life. For most other highway elements, such data are not available and predictions must be made without the support of meaningful information. In the 1960 *Red Book* an AASHO Committee assumed that rights of way had a life of 100 yr and grading and structures 40 yr. Pavement lives were to be based on the results of road-life studies. The 1977 *Red Book* states that the life of facilities ranges from about 5 yr (some traffic signals) to over 50 yr (for earthwork and some bridges). Winfrey suggests ranges of 75 to 100 yr for land for rights of way but 10 to 30 yr (the study period) for damages to property and for buildings; 60 to 100 yr for earthwork; 50

[8]For details and references, see Winfrey op. cit., Chapter 9, and I. E. Corvi and J. U. Houghton, *Public Roads,* Aug. 1971. Procedures such as these are also widely used for other purposes such as predicting the length of human life for insurance purposes.
[9]See I. E. Corvi and B. G. Bullard, *Public Roads,* Dec. 1970.

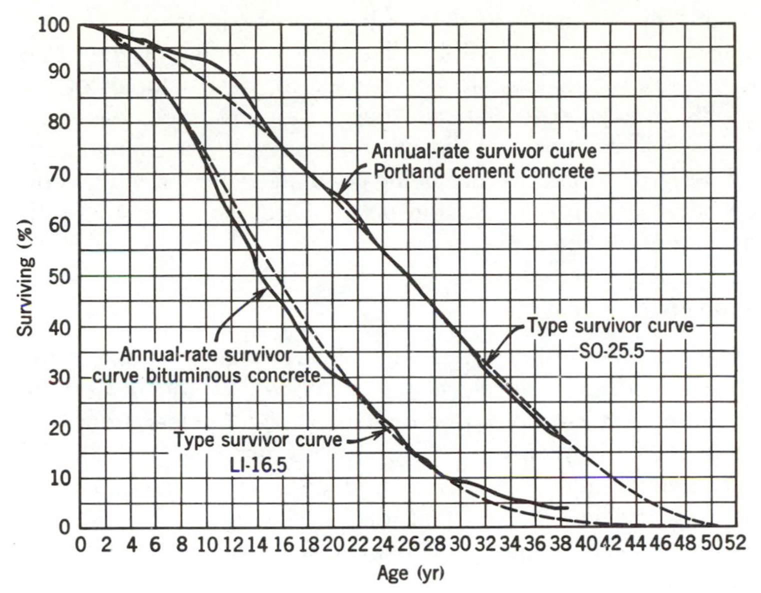

Fig. 4–1. Annual-rate and type survivor curves for bituminous-concrete and portland-cement-concrete surfaces retired 1955–1959—composite data for 26 states and Puero Rico. (Source: R. Winfrey and P. D. Howell, *HRB Record 252*.)

to 75 yr for bridges and major structures; 18 to 30 yr for high-type pavements; and 5 to 20 yr for signs and traffic control devices. Recent experience has indicated that the values (tabular or Winfrey's) for high-type pavements may be too high on routes carrying large volumes of heavy trucks, such as the Interstate System. There is, however, good reason to question assumed useful lives longer than, say, 40 to 50 yr. For, looking back 50 yr, highway design was in its infancy. Also, as will be pointed out in more detail below, if a realistic interest rate is used in the analysis, assumptions of 50 or 100 yr will give almost identical results.

Transportation Agency Costs

Transportation agencies are charged with the responsibility of planning, constructing, maintaining, and operating the facilities under their jurisdictions. Where appropriate, the costs of these activities are included in economy studies. However, many of these agency costs, particularly those for planning and some of the overhead charges associated with other operations, are not affected by proposed investments and they should be excluded from economic comparisons.

The costs per mile to construct highways vary tremendously: from a few thousand to as much as $100 million. Maintenance costs also vary greatly. However, appropriate values are usually available in the records of highway agencies.

For economy studies, the accuracy needed for the costs of rights of way, construction, maintenance, and operation will be different, depending on their use. For example, if long-range projections are being made for needs studies or advanced planning, only average costs per mile might be considered, based on past experiences in similar situations. At the other extreme, if, for example, alternative pavement designs or materials are being compared, the individual cost elements might be developed in considerable detail. Discussions of such estimating procedures are outside the scope of this book.

Interest and Inflation

In economy studies for private business, interest is always treated as one of the costs of invested capital. This is logical, since the money for the investment comes either from withholding earnings from owners or stockholders, from borrowing from others and paying interest on the borrowed funds, or from foregoing other investment opportunities that should produce a return. In the past in the public-works field some have argued that interest should be charged only where borrowing will finance the proposed project. This viewpoint has now largely disappeared under the arguments that (1) capital can and should be productive and (2) interest is a reward or incentive for deferred consumption, as is the case when money is invested in transportation facilities because of anticipated future benefits to society. Incidentally, because no monetary return, as such, occurs from most public investments, Winfrey *(op. cit.)* has proposed that the word *vescharge* be substituted for *interest*.

There is no "right" answer for the appropriate interest rate for transportation and other public works economy studies. First of all, because of persistent inflation, the actual dollar measures of future costs and benefits will be considerably higher than those for the same items at the time of the initial investment. Thus when making economy studies a choice must be made between using constant or uninflated dollars (the prices of all elements at the base year for the analysis) or the "current" or inflated prices which will obtain at the future date. If the constant dollar approach is adopted, the minimum rate for governmental investment should reflect the real cost of capital, which some have estimated as being in the range of 4%. In this is roughly 0.5% to cover risk on low-hazard investments. Higher minimums would be appropriate where risk is higher. On the other hand, if future costs and benefits are priced at their inflated (current) values, minimum rates in the range of 8 to 12% have been suggested, depending on the analyst's estimate of future rates of inflation.[10] Because economy studies only use money as a measure of resource use, the constant dollar approach may be preferable since only differential inflation on future costs and benefits on individual items need to be recognized.[11]

[10]The U.S. Office of Management and Budget stipulates a 10% discount rate for most of the economy studies submitted to it by federal agencies.

[11]For added discussion and references on this complex issue see the new *Red Book,* and Grant, Ireson, and Leavenworth, op. cit.

Discount rates such as those just outlined are minimums, appropriate only for situations where each project stands alone, and the decision is to be "go or no go" on that project alone.[12] For agencies that must ration their limited capital resources among many needs, the discount rate probably will be higher. Analysis for rate of return or with other techniques at several interest rates can be made of each highly desirable project. These, then, are listed in the order of decreasing desirability until the available funds are exhausted. The discount rate at this cutoff point is the appropriate one for the agency in question.

For all situations where capital must be rationed, the discount rate must approximate that at the cutoff point if economy studies are to indicate the optimum investment policy. If the rate is set lower than the cutoff rate, it will unduly favor long-term, capital-intensive projects such as new highways or freeways or mass transit facilities. If it is higher it will favor intermediate-life or short-term investments in items such as traffic control devices or temporary rehabilitation.

Differential inflation among future costs or benefits may at times be a factor. To illustrate, assume that cost or benefit estimates for 10 yr from the base year place the consumer price or other base index at 160, but the construction materials index is 140 and that for labor 180. These factors would be recognized directly in the estimated prices if the current, inflated price approach were used in the study. However, if a constant, uninflated price analysis were employed, the 10-yr-hence material prices would be reduced by 20/160 or 12.5% and labor costs increased by 12.5% over base-year values.

Analysts also must recognize that the decision regarding interest rates has a tremendous influence on the results of economy studies. Assume, for example, that interest rates of 0, 3, 6, and 10%, respectively, are employed for a proposal that has a first cost of $1000 and an estimated life of 30 yr. Then the annual return needed to recover the $1000 in 30 yr, using interest at 0%, will be $33.33; at 3%, $51.02; at 6%, $72.65; and at 10%, $106.08. Again, higher interest rates decrease the importance of probable life adopted for the economy study. To illustrate; at 0% interest, the annual cost doubles if the assumed life is reduced from 100 to 50 yr. At 3%, however, the annual cost increases by only 22% with this change in assumption; at 6% the increase is but 6%; and at 10% the increase is less than 1%.

CONSEQUENCES TO HIGHWAY USERS

Costs of Motor-Vehicle Operation[13]

There is no single correct answer to the question: What does it cost to operate a motor vehicle? First, the elements to be included in cost differ, depending on the purpose of the question and the viewpoint to be taken. Second, although much is already known about the individual elements of motor-vehicle costs,

[12]This is the criterion for federal water projects.
[13]For 1970 and 1971 operating costs for rail-transit and buses, see R. P. Roess, *TRB Record 490*.

many details are still lacking. Third, the vehicles in use are changing markedly; for example, passenger cars are lighter and diesel engines have entered the scene. Finally, because of inflation, motor-vehicle operation costs climb year by year, but inconsistently and differentially.[14] For example, between 1970 and 1975, automobile prices increased 15%, and fuel, tires, and repairs for them 52, 10, and 41%, respectively. For diesel trucks, comparable percentages were 36, 129, 60, and 41. Since gathering basic data is costly and time-consuming, many of these 1975 values were obtained by updating earlier research findings by procedures given in the *Red Book*. This was done with equations that reflected changes in appropriate Consumer and Producer (Wholesale) Price Indexes, published by the Bureau of the Census in *Statistical Abstracts of the United States*.[15]

Updating into the future beyond January 1975 can be done in like manner. An approximation of the results, recommended for preliminary analysis, involves multiplying the January 1975 prices of the *Red Book* by the ratio between the All-Item or Transportation Consumer Price Index for the study date and that for January 1975.[16]

Some operating expenses increase more or less directly with miles driven; in other words, their cost per vehicle-mile is relatively constant. In this classification fall such items as fuel, tires, oil, maintenance and repairs, and that portion of depreciation attributable to wear. Other costs vary mainly with time and are constant for a given period such as 1 yr; or, stated in costs per vehicle-mile, they vary inversely with the number of miles driven annually. Included here are drivers' license and registration fees, garage rent, insurance, and obsolescence, which is the portion of depreciation that results from inadequacy or being out of date. Some items are dependent entirely or in part on speed. The most important of these is the travel time of operator and rider; any charges for these will vary inversely with speed. On the other hand, some of the operating costs that vary primarily with miles driven, such as fuel and oil consumption and tire wear, may also be influenced by speed and other factors such as roadway congestion.

Of the costs mentioned here, *running costs* that vary primarily with mileage or speed are most often affected by highway improvements. It follows that these are of particular concern in highway economy studies, for justification of many improvements depends largely on savings in operating costs to offset proposed

[14]Most of the costs quoted here are for January 1975, since that is the base date for the *Red Book* from which the data have been taken.

[15]An excellent earlier reference dealing with inflation in highway economy studies is R. R. Lee and E. L. Grant, *HRB Record 100*.

[16]For greater precision, the *Red Book* offers a group of equations for computing multipliers to be applied to 1975 data. Those for passenger cars are based on the Consumer Price Index for gasoline, (CPI_F); motor oil, (CPI_O); tires, (CPI_T); repairs and maintenance, (CPI_M); and new automobiles, (CPI_D). Those for trucks are based on the appropriate Producer (Wholesale) Price Indexes except that repairs and maintenance use the Consumer Price Index. A typical equation, that for general operation of automobiles on straight level tangents is

$$M = 0.0017CPI_F + 0.0001CPI_O + 0.0004CPI_T + 0.0016CPI_M + 0.0032CPI_D$$

expenditures. However, care must be exercised to consider only those costs or savings that are relevant to a particular comparison. Stated differently, only those costs or savings that will be affected by the proposal should be included in economy studies. The importance of this distinction is made apparent by the fact that, although the average cost per mile of owning and operating a standard-sized automobile in 1975 was 15.9 cents, (17.9 cents in 1979) incremental or running cost of driving an additional mile in 1975 was 10.8 cents.

The stream of vehicles traversing the highways is a complex of small to large passenger cars, pickups, buses, motorcyles, bicycles, and light and heavy trucks of many body styles and axle configurations. In addition, their ages range from new to more than 30 yr. For economy study purposes, the *Red Book* has reduced this complex of vehicles to three, as follows: $4850 passenger cars of 3500 lb empty and 3800 lb loaded weight; $7500 single-unit, gasoline-powered two-axle trucks with six tires, weighing 12,000 lb (this group includes buses); and $30,000, five-axle diesel, 3-S2 (3-axle tractor, 2-axle semitrailer) combination trucks weighing 54,000 lb.[17] These prices are without tires, which are treated separately. Although this classification is considered sufficiently accurate for studies in the United States and Canada, it probably would be unsatisfactory in many parts of the world where passenger cars, on the average, are much lighter and trucks have quite different conformations and weights.

The motor-vehicle running costs presented on the pages that follow are abstracted or simplified from the *Red Book*. They are dated January 1975 and are primarily for a 3800-lb passenger car operating on pavement. Because of space limitations, data for the other two classes of vehicles have largely been omitted. However, several multipliers have been developed with which to obtain first-approximation costs for trucks from the passenger-car data. (See Table 4–2). For detailed analyses, data should be obtained from the *Red Book* or more recent sources as they are available.[18]

Total running costs for each vehicle type vary with such factors as speed, speed changes, degree of congestion, grade, and curvature. The procedure is to develop the cost associated with each separately so that they may be added together to determine total cost.

Running costs for passenger cars traveling at various speeds on uncongested, level, straight, paved roadways, broken down into five cost elements, are shown on Fig. 4–2. Totals only for the two truck classes also appear. Figure 4–3 gives, for passenger cars, costs for sections of four- and eight-lane freeways, arterials, and two-lane highways. They include the effects of congestion, grades, curves, and speed changes (see below). An example appears on the figure. Multipliers developed from *Red Book* data from which rough costs for the two truck

[17]Some such simplification in vehicle classification is essential, otherwise data requirements and computational difficulty would make the determination of vehicle operating costs far too cumbersome.

[18]Studies far more comprehensive than those underlying the *Red Book* have been made in Kenya and Brazil. Reports on these studies and some of the data appear in *TRB Record 702* and *Synthesis 5* of Transportation Technology Support for Developing Countries, TRB.

TABLE 4–2. Approximate Multipliers for Determining Running Costs for Other Vehicle Classes from Those for Passenger Cars*

	Vehicle Class	
Operating Situation	*12,000 lb Single-Unit Truck*	*54,000 lb. 3-S2 Diesel Truck*
Continuous travel on straight, level roadways	1.9	2.1
Forced flow, queues formed (level of service F)	4	15
Slowing, stopping, and accelerating†	2.4	8.0
Idling†	2.3	2.7
Horizontal curves	2.0	3.2
Positive grades	2.6	3.4
Negative grades	1.5	1.5

*Source: Approximations from the *Red Book*.
†See tables on Fig. 4–7 and 4–8 for more precise values.

classes can be obtained appear in Table 4–2. It is to be observed that costs are lowest for passenger cars at around 35 mph and for trucks at about 20 and 25 mph. Costs are higher when speeds are either lower or higher than these values.

The "energy crisis" which first developed in late 1973 has had a substantial impact on vehicle-running costs which is difficult to evaluate. For example, although the cost per gallon of gasoline has increased substantially, it may, in time, be partially offset by trends toward lighter passenger vehicles and diesel engines in passenger cars and light trucks. On the other hand, there is little chance that heavy truck weights will decrease substantially, so that with them, the full impact of fuel cost increases will be felt. A brief discussion of the individual elements that make up operating costs follows.

MOTOR-FUEL COSTS. Motor fuel consumed per mile traveled varies, for a particular vehicle, with operator skill, engine adjustment, speed, the degree of congestion of the road, the road surface, the grade or slope of the road, road curvature and superelevation, the number and duration of stops, temperature, and elevation above sea level. Between vehicles, it varies with age, weight,[19] and size, the efficiency and adjustment of the engine and transmission, the inefficiencies introduced by smog control devices, if required, and the skill of the operator. Approximations of miles obtained per gallon of fuel at various speeds can be determined from Fig. 4–2 for the typical passenger car by dividing the assumed tax-free cost per gallon of fuel ($0.40 gal[20] for gasoline) by the fuel cost per mile driven. For a passenger car, 21.1 mi/gal is obtained at 35 mph but only

[19]The new *Red Book* indicates that fuel consumption increases at about 0.98 times the increase in vehicle weight.
[20]By mid 1981 this tax-free price has roughly tripled, clearly demonstrating that data given here must be updated.

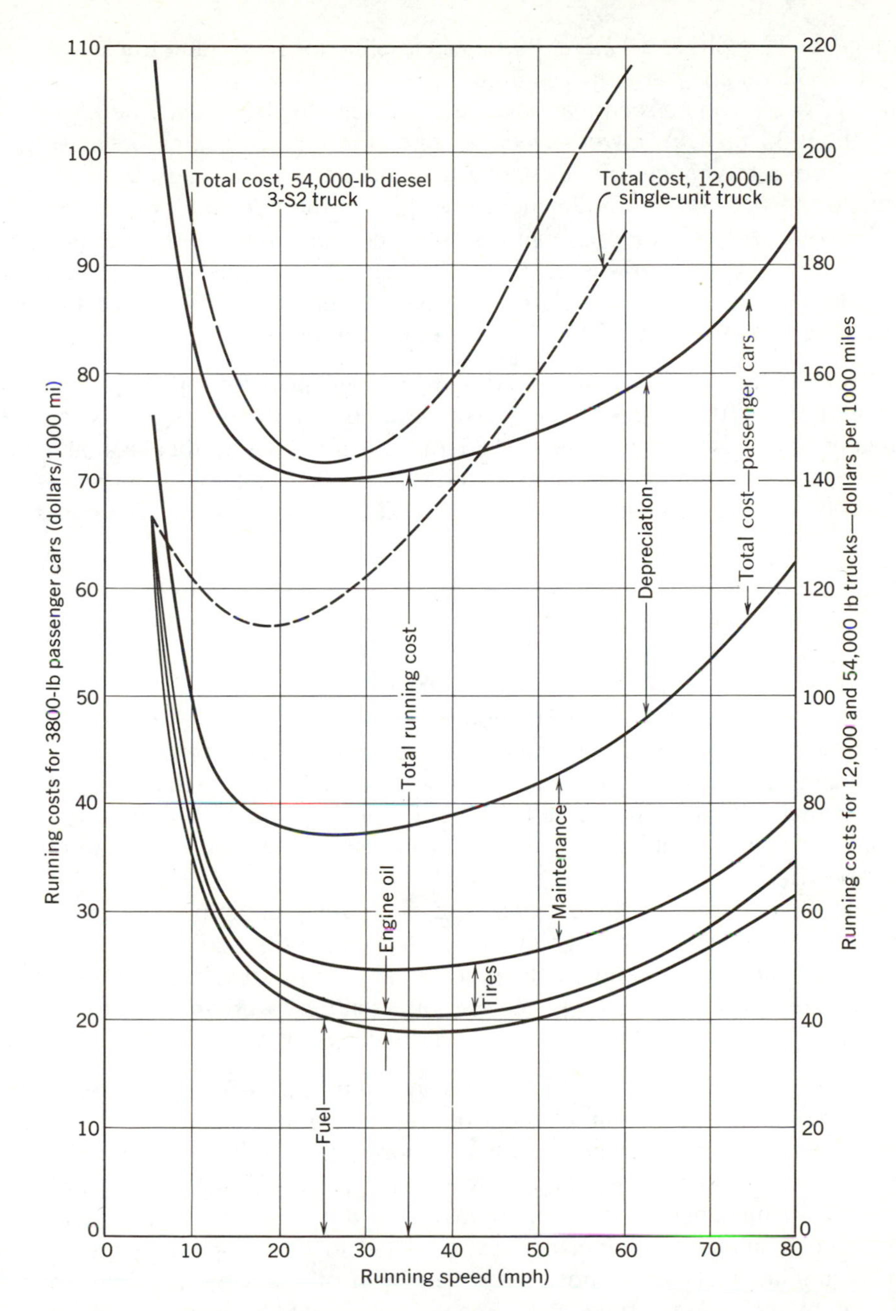

Fig. 4–2. Vehicle running costs on straight level tangents. (Data source—the *Red Book*.)

11.6 mi/gal at 10 mph, and 12.5 at 80 mph. Gasoline for single-unit trucks was priced at $0.33/gal and diesel fuel at $0.31/gal.

From Fig. 4–2 it can be seen that passenger cars take the least fuel when traveling at about 35 mph. At lower speeds, engine and drive efficiency decrease; at speeds above this range, air resistance and internal friction cause gasoline consumption to rise. Where, at 25 mph, one-third of the tractive effort is consumed by air resistance, at 55 mph air resistance accounts for 70% of the total.

Other factors such as slowing and speeding up where traffic is congested, grades, curves, and surface type also affect fuel consumption. Their effects on total costs, but not on fuel consumption alone, are discussed subsequently.

OIL CONSUMPTION. Oil, here priced at $0.90 per quart for passenger cars and $0.44 and $0.40 for trucks, is a small part of operating costs. It is possibly 50% higher at 10 and 70 mph than at 35 mph. It increases further at higher speeds. Data have been accumulated to prove that oil consumption increases progressively as the roadway changes from pavement to loose gravel to unsurfaced.

TIRE COSTS. Tax-free tire prices for this analysis were $32 per tire for passenger cars, and $122 and $290 for the trucks. Figure 4–2 shows that tire costs for passenger cars increase with speeds up to about 45 mph but remain relatively constant beyond that. This is in marked contrast to the earlier situation when tire costs tripled between 20 and 50 mph and again from 50 to 80 mph, an indication of how much passenger-car tires have improved in recent years. On the other hand, for light and heavy trucks, tire costs, although relatively small, increase dramatically with speed; for the former from $0.65 to $12.63 per thousand miles at 5 and 60 mph, respectively; for the latter, from $1.72 to $39.68 at these speeds.

Although the details are not presented here, tire costs predominate in the totals for certain maneuvers such as slowing or stopping and speeding up, travel around curves and corners, and on steep grades and unpaved surfaces. Drivers can reduce them substantially by giving careful attention to driving and to tire inflation and rotation, wheel balance, and overload control.

VEHICLE REPAIR AND MAINTENANCE. As shown by Fig. 4–2, vehicle repair and maintenance is a substantial element of cost for the average vehicle being considered here. For the individual owner, these costs may be either high or low, since they are strongly dependent on weather, road and operation conditions, on the maintenance practices of the owner, and on the age of the vehicle.

Only those repair and maintenance costs related to travel or idling are included in running costs since those associated with off-highway activities are not affected by highway improvements. Repair costs related to accidents are included under accident costs (see below).

As with passenger cars, estimated maintenance costs for trucks increase substantially with speed. For the light truck it goes from 3.1 cents per mile at 5 mph to 5.9 cents at 60 mph. For the heavy diesel comparable figures are 4.6 cents and 8.7 cents.

DEPRECIATION. Depreciation has been defined as "the inevitable march to the junk heap." One part of depreciation is attributable to wear, which occurs largely on the highway; it is primarily mileage-dependent. The second part is attributable to obsolescence or being inadequate or out of date; it occurs whether or not the vehicle is in use and is time-dependent. It seems logical that only the on-highway wear portion of depreciation should be included in highway economy studies, since only it will be affected by highway improvements.

Earlier procedures for assigning depreciation in highway economy studies first determined an average total cost per mile for owning and operating vehicles and then assigned an arbitrary portion, such as 50%, to mileage. More recently, it has been recognized that depreciation can well be dependent on speed and other operating maneuvers such as stopping and starting, all of which affect vehicle wear. Winfrey developed the approach that resulted in depreciation charges such as are shown in Fig. 4–2.[21] He established from data already collected that vehicles which are operated at higher speeds will travel farther each year but have a somewhat shorter life. These effects were largely offsetting for passenger cars. The values for depreciation plotted in Fig. 4–2 were then developed. For passenger cars, these were 3.3 cents per mile for speeds of 40 mph and lower and fell to 3.1 cents at 80 mph. Depreciation charges for the 12,000-lb truck decreased from 4.1 cents per mile at 5 mph to 3.2 cents per mile at 60 mph and for diesels from 4.0 to 2.5 cents per mile at these speeds. The low depreciation value for the diesels reflects the high mileages they are driven before retirement. For heavy trucks he charged substantially larger percentages to mileage depreciation as compared to passenger cars on the basis that such trucks generally are worn out when they are retired from service.

In sum, the only procedure currently available for determining depreciation charges to be used in highway economy studies involves allocating a portion of total depreciation to running costs in an apparently logical manner. Users of such cost figures should be aware of their origins and, where necessary, adjust them in an appropriate manner.

COSTS RELATED TO TRAFFIC VOLUMES. The vehicle running costs shown in Fig. 4–2 are for situations where vehicles travel freely without interference from others. However, as volume increases, these interferences cause not only a reduction in average speed (see Fig. 4–3), but also repeated slowdowns and speedups. At or near capacity (volume to capacity ratio v/c of 1), flow of the traffic stream becomes unstable; it may continue at about 30 to 35 mph under most situations (15 mph on arterials), or it may break down or stop so that traffic operates at even lower speeds under a condition of forced flow (level of service F).[22] Running speeds for passenger cars on representative freeways, arterials, and two-lane highways are given. From the figure can also be read the associated costs and times. An example, including the influence of grades and curves (see below) appears on the figure. Speed and speed change costs for relatively

[21]See Appendix B of the *Red Book* for details of the Winfrey method.

[22]See the discussion of highway capacity in Chapter 8 for added detail.

free and congested conditions are given in the top left of Fig. 4–3. These "volume" costs are in addition to other running costs.

COSTS ASSOCIATED WITH PASSING RESTRICTIONS ON TWO-LANE ROADS. On two-lane roads, passing another vehicle may be impractical because of restricted sight distance or an approaching vehicle. In such instances, the overtaking vehicle must slow to the speed of the one being followed and maintain that speed until there is an opportunity to pass. The reductions in vehicle speed because of these restrictions can be seen on Fig. 4–3 by comparing the 70-mph curves for freeways and two-lane highways. However, Fig. 4–3 also indicates that these speed reductions lower operating costs slightly on high-speed roads.

COSTS RELATED TO GRADES. Figure 4–3 also portrays the changes in running costs for passenger cars due to grades. Since added energy is needed to travel uphill, costs increase rather uniformly with steepness; they are roughly 20% higher for a 6% grade than for traveling on the level. On downgrades flatter than about 2%, the energy demand lowers and costs decrease by roughly the same amount as the increase on upgrades. There is, however, little if any further cost decrease on downgrades steeper than 2%, since braking is required.

COSTS ASSOCIATED WITH CURVES AND CORNERS. Figure 4–3 also gives passenger-car-running costs for traveling on horizontal curves, over and above those of traveling on straight and level tangents. These are given in terms of "degree of curve" D; for radius in feet divide 5730 by D. Figure 4–4 gives parallel data for relatively short 90° corners such as those that occur at intersections and at diamond interchanges. According to the new *Red Book*, two-thirds of these added costs are attributable to added tire wear.

Travel at relatively high speeds on sharp curves or around corners can substantially increase running costs, as comparisons utilizing Figs. 4–3 and 4–4 will show. For example, costs are increased by about three-fourths over those on level tangents if a passenger car travels around a 6° (955-ft radius) curve at 60 mph. There are, of course, limits to speeds on curves and the costs they will generate which are set by superelevation and side friction.[23] The costs associated with speeds appropriate for design purposes are shown by the dotted lines on Fig. 4–3c and Fig. 4–4.

Graphs showing curve and corner costs for other types of vehicles are not given here. However, approximate multipliers will be found in Table 4–2.

SPEED CHANGE (TRANSITION) COSTS. Except on high-type facilities under free-flowing conditions, cars seldom travel along roadways at constant speed. Rather, as they approach sharp curves, narrowed roadways or bridges, or certain other situations, they reduce speed. When possible, they then accelerate to

[23]See Chapters 8 and 9 for a detailed explanation of these relationships.

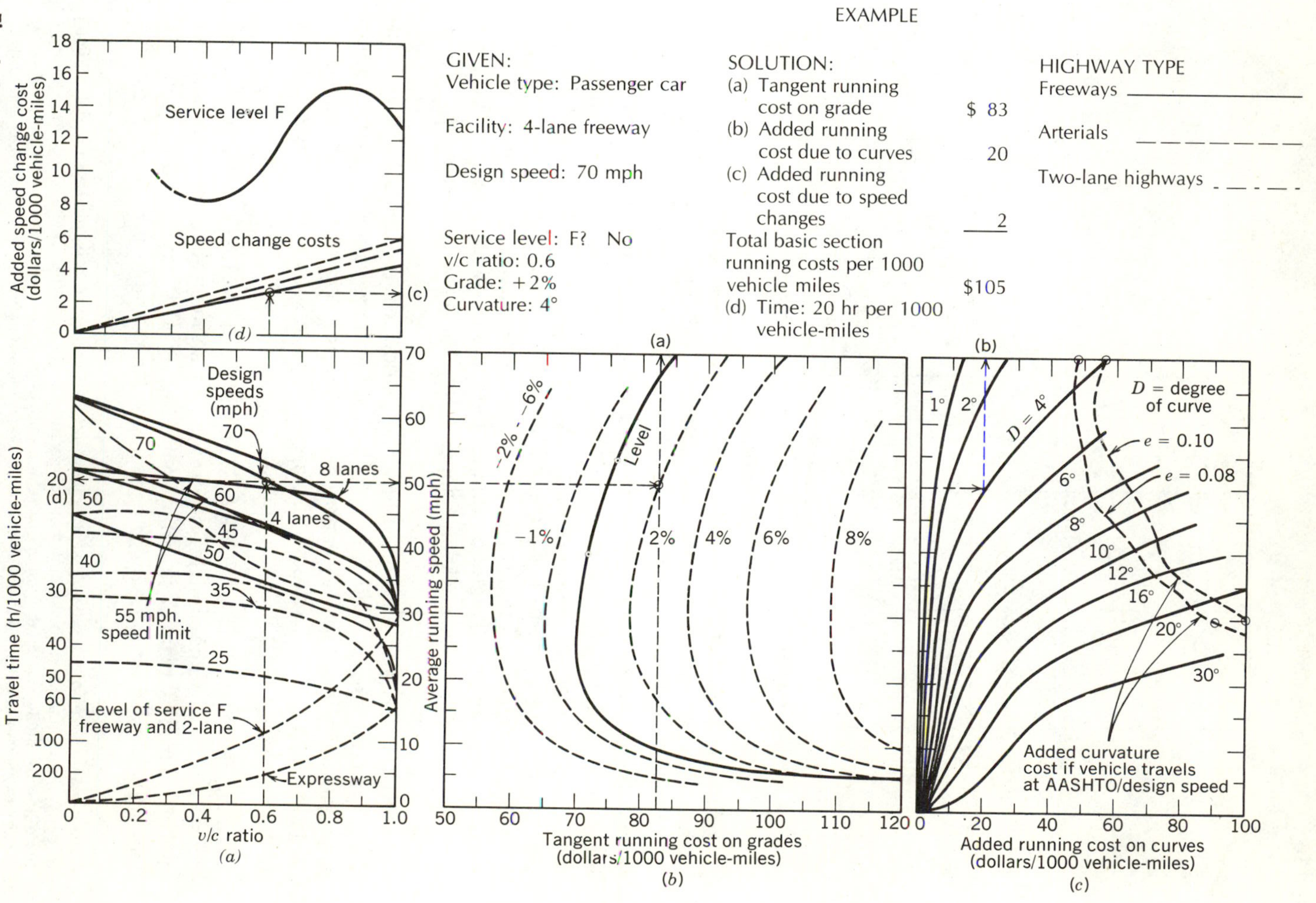

Fig. 4-3. Passenger cars—basic section vehicle-running costs on freeways, arterials, and two-lane highways. (Composite of three graphs in the *Red Book*.)

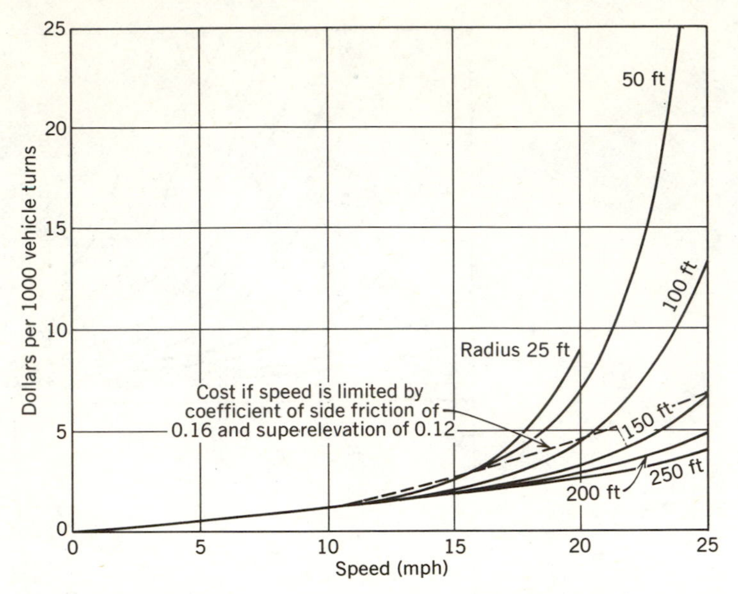

Fig. 4–4. Passenger cars—excess vehicle running costs from turning 90° corners. (Winfrey data updated to 1975 prices.)

higher speeds. These maneuvers increase vehicle operating costs. Dollar values for the cost of these slowings or accelèratings are shown in Fig. 4–5. These are the sum of the costs of both slowing and accelerating. If only one of the maneuvers is carried out, one-half of the value shown will be used.[24]

Figure 4–5 clearly demonstrates that speed changes are costly; to slow from 60 to 30 mph and accelerate to 60 mph, the 1975 cost is $13.60 per 1000 maneuvers. Multipliers for trucks developed from *Red Book*, data are given in Table 4–2 and on the figure.

Speed changes also consume time. The amount varies substantially with situation, among drivers, and with vehicle type; average values for usual but not emergency situations for passenger cars are given in Fig. 4–6. Multipliers from which to obtain times for trucks also appear. The lesser ability of large trucks to change speeds is clearly demonstrated.

INTERSECTION COSTS (STOPPING, IDLING, STARTING). Traffic signals cause a portion of the vehicles on each leg of an intersection to stop and wait and some to slow without stopping. With stop signs, all vehicles on the affected legs stop or nearly stop before proceeding.[25] Nomographs for pricing these ma-

[24]The research necessary to separate accelerating and decelerating costs would be extremely difficult to carry out. It has been speculated that acceleration costs exceed deceleration ones. However, since in most situations equal numbers of vehicles travel in the two directions, an equal number of accelerations and decelerations occur at each affected location and the summed costs and correct. It is recommended that the plotted costs be halved when only the cost of slowing or of accelerating is desired.

[25]This discussion pertains only to the cost aspects of such traffic-control devices. Details on other aspects of their use appear in Chapter 10.

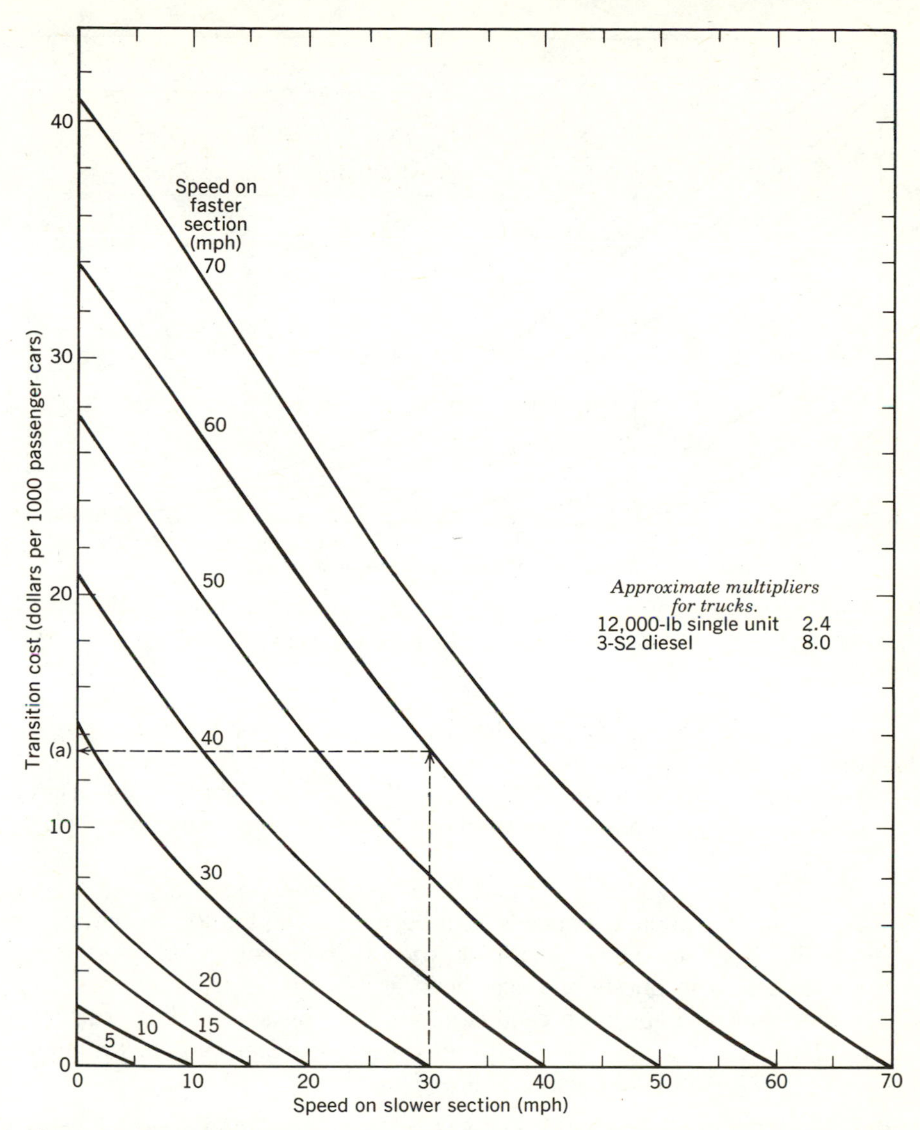

Fig. 4–5. Passenger cars—excess running costs of speed-change cycles (to be added to costs of continuing at constant speed). (Source: the *Red Book*.)

neuvers appear as Figs. 4–7 and 4–8. They give costs and times associated with various degrees of congestion. In addition they provide means to recognize the effects of trucks more accurately than could be done with the approximations of Table 4–2. Illustrations showing how cost and time data are extracted from the figures appear on them.

The data given in Figs. 4–7 and 4–8 are based on observations of many intersections. For intersections controlled by two-phase fixed time signals, results

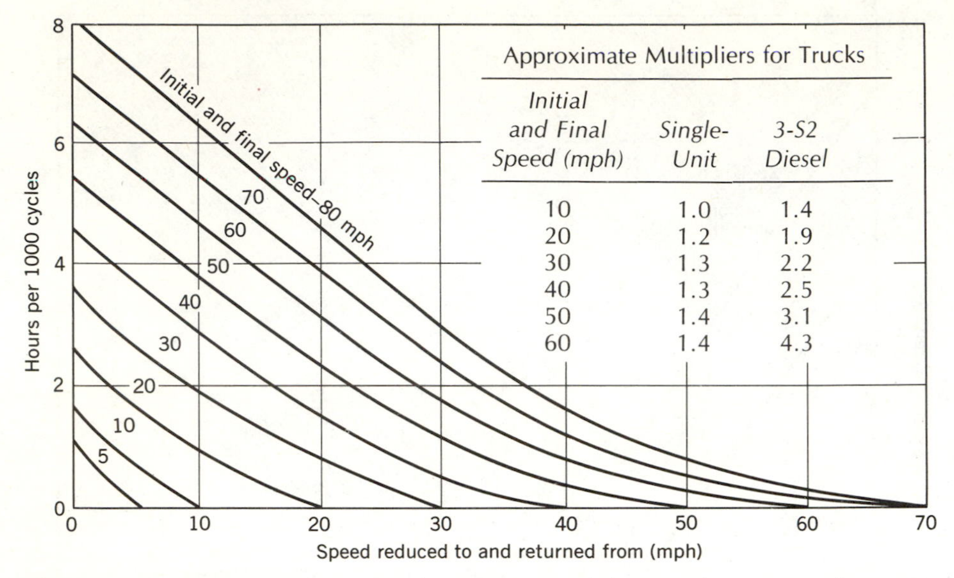

Approximate Multipliers for Trucks

Initial and Final Speed (mph)	*Single-Unit*	*3-S2 Diesel*
10	1.0	1.4
20	1.2	1.9
30	1.3	2.2
40	1.3	2.5
50	1.4	3.1
60	1.4	4.3

Fig. 4–6. Passengers cars—excess hours consumed by speed-change cycles over continuing at constant speed. (Data from R. Winfrey, *Economic Analysis for Highways*.)

for each movement can be obtained directly from the figures. For more complex situations, as when left-turn phases are provided, separate analyses are required. Again, for situations where channelization permits right turns against red without stopping, these can be appraised using cost and time data from Figs. 4–4, 4–5, and 4–6.

For stop signs, the added costs and times over and above driving straight through, for various approach speeds, can be read along the top margin of the graphs on Fig. 4–7. Standing (idling) times must be either observed or estimated for the intersection in question, since they vary greatly with traffic considerations, intersection and approach geometry, and sight distance. Idling costs from the *Red Book* (1975 prices) per 1000 hours are: passenger cars, \$313; single-unit trucks, \$277; and 3-S2 diesels, \$193.[26]

COSTS RELATED TO ROADWAY SURFACINGS. The running costs given so far have been for high-type surfacings. As would be expected, each element of cost increases progressively with changes to lower type pavements, gravel or stone, and earth. A rough approximation of more detailed data offered by the *Red Book* for all vehicle types and running situations is as follows:

1. Gravel or stone surfaces, for passenger cars and single-unit trucks: use given costs multiplied by the factor $(1.05 + 0.92 \times {}^{mph}/_{100})$. For 3-S2 diesels the factor is $(1.15 + 0.83 \times {}^{mph}/_{100})$.
2. Earth surfaces, for passenger cars and light trucks: use given costs multiplied by the factor $(1.10 + 1.84 \times {}^{mph}/_{100})$. For 3-S2 diesels the factor is $(1.30 + 1.66 \times {}^{mph}/_{100})$.

[26]For a study of fuel consumption in urban traffic, see S. L. Cohen and G. Euler, *TRB Record 667*.

EXAMPLE

GIVEN:
Volume: 480 vehicles/hr
Saturation flow: 1600 vehicles/hr
Signal cycle time: 60 s
Effective green time: 30 s
Intersection approach speed: 30 mph
5% single unit trucks
5% 3-S2 combination trucks

SOLUTION:
$\lambda = 30/60 = 0.5$
Capacity of approach = $0.5 \times 1600 = 800$
$\chi = 480/800 = 0.6$
(a) Average stops per vehicle (per signal): 0.71
(b) Stopping delay per signal: 2.5 hrs
(c) Cost of stopping: $10.30

Running costs = 10.30 × 1.42* = $14.60‡
Time
Passenger cars = 2.5 × 0.90 × 1.35* = 3.04 hr‡
Single unit trucks = 2.5 × 0.05 × 1.35* = 0.17 hr‡
3-S2 trucks = 2.5 × 0.05 × 1.35 = 0.17 hr‡

*Adjustment factors for trucks in traffic stream.
‡Values per 1000 total vehicles per signal.

Adjustment Factors for Percent Trucks in Traffic Stream

TIME

Approach Speed (mph)	*Single Unit Trucks (%)*	*3-S2 Combination Diesel Trucks (% in traffic stream)*				
		0	*5*	*10*	*20*	*100*
5–20	0	1.00	1.15	1.30	1.61	4.03
	5	1.07	1.22	1.37	1.67	
	10	1.13	1.28	1.43	1.74	
	20	1.26	1.41	1.57	1.87	
	100	2.31				
21–40	0	1.00	1.25	1.51	2.01	6.05
	5	1.10	1.35	1.60	2.11	
	10	1.20	1.45	1.70	2.21	
	20	1.40	1.65	1.90	2.41	
	100	2.99				
41–60	0	1.00	1.41	1.82	2.63	9.17
	5	1.11	1.56	1.93	2.74	
	10	1.22	1.61	2.04	2.85	
	20	1.44	1.85	2.26	3.07	
	100	3.20				

RUNNING COST

Approach Speed (mph)	*Single Unit Trucks (%)*	*3-S2 Combination Diesel Trucks (% in traffic stream)*				
		0	5	10	20	100
5–20	0	1.00	1.35	1.70	2.40	8.02
	5	1.08	1.43	1.78	2.49	
	10	1.16	1.51	1.86	2.57	
	20	1.32	1.68	2.03	2.73	
	100	2.62				
21–40	0	1.00	1.35	1.71	2.41	8.07
	5	1.07	1.42	1.78	2.48	
	10	1.14	1.49	1.84	2.55	
	20	1.27	1.63	1.96	2.69	
	100	2.37				
41–60	0	1.00	1.35	1.70	2.39	7.96
	5	1.06	1.41	1.76	2.45	
	10	1.12	1.47	1.82	2.51	
	20	1.24	1.59	1.94	2.63	
	100	2.21				

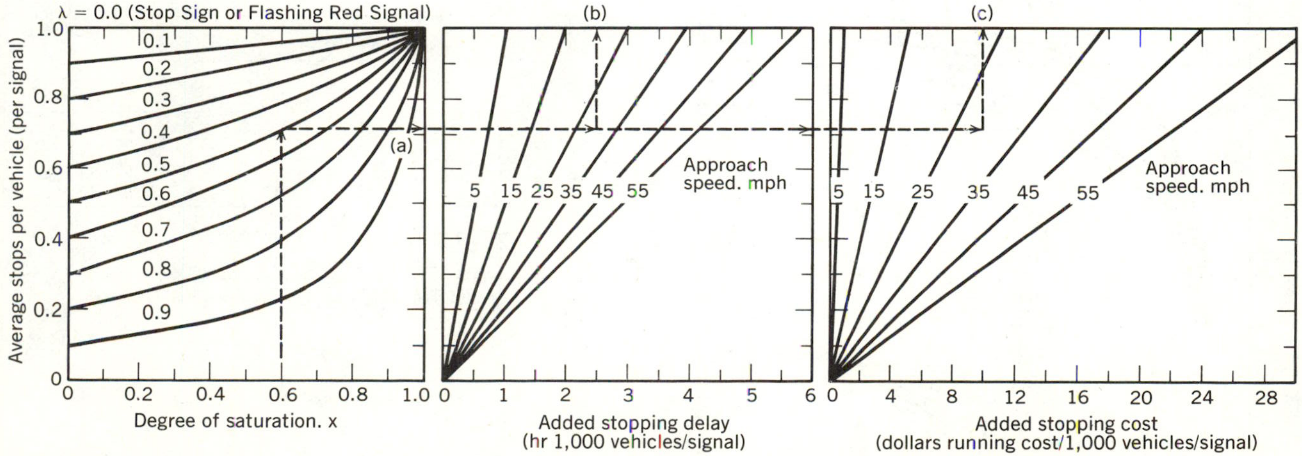

NOTE: Where $\chi = v/\lambda s = v/\text{capacity}$ s = saturation flow v = volume λ = green to cycle time ratio

Fig. 4–7. Costs and times resulting from stopping at intersections—excludes idling. (*Source:* the *Red Book.*)

Adjustment Factors for Percent Trucks in Traffic Stream

Idling Time Factor		*3-S2 Combination Trucks (%)*				
		0	*5*	*10*	*20*	*100*
Single Unit Trucks (%)	0	1.00	1.08	1.17	1.33	2.07
	5	1.07	1.15	1.23	1.40	—
	10	1.13	1.22	1.30	1.47	—
	20	1.27	1.35	1.43	1.60	—
	100	2.33	—	—	—	—

Idling Cost Factor		*3-S2 Combination Trucks (%)*				
		0	*5*	*10*	*20*	*100*
Single Unit Trucks (%)	0	1.00	0.98	0.96	0.92	0.02
	5	0.99	0.98	0.96	0.92	—
	10	0.99	0.97	0.95	0.91	—
	20	0.98	0.96	0.94	0.90	—
	100	0.89	—	—	—	—

EXAMPLE

GIVEN:
$\chi = 0.6$
Capacity = 800
$\lambda = 0.5$
Cycle length: 90 s
5% single unit trucks
5% 3-S2 combination trucks

SOLUTION: Average delay per vehicle: $(a) + (b) = 16.2$s
(c) Idling hours: 4.5 hr
(d) Idling cost: \$1.40
Running costs = \$1.40 × 0.98 = \$1.38
Time
Passenger cars = 0.90 × 4.5 × 1.15* = 4.66 hr‡
Single-unit trucks = 0.05 × 4.5 × 1.15* = 0.26 hr‡
3-S2 trucks = 0.05 × 4.5 × 1.15* = 0.26 hr‡

*Adjustment factors for percent trucks in traffic stream
‡Values per 1000 total vehicles per signal.

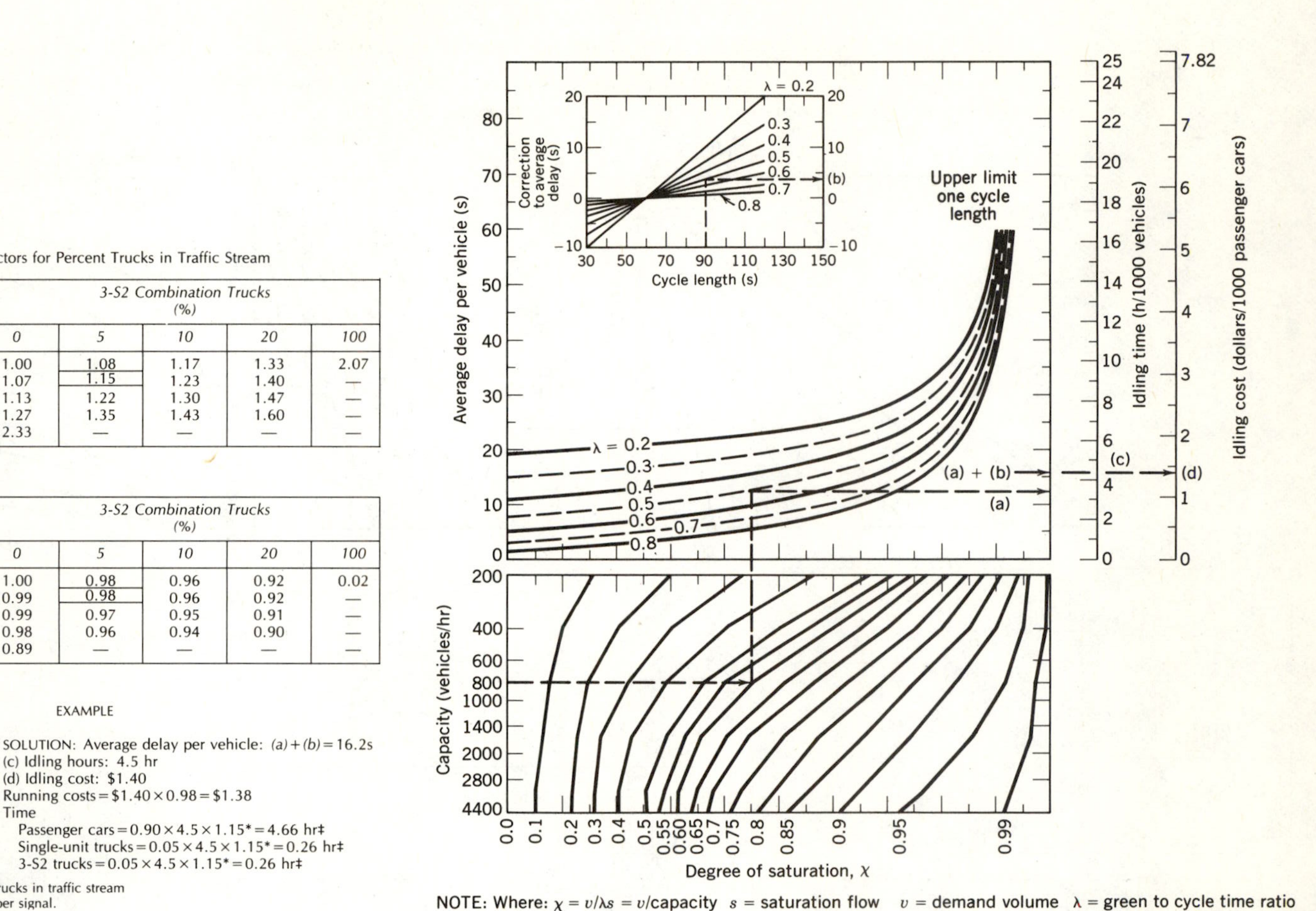

Fig. 4–8. Costs and times resulting from idling at intersections. (*Source:* the *Red Book*.)

3. Low type of pavements. Little if any observed data are available. Use a multiplier greater than 1.0 but less than that for gravel or stone surfaces.

PROCEDURES FOR COMPUTING VEHICLE RUNNING COSTS. The individual cost elements presented here do not overlap: thus they can be added together to determine total running costs in individual situations. The procedure for determining section costs and times is as follows:

1. For the given horizontal and vertical alignment and *v/c* ratio, develop a velocity profile for each operating situation. This profile will have discrete individual blocks with vertical steps between them.
2. For the selected conditions, determine from Figs. 4–3, 4–5, and 4–6 the appropriate running cost elements that are to be combined and compute and sum the numerical values. Where traffic flow varies substantially with time, separate analyses for representative situations may be called for.

For intersections controlled by traffic signals, the procedure is as follows:

1. Based on traffic counts or estimates and using capacity data from Chapter 8 and signal timing procedures from Chapter 10, determine the *v/c* ratio for each signal phase for appropriate time spans.
2. From Fig. 4–7, determine costs and delay times for each signal phase, recognizing its traffic composition. An example is on the figure.
3. From Fig. 4–8, determine idling costs. These, added to the findings from Fig. 4–7, give the total costs and delay times for the intersection.

For intersections on which some or all legs are controlled by stop signs:

1. Based on traffic counts or estimates, determine the number of vehicle stops on each affected leg of the intersection.
2. From the top of the graphs on Fig. 4–7, determine added stopping delay costs for the affected vehicles, including trucks.
3. By observation or estimate, determine average idling times; price these using values given on Fig. 4–8. These, added to the data from Fig. 4–7, give total intersection costs under stop sign operation.

Use of the running cost data is illustrated by the sample economy studies given later in this chapter.

Time Consumption and Time Value

The amounts of time consumed during vehicle travel can be estimated from Figs. 4–3, 4–6, 4–7, and 4–8 or by simple computations. Studies of the value placed on time by travelers in passenger cars and in other transportation vehicles have been discussed in Chapter 3. Suggested values from the *Red Book* for time savings per hour, broken down by family income, the amount of time saved, and for average or business trips, are given in Table 3–4. Proposed dollar values for the wages of truck drivers, including fringe benefits, are $7 for single-unit trucks and $8 for 3-S2 diesels. However, the *Red Book*, in its examples, does not distinguish among time savers; rather, it adopts a value of $3 per vehicle-hour for the entire traffic stream. This approach has not been followed in this book for reasons detailed below.

There is general agreement among transportation economists that a portion of the savings in the wages of drivers and other workers traveling in trucks, buses, vans and other commercial vehicles, and in passenger cars used for business purposes, represents a savings in resources. As such, they should be included in economy studies. Such savings are substantial. For example, Fig. 4–6 shows that stopping 1000 3-S2 trucks traveling at 50 mph takes 5.2 × 3.1 or 16 hr. This is about 1 min per truck. For 1000 trucks per day, the savings would be 16 hr at \$8 = \$128. Annually this would be \$47,000. Capitalized over 20 yr at 6% interest, it represents \$540,000. But it can be argued that such apparent savings actually do not occur, since the many small increments of time that are saved cannot be put to productive use. As an example, Fleischer determined the savings to over-the-road trucking companies resulting from successive completion of segments of the Interstate System. He found that costs were not reduced substantially until enough segments were completed that drivers could make a round trip between terminals. Until this occurred, drivers drove one way in less than 8 hr but were paid for the full shift. The remaining time was not used productively.[27] From this finding, it can be argued that economic gains from time savings may not accrue until some time after the savings begin. On the other hand, it may be argued that even these small increments of time often can be used productively and, furthermore, that the cumulative effects of the many improvements that save time over the life of the project should be recognized.

Transportation improvements also bring time savings to those making nonwork trips for purposes such as commuting, shopping, and recreation. But these savings usually do not directly save resources as do those concerned with work-related activities. These have been referred to in Table 3–1 as quantifiable nonmarket benefits. It is proposed here that these time savings be reported separately from the clearly economic ones. Furthermore, they should be stated in hours and not in money terms. If dollar values are to be given, as some economists propose, then decision makers and not analysts should set the hourly rate.[28]

Accident Costs[29]

The tremendous toll of motor-vehicle accidents not only causes much suffering and misery, but is a waste of economic resources. The National Highway Traffic Safety Administration (NHTSA) placed societal highway accident costs for 1975 at \$37.6 billion. The National Safety Council, using other data and omitting societal costs, gives for 1977 a value of \$30.5 billion made up of \$8.8 billion wage loss, \$2.5 billion medical expense, \$9.1 billion insurance administration,

[27]See G. A. Fleischer, *HRB Record 12*.

[28]This approach is not inconsistent with that of Chapter 3, where planners use money values of nonwork time to determine whether trips will be made and by what mode. Here the aim is different; it is to decide comparative resource use among alternative investments.

[29]This discussion deals with the economic and societal costs of highway accidents as inputs in decisions on investments. Chapter 12 is concerned with other aspects of highway accidents and the efforts under way to decrease them.

and $10.1 billion motor-vehicle property damage. Based on the NHTSA total, other average measures of cost are $175 per inhabitant, $980 per mile of road, $280 per motor vehicle, or 2.8 cents per vehicle-mile driven. All are staggering statistics.

Motor-vehicle accidents commonly are subdivided into those involving fatalities, nonfatal injuries, and property damage only. Average costs for each class are determined separately. Findings of the NHTSA study,[30] shown in Table 4–3, further subdivide nonfatal injury accidents into five classes; they give the number of accidents in each class, including those involving fatalities and property damage only, and the costs assigned to each class. It should be noted that in this and subsequent tables a fatal accident is one in which one or more individuals are killed and possibly others are injured as well. Likewise in an injury accident, more than one person may be injured. The *Red Book* reports averages of 1.17 fatalities and 2.03 injuries per fatal accident and 1.5 injuries per injury accident.

UNIT COSTS OF HIGHWAY ACCIDENTS. It is not possible to find "correct" values to use in appraising the cost of highway accidents. First of all, using average values for arbitrary categories may not be suitable; next, reliable data are lacking and will be difficult and costly to gather; finally, there are disagreements between those who favor the societal-cost rather than the economic-cost viewpoint. Table 4–4 shows some of the values that are in use. It can be seen that variations from low to high are in multiples from two to ten. Given this dilemma, the *Red Book* proposes that decision makers in each agency adopt values that they deem appropriate for their agency's situation.

As Table 4–4 shows, average costs from three categories are generally employed, and these may not be appropriate for comparative accident-reduction cost studies. For example, suppose the study is to contrast accident costs where an intersection is controlled by two-way stop signs or by traffic signals. With this change, accidents at right angles usually decrease but rear-enders increase. It is very unlikely that the severities and costs of either injuries or property damage will be the same for the two situations. Likewise, distinctions would be appropriate in multivehicle collisions between angles of impacts, velocity, and other variables. For single-vehicle accidents, such variables as velocity, obstructions struck, and vehicle behavior could substantially affect costs. It follows that cost comparisons for accident-reduction proposals may be biased using the average cost approach.

As indicated, reliable data on average accident costs even for the few listed classes are lacking. Much of it is old and for different vehicles and traffic conditions. Furthermore, future changes in vehicle design aimed at reducing injury to occupants and damage to vehicles will continue to be made, so that even if up-to-date data were available for today's vehicle mix, it would soon become out-of-date.

The difference between the economic and societal approaches also affects ac-

[30] *Societal Costs of Motor Vehicle Accidents—1975*, National Highway Traffic Safety Administration, Dec. 1976.

TABLE 4–3. Highway Accidents: Annual Number and Societal Costs for the Year 1975*

		NonFatal-Abbreviated Injury Scale†							
Item	*Fatal*	*5*	*4*	*3*	*2*	*1*	*NonFatal Total*	*Property Damage Only*	*Total*
No. of occurrences, (1000)	46.8	4	20	80	492	3400	4000	21,900	25,900
Cost per occurrence ($1000)	287.2	192.2	87.0	8.08	4.35	2.19	3.185 (av)	0.520	1.45 (av)
Total cost ($ billion)	13.44	0.77	1.74	0.65	2.14	7.45	12.75	11.39	37.58

*Source: *Societal Costs of Highway Accidents—1975*, National Highway Traffic Safety Administration.
†Classes of nonfatal accidents are: 1, minor; 2, moderate; 3, severe (not lifethreatening); 4, severe (life threatening, survival uncertain); 5, critical (survival uncertain).

TABLE 4–4. Costs of Individual Motor Vehicle Accidents as Developed by Representative Sources

Source	*Fatal Accidents*	*Injury Accidents*	*Property Damage Only*
National Highway Traffic Safety Administration*	$287,200	$3185	$520
California Department of Transportation†			
Urban	112,000	3500	1000
Suburban	127,000	4000	1200
Rural	142,000	4500	1400
Questionnaire to State Highway Agencies‡	64,000	4800	840
Illinois Accident Study 1958§			
Urban	26,100	5800	970
Rural	34,900	9000	1330

*From *Societal Costs of Highway Accidents 1975.*
†Basic data from *A Report on the Washington Area Motor Vehicle Accident Cost Study* by Wilbur Smith and Associates 1966, as reported in the *Red Book.*
‡From M. Roddin and D. Andersen, *TRB Record 550.*
§Updated to 1970 in *NCHRP Report 133* and to 1975 using the change in the All-Item Consumer Price Index.

cident costs. The economic approach considers net resource use only; the societal-cost approach includes total earnings, without deducting the resources consumed to produce those earnings. It also adds a group of nonmarket, nonquantifiable consequences to the market ones. Table 4–5 is a summary of one proposed set of societal costs. Three of the numerous differences between the societal and economic approaches are as follows:[31]

1. Either approach includes the stream of future incomes which measures lost production brought on by the accident. This stream is discounted to the date of the accident; incomes are computed separately for representative age groups and for men and women, and weighted to reach the average value. The two approaches differ in that the societal one assumes full employment so that all future earnings are included; the economic one at least questions the full-employment assumption. Also, the economic approach uses net earnings; that is, it deducts from future earnings of the individual the resources consumed by that individual. This deduction is disregarded in the societal analysis.
2. The societal approach includes as a loss the home, family, and community "productivity" of the individual. This includes such nonmarket items as home maintenance, household tasks, and training, teaching, and counseling children. Also volunteer activities outside the home are appraised. The economic approach would omit at least some of the home and all the volunteer services on the basis that they cannot be priced and probably would be performed by others.

[31]For added discussion of this complex subject, see *Societal Costs of Highway Accidents—1975,* op. cit. and R. Winfrey, F. W. McFarland and J. B. Rollins, *TRB Record 680.*

TABLE 4–5. Societal Costs of Highway Accidents—1975*

Cost Component	Fatal Accident	Injury Severity (Abbreviated Injury Scale†)					Property Damage Only
		5	4	3	2	1	
Production/consumption:							
Market	211,820‡	126,650‡	55,550‡	1,645	865	65	—
Home, family and community	63,545‡	37,995‡	16,660‡	425	310	20	—
Medical:							
Hospital	275	5,750	2,250	1,095	450	45	—
Physician and other	160	5,520	2,160	525	165	55	—
Coroner-medical examiner	130	—	—	—	—	—	—
Rehabilitation	—	6,075	3,040	—	—	—	—
Funeral	925‡	—	—	—	—	—	—
Legal and court	2,190	1,645	1,090	770	150	140	7
Insurance administration	295	295	285	240	220	52	30
Accident investigation	80	80	70	45	35	28	6
Losses to others	3,685	4,180	1,830	260	130	32	—
Vehicle damage	3,990	3,990	3,960	2,920	1,865	1,595	315
Traffic delay	80	60	60	160	160	160	160
Total	287,175	192,240	86,955	8,085	4,350	2,190	520

*Source: *Societal Costs of Highway Accidents—1975*, National Highway Traffic Safety Administration.
†Classes of nonfatal injury accidents are: 1, minor; 2, moderate; 3, severe (not life threatening); 4, severe (life threatening, life uncertain); 5, critical (life uncertain).
‡All future costs discounted to the present at 7%.

3. In the societal approach, a money value for pain and suffering to the injured individual and deprivation to others often has been included among societal costs. It would be determined by an analysis of awards made in lawsuits. These are not shown in Table 4–5. However, they were excluded only "because of the problem of collecting a statistically valid sample by reading individual cases . . ." In the economic approach, pain, suffering, and deprivation costs would be excluded as nonmarket and nonquantifiable. This is not to say that such social costs should be excluded from the decision-making process. Rather, it is reasoned that they should be weighed by the decision maker as an irreducible.

In sum, it can be seen that there are no "right" unit costs of highway accidents to use in economy studies. Yet the costs are real and substantial and should be included in a study. It would seem that sensitivity analyses using a range of values would often be desirable.

PREDICTING FUTURE ACCIDENT EXPERIENCE. To recognize accident costs in highway economy studies, an estimate must be made of the difference in accident experience between existing and proposed facilities. For the existing road, data should be available from road inventory records; for the proposed replacement, accident experience on roads of similar design is a common basis of prediction. Data such as those in Table 4–4 then are used to assign money values to before and after conditions.

The *Red Book* gives data on the 1973 accident experience on state highways in California. A portion of it is reproduced in Table 4–6. Included in the averages would be high- and low-volume situations. Where access is not controlled, accidents at intersections and those from roadside intrusions are counted; on freeways, interchange incidents have been reported.

Comparisons of the accident rates shown in Table 4–6 clearly reflect the relatively accident-free characteristics of freeways as compared with other rural and urban highways and the substantially higher fatality rate of two-lane rural highways as contrasted with urban and suburban situations, which probably reflects the greater seriousness of accidents at higher speeds. Also, the large variation when ratios are computed among the four accident classifications indicates the distortions in results if any single one of them—for example, total accidents—is chosen to portray the entire experience.

Accident experience for general classes of highways, as given in Table 4–6, sheds little light on the effects on accidents of specific features such as lane and shoulder widths, side slopes, roadside obstructions in marginal areas, and signs and signals. Such data as are available are discussed under these specific topics in Chapters 9, 10, and 12. Unfortunately, the data are sparse and often conflicting.

In predicting accidents, it is also important to challenge the common assumption that higher standards mean fewer accidents. It has been demonstrated that "It seems dangerous, so accidents happen" does not always prove out. Rather, because of driver behavior, the truth may more nearly lie with "It seems dangerous, so accidents do not occur."

TABLE 4–6. Relationships Among Accident Frequency and Highway Types*†

Location and Road Type	Accident Class: Fatal	Injury	Property Damage	Total
Rural				
No access control				
2 lanes	0.070	0.94	1.39	2.40
4 or more lanes undivided	0.047	0.89	1.95	2.89
4 or more lanes divided	0.063	0.77	1.25	2.08
Partial access control				
Divided expressway	0.038	0.44	0.76	1.24
Freeway (full access control)	0.025	0.27	0.49	0.79
Urban				
No access control				
2 lanes	0.045	1.51	3.38	4.94
4 or more lanes undivided	0.040	2.12	4.49	6.65
4 or more lanes divided	0.027	1.65	3.19	4.87
Partial access control				
Divided expressway	0.022	1.08	2.04	3.14
Freeway (full access control)	0.012	0.40	1.01	1.42
Suburban				
No access control				
2 lanes	0.048	1.26	2.56	3.87
4 or more lanes undivided	0.037	1.58	3.31	4.93
4 or more lanes divided	0.030	1.10	2.24	3.37
Partial access control				
Divided expressway	0.060	0.82	1.29	2.17
Freeway (full access control)	0.015	0.32	0.74	1.07

*Source: *The Red Book.*
†Rates per million vehicle miles.

It follows then, that, as with determining the unit cost of accidents, predictions into the future as to their frequency and severity in specific situations is at best inexact.[32]

Other Highway User Consequences

In settled areas in developed nations, land access is already available. Highway expenditures then improve that access by reducing motor-vehicle operating

[32]*NCHRP Report 162* offers several ways to identify accident-prone locations and for evaluating the potential worth of proposed changes.

costs, travel times, and, hopefully, accidents. It then can be argued that the savings in these items represent a full measure of market highway user benefits. To include others would be counting the same benefit twice.

There commonly are (see Table 3–1) several nonquantifiable, nonmarket consequences that highway improvements bring to users. Among the most important is the reduced pain, suffering, and deprivation of those involved in or who are affected by highway accidents. As indicated earlier, reducing the number and severity of such accidents is a major concern of the public, legislative bodies, and highway officials, and it is important that these projected accident reductions be clearly stated so that they will be considered in programming future investments. It is to be noted that federal highway appropriations in recent years have special funds which recognize these noneconomic aspects of highway accidents.

Another nonquantifiable, nonmarket value of highway improvement is a reduction in the strain and discomfort of nonuniform driving. Origin-destination surveys have shown that some drivers choose routes along freeways and expressways in preference to those along conventional highways or streets, even though overall distances are longer and travel times greater on the former. Also, many drivers are willing to use toll roads even though they can reach their destinations in fewer miles and with little time difference on a free but congested route. Thus, there is evidence that drivers impute a money value to the comfort and convenience provided by modern highway facilities.

Some highway economists assert that reduction in strain and discomfort is a nonmarket, nonquantifiable benefit of highway improvement (see Table 3–1) and should not be assigned a monetary value. On the other hand, the 1960 *Red Book* proposed values per vehicle-mile for discomfort and inconvenience in terms of operating conditions as follows: restricted, 1.0 cent; normal, 0.5 cent; free, 0 cent. For loose, unpaved surfaces the assigned value was 0.75 cent. It can be seen, then, that as with time savings to noncommercial vehicles, there is debate as to whether charges for strain and inconvenience are appropriate in economy studies. It is clear that values for them are approximate at best. For this reason they should be computed and reported separately, if they are included at all.

Often, today's highway designs call for separated roadways, extensive landscaping, developing of views, or other means to make driving more pleasant.[33] On the other hand, driving is unpleasant and noisy on many highways and in tunnels or on depressed roadways with vertical walls. Such factors as these, although very important, also are nonquantifiable and noneconomic and any assignment of plus or minus money values would be arbitrary. It has been proposed that, since the benefits of expenditures to improve the driving "climate" cannot be quantified, the costs of providing them also be omitted from economy studies.

Highway improvements also are made to improve access to parks, recreational facilities, and cultural and historical areas or to provide for sightseeing

[33]See Chapters 6, 8, 9, and 13 for added discussions.

and pleasure driving. Again, the benefits of such investments are noneconomic and nonquantifiable, although money values sometimes have been imputed by observing the choices made by travelers. Possibly such dollar measures, even though imperfect, are useful in the decision of whether to undertake a project or in comparing alternative uses of limited funds. But it may be better to state such benefits in nonmonetary terms, such as the number of visitors or visitor-days.

CONSEQUENCES TO OTHER THAN HIGHWAY USERS

Table 3–2 lists a group of market and nonmarket community consequences of urban freeways. This list could also be applied to all highway facilities by excluding the items that are not appropriate. As indicated in Table 3–2, freeways and other controlled-access highways that cut through developed or partially developed areas may have economic consequences; for example, they may change the circulation patterns and costs of local traffic movements and the operating costs of public service agencies such as public transit, police and fire departments, and schools. Thus, if a segment of a community is severed from the remainder and travel to it becomes circuitous, the costs of these services will increase or, in extreme cases, duplication of facilities may become necessary. In such instances the added costs, although difficult to estimate, should be included among the consequences of the proposed improvement. In the overall public viewpoint, it is immaterial which governmental or private body will pay the bill; if the costs will result from the improvement, they belong in the economy study. The same line of reasoning should be used to determine whether a crossover or interchange is needed at all and if so, what its location and conformation should be.

The effect of highway improvements on land values or on the income that land produces raises several difficult and controversial problems. In certain cases, highway improvement unquestionably brings economic gain, as when natural resources or land are made more easily accessible. Here care must be taken to avoid "double counting." To illustrate, a road into an isolated or poorly served area may bring a change in land use which increases both the value of the land and the income from crops or other economic activity. Either of these may be taken as a measure of economic gain, but if both are included, the same benefit has been counted twice. In other situations, land value and business activity may rise along the route of a new highway. However, this rise may be offset or more than offset by depressed land values and reduced business at other locations. Here a "transfer" rather than an economic gain has occurred. Because of these transfer aspects, many highway economists are reluctant to include gains in land value or business activity in economy studies.

That highways, freeways, and transit systems have consequences for those nearby is clearly demonstrated. But these are complex. For example, a study in Chicago indicated that a freeway impacted developing areas far more than mature ones. Again, different land uses are affected differently.[34] In general, industry profits by being in the zone of influence; others such as residences may suffer.[35] In most instances, as with land values, such differential impacts are not quantified in economy studies, since they seldom are measured, may be offsetting, and often can be classified as transfers in the broad public context.

Highways and mass transit systems also have what economists call "distributional effects," which involve redistribution of benefits and costs among those affected.[36]

Freeways and major highways may also bring noise, air pollution, and other ecological impacts to previously unaffected areas.[37] Some of the costs associated with noise can be quantified as when noise barriers are constructed along the highway margin or buildings and homes are insulated to reduce the decibel level. At present, however, many of the other economic costs of noise, air pollution, and other effects cannot be quantified and others fall into the nonmarket area along with such social costs as the pain and suffering associated with highway accidents.[38]

Displacement of people from their homes and the disruption of neighborhoods are among the most frequently cited adverse effects of urban highways and freeways. Recent legislation at both the federal and state levels which provides for compensation for certain displacements may provide at least partial measures for some of the economic and social costs (see Chapter 7). But to date many of the effects are largely unknown as well as unquantified.[39]

At the regional level also, highways and transit may have impacts on development, but these seldom can be quantified.[40] However, some researchers have attributed development to other factors.[41] Another regional effect is that better roads attract tourists, either new travelers or those diverted from other areas. Spending by these visitors for food, lodging, and other purposes bring economic activity, including jobs for local residents. On the other hand, there are associated costs, including the construction and maintenance of roads and parks by government agencies. These economic and financial impacts may appear quite different when appraised from a regional as contrasted with a national viewpoint. Treatment of these topics is outside the scope of this book.[42]

[34]See P. Wang et al., *Transportation Engineering Journal of ASCE,* Aug. 1975.

[35]See F. I. Thiel et al, *Transportation Engineering Journal of ASCE,* Nov. 1976; H. B. Gamble et al., *TRB Record 508*; and C. J. Langley Jr., *TRB Record 583.*

[36]See, for example, J. S. Dajani and M. M. Egan, *TRB Record 516* and A. Chatterjee and K. C. Sinha, *Transportation Engineering Journal of ASCE,* Aug. 1975.

[37]The legal and technical aspects of noise, air pollution, and other community effects of highways are discussed in Chapter 13.

[38]See G. R. Allen, *TRB Record 583* for a proposal for incorporating economic considerations into environmental impact statements.

[39]See A. G. Kriken et al., *TRB Record 528,* for a discussion of methods for identifying impacts on neighborhoods.

[40]See, for example, R. E. Taggart, Jr. et al. and A. M. Gaegler et al., *TRB Record 716.*

[41]For example, D. Levitan, *TRB Record 583,* questions that the construction of Route 128 circling Boston actually was the cause of the outward movement from the downtown area.

[42]For a discussion of this topic in the transportation literature, see D. H. Farness, *TRB Record 490.* There is also extensive discussion of this topic in the water resources literature.

Economy Studies for Transit Systems

Economy studies to establish the feasibility of proposed transit systems, or to make comparisons among systems or components, would be made using the same techniques given here for highways. The difficulties are not in computation but in getting reasonable estimates of first cost, and of annual benefits and operating costs. The *Red Book* gives supporting data, procedures, and an example for appraising bus operations; but limited space precludes doing the same here.

PROCEDURES FOR ECONOMY STUDIES[43]

An economy study must first answer the question: "Why do it at all?" In other words, does the proposed improvement represent an attractive investment when compared with other possible uses of available resources? Where there is only one plan for a particular improvement, a favorable answer clearly indicates that the project is desirable. However., where there are alternative methods for improvement, a second question is in order. It is "Why do it this way?" or "Which of the proposals is the best?" This is answered by finding whether the *increment* of investment between cheaper and more expensive plans also appears attractive. By successively eliminating those proposals that fail either the first or the second of these tests, the best of the lot may be found.

Compound Interest Calculations

First costs, occasional costs such as for resurfacing, salvage values, and annual costs of operation and maintenance lie at different points along the time scale. For economy study purposes, the money values for the different alternatives must be accumulated in a manner that makes them comparable. Common bases of comparison in the highway field are (1) equivalent uniform annual costs, (2) present worth of future expenditures or cost reductions, (3) the interest rate or "rate of return" at which the alternatives are equally attractive, and (4) the "benefit-cost ratio" of the total investment and of each viable increment.

Regardless of which of these methods of comparison is chosen, the same approaches must be used. For example, the first cost of a capital improvement is converted into equivalent uniform annual cost by the formula

$$A = P\left[\frac{i\,(1+i)^n}{(1+i)^n - 1}\right] \tag{4–1}$$

where

i = interest rate per interest period
n = number of interest periods

[43]Derivations of compound interest formulas and procedures for economy studies are spelled out in far more detail in the standard textbooks in engineering or highway economy such as those by Grant, Ireson, and Leavenworth, op. cit., and Winfrey, op. cit. These references give data for either periodic or continuous compounding. Only the former is presented here.

P = present sum of money, in this case the first cost of any particular element of road improvement
A = end of period payment or receipt in a uniform series continuing for the coming n periods, the entire series equivalent to P at interest rate i

For assigned values of i and n, the conversion factor $[i(1+i)^n]/[(1+i)^n-1]$ is constant. It is called the *capital recovery factor* (CR); CRs appropriate for the solution of highway engineering problems are given in Table 4–7.

Salvage value at the end of estimated life is converted to annual cost by multiplying salvage value by the *sinking-fund deposit* factor. This factor is found by subtracting the interest rate from the CR (CR − i). To illustrate, the sinking-fund deposit factor, based on Table 4–7, at 6% interest and 40-yr life is 0.06646-0.06 = 0.00646. The "present worth" of a series of equal annual costs is obtained by using the reciprocal of the CR to make the conversion.

At times, it is also necessary to incorporate a single cost or benefit from a specific year in the future into the study. This can be done by means of the *single-payment present-worth factor* (PW) which is given by the formula

$$P = F\left[\frac{1}{(1+i)^n}\right] \qquad (4\text{–}2)$$

where F = a single sum of money at the end of n periods from the present date which is equivalent to P at interest rate i. Typical values are given in Table 4–8.

At times it may be necessary to find the equivalent of a present sum at some future point in time. An example is where expenditures for construction are made over several years, but it is desired to know the value at the end of the construction period. To make this conversion from P to F, the reciprocal of the PW of Table 4–8 may be employed. For annual cost solutions, the conversion of F to A can be made by first converting F to P and then P to A.

In many cases, traffic volumes and certain annual costs increase over time. An approximate method for recognizing this in economy studies is to average

TABLE 4–7. Capital recovery factors (CR) for Various Lives and Interest Rates.

Life	*Interest Rate*						
(yr)	*0%*	*3%*	*6%*	*8%*	*10%*	*12%*	*15%*
5	0.20000	0.21835	0.23740	0.25046	0.26380	0.27741	0.29832
10	0.10000	0.11723	0.13587	0.14903	0.16275	0.17698	0.19925
15	0.06667	0.08377	0.10296	0.11683	0.13147	0.14682	0.17102
20	0.05000	0.06722	0.08718	0.10185	0.11746	0.13388	0.15976
25	0.04000	0.05743	0.07823	0.09368	0.11017	0.12750	0.15470
30	0.03333	0.05102	0.07265	0.08883	0.10608	0.12414	0.15230
40	0.02500	0.04326	0.06646	0.08386	0.10226	0.12130	0.15056
50	0.02000	0.03887	0.06344	0.08174	0.10086	0.12042	0.15014
60	0.01667	0.03613	0.06188	0.08080	0.10033	—	—
80	0.01250	0.03311	0.06057	0.08017	0.10005	—	—
100	0.01000	0.03165	0.06018	0.08004	0.10001	—	—

TABLE 4–8. Single-Payment Present-Worth Factors (PW) for Various Time Periods and Interest Rates.

Period (yr)	Interest Rate						
	0%	3%	6%	8%	10%	12%	15%
1	1.000	0.9709	0.9434	0.9259	0.9091	0.8929	0.8696
2	1.000	0.9426	0.8900	0.8573	0.8264	0.7972	0.7561
3	1.000	0.9151	0.8396	0.7938	0.7513	0.7118	0.6575
5	1.000	0.8626	0.7473	0.6806	0.6209	0.5674	0.4972
10	1.000	0.7441	0.5584	0.4632	0.3855	0.3220	0.2472
15	1.000	0.6419	0.4173	0.3152	0.2394	0.1827	0.1229
20	1.000	0.5537	0.3118	0.2145	0.1486	0.1037	0.0611
25	1.000	0.4776	0.2330	0.1460	0.0923	0.0588	0.0304
30	1.000	0.4120	0.1741	0.0994	0.0573	0.0334	0.0151
40	1.000	0.3066	0.0972	0.0460	0.0221	0.0107	0.0037

the values for the first and last years. But, because compound interest is involved in cost or benefit calculations, this approach will give equivalent annual amounts that are too high. A correct value can be obtained by computing an *equivalent uniform annual value*. Table 4–9 offers a few multipliers for a convenient way for making this calculation. To illustrate its use, assume that traffic in year 1 is estimated to be 10,000 vehicles per day; it will increase each year for the next 20 yr by 2%, or 200 vehicles, so that the volume in year 20 is 10,000 + 200 × 19 = 13,800 vehicles. The average for years 1 and 20 is then 11,900 vehicles. However, to compute the correct equivalent annual traffic volume for economy studies, multipliers such as those in Table 4–9 should be used. For the example just given, and an interest rate of 8%, the equivalent annual daily traffic volume is the sum of the traffic in year 1 (10,000) and the equivalent uniform annual increase (200 × 7.04). The total is 11,408 vehicles per day.[44] Note that at 0% interest, the total would be 10,000 + 200 × 9.50 or 11,900, the same results computed above by the averaging method.

The values in Table 4–9 can also be employed in economy studies in situations where resource uses such as for maintenance increase uniformly over time. They could also be employed in financial studies.

Annual Cost Method

As already indicated, the annual cost of an element of capital improvement is found by multiplying its first cost by the appropriate CR (given in Table 4–7). The amount so found, if charged at the end of each year for the assumed life span, will exactly repay the initial investment, with interest. The total annual cost of a particular improvement is the sum of all the annual costs of capital

[44]More complete tables of multipliers and another set giving the present worth of such series will be found in Grant, Ireson, and Leavenworth, op. cit. Winfrey, op. cit., offers two sets of tables, one for arithmetic and the other for exponential growth rates; the latter assumes a compounded rate of increase. Also Winfrey's tables begin with year 0 traffic, which calls for a slightly different computational procedure. The *Red Book* offers a graphical solution to find present worth. It uses first and last year's traffic estimates and assumes an exponential growth rate.

TABLE 4–9. Multipliers for Use in Computing a Single Annual Traffic Volume or Cost that is the Equivalent of a Uniformly Increasing Traffic Volume or Cost*

Period (yr)	Interest Rate						
	0%	*3%*	*6%*	*8%*	*10%*	*12%*	*15%*
5	2.00	1.94	1.88	1.85	1.81	1.77	1.72
10	4.50	4.26	4.02	3.87	3.73	3.58	3.38
15	7.00	6.45	5.93	5.59	5.28	4.98	4.56
20	9.50	8.52	7.61	7.04	6.51	6.02	5.37
25	12.00	10.48	9.07	8.23	7.46	6.77	5.88

*These multipliers convert only the gradient portion of an end-of-year series of 0, 1, 2, . . ., $(n-1)$ to an equivalent uniform annual series for n yr.

recovery, the appropriate annual charges for any periodic expenditures, plus annual maintenance and user costs. Annual costs are computed for the existing facility and for each of the proposals for improvement. Other things being equal, that alternative which has the smallest total annual cost is the best choice. A typical solution by the annual-cost method appears in the sample economy studies given later in this chapter.

Results of solutions by the annual cost method are markedly affected by interest rate. Low interest rates favor those alternatives that combine large capital investments with low maintenance or user costs, whereas high interest rates favor reverse combinations.

One use of economy studies is for capital rationing; that is, to establish priorities among projects. For this purpose, the adopted interest rate should be near that earned by the most badly needed improvements; otherwise the result may not indicate the proper choices.

Present Worth of Costs or Present Worth of Benefits-Less-Costs Methods

Present worth economy studies determine, for each alternative, the resources required at the present to cover all costs of the proposed improvement for the duration of the study period. Thus for alternatives of equal life, the plan with the lowest present worth, or, stated differently, that requires the smallest commitment of resources, is the most attractive, other things being equal. If projects being compared do not have equal lives, an added computation is required so that the costs of each cover the same time period.

The results of present-worth and annual costs studies are mutually convertible by means of the CR. It follows that present-worth studies also are markedly affected by interest rate. Low interest rates will favor alternatives having high capital investments and low annual costs, whereas high interest rates will favor reverse combinations.

An alternative to the present-worth method is that of *present worth of benefits minus costs*. Benefits are defined as the differences in costs between alternatives, and are usually computed only for road-user items. With this form of analysis, the best solution is that which shows the highest positive difference between the present worth of benefits and the present worth of costs.

Rate-of-Return Method

The rate-of-return method involves finding the interest rate at which two alternative solutions to an economy problem have equal annual costs or present worths. The first step is to find the rate of return on each proposed investment as compared with the solution that requires the least capital outlay, which often is the status quo. Those plans that fail to show the minimum attractive return on the total proposed investment are discarded as they fail the test "Why do it at all?" Next the rate of return is computed on the *increase* in investment between proposals having successively higher first costs. Any proposal that fails to show the minimum attractive return as compared with the next lower is eliminated from the series before the rate of return on the next increment is computed. This is to answer "Why do it this way?" The alternative of highest first cost that offers more than the minimum attractive return on both total and increment investments is the best from an economy standpoint. A typical rate-of-return solution is included in the sample economy studies.

Proponents of the rate-of-return method claim that it offers certain definite advantages, among which are the following:

1. It lends itself to the "capital rationing" concept of private industry. As in private industry, the funds available for capital improvements for highways usually are limited. By listing proposed projects in the order of descending rates of return it is possible to develop an array of projects in the order of economic desirability.
2. A cutoff point at which all funds are exhausted will be established and this will indicate the minimum attractive rate of return for the particular agency. Thus a standard is established by which to judge all design and administrative decisions that have economy aspects.
3. Rate of return is clearly understood by laymen since it is a part of all business transactions. It follows that explanations to legislators and the public will be far simpler and more direct than if other comparisons are used.

Objections to the rate of return method are:

1. Answers can be found only by tedious trial-and-error when computers are not available, whereas by other methods the answers can be worked out directly.
2. For problems where the stream of annual consequences swings between benefits and costs, the solution may have multiple roots. However, the annual-cost, present-worth, and benefit-cost methods also give equivocal results under these circumstances.[45]

Benefit-Cost Ratio Method[46]

The benefit-cost method, currently favored by some highway engineers and certain others in the public-works field, expresses the comparative worth of proj-

[45]For a more detailed discussion and references see M. Wohl, R. Winfrey, R. S. Leavenworth, H. M. Steiner, and D. R. Bergmann, *TRB Record 731*.

[46]Economists often refer to economic analyses under the heading of cost-benefit studies. This term should not be confused with the benefit-cost ratio, which is one among several analytical methods that can be employed.

ects by the ratio of annual benefits to annual costs.[47] The benefit-cost ratio is expressed by

$$\text{benefit-cost ratio} = \frac{\text{annual benefits from improvement}}{\text{annual costs of improvement}} = \frac{R - R_1}{H_1 - H} \qquad (4\text{–}3)$$

where

R = total annual road-user cost for the basic condition or existing highway, or for that alternative of lower first cost

R_1 = total annual road-user cost for a proposed improvement, or for that alternative of higher first cost

H = total annual highway cost for the basic condition or existing highway, or for that alternative of lower first cost

H_1 = total annual highway cost for a proposed improvement, or for that alternative of higher first cost

The procedure for a benefit-cost economy study, like that for a rate-of-return solution, involves two sets of computations. First the benefit-cost ratios between each alternative and the basic condition are found, and those plans that fail to reach the minimum attractive ratio are discarded. Then the benefit-cost ratio for each increment of added investment is computed, each plan being proved against the preceding acceptable plan. The alternative of greatest first cost that reaches the prescribed benefit-cost ratio (usually 1.0) on both total and increment of investment is the most acceptable on the basis of the assumed interest rate. A typical benefit-cost solution appears in the sample economy studies.

The choice of interest rate can seriously affect the results of benefit-cost solutions. Often, interest charges appear only in the highway-cost portion of the equation; thus an increase in the interest rate increases only the denominator and results in a less favorable benefit-cost ratio. For example, assume that the entire annual cost difference between two alternatives is based on a capital investment of 30-yr life. Then, if the benefit-cost ratio at 0% interest is 3.0, at 3% it will be 2.0, at 6% 1.4, and at 10% 0.94. This means that benefit-cost ratios computed at different interest rates are not comparable. To state the complete results of a benefit-cost solution, both the ratio itself and the interest rate must be given.

The benefit-cost ratio as given by Formula 4–3 does not fit some highway problems. For example, the purpose of many drainage and resurfacing projects is to reduce maintenance costs. User costs (or benefits) are affected very little if at all. In these cases, benefit-cost analyses are meaningless unless benefits are redefined to include reductions in maintenance costs; otherwise there will be no numerator for the equation.

The difficulty of distinguishing between benefits that go in the numerator and

[47]As indicated earlier, Congress stipulated in the Flood Control Act of 1936 that federal agencies dealing with river-basin development make benefit-cost studies. There are no legislative stipulations that require the benefit-cost approach for economy studies for transportation investments.

cost reductions that go in the denominator of the benefit-cost equation restricts its usefulness to "go or no go" decisions when projects of substantially different character are being compared. It can easily be demonstrated that, for any given analysis, the benefit-cost ratio will always be either greater or less than 1 whether particular items are classified as positive or negative benefits or as negative or positive costs. Thus, if the interest rate is suitable, projects showing ratios greater than 1 will be economically viable; those with ratios less than 1 will not be. But the numerical values of the ratio can be substantially altered by the numerator versus denominator decision; this means that the benefit-cost ratio should be used cautiously, if at all, as a device for setting priorities among a list of viable projects.[48]

Sensitivity of Economy Studies

Economy studies for highways are based on forecasts of the cost of such items as construction, maintenance, and vehicle operation; furthermore, they incorporate more or less arbitrary assumptions regarding traffic growth, service life, inflation or differential inflation, and interest rate. It is important that the decision makers be aware of the extent to which moderate changes in the basic forecasts and assumptions modify the results of an economy study. Or stated differently, they should appreciate the *sensitivity* of the findings to changes in the forecasts. Where important decisions are involved, a series of parallel solutions encompassing a reasonable range of assumptions may be warranted. Another approach involves the assignment of probabilities or probability distributions to future events. A detailed discussion of the sensitivity aspects of economy studies is beyond the scope of this book. It can be stated, however, that one of the strongest arguments against unreasonably low interest rates is that they give heavy weight to the uncertain estimates of the more distant future. On the other hand, higher interest rates tend to substantially discount future happenings.[49]

Special Considerations in Economic Analysis

There are many complexities in making highway and transportation economy studies which cannot be discussed here because of space limitations. But three will be outlined briefly in the paragraphs that follow because they raise important questions and have been the subject of considerable attention.

DEMAND CURVES-CONSUMER SURPLUS. Figure 4–9 illustrates by a transportation example the *demand* or *indifference curve,* a conceptual tool widely employed by economists. The principle is that as the cost of a commodity decreases more will be purchased. Referring to curve D_1, for example, if the cost is p_0, V_0 trips will be made. But if the facility is improved so that the cost of the trip is reduced to p_1, then there will be V_1 trips. For the earlier condition, V_0 travelers find that the benefits exceed the costs and make the trip; for the last

[48]See G. A. Fleischer and discussion by R. Winfrey, *HRB Record 383,* for more insight into this issue. Other discussions of cost-benefit approaches include D. G. Haney, *HRB Record 314,* and E. B. Steinberg, *HRB Record 348.*
[49]For further discussion, see Grant, Ireson, and Leavenworth, op. cit., and Winfrey, op. cit.

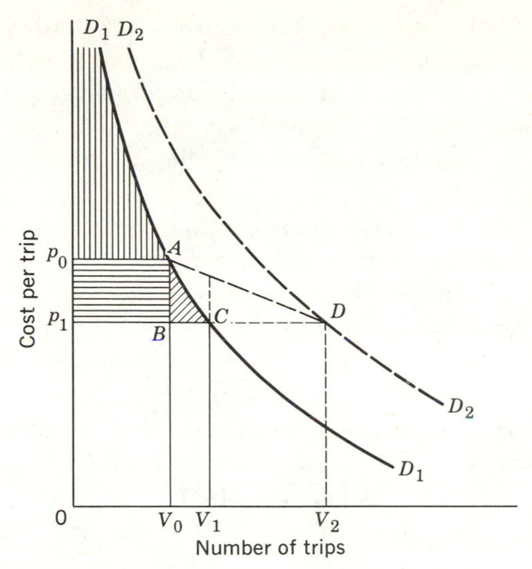

Fig. 4–9. Demand curve illustrating the relationship between trip cost and traffic volume.

traveler V_0, benefits and costs are almost balanced. Traveler $(V_0 + 1)$ does not go since costs exceed benefits. From these choices point A and others on the indifference curve are established.

Indifference curves representing the range of consumer choices have a variety of slopes. If the curve is almost horizontal, demand is said to be *elastic,* that is, small price changes result in relatively large changes in consumption. On the other hand, an almost vertical curve is referred to as *inelastic,* since large price changes are required to produce small changes in consumption.

The demand curve concept is often employed to express the benefits of highway improvements. First, for travelers using the unimproved road, the cost of travel is represented by the area p_0AV_0O in Fig. 4–9. For the same group of travelers, costs after improvements are represented by the area p_1BV_0O. Their savings, then, are given by the area p_0ABp_1 (horizontally hatched on Fig. 4–9). In cases where benefits to future traffic are being forecast, the benefited traffic would be the sum of existing and normal growth volumes.

Another group of travelers are responsible for the increase in traffic $(V_1 - V_0)$ brought by the highway improvement. It can be argued that the benefits that accrue to them can be represented by the approximately triangular area outlined by ACB (diagonally hatched). The idea is that traveler $(V_0 + 1)$ receives almost the full benefit of $p_0 - p_1$, but traveler V_1 gains almost nothing. For forecasts of benefits to future traffic, then, this approach would price the savings to generated and development traffic at half the unit value for existing traffic. In the situation illustrated by Fig. 4–9, all but the V_0 traveler in the "before improvement" condition would be willing to pay more than p_0 to make the trip. This excess in "willingness to pay," called consumer surplus by economists, is shown by the incompletely defined, vertically hatched area above line p_0A. For the "after improvement" condition, the consumer surplus is defined by area p_0ACp_1 plus the vertically hatched area. It follows that the change in consumer surplus (willingness to pay) resulting from the improvement is the area p_0ACp_1. In effect, the benefit to each existing and normal-growth traveler is $p_0 - p_1$, and the average benefit to travelers generated by the improvement is one-half this amount.

In some cases, highway improvements, by raising the level of service, may generate another set of users equal in number to $V_2 - V_1$. There may also be a new demand curve, illustrated by curve D_2. There then is addition consumer surplus consisting of the shaded area above line *CD*. The total increase in consumer surplus (willingness to pay) brought by the improvement is, then, the sum of the shaded area above line *CD* and area p_0ACp_1.

The authors of *Red Book* and many economists argue that the benefits of a highway improvement should be measured by "willingness to pay" and therefore include consumer surplus as just described. Others, including Winfrey, reject the consumer surplus approach on the grounds, first, that "willingness to pay" is an improper way to gauge benefits and, second, it cannot be measured. Instead, they would measure benefits by the change in user costs, with and without the improvement. The issues surrounding demand curves and consumer surplus are complex. A discussion of them is beyond the scope of this book.[50]

REINVESTMENT PROBLEM. As indicated earlier, investments in major highway and transit projects have relatively long lives in the range of 40 yr or more; other projects such as for resurfacings or traffic control may fall in the 10 to 20 yr category. If the relative economy of long-lived and short-lived projects are compared by any of the methods described above, there is the underlying assumption that the identical short-lived project will be done again or that the annual cost reductions resulting from this expenditure can be reinvested in equally attractive projects. Only by examining reinvestment opportunities can an analyst be sure that comparisons are valid. One procedure would be to extend the analysis to include the effects of reinvestment to the end of the comparison period. This might be done at several interest rates to determine how sensitive the result is to them.

SYSTEM VERSUS PROJECT COMPARISONS. Most of the attention to economic analysis for highways has focused on choices among individual projects or on levels of improvement within individual projects. Yet projects do not stand alone; improvements to one part of a highway network invariably affect traffic on others. Furthermore, the sequencing of individual project construction over time likewise affects future traffic flows and the accompanying costs.

To evaluate the economic effects of these interdependencies in overall road or transportation networks requires that the techniques for traffic forecasting and for economic analysis be integrated. Because of its complexity and the number of alternatives that must be examined, only greatly simplified approaches have been attempted to date. Such system-wide comparisons are not commonly

[50]See the *Red Book* and R. Winfrey, *TRB Record 550* for further discussion.

made.[51] Neither is it common to make comparisons among alternative public transportation schemes. Reasons include lack of data, the complexity of the analysis itself, and the fact that the decisions are primarily political and financial.[52]

ECONOMY STUDIES FOR HIGHWAYS IN DEVELOPING COUNTRIES

As indicated earlier, the procedures outlined so far in this chapter apply where there is already access by motor vehicle so that the principal benefits usually accrue to motor-vehicle users. However, in many parts of the world, groups of people in rural areas often have had little economic interchange or social contact with outsiders. Often, travel has been by foot or horseback. Where roads exist, they might be little more than tracks, passable only in good weather.

Today, strong efforts are under way to provide better access to these rural areas. Lending agencies such as the International Bank for Reconstruction and Development (World Bank), the U.S. financed Agency for International Development (AID), and parallel agencies of other nations or groups of nations are sponsoring projects in many developing countries. As a part of these efforts, sponsors commonly require careful planning, with a strong emphasis on economic justification. Several research studies have been carried out and others are under way to develop appropriate design standards and realistic vehicle-operating, construction, and maintenance costs.[53]

The concept of shadow pricing mentioned earlier in this chapter is of particular importance in economy studies for developing nations. Also, inputs for analysis and decision rules can be quite different with roads serving different purposes, as follows:

1. Principal highways or those where access by motor vehicle is reasonably available: For them, as for highways in developed nations, benefits are primarily savings in motor-vehicle operating costs which result from the proposed improvements.
2. Economic penetration roads that provide access to areas currently inaccessible or barely accessible to motor vehicles: For them, benefits are primarily the net economic gain from improved access. Measures could include the value of crops or goods exported from the region or, alternatively, the increased value of the land penetrated by the road.
3. Social roads which provide contacts between affected people and the rest of the country but which probably cannot be justified on economic grounds: Money for such roads would be allocated separately, without being justified under the criteria

[51] See J. W. Spencer, J. H. Shortreed, and D. S. Berry, and S. J. Bellomo and S. C. Provost, *HRB Record 224,* J. Freeman and B. G. Hutchinson, *TRB Record 550,* and the *Red Book* for added discussion.

[52] See, however Gray and Hoel, op. cit., *HRB Record 293 and 459,* and *TRB Record 559.*

[53] C. G. Harral and S. K. Agarwal and several others in *TRB Special Report 160* give details of a World Bank sponsored study in Kenya. A parallel research effort in Brazil costing $19 million is partially reported in *TRB Record 702.* Many analyses of individual projects have been completed. AID and the World Bank have sponsored the development of a computer-based Road Investment Model which is gaining acceptance.

for the other two road classes. In this instance selection among projects could be based on the number of people affected per unit of money spent.[54]

There is a growing realization that, in the past, designers for roads in developing countries have placed too heavy reliance on geometric and pavement design practices from the developed countries. This approach is now being questioned. Another concern has been that all too often roads, once constructed, were permitted to fall into disrepair. It seems clear that in the future construction may be to lower standards utilizing more local materials and with heavy emphasis on labor intensive methods. In addition, maintenance will be given substantially greater attention.[55]

SAMPLE ECONOMY STUDIES[56]

Example 1

PROBLEM STATEMENT. A two-lane rural road follows a circuitous 9.30-mi route. On it are six curves of 90° central angle and 6° sharpness. Total length of these curves is 9000 ft (1.70 mi). The surface is in bad condition and must be renewed. Since a small additional expenditure for grading and structure widening will bring the road to fairly high standards, this cost is included in the "do nothing" alternative called Alternative A. There are two other likely proposals, both of which call for abandonment of the present road and construction along new, largely straight alignments. Alternative B is shorter than the present road by 0.85 mi but requires a sizable expenditure for rights of way, grading, structures, and surfacing. Alternative C is shorter than alternative B by 0.43 mi. It also requires new rights of way and construction. Grading and structure costs are much higher as the terrain is rougher and a river is crossed at a very unfavorable site.

Estimated lives and costs, stated at present, uninflated values, including engineering, are as follows:

[54]Unfortunately, the effects of improved access and contact with the outside world are not always positive. For example, in many countries there has been a movement to the squalor of ghettos in urban areas where the affected people may be even worse off. Also, their cultural values and disciplines may be eroded with serious consequences for them and others as well.

[55]Further discussion of this topic is beyond the scope of this book. *Synthesis 5* of Technology Support for Developing Countries, TRB, is an excellent starting point for further study. There is also a voluminous literature including several books sponsored by AID and prepared by the Brookings Institution, a series of *Occasional Papers* published by the World Bank, and numerous publications by other organizations such as the Transportation and Road Research Laboratory of Great Britain.

[56]Because of space limitations, only three relatively simple economy studies will be presented. These have been designed to illustrate the use of the charts and tables and the various methods of analysis. Numerous others appear in Winfrey, op. cit.; *HRB Record 12, 77, 100, 115, 172, 179, 180, 224, 245, 252, 285,* and *383; TRB Record 528, 550, 572, 737, 747, 751,* and *774. HRB Special Report 92* and *122; NCHRP Report 182; C. W. Dale, ITE Journal,* Apr. 1981, and T. R. Neuman, *Transportation Engineering Journal of ASCE,* Nov. 1979. J. E. Gruver, *TRB Record 490,* summarizes a study of nationwide highway benefits and costs until 1990. Agencies that do economic analyses routinely have them programmed to computers.

Element	Estimated Useful Life (Yr)	Cost in 1981 Dollars		
		Alternative A	Alternative B	Alternative C
Right of Way	40	$ 0	$ 200,000	$ 160,000
Grading	40	100,000	620,000	920,000
Structures	40	60,000	480,000	1,070,000
Surface	20	200,000	940,000	890,000
Total		360,000	2,240,000	3,040,000

The estimated average annual daily traffic for 1981 is 1600 passenger cars and pickups, 200 single-unit trucks and buses, and 100 3-S2 diesel trucks. Forty percent of this volume is evenly distributed over 4 hr; the remainder spreads out evenly over the remaining 20. Traffic is assumed to increase at 2% arithmetic per year for the study period. Average speed over a 24-hr day for all vehicles traveling alternative A will be 40 mph; it will be 45 mph on the new alignments. From a capacity standpoint (see Chapter 8), light trucks are equivalent to passenger cars; one diesel truck is the equivalent of three passenger cars. The small volume of traffic with origins and destinations along the road is ignored. Costs introduced by grades or stops are assumed equal for all plans. Accident hazard is taken as equal on all plans with the longer distances and curves on alternative A being offset by the higher speeds on the others. Incremental maintenance costs are set at $3000/mi/yr in 1981 dollars on all alternatives. Time costs (1975 values) are $7/hr for single unit trucks and buses and $8/hr for 3-S2 diesels. Eighty percent of the passenger car time is classed as nonmarket and will be stated only in hours; the remainder, consisting of work trips, are priced at $4 per vehicle-hour. Since this agency has many unfulfilled highway needs, the minimum attractive rate of return is set at 8%, neglecting inflation. To adjust 1975 costs to 1981, a multiplier of 1.59 (6 yr compounded at 8%) is to be applied.

Some of the assumptions on which this example is based are subject to question. The aim is to illustrate computational method; for this reason the problem has been greatly simplified.

SOLUTION BY ANNUAL COST METHOD (INTEREST AT 8%)

a. *Capital Recovery.* Computation of annual cost of capital recovery (1981 Values—$1000)

Roadway Element		Alternative A		Alternative B		Alternative C	
		First Cost	Annual Cost	First Cost	Annual Cost	First Cost	Annual Cost
Rights of way	CR $(n=40)=0.08386$	0	0	200	16.7	160	13.4
Grading	CR $(n=40)=0.08386$	100	8.4	620	52.0	920	77.2
Structures	CR $(n=40)=0.08386$	60	5.0	480	40.3	1270	106.5
Surface	CR $(n=20)=0.10185$	200	20.4	940	95.7	890	90.6
Total annual cost of capital recovery			33.8		204.7		287.7

b. Vehicle Operating Costs. The volume-capacity ratio for computing speed change costs is as follows.

$$\text{passenger car ADT equivalent (1981)} = 1600 \times 1 + 200 \times 1 + 100 \times 3 = 2100$$

$$\text{passenger car ADT for 20 yr}^{57} = 2100 \left(\frac{1.00 + 1.38}{2}\right) = 2500$$

$$\text{average } v/c \text{ ratio} = \frac{2500}{2000 \times 24} = 0.05^{58}$$

Road User Costs (1975 values) and Times per 1000 Vehicle Miles from Fig. 4–3 and Table 4–2

	Alternative A			*Alternatives B and C*		
	Cost			*Cost*		
Item	*Multiplier*	*$*	*Time (Hr)*	*Multiplier*	*$*	*Time (Hr)*
Free flowing-tangent						
Passenger cars	1.0	72	25.0	1.0	73	22.2
Single-unit trucks	1.9	137	25.0	1.9	139	22.2
3-S2 diesel trucks	2.3	166	25.0	2.3	168	22.2
Curves (added)						
Passenger cars	1.0	20	0	0	0	0
Single-unit trucks	2.0	40	0	0	0	0
3-S2 diesel trucks	3.2	64	0	0	0	0
Speed change (added)						
Passenger cars	1.0	0.3	0	1.0	0.3	0
Single-unit trucks	1.9	0.6	0	1.9	0.6	0
3-S2 diesel trucks	2.3	0.7	0	2.3	0.7	0

[57]Note that traffic volume for computing the *v/c* ratio is the average on the first and last year's traffic. Values are not discounted as are money values.

[58]It is correct to use the average *v/c* ratio to compute costs, since the relationship between it and speed change costs as shown on Fig. 4–3 is a straight line. Averaging would not give exact results for travel times where speeds differed between peak and off-peak conditions.

Estimated Vehicle Operating Costs for 1981 (in 1975 dollars)*

Vehicle and Alternative	*ADT*	*Traffic Condition*									*Total $*
		Freeflowing			*Curves†*			*Speed Changes*			
		Vehicle Miles	*Unit Cost*	*$*	*Vehicle Miles*	*Unit Cost*	*$*	*Vehicle Miles*	*Unit Cost*	*$*	
Alternative A (9.25 mi)											
Passenger cars	1600	5400	72	389	990	20	20	5400	0.3	2	411
Single-unit trucks	200	670	137	92	120	40	5	670	0.6	0	97
3-S2 diesel trucks	100	340	166	56	60	64	4	340	0.7	0	60
Total											568
Alternative B (8.40) mi											
Passenger cars	1600	4910	73	358	0	0	0	4910	0.3	1	359
Single-unit trucks	200	610	139	85	0	0	0	610	0.6	0	85
3-S2 diesel trucks	100	310	168	52	0	0	0	310	0.7	0	52
Total											496
Alternative C (7.97 mi)											
Passenger cars	1600	4650	73	339	0	0	0	4650	0.3	1	340
Single-unit trucks	200	580	139	81	0	0	0	580	0.6	0	81
3-S2 diesel trucks	100	290	168	49	0	0	0	290	0.7	0	49
Total											470

*All data except ADT in 1000.
†Total curve length = 1.70 mi.

Equivalent Uniform Annual Operating Costs with 2% Annual Traffic Increase*

	Vehicle Operating Costs				
		1981 Costs in 1981 Dollars		Equivalent Uniform Annual Costs in 1981 Dollars	
Alternative	*1981 Costs in 1975 Dollars*	*Multiplier*	*$*	*Multiplier†*	*$*
Alternative A	568	1.59	903	$(1+0.0704\times 2)$	1030
Alternative B	496	1.59	789	$(1+0.0704\times 2)$	900
Alternative C	470	1.59	747	$(1+0.0704\times 2)$	852

*All values in 1000.
†Multiplier from Table 4–9.

Time Costs for 1981 Traffic (1975 values)

Item	*Annual Vehicle-Miles (1000)*	*Hours per 1000 Vehicle-Miles*	*Time per Year (hr) in 1000*	*$/hr*	*$/yr (1000)*
Alternative A					
Passenger cars (20%)	1080	25.0	27.0	4	108
Single-unit trucks	670	25.0	16.8	7	118
3-S2 diesel trucks	340	25.0	8.5	8	68
Total					294
Alternative B					
Passenger cars (20%)	980	22.2	21.8	4	87
Single-unit trucks	610	22.2	13.5	7	94
3-S2 diesel trucks	310	22.2	6.9	8	55
Total					236
Alternative C					
Passenger cars (20%)	930	22.2	20.6	4	82
Single-unit trucks	580	22.2	12.9	7	90
3-S2 diesel trucks	290	22.2	6.4	8	51
Total					223

Equivalent Uniform Annual Time Costs with 2% Annual Traffic Increase*

	Time Costs				
		1981 Costs in 1981 Dollars		Equivalent Uniform Annual Costs in 1981 Dollars	
Alternative	*1981 Costs in 1975 Dollars*	*Multiplier*	*$*	*Multiplier*	*$*
Alternative A	294	1.59	467	$(1+0.0704\times 2)$	532
Alternative B	236	1.59	375	$(1+0.0704\times 2)$	428
Alternative C	223	1.59	355	$(1+0.0704\times 2)$	405

*All values in 1000.

Noncommercial Time in Thousands of Hours

Alternative	*Time in 1981*	*Average Time over 20 yr*
Alternative A	$27.0 \times 80/20 = 108$	$108 \times (1 + 1.38)/2 = 129$
Alternative B	$21.8 \times 80/20 = 87$	$87 \times (1 + 1.38)/2 = 104$
Alternative C	$20.6 \times 80/20 = 82$	$82 \times (1 + 1.38)/2 = 98$

Summary of Annual Costs (1981 Prices) and NonCommercial Times for a 20-yr Study Period. Interest at 8% in 1000

	Alternative A		*Alternative B*		*Alternative C*	
Cost Item	*Length (mi)*	*$*	*Length (mi)*	*$*	*Length (mi)*	*$*
Capital recovery		34		205		288
Maintenance	9.25	28	8.40	25	7.97	24
Vehicle operating		1030		900		852
Time		532		428		405
Totals		1624		1558		1569
Average annual noncommercial time		129		104		98

Based on the foregoing analysis, alternative B appears to be slightly better than alternative C from an economy study point of view. Alternative A is less attractive. However, the fact that alternative C will save some 6,000 hr in noncommercial time annually argues for it over alternative B. It also must be remembered that this finding is based on an uninflated interest rate of 8%. At some lower interest rate, alternative C, with its higher capital investment, will be favored. At some higher rate, alternative A becomes most attractive. A brief review of the calculations will also indicate that, because vehicle-operating and time costs are such a high percentage of the total, the results are very sensitive to changes in estimated traffic volume and composition, running speeds, and unit vehicle-operating and time costs. But even with these uncertainties, comparisons such as these are of great assistance in making rational decisions.

SOLUTION BY RATE OF RETURN ON INVESTMENT.

Comparing Alternatives A and B at 12% Interest Rate[59]

alternative A, total annual cost $= (100 + 60) \times 0.12130 + 200 \times 0.13388 + 28 + (903 + 467)(1 + 6.02 \times 0.02) = 1609$

[59]These computations are made in exactly the same manner as shown for the annual cost method, except for changed compound interest factors.

$$\text{alternative B, total annual cost} = (200 + 620 + 480) \times 0.12130 + 940 \times 0.13388 + 25 + (789 + 375)(1 + 6.02 \times 0.02) = 1613$$

This computation shows that, at 12%, alternative A is slightly more attractive. It follows that the rate of return at which alternatives A and B are equally attractive lies somewhere between 8% and 12%. Based on straight-line interpolation or a graphical solution:

$$i = 0.08 + 0.04 \times \frac{1624 - 1558}{(1624 - 1558) + (1613 - 1609)} = 0.118 \text{ or } 11.8\%$$

Comparing alternatives A and C. By a solution following the procedure just demonstrated, the interest rate at which alternatives A and C are identical is 10.0%. Alternative C, then, is better than the basic condition if the minimum attractive rate of return is 10% or less.

Comparing Alternatives B and C. By the same set of computations, it is found that the interest rate at which B and C are identical is 6.7%. Only if the minimum attractive rate of return is 6.7% or lower is the extra investment in alternative C justified on economic grounds. Since 8% was originally chosen as the minimum attractive rate of return, it follows that, from an economic point of view, alternative B is the best among the three choices. This answer agrees with that obtained by the annual cost method.

SOLUTION BY THE BENEFIT-COST METHOD. Annual benefits and costs for computing benefit-cost ratios are found exactly as indicated in the solution by the annual-cost method. To complete the solution these results are substituted in Formula 4–3. Benefits are the differences in road-user costs; costs will be all outlays of the highway agency.

From the summary in the solution by the annual-cost method at 8%, in $1000 units:

R = *annual road-user costs for alternative A = 1562*
R_1 = annual road-user costs for alternative B = 1328
H = annual highway costs for alternative A = 62
H_1 = annual highway costs for alternative B = 230

$$\text{benefit-cost ratio (B vs A)} = \frac{1562 - 1328}{230 - 62} = 1.39$$

In like manner:

$$\text{benefit-cost ratio (C vs A)} = \frac{1562 - 1257}{312 - 62} = 1.22$$

These computations show that, if 8% is the minimum attractive rate of return, either alternative B or C is an improvement over the basic condition.

Comparing Alternatives B and C

$$\text{benefit-cost ratio (C vs B)} = \frac{1328 - 1257}{312 - 230} = 0.87$$

This demonstrated that, at 8% interest, the added expenditure for C over B is not justified. However, if the computation had been made at 11.8%, the benefit-cost ratio would have been 1.0 and either alternative would have been equally attractive. Here, then, is proof of the earlier statement that *both* the benefit-cost ratio and interest rate must be stated in any meaningful decision rule.

Example 2

PROBLEM STATEMENT. A projected recreational road primarily for passenger cars and pickups is to follow a route along a river. At one point, the stream flows around a sharp promontory and abruptly changes direction by 60°. Among the possible locations at this point are the following.

Alignment A. A very sharp (25°-230 ft radius) curve which stays close to the river. With this alignment, vehicles will travel up a constant grade of +2.74%, but will be forced to slow down from 50 to 30 mph and accelerate again. Distances traveled are 3043 ft on the two tangents and 240 ft on the curve. Construction costs will be low since little cut and fill will be necessary. There will be little scar and minimal costs for erosion control.

Alignment B. A flat (2°-2865 ft radius) curve. With this alignment, the 3000-ft-long curve will swing about 430 ft further away from the stream. The road will climb on a +6% grade for 2000 ft and then descend at −3% for 1000 ft. No slowing from 50 mph will be required. Because a deep cut and high fills are needed, the first cost of construction and erosion-control measures will be substantially higher on this alignment.

Given these alternatives:

A. Compute separately the 1975 vehicle running costs and times for 1000 passenger vehicles traveling in each direction on each alignment.
B. Assuming an average daily traffic of 1000 vehicles in each direction 365 days per year, what will be the difference per year in 1975 vehicle running costs and travel times for the two alignments?
C. For the traffic in part B, and an imputed value of time of $3 per vehicle-hour, what will be the difference per year in the dollar value of the time between the two alignments?
D. For the traffic volumes in parts B and C and an economic life of 30 yr for construction and erosion control expenditures, and with interest at 10%, what is the maximum justified expenditure at 1975 prices that should be made for alignment B if: (1) only running cost differences are considered? (2) running and imputed time cost differences are both considered?

SOLUTION—PART A

	Alignment A	Alignment B
RUNNING COSTS PER 1000 VEHICLES 0% GRADE ($)		
At 50 mph (Fig. 4–3)	$\frac{3043}{5280} \times 74 = 43$	$\frac{3000}{5280} \times 74 = 42$
At 30 mph (Fig. 4–3)	$\frac{240}{5280} \times 69 = 3$	
Slowing 50 to 30 mph and return (Fig. 4–5)	$= 8$	
Extra cost on curve	$\frac{240}{5280} \times 100 = 4$	$\frac{3000}{5280} \times 7 = 4$
Total running cost on 0% grade	58	46
FOR UPHILL TRAVEL		
Cost on grades (Fig. 4–3) Different from 0%	$\frac{3043}{5280} \times 11 + \frac{240}{5280} \times 13 = 7$	$\frac{2000}{5280} \times 24 - \frac{1000}{5280} \times 16 = 6$
Total running cost uphill	65	52
FOR DOWNHILL TRAVEL		
Cost on grades	$\frac{3043}{5280} \times (-14) + \frac{240}{5280} \times (-12) = -9$	$\frac{1000}{5280} \times 12 + \frac{2000}{5280} \times (-16) = -4$
Total running cost downhill	49	42
TIME CONSUMED PER 1000 VEHICLES, UPHILL OR DOWNHILL (HR)		
Travel time at constant speed	$\frac{3043}{5280} \times 1000 \times \frac{1}{50} + \frac{240}{5280} \times 1000 \times \frac{1}{30} = 13.0$	$\frac{3000}{5280} \times 1000 \times \frac{1}{50} = 11.4$
Time for speed-change cycle (Fig. 4–6)	1.2	0
Total time	14.2	11.4

SOLUTION—PART B. *Excess annual running costs and travel times, A over B for 1000 vehicles per day in each direction.*

$$\text{excess costs} = (65 - 52) \times 365 + (49 - 42) \times 365 = \$7300$$
$$\text{excess time} = (14.2 - 11.4) \times 365 \times 2 = 2000 \text{ hr}$$

SOLUTION—PART C.

$$\text{cost of excess time at \$3/hr} = 2000 \times 3 = \$6000$$

SOLUTION—PART D. *Maximum added justifiable capital expenditure in alternative B.*

$$\textit{maximum expenditure, based on running costs} = \frac{7300}{0.10608} = \$69{,}000$$

maximum expenditure, based on running costs

$$\text{plus time costs} = \frac{7300 + 6000}{0.10608} = \$125{,}000$$

Example 3

A busy suburban artery currently is carrying 12,000 vehicles per day. A cross street, which carries very little traffic, is a major crossing point for school children. Under pressure from local residents and school parents and officials, four-way stop signs were installed at the intersection. It is now proposed to replace them with vehicle- and pedestrian-actuated traffic signals.

Of the total traffic on the main street, 90% can be considered passenger cars and 10% single-unit trucks. To simplify the analysis, it is assumed that the saturation effects shown in Figs. 4–7 and 4–8, the costs to traffic on the cross street and making turning movements, and delays to pedestrians can be neglected. With the stop signs, all vehicles (supposedly) come to a complete stop from an average speed of 40 mph, wait for 5 s, and return to full speed. Experience with traffic-actuated signals at similar locations indicates that 70% of the vehicles will slow from 40 to 30 mph and then resume speed. The remainder will stop and wait an average of 20 s before proceeding.

Installation of the traffic signals, including incidental striping and signing, will cost $50,000 and have no residual value. Maintenance costs are assumed to be $2000/yr. Anticipated life is 15 yr. Neglect any salvage value for the stop signs; removal costs will offset any salvage value. Currently, the agency finds that its funds are exhausted by projects showing an uninflated rate of return of 10%.

Compute the total cost and time consequences, over and above driving straight through an 40 mph, with and without traffic signals. Make complete computations separately for the two alternatives in order to show the magnitude of the individual elements. Keep total vehicle running costs and time costs for commercial vehicles separately; state noncommercial time costs (all passenger cars) only in hours. Report your findings in terms of:

A. Average benefits from installing the traffic signals, considering:
   1. Vehicle running costs only.
   2. Vehicle running costs plus time savings to commercial vehicles quantified at $8/hr.
   3. Annual time savings to passenger cars, in hours.
B. The net present worth of benefits minus costs for 15 yr of service for conditions 1 and 2 stated above.
C. Assume that traffic in the general area and at the intersection will increase at the rate of 3% arithemetic per year from that of the first year of 12,000 per day including the light trucks. This increase will continue for the following 14 yr.

SOLUTION—PART A

Annual Vehicle Running Cost Increases over Driving Straight Through in Dollars

	Stop Signs	Traffic Signals
Passenger cars (10,800/day)	$	$
Cost of stops (40–0–40) (Fig. 4–5)	$\frac{10,800}{1000} \times 365 \times 20.7 = 81,600$	$\frac{10,800 \times 0.30}{1000} \times 365 \times 20.7 = 24,500$
Costs of speed-change cycles (40–30–40) (Fig. 4–5)		$\frac{10,800 \times 0.70}{1000} \times 365 \times 3.8 = 10,500$
Costs of idling (Fig. 4–8)	$\frac{10,800}{1000} \times 365 \times 0.40 = 1600$	$\frac{10,800 \times 0.30}{1000} \times 365 \times 1.70 = 2,000$
Total for passenger cars	83,200	37,000
Single-unit trucks (1200/day) (using data from above and multipler from Table 4–2)		
Total annual running costs	$\frac{1200}{10,800} \times 83,200 \times 2.4 = 22,200$	$\frac{1200}{10,800} \times 37,000 \times 2.4 = 9,900$
Total annual vehicle running costs	105,400	46,900
Annual savings in running costs with traffic signals, all vehicles		58,500

Annual Time Increases over Driving Straight Through in Hours

	Stop Signs	Traffic Signals
Passenger cars (10,800/day)	hrs	hrs
Stops (40–0–40) (Fig. 4–6)	$\frac{10,800}{1000} \times 365 \times 4.5 = 17,900$	$\frac{10,800 \times 0.3}{1000} \times 365 \times 4.5 = 5,300$
Speed changes (40–30–40) (Fig. 4–6)		$\frac{10,800 \times 0.7}{1000} \times 365 \times 0.5 = 1,400$
Idling (Fig. 4–8)	$\frac{10,800}{1000} \times 365 \times 1.4 = 5,500$	$\frac{10,800 \times 0.3}{1000} \times 365 \times 5.6 = 6,600$
Total for passenger cars	23,200	13,300
Total annual saving in passenger car time		9,900
Single-unit trucks (1,200/day) (using data from above and multiplier from Fig. 4–6)	hrs	hrs
Stops and speed changes	$17,700 \times \frac{1200}{10,800} \times 1.3 = 2,600$	$(5300 + 1,400) \times \frac{1200}{10,800} \times 1.3 = 1,000$
Idling	$5500 \times \frac{1,200}{10,800} = 600$	$6600 \times \frac{1200}{10,800} = 700$
Total commercial time	3,200	1,700
Cost of commercial time at $8/hr	$25,600	$13,600
Annual savings in commercial vehicle time costs with traffic signals		$12,000

Summary of annual savings with traffic signals

$58,500 + $12,000 = $70,500 plus 9900 hr of passenger car time

SOLUTION—PART B. Present worth of 15-yr savings at 10%:

vehicle running costs only $= \frac{\$58,500}{0.13147} = \$445,000$

vehicle running costs + plus commercial vehicle time costs $= \frac{\$70,500}{0.13147} = \$536,000$

Present worth of agency costs:

$\$50,000 + \frac{\$2,000}{0.13147} = \$65,000$

Net present worth of savings:

savings based on running costs = \$445,000 − \$65,000 = \$380,000

savings including commercial vehicle time costs = \$536,000 − \$65,000 = \$471,000

SOLUTION—PART C

annual traffic increase = 12,000 × 0.03 = 360 vehicles

The equivalent uniform increase in traffic volume for n = 15 yr, i = 10%, (see Table 4–9) = 360 × 5.28 = 1900. Thus, the total equivalent traffic volume = 12,000 + 1900 = 13,900, an increase of 15.8%.

Net present worth of savings, including the traffic increase:

savings, based on running costs only = 445,000 × 1.158 − 65,000 = \$450,000.

savings, including commercial vehicle time costs = 536,000 × 1.158 − 65,000 = \$556,000.

ECONOMY STUDIES FOR PUBLIC TRANSPORTATION

The principles involved in economic evaluation of public transportation alternatives, referred to by UMTA as alternatives analysis, are the same as those for highways given here. However, the problems are far more complex and input data are scarce. In addition, many of the cited benefits fall into the nonmarket, nonquantifiable categories for which money measures may be inappropriate. For these reasons plus space limitations, the economics of public transportation is not explored here.[60]

SUFFICIENCY RATINGS—DEFICIENCY RATINGS—PRIORITY INDEXES

Sufficiency Ratings

Sufficiency ratings are used by some highway and street agencies to establish priorities for road improvement. Each segment of the particular highway system is rated on the basis of its efficiency, safety, and service, and possibly on its

[60]An excellent reference for those who wish to study this topic is Part III of *Public Transportation* by Gray and Hoel, op. cit. See also *TRB Record 559* and *751* and R. E. Skinner, Jr. and T. B. Deen, *Transportation Engineering Journal of ASCE,* May 1978. The *Red Book* examines the costs and benefits of bus transportation and gives sample analyses for reserved lane bus transit and for bus subsidy.

environmental impact. Results are weighted to recognize traffic volume. Roads in perfect condition have a sufficiency rating of 100% (or some other value); deficiencies of any kind cause the rating to drop. A tabulation of all projects in the order of ascending sufficiency ratings forms a priority schedule for road improvement. Such a list gives administrators a basis for allocating funds. It also provides an effective basis for resisting the demands of pressure groups that their favored projects be completed first.

At present there is no uniform method for presenting needs to federal and state legislative bodies. It has been proposed that sufficiency ratings be used for this purpose. Each administration would establish a minimum tolerable sufficiency rating for its system and estimate the cost of bringing all roads at least to that level. This would establish a statement of needs in terms understandable to legislative bodies and the public.

Economists recognize that sufficiency ratings are not economy studies based on cost comparisons. Instead, the sufficiency rating measures the quality of a road by an arbitrary assignment of weights to certain of its characteristics (see below). A change in selection or weighting of these characteristics affects the final sufficiency rating so that ratings normally are not comparable between agencies. Even so, as a start toward economy studies or as a means for selecting a relatively few projects for detailed analysis, sufficiency ratings have merit because they compel a periodic and orderly appraisal of a highway plant. On the other hand, sufficiency ratings are not money-based and therefore cannot answer the basic question, "Are the proposed expenditures the best use of public funds?"

The sufficiency rating was first developed by the joint efforts of the Arizona Highway Department and the U.S. Bureau of Public Roads and has been used for 40 yr. Roadway characteristics considered in the Arizona plan, and the weights assigned to each, are as follows:

	Number of Points
Condition	
Structural adequacy	17
Anticipated remaining life	13
Maintenance economy	5
Condition total	35
Safety	
Roadway width, or marginal friction	8
Surface width, or medial friction	7
Sight distance, or intersectional friction	10
Consistency	5
Safety total	30
Service (expressed in terms of dispatch and ease of driving)	
Alignment (dispatch)	12
Passing opportunity (dispatch)	8
Surface width (ease)	5
Ride quality	10
Service total	35

Rules have been set for evaluating each of these factors to assure uniformity in rating various roads. Final results are weighted, a set of special graphs being used to give priority to roads carrying large volumes of traffic.[61]

Deficiency Ratings

Deficiency ratings offer another noneconomic way for assessing the needs and priorities for highway improvement. With them, acceptable or tolerable standards for rural and urban situations for capacity and geometric features such as lane and shoulder width, alignment, and surface type are set for various traffic volumes and terrains. Weighting may be given to accident experience. Installations not meeting some accepted standard are classed as deficient. This information, combined with estimates of the cost of correcting the deficiencies, then offers a basis for establishing both needs and priorities.

Priority Indexes

Numerous other schemes for setting priorities for highway improvements, some more complex than sufficiency ratings and others bordering on opinion polls, have been advanced. The difficulty with them is that, because they consolidate all the factors into a single number or rating, they may disguise rather than clarify differences important to decision making.[62]

ENERGY FOR TRANSPORTATION

Introduction

Of the total energy consumed in the United States, one-fourth goes to operate transportation facilities. An additional 17% is used for materials and manufacture of transportation vehicles and to construct, rehabilitate, and maintain the supporting physical plant. Another dimension is that transportation consumes 53% of all the strategically important petroleum products and depends on it for 97% of its energy. Of this, automobiles use 52%, trucks 22%, buses 1%, aviation 12%, railroads 3%, pipelines 6%, and water 4%. From these statistics it can also be seen that vehicles traveling on the highways use three quarters of the total.[63] It follows that highway-based transportation in the United States and much of the rest of the world is critically dependent on petroleum. In sum, the "energy" problem must loom large in all transportation planning and decision making. Trade-offs will be required and they will not be easy ones. As one observer has noted, the problem has four facets: economic or resource use, de-

[61]For a similar application in Kentucky, see C. V. Zeeger and R. L. Rizenbergs, *TRB Record 698*.
[62]Discussions of or procedures for rating schemes will be found in *NCHRP Reports 122, 156,* and *217. NCHRP Synthesis 48; HRB Record 238* and *305; TRB Record 490* and *680;* and *Transportation Engineering Journal of ASCE,* Jan. 1973. A priority rating scheme for city street improvements is proposed by G. R. Thiers et al. in *HRB Record 348*. Possible noneconomic measures for the quality of traffic service appear in *HRB Special Report 130*.
[63]See *NCHRP Synthesis 43* for the source of this and much other data.

pendence on foreign sources, depletion of local resources, and social costs in the environmental, health, and safety areas. Attacking it on any one of these fronts has adverse consequences on all the others.

Alternative Liquid Energy Sources

In the short run, energy for highway-based transportation will continue to come from internal-combustion engines burning primarily liquid hydrocarbons derived from petroleum. Three substitute sources offer promise in either the immediate or more distant future.

Alcohol, produced by fermenting feed grains, is being added to unleaded gasoline to produce gasohol. In a 90% gasoline, 10% alcohol mixture it serves both as an extender and to improve the octane rating. In 1979, gasohol production was 80 million gal; estimates are that with a crash program, production by the mid-1980s could be 1.8 billion gal, so that one-third of the unleaded supply could be gasohol. An alternative is to modify engines to run on alcohol alone. Furthermore, feed grain is but one biomass from which alcohol can be produced. In the longer term, other agricultural residues such as bagasse, corn stalks, and distressed crops, coal, ocean-grown plants, and forest and municipal waste also may be feedstocks. Gasohol, unfortunately, may add to air pollution, since evaporative losses may be 50 to 100% greater than for gasoline.

The conversion of coal to liquid fuel is already a developed process in South Africa and pilot plants to produce gasoline from oil shale have been operated successfully in the United States for years. Developing these sources will require a long-time effort, with many economic, environmental, and other difficulties to be overcome.

In the United States, research on new liquid energy sources is being sponsored by both government and industry, but production is being left to the private sector. However, massive subsidies or tax advantages have been proposed to encourage rapid progress in these areas, so that the problem is political as well as technical.[64]

The Vehicle and Energy

After the Arab oil embargo of 1973–1974, Congress in 1975 passed the Energy Policy and Conservation Act. This act, and subsequent regulations, set standards for fuel consumption of increasing stringency for new passenger cars. Manufacturers who failed to meet these requirements were to be penalized. Coupled with these requirements has been a shift by buyers to more fuel-efficient vehicles, either of lighter weight or diesel powered. It follows that as the motor-vehicle fleet is renewed, fuel consumption per mile will decrease. Unfortunately, these gains will not be fully realized. For example, pollution-reducing

[64]*National Geographic,* Feb. 1981, offers an excellent summary of the energy dilemma. Also, one among many reviews of the synthetic fuel field is given by G. Dallaire in *Civil Engineering,* Jul. 1980.

devices, also stipulated by law, make engines less efficient.[65] Again, a smaller quantity of the unleaded gasoline required for pollution control can be produced per barrel of crude oil. Both of these increase energy consumption. And, as pointed out in Chapter 12, passenger-car weight reduction may mean an increase in accident severity.[66]

Battery-powered electric vehicles have been in use on a limited scale for 80 yr and recent research has produced a prototype that can travel 70 mi at 35 mph before recharging. Cost, at 1979 prices, is $6500 or 18 cents per mile. Such vehicles offer an alternative to gasoline powered ones, particularly for short commutes and local trips and as a replacement for a second car. To date, the difficulty is the lack of an alternative to the heavy and costly lead-acid battery.[67] Favoring electric vehicles, is the fact that, while conventional automobiles use scarce liquid fuels, electricity to power them can be generated from coal, oil, the atom, and possibly solar cells or other energy-capturing devices. To date, however, there is little evidence that the automobile industry is tooling up to mass produce them. Among other suggested mobile power sources are hydrogen and flywheels.

Alternatives to Automobile Transportation and Improvements in It

Energy conservation based on present-day technology will take several forms. Among them are the following:

1. Greater use of public transportation (see Chapter 3).[68]
2. Reduction in travel by making urban areas more compact by controlling outward growth and encouraging close-in redevelopment (see Chapter 3).
3. Improved traffic flow and decreased congestion by strategies referred to as TSM. (See Chapter 3.)

Some of the energy conservation measures just mentioned have already come into play. These, along with substantial increases in the price of fuel and the difficulty in obtaining it, and higher costs of vehicles and their maintenance brought a 7% drop in fuel consumption from 1978 to 1979 and an added 6% drop from 1979 to 1980.[69]

[65]See W. D. Glauz et al., *TRB Record 772* for projected changes in vehicle characteristics. R. J. Tabaczynski, *TRB Special Report 169,* claims that, because of the complexities of emission control, it will not be possible to eliminate such devices if present standards are maintained.

[66]D. J. Kulash and C. Difiglio and J. P. Stucker et al. in *TRB Record 648* and *689,* respectively, present means for modeling such effects.

[67]Many reports on electric cars have appeared in the popular and technical press. See, for example, W. Hamilton, *Electric Cars,* McGraw-Hill Book Co., New York 1980, and *TRB Record 605.*

[68]For comparisons of energy use by mode see *Transit Fact Book,* American Public Transit Assn. However, research to date does not indicate dramatic overall energy savings by shifting travelers to public transportation or building new transit facilities. For example, P. S. Shapiro and R. H. Pratt, *TRB Record 648,* estimate a 1 to 3% saving if transit ridership is doubled or tripled. C. A. Lave, ibid., and Lave and D. L. Boyce et al., *TRB Record 689,* see a net energy loss by building rail transit or people movers although several discussors disagree. W. B. Tye et al., *TRB Record 689,* argue that stopping federal aid for urban highways will not bring energy savings.

[69]Among the many additional references are *TRB Record 552, 561, 567, 571, 592, 599, 601, 648, 689, 707, 710* and *764,* and *TRB Special Report 166, 169, 179* and *191,* and *Energy Impacts of Urban Transportation Improvements,* ITE.

Transportation Agency Use of Energy

Producing materials for and carrying out construction, operation, and maintenance of transportation facilities account for an estimated 1 to 2% of total national energy consumption. A conspicuous example of such activities is highway and street lighting for which highway and public safety and crime prevention are the principal claimed advantages (see Chapter 10). Today, as energy costs soar, both the need for and levels of such lighting are being challenged. In many other areas also, administrators have become increasingly conscious of the energy aspects of different construction and maintenance strategies and procedures.

Analysis of energy consumption for construction and maintenance begins with materials; for example, how much is needed to produce a unit amount of cement, asphalt, steel, or aggregate. To these are added summations of the energy employed in individual steps in the construction or maintenance activity itself. Table 4–10 lists approximate total energy requirements for a few typical construction processes, including that devoted to producing the materials.

Some of the items in Table 4–10 can be used to illustrate the effects of scarcity and costs on highway practices. For example, roughly 45% of the energy required to produce and place asphalt concrete goes to drying and heating the aggregate. Already, certain highway agencies are reducing dryer temperatures (see Chapter 19) or substituting emulsified asphalts because, with them, drying and heating the aggregate are not required. Again, because so much energy is required to produce portland cement, "lean" concrete of reduced cement content or flyash extenders as a substitute for some of the cement are being employed.

Table 4–11 is a short list of energy requirements for carrying out a few surface maintenance operations. From data such as these it is possible to evaluate the energy effects of different strategies; for example, to compare seal coats with overlays or hand-labor and machine methods.

Another use of energy data could be to develop an "energy" or "liquid fuel" budget, broken down by maintenance operation. Data supplied by one agency

TABLE 4–10. Total Energy Requirements for Certain Construction Operations—Materials in Place*

Item	*MJ/m³†*
Earth excavation	82
Rock excavation	106
Crushed stone base	550
Cement-stabilized base	1500
Asphalt concrete	1400
Portland cement concrete-jointed, unreinforced	2800

*Abstracted and averaged from J. A. Epps et al., *TRB Record 674*.
†One megajoule per cubic meter (MJ/m³) = 724.6 Btu/yd³.

TABLE 4–11. Energy Requirements for Typical Surface Maintenance Operations*

Item	*MJ/m²*†
Surface patch, hand method	2.1
Surface patch, machine method	3.5
Dig out and repair, hand method	4.5
Dig out and repair, machine method	10.1
Chip seal, partial width	1.5
Chip seal, full width	5.2
Slurry seal	1.7
Asphalt concerete overlay, 5 cm (2 in.) thick	71

*Abstracted and averaged from J. A. Epps et al., *TRB Record 674*.
†One megajoule per square meter (MJ/m^2) = 792.4 Btu/yd^2.

indicated that 15% of its energy consumption was for roadside mowing. Knowing this and similar facts for other operations, administrators could, in the case of an energy shortage, reduce or eliminate this or other less-essential operations.[70]

Transportation of Energy Materials

An often overlooked aspect of the energy problem is that of transporting energy materials to the point of consumption or conversion. For example, energy in coal or oil shale is abundant in the ground in the United States. But it must be transported to the conversion plant, generating station, industrial user, or direct consumer. In addition, wastes must be transported to environmentally acceptable disposal areas. Similarly, the raw materials for biomass conversion to alcohol must be transported to the distillery and both the alcohol and waste from the process delivered for use or disposal. It follows that as energy sources change, the transportation system will likewise have to change. This will involve both new transportation facilities and expanding, modifying, or rehabilitating existing ones. There will be many problems, given that the affected transportation modes are in both private and public ownership, often in competition, and may be financed, subsidized, or regulated by federal and state legislative and administrative bodies. One problem accompanying these shifts which has already surfaced has been the serious deterioration of pavements and railroad roadbeds from increased coal hauling. A 1980 estimate places the cost of bringing coal-haul roads to tolerable standards at $20 billion.[71]

PROBLEMS

4–1*A*. Determine the tax-free price for gasoline in your local area. Then, using this price and values scaled from Fig. 4–2, plot a curve for fuel

[70]References on energy in highway construction and maintenance include *TRB Record 674* and *TRB Special Report 166*.
[71]Further exploration of this complex problem cannot be done here. For added information, see *National Energy Transportation Study*, DOT and DOE and *TRB Circular 216*.

cost versus speed. Note: The tax-free price for gasoline used in developing Fig. 4–2 is 40 cents per gallon.

4–1*B*. Work Problem 4–1*A*, but plot the curve with local currency values as the ordinate and speeds in kilometers per hour as the abscissa.

4–2. From the latest edition of *Statistical Abstracts of the United States* (available in any library) or other source find the All-Item Consumer Cost Index. (This was 100 in 1967 and 156.1 in 1975 when the costs in the *Red Book* were established.) Divide this new value for the index by the *Red Book* index to obtain a multiplier. Then with this multiplier and values scaled from Fig. 4–2, plot a new curve for passenger car operating cost versus speed.

4–3. Repeat the process called for in Problem 4–2, but determine the multiplier by using the separate indexes for gasoline, motor oil, tires, repairs and maintenance, and new automobiles and the equation given in the textbook for general operation on straight, level tangents. Compare the results obtained by the two methods.

4–4. Assume that the 55 mph maximum speed limit is retained and observed. Then, for a rural freeway which flows freely at 20,000 passenger vehicles per day:

a. Using data scaled from Fig. 4–2, compute the reduction in fuel cost per mile per year compared to that for an average speed of 65 mph.

b. Using the 1975 costs given in Fig. 4–2, and a tax-free gasoline price of 40 cents per gallon, compute the annual fuel savings per mile.

c. Convert the 1975 costs obtained for part *a* to current prices using the present tax-free price of gasoline.

4–5*A*. Because residents complained that vehicles on the main road were passing through an intersection too fast, local authorities added signs which changed intersection control from two-way to four-way stops. As a result, 1000 passenger vehicles per day on the main road now come to a stop from 40 mph, pause for 3 s, and resume speed again. The operation of cross and turning traffic is little affected. For this situation:

a. Compute the added annual costs to motorists in 1975 dollars.

b. Update the 1975 costs to current ones using changes in the All-Item Consumer Cost Index from the January 1975 value of 156.1.

c. Compute the annual added hours of time consumed.

4–5*B*. Work Problem 4–5*A* in currency and speed units appropriate for your country.

4–6*A*. During peak hours, a six-lane freeway, designed for 70 mph, usually flows at an 80% volume-capacity ratio. For this base condition, and for passenger cars only, determine from Fig. 4–3:

a. Average traffic speed.

b. Running cost per vehicle-mile in 1975 prices.

c. If a disruption in flow occurs and forced flow develops at an 80%

volume-capacity ratio, what then is (1) traffic speed and (2) running cost per vehicle-mile in 1975 prices?

4–6*B*. Solve Problem 4–6*A* in units appropriate for your country.

4–7. Using Fig. 4–3 and Table 4–2, solve Problem 4–5*A* for single-unit and 3–S2 trucks, assuming 100 of the former and 25 of the latter.

4–8. Using Fig. 4–3 and Table 4–2, solve Problem 4–6*A* for single-unit and 3–S2 trucks.

4–9. By reconstructing a section of road crossing a divide, the up and down grades can be reduced from 6% to 3%. The length of the road remains approximately the same; it rises for ½ mi and falls for ½ mi. For this situation:

a. What is the average saving in cost at 1975 prices to passenger cars for a single passage at 50 mph?

b. If the average daily traffic is 2000 passenger cars, what is the annual savings at 1975 prices? At current prices?

c. If the economic life of the proposed improvement is taken as 30 yr and 10% is an appropriate interest rate, what expenditure for the improvement is justified at 1975 prices? At current prices?

4–10*A*. Vehicles are now operating on a level, gravel, two-lane, farm-to-market road at 35 mph. The road must either be resurfaced with gravel or be paved. In the latter case, speeds will increase to 45 mph. Gravel surfacing will cost $10,000/mi. at 1975 prices and last 5 yr; base course and road-mix bituminous surfacing will cost $40,000/mi and 1975 prices and last 15 yr. Assume no salvage value. For these alternatives:

a. What saving in operating costs per vehicle-mile for passenger cars, at 1975 prices, results from paving the road?

b. What average daily passenger-car traffic is required to justify the pavement, based solely on savings in operating costs? Assume that maintenance costs are the same for either alternative. The road agency has many demands for its very limited funds, so that 10% interest seems appropriate.

c. How many hours are saved per year per vehicle per mile?

d. Assuming that half the vehicles are traveling on business purposes and that the time savings can be priced at $5 per vehicle-hour, what average daily traffic is then needed to justify the investment?

4–10*B*. Work Problem 4–10*A* for conditions that seem appropriate for your country. Use the vehicle-operating costs from the textbook but develop or assume construction and time costs that are appropriate.

4–11*A*. Two major highways now intersect at grade, and traffic is controlled by fixed-time signals that operate 24 hr a day. Average daily traffic on highway *A* is 10,000 passenger vehicles and that on highway *B* is 4000. There is relatively little left-turn traffic and a diamond interchange like that shown in Fig. 9–17*a* offers a satisfactory plan for grade separation.

Observations taken over a typical 24-hr period show that 50% of the vehicles on highway *A* and 15% of those on highway *B* pass through the intersection without stopping or making an appreciable change from their usual speed of 40 mph. The remaining cars are stopped by the signal. Average standing delay to those stopped on highway *A* is 20 s and on highway *B* 30 s.

a. What are the additional operating costs at 1975 prices, without time costs, for each passenger vehicle stopped *(1)* on highway *A*? *(2)* on highway *B*?

b. What are the time costs at $4.00/vehicle-hour, for each vehicle stopped *(1)* on highway *A*? *(2)* on highway *B*?

c. What are the annual stopping costs at the intersection at 1975 prices *(1)* without time costs? *(2)* with time costs of $4.00/hr applied to the 20% of the vehicles whose occupants are engaged in economically productive activities?

d. What investment in a grade separation structure at 1975 prices is justified *(1)* without considering time costs? *(2)* considering time costs? An appropriate interest rate is 8% and 30 yr a reasonable useful life. Assume that the costs and times for turning vehicles are the same for either alternative and that the gains from improved geometrics offset the costs of added distances and elevation changes with the diamond.

e. Update the answer of part *d* using the current value of the Construction Cost Index of FHWA obtained from the most recent cost report from *Engineering News-Record* or from *Highway Statistics*.

4–11*B*. Work Problem 4–11*A* for an approach speed of 65 km/h. State the results in the currency of your country.

4–12. Work Problem 4–11*A* assuming that the total traffic on highway *A* is 20,000 vehicles and on highway *B* 8000; traffic composition is 84% passenger cars, 10% single unit trucks, and 6% 3-S2 diesels. Assume that time costs for passenger cars are $4/hr, for single-unit trucks $7/hr. and for 3-S2 diesels $8/hr. All trucks and 20% of the passenger cars are engaged in economically productive activities.

4–13. A heavily traveled suburban traffic artery carries 30,000 vehicles per day. It is proposed to reconstruct it to freeway standards. Assuming that accident costs and rates on the existing and proposed facilities correspond to those shown in Table 4–4 (California) and Table 4–6, respectively:

a. Compute the anticipated annual number of fatal, injury, and property-damage accidents per mile for each, and the differences between them.

b. Compute the 1975 anticipated costs and cost differences per mile for each accident class.

c. Based on these cost differences, compute the investment per mile in 1975 dollars which could be justified solely by reductions in accident costs. Assume a 20-yr study period and a 10% interest rate.

d. Assuming that traffic volumes will increase each year by 2% of the base year and that accident frequency per vehicle-mile will not change, what investment in 1975 dollars is justified?

4–14. Obtain the answers called for in Problem 4–13 for an urban facility carrying 40,000 vehicles per day.

4–15. Work through Example 1 in the text by the annual cost and benefit-cost methods using an interest rate of 12%.

4–16. Revise Example 1 in the text on the assumption that the average daily traffic in all vehicle categories will increase at a 2% arithmetic rate until the end of year 20 but remain at that level for the second 20-yr period.

4–17. Introduce the effect of accident costs into Example 1 of the text. Assume that the accident experience of the present road is that shown in Table 4–6 but that the safety features of the new alignments will reduce accident rates by 20%. Use the California cost per accident from Table 4–4.

4–18. Change the basic data of Example 1 of the text to read as follows:

Element	*Estimated Useful Life (yr)*	*Cost*		
		Alternative A	*Alternative B*	*Alternative C*
Right of way	40	$ 0	$ 220,000	$ 180,000
Grading	20	80,000	600,000	900,000
Structures	20	50,000	450,000	1,090,000
Surfaces	10	230,000	970,000	920,000

Solve this problem using 1975 or updated vehicle-operating costs, one method or several methods, and the interest rate and one or more of the assumptions suggested below, all as designated by the instructor.

a. Assuming that traffic volume remains constant.
b. Assuming that the average daily traffic increases at an arithmetic rate of 3% per year for 10 yr but remains constant thereafter.
c. Assuming that the accident experience of the different alternatives will be as indicated in Table 4–6.
d. Assuming that the average daily traffic increases as indicated in part *b* above and that the accident experience differs among alternatives as outlined in Problem 4–17.

4–19. Two six-lane urban traffic arteries controlled by traffic signals now carry a total average daily traffic of 50,000 vehicles, of which 15% are single-unit trucks and 8% 3-S2 diesels.
Running speed is 30 mph between stops which average 2½ per mi. Average standing time at each stop is 30 s.
An eight-lane freeway costing $12,500,000/mi, at 1975 prices, including interchanges and connections, is planned to take through traffic from these streets. Half the cost is for rights of way having a

40-yr assumed life; the remainder is for grading, structures, and pavement with an assumed life of 20 yr. The freeway will divert 75% of the traffic from the present arteries. Speed on the freeway will be 55 mph, delays will be negligible. Maintenance and administrative costs for the freeway and streets will be $15,000/mi/yr more than for the streets alone. Assume that operating conditions on the street do not change with the decrease in traffic.

a. Compute, for passenger cars, single-unit trucks, and 3-S2 diesels the average saving in operating costs per vehicle-mile at 1975 prices. *(1)* Omit time costs from your answer. *(2)* Include time costs at $4.00, $7.00, and $8.00/hr in your answer.

b. Compute the total annual savings in operating costs per mile of freeway. *(1)* Omit time costs. *(2)* Include time savings for 10% of the passenger cars and for all trucks.

c. Compute the difference in annual costs, including road-user costs, between continued operation on the streets and operation if the freeway is constructed. Employ Example 1 as a guide. Interest is at 10%. *(1)* Omit time costs. *(2)* Include time costs for 10% of passenger cars and for all trucks.

d. Compute the rate of return on the investment in the freeway for assumptions *(1)* and *(2)* in part *b*.

4–20. Solve Problem 4–19 assuming that the average daily traffic of 50,000 vehicles traveling streets and freeway will increase by 1200 vehicles per year for the first 15 yr and remain constant thereafter.

4–21. Solve Problem 4–19, but include the expected reduction in accident rates and costs shown in Tables 4–6 and 4–4 (California).

4–22. Solve Problem 4–19 considering both the traffic increase described in Problem 4–20 and the reduction in accident costs outlined in Problem 4–21.

4–23. As directed by the instructor, investigate the sensitivity of the answer for one of the examples or earlier problems in this chapter to a change in one or more assumptions. For example, assume that the cost estimates are too high or too low by 10%. Again, assume that traffic increases at a rate of 2% per year faster than was assumed in the original situation.

4–24. Rework Example 2, assuming that alignment *B* climbs uniformly on a +3% grade for its 3000-ft length. Use a 10% interest rate.

4–25. As a class project, secure from the local office of a highway agency layout drawings, construction-cost data, and projected traffic volumes by individual maneuvers for a proposed or recently completed grade-separation structure. Using them and appropriate cost and time data from this book, appraise the economic worth of the project as compared with the "before" condition.

4–26. As a class project, study the relative economy of two or more appropriate traffic-control arrangements for a busy intersection near your campus. If turning maneuvers represent a substantial portion of the traffic, the costs and times of their slowings, stoppings, and turnings should be included in the analysis. Example 3 may be helpful as a guide.

4–27. Explore the short- and long-term energy conservation programs of the local highway, street, or transit operation near your college. Then prepare a written evaluation and recommendations for changes.

5 FINANCING HIGHWAYS AND PUBLIC TRANSPORTATION

The highways and streets on which automobiles, trucks, and buses travel are almost entirely provided, maintained, and operated by government as one of its primary functions. Just as with other governmental services, the funds to pay for them are raised through taxes levied by the various legislative bodies; expenditures in 1979 totaled about $37 billion (see Fig. 1–1). From 1921 through 1980, over $290 billion was collected and expended in the United States to construct this road system. In 1979, total annual capital outlays for highways and streets were $14.8 billion, or about $60 per person[1] with a somewhat larger amount for other highway purposes such as maintenance, administration, policing, and debt retirement.[2]

In contrast to the public financing of highways and streets almost from their inception, support for transit from government is fairly recent. Before World War II, transit facilities except in a few large cities were privately owned; almost all, private and public, operated at a profit. Ridership in 1940 was 10.5 billion trips a year. Following a bulge to 19 billion riders in World War II, the number fell to 5.3 billion in 1972, but had risen to 7.6 billion in 1978, an upsurge that can be expected to continue. Even so, to keep transit operating, governmental agencies have taken over many of the larger transit systems; as of 1978, they operated 48% of the systems that accounted for 91% of the passenger miles. Subsidies from federal, state, and local governments paid 47% of transit operating costs and provided large sums for equipment and capital improvements as well. In addition, some 67% of the passengers rode in buses that traveled on public highways.

Throughout the automobile age, funds for highway improvement have not kept up with the demands for road improvement. In early years these demands were for all-weather or paved surfaces to get traffic out of the mud. After World

[1]Real-dollar capital expenditures per person for highways have decreased over time when account is taken of population increase and inflation. For example, 1960 expenditures, stated in 1979 dollars, totaled about $90 per person.

[2]The source of these statistics and many others that will be cited is *Highway Statistics* published annually by the Federal Highway Administration. There is also a *Summary to 1975*.

War II until the late 1960s or early 1970s the major thrust in both rural and urban areas was on safe, free-flowing arteries to relieve congestion, ease driving tensions, and reduce the toll of motor-vehicle accidents. Currently the emphasis is shifting toward reconstruction and maintenance of existing roads which are deteriorating and toward decreasing congestion and smoothing traffic flow on major arteries by removing bottlenecks and upgrading traffic control.

Viewing highway financing in the United States over time, the first major shifts came about 1920. They were from local autonomy to federal aid on a matching basis for major routes, and state augmentation of property taxes with revenues collected from highway users. A second series of changes occurred after World War II, when Congress in 1956 shifted federal aid from the general fund to substantially increased federal participation by funding the Interstate System on a 90% federal, 10% state basis. State contributions from highway-user revenues also grew substantially. Recent Federal-Aid Highway Acts, successive Urban-Mass Transportation Acts, and the Surface Transportation Assistance Act of 1978 mirror increased emphasis on mass transportation at federal, state, and local levels. At the same time, the resources devoted to highways have decreased when account is taken of inflation.

This chapter first presents views regarding the financial responsibility of the various beneficiaries of transportation. It then describes past and current sources for and distribution of highway and other transportation funds and how they were and are expended. There follows a discussion of financing methods which permit acceleration of improvements. The final section treats proposals for using the taxing power as a means for controlling motor vehicle use.

DISTRIBUTION OF TRANSPORTATION COSTS AMONG GOVERNMENT, USERS, AND OTHERS[3]

Roads serve multiple functions. They make it possible for government to render various essential services, supply the avenues of intercommunity mobility for persons and goods, facilitate these movements within each neighborhood, and give access to land and dwellings. However, different classes of roads are devoted more to some of these functions than to others. At one extreme would be a rural freeway on the Interstate System. Its most important functions are to provide essential governmental services and intercommunity mobility. At the other extreme is the *cul-de-sac* (dead-end street) of the modern residential subdivision or a road ending at a farm gate. In both instances, the sole purpose is to give access to land and dwellings. Although the main functions of most roads and streets cannot be so clearly distinguished, some logical theory for classifying

[3]The study that best addresses the many complex aspects of this subject is *The Final Report of the Highway Cost Allocation Study, House Documents 54 and 72,* Eighty-seventh Congress, First Session (1961) and the *Supplementary Report, House Document 124,* Eighty-ninth Congress, First Session, 1965. Less elaborate studies were made in 1969 and 1975. See also, Robley Winfrey, *Economic Analysis for Highways,* International Textbook Co., Scranton, Pa., 1969. References to more specific topics are given at appropriate places below.

each road segment by its use may be possible. These might include its more important or *predominant use,* or alternatively, its *relative use*.

Many economists support the notion that financial responsibility for transportation should be assigned among the beneficiaries on the basis of the benefits each receives. The first step in implementing this "benefit" theory would be to classify each road segment on the basis of its functions. Then, similar roads would be classified into systems (see Chapter 2); the responsibility for them would be given to appropriate governmental agencies; and the agencies would assess proper charges against the various beneficiaries. Following this line of reasoning, financial responsibility might reasonably be distributed as follows:

1. Responsibility of the federal government would be limited to payment for activities designed to serve broad national objectives. Road needs for mail deliveries and national defense would clearly be among these. Without question the federal government would supply a sizable portion of the financing to complete the rural Interstate System and would continue to participate in the federal-aid primary system. The federal role in major urban highways and in urban and rural public transportation also could fall in this category since it aims to increase the efficiency of the transportation system, reduce demands for energy, decrease air pollution, and provide transportation for the disadvantaged. Continued appropriations for certain rural roads now in the federal-aid secondary system probably would be questioned, while federal aid for local roads would be ruled out. Those who suggest such limits on federal participation often argue that federal appropriations should be from general taxes and that user taxes should be collected only by the states.
2. Highway-user responsibility would be limited to highway-user benefits resulting from improved interarea or interneighborhood mobility. It would include, as a minimum, a major share of the costs of principal state highways, other rural roads connecting towns, and major urban arterials.
3. Local governments would assume responsibility for local arterials, roads, and streets within each neighborhood, and for facilities providing access to land and dwellings.[4] The local arterials would be financed and public transportation subsidized largely by general taxes, including a sales tax, and purely local streets would be paid for by direct assessments against the property served or by creating improvement districts. Only where there was marked use of a road by interneighborhood traffic would user taxes be devoted to its improvement.

The scheme for transportation support outlined in the preceding paragraphs relates payments through taxes to "benefits anticipated or received." It has also been proposed that "costs caused or occasioned" by various beneficiaries are a more suitable measure of financial responsibility. Yet another approach is to employ "willingness to pay" criteria.[5] To implement any of these theories calls for, among other things, suitable and acceptable measures of benefits, costs, and willingness to pay. Unfortunately, data for them are not available.

Earmarking taxes on the basis of benefits, costs, or willingness to pay is com-

[4]C. A. Steele and T. R. Todd, *HRB Record 20,* suggest that neighborhoods might have radii ranging from 3 mi for sparsely settled rural areas down to ¼ mi for central cities, downtown.
[5]See, for example, G. J. Roth, *TRB Record 731.*

mon practice. It is reported that there are some 800 such funds in the federal government alone.

Some students of government state that this "linkage theory" is wrong in principle. They argue that this procedure usurps the right and duty of legislative bodies to levy taxes and appropriate funds in the best interest of their constituents. In their view, taxes of all sorts should go into a general fund from which appropriations are made for all purposes. It is noteworthy that this "general fund" approach to highway financing was followed by the federal government until 1956, at which time highway-user and motor-vehicle excise taxes were placed in a separate Highway Trust Fund. Today the inviolability of that fund is being challenged, particularly by supporters of public transportation.

Actually, financing in the United States today, with notable exceptions, roughly follows the "benefit" theory outlined above. Furthermore, a scheme such as that outlined provides a valuable frame of reference from which to judge present and proposed methods for raising and distributing funds.

HIGHWAY USER TAXES ON MOTOR VEHICLES

The subject of a "fair" distribution of highway costs among vehicle classes and other beneficiaries has been argued for 50 yr. In this ongoing controversy, trucking interests maintain that they are now paying more than their fair share. Furthermore, they strongly urge legislation to permit longer and heavier trucks on the highways as a means of reducing costs. On the other hand, some attribute much highway damage and congestion to large or overweight trucks and urge increased truck taxes, prohibition of further size or weight increases, or even that these limits be lowered. Railroad and "water" interests strongly advocate increased truck taxes "to remove the subsidy that truckers now enjoy." They contend that trucks are given a competitive advantage because they do not pay their just share of the costs of the highways over which they travel, and furthermore that railroads must pay taxes on the roadbeds over which they operate while trucks roll over publicly owned highways which are tax exempt. At the same time, much of the motoring public has become alarmed by congestion of roads and streets and by fear of truck-induced accidents.[6] They also advocate restrictions on trucks. As would be expected, proponents on each side are continuously pressing state legislatures and Congress for changes in the tax structure and size and weight limits. This question came to the forefront again in the late 1970s when serious pavement deterioration developed on the Interstate System. Following a study by its Budget Office, Congress, in the Surface Transportation Assistance Act of 1978, mandated a 4-year study on the subject. This has been limited to the federal-aid system, considers only motor vehicles, and is to be on a "cost occasioned" basis, which means that benefits will not be considered. Thus many of the broader issues will not be addressed.

[6]There is a basis for these concerns. Mileages traveled by heavy trucks tripled between 1956 and 1977.

At the outset it must be recognized that laws setting vehicle taxes and laws for tightening or loosening size and weight limits are a legislative responsibility and as such are subject to all the pressures and maneuvering that accompany all controversial political issues. Furthermore, engineering analysis can provide only partial answers, and often these are based on assumed rather than known conditions. It follows that political strength rather than rational findings based on research is the major control over tax schedules and weight limits for trucks. This is demonstrated by Table 5–1, which shows the level and wide divergence of taxes on a few classes of vehicle in the 11 western states.

Proponents of different tax levels advance various theories to support their views. Currently favored are the "increment cost" theory, a "consumption rate" theory, and an "economic cost" theory. Also advocated are the "theory of differential benefits," the "ton-mile" theory, and the "unit-vehicle-operating cost" theory. Each of these is described briefly below.

To implement the tax level developed under any tax theory usually calls for an additional or a *third structure* tax on trucks and for-hire buses. These are added onto the road-user and license taxes applied to passenger vehicles. They take a variety of forms. Included are a "gross-receipts" or "passenger-mile tax" on contract haulers, fuel surcharges, or mileage or weight-distance assessments.

None of these theories of truck taxation or the third structure tax to implement it can produce a total tax burden that seems fair in the eyes of all concerned, and the process of enacting any of them into law is accompanied by long and acrimonious debate.[7]

Theories of Motor-Vehicle Taxation

INCREMENT COST THEORY. The increment cost theory, sometimes called the differential cost theory, takes a "cost occasioned" view. It assumes that highways are constructed for the joint use of passenger cars and trucks. It seeks to distribute equitably the cost of a basic road suitable for passenger cars among all classes of users, but to assign to heavier or larger vehicles all costs for which they are solely responsible.[8]

Under the increment cost theory, costs are placed in two categories: (1) *joint (or common) costs,* allocated among all vehicles, and (2) *size and weight costs,* assigned only to heavier or larger vehicles.

1. Joint costs, which for new freeways may be half or more of the total, are distributed among the various weight classes on some reasonable basis. For example, it has been

[7]For added detail see *The Role of Third Structure Taxes in the Highway User Family,* Bureau of Public Roads, 1968. See also K. Bhatt et al. and R. McGillivray et al., *TRB Record 680* for other approaches and discussion.

[8]The selection of the name "increment-cost theory" is unfortunate since the principle of increment or added costs is followed only partially. For a true increment-cost study, it would be assumed that the passenger car was the basic vehicle to which all costs of a basic road system would be assigned. Trucks would be held responsible for costs occasioned solely by their presence, such as added capacity, wider lanes, or stronger pavements and bridges. The cost assignments obtained by this method would represent the level at which each heavier weight group was barely paying its way.

TABLE 5–1. Annual Fees Assessed in 1978 against Average Passenger Cars and 3-S2 Trucks Doing Intrastate Commercial Hauling in the Eleven Western States (Dollars per Vehicle)

Vehicle	Annual Mileage	Fees		
		Lowest	*Highest*	*Average*
Passenger car*	10,200‡	60	128	81
3-S2 truck†	30,000	1395	4493	2830
3-S2 truck†	50,000	2101	6100	3611

*Total is the sum of (a) total vehicle tax collected divided by the number of passenger cars and (b) state tax per gallon times gasoline consumption computed at 14.3 mi/gal.
†Data from L. Henion, Oregon Dept. of Transportation.
‡National average annual mileage.

proposed that highway administrative expense and the costs of rights of way, utility adjustments, roadside development, and traffic and pedestrian service be allocated on a basis of "vehicle-miles traveled" disregarding vehicle size and weight.[9] Furthermore, the assignment to passenger cars of grading and drainage costs would also be on a vehicle-mile basis, except that on intermediate- and high-type roads, passenger-car mileage would be weighted at about 90% of that for trucks. There are many who argue that a cost distribution such as that just described is far too favorable to trucking interests. An extreme view in the other direction would be to allocate joint costs on a ton-mile basis.

2. Size and weight costs such as the expense of greater pavement and base thicknesses, stronger structures, additional maintenance, and additional capacity demands beyond those for passenger cars are assigned among the heavier vehicles, on a cumulative basis. Similarly, any added geometric costs such as wider lanes or flatter grades to ensure safe and free operation would be chargeable to the larger, heavier classes.

The increment-cost theory cannot be conclusive. In the first place, a more or less arbitrary distribution of joint costs is made among the various classes of vehicles; and this means that subjective judgment, not engineering analysis, is used. Secondly, data are incomplete concerning the relationship of vehicle weight and size to highway design features and their costs.

CONSUMPTION RATE THEORY. This theory argues that, with the highway system largely in place, the principal concern is to rehabilitate it as it wears out. The problem then becomes one of allocating these "wearing out" costs equitably among the vehicle classes in terms of the damage they do. There would be few joint costs as defined above. However, if some of the deterioration could be attributed solely to weather or age, these would be classed as joint costs to be allocated arbitrarily. If it is assumed that pavement distress is the major rehabilitation item, it follows that under this theory heavy trucks would carry most of the costs.[10]

[9]See *Supplementary Report of the Highway Cost Allocation Study, House Document 124,* Eighty-ninth Congress, First Session, 1965.
[10]The relative effects of various axle loads on pavement deterioration are discussed in Chapter 19 to 21.

ECONOMIC COST THEORY. This theory assigns all economic costs of highway use equitably among the various vehicle classes. Included would be those of highway deterioration, congestion, environmental effects, and accidents. It can be seen that this theory extends the consumption rate theory by adding social costs.

THEORY OF DIFFERENTIAL BENEFITS. A theory of differential benefits for allocating tax responsibilities among vehicle classes also has been investigated. With it, the services that highways render are measured in terms of the savings resulting from road improvement. By analysis of the current or proposed improvement program, savings to each class of users would be computed, such items being considered as reduction in distance, improved surfacing, reduction in rise and fall, improved grade and alignment, and reduced impedance. Taxes would be set in proportion to these benefits. Many of the data for and techniques of highway economy studies could be utilized directly in making tax studies under this theory.

THE TON-MILE THEORY. The ton-mile theory of truck taxation makes no attempt to relate vehicle size and weight to the cost of the highway over which the vehicles travels or to the benefits accruing to different classes of highway users. Rather it assumes that taxes graduated to reflect the combination of gross weight and distance give a fair charge for the use of the highway. Thus, the charge against a truck and load weighing 20 tons and traveling 1 mi would be ten times that levied against a 2-ton passenger car traveling the same distance. In the view of many analysts, the ton-mile theory sets an upper bound on truck taxes, just as a fixed charge per mile, regardless of vehicle size and weight, would set a lower bound.

The ton-mile theory should not be confused with third-structure taxes based on weight-mileage charges. These, which are in effect in many stages, can be employed to implement any tax theory or to develop a tax structure unrelated to any theory.

Studies have been made in a number of states in which the gross ton-mile theory has been wholly or in large part the basis for a recommended allocation of motor-vehicle taxes. Several states have tried the ton-mile tax and abandoned it. The principal argument for repeal was that it could not be administered fairly, since determination of the tax depended solely on the accuracy and honesty of the taxpayer's records.[11]

THE UNIT-VEHICLE OPERATING-COST THEORY. The unit-vehicle operating-cost theory proposes a tax based on operating costs. As with the ton-mile theory, it is a charge for use of the highway, and does not consider highway costs or benefits. It has been established that vehicle operating costs increase with the weight of the vehicle but that the cost per ton-mile of pay load decreases. It follows that, if the tax on passenger cars were fixed at some designated level,

[11]Spokesmen for the trucking industry strongly oppose the ton-mile tax and advocate a tax based on gross freight revenue instead. See A. C. Flott et al., *TRB Record 577*.

truck taxes set by the operating-cost theory would be lower than if the ton-mile theory were used.

SOURCES OF FUNDS FOR TRANSPORTATION

Early Highway Financing

Before about 1890, public roads and streets were financed almost entirely by local property taxes.[12] In the northeastern states the road tax unit was the town, elsewhere the county. Commonly, the tax rate was low. In New England it was fixed at town meetings; elsewhere, county boards of assessors usually set the rates. Often taxpayers "worked out" their tax on the road. When money was borrowed by a town or county bond issue, a vote of the people, frequently a two-thirds affirmative majority, was required.

After 1890, state governments began financing state highways and providing aid for county roads. They often initiated bond issues which were then authorized by popular vote. In populous states, taxes for state highways largely fell on the cities where there were high concentrations of property values. Thus, in 1915 Boston paid an estimated 40% of the Massachusetts state highway bill and New York City 60% of the New York total. These early state or state-aid highways did not extend into the cities, and city streets were paved and maintained wholly by local property taxes. For many years after modern rural highways began to develop, city streets were presumed to be adequate. Thus, they at first were denied any federal aid whatever.

Up until the late 1910s or early 1920s hauling over country roads was exclusively by horse-drawn vehicles which could transport only limited loads. Grades, mud, and sand were severe handicaps. Early bond issues for county road improvements naturally were used largely for macadam and gravel surfaces. Administration of many of these projects was poor, and as a consequence some of the money was wasted.[13]

Present-Day Sources of Highway Funds

Figure 5–1 shows by sources the annual nationwide highway revenues since 1921. It clearly reflects the road-building surge of the 1920s, the limitation in highway improvement during the depression except for relief activities, and the cutback during the World War II years. Following that war and through 1972, federal aid and state-collected, highway-user revenues increased substantially and continually, both in amounts and purchasing power, influenced by the in-

[12]An exception was a federal appropriation for the Cumberland Road in 1806. In the decades that followed, President Madison's belief that federal spending for internal improvements was unconstitutional prevailed and federal aid for roads ceased. Not until 1912 did the federal government again participate to any extent in financing road improvements.

[13]Investigations by the Office of Public Roads revealed that county taxpayers often were burdened unnecessarily with taxes for long-term sinking-fund bonds. It advocated a shorter-term serial type by which service charges annually were reduced by beginning retirement of the bonds as soon as the new roads were built.

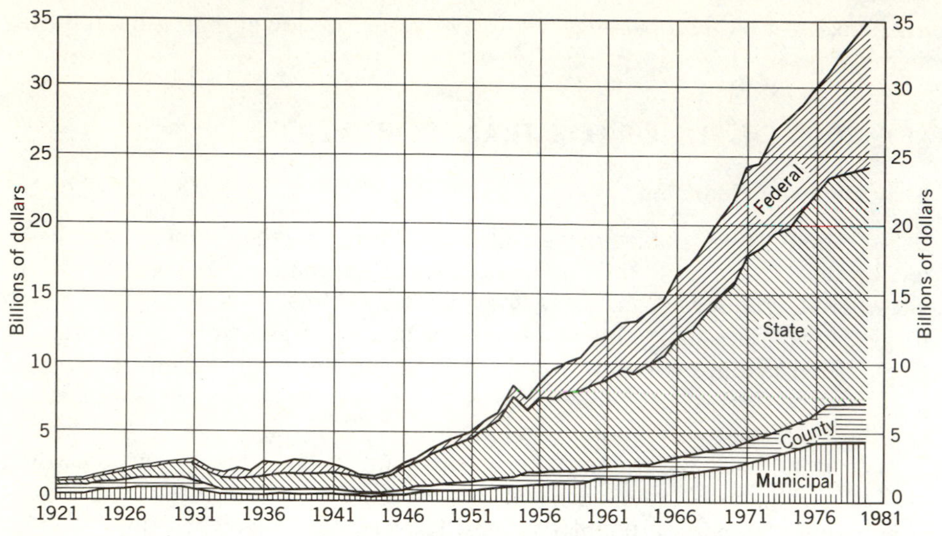

Fig. 5–1. Total receipts for highways, by governmental unit, 1921 to 1979.

creased number of motor vehicles and an accompanying increase in fuel use (see Fig. 1–1), and by substantial increases in state taxes per gallon on motor fuel. Since 1972, although dollar revenues have increased, the high rate of inflation has caused purchasing power to decrease so that the resources for highway construction, maintenance, and operation have fallen drastically. This is demonstrated by the dashed line on Fig. 1–1 which converts annual expenditures to 1977 dollars, using indexes based on construction, maintenance, and management costs.

Federal revenues devoted to highways took a substantial jump beginning in 1957 after Congress created the Highway Trust Fund, primarily to finance the Interstate System. Federal spending has continued to increase since. County and local revenues also have increased, but at a far less rapid rate, primarily because local governments were successful in tapping highway-user sources. As already noted, however, these later funding increases were far less spectacular than the graph of dollar values might indicate, since it does not recognize the continuing effects of inflation. It is also worth noting again that 1979 vehicle registrations were 6.0 times 1930 values and 3.2 times those in 1950, while expenditures converted to constant dollars were 2.0 and 1.6 times as great, respectively.

Brief discussions of the individual major sources of revenue are given in the paragraphs that follow.[14]

FEDERAL SUPPORT FOR HIGHWAYS. By 1910 state aid to accelerate rural highway construction had in some form spread to many states. There naturally

[14]See *Highway Statistics,* op. cit. A few details from the various legislative acts also are summarized below, but their complexities can best be appreciated by reading the legislation itself.

arose the question of how the federal government similarly might aid this movement for better rural roads, especially along mail routes. Several bills were introduced in Congress; in 1912 one became law, appropriating $500,000 to be used on post roads in states whose governors accepted the act. The principal stipulation was that the state (or a subdivision) should match the federal dollars two for one. The response was not enthusiastic. In 1916 the first federal-aid road law was passed. It carried $5 million for the first year with an increase of $5 million each year for 5 yr—$75 million in all. This grant was greatly increased by 1919 by an emergency appropriation of $200 million, largely as a re-employment measure for veterans of World War I.

In 1921, Congress passed the Federal-Aid Highway Act which, supplemented and amended numerous times, has been the basic federal law. In 1978 its name was changed to the Surface Transportation Assistance Act to reflect the inclusion of public transportation funding assistance. These legislative acts established the premise, still followed today, that the nation as a whole has a stake in highways and other forms of transportation. Under this act (through 1982) approximately $120 billion of federal aid to the states has been authorized.

The federal program in effect in 1981 is shown in Table 5–2. It details the expenditures, by category, authorized in the 1978 act. Some of the more important among this bewildering array are discussed briefly below. But even a cursory examination of the items listed and the amounts appropriated for them will clearly demonstrate how the Congress uses federal appropriations to direct the nation's transportation programs.

As of mid 1981, future federal appropriations were under debate. The Reagan administration proposed, for the fiscal years 1981 through 1986, retaining the Highway Trust Fund and continuing the funding for Interstate completion and the Federal-aid Primary System. Amounts for the Interstate 3R program would be increased, but the Federal-aid Secondary and Urban programs would be phased out. Amounts for some of the Title II and III programs would be drastically reduced. Legislation proposed in the House of Representatives, which is controlled by the Democrats, would extend the present levels of expenditure for 1 yr. It is almost certain that legislation representing a compromise will be enacted.

Until 1956 federal aid and other appropriations for transportation came from the general fund of the federal government. In that year Congress established the Highway Trust Fund made up of proceeds from a group of highway-user taxes (see below). This tax plan originally was to operate for 16 yr but has been continued. Its intent is to meet authorizations for the Interstate and federal-aid primary and secondary systems, and what is now designated as the urban system. Also, as shown in Table 5–2, a number of other highway and a variety of highway-safety activities are now supported with trust fund money. At the same time, Table 5–2 shows that Congress has chosen to continue to appropriate general fund money to support or partially support several other highway-related programs that it deems to be in the national interest but not so directly related to highway users. It should be understood that all the "authorized" expenditures

TABLE 5–2. Financing Provisions of the Federal Surface Transportation Assistance Act of 1978

TITLE I AUTHORIZATIONS (IN MILLIONS)

	Fiscal Year 1979	*Fiscal Year 1980*	*Fiscal Year 1981*	*Fiscal Year 1982*	*Four-Yr Total*
*Interstate Completion	3250	3500	3500	3200	13450
*Interstate 1/2% minimum	125	125	125	125	500
*Interstate 3R	175	175	275	275	900
*Federal-Aid Primary	1550	1700	1800	1500	6550
*Federal-Aid Secondary	500	550	600	400	2050
*Federal-Aid Urban	800	800	800	800	3200
*Forest highways	33	33	33	33	132
*Public Lands Highways	16	16	16	16	64
*Economic Growth Centers	50	50	50	50	200
*Urban high Density	85	—	—	—	85
*Bridges on Dams	15	—	—	—	15
*Overseas Highways	9	—	—	—	9
*Integrated Motorist Systems	1.5	2.5	26	—	30.0
*Access Control Demos	10	20	—	—	30
*Bypass Highway Demo	5	25	20	—	50
*Bloomington Bridge Els	0.2	—	—	—	0.2
*Multimodal Concept	9	—	—	—	9
†Bicycle Program	20	20	20	20	80
†Railroad Relocation Demo	70	90	100	100	360
*Carpooling and Vanpooling	4	10	1	—	15
†Great River Road	35	35	35	35	140
†Safer Off-system Roads	200	200	200	200	800
‡Billboard Control	30	30	30	30	120
‡Admin. Expenses (HBA)	1.5	1.5	1.5	1.5	6.0
Territories					
‡Virgin Islands	5	5	5	5	20
‡Guam	5	5	5	5	20
‡American Samoa	1	1	1	1	4
‡Northern Marianas	1	1	1	1	4
‡Access Highways	15	15	15	15	60
‡Northeast Corridor Demo	45	40	—	—	85
Appalachian Project	1.8	—	—	—	1.8
‡Forest Devel. R & T	140	140	140	140	560
Rail Cross Study	—	0.35	—	—	0.35
‡Park Roads & Trails	30	30	30	30	120
*Parkways	45	45	45	45	180
‡Indian Roads	83	83	83	83	332
*Public Lands Devel. Roads & Trails	10	10	10	10	40
Gasohol Study	1.5	—	—	—	1.5
Total Funding	7377.5	7758.35	7967.5	7120.5	30223.85

*Funded out of Highway Trust Fund
†Funded out of Highway Trust Fund and general fund.
‡Funded from the general fund.

TITLE II AUTHORIZATION (IN MILLIONS)

	Fiscal year 1979	*Fiscal year 1980*	*Fiscal year 1981*	*Fiscal year 1982*	*Four-yr Total*
*Sec. 402: NHTSA	175	175	200	200	750
*Sec 403: NHTSA	50	50	50	50	200
*Sec. 402: FHWA	25	25	25	25	100
*55 mph Enforcement	50	67.5	67.5	67.5	252.5
*Sec. 403: FHWA	10	10	10	10	40
*Bridge Replacement	900	1100	1300	900	4200
*Pavement Marking	65	65	65	—	195
*Hazards Elimination	125	150	150	200	625
*Schoolbus Driver Training	2.5	2.5	2.5	2.5	10
Innovative Grants	—	5	10	15	30
*RR-Highway Crossings	190	190	190	190	760
*Highway Safety Education	16	—	—	—	16
*Accident Data	5	5	5	5	20
Totals	1613.5	1845.0	2075.0	1665.0	7198.5

*Funded out of Highway Trust Fund

TITLE III (MASS TRANSIT)—GENERAL FUND (IN MILLIONS)

	Fiscal year 1979	*Fiscal year 1980*	*Fiscal year 1981*	*Fiscal year 1982*	*Fiscal year 1983*
Section 3—Discretionary	1,375	1,410	1,515	1,600	1,580
Section 5—Formula	1,515	1,580	1,665	1,765	
Basic Tier	(850)	(900)	(900)	(900)	
Second Tier	(250)	(250)	(250)	(250)	
Commuter	(115)	(130)	(145)	(160)	
Bus	(300)	(300)	(370)	(455)	
Rural Assistance	90	100	110	120	
Miscellaneous	90	95	100	105	
Bus Terminals	40	40	40	40	
Intercity Operations	30	30	30	30	
Transportation Institutes	10	10	10	10	
Waterborne Demonstration	20	5	—	—	
Total General Fund	3,170	3,270	3,470	3,670	1,580
	Total Funding $13,580,000,000 (four year authorization)				

shown in Table 5–2 may not be made or that Congress may later modify the 1978 law in "appropriations" legislation.

A detailed discussion of all facets of federal aid and of the many conflicts and compromises that accompanied each federal highway or mass transportation act is beyond the scope of this book. A few pertinent recent clashes will be described below under specific topics. Studied carefully, the acts themselves present classic examples of the legislative maneuvering and give-and-take that underlie the initiation and continuation of any large-scale public works undertaking.

FEDERAL-AID TITLE 1 (HIGHWAY) PROGRAMS. As shown in Table 5–2, authorizations of federal-aid funds for construction are by systems[15] and by fiscal year, with fiscal years running from October 1 through September 30. For example, fiscal year 1979 ends September 30, 1979. From each authorization, a sum not to exceed 3¾% is first set aside to administer the program and to support the administrative, planning, and research activities of the Federal Highway Administration (FHWA). The remainder is apportioned among the individual states, provided they meet certain "matching" provisions. An incomplete summary of apportionment and matching provisions, by systems, follows:

Interstate. Apportionment of Interstate funds among the states since 1960 has been proportional to the estimate of the cost of completing each state's Interstate System. The matching ratio, with exceptions to be noted later, is 90 to 10, that is, each $90 of federal aid must be matched with $10 of state money. The small minimum Interstate budget item shown in Table 5–2 assures that each state will receive a small sum of Interstate money. Concern over deterioration of the system has brought the Interstate 3R (Resurfacing, Restoration, Rehabilitation) appropriation. Three-fourths of this is apportioned among the states on the ratios of Interstate lane miles that are over 5 yr old, and one-fourth on vehicle miles of travel on the system. Matching ratio is 75–25.

At the time of the 1973 act, attempts to give the states the option of using Interstate money from the Highway Trust Fund for mass transportation were defeated. However, states are permitted until September 30, 1983, to be credited with Interstate apportionments if they are used for other approved transportation purposes, with the money coming from the Federal General Fund. Also, until September 30, 1983, states are permitted to substitute new urban Interstate routes for those deleted. The matching ratio for these is 85–15. These substitutions must be under contract by September 30, 1986. As of 1978, no new mileage could be added to the system.

Federal-Aid Primary. There are two formulas for distributing federal-aid primary funds. The first has been the same since 1916. Under it, two-thirds of the authorization is divided into thirds; these thirds are distributed in proportion to relative population (today rural population only), the state's area, and miles of rural-delivery and intercity mail routes. The remaining one-third is distributed in proportion to urban population. The matching ratio, with exceptions to be noted, originally was 1 to 1; that is, each $1 of federal aid was matched with $1 of state funds. Under the 1970 act, however, the ratio was changed to 70 federal and 30 state, and in 1978 to 75 federal and 25 state.

Federal-aid Secondary. Distribution among the states follows the original federal-aid formula; one-third divided by rural population, one-third by area, and one-third by postal mileage. The 75–25 matching ratio and its exceptions apply.

[15]These systems are described in Chapter 2.

Urban System. As indicated in Chapter 2, Congress in 1944 set aside funds to extend the federal-aid primary system into urban areas. In 1970, it established a separate urban system. For it also the matching ratio is 75–25.

For reasons such as a lack of matching funds or disputes as to whether certain highways should be built, some states may not use their full portions of allocated federal-aid funds. Commonly, these are made available to other states. Also, to provide flexibility, individual states are permitted to carry over unused funds for an additional year (or years) and to transfer 50% of their allotments between primary and urban or primary and secondary systems.

Other Title 1 Programs. Congress also makes a variety of special-purpose appropriations, as shown in Table 5–2. Space limitations preclude detailed discussion here. Some of these are from the Highway Trust Fund, some from the general fund; or their costs are shared between the two. Some are for aid to the states or their subdivisions; for example, "safer off-system roads." Others are for federal projects such as Forest, Park, and Indian roads. In several instances "linkage" is clearly evidenced. To illustrate, funds for forest highways, which are major arteries, come from the Highway Trust Fund while those for forest development roads and trails and park roads and trails, which serve purposes such as timber access, conservation, and recreation, are financed from the general fund.

TITLE II PROGRAMS FOR HIGHWAY SAFETY. As shown in Table 5–2, federal funds for a variety of highway safety programs, some at the federal level and some to support state or local efforts, with or without matching, have been provided. All are financed through the Highway Trust Fund. Five percent can be retained by the appropriate federal agencies to administer the programs. Seventy-five percent of the money is allocated to the states in proportion to population; the remainder in proportion to public road mileage.

Title II Section 402. These funds are to finance approved programs at the state and local levels "designed to reduce traffic accidents and deaths, injuries, and property damage resulting therefrom." Responsibility for them is placed squarely on each governor. As indicated, the larger share is administered by NHTSA, but some through FHWA. A wide variety of programs is permitted. Many of these programs are discussed in Chapter 12.

Title II Section 403. These funds are for research in highway safety. Again, NHTSA and FHWA each administer a portion of the funds.

Title II Enforcement of the 55 mph Speed Limit. These funds are to be employed to increase the effectiveness of state enforcement of this national law. Not only are these funds made available, but states can lose 5 to 10% of their federal-aid primary, secondary, and urban funds if enforcement does not reach increasingly stringent standards for measuring enforcement. States that meet the standards can receive a bonus of 10% of their safety grant program apportionment. An interesting side effect of this provision is that, although legislatures in certain western states have been pressured to de-emphasize the 55 mph speed limit, they have refused to do so because of the financial consequences. This discontent was reflected in the 1980 platform of the Republican party which calls for repeal of this provision of the law.

Title II Bridge Replacement and Rehabilitation. These funds have recently been made available in large amounts on an 80–20 matching basis with the realization that many bridges are seriously deficient in geometry, strength, or both. Because many of them are not on one of the federal-aid systems, the law provides that between 15 and 35% of a state's apportionment be for "off-system" projects. Also, $200 million annually is for major projects costing over $10 million. These are selected by the Secretary of Transportation.

Railroad-Highway Crossing Projects. These projects have been supported by federal grants for many years. The 1978 act apportions funds equally to "on-system" and "off-system" projects and among the states as follows: 25% each by the federal-aid secondary and urban formulas and the remaining 50% in proportion to the number of rail-highway grade crossings in each state. Emphasis appears to be shifting from costly grade separation to warning devices and gates.

Other Safety Programs. These include Pavement Marking and Hazard Elimination with the former to be consolidated with the latter in fiscal year 1982. And there are certain other special-purpose amounts.

TITLE III PUBLIC (MASS) TRANSIT ASSISTANCE. The financial difficulties of publicly owned and private mass transportation systems (see Chapter 3) first brought congressional action with the Housing Act of 1961 followed by the Urban Mass Transportation Act of 1964. As of May, 1977, $4.5 billion had been distributed for new rail facilities and $0.8 billion, $0.7 billion, and $0.2 billion to modernize heavy, commuter, and light-rail facilities, respectively. Total capital grants through 1978 were $8.4 billion. The 1974 act also redefined "highway purpose" to permit the use of federal-aid highway funds to support transit (see below). Table 5–2 indicates the 1978 aid program. Major elements of it, all from the general fund, are described briefly as follows:

Title III, Section 3 Public Transit Discretionary Grants. These grants, made by the Secretary of Transportation, are for individual large projects. Examples of such projects already under way are rail-transit facilities in cities such as Washington, D.C., Atlanta, Baltimore, Boston, and Miami, and people movers similar to the one in Morgantown, West Virginia (see Chapter 3). The act also permits the secretary to make long-range commitments for major projects, but these are subject to the appropriations process. Of the total, at least $350 million annually is for modernization of existing facilities. Matching ratio is 80–20. In the 1980 Appropriations Act, Congress increased the Section 3 funding from the amount shown in Table 5–2 to $2.19 billion for fiscal year 1981.

Title III, Section 5 Public Transit Formula Grants. These grants are distributed on the basis of urban population. The basic and second tiers may go either for capital improvements or operating assistance with 85% of the second-tier funds restricted to urbanized areas with populations over 750,000. Commuter rail and bus funds are solely for capital grants. For capital grants, the matching ratio is 80–20; for operating assistance it is 50–50.[16]

Other Title III Grants. These grants reflect specific concerns. For example, the Rural Assistance Authorization is aimed primarily at providing transportation for the disadvantaged who live in the country. The Intercity Operations item is to prop up the intercity bus companies, just as Amtrak has been an attempt to maintain intercity rail passenger service. Matching ratios and distribution provisions for each reflect congressional judgments of need and equity.

The federal programs for assisting public transportation have rightly or wrongly been criticized on a number of scores. One is that they are too little and too late; the counter charge is that they are wasteful. Another is that the large block grants have been made on political grounds rather than for carefully planned and economically justified proposals. A third is that they have been inconsistent, vacillating among heavy rail, light rail, buses, and people movers. These specific issues are too complex to discuss here.[17]

SOME SPECIAL PROVISIONS IN THE FEDERAL-AID ACTS. Congress, in writing the federal-aid acts, attempted to recognize certain unusual situations. The matching requirement is one of these. It provides exceptions to the "matching provision" for the so-called public land states where more than 5% of the land is controlled by the federal government. Alaska, Arizona, California, Colorado, Idaho, Montana, Nevada, New Mexico, Oregon, South Dakota, Utah, Washington, and Wyoming are the principal beneficiaries, although several other states qualify for small credits. These states, for Interstate matching, have the 10% reduced by the ratio of federal lands, excluding national forests, parks, and monuments, to the total land area, with the limitation that federal participation may not exceed 95%. For the other systems, the 25% state matching requirement is reduced by 25% of the ratio of federal land area, including national forests, parks, and monuments, to total land area. Of the contiguous states, Nevada has the most favorable ratio; there the federal government controls 70.1% of the land under the Interstate definition and 88.4% under the other formula. Its matching ratio is 95–5 for both categories.

Payments to states are also affected if they do or do not follow federal standards for control of air pollution and for billboard control on nonurban and nonindustrial sections of the Interstate and primary systems. Again, as mentioned, there are the bonus and penalty provisions of the safety sections of the act.

[16]There has been concern in the Congress that federally subsidized operating grants which cost local communities very little would be almost entirely swallowed up in wages with little improvement in service. In earlier years there were no grants for operating subsidies. The 50–50 ratio now authorized is intended to provide assistance while still pressuring local officials to resist excessive wage demands.

[17]For further discussion, see among other sources Wilfred Owen, *Transportation for Cities*, Brookings Institution, Washington, D.C., 1977, *TRB Special Report 177*, and *TRB Record 759*.

OVERALL INTENTIONS OF THE FEDERAL-AID HIGHWAY PROGRAMS. A review of the provisions of federal-aid legislation demonstrates that its main aim is to help the states develop a well-planned and effective system of free roads and to facilitate their use in the national interest. A few salient illustrations of these purposes are as follows:

1. Federal and federal-aid funds may be used for planning and research. These activities are specifically authorized at the federal level and encouraged in the states by stipulating that up to 1½% of federal aid can be used for these purposes without matching with state funds.
2. Federal-aid funds cannot be used for routine operation, maintenance, or administration of the states' highway systems. Furthermore, the states must assume responsibility for maintenance of completed federal-aid projects at the hazard of losing a portion of their federal-aid funds if maintenance is neglected. A recent exception to this policy is the 3R program to resurface, restore, and rehabilitate the Interstate system. As maintenance assumes increasing importance, further shifts of federal aid to it are likely.
3. Federal aid may be used in all the steps required to initiate and complete a construction or reconstruction project. From the beginning, these included not only supervising, inspecting, and building a project, but also locating, surveying, and mapping. As situations have changed and national purposes have expanded, the definition of construction has been repeatedly extended to include planning, acquisition of rights of way, beautification, highway-related facilities to improve mass transit,[18] and constructing or reconstructing replacement housing and paying moving expenses for families or businesses. It should be noted that all of these are designed to assist the states or local governments financially, but not to take over primary responsibility for roads or transit.
4. In the beginning the states were required to pay for maintenance and to provide rights of way from their own funds. The aim was to protect federal-aid funds since maintenance and right-of-way acquisition offered fertile fields for favoritism and graft; manipulations in this area are extremely difficult to detect and control. Not until 1944 did Congress recognize that rights of way were expensive, particularly in urban areas, and permit federal participation in right-of-way costs of specific projects. However the present law not only allows full federal participation in right-of-way costs, but also permits the use of federal funds for advance acquisition, provided only that actual construction follow within 10 yrs. As indicated earlier, only recently has federal aid been available for maintenance under the 3R program, and only on the Interstate System.
5. Federal aid may not be used to construct toll roads. In this provision, Congress has retained its traditional view that all federal-aid roads should be free roads. Exceptions have been made which permitted the use of federal-aid funds to improve publicly owned toll roads on the Interstate System which had only two lanes, and for approaches to toll bridges or tunnels if there is no feasible alternative route. However,

[18]One purpose stated in the 1970 Federal-Aid Act was "to encourage the development, improvement, and use of mass public transportation systems operating motor vehicles on highways." It stipulated that federal-aid highway funds could be used to share in the cost of the construction of exclusive or preferential bus lanes, highway traffic control devices, bus passenger loading areas and facilities, including shelters, and fringe and transportation corridor parking facilities. Later acts, beginning in 1975, permitted the use of some urban funds for rolling stock for rail facilities. In the 1978 act, however, equipment purchases are from transit appropriations.

revenues over and above operating expenses must go to paying off the debt and when that is done, the facility must be made toll free.

OTHER FEDERAL PROGRAMS AFFECTING TRANSPORTATION. As of 1975, there were 448 federal grant programs to state and local governments. These included 45 in natural resources and energy, 103 in community and economic development, 258 in human resource development, 23 in environmental protection, 14 in public safety, and 5 in policy coordination. All of these have transportation implications. More specifically, each of the two "development" grant categories had 45 programs for transportation facilities or services. In addition, many other federal activities such as management of federal lands, site locations of federal employment and of contractors dealing with the federal government, and federal credit and regulatory programs have direct transportation implications. It is charged that coordination among these many activities and the direct federal-aid programs to support transportation is largely lacking.

FUTURE ISSUES IN FEDERAL FINANCING FOR TRANSPORTATION. There are many areas in which basic disagreements have and will continue to plague Congress as it enacts new and modifies old financial legislation. Among these are:

1. Will the role of the federal government continue to be that of assisting state and local governments to finance surface transportation or will it assume more direct administrative and management roles?
2. With whom at state and local levels will the federal government deal? With federal aid for highways, relationships have been at the state level, between the Federal Highway Administration and state highway or transportation departments. In mass transportation, many federal relationships are with representatives of urban areas. How will these relationships be modified as highway and transit functions are merged?
3. How may the level of federal aid for transportation and the division between highways and public transit be altered? This question is closely related to the continuation of the Highway Trust Fund and whether or not revenues from highway users will remain largely earmarked for highways.[19]
4. How may the division of federal aid among the states and between rural and urban areas change? For example, will the formulas for distribution which weigh such factors as population, area, and road mileage be modified substantially? Also, how will federal funding affect the various state and local agencies?[20]

Federal Taxes on Motor Transport

Federal taxes on motor transport are of long standing. Manufacturer's excise taxes on motor vehicles were first imposed in 1917 and, except for the years 1928 to 1932, have been continuously in effect on some vehicles. A federal

[19]The 1979 report *National Transportation Policies through the Year 2000* recommends retention of the Highway Trust Fund and consideration of a transit trust fund.
[20]For the effects on county road programs, see R. A. Larson, *Transportation Engineering Journal* ASCE, Jan. 1979.

gasoline tax was instituted in 1932. Until 1956, these revenues went into the general fund, and from it a portion was apportioned to highways. For example, excise taxes paid by highway users in the calendar year 1955 totaled $2.7 billion; federal aid for the year 1954 to 1955 was $575 million and for 1955 to 1956, $875 million. In 1956, as indicated earlier, a separate Highway Trust Fund was created. Under 1978 legislation, its life is extended through September 1984. Into it go, as of June 1981, proceeds from federal taxes approximately as follows: 4 cents per gallon on motor fuel and 6 cents on lubricating oil; a 10% tax on the manufacturer's sale price on buses (except school buses), trucks, and trailers; 8% of manufacturer's sale price on automobile parts and accessories, 10 cents per pound on highway tires, and inner tubes, and $3 per year per 1000 lb on trucks weighing over 26,000 lb.[21]

State and Local Motor-Fuel and Other Road-User Taxes

The Federal-Aid Act of 1916 required that federal funds be matched with state funds. Many states issued bonds, usually based on property taxes, to finance their shares. In the succeeding years the total state bonded indebtedness for highways increased, and at the end of 1979, with toll-facility obligations included, totaled $18 billion. The 1919 session of the Oregon Legislature, however, decided to tax gasoline for highway construction. A 1½ cents per gallon tax became effective in February of that year. It was an epoch-making law which proved so successful that similar laws forthwith were adopted by other states. By 1926 all the states had a highway motor-fuel tax.

There have been many changes in the state tax rates on motor fuel, but no state has abandoned this popular method for raising highway revenue. The tax per gallon on gasoline varies from state to state; in 1979 it ranged from 6 to 11 cents.[22] Some states had a different tax per gallon, usually higher, on diesel fuel, although Vermont and Wyoming had no such tax. State motor-fuel receipts in 1979 were $9.8 billion.[23] In a few states, the counties or cities are authorized to levy motor-fuel and other highway-user taxes, but the sum collected is relatively small, some $221 million in 1978.

As noted, motor fuel taxes have traditionally been set at a price per gallon. With inflation and other factors substantially increasing prices, the percentage of the total devoted to taxes has decreased markedly. This has brought efforts to increase the tax per gallon, to set the tax as a percentage of the sale price, or to substitute or add a percentage sales tax. Proposals to temporarily or permanently exempt new fuels such as gasahol from taxes to encourage their production and use is seen by some as another serious threat to highway-user income.

Another force that has and will continue to affect fuel-tax revenues is legisla-

[21]This overview of taxes omits many details, including certain exemptions. For further detail, see the legislation itself or the appropriate issue of *Highway Statistics*.

[22]Tax increases were enacted in 1978 in 5 states and in 1979 in 11 states.

[23]A state by state analysis of motor-fuel and other state and local highway-user taxes will be found in *Highway Statistics*. See also *NCHRP Synthesis 62*.

tion to force automobile manufacturers to produce more efficient automobiles. This has taken two forms: (1) a requirement that the average new vehicle meet ever tightening efficiency requirements and (2) high taxes on "gas guzzlers," possibly accompanied by a tax credit for more efficient models. It seems that the Congress agreed with this notion when it chose to control vehicle size rather than to rely on very high fuel costs as its conservation weapon.

There have been careful studies to determine the effect of increased prices, including taxes on the consumption of gasoline. The original reason was to predict their influence on revenues[24]; today it is to see if they would bring fuel conservation. One analysis, made after the oil embargo and price increases of 1973 and 1974, found that a 35% increase in price brought a 7% decrease in travel. However, this was for commuters in the New York City area for whom carpooling and transit were available alternatives.[25] Effects of price increases, shortages, and tax increases in the late 1970s and the early 1980s on both national and local consumption are still to be evaluated.

Motor-vehicle registration and license fees began in New York about 1900. Progressively they developed separate fee schedules for trucks, buses, trailer combinations, and so forth, as illustrated by Table 5–1. In most states such fees, as well as license taxes for motor carriers, now are applied primarily to highway construction and upkeep. These fees, for 1979, totaled more than $3.5 billion.

Many states charge fees for drivers' licenses. Often this income goes to operate the state highway patrol. As indicated earlier, a few states level a third structure tax in addition to a license fee and fuel tax against trucks.

In 1979 the total state-collected user revenue was about $15 billion, counting toll road and bridge tolls of $1.2 billion. The average weekly state road-service bill (1979) was about $2 per vehicle, and federally collected taxes added about $1 more.

Property Taxes for Roads

Property taxes and other general fund revenues have always provided a substantial share of the revenues raised by local governmental units. In earlier times, when roads were a local concern, practically all financing came from them. Some of the funds so gained were used directly for construction or maintenance; others went to pay interest and principal on bonds for roads already constructed. Beginning about 1930, at the start of the depression, both county and city collections for road purposes fell off sharply (see Fig. 5–1). Since that time many local governmental agencies have been successful in gaining state-collected user taxes to meet some of their roads needs. The $2.5 billion of local rural and urban revenues collected and available for road purposes in 1978 were 3.5 times those gathered in 1950. But they are roughly equal to the 1950 total when a correction is made for change in the highway construction cost index. State governments have made little use of property taxes to finance high-

[24]See, for example, *Public Roads*, Mar. 1949.
[25]See E. J. Lessieu and A. Karvasarsky, *TRB Special Report 153 and NCHRP Reports 229* and *231*.

ways. In 1921, only 5% of the funds came from this source and no such income was reported separately in 1979.

Bond Issues for Highways

As mentioned earlier, borrowing to finance road improvement was common practice in the early years of highway improvement. As a rule, the issues consisted of general obligation bonds to be repaid from property taxes. After the boom period of the 1920s, bond financing was less common, largely because of the depression and World War II. Following World War II, however, the urgent need for highway funds resulted in many bond issues, to be paid generally from user-tax funds or from tolls. State borrowings during the peak year 1954, just before passage of the Interstate Highway legislation, totaled $2.3 billion. In 1979 they were $1.0 billion.

At the end of 1979 the principal outstanding on state-sponsored projects, including toll roads, totaled $18 billion. Outstanding obligations for local rural and local urban highways and streets were, at the end of 1978, $2.0 billion and $5.2 billion, respectively.

Many city streets, particularly in residential areas, have been paved under the provisions of street-improvement acts. Owners are assessed for the cost of paving the street in front of their properties, plus a prorated share of the costs of improving adjacent intersections. At times the city also may provide some aid. The sale of bonds guaranteed by liens against the property provides funds for these improvements.

EXPENDITURE OF HIGHWAY FUNDS

Federal Funds

As indicated above, federal appropriations for highways are earmarked for specific purposes. Actual expenditure of the funds, including selection of the individual projects and their design and construction, is under the direction of officials of the states, sometimes in consultation with local officials, but is subject to review by the Federal Highway Administration. However, the states must allocate to each chosen project the stipulated matching amount, which in effect permits federal officials to exert some degree of control over these funds as well. This combination of state and federal authority has been beneficial in many ways. On the one hand, it has left control and operation of the highway system in state hands. On the other, it has provided guidance to the state highway departments. For example, it has aided in introducing uniformity in highway practices over the nation. Probably the most important single result of federal review of expenditures has been an impetus toward the creation of integrated highway systems of limited mileage. There are also many instances in which support of the FHWA has aided state highway officials in resisting demands for the dissipation or unwise use of funds.

State Highway Expenditures

In all but five states the state highway system consists of a limited mileage of the more heavily traveled roads and streets.[26] Support for this system has always come mainly from highway user taxes and federal aid. Expenditures in 1979 totaled $22.1 billion, of which 60% was for rights of way and for capital outlay for construction and reconstruction, 20% for maintenance and traffic service, 7% for administration and research, 8% for highway police and safety, and 5% for interest and repayment of indebtedness.

Expenditures for Local Rural Roads

Local rural road agencies have several sources of funds. Of the $5.6 billion spent in 1978, 19% was from property taxes, 24% from the general fund and other local imposts, 41%, largely highway-user taxes, from the state (but not in four states), 10% from various special federal sources, and 6% from borrowing. The percentage of state support for local rural roads varies widely among states. For example, in Indiana in 1978, income from local taxes was only 30% of that contributed by the state government while in Kansas income from local sources was almost three times the state's contribution. In a given state, the percentages will vary from county to county, depending on the tax assessed by each county unit and the procedure for allocating state funds.

Direct expenditures for local rural roads were divided as follows: 27% for rights of way, construction, and reconstruction; 52% for maintenance; 11% for administration; 7% for interest and debt repayment, and 3% to other road agencies. The shift in emphasis from construction to maintenance, as compared to state highways, is to be expected when the relative mileages, degree of improvement, and traffic volumes are considered.

Expenditures for City Streets

Funds for city streets (excluding federal and state highways within the cities) come mainly from local sources. In 1978, 13% of the $7.4 billion was from property taxes, 48% from the general fund and similar sources, 27% from other governmental units, and 12% from borrowing. Before 1931, state aid for city streets was very limited. Since that time, the states have taken over the responsibility for some major arteries by incorporating them in the state system and have given funds for other street work, but the level is low compared to that for local rural roads. In 1978, three state governments gave no support whatsoever for city streets; in 29 others, the amount was less than $6 million annually. In only three cases did state contributions exceed local ones.

In 1978, 27% of the expenditures were for rights of way and construction or reconstruction, 35% for maintenance, 23% for administration, 13% for debt repayment and interest, and 2% went to other agencies.

[26]In Delaware, North Carolina, West Virginia, and almost all counties of Alabama and Virginia, the state highway department administers and finances local rural roads.

The Pattern of Highway Revenues and Expenditures

Among the individual states, mileages in the state highway system vary from 6 to 60% of the total road and street mileage. The remainders, then, might be classified as local rural and urban facilities. Furthermore, it has just been stated that the division of highway user funds between state highways and local roads and streets differs widely among states. It follows that any attempt to develop a pattern from national averages is dangerous. Even so, such an attempt may be useful. Figure 5–2 shows, at 5-yr. intervals, the financial situation for agencies that collected and disbursed highway funds. It can be seen that ever since 1921 local road systems have been subsidized by state funds, in the main from highway user levies. Federal subsidy of local roads and streets has been relatively small.

Figure 5–2 shows that, beginning in the early 1920s and until the end of World War II in 1945, local governments had continuing success in wresting highway-user funds from the state highway departments. This shift in support for local roads was particularly marked during the depression decade 1930–1940 when property tax revenues fell markedly and highway-user taxes picked up the slack. By the late 1940s the state highway systems that contained the more heavily traveled roads and streets were in serious distress. Since that time further encroachments by local governments on highway user revenues generally have been modest. It would seem doubtful, however, that local governments again would assume primary financial responsibility for their road and street systems. They are faced with too many other pressing needs for the funds at their disposal.

It is difficult, if not impossible, as yet to appraise the effect of the large increases in federal funds for highways on the overall pattern of highway financing. To date, the bulk of this money has gone for the Interstate System. With its completion scheduled for the mid-1980s, local governments may seek to make new inroads into federal and state highway-user funds, particularly for mass transit.

Diversion of Highway-User Taxes to Nonhighway Purposes

Beginning in 1934 the successive Federal-Aid Highway acts have read:

> Since it is unfair and unjust to tax motor-vehicle transportation unless the proceeds of such taxation are applied to the construction, maintenance, or improvement of highways . . . .

To carry out this principle, the law goes on to stipulate that federal aid "be extended only to those states that use at least the amounts provided by law on June 30, 1935" for highway purposes. Any state that, after that time, increased its percentage diversion of highway-user funds would lose one-third of its Federal aid. By 1969, 28 states also prohibited diversion of highway-user funds by constitutional amendment. Thus, it would appear that, up to that time, there was preponderant legislative and public support for the "linkage theory" that

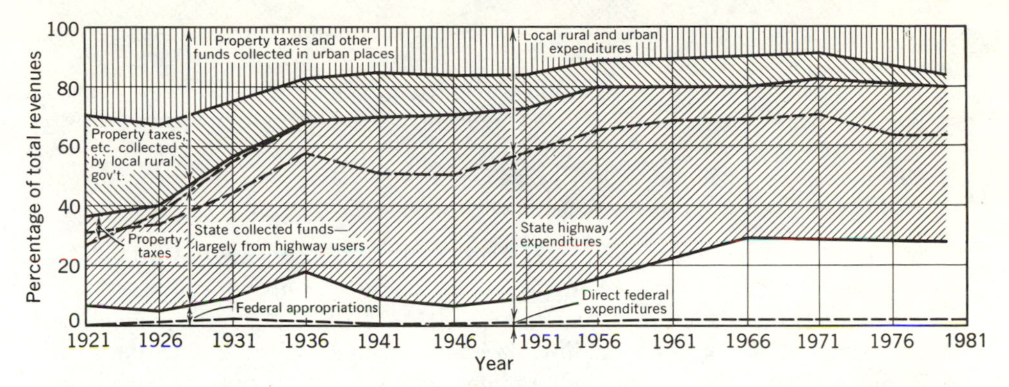

Fig. 5–2. Approximate percentage contributions to and expenditures from highway funds by the various levels of government, 1921 to 1979 (relief funds and toll road data excluded).

ties highway-user taxation to highway expenditures. Since then pressures have developed to repeal antidiversion amendments, particularly in populous states where urban mass transit is in distress or where new transit schemes are being considered.

Diversion of highway-user taxes to nonhighway-user purposes in 1979 was $1.7 billion or 10% of total income from that source. Another $580 million went to mass transit. However, it has been argued that some of these are not diversions. For example, some state governments collect an "in-lieu" tax on each vehicle along with the license fee. This tax is in lieu of the personal property tax on the vehicle and in many cases was instituted to guarantee collection of the personal property tax. It then might be argued that the governmental units that had previously received the personal property tax are entitled to some or all of the in-lieu tax and that assigning it to them cannot be classed as a diversion. However, others claim that personal property taxes on vehicles are highway-user taxes. Thus, diversion itself is still to some degree undefined.

METHODS FOR ACCELERATING HIGHWAY AND PUBLIC TRANSPORTATION IMPROVEMENTS

Most of the expenditures for highways and public transit are on a "pay-as-you-go" basis which links expenditures to revenues. To move forward more quickly agencies may borrow funds and repay them later. Principal arguments favoring this view are that (1) the need is immediate, whereas pay-as-you-go financing means slow progress, and (2) based on past history, pay-as-you-go funds will be dispersed over the entire road and street system and not concentrated on the principal arteries where the situation is most critical.[27] The arguments against financing by borrowing are:

[27]The argument is often advanced that money is saved by borrowing because of the increase of construction costs over time. In terms of financing, this argument is true. In terms of resource use, this argument holds only for any differential increase in construction costs over and above the overall inflation rate.

1. That borrowing may impair or use up the credit rating of governmental agencies and thus block other and more important improvements;
2. That if future income is committed to pay for past improvements, funds will not be available to provide for changed traffic requirements, and even to maintain the existing system;
3. That with large amounts of money available, there is a temptation to overbuild; and
4. That paying interest is a waste of the public's money.[28]

Two basic approaches to accelerated financing are (1) borrowing with repayment guaranteed by general taxation or by giving first claim on all the agency's future revenues or (2) limited guarantees, where repayment comes solely from the earnings of the specific facility. In many instances, these two basic approaches are combined in an attempt to gain the advantages of both.[29] The New York Thruway offers one of many interesting examples of this approach. It is a toll facility and its bonds will be paid off from them; yet borrowing was at a very favorable rate because the debt is underwritten by the general faith and credit of the state of New York.

Financing by Borrowing

Borrowing to finance highways was common practice even before 1900. In early years, as already noted, almost all money came through the issue of general obligation bonds guaranteed by and repaid from the overall income of the issuing agency. Only in fairly recent times, and largely since World War II, have "limited obligation" bonds also been used. Commonly these are to be repaid from road-user taxes. In 1979, 18 state governments borrowed $1.0 billion through general and limited obligation bonds. For local rural units, in 1978, the sum was $355 million and for local urban units, $920 million.

Toll Roads and Bridges

Many early American highways were built with private funds as toll roads. However, private investment had little part in the rapid improvement of our highway system, which began with the coming of the automobile. Not until 1940, with the opening of the Pennsylvania Turnpike, did financing based on tolls reappear to contest the free-road concept. In 1979, toll roads were in service in 18 states. Receipts totaled $1.5 billion.

Although toll roads had little acceptance before World War II, many toll bridges and tunnels were constructed, some by agencies created by state and

[28]An example of this line of reasoning is as follows: It is proposed to compress a 20-yr $1 billion program into 10 yr. This requires borrowing $100 million at the beginning of each year for 10 yr. With interest and borrowing costs at 7%, repayments over 20 yr will total $1.42 billion: therefore $420 million, or 42% of the funds, will be wasted. This argument is invalid in at least two regards. First, it ignores the fact that, if the plan is economically justified, the attendant gains from the improvement will more than offset the interest charge. Second, in totaling dollar amounts along the time scale to determine the total interest charges, the analyst is violating fundamental principles of economic and financial analysis.

[29]See the annual issues of *Highway Statistics* for details of individual plans for highway borrowings and toll road financing for that year.

local governments and others by private interests operating under franchise. For the most part these ventures enjoyed a complete monopoly, as the motorist often had no alternative to the toll facility. Revenues totaled $573 million in 1979. In addition, a few states operated ferries, for which the 1979 revenues were $10 million.

Toll-road activities reached their peak in 1954 when more than $2 billion in bonds for them were sold. After that, the imminence of 90% federal financing for the Interstate System slowed activities markedly. State officials were reluctant to build toll roads to be paid for by local users if federal money would soon be available. Even today, however, the toll-road question has considerable significance, although there is far less controversy than formerly.

As indicated earlier, there are many alternatives for financing toll roads. For example, consider the original Pennsylvania Turnpike. Of its $71,500,000 total cost, $29,250,000 was from a free grant of the Public Works Administration. Likewise, the easterly and westerly extensions did not need to be self-supporting, since the revenues of the entire turnpike were pledged to pay for the newer sections. The Westchester County parkways in New York were originally built to be public freeways, but authorization to impose tolls was given by the legislature. In this instance, it was necessary to rebate federal aid before tolls were imposed. Another illustration is the Denver-Boulder Turnpike in Colorado. By legislative action, state user tax revenues were pledged if needed to make up 30% of a reasonable sinking-fund reserve to repay authorized bonds. For the Connecticut Turnpike, the first issue was supported solely from revenues; but, because of increased construction costs and rising interest rates, it was necessary that the state legislature support later issues with gas-tax funds. In 1978, $18.4 billion in toll road bonds were outstanding. On these, $3.4 billion were supported only by revenues, $4 billion carried limited obligation, and $4 billion were guaranteed by the states' full faith and credit.

Today, then, any controversy over toll roads revolves around a basic public issue: Should or should not government provide a free road system? In some states the answer is "yes," in others "no."

Arguments favoring toll roads center around the urgent need for limited-access express highways. Where these facilities cannot be provided as free roads, motorists appear willing to pay tolls of 1 cent per mile more in order to use them which, in terms of gasoline tax, is about 15 cents per gallon. Government, then, is meeting a public demand by the only available means. There are two other claimed advantages. This first is that, because the projects should be self-liquidating, only those roads that are economically justified will be constructed, and all possible economies will be effected in design, construction, and operation. The second argument is that toll roads will divert part of the traffic from the generally parallel free routes and relieve their congestion.[30]

In certain instances, toll roads may introduce a degree of equity into highway taxation. For example, Connecticut and Rhode Island are "corridor" states, serving New York drivers going to Boston or Maine. Unless tolls are charged,

[30]For a presentation favoring toll roads, see *Civil Engineering* Aug. 1951.

they will pay little of the cost of the roads they travel. It should be observed, however, that this argument partially loses its force where roads are on the Interstate System paid for largely by federal highway user taxes.

Many engineers and public officials are opposed to toll roads. Their principal arguments may be summarized as follows:

1. Toll roads do not offer a nationwide solution to our highway needs and may slow a solution to them.
2. It is difficult to fit toll roads into an overall highway system. For economical toll collections and to get revenues from all users, access and egress must be limited to a few points where traffic volumes are large. This means that, with few exceptions, toll roads can effectively serve only long-haul traffic; other provision must be made for short-haul vehicles. This argument applies with particular force to roads approaching or within metropolitan areas where a predominant portion of the traffic is making short trips.
3. The first cost of toll roads is higher than that of free roads. Interchanges for free roads are simpler, and travel distances are shortened, as traffic does not have to pass through toll stations. Also, crossings at grade might be permissible on free roads where traffic volumes are low and site conditions favorable; on toll roads grade separations are required at all crossings.
4. Toll collections are costly. Reported costs ranged from 4.3% of tolls for the New Jersey Turnpike to 12.1% for the Colorado (Denver-Boulder) facility. To this should be added the costs of stopping which, at 3 cents for each of 50,000 daily vehicles, would total $500,000/yr.
5. Toll roads increase the overall cost of motor transportation. Access to and egress from the toll road is often inconvenient and circuitous in order to route vehicles through the toll gate. In addition, there are cases where construction of a new free road will permit abandonment of the old one. Both toll road and old free road must be kept in service, with the free road at lower standards to protect the toll road's financial position.
6. Toll roads may introduce inequitable taxation. Motorists pay tolls and, in addition, are assessed motor-fuel taxes to pay for free roads they do not use. In at least one case, that of the Massachusetts Turnpike, rebates of motor-fuel taxes have been provided to remove this inequity.

ROAD AND OTHER TRANSPORTATION PRICING

Economists propose that more "economically efficient" use can be made of highways, public transportation, and appurtenances such as parking by imposing graduated user charges. These "efficiency tolls" would be set at high levels during periods of high demand so that only those willing to pay a high price would use the facilities. With volumes lowered, less capacity would be needed and congestion would be reduced, resulting in a saving in resources. In theory, the price for using a facility during the period of potential congestion would be set at such a level that the marginal user (the last to join the traffic stream) would pay personal travel costs plus the added cost that the user's presence imposes on all others already traveling the route or entering the area. In practice, the same charge would be assessed against every user, with the assessment set to hold traffic volume and congestion to a desired level.

Although road and other transportation pricing would produce revenue for improvements or other governmental uses, this is not its primary purpose; rather the income would be a desirable by-product accompanying the main objective of more efficient transportation. It can be argued, then, that similar results might be obtained by subsidizing or even paying people to use mass transportation or to join carpools. The real world problem with this stratagem is that it would call for added spending by already financially pressed governmental agencies.

A large-scale road-pricing scheme in Singapore has been gauged to be quite successful.[31] The plan requires a supplementary license of substantial cost to bring an automobile into the downtown area during the morning peak period. Parking charges also have been increased. Among the results were a reduction of 75% in the morning peak traffic with a significant shift to buses and a reduced trip rate by car owners into the affected area. Afternoon traffic peaks were not greatly affected. Public attitudes were favorable. It must be recognized that the Singapore situation is highly suitable for such a scheme, in that the population is largely concentrated on a single island with its primary business district on the southern tip where it can be easily cordoned off.

In many urban areas, strategies such as barring passenger cars from certain areas or streets, developing pedestrian malls, reducing the number of parking spaces, or placing high or differentiated charges on parking have been employed. These seem primarily intended to reduce automobile use and the attendant congestion or to encourage a shift to public transportation. But there is little evidence that efforts have been made to quantify the resulting economic gains.

Implementing road pricing strategies in large urban areas would be difficult. From a theoretical viewpoint, given the variable nature of traffic demand over time and little knowledge on the relationship between cost and demand for road use and between volumes and congestion, setting a rate structure to produce the desired result would be difficult. Furthermore, the mechanics of and costs associated with collecting the money may present serious complications.

It has also been found that where the users are affluent such schemes are ineffective. Finally, from a political point of view in a democratic society, overcoming inertia and gaining acceptance of the potentially high fees to be assessed becomes difficult and slow.

It can also be argued that road and transportation pricing to produce economic efficiency is undesirable. Congestion is only one of a number of ills and correcting it might have serious and unexpected side effects. For example, increasing the costs of going into or through a given area or of doing business there could cause these activities to shift to other areas. There is also the added concern that "ability to pay" may not be a suitable measure of social costs to be applied in deciding priorities on the use of a public facility. One of the illustrations that "ability to pay" might be a poor measure is that socially desirable ends such as giving the less affluent access to the area for work or other useful purposes would be prejudiced.

None of the points presented here are to argue that road and transportation pricing for congestion control should not be given serious consideration. On the

[31]See E. P. Holland and P. L. Watson, *Transportation Engineering*, Feb. 1978.

other hand, what on first glance may appear to be a simple solution to the congestion problem could offer serious difficulty in application and have many unforeseen consequences.[32]

PROBLEMS

5–1. For the state in which your college is located, compare the actual responsibilities for highway financing with the theoretical proposal outlined in the text.

a. Direct your study particularly at the state highway system.

b. Direct your study particularly at local road agencies.

5–2. Compare the taxes for the various classes of vehicles in the state where your college is located with the average, lowest, and highest rates listed in Table 5–1.

5–3. By contacting appropriate officials in the state or community where your college is located, find out how federal-aid urban funds are distributed between highways and mass transportation and how they and urban mass transportation funds are distributed among urban areas in the state.

5–4. What sums of money, by sources, are available annually for highway purposes in the state in which your college is located? How, by amounts, is this money distributed among state highways, local rural roads, and city streets? How much is this per mile of road in each system? (Source: *Highway Statistics*.)

5–5. Extend Figs. 5–1 and 5–2 through the last year of record.

5–6. Obtain the most recent operating and financial statement of the public transportation agency in the community where your college is located. From it determine:

a. The number of trips and the trips per resident per year and the average fare paid.

b. The sources and amounts of income. What amounts and percentages come from fares, other revenues, and from federal, state, and local subsidies? How do these percentages compare with national averages?

5–7. For the last year for which data are available, determine the principal outstanding and the annual payments made on highway and street bonds in the state where your college is located. Break your report down by levels of government. (Reference: *Highway Statistics*).

5–8. How much, if any, user-tax revenue is diverted to nonhighway purposes in the state in which your college is located? (Reference: *Highway Statistics*).

5–9. Charges for parking to limit automobile use are one form of road pricing. Find out how they are used on or near your college campus. If they are, determine their effectiveness and how the collected funds are used.

[32]No attempt has been made here to develop the economic theory presented by proponents and opponents of road pricing. Many useful references will be found in *HRB Record 296, 314, 348; TRB Record 494, 528, 574, 731, and 767,* and *TRB Special Report 181.*

6 COMPUTATIONS, SURVEYS, AND PLANS

To plan, locate, design, construct, operate, and maintain highway or public transportation facilities requires the collection, processing, and reporting of vast amounts of data. Today, in most agencies, data processing and reporting are done largely with computers. Their overall role is discussed at the beginning of this chapter. Following is a brief description of the planning and decision-making role of computer graphics, which combines computer-generated data or analysis with display on a picture tube. Remote sensing, which is the use of aerial photographs and radar and similar devices for map-making and other purposes, is then presented. Next, the principles and practices of highway and transit location in rural and urban situations and at bridge sites are outlined. Finally, highway plans and specifications, the drawings and written documents employed to describe projects in detail, are illustrated.

COMPUTATIONS

Engineering approaches to almost every highway and urban transportation problem are based on the results of computations ranging from simple to complex. Until the 1950s these were done by calculating machine or slide rule. Often routine calculations occupied a major share of the time of the engineer. Today, however, the computer and its appurtenances have brought great change. Almost all agencies and engineering consultants have computers or have access to them.

As noted throughout this book, computers can do the analyses and data processing for many problems. Examples include planning activities such as projections and statistical studies of traffic and transit ridership, economic analysis, financial programming, geometric, bridge, and pavement design, and maintenance and pavement management. In addition, scheduling for design and construction and computation of earthwork and other quantities both for planning and payment to contractors are done by computer.

Computers are coupled with stereo-plotters in mapmaking and location practices. And computer-based interactive graphics has become a valuable tool in bringing problems into focus, not only in technical areas but in management decision making and situations involving citizen and community participation.

On the management side, budgets, payroll, accounting, and many other activities are computer based.

Computers offer many advantages. They do quickly many routine, time-consuming calculations, making it possible to evaluate more alternatives. When correctly programmed, they are less prone to computational errors. On the other hand, reliance on them can be a trap for the uninformed or lazy who assume that the stipulated program inputs are appropriate. Too often, also, there is a tendency to reduce analyses that need professional attention to computer programs. Accepting such results will often result in poor decisions.

The subject of computer selection and programming is far too complex to treat here. The "hardware" situation is changing very rapidly under such forces as increased use, which brings rapidly decreasing costs, developments such as programmable handheld or desk-top computers, remote terminals or small machines coupled with large central ones and the availability of on-line access, time-sharing, editing capabilities, and interactive graphics. On the "software" side, an abundance of programs are available to solve almost all the repetitive highway and urban transportation problems, so that there is often little point in preparing new ones. The difficulty is to find the most appropriate one for a given purpose. They are scattered among many public agencies, computer manufacturers, computer service organizations, and consultants; sometimes they are privately owned. Then, even if an appropriate program is located, it may be written in any one of several languages or be fitted to a particular generation, make, or model of computer so that adapting it may be arduous and time-consuming. Even with all these difficulties, planners and engineers, at least in the developed nations, must recognize that computers are essential tools that must be understood and used with judgment.

There are many sources of information about available computer programs. Among them are the Implementation Division of the Offices of Research and Development, FHWA, Washington, D.C. 20590, and the AASHTO publication, *Computer Systems Index.*[1]

COMPUTER GRAPHICS—INTERACTIVE COMPUTING

Computer graphics involves projecting information stored or developed by a computer on a picture tube. Displays can be words, tabular materials, or graphs in a variety of conformations. It is possible to query the computer, have it do a previously programmed analysis, and display the new findings. Also, using the equivalent of a light pen, graphs or diagrams can be altered by moving control points around on the screen, after which the computer can be directed to calculate the effects of the change. In turn the new answer or graph will be displayed. A high-speed printer can be placed on line with the display so that inputs and results can be recorded. It is clear that as computer hardware and

[1]*NCHRP Synthesis 55* summarizes computer use in highway and urban transportation situations. References on specific applications appear in the discussion of individual topics.

software become less costly and more sophisticated, computer graphics will come into increasing use.

One among many applications of computer graphics is to display a motorist's view of a highway so that alternative treatments of such features as transitions from fills to cuts can be seen. There are many uses associated with route location, traffic and transportation planning, and accident analysis. One among the programs now operating is for transit route selection and operation. A number of alternative bus routes and schedules to a destination are selected and displayed. For each, impedances to travel by walk-and-ride, park-and-ride, and drive-yourself modes are evaluated in cost and time measures. From this, the most favorable routings and scheduling can be selected.[2]

Another technique that might be classed as interactive graphics without a computer is called environmental simulation. With it a remotely controlled television camera passes through a small-scale physical model of an area or route and records the passage on videotape. The playback gives impressions similar to those experienced by a walker or vehicle occupant.[3]

REMOTE SENSING[4]

Photogrammetry, often called "remote sensing," is defined as the science or art of obtaining measurements by means of photography. Quite commonly it is construed more broadly to encompass procedures for photointerpretation and for converting single photographs into composite ones (called orthophotographs or mosaics) and into maps. Photogrammetry in this broader sense, and particularly as based on aerial photographs, is today a basic working tool of the engineer. Applications appear not only in location, but also in planning, geometric design, rights of way, traffic studies, drainage, soil classification and identification, earthwork measurement, materials location, and pavement condition surveys.

Other forms of remote sensing are helpful. SLAR (side-looking radar) which cuts through clouds, infrared imagery, and multispectral imagery, are among them. To date, information obtained from remote-sensing satellites such as Landsat has not provided the detail needed for highway purposes, except possibly for hydrologic studies. Proposed new satellites offering greater resolution may alter this situation.[5]

[2]A few among the many applications of computer graphics are reported in *TRB Record 455, 491, 553, 559, 574, 619, 657, 677,* and *729.*

[3]See D. Appleyard and K. H. Craik, *TRB Record 617.*

[4]Since remote sensing usually is not a part of transportation or highway courses only a brief discussion of its implications to those areas is included here. There are several textbooks on the subject as well as a *Manual of Remote Sensing* and *Manual of Color Photography* of the American Society of Photogrammetry. Its magazine *Photogrammetric Engineering and Remote Sensing* is also excellent. Many articles also appear in the *Journal of Surveying and Mapping* of ASCE and the *Proceedings* of the American Congress of Surveying and Mapping. Recent publications covering some of the applications of remote sensing to transportation include *HRB Record 201, 270, 319, 375, 421, and 452, HRB Special Report 102, Public Roads,* Oct. 1970, and *Vol. XIII* of the National Association of County Engineers' Action Guide Series.

[5]See, for example, K. A. Godfrey, Jr., *Civil Engineering,* July 1979.

Mapping by Photogrammetric Methods[6]

Aerial photography and the preparation of orthophotographs (mosaics made from corrected photographs) and maps may be done in part or in whole by the transportation agency or by contract between private companies and the individual agencies. If done by contract, the agreement generally stipulates the "results to be obtained" and leaves the "manner and method" of specific photographic and photogrammetric equipment and procedure to the company. To illustrate, a common specification for accuracy of topographic maps states that 90% of the elevations be correct within one-half of a contour interval and the remainder within one contour interval. Another requirement is that 90% of the planimetric features be positioned on the map within one-fortieth of an inch of correct location—the rest within one-twentieth of an inch. Incidentally, these specifications clearly demonstrate the high degree of accuracy that can be obtained.[7]

Vertical aerial photographs taken with the camera pointed nearly straight down are the most useful for highway mapping purposes. The country to be covered is photographed in parallel runs with the individual pictures lapped both in the direction of flight (endlap) and between successive runs (sidelap). For stereoscopic uses, endlap must be greater than half the picture width (possibly specified as not less than 55% nor more than 65%) in order that the center (principal point) of one photograph is included in both adjacent photographs. Sidelap should average 25%, with percentages less than 15 or more than 35 unacceptable. Alternatively, overlap can be stated in terms of acceptable base to height and width to height ratios. Selection of the height from which photographs are to be taken depends on the uses which they, or the maps to be made from them, are to have. For mapmaking purposes the variables include the focal length of the aerial camera, the desired combination of map scale and contour interval, and the ratio of map scale to photograph scale. The latter is, in turn, a function of the stereoscopic projector used for map making.

Several instruments of varying complexity are available for converting data from aerial photographs into maps. These include the multiplex, (largely obsolete today), the Kelsh and Balplex stereoscopic plotters, the Wild autograph, the Kern PG2, and the Zeiss stereoplanigraph. With some, plotting is done directly under the projectors; with others, such as the Wild autograph, the stereoscopic instrument controls a remote coordinatograph mounted on a separate large table. All utilize the concept that when the area common to a pair of matched photographs is viewed through a stereoscope, the topography is seen in relief. It is possible with any of these instruments to produce an accurate map showing all natural and artificial features. Also, contours may be drawn or spot elevations determined. Only those features that cannot be identified on the photographs must be located by ground measurement. Difficulties will of course be

[6]See C. D. Burnside, *Mapping from Aerial Photographs,* Wiley, New York, 1979, for added detail.
[7]Detailed specifications have been published by the Federal Highway Administration as *Specifications for Aerial Surveys and Mapping by Photogrammetric Methods for Highways*.

encountered in attempting to map heavily wooded areas where the ground is not visible in the photographs. Even in this instance, mapping is possible by taking photographs when the trees are bare, by setting ground elevations by using estimated tree heights, or having field crews determine details.

Ground control to give scale and elevation to the maps is done by successively tying together a chain of geometric figures defined by marked points on the ground or by dropping it from high-flight photography. Other points to be incorporated into ground surveys also are marked for easy identification on the photographs.

Accuracies of ground control triangulation networks or traverses must be in keeping with the scale of the aerial photographs. It has been generalized that third-order triangulation is satisfactory in rural areas and second-order in urban locations. For these, maximum errors in distance are 1 to 5000 and 1 to 10,000, respectively. Sufficiently accurate angular measurements are easily obtained with modern theodolites. Distance measurement is usually carried out by an electronic distance measuring device (EDM) which employs visible or infrared light beams, microwaves, or laser light.

Traditionally, map making from aerial photographs has been done by a skilled operator. However, analog computer devices have now been developed which determine the locations and elevations of points and direct the operation of the map plotter. Much of the lettering detail such as style and size can be computer directed.[8]

Photogrammetric techniques have been coupled with computer and digitizer to produce *digital terrain models*. With them the horizontal and vertical positions of the ground surface or other topographic features are transferred directly from matched aerial photographs to a computer data bank. The information can then be recalled and the computer programmed to develop profiles, cross sections, cut-and-fill earthwork quantities, and, as mentioned, motorists' views of the road.[9]

Given aerial photographs and computer-recorded data based on them, separate maps giving specific kinds of information can be plotted easily. Among the concerns served could be highways, drainage, utilities, housing, land use and zoning, and property assessment.[10]

For many parts of the United States, excellent and accurate small-scale maps have already been made by the U.S. Geological Survey. This agency utilizes aerial photographs as its basic source of data. Field surveys are only for marking for ground control and for filling in detail that cannot be gained from the photographs. Scales for the finished maps are as follows: older ones for rural areas 1 to 62,500 or approximately 1 mi to 1 in.; for recent rural and present urban areas, 1 to 24,000 or about 2000 ft to 1 in. New maps will be 1 to 25,000 and

[8]See, for example, K. A. Godfrey, Jr. *Civil Engineering*, Feb. 1977.
[9]See, for example, articles by D. A. Maxwell and C. E. McNoldy, *HRB Record 319* and M. M. Thompson, *Journal of the Surveying and Mapping Division of ASCE, Sept. 1977*.
[10]See V. W. Cartright and J. P. Alessandri, *Civil Engineering*, Nov. 1976.

metric based. State-by-state lists showing the areas that have been mapped are available on request from the Geological Survey.

Orthophotographs, Mosaics, and Overlays

Orthophotographs, aerial photographs corrected for scale and tilt, serve many engineering and other purposes. When the center portions are skillfully matched and copied they give the appearance of a single photograph and show far more detail than maps. Less frequently today, the uncorrected or partially corrected photographs are combined to form a "mosaic." Either the combined orthophotographs or mosaics are useful for public presentations or similar purposes. However, the assembled orthophotographs are far better for engineering and right-of-way purposes where exactness is important.

At times situations arise where alternative locations involve trade-offs between competing land uses or impacts such as developments already in place, adverse soil conditions, parks, open space, scenic views, historical sites, or numerous other controls. In such instances, it may be helpful to develop a series of transparent drawings from the aerial photographs or orthophotographs. These would show each of these influences separately. These transparencies then can be overlaid so that the impacts of the various factors, taken in different combinations, can be examined. As with mosaics, such exhibits will be particularly helpful when describing the alternatives to interested individuals or groups.[11]

Color Aerial Photography

Color film has now been developed which has excellent resolution, sensitivity, and metric stability. Also aerial camera lenses are now color-corrected and distortion-free. Today, then, aerial photographs in color offer the highway engineer a far better tool than black and white alone. Far more detailed and precise information can be gained on, for example, traffic and parking studies, land uses, geological conditions and material sources, and surface and subsurface drainage. Photographing the same area with several different films, such as natural color, color positive, color infrared, black and white panchromatic, and with radar sensors also will provide added insights.

Oblique Photographs

Oblique photographs, either individually or in stereoscopic pairs, can be used to advantage for special situations. Among these are illustrations for right-of-way and public-hearing purposes, and for special studies either where the ground is almost flat, or where cliffs are so steep that vertical photographs do not provide sufficient detail.[12]

[11]See, for example, I. L. McHarg, *HRB Record 246*.
[12]See, for example, H. W. Smedes et al., *TRB Record 594*.

PRINCIPLES OF HIGHWAY LOCATION

Engineering played little part in early highway locations and designs. For example, of the 2½ million mi of rural highways that had accumulated by 1890, many in the east had been positioned by the successive development of earlier trails. In the vast Mississippi valley area, and later in the settled areas of the west, roads commonly followed the north-south and east-west section and township lines.[13] With state aid and state highway construction in the 1890–1920 decades came better road location and design. Then, with the development and increased use of the motor vehicle, alignments were improved and grades flattened. Even so, before World War II, rural highway location and design in settled areas involved mainly higher standards for the width, line, and grade of existing roads. In urban areas the primary concern was with street realignment and widening and with subdivision layout. Only in the sparsely settled portions of the far west and Pacific states and certain mountainous areas were major new locations made over long distances and to consistent standards. Surveys for them were carried out primarily on the ground using traditional techniques based on the transit, level, and tape.

Since World War II, highway location and geometric design practices have been revolutionized. First, the principle was established that access to major highways facilities must be controlled to protect them from encroachments by land-use activities; this forced the adoption of new locations for many major arteries in both rural and urban areas. For example, 85% of the Interstate System lies on new locations. Second, new techniques for computation, mapping, and location have largely supplanted traditional methods, at least for the larger agencies.

For new major highways, locations must blend curvature, grade, and other roadway elements to produce an easy-riding, free-flowing traffic artery that has high capacity and meets exacting safety standards while minimizing disruption to historic and archaeological sites and to community, industrial, business, residential, scenic, and recreational developments. Furthermore, as indicated elsewhere in this book, economic and environmental impacts must be evaluated. The same principles, but on a more limited scale, must be applied when existing highways are being upgraded.

Before any highway improvement is begun, tentative decisions regarding controlling or minimum design speed, roadway cross sections, and maximum grade must be made. These, to be sound, must rest on reliable estimates of the amount, character, and hourly distribution of traffic, coupled with knowledge of the area to be traversed, economic and community factors, and the available

[13] Among the exceptions was the old National Pike from Cumberland, Md., to Ohio and later to St. Louis. For this the act of Congress in 1806 authorized the employment of a surveyor at $3 per day, with assistants. Another exception was the North-Western Turnpike from Winchester, Va., to Parkersburg, W. Va., for which the original act of the Assembly of Virginia named an impassable mountainous route. Later Claude Crozet was employed as locator; his location, on which original construction was completed in 1848, is today largely used by U.S. Route 50.

funds. Then, as planning progresses, choices between possible routes and decisions regarding design alternatives must be selected. These often are made with the active participation of local officials and community groups.

LOCATION SURVEYS IN RURAL AREAS

Traditionally, rural highway location practice has been field oriented, that is, the bulk of the location party's time and effort went to measurement and observation "on the ground." Reconnaissance of the area was the first step; the locator, using available topographic maps and sometimes an airplane, explored the area. His aim was to search out feasible routes and determine such primary controls as mountain passes or suitable river crossings and to locate major obstacles such as steep slopes or marshes. Reconnaissance of feasible routes came second; each of these was covered on foot and rough measures of relative length, difficulty, and cost were taken. Where rate of climb was critical, slopes were measured with an Abney hand level or some comparable instrument. Often the locator "flagged" out the line as he went to establish control points for more detailed surveys that might follow. The third step was for the survey party to run in the preliminary of *P* line, or, where necessary, two or more alternate lines. Distances and angles commonly were measured by transit and taping methods; a profile was taken by differential leveling. Topographic features and often contours were "hung onto" the *P* line. In the office, after the *P* line data were plotted, the engineer laid out the final location or alternate locations by study of the maps and profiles. Finally this *L* line was staked on the ground and profiles, cross sections, and drainage were taken for it.

Figure 6-1 outlines modern location practice based on photogrammetric techniques. The parallel to traditional methods as outlined above is striking; each contains the succeeding steps of area and route reconnaissance, preliminary-line survey, and final location. The difference is that the new method is "office" oriented. Field work before the final location is ready for staking is primarily devoted to the aerial survey and ground control for it, to checking out obscure or incomplete data on the photographs or maps, and to soil surveys and subsurface exploration.

Most highway surveys are tied into the state plain coordinate systems developed by the National Geodetic Survey as an adjunct to its nationwide triangulation network. These master coordinate systems provide a firm base and a check not only for ground control surveys, but also for conventional surveys for alignment and grade and for property description.

Preliminary Reconnaissance

At the reconnaissance stage, the engineer's task is, by cut and try, to determine which routes deserve further study. Terminals of the road and intermediate points through which it must pass form the primary controls. A unique bridge site or single mountain pass also may become a primary control if no alternative exists. Likewise, for scenic highways, the positions of timbered areas, waterfalls,

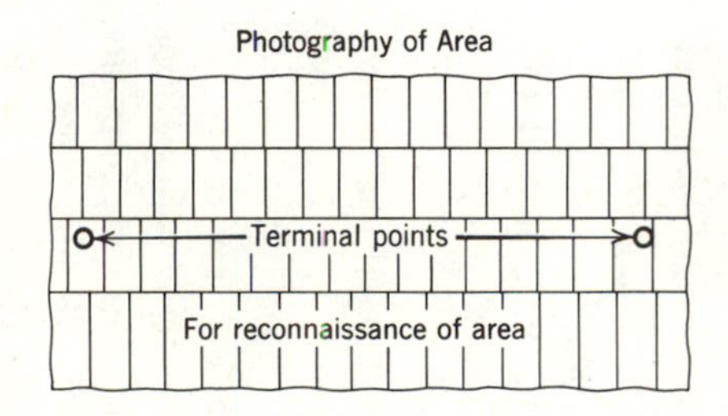

(a)

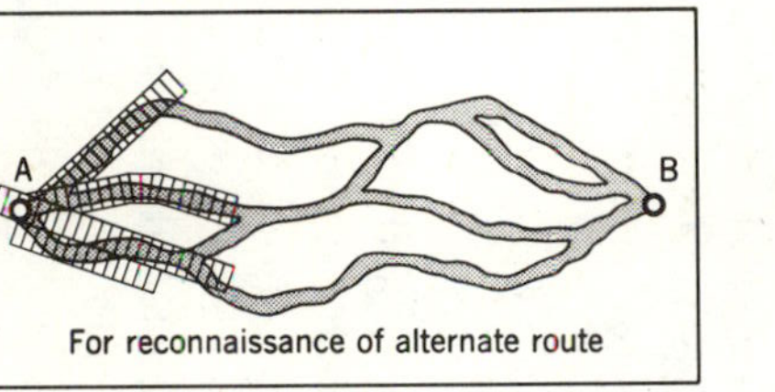

(b)

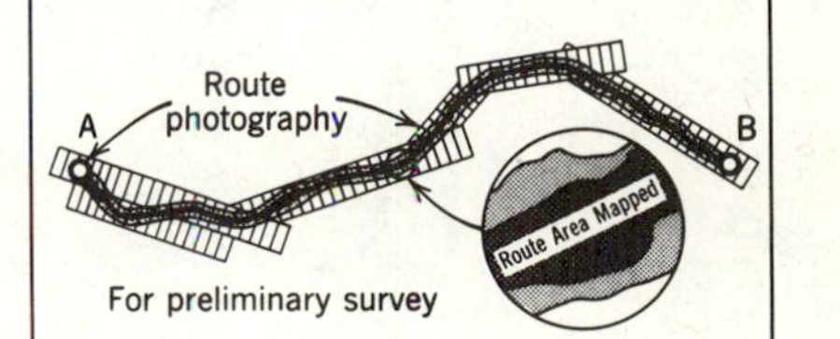

(c)

FIRST STAGE

Reconnaissance survey of the entire area between terminal points:

1. Stereoscopic examination of small-scale aerial photographs of the area, supplemented by available maps
2. Determination of controls of topography and land use
3. Location of feasible routes on the photographs and maps

Scale: 1500–2500 ft/in.

SECOND STAGE

Reconnaissance survey of feasible routes:

1. Stereoscopic examination of large-scale aerial photographs of each route
2. Determination of the detailed controls of topography and land use
3. Preparation of route maps by photogrammetric methods when necessary
4. Location and comparison of feasible routes on photographs and maps
5. Selection of the best route

Scale: 800–1000 ft/in.

THIRD STAGE

Preliminary survey of the best route:

1. Preparation of large-scale topographic maps using route photographs and photogrammetric methods, or

Preparation of large-scale topographic maps by ground surveys, guided by best route location made on photographs in second stage

2. Design of the preliminary location:

a. Using topographical dimensions of the large scale map or computor-assisted interactive graphics

b. While stereoscopically examining the route photographs

3. Preparation of highway construction plans

Scale: 100–200 ft/in.

FOURTH STAGE

Location survey staking of the right of way and of the highway and structures for construction

Fig. 6–1. Progressive stages in rural highway location based on photogrammetric methods.

lakes, and other attractions may be primary controls. Small settlements which would be ignored or purposely bypassed in locating principal highways may be primary controls for secondary roads. Drainage systems, mountain passes, low points in ridges, or swamps in low country, often form secondary controls. Cost factors such as favorable or unfavorable soil conditions, the numbers and sizes of structures, and the amount of excavation and embankment required for satisfactory alignment and grade likewise can be classified as secondary controls.

In mountainous country with well-defined summit ranges, there is usually a suitable pass with possible approaches following along the drainage on both sides. The least expensive and frequently the straightest line may lie just above high water in the streams. Often, however, the rise of the valley or canyon may exceed the maximum permissible grade. Then, if the stream grade is to be followed, extra roadway length must be gained on adjacent mountain spurs, in side canyons, or with switchbacks.

In snow areas, locations should, if possible, be confined to slopes exposed to the sun to avoid icing of the roadway and to ease snow-removal problems. Likewise, the spotting of areas where drifts form, snowslides occur, and the snow melts late is extremely important and may require separate snow surveys in winter or early spring.

Reconnaissance Survey of Feasible Routes

The preliminary reconnaissance will have established primary and secondary controls for one or more feasible routes and will have fixed each location within a band of limited width, possibly within a few hundred feet. The second stage (see Fig. 6-1) is to set the position of alternate routes quite closely by establishing all control points and fitting tentative vertical and horizontal alignment to them, and by roughly estimating their relative costs. Photogrammetry and computers permit more alternatives to be examined than could be done in the field.

Preliminary Location

After the preferred location has been established within a reasonably narrow band, location on the ground (the traditional method) calls for a preliminary or *P* line which follows as closely as possible to the apparent position of the final center line. Survey data are reduced to maps and profiles, often plotted to a scale of 100 ft to 1 in. Topography and possibly contours for a strip 200 to 800 ft wide usually are shown.

The preliminary survey based on photogrammetric methods (see Fig. 6-1) is almost entirely an office operation. Maps to proper scale, often 200 ft to 1 in., are prepared from aerial photographs and the location laid out on them.

Procedures for rural multilane facilities are somewhat different. For them, the most advantageous location may result if the roadways for opposing directions are designed as separate highways. For example, they may occupy opposite walls of a canyon or go on opposite sides of streams or small hills. Median width will be variable, and alignment and grade line will be quite different for the two. Often this dual location is cheaper than for a constant cross section; driving is more pleasant and less monotonous since the view and noise of the opposing roadway is minimized. With care, headlight glare is almost entirely eliminated.

Final Location

Final location is essentially the fixing of the details of the projected highway. It offers opportunity for small shifts of the line and adjustments in grade. At this time, final horizontal and vertical positioning of structures, channels, and other drainage facilities is set. Particular attention should be paid to coordinating horizontal and vertical alignments. For example, beginning or ending a horizontal curve within the limits of a vertical curve over a crest should be avoided, since drivers cannot see the change before they reach it.

As mentioned above, computer-based techniques have been developed which display the road ahead on a cathode-ray tube. These will enable the designer to have a driver's eye view and adjust the design to make it aesthetically pleasing.

Following final location, whether done in field or office, sufficient curvature, tangency, and other control points must be carefully referenced on the ground to permit easy location of the line during all phases of construction. Likewise, benchmarks must be set at relatively close intervals and in positions free from disturbance by construction activities. Also, directions of all property lines, distances to property corners, and the locations of buildings, fences, and other improvements must be established accurately. On these will hinge the future property acquisitions and settlements made by the right-of-way agents. Large-scale topographic maps or other special surveys at bridge and structure sites are also needed. Soil surveys and foundation explorations for structures will be made, probably by specially trained and equipped crews.

Sometimes a field survey party is called upon to bound stream basins and to gather other information from which stream flows can be predicted. Where good topographic maps such as the U.S. Geological Survey quadrangle sheets are available, the large areas can be traced on them, and only the small areas adjacent to the road must be surveyed. Alternatively, it may be possible to trace these boundaries on matched aerial photographs.

With field survey methods, earthwork quantities, both for estimating purposes and for payment to contractors, have been based on cross sections taken either during the final location survey or just preceding construction. It is now common practice to carry out both of these functions by photogrammetric-computer methods. Even the locations of "slope stakes," which mark the limits of the roadway prism, can be predetermined without conventional field surveys, thus reducing the field surveying time substantially.[14]

A location survey across the almost impassable swamps of the "Darian Gap" in the Pan American Highway on the Panama-Columbia border employed many advanced techniques in surveying and soil exploration.[15]

[14]See, for example, L. L. Funk, *HRB Bulletin 228,* and V. H. Schultz, *HRB Record 375.*
[15]See A. F. Ghiglione, *HRB Record 299,* and L. F. Delwig and C. Burchell, *HRB Record 421,* for added detail. For a description of the U.S. Forest Service computer-aided location and design system, see T. A. George, *TRB Special Report 160.*

Locating Roads in Recreational and Scenic Areas

The primary purpose of recreational and scenic roads is different than that for other highway facilities. Thus, although the engineer follows the step-by-step location process outlined above and also must adopt suitable design standards as outlined in Chapters 8 and 9, the facility that results probably will be different. To illustrate, a circuitous rather than direct route may be appropriate in order to provide access to lakes, streams, campsites, or scenic views. Again the alignment may have more curves, possibly to follow a stream, lake, or cliff top, or to encourage more leisurely driving. A narrower roadbed, sometimes placed under a canopy of trees, is another stratagem to discourage fast driving.

Particular attention should be exercised to relate scenic views to the need for driver attention to the road. For example, a spectacular view of a mountain from the road might be developed on a straight section by carefully directing the alignment. On the other hand, similar visibility should be avoided on curves. Again, very wide shoulders, turnouts, or parking areas should be provided where motorists may wish to stop; on other sections of the road, trees or planting to block the side view will discourage stopping on the roadbed.

Locations often can be set to develop "explosive views." For example, in Yosemite National Park the first view of Yosemite Valley on the south entry road comes as the motorist emerges from a tunnel. In this instance, an adjacent parking area has been developed so that motorists do not stop in the road. Sometimes, as at the south rim of the Grand Canyon, the road is set in a forest somewhat back from the rim, and short stub roads lead to parking adjacent to view points. With such an arrangement, the view and the road are not competing for the driver's attention. Locations should also be sited so that the road is not disruptive to the area. Scars from cuts and fills should be minimized. Again, preferred locations would lie at the edge of a meadow rather than across its center.

Plans at both federal and state levels have been developed for an extensive system of scenic roads. Guides putting forth detailed recommendations for location and design are available.[16]

HIGHWAY AND TRANSIT LOCATION IN LARGE URBAN AREAS[17]

General Principles

Before World War II, the typical American city contained a central business district with large stores and offices and often the cultural and civic centers. Adjacent to it along railroad lines or waterways had developed commercial and in-

[16] *A Proposed Program for Scenic Roads and Parkways* prepared by the U.S. Department of Commerce for the President's Council on Recreation and National Beauty, proposes a national program, a method of financing, and offers an excellent design guide. Another design guide is *A Guide for Highway Landscape and Environmental Design*, AASHTO.

[17] The report, "Interregional Highways," *House Document 379*, Seventy-eighth Congress, Second Session (1944), even today offers one of the best discussions of this subject on pp. 53–74. *The Free-*

dustrial districts. These merged into secondary business areas which in turn changed into neighborhoods of mixed land uses and rundown commercial buildings and houses. These were referred to as "blighted areas." Beyond these were the newer residential developments. With the coming of the automobile, these often extended far beyond the city limits in the form of widely scattered subdivisions. Vehicular travel in this typical city was heaviest along radial routes leading from the rural areas into the business and industrial areas. Volumes usually increased substantially with proximity to the central areas. Widening these radial arteries to relieve congestion was done at great expense but with little success. This procedure failed because it attempted to combine the incompatible functions of business or residential street and traffic artery in a single facility. In sum, the developed portions of the average large city resembled a rimless wheel, with central commercial, industrial, and older residential areas forming the hub and the finger-like developments along the highways making the spokes.

After World War II, favorable locations for the early radial urban freeways often were found between these outstretched spokes or fingers. Right-of-way costs were lower and the number and complexity of grade-separation and interchange structures reduced. In other instances, the best locations fell along railroad lines, valleys, or water courses, where the normal subdivision pattern had been broken and cross streets were widely spaced. Because traffic volumes increased toward the downtown, more lanes generally were needed close in. The effect of this easy access to the downtown area along radial freeways was to accentuate the "flight to the suburbs" discussed in Chapter 3. In addition, the freedom provided by the automobile made residential development in the open areas between the spokes of older urban developments attractive. In turn, shopping and service facilities, and later light industries that needed skilled labor, followed the residential movement. All these, in turn, created substantial circumferential traffic which was accommodated, or may in the future be accommodated, by the construction of major circumferential traffic arteries and sometimes by "beltways" designed to freeway standards.

Radial freeways seldom penetrated into the heart of a central business district (CBD) where property values were high. Rather they connected to a close-in circumferential loop that circled the city's heart. Often this was located in the "blighted area." Vehicles bound to the CBD were to travel around this ring road, leaving it at street connections near their destinations.

A radial and circumferential freeway and expressway system to fit an automobile-oriented development pattern is typified by past and proposed developments in Houston, Texas. Plans for its closer-in areas are shown in Fig. 6-2.

---

way in the City, a report to the Secretary of Transportation prepared in 1968 by an eminent group of landscape architects, urban designers and planners, architects, and engineers is an excellent and authoritative treatise on the principles of urban freeway planning, location, and design. Chapter titles, which in themselves are an excellent summary of problem areas, are (1) Comprehensive Planning and Community Values, (2) The View from the Freeway, (3) Location of the Freeway, (4) The Roadway, (5) Highway Structures, (6) Multiple Use of the Corridor, and (7) The Systems Approach.

Population of the metropolitan area is roughly 2.5 million. As of 1979, about 370 of a projected 700-mi system was in place. In this case, there are no topographic barriers to distort a symmetrical pattern. Planned are 11 radial freeways or expressways that bring traffic from outlying areas. Most of them end less than a mile from the downtown center, where they connect to a freeway loop as well as to the streets. In addition to the downtown loop, there are two other circumferential routes having diameters of roughly 10 and 25 mi. As shown in Fig. 6-2, there is a well-developed grid of arterials serving the areas between the freeways and expressways.

One among many difficulties in laying out urban freeways is that of connections to city streets. To provide both off and on ramps to every street usually is out of the question. One reason is the extreme cost. Another and more compelling argument is that such close spacing of connections impairs the free-flowing and accident-free characteristics of the freeway. As will be demonstrated in Chapter 8, the clear distance preceding each connection should be great enough that drivers can be alerted and given time for appropriate action. Spacings as great as several ordinary city blocks are required where the design speed of the freeway is high. Often the best solution is to change the flow on the adjacent city streets to one-way and to provide connections to appropriate pairs of them. Where large volumes must be carried by single ramps, good practice calls for an added lane on the freeway solely for these vehicles.

Where an existing city development is on a gridiron pattern, the ideal right-of-way width for a depressed freeway is a full city block. This provides amply for all future needs and for on and off ramps. Grade separation for cross streets can be provided without disrupting established development in adjoining blocks. Rearrangement of underground utilities that are cut by freeway excavation is usually less troublesome at midblock than near intersections. Space remains for developing parks and recreational areas. Then, too, the existing streets along the boundaries serve local traffic and provide full access to fronting property. Unfortunately, rights of way this wide often cannot be taken, either because funds are not available or because there is objection to the removal of so much property from the tax rolls. In some cases, rights of way for freeways or expressways have been gained by widening the alleys that bisect main blocks or by taking a half-block including one street and one alley. Alley frontage is of much lower value than street frontage, which makes the overall cost of such rights of way less.

In many instances, freeways have been placed on fills or continuous structures above ground level. These designs largely eliminate interference with normal street traffic and require little rearrangement of underground utilities. On the other hand, providing width for the roadbed and the fill slopes takes valuable property and is objected to on aesthetic grounds. Continuous structures are extremely expensive, and provisions for access and egress are both expensive and troublesome. Settlements with owners of adjacent property who claim damage to the value of their properties from noise, fumes, and loss of privacy (see Chapter 7) may also be high.

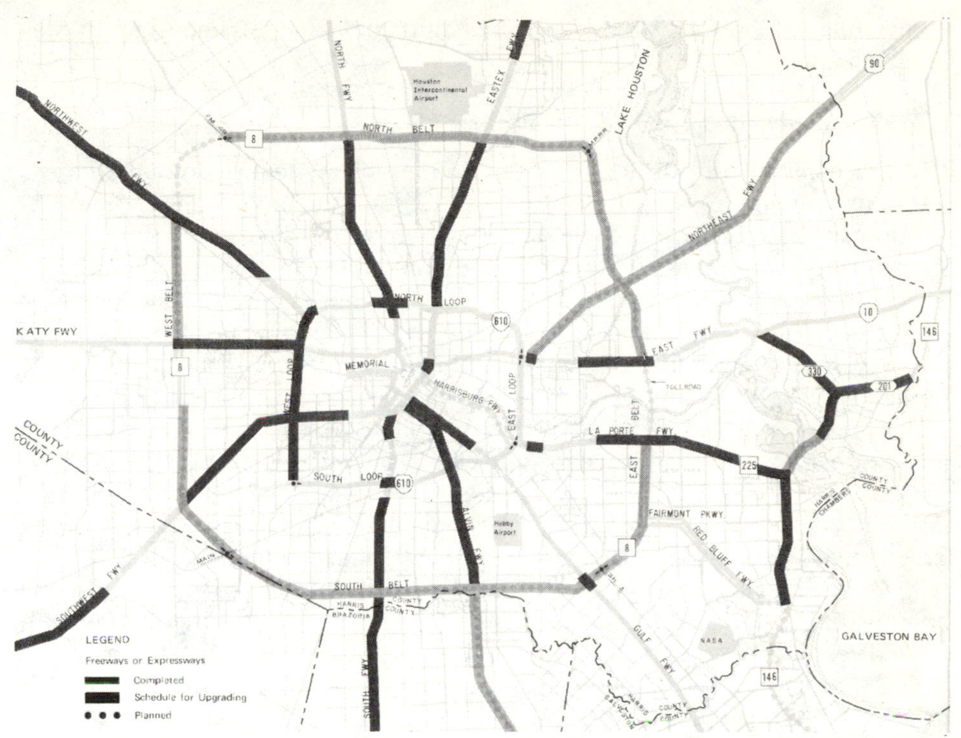

Fig. 6–2. Existing and proposed freeway and expressway system for the metropolitan area of Houston, Texas, simplified from map dated Jan. 1, 1979. (Courtesy, Houston-Galveston Regional Transportation Study.)

It is claimed by some that elevated highways form a barrier much like that created in earlier days by railroad tracks. They assert that on one side normal development will take place but that property on the "other side of the tracks" will decrease in value. They also declare that elevated roads are out of place except in industrial areas or in locations where it is desired to separate one kind of land use definitely from another. Little recent factual data are available to support or refute these arguments, although a 1960 study by the Special Freeway Study and Analysis Committee of the AASHTO noted that, for elevated freeways constructed on fills, detrimental effects were not apparent.

The location of arterial highways in the vicinity of small communities raises a quite different problem, since a large percentage of the traffic will not stop at all; possibly 40% will stop in a community of 5000. Here, then, it is often desirable to swing entirely clear of the community rather than to pass between a segment of the residential area and the business district. Rights of way for the close-in route will be more expensive, and if the bypass is to be developed as a freeway, more grade separations will be required to accommodate local movements, which means greater first cost. Often local business interests will oppose the complete bypassing of their community. Yet much evidence is accumulating to show that business is not hurt and often is improved by this arrangement.

Coordinating Highway and Transit Facilities

Given a completed or planned freeway and arterial system, many urban areas are now planning to integrate it with a public transportation system. The problem in each case is to choose among the many possible alternatives for both the short and long run, considering potential use and financial constraints. All these approaches cannot be discussed here. Rather, some of the alternatives being evaluated for implementation in Houston, Texas, beginning in the early 1980s will be used as illustrations. Involved are highway, planning, and public transportation agencies as well as groups of public officials and citizens.

As shown in Fig. 6-2 and described earlier, Houston has an extensive freeway and expressway system in place or projected. This is designed to serve the central business district which lies within the inner ring, a highly developed area laid out in a 19 X 13 block grid. Individual blocks are about 400 ft between street centers. Completion of the highway network and some form of advanced public transportation scheme seems almost assured. In August 1978, area voters approved a 1% limited sales tax for transit; and income from this, coupled with anticipated federal and state subsidies, provides a reasonable financial base.

The short-term approach in Houston is to greatly increase commute-bus use. Of the 160,000 people employed in 1979 in the CBD, only 1500 enter on express buses traveling the freeways and expressways. Planned are some 18,000 park-and-ride spaces to be in place by 1983 on owned and leased land. It is anticipated that the resulting large shift to transit will substantially reduce private automobile use with an accompanying lessening of congestion, fuel use, and air pollution.

Planning for the long run envisions, by 1995, 55,000 commuters in a work force of 240,000. Studies have examined both the commuter system and circulation in the CBD. For commuter service, the first step was to identify those traffic corridors that justify a facility completely separate from existing freeways. For them, an alternatives analysis[18] will lead to a choice between fixed guideways (rail) and busways (see Chapter 3).[19]

For the rail commuter alternative, terminals adjacent to or within the CBD were considered. Joint development (see below) would be a feature. Such terminals were among the alternatives for an all-bus commuter system. Passenger movements to and from these terminals could be by people movers (see Chapter 3) or shuttle buses on a transit mall and possibly other streets.

With an all-bus or largely bus system, an alternative to terminals is to route commuter buses on reserved lanes throughout the CBD and possibly along a transit mall. With this arrangement many passengers would walk to their desti-

[18]Alternatives analyses are called for by UMTA as a step toward federal funding for public transportation projects.
[19]In late 1980, a 13-mi rail line was chosen to serve the southwest corridor.

nations or use a system of shuttle buses. All of the alternatives considered for Houston involve conversion to a transit mall of a street running the long dimension near the middle of the CBD.

Space limitations permit only this brief sketch of the planning for transportation in Houston. Even so, it illustrates the complexities to be faced in planning public transportation in urban areas.[20]

Joint and Multiple Use Concepts

In the past, highway and transit rights of way and the space above and below them have been reserved almost exclusively for agency purposes. One reason was that joint ownership or easements to permit occupancy of space above, below, or alongside them created many legal complexities.[21] Another is the fear of loss of life or property in case of vehicular accidents on the facility. For example, there have been instances where trucks with cargos of flammable liquids ignited after collision so that developments below elevated structures or above depressed or ground-level ones were damaged. Parallel concerns affected public transit agencies.

In contrast to many other countries, political and public interest concerns in the United States have made joint development and multiple use between private enterprise and governmental agencies particularly difficult. Elected or appointed officials or agencies who become involved in such arrangements have been charged with giving away public assets or favoring special interests. Among the difficulties are placing a suitable value on the public's share.[22] As a consequence, some of the more cautious have avoided being involved in such schemes. Other agencies have kept the developments in-house. Today, however, pressures to integrate all transportation modes and various associated activities are counterbalancing these less venturesome approaches. For example, federal-aid highway legislation now permits the states to develop agreements for the use of air space above or below Interstate highways. Also, urban transit discretionary grant and loan funds can be applied to investments which enhance urban economic development or incorporate private investment, including commercial and residential development.

One form of multiple use involves shared rights of way between highway and transit agencies. Busways or rail facilities have been placed in the median or alongside freeways. Recent examples include the Congress Street and Dan Ryan Expressways in Chicago, certain sections of BART in the San Francisco Bay Area, and I-66 in northern Virginia. At times, the median has been expanded to accommodate transit stations and parking for passengers. One difficulty with such arrangements is the high cost of grade separations when highway and transit alignments diverge.

[20]For a detailed street layout of downtown Houston and estimates of the use of a shuttle service, see J. J. Hinkle, *Transportation Engineering Journal of ASCE*, Jul. 1980.

[21]See the discussion of value capture in Chapter 7.

[22]See *NCHRP Report 142* and *Special Studies on Highway Law, Vol. 2, NCHRP*, for approaches to this problem.

A number of suburban joint-development schemes involving park and ride for transit have been proposed and some have been implemented. These recognize that peak demand for parking at outlying shopping centers, churches, theaters, and other cultural and recreational centers are at night or on weekends and that the space is underused and available for commuters on weekdays. There are instances where shopping centers have permitted free use of parking space as a means of attracting commuters as customers. In other cases, a cost-sharing arrangement has been worked out between the transit agency and the private property owner or participating public entity. Unfortunately, even such mutually advantageous schemes may come under public scrutiny and bring charges of favoritism or patronage.

Several downtown joint-development approaches have been proposed and some have been implemented. For example, private developers have constructed stores or offices on transit-agency property adjacent to or in the air space over a rail or bus station or terminal. One scheme envisioned two terminals near the opposite fringes of the downtown, connected with shuttle buses or a people-mover. Each terminal incorporated a shopping area and office tower. Revenues to the public agency would permit it to capture the increased land value.[23]

Surveys for Urban Highways

In general, the sequence of area and route reconnaissance and preliminary and final location is followed for urban as well as rural highway surveys. On urban projects, however, procedures are much less uniform and fixed. For example, there may be prior surveys and maps made for property location, street improvement, utilities, or other purposes that will furnish much of the information that in rural areas must be gathered in the field or from aerial photographs. In many instances these data may be complete and accurate enough that no preliminary survey is required. Thus the finished location can be developed largely from the results of earlier surveys.

Photogrammetric methods are particularly effective for urban highway location since data gathering and inspection on the ground are so difficult. The same four steps as for rural locations apply, namely: area reconnaissance, reconnaissance of feasible routes, preliminary survey, and final location. The principal difference is in the scale of the aerial photographs and drawings. As a generalization, photographs, mosaics, and maps for urban projects have roughly double the scale of their rural counterparts because greater detail is required. As rough approximations, photograph scales for preliminary reconnaissance might run 800 to 1000 ft to 1 in.; and for route reconnaissance 400 to 500 ft to 1 in. For preliminary surveys and final detailed designs, photographs producing maps

[23]For some of the earlier proposals for joint development see *The Freeway in the City*, op. cit.; *A Policy on the Design of Urban Highways and Streets*, AASHTO; and *HRB Special Report 104*. Later reports appear in *TRB Record 528, 565,* and *634,* and *TRB Special Report 183*. R. G. Arbogast et al., *Transportation Engineering Journal of ASCE*, Sept. 1980, presents a method for optimizing station location and M. D. Rivkin, *TRB Record 634* gives several specific examples.

to scales of 40 to 100 ft to 1 in. often are appropriate. Underground installations for utilities such as water, sewer, gas, electricity, and telephone often require special investigation. At times, careful field surveys are required, particularly in older, established areas for which maps may be inaccurate or, in some instances, nonexistent.

Each step of urban highway location calls for careful appraisal of many factors; there appears to be no substitute for the cut-and-try approach by an experienced and qualified engineer. Repeated consultation with local governmental officials, both professional and political, and possibly with citizens' groups is a must (see Chapter 3).

Just as in rural areas, location surveys, if done by traditional methods, will consist of staking and referencing the centerline, taking profiles and cross sections, and determining the locations of all property monuments and culture. Surveys on new rights of way through built-up areas will be complicated by the many obstructions. Often it will be impossible to run a continuous centerline until the right of way is cleared for construction, and all earlier work must be done by using offset lines and improvising in other ways. Then, too, seldom is an urban survey made in a location free from conflict with motor vehicles on the existing streets, which occasions many delays and is at times actually dangerous to the personnel involved.

BRIDGE LOCATIONS

Since the purpose of bridges as well as highways is to convey traffic, the location and positioning of all but the most important structures should be subordinate to general alignment and grade. There have been many instances where sharp turns at the approaches and generally tortuous alignment have resulted because the most favorable bridge site was the sole criterion for the location. Sometimes favorable alignment has been sacrificed merely to provide a cheap right-angle crossing of a small stream. Today the general policy for all but minor roads is to determine the proper highway location and furnish structures for it. This, of course, results in more expensive crossings, for skew bridges cost more than right-angle ones, and the introduction of horizontal and vertical curvature into large bridges creates serious design and construction problems. However, the end result is a better roadway.

If traffic would be nearly equally served at several sites, cost will determine the location of the bridge. Foundation conditions for piers and abutments will seriously affect costs, but will not always be determinative. Obviously the cost of approaches is important, and the combined cost of the bridge and its full approaches must be determined before the crossing site is selected. The engineer must recognize these basic considerations and place the line accordingly.

When the approximate location of the bridge is fixed, there must be a complete and extensive report and a special survey for the site, supplemented by sketches and full-scale maps and profiles. The bridge-survey report should include accurate data on the channel or waterway for all stages of water, the foundation conditions, and the stream character. Data on adjacent structures on the stream, particularly their waterway openings, is especially important.

In certain instances, waterway conditions even for minor structures may be improved greatly through slight modifications in location. As an illustration, consider Fig. 6-3. The submitted or original alignment involves (1) a skew crossing which is in itself undesirable from a standpoint of first cost because of the increased length of barrel and more elaborate abutments and piers, and (2) the construction of a long wing at A to eliminate the tendency of the stream to wash the fill during flood periods. Also, the stream encroaches so near the shoulder of the road at B that it is quite possible that even the length of wing shown may not always prove adequate during a succession of future flood periods, and riprap or some other form of bank protection may be required. This tendency to cut behind and under wing walls is the source of a great deal of serious trouble in the maintenance of small bridge structures. The channel change (shown dotted in Fig. 6-3) might be a way out of these difficulties, but adds first cost and may not be cheaper than the skewed culvert and long wing. Moreover, the channel change affects the natural watercourse, and the builder at once becomes liable for any property damage resulting from overflow and erosion which may take place at C.

These undesirable features may be eliminated by the revised alignment shown in Fig. 6-3. This method seems logical but for the fact that the new alignment throws into the hill at D, thus involving more excavation. These considerations must be balanced against the undesirable waterway features enumerated above. The solution will obviously depend upon the exact conditions disclosed by careful surveys and estimates along both lines.

At major stream crossings, the bridge structure itself becomes increasingly important and tends more and more to outweigh considerations of cost for the approach alignment. Under such conditions, the location should take advantage of any narrow neck or point of constriction in the waterway and of particularly favorable foundation conditions. Also, where possible, skewed structures should be avoided. Their superstructure is generally more expensive, since it demands greater floor area which means more material; also the cost of fabrication and concrete forming usually is higher. The piers and abutments will be increased in length, and the wing walls on abutments will be longer on two corners.

HIGHWAY PLANS

Plans and specifications are, in effect, the instructions under which highways are built. Where the work is done by a contractor, they are an integral part of the contract between contractor and highway agency and as such are legal documents. In general, the plans contain the engineering drawings of the project whereas the specifications present the written instructions.

After the final location for a given project has been completed, including the field soils investigation, a complete and detailed scheme for the road is worked out by specialists in the fields of geometric design, traffic, drainage, erosion

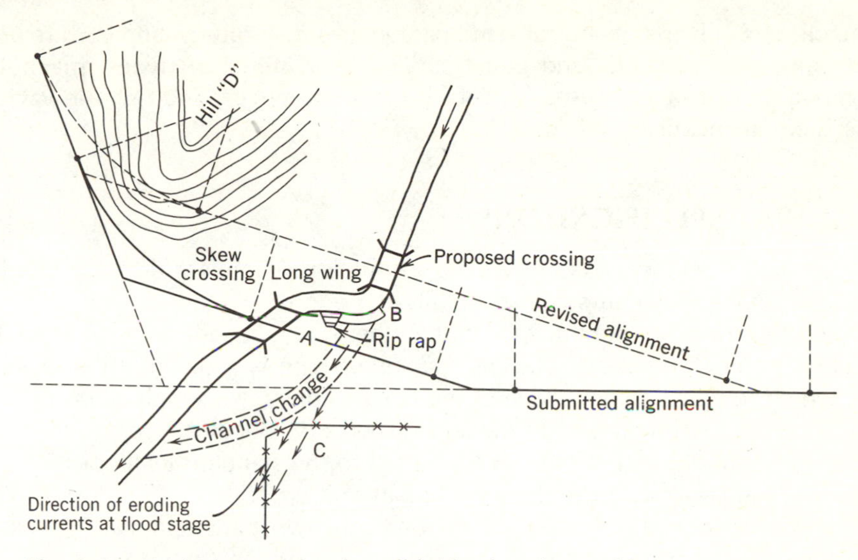

Fig. 6–3. Typical example of small bridge location problem.

control and roadside development, structures, soils, and pavements.[24] All the dimensional features and many other details of each final design are recorded on a series of drawings commonly referred to as plans. Figure 6-4 shows one sheet of plans for a typical two-lane rural highway in open country. In this case, the upper half of the sheet is devoted to a "plan" view showing horizontal alignment, right-of-way takings, arrangements for handling drainage, and many other features. The lower half is the "profile" on which are plotted elevations of the original ground surface along the roadway centerline and the vertical alignment or "grade line" for the road. The vertical scale of the profile usually is exaggerated five to ten times. Sometimes profiles and other details of drainage channels or connecting roads and ramps also are placed here. Many agencies list the estimated earthwork quantities for each 100 ft station or other interval along the bottom of each sheet along with estimated overhaul (see Chapter 15) as they are often needed by contractors or field engineers. Other sheets of the plans show roadway cross sections fitting every situation in the entire project. Also included will be sheets of drawings for all structures and roadway appurtenances.

Nearly all highway agencies use standard-size drawings about 36 in. wide and 22 in. high. Some reduce the final plans during the blueprinting process to make for easier handling in the field. Some states have reduced the number of drawings to be supplied for each project by providing "standard drawings" for certain elements that appear repeatedly. Engineers, contractors, and other interested parties are supplied with them or permitted to purchase them. A partial list of subjects covered by the standard drawings would include pipe culverts,

[24]For a discussion of techniques for managing these and other preconstruction activities, see *TRB Record 742*.

concrete box culverts, guard rails and parapets, curbs, gutters, and curb returns, sidewalks, drainage inlet and outlet structures of numerous types, manholes, rip'rap and other devices used for bank protection, fences, and right-of-way and other permanent survey markers.

HIGHWAY SPECIFICATIONS

As indicated, specifications are written instructions. Careless or loose wording can result in the use of improper materials and poor workmanship and, at times, extra cost to owner or contractor. On the other hand, specifications that are too exacting result in higher costs. Specification writing is a difficult and exacting task which requires a knowledge of the law of contracts as well as of highway practices. Detailed treatment of the legal aspects of specifications is outside the scope of this book, and the reader is referred to the standard textbooks in specifications and contracts or to the legal literature itself.

Highway specifications often are divided into two parts: standard specifications and special provisions. The standard specifications apply to every project constructed by the agency and treat subjects that occur repeatedly in the agency's work. The special provisions cover subjects peculiar to the project in question and include additions or modifications to the standard specifications. The special provisions issued by some highway agencies include copies of all the documents required for securing competitive bids and for the contract. This is helpful to both contractor and construction engineer.

Specifications often are subdivided in another manner. One portion, called the general clauses, deals primarily with bidding procedures, award, execution, scope, and control of the work and other legal matters. The other gives details regarding materials, manner of executing the work, and how pay quantities are to be measured.

It is common practice to incorporate many items into the specifications by reference. For example, standard specifications rarely duplicate the details of specifications or test procedures for highway materials. Instead, a statement is included that the material shall conform to the requirements set out in the appropriate standard of the American Association of State Highway and Transportation Officials or the American Society for Testing Materials.

The highway agency is represented during construction by a resident engineer and the engineer's staff of engineers, surveyors and inspectors. Their responsibility includes inspection of the contractor's work, making design adjustments to fit actual conditions, measuring and computing the quantities for which the contractor will be paid, and usually surveying layout work. This topic is discussed further in Chapter 15.

PROBLEMS

6-1 Investigate and prepare a brief report on the computer hardware, software, and its uses (including computer graphics) in the local highway, street, or public transportation agency.

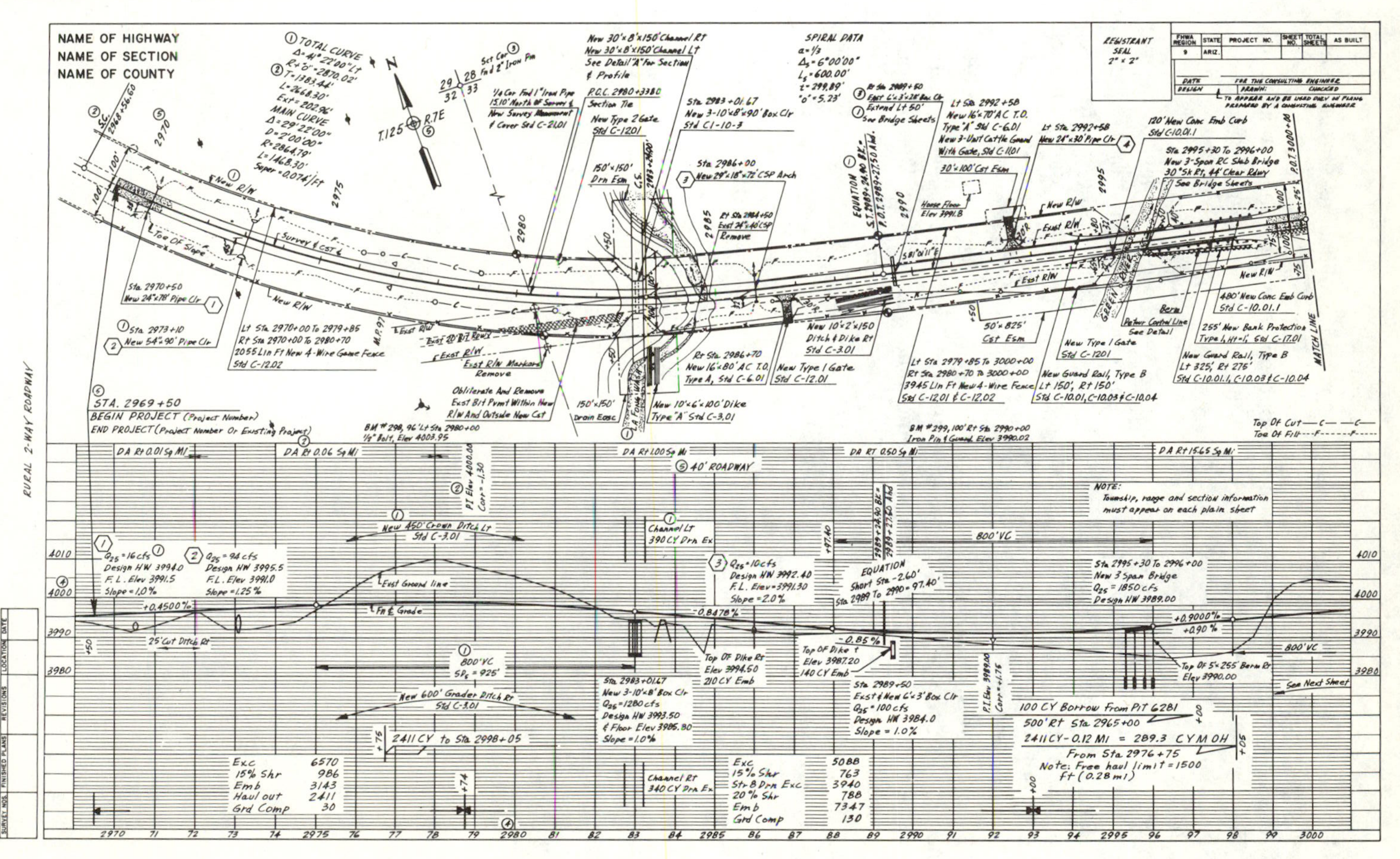

Fig. 6–4. A typical sheet of plans for a two-lane highway. (Courtesy Arizona Department of Transportation.) Please refer to insert at back of book for an enlargement of plan.

6-2 Find out what remote sensing equipment and applications are used by your state transportation agency.

6-3 Assume that you are representing the appropriate agency at a public hearing to explain why a bypass is to be constructed around a small town. Prepare a brief written statement outlining the advantages to motorists and to the town.

6-4 On a map of your college campus or of a nearby community (as designated by the instructor), determine the approximate location of a suitable route for either a bypass or a circumferential road system to remove through traffic from the congested contral area. Reconnaissance in the field or a study of aerial photographs is highly desirable if time and facilities permit. Defend your selection with a brief written statement.

6-5 Submit a freehand sketch showing a plan view of a badly located or otherwise inadequate bridge or culvert in the area near your college campus. On the same sketch, show the realignment that you propose. Defend your proposal with a brief written statement.

7 LEGAL PROBLEMS IN TRANSPORTATION

During the earlier decades of this century, most rural roads were constructed along section lines or other dedicated ways. The few additional rights of way were usually secured cheaply; often property owners were eager to supply land in return for the benefits to be gained from road improvement. In cities, rights of way for highways and street cars were available along dedicated streets. Conditions are far different today. Major highway improvements require that existing rights of way be widened or new ones acquired. Transit also demands a strip of land and space for stations. Entire residential or business properties may be swallowed up, or, with partial takings, the value of the remainder may be reduced. Again, access to abutting lands or nearby property may be impaired, and light, air, view, and quality of the environment affected. In addition, the feeling is widespread that property owners are entitled not only to just remuneration but to the largest settlements that they can collect. For these reasons, rights of way now are one of the major costs of most highway and major transit improvements. To illustrate, in 1977 the state highway departments alone spent more than $800 million or 9% of their capital outlay for rights of way. On many urban freeways, expenditures for property or for damages have equaled or exceeded those for construction, and on one, in Los Angeles, they averaged $50 million per mile.

Traditionally, public agencies have acquired or retained only lands needed for rights of way. However, recent practice, particularly by urban transit agencies, has been to acquire additional property either to develop income from it or to capture the increase in value for the public. Such ventures introduce many legal problems.

The Federal Highway Act of 1968 brought another set of problems in right-of-way acquisition and land use. This act required that, on projects involving federal aid, occupants of housing and businesses displaced by a right-of-way taking must be relocated or compensated before construction could begin. To comply with the requirements of this law, the states have enacted parallel legislation. Today, then, highway and public transportation agencies have become deeply involved in administering relocation programs.

Concern over the environment, as expressed by Congress in the National Environmental Policy Act of 1969 and by parallel state legislation (see Chapter 13) has delayed many highway and public transportation projects and has raised many difficult legal questions.

Lawsuits against highway and transportation agencies and their employees resulting from negligence in design, construction, operation, and maintenance which results in injury to users have brought many complications. Conflict of interest also has received increasing attention. Finally, administering legislation regarding equal rights, wages and hours, and employee safety both within the agency and in its contract operations bring other management and legal problems.

This chapter deals primarily with the legal aspects of the first four concerns mentioned above, namely: right of way and related problems, displacement and relocation, agency and employee liability, and conflict of interest. The legal and other aspects of impacts on the environment are treated in Chapter 13. A discussion of the many legal and administrative problems associated with equal rights, wages and hours, and employee safety both within the agency and in its contracting operation is outside the scope of this book.

THE LEGAL SETTING

The rights and powers of various levels of government to take land for highway or public transportation purposes, the procedures for acquiring property by purchase or by condemnation, and the settlement of claims by property owners and others against the government and its employees are prescribed by the constitutions and laws of individual states. Rulings on disputed cases, generally based on interpretations by appeal or supreme courts, establish the bounds of governmental privilege and obligation. Where a dispute over a law or regulation arises that is not covered by an existing court ruling, action to establish precedent is necessary. On occasion, a court will reverse an earlier decision, and this changes the basis for future settlements on the point in question. Sometimes rulings with seemingly contradictory findings appear in the same jurisdiction, which introduces added confusion. Since each state is sovereign in such matters, the laws and court rulings differ among them, making it hazardous to predict settlements in one jurisdiction on the basis of those in another.[1] Furthermore, the powers and rights delegated by the state legislature may differ among states and with those given to counties or cities, and this is a further complication.

Legal decisions are greatly influenced by precedents set by earlier court rulings. Findings of courts in other jurisdictions, particularly the appellate divisions, carry weight although they are in no way binding. At times opinions as old as 50 or more years may hold the governing authority. Because precedence is weighed so heavily, there is a certain timelessness about the law that those from the fastmoving technical world have difficulty in understanding. Engineers

[1]This idea is illustrated by the term "weight of authority," which appears frequently in legal literature. Weight of authority says, in effect, that more of the courts have decided one way than the other on a particular point.

often fail to recognize this and other differences between the legal and technical approach to problems and in so doing put themselves at a disadvantage.

It should be clear, therefore, that engineers must seek competent legal advice when dealing with legislative and judicial problems. Certainly the general statements presented here cannot be taken as authoritative.[2]

AUTHORITY AND RIGHTS IN TRANSPORTATION SITUATIONS[3]

As indicated earlier, governmental agencies have responsibility for highways and are assuming it for most public transportation. This in turn vests in the appropriate agency the authority necessary to accomplish that task. This includes the right to acquire rights of way and to make settlements with those affected by property takings or other actions of the agency.

The right of government to take private property for public purposes or to otherwise cause individuals to suffer for the public good is firmly established in American law. In some instances, these losses are compensated in full, in others only in part, and in others not at all.[4] In the United States, for those transportation agencies that are creatures of government, right-of-way law is cast within this framework of public versus private rights and privileges.

The individual state constitutions and statutes, as interpreted and restricted by the courts, set powers such as those to locate and construct facilities, to acquire rights of way, and to determine and pay damages, sometimes but not always including attorneys' fees. A discussion of the more general legal concepts follows.

Authority To Locate, Relocate, Plan, Construct, and Operate Transportation Facilities

Until recently, the courts have almost universally held that the state legislature, or by delegation the designated agency, has the vested power to locate or relocate facilities.[5] This includes authority to set standards of improvement including line, grade, right-of-way widths, and the need for access control. Furthermore, the courts generally have refused to review such actions, unless it can be shown that they are so fraudulent, arbitrary, discriminatory, or unreasonable as to deprive property owners of their constitutional guarantee of due process of law. A corollary to this delegation of power is that engineering or administrative

[2]NCHRP has sponsored a series of authoritative studies on highway laws which provide the basis of much of the discussion that follows. They have been published in the *NCHRP Research Results Digest* and *TRB Circular* series. Many of these also are available in three volumes under the title *Selected Studies in Highway Law*, hereafter referred to as *Selected Studies*.

[3]The FHWA publication *Highway Condemnation Law and Litigation in the United States*, 2 vol., is an excellent reference and cites many cases.

[4]Drafting individuals into military service at low pay and with hazard to life and limb was a vivid illustration of these principles.

[5]In some states the legislature stipulates general routes and their end points and leaves exact location and other details to the agencies. In others, the legislature retains considerably more control.

decisions on location, design, construction, or operation, and on the property needed for them usually cannot be challenged in the courts.[6]

Rights Related to Traffic Flow

The courts have almost universally ruled that property abutting a road or other transportation route has no right to a continuation or maintenance of traffic in front of it. Accordingly, no compensable damage usually occurs when an improvement or change diverts the main flow vertically, horizontally, or to another route. Neither is there a cause of action if traffic flow increases.

On first glance, this viewpoint may seem to create an unwarranted hardship on the property owner. However, it can be reasoned that the traffic flow was created by governmental action in building the present facility and not by any act of the property owner. As a practical matter, to permit recovery of damages by property owners when traffic is diverted from or to their properties would almost preclude major construction or traffic-control measures.

Public Versus Property Rights

The legal right of the public to convenient and safe travel over public roads and streets is firmly established in the United States. On the other hand, property also has certain legal rights, among them the rights to reasonable access and egress to existing highways, to a reasonable view of the property from the highway and of the highway from the property, and to a flow of light and air to the property. These rights of property (and not of a person as such) are protected by the "due process" clauses of the state and federal constitutions. They can be taken only for public purposes and then only under the police power or some other constitutional provision, or by payment of just compensation, with final determination in disputed cases made by the courts.

Many highway improvements, particularly freeway type facilities involving access control and grade separation, bring these public and property rights into direct conflict. In each state the courts must, over time, interpret the constitution and classify public and private rights into four general groups.[7]

1. Where public rights to use of the highway or transit facility are so paramount that no compensation is due property owners when they suffer damage because of an improvement or change.
2. Where private rights are infringed, but compensation is denied because the infringement results from proper exercise of the "police power," which is government's constitutional right to legislate to protect the public health, safety, and morals. Some courts have referred to the police power as "regulation" in contrast to "taking" under eminent domain. Traffic control by signs or signals, setting up one-way street patterns, or designating freeway lanes or ramps for high-occupancy vehicles are exam-

[6]As indicated, there are exceptions; several of these are reported in *NCHRP Research Results Digest No. 6*. Also, the courts are permitting challenges to administrative decisions on environmental issues. (See Chapter 13 for further discussion.)
[7]See *Selected Studies*, op. cit. Vol. 1, for added detail.

ples of proper exercises of the police power. Control of junkyard location and of billboards are others. (See Chapter 13.)

3. Where private rights are infringed because property is taken under "eminent domain," which is the power of government to take private property for public use, with or without the owner's consent. In this instance, the property owner is entitled to payment for the property taken. If only a portion of a property is taken, the owner may also be entitled to damages to the remainder because the use, access, light, air, view, or environment of this remainder is impaired.
4. Where no property is taken but private rights such as access, view, light, air, or the environment are damaged as a result of the improvement. Here the right to compensation varies among states and with individual situations.

PROPERTY ACQUISITION

Procedures for Acquiring Property

Transportation agencies, along with many other public bodies or quasipublic corporations such as railroads and public utilities, are vested with the power of eminent domain. If owners are not satisfied with the payment offered for their properties, they have recourse to the courts to secure just compensation. Agencies try to avoid using eminent domain when at all possible. Instead, property is obtained by direct negotiation with the owner. Condemnation is resorted to only when owners refuse to sell or when their demands are unreasonable.[8]

Procedures for acquiring property are far from uniform among the various agencies. Many of the methods are extremely cumbersome and time-consuming. Some agencies use different techniques under different circumstances. In any event, the evidence is strong that in many jurisdictions efficiency could be greatly improved by a drastic revision of present laws and practices such as wider use of the discovery procedure described in the next section.

Right-of-way valuation is far from an exact science. In some cases, values set by the appraisers for the claimant and the condemning agency will differ several times over. These divergences result because of honest differences of opinion, through differences between or misapplications of the methods used to determine payment or damages (see below), and, at times, because appraisers and lawyers submit inflated or unreasonably low valuations. Strong efforts are being made to develop professional approaches and disciplinary rules for appraisers.[9]

Among the complexities in determining the proper settlement for the rights or damages to a given property is the time at which the valuation is made. Often the announcement that a facility may or will be built on a given alignment will bring appreciable changes in value, either up or down. Generally, the date of valuation is that of the actual taking of the property. Even so there are many

[8]The California and Minnesota Highway Departments reported in the early 1970s that direct settlement was made with the owner in 97% and 58%, respectively, of their cases.

[9]See *NCHRP Report 126* for a detailed discussion of the reasons for divergence in valuations and the Code of Ethics and Standards of Professional Conduct of the American Institute of Real Estate Appraisers.

ramifications to this date setting and to how values change before and after that date. They are too complex to be discussed here.[10]

Zoning regulations or prospective changes in them also can have substantial effects on the value of property. Evidence regarding these matters generally is admissible in eminent domain proceedings and adds yet another complexity to an already difficult problem.[11]

Court proceedings involving property valuation are conducted under the adversary system where each side, through its attorney and expert witnesses, presents its views in the most favorable light possible. Cross examination of witnesses by the opposing attorney attempts to discount or discredit their testimony. Depending on circumstances and the procedures of a particular court system, value determination may be made by the judge alone or by a jury.

Appeals to higher courts usually are restricted to matters of law, such as a trial judge's ruling on admissibility of evidence or instructions to a jury.[12]

Right-of-way acquisition and damage settlement in urban areas are time-consuming operations. There are instances where 40 separate property settlements were made in each block. In such cases, right-of-way acquisition should begin several years before the time scheduled for construction.

With the exception of certain appraisal work, right-of-way acquisitions and settlements are more a legal than an engineering matter. Even so, engineers, although short of knowledge of the legal side of the right-of-way problem, should understand the procedures and rules under which property is obtained. Furthermore, since they must design and construct facilities at the lowest overall cost, they should know something of the legal factors that so often control property costs and damage settlements. The paragraphs that follow offer a brief introduction to these subjects.

Discovery Procedure

Discovery is a judicial process permitted in most states as a means for expediting settlements or speeding up trials. Under it, one party must furnish pertinent information to the other before the case comes to trial. In general, each side is required to produce facts asked for by the other at pretrial conferences. There is disagreement among court systems as to whether privileged information such as the results of special studies or statements by witnesses must be disclosed. The trend appears to be toward requiring more disclosures.[13]

Immediate Possession of Rights of Way

Where property must be obtained by condemnation, the procedure may be long drawn out. This is particularly true when the findings of the trial court are reviewed by the higher courts. There are many instances where completion of an

[10]See *NCHRP Report 114* and *NCHRP Research Results Digest No. 11* and *45* for added detail.
[11]See *NCHRP Research Results Digest No. 22* for further detail.
[12]For a discussion of courtroom strategies, see *NCHRP Research Results Digest 47* and 111, or *Selected Studies*, op. cit., Vol. 1.
[13]See *NCHRP Report 87* or *Selected Studies*, op. cit., Vol. 1. for more detail.

improvement has been delayed for several years because of tangles over rights of way. At times, property owners have forced exorbitant payments by the threat of litigation. To avoid these losses in time and money, many states have enacted legislation permitting transportation agencies to take almost immediate possession of needed land. Details of these arrangements vary from state to state. In general, they require that taking the land first be justified to the proper court. Next, money or bond equal to or greater than the appraised settlement is deposited with the court for the landowner. The agency is then free to begin construction while the final settlement is worked out.

For the Interstate System, Congress made available to the states the legal resources of the federal government to acquire early possession of rights of way, including access rights. As a further aid, a federal revolving fund permits acquisition of rights of way for 10 yr in advance of construction.

Payment for Acquired Property[14]

Fair payment for rights of way would, on the one hand, provide just compensation for the property rights that were taken and, on the other, represent a proper and fair use of public funds. However, it is often impossible to provide full compensation to property owners, since public officials are prevented by law from paying for certain losses that a property owner may suffer. For example, the courts generally have ruled that the owners of business property taken for public purposes cannot be compensated for such things as loss of goodwill, inability to locate an acceptable substitute location, loss of profit caused by moving or by interruptions during construction, or the cost of moving goods or other personal property.[15] This is an illustration of the basic principle that government is not required to make payment for all injuries that it imposes on persons or property.

Payment When Entire Property Is Taken

When property is taken in its entirety, just compensation is usually based on the *market value* of the property. AASHTO defines market value as "the highest price for which property can be sold in the open market by a willing seller to a willing purchaser, neither acting under compulsion and both exercising reasonable judgment, and both being fully aware of the highest and best use to which the property can be put."[16] This definition would permit land value to be set at the "highest and best use" which is defined as "the most productive use, reasonable but not speculative or conjectural, to which property may be put in the near future." Thus a vacant parcel of land located in an industrial area, but currently being farmed, could be priced as industrial land. On the other hand,

[14] *NCHRP Report 104* is an excellent source of detailed information on many aspects of this topic. For residential takings, see *NCHRP Report 107*.

[15] See the section on "displacement and relocation problems" for certain exceptions.

[16] This definition and others that follow are from *AASHTO Highway Definitions*.

farmland at some distance from town will be paid for as such, and not on the basis that it might, in the future, become business or commercial property.

Market value may be approached in several ways, including the following:

1. *Market-data approach.* Comparison of property being appraised with similar properties being sold or listed for sale.
2. *Replacement cost new less depreciation approach.* The market value of the land is determined by the study of market data; to this is added the estimated cost of replacing the improvements today less the proper depreciation from all causes. On improved property, this method supplements the market-data approach. An alternative within this method may be to determine the cost of the land and add to it the cost of moving the buildings and other improvements to another site.
3. *Income approach.* For types of property where income is important, the actual and fair earnings of the property are compared to earnings of comparable properties that have been sold, and from this an estimate of value is obtained. The courts have generally ruled that possible future use and profits, if only hypothetical or speculative, cannot be considered as evidence of value.
4. *Historical cost new less depreciation approach.* Actual cost is generally the best evidence of value for both land and improvements of new property. In older properties, historical cost becomes less and less important.

An appraisal to determine market value, based on one or more of the approaches outlined here, is the first step toward securing property for rights of way. Armed with this appraisal, a negotiator attempts to make a settlement with the property owner. If agreement cannot be reached, condemnation proceedings are instituted in the courts.[17]

Payment for Partial Takings

When only a portion of a property is to be acquired, payment is usually set as the difference in value of the property "before and after taking." Part of the settlement is for the property actually acquired and the remainder for "severance" damages. Under certain circumstances, "benefits" to the remaining property are offset against damages in reaching the final settlement.

Value of property taken may be determined by the approaches outlined in the preceding section. For farmland, a price per acre may be appropriate. For business property, where value per unit of area decreases from front to rear of the property, "depth factors" are sometimes used as a guide.

AASHTO defines severance damage as "loss in value of the remainder of a parcel resulting from a partial taking of real property." It occurs, for example, if a remaining piece of farmland is too small to be tilled economically or if a lot in a business or industrial area is shortened to a point where the enterprise operates less efficiently. Severance damages may also include losses resulting from

[17]For more detail on these approaches see *NCHRP Research Results Digest 41, 42, 47,* and *54,* or *Selected Studies,* op. cit., Vol. 1. Valuation procedures for special situations appear in the following *NCHRP Reports: No. 92,* special purpose properties; *No. 94,* trade fixtures; *No. 107,* residential; *No. 112,* junkyards; and *No. 142,* airspace; or in *Selected Studies,* op.cit., Vol. 2.

impairment of access, light, air, and view, or quality of the environment (see below).

Benefits to adjoining property and to surrounding areas often result from improvements in transportation. For legal purposes, these are grouped into two classes, "general" and "special." AASHTO defines a general benefit as "advantage accruing from a given highway improvement to a community as a whole, applying to all property similarly situated." An example would be a general rise in property values in an area made accessible by a new facility. Special benefit is defined as "advantage accruing from a given highway improvement to a specific property and not to others generally." It is described as anything that adds to convenience, accessibility, and use of property. Examples of special benefits are: opening a street giving additional frontage or making a corner lot; opening a new road or changing the grade of an existing road, making the property more accessible or providing ingress and egress where none was provided before; and improving the road fronting a property by grading, paving, or drainage. Note that in each of these examples a particular piece of property benefits, as contrasted to benefits accruing to the community in general.

There is evidence that transportation improvements usually increase land values of the area penetrated. One early study found that the median price per acre of remainders increased by 38%. In 26% of the cases, price per acre fell; in 74% it increased. In 34% of the cases, the value tripled.[18]

The extent to which state legislatures have permitted benefits to offset damages varies widely. The provisions in a majority of states are either that (a) special benefits but not general benefits can be offset only against severance damages to the remainder or (b) special but not general benefits can be offset against both the value of the taking and severance damages to the remainder.[19]

SETTLEMENTS WHEN NO PROPERTY IS TAKEN

An improvement or change in a transportation facility sometimes impairs the usefulness of lands abutting a proposed location or in the area penetrated. Injury from this cause is called "consequential damage," defined by AASHTO as "loss in value of a parcel, no portion of which is acquired, resulting from a highway improvement." A common cause of consequential damage is impairment of access, light, air, and view, or quality of the environment (see below). Another is delay in the condemnation process. Often suits for consequential damages are referred to as "inverse condemnation," because the property owner sues the governmental agency.[20]

Whether consequential damages must be paid depends primarily on the constitution of the state in question. Over half the states require compensation for damage to property caused by public improvement. Here, the courts would be

[18]See G. V. Broderick and F. I. Thiel, *Public Roads*, June 1964, and F. I. Thiel ibid., Apr. 1967; also Chapter 4.
[19]For a detailed analysis see *NCHRP Report 88*.
[20]See *Selected Studies*, op. cit., Vols. 2 and 3 and *NCHRP Research Results Digest 119*.

required to entertain suits for consequential damages. In the remaining states, the wording is that compensation is due only if private property is taken for public purposes; this indicates that only where there is a taking can damages be awarded. The general rule just stated does not always apply. There are even instances in which the courts have taken a viewpoint opposite to their state's constitution.

EASEMENTS

An *easement* is defined by AASHTO as "a right to use or control the property of another for designated purposes." Thus an easement does not call for any taking of property, as such. But it is common practice for agencies to acquire easements for drainage facilities, planting, to provide for sight distance, or for space needed for cut or fill slopes. Particularly in rural areas, easements usually cost far less than to acquire the property outright.

The *scenic easement*, as defined by AASHTO, provides "an easement for conservation and development of roadside views and natural features." Another definition is "a conveyance of those ownership rights in property which will permit a public body to effectively preserve (protect and restore) the scenic beauty of the property when viewed from public lands, reserving to the grantee all other beneficial interests." Following the Highway Beautification Act of 1965, legislation was enacted in most of the states to permit the use of the power of eminent domain to acquire scenic easements where property owners refused to grant them. Expenditures to cover the costs were also authorized.[21]

RIGHTS OF ACCESS, LIGHT, AIR, VIEW, AND QUALITY OF ENVIRONMENT[22]

Control of Access

In the past, business, commercial, and other developments along conventional highways and streets often have practically destroyed their usefulness as traffic arteries. As an early illustration, Garvey Road, which carried U. S. Highways 60 and 70 east from Los Angeles for 30 mi, was built in the middle 1930s through open fields, away from the then developed towns. Access from adjoining property was not controlled. Within 1 yr, congestion caused by roadside development had decreased the capacity by about 25%. Within 4 yr, ribbon development was almost continuous, and speed zoning was in effect over much of the road's length. Today, Garvey Road is a city street and through traffic moves on a parallel freeway. Similar examples may be found throughout the

[21]See *NCHRP Report 56* for a detailed presentation and examples.

[22]Because each state is sovereign and its constitution and laws are different, and because the facts among situations vary widely, court findings in this area lack consistency. It follows that a brief presentation can, at best, summarize the areas of concern and highlight a few instances. E. B. Hatch, in *HRB Research Circular No. 75* offers a summary and cites many specific cases.

country. It is almost universally true that urban and suburban highways built without access control soon become ineffective as traffic carriers.

Control of access is defined by AASHTO as "that condition where the right of owners or occupants of abutting land or other persons to access, light, air, or view in connection with a highway is fully or partially controlled by public authority." Under this principle, agencies are free to select conditions under which access is permitted. Entrances, exits, and crossings for existing roads and outlets for abutting property may or may not be provided, at the discretion of the agency. With this authority, it is possible to prevent ribbon development with its attendant evils. Furthermore, access, egress, and crossing facilities can be located and designed primarily to meet traffic and other engineering considerations rather than to satisfy local demands. Stated differently, it can be said that the controlled access device insulates the road from its environment and, at the same time, liberates the environment from the road. Both profit. Transportation service becomes optimum and the community becomes a better, cleaner, and safer place in which to live.

Except for toll facilities, control of access is of recent origin; it developed after World War II. The stipulation that access control must be acquired for the Interstate System brought the concept into widespread use. All states now permit highway agencies to acquire access rights from abutting property owners.[23]

Payment for Access Rights

The rights of access to property are as fundamental as the rights of ownership. It follows that if an agency takes away access rights to an existing road, the abutting property owner is entitled to damages. Where a portion of the property is taken, impairment of access is a part of severance damages; where access rights but no property are taken, impairment of access alone can be compensable. There are exceptions, as where access is restricted to reduce accidents, for this is an exercise of the police power and as such, noncompensable.

Abutting property owners are not necessarily entitled to access at all points on their property, and as long as a "suitable" means of access remains they have suffered no legal injury. In some cases it has been ruled that access equal to that before the taking is a proper measure of what is suitable. Of course damages are granted if all access is blocked.

Whether the substitution of access to frontage roads for access to through highways is compensable has been answered differently by different courts. Although closely related to the concept of circuity of route (see below) the courts have sometimes considered it separately.

Where access to nonabutting property is impaired, provisions in the constitution of the particular state as to payment of consequential damages, as discussed earlier, apply. Even where consequential damages are permitted, it must be demonstrated that the damage is special to the subject property rather than general. Furthermore, in some instances, the courts may rule that the damage

[23]Recent references include *NCHRP Report 93* and *121*.

results from a proper exercise of the police power or from circuity of route (see below) and is noncompensable.

Payment for Loss of Light, Air, View, and Quality of the Environment

Numerous claims for loss of light, air, and view have been occasioned by grade-separation structures and other improvements where the roadway grade was altered. Where a portion of a property is taken, awards for impairment of light, air, or view are included with other severance damages. The question of what constitutes impairment is still largely undefined.

When no property is taken, suits for damages for impairment of light, air, and view are not admissible in those states where the constitution or courts bar the payment of consequential damages. In certain other states, the courts have ruled that public rights to the use of the highway transcend private property rights so that no compensation need be paid; others have taken the opposite view. To illustrate, certain courts have held that compensation need not be paid abutting owners when the change involved a legitimate street use. In particular cases, the construction of bridges, viaducts, and their approaches, and other changes in street or highway grade have all been classed as proper street uses. On the other hand, in Iowa, the Supreme Court ruled for the property owners in a case where a viaduct constructed in the street as part of a railroad grade separation left a narrow and circuitous means of access and also impaired light, air, and view. The court stated, in effect, that abutting owners were entitled to compensation for the destruction, substantial impairment, or interference with their rights of access, light, air, or view by any work or structure built for the improvement of the highway.

Lawsuits against highway agencies have been brought where a highway diminished the quality of the environment. In a New York case, severance damages to a remaining property were awarded where a 20 ft fill blocked a "beautiful view of forest and mountain." The appeal court ruled that privacy and seclusion, view, and traffic noise, lights, and odors could properly be introduced as evidence of damage. In other states such awards have been disallowed.[24]

CIRCUITY OF TRAVEL—CUL-DE-SACS

Freeways involve separated roadways, frontage roads, and traffic interchanges that almost always lengthen the route to or from some parcels of abutting or nearby property. One-way streets, U-turn prohibitions, and many other traffic-control measures have a similar effect. Many courts have held that this "circuity of travel" may be imposed under the police power and therefore is noncompensable unless public officials have acted in an arbitrary manner.

Another line of reasoning is that circuity of travel does not substantially impair ingress and egress and that any interference that does result is one of the prices

[24]See *HRB Highway Research Circular 71* and *87*. The right to compensation for impairments to the quality of the environment is discussed in Chapter 13.

of improvement shared in common with the general public. It follows that the courts will seldom award damages unless access to the particular property in question is substantially impaired as compared with other property in the vicinity.

Limited-access highways or at-grade transit facilities often cut across and block off existing roads and streets. Property that earlier had access to cross streets in either direction is placed on a dead-end street, commonly called a *cul-de-sac*. In numerous instances, owners of property so affected have sued for damages. In a majority of cases, the courts have denied such claims when adequate means of access were left. Often the reasoning has been that the principal factor was mere "circuity of route" which commonly is not compensable, as explained in the preceding paragraph. In other cases, the ruling was based on the argument that the injury, if any, differed only in degree but not in kind from that suffered by the public in general. Another court held the damages to be *damnum absque injuria*, which means injury without violation of a legal right. In some states, however, the courts have granted damages when property was placed on a *cul-de-sac*, so that generalization is not possible.

ACQUISITION OF LAND OTHER THAN FOR RIGHT-OF-WAY—VALUE CAPTURE

The taking of property under eminent domain is not limited to rights of way of a definite width. For example, in most states, an agency may acquire any real property which it considers necessary to fulfill its purpose. This may include quarries; borrow pits for fill materials, sand, or gravel; sites for offices, shops, and storage yards; space needed to accommodate cut or fill slopes, drainage channels, and access roads; park sites; and areas needed to protect scenic views.

In some states the statutes also permit taking small fragments that have little value. This authority may extend further, so that a whole property may be taken when its cost is less than that of the partial taking plus severance damages to the remainder or when the remainder becomes land-locked because of a partial taking. In these cases, legal means are also established for the sale of the excess to private owners, usually by auction.

The taking of continuous strips of marginal land or that adjacent or near to interchanges or terminals is permitted by the constitutions of some of the states. Acquisition may be for one or more of four purposes: (a) to provide space for future transportation facilities, if needed; (b) to protect the facility from objectionable or unsightly developments; (c) to provide for joint development (see Chapter 6); and (d) to capture the increase in value of the land for the public. In certain other states, the legislatures have passed statutes that permit the acquisition of lands for these purposes, but these, unlike constitutional provisions, must be tested in the courts.

In general, the courts have upheld provisions serving the first two purposes, but are divided on the others. The "recoupment" or "value capture" approach, which provides for public rather than private gain from increases in land value has been little used in conjunction with free highways in the United States although it has been followed in some foreign countries. Toll-road authorities have followed it by acquiring land on which to place facilities such as service stations and restaurants.[25]

Value recapture can be an element where added land is acquired at transit terminals or stations. Through joint development (see Chapter 6) air rights can be sold or leased or space can be rented to a variety of enterprises.

DISPLACEMENT AND RELOCATION PROBLEMS[26]

As indicated earlier, payments for land and improvements taken and for severance or consequential damages are made to the "property itself" and not to owners or tenants. With this approach, many costs and inconveniences to individuals and businesses associated with property takings went uncompensated, although this was clearly possible under the federal Constitution. For example, the U.S. Supreme Court has stated that " . . . while the legislature was powerless to diminish the constitutional measure of just compensation (as applied to property under the Fifth Amendment) we are aware of no rule which stands in the way of an extension of it, within the limits of equity and justice, so as to include rights otherwise excluded."[27] The Supreme Court pointed out that, under the Constitution, the basis for such payments to individuals is the power of Congress to determine whether claims upon the public treasury are founded on moral obligation and the principles of right and justice.[28]

Congress, then, was on firm ground when in the 1962 Federal Highway Act it provided that before a state's federal-aid plan would be approved the highway department must give satisfactory assurance that relocation advisory assistance would be provided for displaced families. The act then redefined the term "construction" to permit the use of federal-aid funds to pay moving expenses for families and businesses to the extent authorized by state law, and within prescribed maximums. However, Congress recognized that some state constitutions prohibited such uses of public funds and made the provision to pay relocation costs permissive.[29]

In the 1968 act, Congress made relocation assistance mandatory on July 1, 1970, with all federal aid cancelled for those states that failed to comply. That act stipulated that:

[25]See *NCHRP Research Results Digest No. 42* or *Selected Studies*, op. cit., Vol. 2 for a detailed discussion of and references on the legal aspects of excess and substitute condemnations.

[26]For more detail, see, among other sources, *NCHRP Research Result Digest 3, 32,* and *40*, and *HRB Research Circulars 69, 95,* and *114*. *TRB Special Report 192* is a 1980 assessment of relocation assistance and offers recommendations for its improvement.

[27]*Mitchell* v. *United States*, U.S. 341 (1925).

[28]*Joslin Mfg. Co.* v. *Providence* 262 U.S. 668 (1923) and *Mitchell* v. *United States*, op. cit.

[29]The courts in several states had ruled earlier that reimbursement to privately owned public utilities for moving facilities from rights of way would violate the constitutional prohibition against using public funds to compensate private parties. See *NCHRP Research Results Digest No. 3* for court cases.

The Secretary shall not approve any project which will cause the displacement of any person, business, or farm operation unless he receives satisfactory assurances [that] . . .

1. Fair and reasonable relocation and other payments shall be afforded to displaced persons . . .
2. Relocation assistance programs offering [certain stipulated] services . . . shall be afforded to displaced persons.
3. Within a reasonable period of time prior to displacement there will be available, to the extent that can reasonably be accomplished, in areas not generally less desirable in regard to public utilities and public and commercial facilities and at rents and prices within the financial means of the families and individuals displaced, decent, safe, and sanitary dwellings, as defined by the Secretary, equal in number of and available to such displaced families and individuals and reasonably accessible to their places of employment.

In 1970, Congress passed a Uniform Relocation and Assistance and Land Acquisition Policies Act which replaced the relocation portion of highway legislation. Its aim was to bring uniform policies among all federal and federally assisted programs. This act carried forward the concepts of the 1962 and 1968 acts enumerated above. All the states have now enacted satisfactory legislation or, by various strategems, complied.

A brief abstract of some of the features of federal relocation legislation includes:

A. *Payment to be made to displaced persons.*
   1. Persons displaced will be reimbursed for actual reasonable expenses in moving themselves, their families, businesses, farm operations, or other personal properties.
   2. In addition, they will be reimbursed for the actual direct losses of tangible personal property as a result of moving or discontinuing a business or farm operation.
   3. In addition, they will receive actual reasonable expenses in searching for a replacement business or farm. In lieu of payment based on an actual accounting, a person may choose a moving allowance of $300 and a relocation allowance of $200. Businesses or farms may choose to receive a fixed sum equal to annual net earnings of not less than $2500 nor more than $10,000.

B. *Provisions for replacement housing for homeowners.*
   To homeowners of over 180 days standing, payments not to exceed $15,000 will be made based on the following:
   1. The amount, which when added to the acquisition cost of the present dwelling, will provide a comparable replacement dwelling which is decent, safe, and sanitary, adequate to accommodate such displaced person, reasonably accessible to public services and places of employment and available on the private market. (Detailed specifications are provided as to what constitutes decent, safe, sanitary, and adequate housing.)
   2. The amount required for increased interest costs.
   3. Reasonable expense for title searches, recording fees, and closing costs.

   Homeowners with less than 180 days ownership receive lesser sums.

C. *Replacement housing for tenants with at least 90 days occupancy.*
Payments may be either:
  1. The amount necessary to enable such displaced person to lease or rent for a period not to exceed 4 yr comparable housing (as defined in B above), but not to exceed $4000.
  2. The amount necessary to enable a person to make a down payment on the purchase of comparable housing. This amount shall not exceed $4000; if the amount exceeds $2000, the person must equally match the remainder.

D. *Last resort housing by federal agency.*
If the project cannot proceed to actual construction because comparable sale or rental housing is not available, the agency is authorized to use project funds to provide it. Authority is given to construct new housing or to reconstruct, rehabilitate, or move existing housing. States may contract with local governmental housing authorities to carry out this plan.

A program as complex as this, and which often involves dealing with the disadvantaged, is bound to generate many complaints and charges that the agencies involved are arbitrary and unfair. To deal with this problem, the law provides that the head of the state agency must review complaints. In turn, the individual states have set up appeal procedures. Some are administrative, others provide for review boards. Whether further appeal through the courts is possible is not clear, since the act is silent on that matter. *HRB Research Results Digest 40* indicates that a state law prohibiting appeals probably would stand. Without it, appeals might be possible.

Agencies must follow the 1970 replacement act or they are effectively denied federal funds. The act stipulates that no person shall be required to move until the federal agency head is satisfied that replacement housing is available. This would stop construction. Thus, like it or not, all transportation agencies using federal grants have been thrust into the world of real estate since they must both advise and provide replacements for housing, businesses, and farms. And the problem is a large one. In 1978, federally funded relocation assistance cost $37 million and involved 4700 dwellings and 13,000 persons.

Relocation has many impacts and brings many social problems that are far too complex to discuss here.[30]

SHARING RIGHTS OF WAYS WITH UTILITIES

Traditionally, overhead utilities such as electric-power and telephone lines have been placed along the edges of conventional highways and streets inside the right of way. Also mains for water, sewer, and gas and their service connections occupy underground space. In some instances there are also pipelines or other accommodations for telephone or power cables, petroleum products, steam, drainage, and irrigation. Even where these utilities do not run along the highway or street, they must cross at intervals, so that joint use of land is involved. In the

[30]For a detailed study of the issues in and effects of relocation, and an extensive bibliography, see J. E. Burkhardt, *TRB Record 716*.

United States, it has been generally accepted that this joint use of land is in the public interest and is to be encouraged. On the other hand, if improvement to the highway or street is to be undertaken later, some or all the utilities must be relocated or modified. This can be very costly.

The question as to who pays for utility relocation is particularly troublesome and controversial. In most instances, the utility is on the right of way under an easement granted by the highway authority, and its terms may require that the utility company pay the costs of moving or relocating its facilities. Whether or not the utility companies are liable for these costs, they are reluctant to pay them and often resort to legal or political means to gain reimbursement. In the case of privately owned utilities in particular, reimbursement raises many issues, one of them being the fact that payment of public funds to private parties may be unconstitutional.

Congress has stipulated that federal-aid funds can be used to pay relocation costs except when the payment to the utility violates the law of the state or a legal contract between the utility and the state. Furthermore, utilities were so defined as to include those owned publicly, privately, or cooperatively. All but 11 state legislatures have passed laws granting reimbursement; in some instances this has been limited to all or some portion of the federal-aid systems.[31]

To reduce utility relocation problems in the future, some highway officials propose restrictions on placing them in the rights of way of new major facilities. Means for dealing with this problem are outlined in *A Policy on Accommodation of Utilities on Highway Rights of Way,* AASHTO, and *NCHRP Synthesis 34*.

RESERVING RIGHTS OF WAY FOR FUTURE CONSTRUCTION

Almost all agencies select locations and prepare tentative plans several years before funds become available to program right-of-way acquisition and construction. During this period, buildings and other improvements often are constructed on the property. The result can be economic waste as well as high right-of-way costs. Early purchase of rights of way is often the ideal solution to this problem, but funds for that purpose are usually lacking. In most of the states there is statutory authority to acquire lands early. Some laws are intended to head off land improvement; a few are to alleviate hardships to landowners; but most serve both purposes. Some states have special budget items or rotating funds for such programs. And, as noted earlier, federal-aid funds can be used for up to 10 yr before construction.

Another basic approach to the problem is to acquire *development rights.* In this instance, land is not purchased; rather, owners are paid to observe restrictions as to what developments or structures they can put on the property. In this case, as with outright purchase, some funds must be advanced.

Yet another method or group of methods for reserving rights of way stems from the police power. In 1926, the U.S. Supreme Court approved the concept

[31]For further detail, see *HRB Special Report 91* and *TRB Research Results Digest 68* and *116*.

that a community could act *before* private land use injured the public interest. Today, this concept has been expanded to cover land uses that are incompatible, or that will create a nuisance or adversely affect the public health, safety, and welfare. Three of these uses of the police power are the *official map procedure, zoning,* and *subdivision controls.*

Under the official map procedure, the agency determines the location and files an official map with an appropriate official. Thereafter no structure or other improvement is permitted on the property. Enforcement is by several means such as (a) denial of compensation for the improvements when the land is acquired; (b) requiring that unauthorized structures be removed within a stated period; or (c) declaring that making improvements is a misdemeanor. In some states a time limit such as 1 or 2 yr makes the procedure suitable only for emergency situations.

Zoning vests in local government the power to regulate land use in terms of permitted developments in certain areas. It also can place limitations on such features as building height, land coverage, and structure setbacks. Subdivision controls, which require planning commission or other approval before land development or redevelopment or urban renewal are permitted, are a parallel to zoning. By these measures, local officials can, if they so desire, protect rights of way needed in the future. It is to be remembered that, as with all recourse to the police power, the courts will not tolerate its unreasonable or capricious use. Adequate plans and some degree of assurance that the plans will be carried out are essential. Neither can the police power be employed to reduce costs where a taking under eminent domain is appropriate.[32]

LAND-USE CONTROL AT FREEWAY INTERCHANGES

As pointed out earlier, access control protects freeways from the encroachments of roadside activities. However, in many instances, unordered development has occurred on cross-roads and service roads adjacent to freeway interchanges. At times, sufficient traffic has been generated to impair operation of the interchange and to congest the on and off ramps of the freeway proper. It follows that the highway agency has a stake in seeing that local communities control these developments by realistic zoning. In many states, local communities are now required to institute such controls. For example, in California, the local governing body must impose zoning controls on all property within a 1-mi radius of any interchange on the rural Interstate System.

LIABILITY ACTIONS AGAINST TRANSPORTATION AGENCIES AND THEIR PERSONNEL

A "tort" has been defined as a negligent or illegal act or neglect of duty (except breach of contract) by one party which results in injury to other persons or dam-

[32]*NCHRP Report 31* is an excellent source of detailed information.

age to their property. It is under tort law that claims between private parties over, for example, automobile accidents, are settled. It would also include product liability suits against automobile manufacturers and others when a defect results in a loss or damage and third-party claims for injuries or property damage at construction or maintenance sites. Until the coming of workman's compensation insurance, suits against employers for injury in the work place came under tort law.[33]

Under English common law as it was followed in the United States, agencies and their employees were in the past protected from tort actions under the doctrine of "sovereign immunity." This is the concept that the sovereign can do no wrong; he can be sued only with his permission. Erosion of this doctrine began in the 1930s; by 1977, legislation at the federal level and in all but 14 states had created legal mechanisms which permitted suits for "defective" highways. Four other states had established boards with which claims could be filed. In other instances, the courts have declared "sovereign immunity" unconstitutional.

Grounds for claims differ; in some jurisdictions the existence of a defect is sufficient to permit recovery; in others it must be proved that there was knowledge of the defect and that it had not been corrected or that appropriate warnings had not been posted. Claims are sometimes barred against "discretionary" acts such as design decisions but permitted against "ministerial" ones involving, for example, traffic control or maintenance.

The courts offer governmental agencies and their officials certain protections from tort claims that are not available to private parties. First of all, because of the constitutional separation of powers, they are reluctant to intrude into administrative decision making. Secondly, they recognize that public officials are often required by law to make judgments or choices that leave opportunity for injury and should not be hindered from doing so. For example, given limited resources, officials must decide which among numerous potentially dangerous situations should be corrected. Finally, they recognize that the threat of unlimited liability will deter able persons from accepting employment with governmental agencies. Even so, it must be recognized that decisions are made in the courts and by judges and juries who may not fully understand or accept budgetary or technical arguments. Furthermore, there is an increasing tendency to attach liability strictly to the agency and its representatives and to ignore the contributory negligence of the injured party.

Many of the tort lawsuits brought against both private parties and governmental agencies and their employees are under "contingency" contracts between the plaintiffs and their attorneys. Under them, lawyers carry all of almost all of the costs of the legal action; they share in the award, if any, at a preagreed rate. This share often is about one-third if the settlement is out-of-court and possibly 40% if the case is tried.

In many countries contingency contracts for legal services are not permitted and their pros and cons are vigorously argued in the United States. A "pro" ar-

[33]Workman's compensation and liability laws in general are outside the scope of this book. The reader is referred to legal textbooks for added detail.

gument is that only with them is it possible for persons of limited means who suffer injury to press their claims. A "con" response is that they create an "ambulance chasing" climate where attorneys seeks out clients and make it advantageous for them to seek exorbitant settlements. Recent legislative actions in some states such as those creating "no-fault" automobile insurance and limits on the size of attorneys' fees are responses to some of these excesses.

Several examples of situations creating agency and possibly personal liability of public officials and agency employees or consultants are cited elsewhere in this book. The point of this brief discussion is to bring out another situation in which professionals may be vitally concerned in legal matters.[34]

CONFLICT OF INTEREST

Concerns that public officials or employees of public agencies might use their positions for personal gain or political advantage have always been high in the United States. The earliest legislative efforts were directed at preventing bribes, extortion, or embezzlement. Later came actions to control political favoritism and the spoils system in the award of contracts and jobs; unfortunately such abuses continue in a few agencies.[35]

More recent legislation is aimed at controlling acceptance of other favors, such as loans, discounts, or services, taking outside employment with those dealing with the agency, or holding interests in property that might be purchased or leased. Today there are even more stringent requirements which attempt to force officials to disclose or divest themselves of financial interests whose value might be affected by their actions as public servants. It can be seen, then, that those involved in administering transportation agencies are expected to avoid even the suggestion of many forms of conflicts of interest.[36]

PROBLEMS

7–1. Define, in your own words: weight of authority, police power, eminent domain, market value, severance damage, special benefit, general benefit, easement, excess condemnation, tort, conflict of interest.

7–2. In general terms, for what losses is a property owner compensated in the following situations? What losses go uncompensated? What claimed losses might be compensable in some states but not in others? Explain each answer briefly.

a. A new highway diverts 75% of the traffic from the road in front of a service station so that business declines very substantially.

b. The property occupied by a large department store is taken for high-

[34]For added detail see, among other sources, *NCHRP Research Results Digest 79, 80, 83,* and *95* or *Selected Studies, Vol. 3;* S. I. Pivnik, *Transportation Engineering Journal of ASCE,* Jul. 1979; D. C. Oliver, *Traffic Engineering,* May 1977; and *Civil Engineering,* May 1979.
[35]See, for example, *Engineering News-Record,* Jan. 25, 1979.
[36]See *NCHRP Research Results Digest 109* for details.

way purposes. The store is moved to a nearby site, but business is interrupted for a month during the move and a large number of customers begin trading with other stores.

c. An elevated freeway passes through an area at the midpoint of a long block. The street is dead-ended at the freeway, creating a *cul-de-sac*. A large tower apartment development is affected by this situation as follows. The land occupied by one of two tower apartments plus most of the garden for the entire development is taken. Automobile and pedestrian access from the street and the parking area to the remaining apartments is not impaired. The front windows of the remaining apartments, which formerly faced on the garden, now look out on the freeway.

d. Same situation as *c*, except that the freeway skirts the development and no land nor buildings are taken.

e. A storage yard for raw materials at the rear of a manufacturing plant is taken. The factory building itself is not disturbed.

7–3. Prepare answers to the following questions:

a. What is control of access?

b. In general, must property owners be compensated when access rights to an existing road or street are taken? Are there exceptions to the general rule?

c. What have the courts ruled as to compensation for access rights to new controlled-access highways constructed on entirely new locations?

7–4. In general, can property owners collect damages when a highway improvement lengthens the route of access to or egress from their properties? Explain briefly.

7–5. Assume that you are a representative of the local highway agency. You are to advise affected residents regarding the compensation they will receive because a right-of-way taking for a highway forces them to move. Prepare your statement for the following people:

a. A homeowner of 5 yr standing.

b. A tenant who has lived in a house for 2 yr.

7–6. Through an interview with an official of the local road, street, or transit agency, assess the seriousness of the problem of tort liability lawsuits to the agency and its employees. Also find out what provisions, if any, are made to protect employees from such suits.

8 DRIVER, VEHICLE, TRAFFIC, AND ROAD CHARACTERISTICS

Motor vehicles travel the highway under the control of individual operators. If highways are to be designed and operated on a rational basis, the abilities and limitations of driver, vehicle, and road individually and in combination must be appraised and taken into account. Additionally, the ability of various highway types to accommodate traffic and do so within reasonable limits for congestion and safety must be estimated so that the road is properly sized and detailed.

This chapter and the two that follow form a unit. The first (Chapter 8) is concerned with the fundamental mechanical and behavioral characteristics of driver, vehicle, traffic, and road as they affect the demand for and efficiency and safety of highway travel; the second (Chapter 9) deals with the actual design of highways, streets, and their appurtenances in terms of these characteristics; and the last (Chapter 10) discusses their operation, and the regulation and control of vehicles traveling over them. The topic of highway accidents is treated separately in Chapter 12.

Many of the covered topics have been approached theoretically under the general classification of "traffic flow theory," which is defined as "the description of traffic behavior by the application of the laws of physics and mathematics."[1] Even so, present day practice draws heavily on observed behavior.

In this and the following chapter, repeated reference will be made to "AASHTO policies." These represent standards agreed to by the state highway or transportation departments and the Federal Highway Administration and may be considered authoritative. They have been developed over many years. Most of them are now being consolidated in a single publication titled *A Policy on Geometric Design of Highways and Streets,* hereafter referred to as *A Policy on Geometric Design*. As of mid 1981, this document was in the review-draft stage and had not been formally adopted.

[1]*TRB Special Report 165*, a monograph by D. L. Gerlough and M. J. Huber, brings together in one volume the research till 1975. More recent publications include *TRB Record 456, 509, 533, 567,* and *596*.

DRIVER CHARACTERISTICS

In any discussion of human beings as vehicle operators, it must be recognized at the outset that there is no such thing as an average driver or an average driving condition. First, drivers in a single age group have far different abilities to see, process information, judge, and react. These abilities may change under such effects as fatigue, frustration, and monotony. Then the age spread of drivers ranges from 16 or less to 80 or more, and abilities change in differing degrees as age increases; at the same time older drivers may compensate for some deficiencies by experience and self-pacing, and by being more cautious. Such differences as those between daylight, dusk, and dark and between good and foul weather bring further complications. To some degree, driving characteristics are different between the sexes. There is also substantial evidence of a correlation between social adjustment and driver behavior. Thus highway designers must weigh existing knowledge of all these diverse factors if they are to make sound decisions where driver ability and behavior are factors. Then their objective will be, insofar as possible, to meet the expectations of this diverse group.

Certain driver characteristics have particular influence on susceptibility to accidents; they are discussed further in Chapter 12. Among these are general behavioral patterns, age, and the effects of alcohol and drugs.

The Human Mechanism

After a person's eyes detect and recognize a given situation, a period of time elapses before muscular reaction occurs. This period, called *decision and response initiation time* is appreciable, and differs between persons. It also varies for the same individual, being increased by fatigue, drinking, and other causes. The average time for a simple eye-to-finger reaction, which requires depressing the finger after a light flashes, is about ⅜ s. Eye-to-foot reaction called for in depressing the brake pedal of a motor vehicle requires a longer period, ⅔ s for the median driver but 1.0 s for the 85th percentile. The subjects of these tests were expecting the signal to which they were to respond; often the motor-vehicle operator is not. Thus a preceding *detection and recognition time* to cover this lack of expectation may be needed. A minimum value falls between 0.2 and 0.3 s; under average road conditions the interval is in the order of 1.5 s. As will be shown below, the AASHTO brake reaction time for stopping has been set at 2.5 s to recognize these factors.[2]

Often drivers face situations much more complex than those requiring a simple response such as steering adjustments or applying the brakes. Then, external stimuli, operator sensing, perceiving, and judging, and vehicle response all are involved. This process is diagramed in Fig. 8–1. The right-angle intersection with vehicles approaching from all directions is a common example. It involves directing vision successively to each leg of the intersection, since the eye can

[2]See *A Policy on Geometric Design* for added detail.

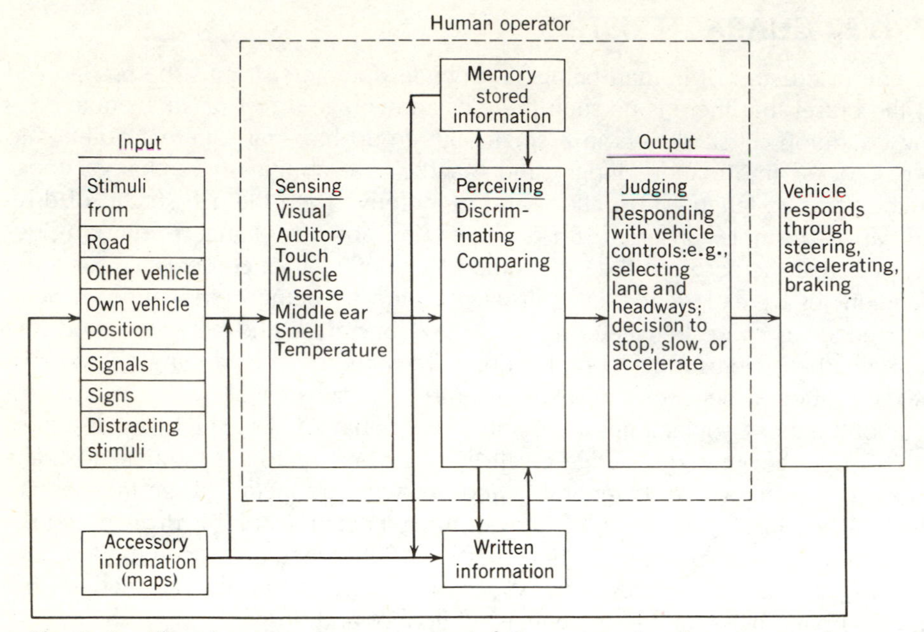

Fig. 8–1. Flowchart showing external stimuli, operator involvements, and vehicle response in driving decisions. (After T.W. Forbes.)

see only a 20° arc clearly. The median, alert driver requires about 0.5 s to observe and process each bit of information; this interval is about 0.7 s at the 85th percentile. After all bits are observed and processed, 0.7 and 1.0 s, respectively, are required to decide what to do and how to react. If the situation is unexpected, the median driver takes 30% and the eighty-fifth percentile driver 50% longer to carry out the process. Thus, in complex situations, several seconds elapse before any of the controls of the vehicles are changed. Recognition that complex decisions are so time-consuming leads to an axiom in highway design; it is that drivers should be confronted with only one decision at a time, with that decision simple rather than complex. As another instance of judgment time, the decision of a driver to pass on a two-lane highway requires at least a second and may take several seconds.

To make the complex driving task easier and less hazardous, highway designers use the concept of positive guidance. Upcoming situations are communicated to the driver through a variety of information carriers which include alignment, sight distance, cross-sectional features, pavement markings, and a variety of signs and signals.[3]

A substantial amount of research on humans as mechanisms for driving has been carried out. For example, devices that record head and eye positions and movements have been widely used in both road and simulation testing situa-

[3]See, for example, *A User's Guide to Positive Guidance,* FHWA, June 1977, and J. E. Leisch, *TRB Record 631*. These strategies are discussed further in Chapters 9, 10, and 12.

tions.[4] One among the many findings was that drivers traveling along a road employ a series of visual fixations; when the test driver's vision was restricted to a small aperture, 96% of the fixations included the road edges or centerline markers. Also, it was found that drivers are not constantly alert; they have a continuous series of "blind" intervals of 1 to 2.5 s or more between fixations.[5] Another finding was that many drivers cannot accurately appraise the closing rate of the opposing vehicle or the distance available for passing, particularly at high speeds. For instance, at 50 mph the estimates of 78% of those tested were so much in error that passing would have been dangerous.[6] Tests of drivers deprived of sleep for 24 hr showed, among other differences, that the point alongside the road on which the eye was focused was reached 3 s earlier than when the driver was rested.[7] This argues convincingly that the instinctive attempts of drivers to compensate for fatigue may lead to unsafe driving practices. There are, of course, many other characteristics of the human mechanism that affect driver performance, but those mentioned here give some notion of its complexity.[8]

Other Driver-Related Characteristics

Self-interest, particularly fear of possible accidents, controls many driver decisions. Operating speeds are ordinarily set instinctively at the highest level at which drivers feel that their information-gathering, data-processing, and reacting abilities match the situation, so that they feel secure. The decision to pass another vehicle or to conduct other maneuvers, or to observe or ignore a warning sign, is similarly affected. Since different drivers have varying impressions of their abilities and exercise different degrees of caution, the speeds at which they drive will not be the same. Operators also tend to follow the path that involves the least change of speed and least discomfort. The "cutting of corners" by motorists who approach intersections too fast and do not slow down is an illustration; setting the brakes when confronted with any danger is another involuntary response.

Drivers almost invariably veer away from fixed objects like walls or railings set close to the roadway edge. Likewise they stay at considerable distances from unpaved, depressed, or rough shoulders. If these conditions exist, the effective

[4]See, for example, D. A. Gordon, *HRB Record 122,* W. W. Senders et al., *HRB Record 195,* T. H. Rockwell et al., *HRB Record 247,* and articles in *HRB Record 364* and *440.* For a discussion of simulation see and R. W. Allen et al., *TRB Record 706* and *739,* and K.M. Roberts, *Public Roads,* Dec. 1980.

[5]See D. A. Gordon, op. cit.

[6]See D. A. Gordon and T. M. Mast, *HRB Record 247.* Articles by E. Farber, C. A. Silver, and D. Landis, ibid., discuss other aspects of driver behavior in passing maneuvers.

[7]See N. A. Kaluger and G. L. Smith, Jr., *HRB Record 336.*

[8]T. W. Forbes reported some of the earliest authoritative findings on driver characteristics in *Transactions,* ASCE, 1941. See also *Human Factors in Highway Traffic Safety Research,* T. W. Forbes, Editor, Wiley, New York, 1972. Among recent references are *HRB Record 247, 292, 349, 364, 414,* and *464; TRB Record 520, 530,* and *623; TRB Special Report 165; NCHRP Report 123,* and D. A. Gordon, *Public Roads,* Mar. 1981.

width of the roadway and its traffic-carrying capacity are reduced by the driver's natural reactions.

Driver decisions regarding *"gap acceptance"* are a principal control over highway and intersection capacity and safety. Gaps (or headways) are commonly measured in seconds. Examples affecting vehicle operation include the headways between leading and following vehicles, the gap between succeeding vehicles into which another will enter when changing lanes or entering a traffic stream at an "on" ramp, or the acceptable openings to vehicles entering or crossing traffic lanes at intersections. Data on headways in traffic lanes are given later in the discussion of capacity. Another example involves the acceptable gaps in traffic on a two-way through street for drivers entering from the third leg of a T intersection controlled by a stop sign. One study found that where the near lane must be crossed to make a left turn into the far lane, only 8% of the drivers would accept a 2-s gap. Of the drivers offered a gap of 4 s, 15% accepted; gaps of 6 and 8 s were accepted by 71% and 87% of the drivers, respectively. These and many other results clearly indicate the differences among drivers.[9]

Fear of arrest and punishment also affects driver behavior. To illustrate, it was found that drivers reduced their speed by 5 mph when a highway patrol car replaced an orange station wagon alongside the road. In heavily policed urban areas excessive speed and other reckless practices occur less frequently than on suburban or rural roads where the chance of detection is remote. For the same reason, traffic-control devices are more effective in urban than in other areas; likewise designs used may differ between urban and rural situations.

The effects of traffic regulations on driver behavior may at times produce unanticipated results. For example, imposing minimum speed limits on freeways was expected to increase the use of the outer lanes and decrease travel times; actually the opposite happened.

Driver attitudes and behavior are a strong contributor to highway accidents. This topic is discussed in Chapter 12.

Dual Mode Systems for Vehicular Control

Schemes that combine electronic controls on main traffic arteries with driver control elsewhere have been proposed, particularly for mass-transit vehicles. They remain largely in the experimental stage. One scheme is a "guidance" system whereby vehicles position themselves over or alongside wires imbedded in the pavement, thus staying in the center of the traffic lane. Another is a detector-warning system that regulates vehicle spacings. Both would operate far more quickly than the perception-reaction sequence of the driver. Thus an effective electronic system could increase lane capacity by permitting closer vehicle spacing or, conversely, give a greater factor of safety. Most certainly the strain and effort of driving could be reduced.[10]

[9]See N. G. Tsongos and S. Wiener, *Public Roads,* Apr. 1969. Recent reports on gap acceptance appear in, among other places, *HRB Record 195* and *409* and *TRB Record 567* and *667*. *TRB Circular 212* shows design values for several conditions.

[10]For more detail see K. W. Olson et al., *HRB Record 275;* R. E. Fenton et al., *HRB Record 344; NCHRP Report 51;* and *TRB Special Report 170.*

VEHICLE CHARACTERISTICS

Almost all highways carry both passenger automobiles and trucks and design standards must be set to meet the requirements of both. Passenger car features such as driver's eye height and behavior at high speeds provide one set of criteria (see Chapter 9). In the past changes in some of these have forced revision of design standards. There has recently been a shift toward smaller, lighter, and lower vehicles because of high first and fuel costs, environmental considerations, and possible fuel rationing. Further changes are to be expected in the years ahead.[11] On the other hand, the size, weight, and other characteristics of legally permitted trucks govern other standards such as those for lane width, vertical clearance, and pavement and bridge loadings. The application of these standards to particular design situations is discussed at appropriate places throughout this book.

In each state the legislature prescribes the limits for each characteristic of the trucks operating there, although local authorities sometimes permit greater loads or interpose more severe restrictions on particular roads or streets. Table 8–1 is a simplified listing of some of the more important size and weight features for which maximums are prescribed by law and gives the extreme upper and lower values set by the individual states.

As is to be expected, the desire of the trucking industry for more economical operation through the use of wider, longer, and heavier vehicles has brought demands for upward revisions of existing limits. Changes, particularly greater uniformity in some of these standards, probably are desirable but will come only after careful study of the consequences. Possibly the strongest arguments against raising the recommended requirements are that many roads and bridges designed to meet existing standards would have to be strengthened and that there is a large mileage that is substandard even for the vehicles now permitted.[12]

Occasionally, loads with dimensions or weights exceeding the legal limits set by each state are carried over their highways under permits granted by an appropriate agency. *NCHRP Synthesis 68* discusses this problem in detail and gives the procedures and regulations of the individual jurisdictions as of 1978.

HIGHWAY TYPES

All state highway systems and most of the local highway and street systems encompass several types or classes of highways. At one extreme are high-speed, high-volume facilities carrying through traffic, with no attempt made to serve abutting property or purely local traffic. At the other are local rural roads or streets that carry low volumes, sometimes at low speeds, and with a primary function of land service.

[11]See D. J. Kulash and C. Difiglio, *TRB Record 648* and J. P. Stucker et al., *TRB Record 689*.
[12]Methodologies for making cost/benefit analyses as a measure for the effects of changes in vehicle sizes and weights appear, along with an extensive bibliography, in *NCHRP Reports 141* and *198*. The reports state, however, that the results of such analyses do not offer conclusive arguments as to the desirability of retaining or raising present standards.

TABLE 8–1. Summary of State and Other Limitations on Truck Dimensions, Axle Loads, and Weights

Item	*AASHTO Recommendations**	*State Limits†*		*Interstate Limits*
		Maximums	*Minimums*	
Width (in)	102	102	96	96
Height (ft)	13.50	14.00	12.50	—
Length (ft)				
Single-unit truck	40	60	35	—
Tractor-semitrailer	55	85	55†	—
Any combination	65	85	55†	—
Axle load (lb)				
Single axle	20,000	22,400‡	18,000§	20,000
Tandem axles	34,000	40,000‡	32,000§	34,000
Maximum gross wt. (lb)	86,500	139,000	73,000§	80,000

*AASHTO recommendations published in 1974.
†From *NCHRP Report 198*. See R. D. Layton and W. G. Whitcomb, *TRB Record 687* for a geographic distribution of these limits.
‡These limits are common along the eastern seaborad.
§Because these limits are in effect in a north-south tier of states from Lake Michigan to the Gulf of Mexico, they are, in effect, applied to all east-west cross-country trucking.

Definitions for the various types of highways (and for many other highway terms) were prepared in 1968 by a (then) AASHO Special Committee on Nomenclature. These are published as *AASHO Highway Definitions*. Some of these are as follows:

Expressway. Divided arterial highway for through traffic with full or partial control of access (see below) and generally with grade separations at major intersections.

Freeway. Expressway with full control of access.

Parkway. Arterial highway for noncommercial traffic, with full or partial control of access, and usually located within a park or a ribbon of parklike developments.

Control of access. Condition where the right of owners or occupants of abutting land or other persons to access, light, air, or view in connection with a highway is fully or partially controlled by public authority.

Full control of access means that the authority to control access is exercised to give preference to through traffic by providing access connections with selected public roads only and by prohibiting crossings at grade or direct private driveway connections.

Partial control of access means that the authority to control access is exercised to give preference to through traffic to a degree that, in addition to access connections with selected public roads, there may be some crossings at grade and some private driveway connections.

The other highway types lack the feature of access control. They include:

Major street or major highway. Arterial highway with intersections at grade and direct access to abutting property, and on which geometric design and traffic-control measures are used to expedite the safe movement of through traffic.

Through street or through highway. Every highway or portion thereof on which vehicular traffic is given preferential right of way, and at the entrances to which vehicular traffic from intersecting highways is required by law to yield right of way to vehicles on such through highway in obedience to either a stop sign or a yield sign, when such signs are erected.

Local road. Street or road primarily for access to residence, business, or other abutting property.

Precise definition for other highway types such as arterial, belt, bypass, radial, frontage also are given. In general the label explains the function.

A more detailed discussion of the location and design practices appropriate for the various types of highways is given elsewhere in this book. It should be pointed out here, however, that the freeway, as typified by the Interstate System, represents the highest type of highway facility; for all other classes certain important advantages must be sacrificed. In capsule form, these advantages include the following:

Capacity. On freeways the absence of intersections or crossings at grade and the elimination of marginal friction through access control permit unrestricted, full-time use by moving vehicles, rather than restricted, part-time flow.

Reduced travel time. On freeways, time losses from stopping and waiting at intersections are eliminated. In addition, most of the conflicts that contribute to accidents are eliminated, except under unusual circumstances. Drivers normally can and will travel at higher and sustained speeds.

Safer operation. On freeways, elimination of conflicts at intersections and along both margins of the roadway and the barring of pedestrians from the right of way usually bring substantial reductions in accidents.

Permanence. Access control along freeways prevents the growth of businesses or other activities along the roadway margin. Without access control, these activities generate unordered traffic and parking. In a short time, capacity is greatly reduced and accident potential is substantially increased.

Reduced operating cost, fuel consumption, air pollution, and noise. Smoother operations and fewer stops reduce fuel consumption and other operating costs. Reduced fuel consumption in turn reduces air pollution. Smoother operation with fewer stops also greatly reduces noise, particularly that from trucks.

TRAFFIC FLOW CHARACTERISTICS[13]

Traffic flow or volume on a highway is measured by the number of vehicles passing a particular station during a given interval of time. In many instances, traffic is stated as the "average annual daily traffic," commonly called the

[13]Techniques for predicting future traffic flows on various arteries and for assigning it between alternative routes are discussed in Chapter 3. The presentation here deals with variations in present or future flows by month, day of week, and time of day.

AADT or ADT if the period is less than a year. Again, volume may be stated on an hourly basis, such as the "hourly observed traffic volume" or the "estimated 30th-hour volume," which is commonly used for design purposes (see below). Some agencies now use volumes for 5-min intervals to distinguish short peak movements.

Traffic flow at a given location depends on numerous factors peculiar to that site. As would be expected, it varies by hours of the day, days of the week, and months of the year. Likewise, its character changes; for example the percentage and kind of trucks is a function both of time of day and the contributing area. Figure 8–2 shows traffic flow for a representative major rural highway at some distance from a large urban area. Among its characteristics are: balanced movement in two directions throughout the day; absence of sharp morning and afternoon peaks; and relatively uniform volume of heavy trucks. Greater daily volumes are indicated for the weekends, with smallest flows midweek. The predominant summer traffic surge is also apparent. Figure 8–3 typifies the flow on a major urban traffic artery. Heavy volumes of automobiles and buses carrying commuters move westbound (toward San Francisco) on five lanes on the upper deck of the bridge in the morning and reverse their directions on five lanes on the lower deck in the afternoon. There is a decided falloff in both in the middle of the day. In contrast, heavy truck flows generally begin later in the morning, are relatively constant through the midday period, and fall off somewhat in the late afternoon. In this instance, division of traffic to upper and lower decks does not permit reversals of direction in some of the lanes, as is done on certain other facilities.

At the time the traffic count of Fig. 8–3 was taken, about 350 buses traveled the Bay Bridge westbound in the morning peak hour, carrying about one-half the persons crossing. To make bus travel more attractive, transit time had been shortened by about 5 min by assigning a separate lane through the toll plaza and priority access to the bridge.[14] But this number of buses is not sufficient to justify assigning a lane of the bridge to them, since the capacity of a lane is about 1200 buses per hour. Car pools of three or more persons also are given this favored treatment. Another innovation has been to alternatively feed traffic from the 17 toll-plaza lanes onto the five bridge lanes by means of traffic signals. In this way, congestion at merging points is eliminated and the bridge flows at near capacity.

Figure 8–3*b* does not completely show the unbalance in bus use between peak and off-peak hours. At peak hours in the direction of major flow, buses carried about 40 passengers each; at off-peak hours occupancy was below 20. Occupancy of automobiles, exclusive of carpools, was in the range of 1.2 to 1.3 persons per vehicle.

It is not economically sound to have a highway congestion-free every hour throughout the year. However it has been established that for many hours each year the traffic volume approaches that of the 30th heaviest hour, which is the

[14]See Chapter 10 for a discussion of priority lanes.

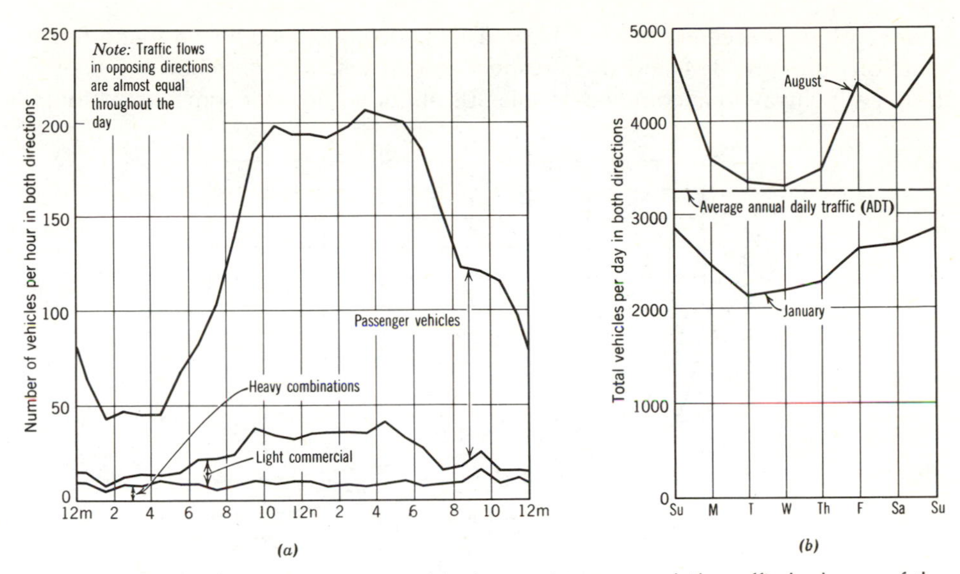

Fig. 8–2. Traffic flow on a major rural highway. *(a)* average daily traffic by hours of the day; *(b)* average daily flow in high and low months.

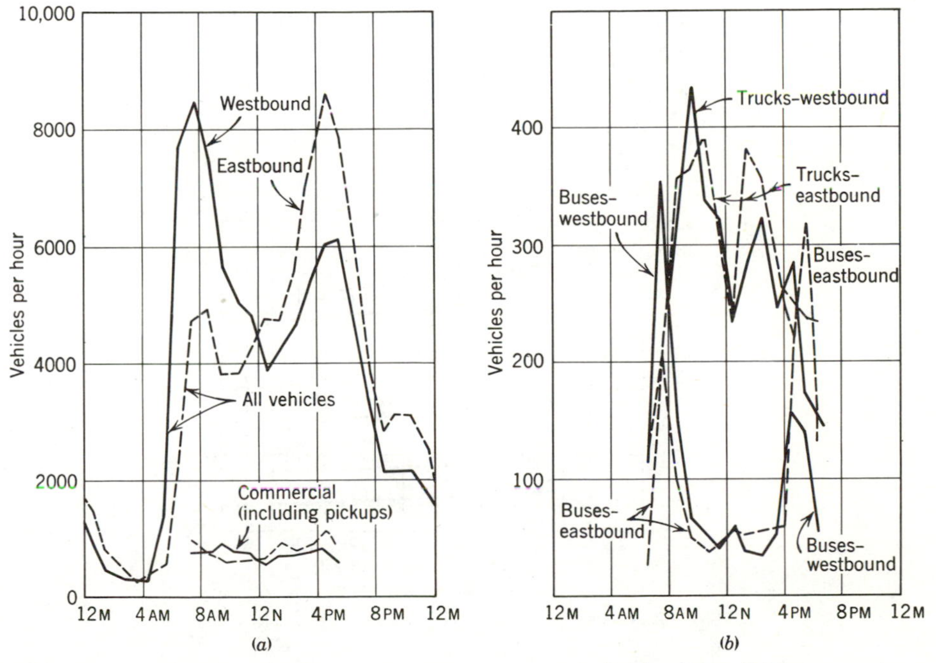

Fig. 8–3. Weekday traffic flow on the San Francisco–Oakland Bay Bridge, a major urban traffic artery; *(a)* all vehicles and *(b)* heavy trucks and buses.

hourly volume exceeded only 29 hr/yr. This is demonstrated for rural and urban situations by Figs. 8–4 and 8–5, respectively. Thus, it is common practice to design a highway to accommodate this 30th-hour volume for some stated future

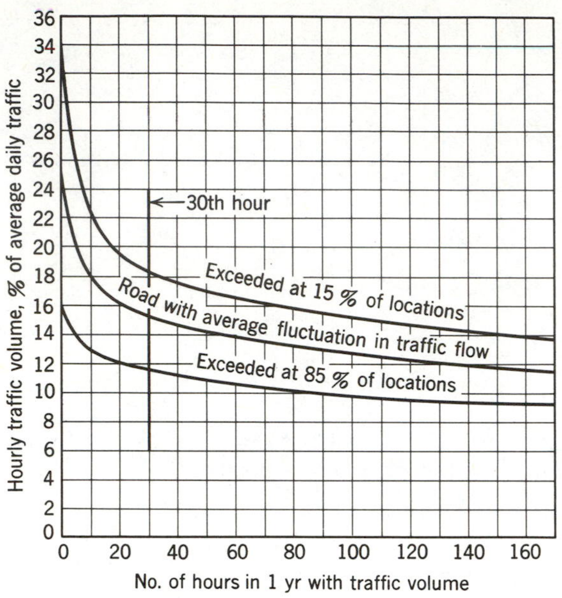

Fig. 8–4. Relationship between peak hourly flows and annual average daily traffic on main rural highways. (Source: *A Policy on Geometric Design*.)

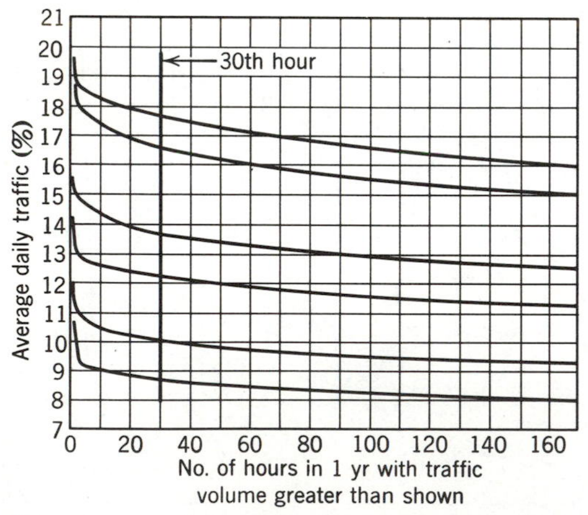

Fig. 8–5. Relationship between peak hour flows and average daily traffic on city streets. Data from six locations in Cincinnati, Ohio. (After A. M. Vorhees, *HRB Bulletin 203*.)

year. For most situations, designs based on estimated 30th-hour volumes 20 yr after plans are approved, or, alternatively, after construction is completed, are considered to be appropriate.[15]

As shown by Figs. 8–4 and 8–5, the ratio between 30th-hour volume and ADT is far from constant for all roads; actually, recorded extremes have ranged from 0.08 to 0.38. For rural locations the average is 0.15; the irreducible minimum appears to be about 0.095; lower percentages would be a definite indication that traffic desires are being suppressed. There was evidence in the past to indicate that the ratio for individual stations remained relatively constant through the years, but later studies do not entirely support this view. In any event, it is clear that each individual highway requires special analysis.

HIGHWAY CAPACITY AND LEVELS OF SERVICE

Introduction

Earlier sections of this book have discussed methods for measuring current traffic flows and for predicting traffic demands over time on existing or proposed highway facilities. With a knowledge of "capacity," then, it is possible to relate these current or predicted traffic volumes to the quality of the service that is provided by an existing highway or to "size" proposed facilities to meet stipulated criteria for levels of service. It follows that capacity is one of the most important factors in highway design and operation. As such, it has been the subject of intensive theoretical analysis and field observation.

The results of the many studies of highway capacity and of the relationship between traffic volume and the quality of traffic flow were brought together in 1965 in the authoritative *Highway Capacity Manual*.[16] In general, *The Manual* presents the findings of many field observations, although it acknowledged the growing contribution of traffic flow theory. It notes that the difficulty with such theoretical approaches is that "the traffic stream is not homogeneous with regard to either drivers or environmental factors, and cannot, therefore, be precisely reproduced." For example, in the case of urban freeways, flow is seldom truly "ideal" because of the presence of nearby on or off ramps, changes in the number of lanes, or the tunnel effect of bridges. Such factors make it extremely difficult to express the true conditions in mathematical terms.

Since *The Manual* was issued, research on traffic flow has continued on many fronts. A revised edition is planned for the mid-1980s. Some of the findings of research and a number of revised approaches were made available in *TRB Circular 212* titled *Interim Materials on Highway Capacity*, 1980. It is not a revision of the entire *Manual*, rather it treats the following specific items in detail: signalized and unsignalized intersections; transit; pedestrians; and freeways.

[15]See *A Policy on Geometric Design* for a detailed discussion.

[16]*HRB Special Report 87*, hereafter referred to as *The Manual*. Similar manuals also are available in, for example, Great Britain, Australia, and Sweden. Some sections of *The Manual* are updated in *TRB Circular 212*.

Like *The Manual,* it relies almost entirely on observed data rather than on theoretical approaches.

The brief presentation on capacity which follows is drawn from *The Manual, TRB Circular 212,* and reported research. The aim is to give readers an understanding of this complex subject; for details they must go to the references.

Capacity Defined

A generalized definition of capacity is: *The capacity of any element of the highway system is the maximum number of vehicles which has a reasonable expectation of passing over that section (in either one or both directions) during a given time period under prevailing roadway and traffic conditions.* A sampling of capacities for modern highway elements is as follows:

Facility	*Capacity in Passenger Cars*
Freeways and expressways away from ramps and weaving sections, per lane per hour	2000
Two-lane highways, total in both directions, per hour	2000
Three-lane highways, total in each direction, per hour	2000
Twelve-foot lane at signalized intersection, per hour of green signal time (no interference and ideal progression)	1800

Close examination of the individual terms of the definition for capacity is valuable as a means of putting the overall concept into perspective. Consider, for example:

1. *Maximum.* Stated values for capacity indicate the maximum volume at which traffic flow is likely to continue without breakdown and serious congestion. It follows that, at capacity, the quality or level of service is far from ideal.
2. *Number of vehicles.* Capacity usually is stated in passenger cars per hour; trucks and buses in the traffic stream can decrease the stated capacity substantially.
3. *Reasonable expectations.* Values for capacity cannot be determined precisely because of the many variables that affect traffic flow, particularly at high volumes. Thus, actual capacities for apparently similar conditions may vary substantially. Stated differently, assigned values for capacity represent probabilities rather than certainties.
4. *One direction versus two direction.* On multilane facilities, traffic in one direction flows independently from that in the other. On the other hand, on two- and three-lane roads, there are interactions between traffic in the two directions and these affect traffic flow and capacity.
5. *A given time period.* Traffic volumes and capacity are usually stated in vehicles per hour. However, because flow does not vary uniformly with time, volumes and capacities sometimes are stated for a shorter period, such as 5 or 15 min. More commonly, this variation within an hour is expressed by a *peak hour factor* (PHF). This factor, which is always less than or equal to 1, is the quotient of the hourly volume

divided by the shorter-period volume multiplied by the number of periods in an hour. For example, if the hourly volume is 1500 and the highest 5-min volume is 150, the PHF equals 1500 divided by 150×12 or 0.83.

6. *Prevailing roadway and traffic conditions.* Prevailing roadway conditions include physical features that affect capacity such as lane and shoulder width, sight distance, and grades. Prevailing traffic conditions reflect changes in the character of the traffic stream; for example, changes over time in the percentage of trucks and buses as illustrated in Figs. 8–2 and 8–3. Physical or traffic conditions might also reflect the fact that the demand at a given point along a roadway might not be related to capacity at that point, but rather to lower capacity at some point upstream or downstream.

Not included in the definition for capacity but of importance in capacity studies are ambient conditions which are primarily weather-related. These include rain, snow, ice, fog, smog, or wind.

In treating capacity, *TRB Circular 212* divides freeways into components: basic freeway segments and those in the zone of influence of weaving areas and ramp junctions. Capacities of expressways, multilane highways, and two- and three-lane facilities also have the two components: basic and those in the zone of influence of intersections. Each of these is treated separately below.

Speed-Volume-Capacity Relationships for Basic Freeway and Multilane Highway Segments

A knowledge of the relationships among speed, volume, and capacity is basic to understanding the place of capacity in highway design and operation. Figure 8–6, which gives such a relationship for a single freeway or expressway lane, is used for illustrative purposes.

If a lone vehicle travels along a traffic lane, the driver is free to proceed at the design speed.[17] This situation is represented at the beginning of the appropriate curve at the upper left of Fig. 8–6. But as the number of vehicles in the lane increases, the driver's freedom to select speed is restricted. This restriction brings a progressive reduction in speed. For example, many observations have shown that, for a highway designed for 70 mph, when volume reaches 1900 passenger cars per hour, traffic is slowed to about 43 mph. If volume increases further, the relatively stable normal-flow condition usually found at lower volumes is subject to breakdown. This zone of instability is shown by the shaded area on the right side of Fig. 8–6. One possible consequence is that traffic flow will stabilize at about 2000 vehicles per hour at a velocity of 30 to 40 mph as shown by the curved solid line on Fig. 8–6. Often, however, the quality of flow deteriorates and a substantial drop in velocity occurs; in extreme cases vehicles may come to a full stop. In this case the volume of flow quickly decreases as traffic proceeds under a condition known as "forced flow." Volumes under forced flow are shown by the dashed curve at the bottom of Fig. 8–6. Reading from that curve, it can be seen that if the speed falls to 20 mph, the rate of flow will drop to 1700 vehicles per hour; at 10 mph the flow rate is only 1000; and,

[17]The variability in speeds chosen by individual drivers and the effect of speed limits are discussed in Chapter 9.

of course, if vehicles stop, the rate of flow is 0. The result of this reduction in flow rate is that following vehicles all must slow or stop, and the rate of flow falls to the levels shown. Even in those cases where the congestion lasts but a few seconds, additional vehicles are affected after the congestion at the original location has disappeared. A "shock wave" develops which moves along the traffic lane in the direction opposite to that of vehicle travel. Such waves have been observed several miles from the scene of the original point of congestion, with vehicles slowing or stopping and then resuming speed for no apparent reason whatsoever.

The speed-volume relationships for traffic flow on the basic sections of multilane highways may be summarized briefly as follows. Beginning with the appropriate design speed curve on Fig. 8–6, progress from the upper left to the right down the slope of the speed-volume curve. At the nose, the flow may stabilize; on the other hand it may break down. In this case speed-volume relationships follow the dashed curve downward to the left to the origin. Stated differently, this progression passes through the zones of normal flow, unstable flow where actual behavior of traffic is highly unpredictable, and possibly into forced flow where volume is substantially reduced.

Effects of the imposition of speed limits of 60, 50, and 40 mph are suggested by the dotted lines on Fig. 8–6. A 55-mph curve could also be drawn midway between the 60 and 50 mph dotted curves to reflect the effects of the federally

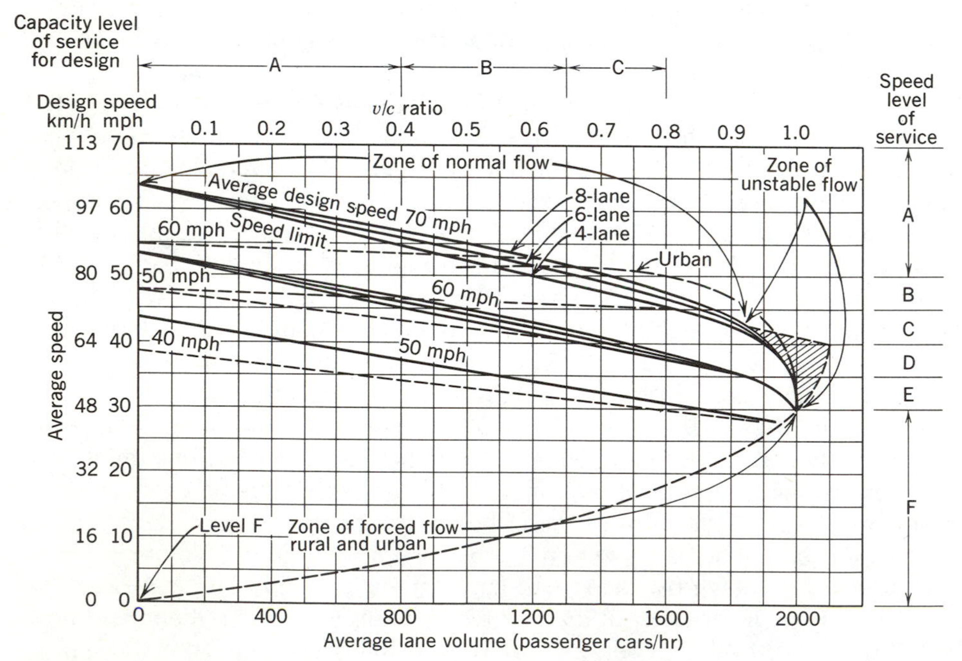

Fig. 8–6. Typical relationships between volume per lane and average speed in one direction of travel under ideal uninterrupted flow conditions on freeways and expressways. Except as noted, curves apply to rural conditions.

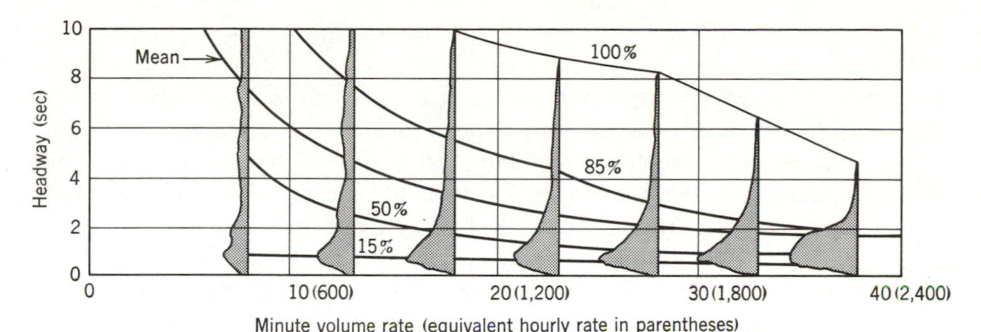

Fig. 8–7. Headway distribution related to traffic flow in the outer lanes of the Ford Expressway, Detroit.

imposed 55-mph limit, but this is conjectural since the level of enforcement varies so widely.

Vehicle spacing, or its reciprocal, traffic density, probably have the greatest effect on capacity since it generates the driver's feeling of freedom or constraint more than any other factor. Studies of drivers as they follow other vehicles indicate that the time required to reach a potential collision point, rather than vehicle separation, seems to control behavior. However, this time varies widely among drivers and situations. Field observations have recorded headways (time between vehicles) ranging from 0.5 to 2 sec, with an average of about 1.5 s. Thus, the calculated capacity of a traffic lane based on this 1.5-s average, regardless of speed, will be 2400 vehicles per hour. But even under the best of conditions, occasional gaps in the traffic stream can be expected, so that such high flows are not common. Rather, as noted, they are nearer to 2000 passenger cars per hour.

Figure 8–7 shows frequency distributions of observed headways on the right-hand lane of an urban freeway for seven different traffic volumes. The lines running slightly downward toward the right indicate, for each traffic volume, the percentage of vehicles traveling at headways less than a given value. At the lower volumes, as shown on the left side of the graph, some of the headways are, as expected, large. But at the maximum volume shown in the figure, half of the headways are less than a second and only 15% are greater than 2 s. It is clear that drivers traveling under such conditions would feel greatly constrained.

The observed headways plotted in Fig. 8–7, which are typical for many American freeways, provide theoretical support for the common statement that a substantial percentage of drivers follow preceding vehicles too closely to be safe. For example, Fig. 8–7 shows that, at all volumes, headways for 15% of the vehicles are less than 0.8 s. If, for example, the traffic stream were flowing at 50 mph at this headway, center to center spacing would be 59 ft. Allowing 20 ft for the length of a passenger car, the rear-to-front spacing would be 39 ft and the time gap between vehicles only 0.53 s.[18] As was pointed out earlier, detec-

[18]It is readily apparent that this behavior clearly violates the "safe driving" rule of the National Safety Council of "one clear car length for each 10 mph which, at 50 mph, is approximately 100 ft.

tion, recognition, decision, and response initiation time for alerted drivers range from 0.5 to more than 1 s. Computations would show, then, that at the very least, one in six drivers could avoid striking the vehicle in front only by swerving out of the lane, if for some reason the forward driver suddenly applied the brakes. The mitigating factor is, of course, that drivers can see some distance ahead by looking through or around the vehicles ahead. Thus, they can anticipate what the leading drivers will do. But what about drivers who follow trucks at the same distance with their foward line of sight completely blocked?

The speed-volume-capacity relationships for basic segments described above assume no change in number of lanes and no incidents that interrupt traffic flow. When they do occur, queues form and speeds are drastically reduced. Getaway flow rates from geometric obstructions such as a decrease in the number of lanes have been found to be about 2000 passenger cars per hour for each downstream lane. At traffic-incident locations, the rate varied from 1100 to 1900 vehicles in the remaining lanes, with 1500 considered a reasonable value.[19]

Speed-Volume-Capacity Relationships for Basic Two- and Three-lane Highways

Figure 8–8 shows speed-volume relationships for two-lane, two-directional roadways. At low volumes where there is little interference from other vehicles, drivers proceed at their desired free-flow speeds. But, as volumes increase, interference between vehicles causes speed to fall as shown in Fig. 8–8. And, as with a lane on a multilane facility, if serious congestion develops, speed falls sharply and the forced flow condition develops as shown by the dashed line at the bottom of the graph.

Figure 8–8 shows that capacity for a two-lane highway totals 2000 vehicles per hour for both directions, which is only half that of two lanes in the same direction. This can be explained as follows. Given all the flow is in one direction on a two-lane road, vehicles can keep one lane filled by immediately passing into the gaps that form. And this single lane can then carry the same number of vehicles as a lane of a multilane facility. However, when passing is restricted by vehicles from the opposite direction, the spaces between vehicles cannot be filled. Instead, queues of vehicles form in each direction behind slower-moving vehicles until the spaces between queues become long enough to permit passing. After a period during which passing is possible, new queues form. As already indicated, total capacity for both directions is 2000 vehicles per hour. This can be 2000 vehicles in one direction with none from the other, 1000 from each, or any other combination adding to 2000.

With three-lane roads the center lane is used for passing in either direction. Under ideal conditions, vehicles can completely fill both outside lanes by passing in the center one. Thus, capacity can reach a total of 4000 vehicles for both directions. In the United States, capacity of three-lane roads is only of concern

[19]See J. A. Lindley and S. C. Tigor, *Public Roads,* June 1979.

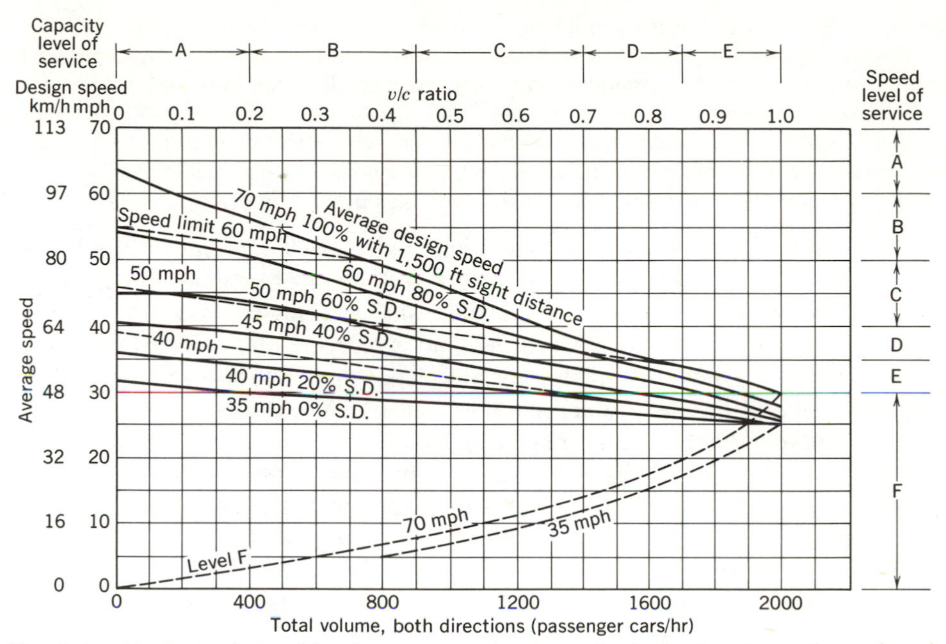

Fig. 8–8. Typical relationships between total volume for both directions of travel and average speed under ideal uninterrupted flow conditions on two-lane highways. (Source: *The Manual*.)

for existing highways since new ones are seldom built because of poor accident experience.

The "Level of Service" Concept

As indicated in the discussion of the relationships of speed, volume or density, and vehicle spacing, operating speed goes down and driver restrictions become greater as traffic volume increases. "Level of service" is commonly accepted as a measure of the restrictive effects of increased volume. Each segment of roadway can be rated at an appropriate level, A to F inclusive, to reflect its condition at the given demand or service volume. Level A represents almost ideal conditions; Level E is at capacity; Level F indicates forced flow.

The two best measures for level of service for uninterrupted flow conditions are operating or travel speed and the ratio of volume to capacity, called the *v/c* ratio. For two- and three-lane roads sight distance is also important.

Abbreviated descriptions of operating conditions for the various levels of service are as follows:

Level A. Free flow; speed controlled by driver's desires, speed limits, or physical roadway conditions.

Level B. Stable flow; operating speeds beginning to be restricted; little or no restrictions on manueuverability from other vehicles.

Level C. Stable flow; speeds and maneuverability more closely restricted.
Level D. Approaches unstable flow; tolerable speeds can be maintained but temporary restrictions to flow cause substantial drops in speed. Little freedom to maneuver, comfort and convenience low.
Level E. Volumes near capacity; speed typically in neighborhood of 30 mph; flow unstable; stoppages of momentary duration. Ability to maneuver severely limited.
Level F. Forced flow, low-operating speeds, volumes below capacity; queues formed.

A third measure of level of service suggested in *TRB Circular 212* is traffic density. This is, for a traffic lane, the average number of vehicles occupying a mile of lane at a given instant. To illustrate, if the average speed is 50 mph, a vehicle is in a given mile for 72 s. If the lane is carrying 800 vehicles per hour, average density is then 16 vehicles per mile; spacing is 330 ft, center to center. The advantage of the density approach is that the various levels of service can be measured or portrayed in photographs.

The Manual proposed adjusting levels of service by a peak hour factor (PHF) to reflect the fact that peak flow rates for short intervals are higher than the hourly average. However, *TRB Circular 212* proposes that level of service be stated separately for each representative period of time.

Appropriate divisions among levels of service for design purposes for multilane highways are shown along the top of Fig. 8–6. *TRB Circular 212* shows considerably more detail, with slightly different *v/c* ratios and volumes per lane designated as break points for four-, six-, and eight-lane facilities. Division points among levels of service for two-lane roads are shown across the top of Fig. 8–8.

The 1950 edition of *The Manual* offered only one criterion by which to judge the level of traffic service. This was "practical" capacity which was defined as the maximum number of vehicles that can pass without *undue* delay, hazard, or restriction on the driver's freedom to maneuver. As an example, for a two-lane rural road, practical capacity was set at 900 vehicles per hour, giving a *v/c* ratio of 0.45. If traffic at (say) the 30th highest hour exceeded this figure, the traffic service offered by the road would be classed as unacceptable. The present "level of service" concept provides several judgment points for the service offered by the road.

There are close parallels between the concept of "level of traffic service" and "sufficiency and deficiency ratings" (see Chapter 4). All require orderly, systematic assessment of individual facilities and provide a basis for comparison between all facilities in a system. Also, they provide a means for establishing some "tolerable level of service" which will aid in formulating a financial program to bring all facilities to the selected level over a stated period of years. But none of them, taken alone, incorporate economic comparisons.

Factors That Reduce Capacity and Level of Service of Basic Sections

The capacities cited earlier are for "ideal" conditions for roadways carrying uninterrupted flow. These include 12-ft lanes, shoulders at least 6 ft wide, flat grades, unrestricted sight distance, and no trucks or buses. If these conditions are not fulfilled, capacity is reduced. The conditions bringing reductions and their effects on capacity are given in the paragraphs that follow.

EFFECTS OF REDUCED LANE WIDTH AND EDGE CLEARANCE. Narrow lanes and narrow shoulders or other restrictions on edge clearance reduce capacity. Some of the data from *TRB Circular 212* for freeways and from *The Manual* for other multilane and two-lane facilities are given in Table 8–2. To illustrate their use: If vertical obstructions are placed at both edges of the 24-ft roadway of a four-lane freeway, capacity per lane will be 81% of 2000 or 1620 vehicles per hour. Again, if an 18-ft-wide bridge, curb to curb, remains when a two-lane highway is rebuilt, its capacity will be 0.58×2000 or 1060 vehicles per hour.

EFFECTS OF HORIZONTAL OR VERTICAL ALIGNMENT. Sharp horizontal curves cause vehicles to travel more slowly because they create dynamic forces to which the driver reacts. Short vertical curves over crests or obstructions to vision on the inside of horizontal curves likewise should cause vehicles to slow down.[20] At lower traffic flows, these real or desirable slowings lower the level

TABLE 8–2. Combined Effect of Lane Width and Restricted Lateral Clearance on Capacity and Service Volumes of Divided Freeways and Expressways and Two-Lane Roads with Uninterrupted Flow*

	Adjustment Factor for Lane Width and Lateral Clearance†							
	Obstruction on One Side of Roadway or Roadways				*Obstructions on Both Sides of Roadway or Roadways*			
Distance from Traffic Lane Edge to Obstruction (ft)	*12-ft Lanes*	*11-ft Lanes*	*10-ft Lanes*	*9-ft Lanes*	*12-ft Lanes*	*11-ft Lanes*	*10-ft Lanes*	*9-ft Lanes*
Four-lane divided freeway, one direction of travel								
6	1.00	0.97	0.91	0.81	1.00	0.97	0.91	0.81
4	0.99	0.96	0.90	0.80	0.98	0.95	0.89	0.79
2	0.97	0.94	0.88	0.79	0.94	0.91	0.86	0.76
0	0.90	0.87	0.82	0.73	0.81	0.79	0.74	0.66
Six- and eight-lane divided freeway, one direction of travel								
6	1.00	0.96	0.89	0.78	1.00	0.96	0.89	0.78
4	0.99	0.95	0.88	0.77	0.98	0.94	0.87	0.77
2	0.97	0.93	0.87	0.76	0.96	0.92	0.85	0.75
0	0.94	0.91	0.85	0.74	0.91	0.87	0.81	0.70
Two-lane highways								
6	1.00	0.88	0.81	0.76	1.00	0.88	0.81	0.76
4	0.97	0.85	0.79	0.74	0.94	0.83	0.76	0.71
2	0.93	0.81	0.75	0.70	0.85	0.75	0.69	0.65
0	0.88	0.77	0.71	0.66	0.76	0.67	0.62	0.58

*Sources: *TRB Circular 212* for multilane, *The Manual* for two-lane.
†Note, however, that traffic flow is unaffected by high-type median barriers.

[20]Unfortunately, drivers generally do not slow down for such restrictions on sight distance.

of service. However, they have but a small effect on capacity, since speeds are relatively low in any case when the highway is operating near capacity. The vertical scales on the right side of Figs. 8–6 and 8–8 can be used to determine the levels of service associated with lower design speeds. For example, it can be seen on the vertical, right-hand scale of Fig. 8–6 that for freeways, a design speed of 50 mph can barely reach level of service A. A design speed of 40 mph is at the lower limit of level of service C. The same measures for two-lane roads are given on the right-hand side of Fig. 8–8. Notice the part played by "percentage of length with passing sight distance less than 1500 ft." The 1500-ft criterion was chosen because field observations have shown that when sight distance is less than this drivers are reluctant to pass and the average speed of the traffic stream falls.

EFFECTS OF COMMERCIAL VEHICLES. Trucks[21] require substantially more highway capacity per vehicle than do passenger cars. A truck in the traffic stream may have the effect of from 2 to more than 100 passenger cars, depending on the particular situation. Buses likewise use more capacity than passenger cars. However, because of their relatively higher performance characteristics, the range of multipliers is from only 1.6 to 12.

Generalized passenger car equivalents for trucks and buses for the major classes of highways are given in Table 8–3. As noted, these values apply over substantial lengths of roadway. The equivalents increase as the terrain becomes rougher because there are more curves and a rolling grade line. Also, the effect on two-lane roads is greater than for multilane facilities at the lower levels of service, since, without a separate lane for slower-moving vehicles, traffic tends to pile up behind them.

Care must be exercised in computing the effects of commercial vehicles on capacity or level of service since, as mentioned, the peaks of passenger car and truck traffic do not coincide. In some instances, the difference is striking. For example, trucks contribute about 15% of the trips in central business districts, but only 5 to 8% during the peak hour.

EFFECT OF GRADES. Brakes are assisted by gravity on upgrades and opposed by gravity on downgrades. On uphill stretches safe or desirable vehicles spacings can therefore be smaller, which leads to increased capacity. However, with grades often go restricted sight distances which, as already indicated, decrease capacity.

Speeds of passenger cars do not change with 3% upgrades and are not greatly affected by upgrades as high as 6 or 7%. However, those of trucks are markedly affected. Bus speeds are influenced to an intermediate degree. The effects of grades on the speed of trucks operating uphill or downhill are illustrated by Fig. 8–9. Distance-speed relationships for an approach speed of 50 mph are shown by solid lines; those for low approach speeds by dotted ones. As mentioned

[21]A truck is defined as a vehicle designed for the transportation of cargo rather than passengers and having dual tires on one or more axles or having more than two axles. Included are truck tractors with trailers and semitrailers. Excluded are two-axle, four-tired vehicles, which have operating characteristics similar to passenger cars.

TABLE 8–3. Average Generalized Passenger Car Equivalents of Trucks, Buses, and Recreational Vehicles on Freeways, Expressways, and Multilane and Two-Lane Highways over Extended Section Lengths (Including Upgrades, Downgrades, and Level Subsections)*

		Equivalent, E, for:		
Level of Service	*Vehicle Type*	*Level Terrain*	*Rolling Terrain*	*Mountainous Terrain*
Freeways, expressways, and multilane highways				
A		Widely variable; one or more trucks have same total effect, causing other traffic to shift to other lanes. Use equivalent for remaining levels in problems.		
B–E	Trucks	2	4	8
	Buses	1.6	3	5
	Recreational	2	3	4
Two-lane highways				
A		3	4	7
B and C	Trucks	2.5	5	10
D and E		2	5	12
All levels	Buses	2	4	6

*Sources: *TRB Circular 212* for multilane, *The Manual* for two-lane.

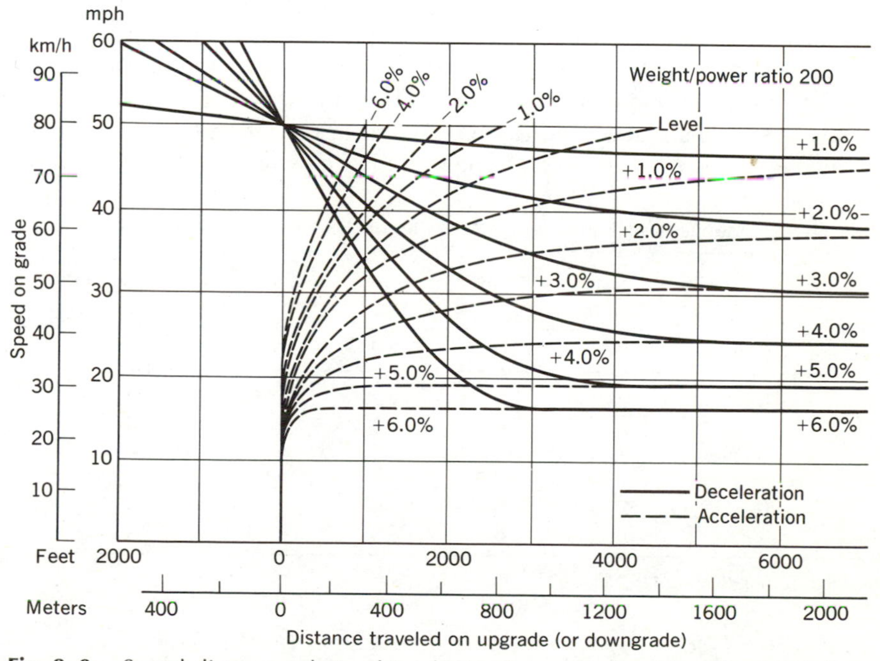

Fig. 8–9. Speed distance relationships for trucks on grades. Note that by shifting the horizontal scale the diagram can be used for other approach speeds. (Source: J. E. and J. P. Leisch, *TRB Record 631*.)

above, capacity of the road is not seriously affected until speeds fall below 30 or 35 mph. Assuming that the curves for Fig. 8–9 fit the vehicles being considered, it follows that capacity is not reduced substantially on grades flatter than 3% or 5% grades less than about 1500 ft long. But to maintain full capacity on 5% grades longer than this, a lane for trucks must be added to the upgrade side of the road. If, however, the aim is to permit passenger vehicles to maintain full speed, the lane must begin near the base of the grade and continue past the summit.[22]

Table 8–4 is a sampling of data on passenger car equivalents of trucks on grades on freeways. It gives the highest values offered by the reference; reasonably accurate multipliers result from interpolation. The table reflects clearly that trucks on longer and steeper grades substantially reduce freeway capacity with the effect more pronounced on four-lane facilities. It also shows that, for steeper grades the equivalents decrease substantially with truck percentage, indicating that they form queues in the outside lane.

Figure 8–10 is representative of data for passenger car equivalents of trucks on sustained grades on two-lane highways. It can be seen that a relatively small number of trucks has significant effects not only on capacity but on passenger car operation as well, unless a climbing lane is added.[23]

Weaving Sections

Weaving is defined as the crossing of two or more traffic streams traveling in the same general direction along a significant path of highway without the aid of traffic signals. It can entail intense lane-changing maneuvers as vehicles from entrance legs move into lanes appropriate to their exit legs. The traffic circle, rotary, or British "roundabout" (see Fig. 9–14) can be considered to be a series of weaving sections placed end to end. Again, at partial and full cloverleaf interchanges (see Fig. 9–17) vehicles entering the freeway must weave with those

TABLE 8–4. Passenger Car Equivalents of Trucks on Freeway Upgrades*

Grade (%)	Length (mi)	Four Lanes			Six or More Lanes		
		Percentage Trucks			Percentage Trucks		
		2	10	20	2	10	20
2	0–¼	2	2	2	2	2	2
2	>1½	7	4	3	7	4	3
4	0–¼	7	4	4	7	4	3
4	>1	17	9	9	13	8	8
6	0–¼	9	6	6	10	5	5
6	>½	28	18	18	20	14	14

*Abridged from *TRB Circular 212*.

[22] *TRB Circular 212* and *A Policy on Geometric Design* give detailed procedures for positioning climbing lanes.

[23] Recent research reports on trucks on grades appear in *NCHRP Report 185* and *TRB Record 615, 631,* and *699*.

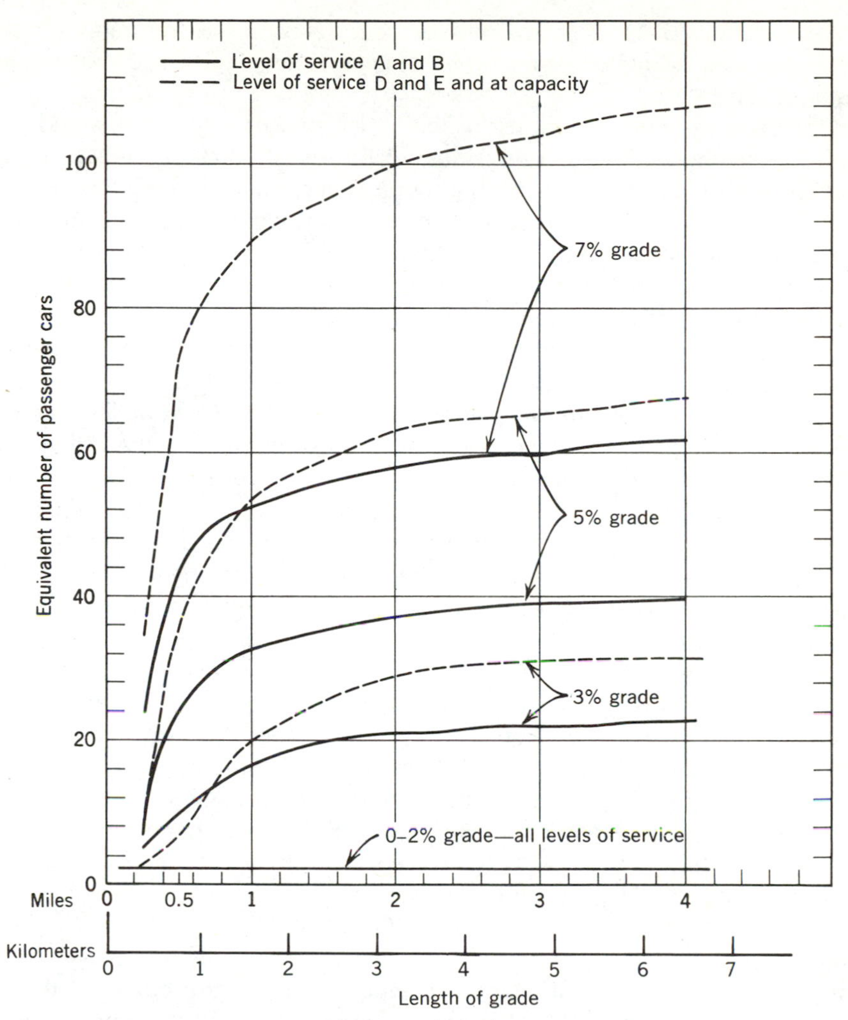

Fig. 8–10. Passenger-car equivalents of trucks on two-lane highways on specific individual subsections of grades. (Based on *The Manual*.)

leaving it. In the more elaborate cloverleaf with collector-distributor roads (see Fig. 9–19) this weaving is on a roadway physically separated from the through lanes. However, *TRB Circular 212* classifies on ramps followed closely by off ramps as ramp junctions (see below) rather than weaving sections unless an auxiliary lane is added.

The roadway length affected by weaving sections or ramp junctions and therefore not in the basic sections mentioned earlier can be substantial. Under congested conditions, the effects of off ramps have been observed 6250 ft upstream. *TRB Circular 212* gives no single figure, seeing it as a matter of engineering judgment. The *Manual on Uniform Traffic Control Devices* possibly sees this approach zone for freeways as more than a mile long, since signing begins there (see Fig. 10–1). *A Policy on Geometric Design* sets the minimum overall length of lane additions, lane drops, and auxiliary lanes at 2500 ft from the point of separation.

One set of recommendations for the length of weaving sections and number of lanes in the section covering most of the usual situations is shown in Fig. 8–11. In this instance, V_1 is the hourly passenger car volume going through on the freeway, V_2 is the nonweaving volume that enters the upstream ramp and leaves the downstream one, and W_1 and W_2 are, respectively, larger and smaller weaving volumes. A graph for determining a weaving factor k and an equation for setting the number of lanes appear on the figure.

A procedure for designing right-side weaving sections is shown as Fig. 8–12. It involves four paths: (1), vehicles proceeding straight through on the freeway; (2) vehicles leaving the freeway at a downstream exit; (3) vehicles entering the freeway at an upstream point; and (4) vehicles entering on the right upstream but leaving on the right downstream. These movements are shown in the plan view in the upper right of Fig. 8–12. The remainder of the figure is a nomograph from which can be read, for stated numbers of passenger cars in each movement and stated levels of service for through and weaving vehicles (a) the required length of weaving section and (b) the required number of lanes in the weaving section.

An example illustrating the use of Fig. 8–12 is as follows. Approaching lanes of freeway, 4(N_b), carrying 5100 passenger cars per hour, 600 of which (W_2) leave by the off ramp; 1100 enter the section from the on ramp, making the total entering volume V, 6200. Of those from the on ramp, 950 (W_1) enter the freeway, the other 150 leave by the off ramp. Desired levels of service for both freeway traffic and weaving are on the border between C and D. The solution is traced through Fig. 8–12 with arrowed lines. Reading down on the diagram at the lower left shows a minimum weaving length of 1300 ft. Following the upward, right, down, and right lines to the graph for N_b equals 4, the line separating levels C and D, gives 5.2 or 5 lanes in the weaving section. With the same inputs, Fig. 8–11 will give a weaving-section length, based on the collector-distributor curve, of 1000 ft and the number of lanes as 4.7 or 5.

On the lower left-hand diagram of Fig. 8–12 are dotted lines labeled "lane inbalance weaving sections." An example of such an unbalance would be two weaving lanes in the weaving section but only one exit lane. In such cases, the solution calls for a longer weaving section to permit drivers to adjust.

In the references cited on Fig. 8–12 is a nomograph for two-sided weaving sections where the entrance ramp is on one side of the freeway and the exit ramp on the other. The nomograph is not given here, since such conformations are unusual.

TRB Circular 212 also presents another approach and a series of equations and graphs for analyzing weaving sections, including those having more complicated layouts. Because of space limitations, they cannot be offered here. Neither can recent studies of weaving sections and ramps reported by R. P. Roess et al. in *TRB Record 772* be summarized here.

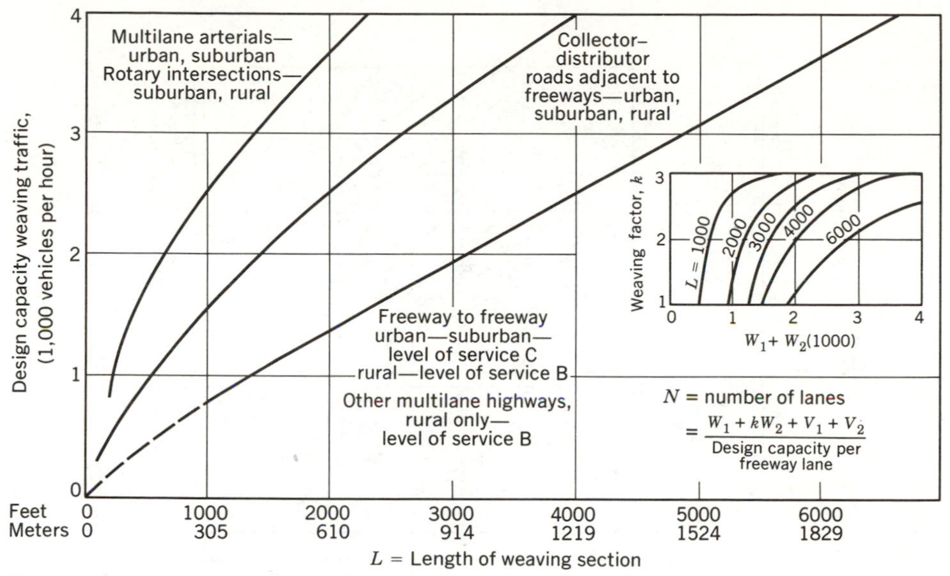

Fig. 8–11. Design capacity of weaving sections. (Based on *A Policy on Geometric Design*.)

Ramps and Ramp Junctions

Ramps transfer traffic from freeway to freeway or provide access to and egress from freeways and some expressways. As noted above, those connections having added lanes are classed as weaving sections.

For capacity and design studies, three ramp elements must be considered separately. These are the ramp junction, the ramp proper, and the connection with other facilities, be they another artery or local highways or streets. Today, another element, on-ramp metering or other forms of control have been added. These are discussed in Chapter 10.

Almost all ramps on modern facilities join to or depart from the right-hand lane (lane 1). Thus, ramp capacity is directly dependent on the capacity of that lane after through vehicles using it have been accommodated. *TRB Circular 212* indicates that this can be a particularly serious problem on four-lane freeways carrying heavy traffic. For example, if through traffic on the freeway is 2000 passenger cars per hour, about 400 stay in the outside lane; with 3500 to 4000 going through, 1600 use it. These vehicles would greatly restrict access to the freeway from on ramps or from it to off ramps. With six-lane freeways, only 6% stay in the outside lane at volumes less than 3500; at 5000 vehicles, only 16% carry through in the outside lane. With eight-lane freeways, 10% or less follow this path at 6500 through vehicles so that the outside lane is largely clear.

The number of vehicles that can enter the outside lane from an on ramp is also strongly dependent on ramp entry layout. Data cited in *TRB Circular 212* show, with an acceleration lane 1000 ft long and a flat 2° convergence angle,

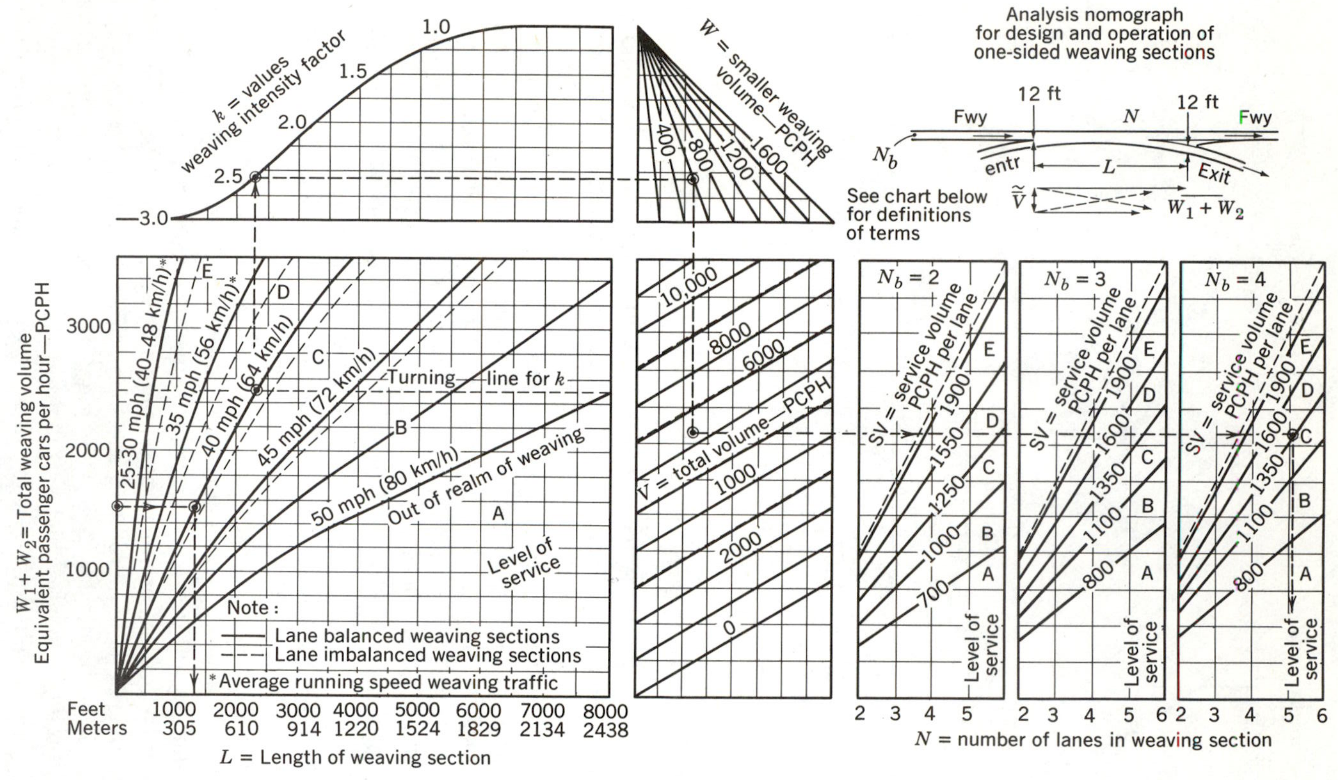

Fig. 8–12. Nomograph for design and analysis of weaving sections, one-sided configuration. (After J. E. Leisch, *ITE Journal*, Mar. 1979 and *TRB Circular 212*.)

about 97% of the gaps can be filled. If acceleration length is short, 400 ft, but the convergence angle is 2°, 32% enter. In the extreme case of a 400-ft acceleration lane and a 10° convergence angle, only 8% of the theoretically possible entries can be made.

Table 8–5 gives checkpoint values for the limits of the various levels of service for isolated freeway on and off ramps. These are for high standards of ramp configuration (see Fig. 9–16). If capacity above 1500 passenger cars per hour is needed for a ramp junction or ramp proper, *TRB Circular 212* recommends adding a lane to the ramp junction entry or exit and to the ramp proper. That publication gives nomographs for the conditions covered by Table 8–5 and also for combinations of two adjacent on ramps, two adjacent off ramps, adjacent on ramps followed by adjacent off ramps and vice versa, and major merges and diverges.

Capacities and levels of service for ramps proper leading to and from ramp junctions are given in Table 8–6. It should be noted that if level of service C or higher is desired, volumes less than the 1500 passenger cars per hour cited earlier are required. It is also recommended that ramps longer than 1000 ft be two-lane to permit passing, or in situations where queues are expected to form or the ramp is on a steep grade with minimal geometry.

Capacity and level of service for the ramp-highway or street interface must also be appraised for each situation. If control is at an intersection, the procedures outlined below for them would be followed.

Overall Capacity and Level of Service for Freeways

From estimated traffic volumes and the data presented above as augmented by the references, it is possible to determine the level of service offered by the basic freeway, weaving sections, and ramps and ramp junctions. Alternatively,

TABLE 8–5. Limiting Volumes in Passenger Cars per Hour for Freeways and on and off Ramps for Various Levels of Service and PHF of 1.00*

Level of Service	*Freeway Volume in One Direction†*		*Checkpoint Volumes*		*Weave Volumes‖*
	Four-Lane	*Eight-Lane*	*Merge‡*	*Diverge§*	
A	1600	3280	>750	>800	>500
B	2500	5400	1200	1300	700
C	3400	6800	1500	1650	1300
D	3850	7700	1800	1900	1550
E(capacity)	4000	8000	2000	2000	2000
F	———	Highly variable	———	———	———

*Source: *TRB Circular 212*.
†For 70-mph freeway design speed.
‡Lane 1 volume plus ramp volume for one-lane on ramps.
§Lane 1 volume immediately upstream of off ramp.
‖Weave volumes between on ramp, off ramp pair per 500 ft of length.

TABLE 8–6. Approximate Service Volumes for Ramps in Passenger Cars per Hour (PHF = 1.00)*†

Level of Service	*Ramp Design Speed (mph)‡*				
	20	*20–30*	*30–40*	*40–50*	*≥50*
A	NA	NA	NA	NA	700
B	NA	NA	NA	1000	1050
C	NA	NA	1125	1250	1300
D	NA	1025	1200	1325	1500
E	1250	1450	1600	1650	1700

NA—This level of service not achievable at this design speed.
*Source: *TRB Circular 212*.
†For two-lane ramps multiply tabular values by 1.7 for <20 mph; 1.8 for 20–30 mph; 1.9 for 30–40 and 40–50 mph; 2.0 for ≥50 mph.
‡For metric conversion: 1 mph = 1.6093 km/h.

given a stated level of service and traffic volumes, each of the elements can be designed to meet this standard.

In establishing levels of service for design purposes, each agency must establish its own guidelines. These have two elements. The first is the traffic volume to be served; for example that for the 30th highest hour in the design year, adjusted to recognize the peak hour factor (PHF). For freeways it is commonly based on a 5-min interval. For large metropolitan areas, this short-time peaking may exceed the hourly flow by 5 to 15% (PHF of 0.94 and 0.87, respectively); in small urban communities it may reach 40% (PHF of 0.71). Arguing for this upward adjustment in design traffic volumes is the fact that a short period of unstable or forced flow and the capacity loss it brings may result in the freeway being clogged for an hour or more. The second step is to proportion the basic freeway, weaving sections, ramps, and ramp junctions to accommodate this flow at the designated level of service using appropriate data.

For basic sections of two- and multilane highways between intersections, the same steps would be followed.

STREET CAPACITY AND LEVELS OF SERVICE

Signalized Intersections

Intersection capacity, rather than that of the street itself, usually determines how many vehicles can be accommodated. Between intersections, the street is alternatively heavily loaded and largely unoccupied. At these intersections, control by traffic signals is common; without them traffic would become almost completely snarled.

As indicated earlier, the capacity of a 12-ft traffic lane is 2000 vehicles per hour. This requires operating speeds of 30 mph or more. For intersection approaches this capacity per hour of green time could only be achieved with perfectly coordinated signals. In almost all cases, however, capacity is reached when the green intervals are fully utilized by waiting vehicles. Many observations were combined to establish that where traffic is not impeded by turning movements, pedestrians, or other interferences, each 12-ft lane will pass automobiles through the intersection at an average rate of one for each 2.1 s of green at a speed of 10 to 15 mph. From this comes a volume of 1700 vehicles per hour of green per 12 ft width.

The average value of 2.1 s per vehicle oversimplifies actual conditions. Rather, the time interval for the first vehicle is about 3 s, but it decreases progressively.[24] Furthermore, drivers encroach on the yellow (caution) signal phase, giving an increase in capacity. Summing up, *TRB Circular 212* assigns a value of 1800 passenger cars per hour for the capacity of a 12-ft lane.

Based on many observations of traffic behavior at intersections, *The Manual* presents graphs and accompanying correction factors for the more usual conditions. Two graphs based on it appear as Figs. 8–13 and 8–14. Multipliers to recognize peak hour factor and location have been tabulated on the figures. It should be noted that the plotted capacities assume 10% each of left and right turns and 5% through trucks and buses, but no local transit buses. Although some of the specifics in these figures may be challenged, they are given here to provide a basis for further discussion.

Factors Affecting Street Capacity and Levels of Service

The Manual divided the factors that affect intersection capacity and levels of service into four categories. These are (1) physical and operating, (2) environmental, (3) traffic characteristics, and (4) control measures. Each of these, in turn, has two or more subcategories. These are discussed in the paragraphs that follow, and also in Chapter 10.

PHYSICAL AND OPERATING—WIDTH OF APPROACH. For one-way streets, *The Manual* stated the capacity of an intersection approach in terms of widths, measured from face of curb to face of curb. For two-way streets, width is that from face of curb to the division line between opposing vehicles or the median. It has been argued that number of lanes is a better measure than width and that lane markings have an effect on capacity. In any event, the width measure seemed best to correlate with observed capacities and traffic behavior and was adopted.

PHYSICAL AND OPERATING—PARKING CONDITIONS. *The Manual* gives different values for streets with or without parking. Where parking is permitted within 250 ft of the intersection, the street is usually classified as "with parking" since capacity is substantially affected. Exceptions are made for approaches that have only a small percentage of total "green" time at the intersection.

The influence of parked vehicles on the effective width of the roadway usually is substantially greater than the space occupied. Driver fear of sudden ma-

[24]See, for example, G. F. King and M. Wilkinson and discussion by D. S. Berry, *TRB Record 615*.

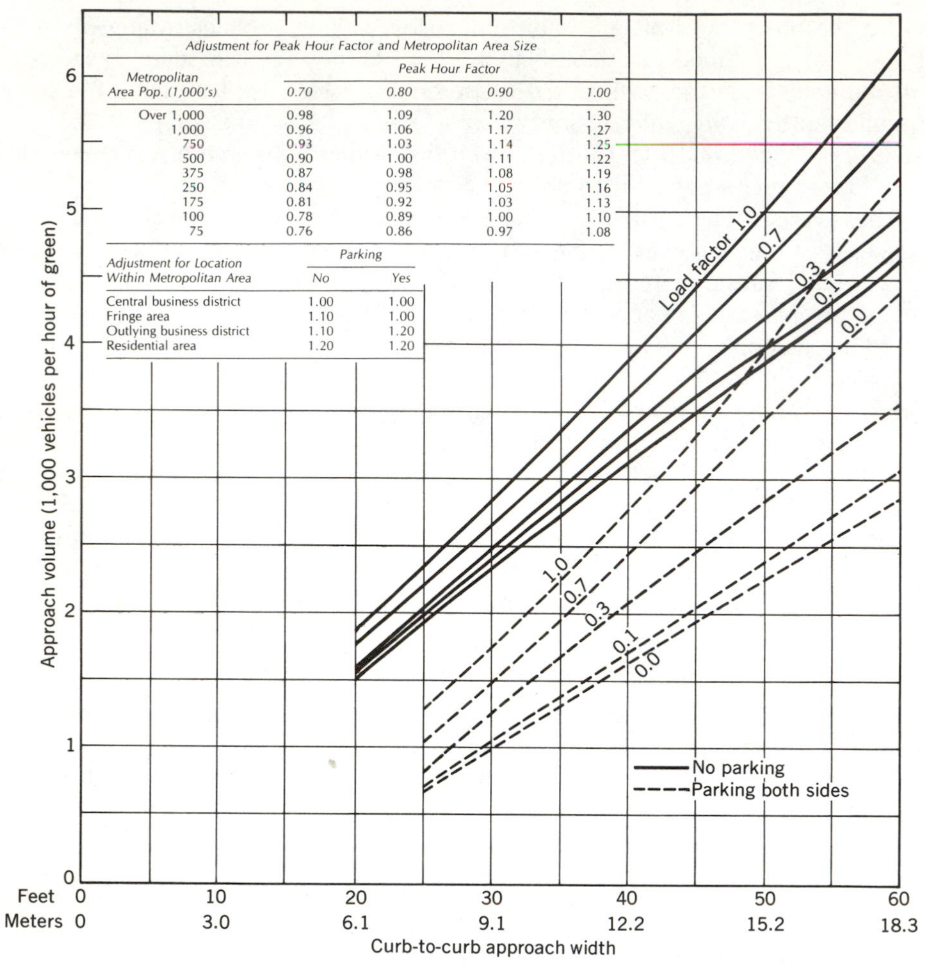

Adjustment for Peak Hour Factor and Metropolitan Area Size

Metropolitan Area Pop. (1,000's)	Peak Hour Factor 0.70	0.80	0.90	1.00
Over 1,000	0.98	1.09	1.20	1.30
1,000	0.96	1.06	1.17	1.27
750	0.93	1.03	1.14	1.25
500	0.90	1.00	1.11	1.22
375	0.87	0.98	1.08	1.19
250	0.84	0.95	1.05	1.16
175	0.81	0.92	1.03	1.13
100	0.78	0.89	1.00	1.10
75	0.76	0.86	0.97	1.08

Adjustment for Location Within Metropolitan Area	Parking No	Yes
Central business district	1.00	1.00
Fringe area	1.10	1.00
Outlying business district	1.10	1.20
Residential area	1.20	1.20

Fig. 8–13. Urban intersection-approach service volume, in vehicles per hour of green signal time, for one-way streets. Conditions: Right turns 10%; left turns 10%; trucks and through buses, 5%; local transit buses, none. (Note: Factors are for no-parking conditions; those for parking conditions differ slightly.) (Source: *The Manual*.)

neuvers or opening doors may cause a loss of 12 to 14 ft of street width, except where streets are narrow and volumes high, where the effect is less severe.

PHYSICAL AND OPERATING—ONE-WAY VERSUS TWO-WAY STREETS. As discussed in more detail in Chapter 10, differences in the operating characteristics of one-way as contrasted with two-way streets strongly affect capacity, as shown by Figs. 8–13 and 8–14. However, direct comparisons of capacities are not sufficient, since an overall analysis may show two-way operation to be more satisfactory in some cases.

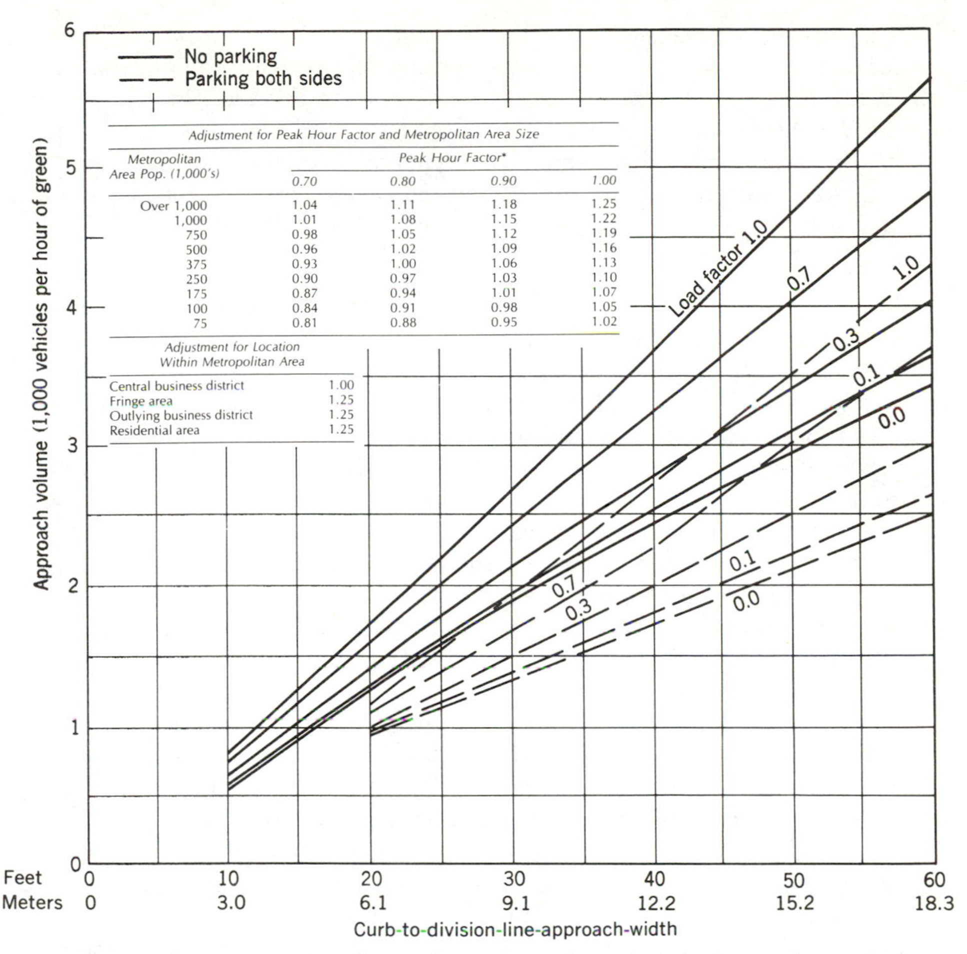

Adjustment for Peak Hour Factor and Metropolitan Area Size

Metropolitan Area Pop. (1,000's)	*Peak Hour Factor** 0.70	0.80	0.90	1.00
Over 1,000	1.04	1.11	1.18	1.25
1,000	1.01	1.08	1.15	1.22
750	0.98	1.05	1.12	1.19
500	0.96	1.02	1.09	1.16
375	0.93	1.00	1.06	1.13
250	0.90	0.97	1.03	1.10
175	0.87	0.94	1.01	1.07
100	0.84	0.91	0.98	1.05
75	0.81	0.88	0.95	1.02

Adjustment for Location Within Metropolitan Area	
Central business district	1.00
Fringe area	1.25
Outlying business district	1.25
Residential area	1.25

Fig. 8–14. Urban intersection-approach service volume in vehicles per hour of green signal time, for two-way streets. Conditions: Right turns, 10%; left turns, 10%; trucks and through buses, 5%; local transit buses, none. (Note: Factors are for no-parking conditions; those for parking conditions are slightly different.) (Source: *The Manual*.)

ENVIRONMENTAL CONDITIONS—LOAD FACTOR. Load factor offers a means for quantifying level of service by measuring the utilization of an intersection approach roadway during 1 hr of peak traffic flow. Specifically, it is the ratio between the number of green phases that are fully utilized and the total number of green phases available. Thus, a load factor of 0.0 means that no green phase during the hour is fully loaded (level of service A); a load factor of 0.3 reflects good operating conditions (level of service C); a load factor of 0.7 approaches unstable flow (level of service D); and a load factor of 1.0 is unstable flow (levels of service E or F).

There has been dissatisfaction with "load factor" as an indicator of level of service because it does not well define situations involving traffic-actuated or coordinated signals or where the intersection is oversaturated. Recent findings comparing observation by several manual methods and time-lapse motion pictures have demonstrated that manual observations of approach delay (approach time minus approach free flow time) is an effective measure of intersection efficiency and thus of level of service.[25] Another alternative is discussed below under critical movement analysis.

ENVIRONMENTAL CONDITIONS—PEAK HOUR FACTOR (PHF). As noted, the peak hour factor recognizes that traffic flows do not remain constant for a full hour. For capacity and level of service analysis for intersections, the PHF is commonly based on a 15-min period. Observed extremes range from 0.47 to 1.0, with those between 0.85 and 0.90 most common. Major arteries in large metropolitan areas will usually have high PHFs while side streets serving activities in a small community will have low ones. Often they are determined separately for each leg of an intersection.

Multipliers to be applied to the plotted values for capacity on Figs. 8–13 and 8–14 to recognize metropolitan area population and location in the area are given in the tables on the figures. The multipliers related to population recognize that drivers in large cities are more skilled in using the streets than those in smaller towns; those for location indicate that interferences from pedestrian conflicts and other causes are usually more severe in the central business district than in outlying areas. Also, tabular values less than 1.0 recognize the need for a safety factor so that occasional long traffic backups will be avoided.

The Manual gives added graphs, not reproduced here, for parking on one side of the street. Approximations can be obtained by interpolating between the no-parking and parking values on Fig. 8–13. Also the no-parking values on Fig. 8–14 can be used to get approximate capacities for rural intersections where parking is prohibited on the traveled way. For average traffic conditions the PHF is about 0.70 and an adjustment factor of 1.00 is recommended; for recreation routes at peak flows (PHF 1.00) the recommended adjustment factor is 1.40.

TRAFFIC CHARACTERISTICS—TURNING MOVEMENTS. As would be expected, turning movements can have a substantial effect on capacity. *The Manual* offers procedures for recognizing them, and the reader is referred to it for details. Some of the influences recognized in *The Manual* are:

1. Effect on capacity per turning vehicle decreases as the number of turning vehicles increases.
2. On two-way streets, the effect of left-turning vehicles is related to the number of opposing vehicles.[26]
3. The effect of turning movements on capacity is dependent on conflicting pedestrian flows.

[25]See H. Sofokidis et al., *HRB Record 453* and "A Technique for Measurement of Delay at Intersections," *Reports No. FHWA RD 76-135* and *76-137* for added detail.

[26]See D. B. Fambro, *TRB Record 644,* for recent data.

4. Turning vehicles cause a relatively greater reduction in capacity on narrow streets than on wide ones.
5. Wider cross streets have increased capacity because left turns can be made more easily, given more space and increased maneuvering speed. The effect of cross-street widths on right turns is variable, depending on such factors as turning radius and pedestrian movement.
6. Provision of separate left-turn lanes, possibly with a separate left-turn signal phase, will have a marked effect on capacity and requires special analysis.[27]

TRAFFIC CHARACTERISTICS—TRUCKS AND THROUGH BUSES. Trucks and through buses which are not scheduled to stop in the vicinity of the intersection reduce capacity because they occupy more space and have slower acceleration rates than passenger cars. Often light trucks are considered to be passenger cars and heavier trucks or buses as two passenger cars. Higher multipliers are considered appropriate where trucks are preponderantly large and heavy.

TRAFFIC CHARACTERISTICS—LOCAL TRANSIT BUSES. *The Manual* and *TRB Circular 212* give data on the effects of local transit buses which have scheduled stops near the intersection on intersection approach capacity. Important factors include the following:

1. Increased volumes of buses reduce capacity in proportion to their number.
2. Effects of buses on capacity are greater in the more congested central business district.
3. Percentage reduction in capacity is inversely proportional to available street width.
4. Location of the bus stop markedly affects capacity. Near-side locations are generally favorable to faster bus operation, since some unloading and loading can coincide with waits for the signal to turn green. But, particularly where parking is banned, they can bring a substantial reduction in approach capacity. For example, 60 buses per hour, loading in the central business district on a 24-ft approach roadway with parking prohibited, will reduce capacity by almost 30%. On the other hand, far-side loading, particularly when combined with heavy turning movements, with or without parking, will not have a substantial effect on capacity. The effects of midblock locations generally require special analysis.

In numerous cities around the world, traffic lanes are reserved for the exclusive use of buses and possibly taxis. In these circumstances, approach width would be reduced accordingly in making capacity and level of service calculations for the remaining vehicles.

TRB Circular 212 offers a *critical movement analysis* approach for determining capacity and level of service for intersections that is in widespread use. It provides a standard form on which a nine-step "planning" analysis can be made. A reproduction of that form, with a typical solution, is given as Fig. 8–15. A brief explanation of the steps is as follows:

Steps 1 and 2. Determine intersection geometry and estimated traffic volumes.

[27]For recent studies of the effect of left turns see E. Fellinghauer and D. S. Berry, *TRB Record 489*, and C. J. Messer and D. B. Fambro, *TRB Record 644*.

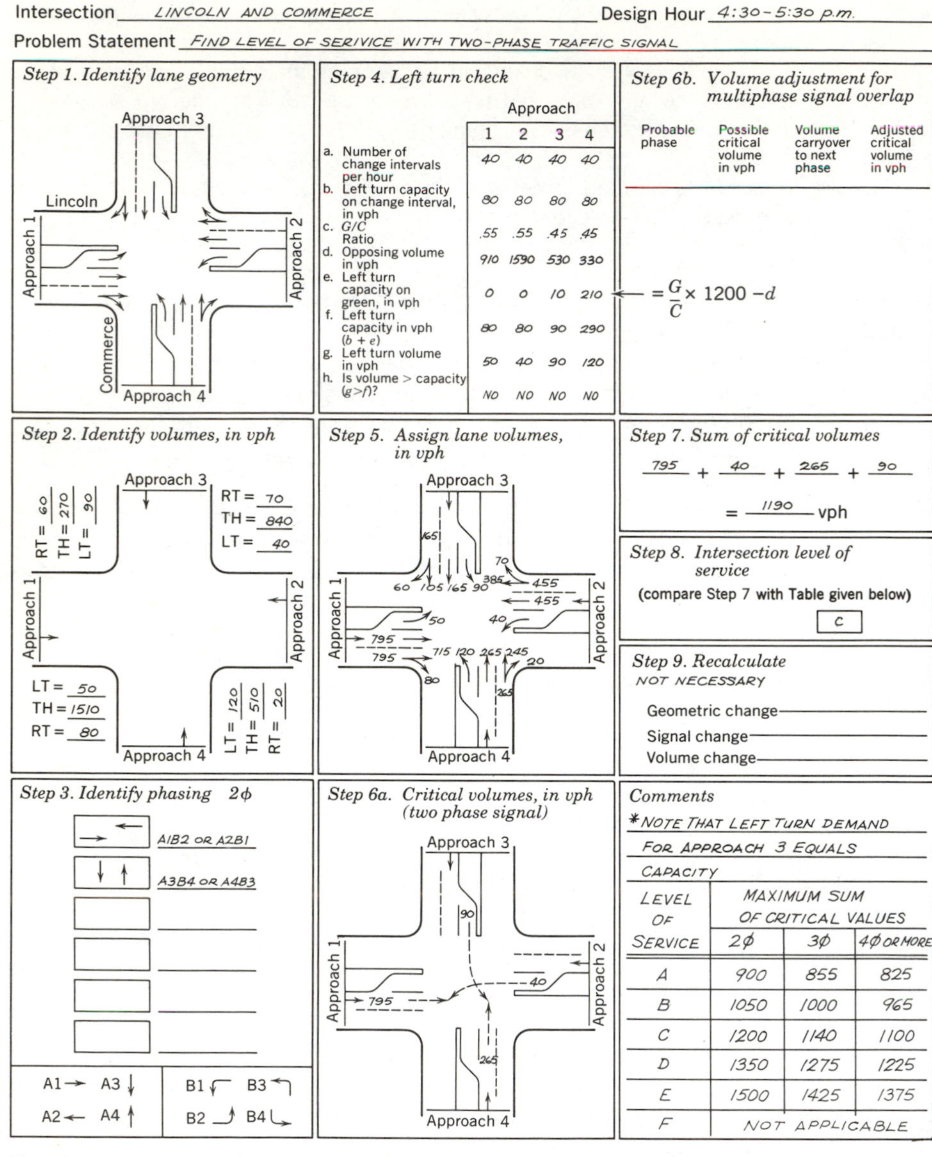

LEVEL OF SERVICE	MAXIMUM SUM OF CRITICAL VALUES		
	2ϕ	3ϕ	4ϕ OR MORE
A	900	855	825
B	1050	1000	965
C	1200	1140	1100
D	1350	1275	1225
E	1500	1425	1375
F	NOT APPLICABLE		

Fig. 8–15. A typical critical movement planning analysis. (Adapted from *TRB Circular 212.*)

Step 3. Assign signal phasing. In this case timing is two-phase, with left turns through opposing traffic.

Step 4. Make left-turn check:

a. Set length of signal cycle—1.50 min or 40 per hour.

b. Determine left-turn capacity *(c)* during signal change interval. This is two vehicles per cycle or 80 per hour.

c. Assign cycle split—55% to horizontal and 45% to vertical movement.

d. Record the number of vehicles opposing the left turns.
e. Determine left-turn capacity during green interval by formula shown.
f. Compute left-turn capacity $(b+e)$.
g. Record left-turn volume.
h. Determine if volume exceeds capacity. Approach 3 is at capacity, others have capacity to spare.

Step 5. Assign lane volumes: all left turns to left-turn lanes and remaining approach movements equally to other lanes.

Step 6a. Determine critical volumes. For the horizontal movement, the larger of the sums of the two conflicting movements are one lane of A1(795) + the opposing left turns A2(40). For the vertical movement, the larger of the sums is A4(265) + the opposing left turns A3(90).

Step 6b. This computation does not apply to two-phase signals but to more complex signal phasings, for example where left-turn movements B1 and B2 have a separate phase and can, with traffic actuation, delay either movement A1 or A2.

Step 7. Sum the critical volumes. The total is 1190.

Step 8. Determine level of service (see table at lower right of Fig. 8–15). Since 1190 is less than 1200, the level of service is C.

A parallel critical movement computation for "operations and design" is also given in *TRB Circular 212*. It corrects volumes for trucks and buses, applies a peak hour factor, and provides for turn adjustments. Because of space limitations, neither that analysis nor a more complex "planning" analysis can be offered here.

Unsignalized Intersections Controlled by Two-Way Stop or Yield Signs

A widely-used procedure for determining the capacity and level of service of unsignalized intersections when a cross or T street is controlled by stop or yield signs is offered in *TRB Circular 212*. The conflicts among traffic entering from or into the controlled streets with traffic on the uncontrolled street are detailed separately for each movement. Then, given the sum of traffic in all conflicting streams and observed values of the critical gaps that drivers accept for a particular maneuver, the potential remaining capacity is determined. Further reductions are applied for impedances where there is more than one conflict and where more than one movement must share a single lane. Finally, a "reserve" or unused capacity for each movement is found and this is related to level of service. Reserve capacity of over 400 vehicles per hour is assigned level of service A, values less than 0 are level of service E, extreme congestion. The analysis of each movement utilizes a two-page computation form, tabular data on critical gaps ranging from 5 to 10 s for high and low speeds and two and four lanes on the main street, and two graphs. Space limitations preclude presenting these details here.

Unsignalized Intersections Controlled by Four-Way Stop Signs

With four-way stop signs, vehicles from all legs of an intersection have equal access to the conflict area. Capacity is set by the number of vehicles that can pass through. Values recommended by *The Manual* are given in Table 8–7.

TABLE 8–7. Examples of Total Capacities of Intersections Controlled by Four-Way Stop Signs*

Intersection Type	*Two-Lane by Two-Lane*					*Two-Lane by Four-Lane*	*Four-Lane by Four-Lane*
Demand split	50:50	55:45	60:40	65:35	70:30	50:50	50:50
Capacity (vehicles per hour)	1900	1800	1700	1600	1550	2800	3600

*Source: *Highway Capacity Manual.*

OTHER CAPACITY AND LEVEL OF SERVICE SITUATIONS

Bus Lanes

Provision is being made increasingly for exclusive lanes on freeways for buses, either by new construction or reserving lanes for them. For uninterrupted flow, *TRB Circular 212* proposes that a bus is the equivalent of 1.6 passenger cars. This would give a headway of 2.9 s and a theoretical capacity of 1250 buses per hour. Assuming typical transit bus capacities ranging from 55 to 85 passengers, including standees, and articulated ones accommodating 110 to 172 (European), theoretical lane capacities would be respectively 69,000, 106,000, 137,000 and 217,000 passengers per hour. At the limit of level of service C (1000 buses per hour) and with all seats filled but no standees (36, 53, 69, and 48 passengers, respectively), persons per hour would be 36,000, 52,000, 70,000, and 48,000 passengers per hour.

Capacities of buses on city streets vary widely with such factors as near- versus far-side loading, bus loading and unloading arrangements, passenger hand baggage, and fare collection arrangements. *TRB Circular 212* provides data from which capacities can be calculated.

Pedestrians

Provision for pedestrians as well as for motor vehicles is required in urban areas and for access to and egress from places where people assemble. Walking and standing patterns are complex and affected by obstructions, by purpose such as hurrying commuters or leisurely shoppers, and by the possible mixture of ages. Many details are offered in *TRB Circular 212*. Among them are capacities for various levels of service. For walkers, these range from 40 ft^2 per person and a flow rate of 6 persons per minute per foot of clear width for level of service A to 6 ft^2 and a flow of 25 persons per minute for level of service F. For standing, the range is from 13 ft^2 per person and a spacing of 4 ft for level of service A to 2 ft^2 and direct physical contact with others at level of service F.

SIGHT DISTANCE

For safe vehicle operation, a clear line of sight of suitable length must be provided along the road. For each situation, the minimum safe sight distance can be computed by the principles of dynamics, using appropriate multipliers or coefficients to recognize characteristics of driver, vehicle, road, or their combined effects. Thus, the equations that will be given are fundamental; however, the results obtained are of necessity based on measured or observed performance of driver, vehicle, or road.

The concept of safe sight distance has three facets—*stopping* or *'nonpassing'*, *passing*, and *decision*. These are discussed separately here. Their applications to the design of vertical and horizontal alignment are treated in Chapter 9.

Stopping (Nonpassing) Sight Distance

At times large objects may drop onto a roadway and will do serious damage to a motor vehicle that strikes them. Also, pedestrians or animals may enter the traveled way. Again, a car or truck may stop in the traffic lane in the path of following vehicles. In such instances, proper design requires that such hazards become visible at distances great enough that drivers can stop before hitting them. Furthermore, it is unsafe to assume that the oncoming vehicle may avoid trouble by leaving the lane in which it is traveling, for this might result in loss of control, leaving the traveled way, or collision with another vehicle.

Stopping sight distance is made up of two elements. The first is the distance traveled after the obstruction comes into view but before the driver applies the brakes. During this the vehicle travels at its initial velocity. The second distance is consumed while the driver brakes the vehicle to a stop. Expressed in formulas, the distances covered are:

detection, recognition, decision, and response-initiation distance (ft) $= tv_t = 1.47tV$ (8–1A)

detection, recognition, decision, and response-initiation distance (m) $= t\ m/s$ (8–1B)

braking distance (ft) $= v^2/2g_t f = V^2/30f$ (8–2A)[28]

braking distance (m) $= \dfrac{(m/s)^2}{2g_m f} = \dfrac{(m/s)^2}{19.6f}$ (8–2B)[28]

where

v_t = initial speed, feet per second
V = initial speed, miles per hour
m/s = initial speed, meters per second
t = detection, recognition, decision, and response initiation (brake-reaction) time.
g_t = acceleration of gravity, 32.2 feet per second squared
g_m = acceleration of gravity, 9.80 meters per second squared
f = coefficient of friction between tires and pavement

[28]Derivation of these formulas will be found in any standard textbook in engineering mechanics.

Table 8–8 gives the assumptions underlying and computations for minimum stopping sight distance. It will be seen that for each design speed two values for "Assumed Speed for Condition" are offered. Computations based on these assumed speeds are carried through using Formulas 8–1A and 8–2A and assumed braking reaction time and coefficients of friction to give two stopping sight distances. The high values, which are based on assumed speed equaling design speed, are recommended for freeways and other high-volume facilities; the lower ones are acceptable for less-heavily traveled roads and streets. The lower values were used in all designs until 1971, because it was assumed that drivers slowed down when pavements were wet. However, this behavior is not borne out by recent observations.

The values for stopping sight distance given in Table 8–8 are based on passenger car performance but are considered appropriate for trucks also. Observations have shown that trucks have poorer braking abilities with stopping distances 50% greater at 30 and 100% greater at 70 mph. Offset against this is the greater eye height of truck drivers which permits them to see over preceding vehicles and also gives longer sight distance over crests. An exception is where downgrades are combined with limited horizontal sight distance; in this instance, provision of sight distances greater than those of Table 8–8 are recommended.

In contrast to common design practice, the values for stopping sight distance contain almost no margin of safety. As pointed out earlier, a 2.5-s detection, recognition, decision, and response initiation time for the nonalerted, poorer-than-average driver represents a minimum figure. Furthermore, the assumed coefficients of friction are far from conservative, as is demonstrated in the discussion of skid resistance which follows.

Formulas 8–2A and 8–2B for braking distance assume a level roadway. If the vehicle is traveling uphill, braking distance is decreased, for gravity forces aid in slowing the car. For downhill operation, braking distance is increased. Braking distance on slopes is expressed by the approximate formulas

$$\text{braking distance (ft)} = \frac{V^2}{30(f+g)} \qquad (8\text{–}3A)$$

$$\text{braking distance (m)} = \frac{(m/s)^2}{19.8\,(f+g)} \qquad (8\text{–}3B)$$

where

g = longitudinal slope of the roadway, or % grade/100

Uphill grades are positive (+) and downhill grades, negative (−). For flat grades and low speeds, braking distance is little influenced by this correction. For example, a design speed of 30 mph, and a 6% grade, braking distance changes only 20 ft. For 60 mph and a 6% downgrade, however, the change is substantial, amounting to 110 ft.

TABLE 8.8. Minimum Stopping Sight Distances (Wet Pavements)*†

Design Speed (mph)	Assumed Speed for Condition (mph)	Brake Reaction		Coefficient of Friction f	Braking Distance on Level (ft)	Stopping Sight Distance	
		Time (s)	Distance (ft)			Computed (ft)	Rounded for Design (ft)
20	20–20	2.5	73– 73	0.40	33– 33	106– 106	120– 120
30	28–30	2.5	103–110	0.35	75– 86	178– 196	200– 200
40	36–40	2.5	132–147	0.32	135–167	267– 314	275– 325
50	44–50	2.5	161–183	0.30	215–278	376– 461	375– 475
60	52–60	2.5	191–220	0.29	311–414	502– 634	525– 650
65	55–65	2.5	202–238	0.29	348–486	550– 724	550– 725
70	58–70	2.5	213–257	0.28	400–583	613– 840	625– 850
75	61–75	2.5	224–275	0.28	443–670	667– 945	675– 950
80	64–80	2.5	235–293	0.27	506–790	741–1,083	750–1,100

*Source: *A Policy on Geometric Design.*

†*The metric conversion units are 1 mph = 1.6095 km/h; 1 ft = 0.3048 m.*

Skid Resistance

The subject of skid resistance, and in turn, of stopping coefficients of friction, is extremely complex. It has been and continues to be the object of intensive research.[29] Variations occur with many factors, among them whether the pavement is dry, wet, muddy, or icy, pavement surface texture and roughness, whether or not the vehicle wheel is rotating (incipient sliding) or locked (sliding), with vehicle speed, with tire inflation and tread design (including grooves), with tire wear as it affects smoothness, and with the temperature, viscosity, adhesive, and energy-absorbing (hysteresis) characteristics of the tread rubber. The degree of variation with some of these factors is illustrated by Figs. 8–16 and 8–17, which represent typical research findings. As shown, skid resistance can be stated several ways. One is as the coefficient of friction, the ratio between parallel and normal forces. Another is as skid number which is 100 times total parallel forces divided by total normal forces. Yet another gives two coefficients, "adhesion" (see Fig. 8–16) and "hysteresis" or damping losses in the rubber in the tire which sum to the coefficient of friction.

Performance of brakes is a primary control in setting a design coefficient of friction. As indicated by Fig. 8–16, it is far lower when a tire skids than when the wheel is turning. Since, in emergencies, most drivers instinctively set the brakes and lock the wheels, sliding coefficients must be used in design unless

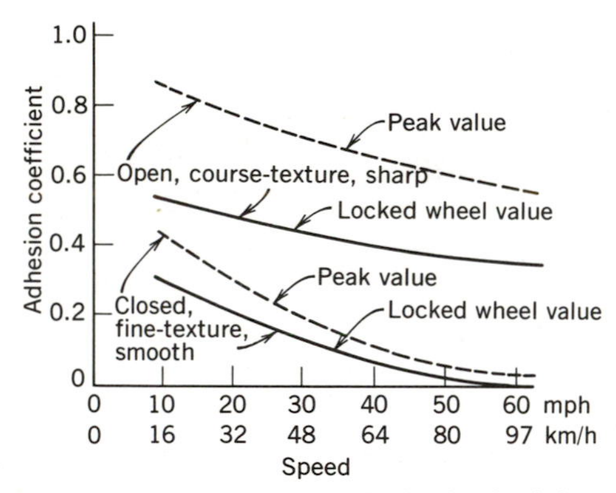

Fig. 8–16. Peak and locked-wheel (sliding) adhesion coefficients for wet fine-textured smooth and open-textured sharp aggregate pavement surfaces (Source: G. G. Balmer, *TRB Record* 666.)

[29]*TRB Record 621, 622, 623,* and *624,* totaling 851 pages, summarize some 50 yr of worldwide skid resistance research until 1977. Among other sources are *Guidelines for Skid Resistant Pavement Design,* AASHTO and *TRB Record 523, 584, 602, 712, 715,* and *777*. Only the briefest sketch of these findings can be given here. However, references to specific studies on certain aspects of the problem appear below. Among the earliest is *Iowa Engineering Experiment Station Bulletin 120,* 1934, by R. A. Moyer, a pioneer in the field.

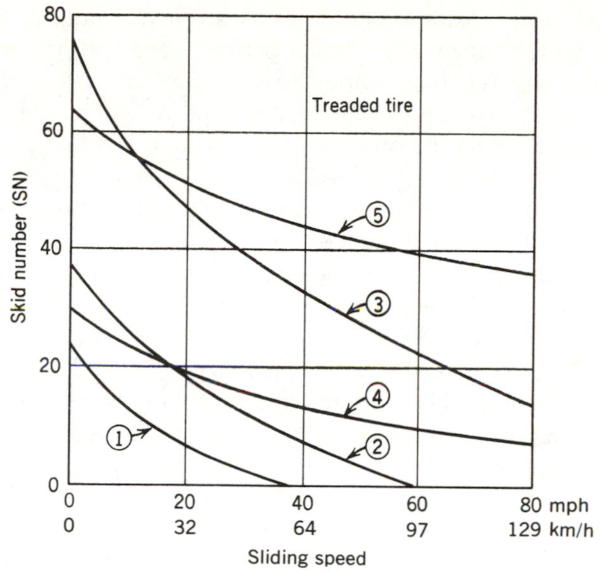

Fig. 8–17. Skid numbers for various surface types (wet) and vehicle speeds. *Surface Types*: A *Smooth*—bleeding asphalt, highly polished stone, asphalt or cement concrete; B *Fine-Textured, Rounded*—worn stone or silica sand of fine gradation; C *Fine-Textured, Gritty*—new silica sand or metal carbide epoxy; D *Coarse-Textured, Rounded*—polished slag or limestone or uncrushed gravel of large size; E *Coarse textured, Gritty*—new slag of large gradation possessing large- and small-scale macroscopic roughness or limestone surfaces with more than 10% sand-sized siliceous material. (Source: *NCHRP Report 37*.)

or until braking systems that prevent locking of the wheels are standard equipment. Although nonlocking brakes for passenger cars have been developed, to date they have not been made available because of their greater cost.[30] Nonlocking brakes for trucks, controlled by an on-board computer, also have been demonstrated. An added advantage of nonlocking brakes is that, with them, drivers retains the ability of steer which means that they may be able to change a vehicle's path or prevent it from spinning out of control. At times high brake temperatures resulting from stops at high speeds or continuous or repeated application may cause the brakes temporarily to "fade," so that even the locked-wheel stopping force is not developed. In other instances wet brakes may not engage satisfactorily.

Seasonal and weather factors have strong influences on skidding behavior.[31]

[30]See N. H. Pulling, *TRB Record 623*.
[31]See J. M. Rice, *Public Roads*, Mar. 1977 for an excellent summary. Also, see B. L. Elkins et al., *TRB Record 777*.

Rain is a major factor. Skidding coefficients are lower on wet than dry pavements because water impairs the contact between tire and pavement. To establish this contact, the tire must wipe the water away. It follows that a well-designed tread, grooved to provide ready escape of water, will perform better than a poorly designed or smooth one. Also, a thick layer of water from heavy rainfall which is not completely wiped away may result in extremely low coefficients of friction, particularly at high speeds, a phenomenon referred to as hydroplaning. Under these circumstances, contact between tire and road does not develop. Hydroplaning is primarily a high-velocity phenomenon. It has been found that gritty, sandpaper-like surfaces produce suitable frictional coefficients at speeds lower than about 40 mph, but that at higher speeds, drainage potential produced by aggregate of ¼ in. or larger size is needed.[32] Furthermore, during long periods of dry weather, a film composed largely of unburned crankcase oil and rubber may form on the pavement surface. This can greatly reduce the coefficient of friction when the pavement is first wet by rain. It has been found that coefficients of friction increased by 20 to 60% after the traffic film was removed by the action of heavy rains coupled with scour from traffic and sand.

Research also has established that surface roughness and wearing characteristics of the aggregate in both bituminous and concrete pavements are critical in developing and maintaining high coefficients of friction. Pavements constructed of limestones have proved particularly troublesome since the aggregate sometimes polishes under traffic. On the other hand, sandstones and certain granites and trap rocks undergo what is termed "coarse wear" as individual, loosely bonded grains are plucked away. Other aggregates contain minerals of different hardnesses which wear differentially. Pavements containing these retain high coefficients of friction. Presently, some highway agencies are specifying that nonpolishing aggregates be used in the surface layer of pavements. In addition, slick pavements are being resurfaced or seal coated with nonpolishing materials (see Chapter 19), or are being grooved (see Chapter 20). This procedure may not greatly increase skid resistance, but is effective in reducing uncontrolled skidding.

Bituminous pavements that "bleed" from an excess of binder can become extremely slick since repeated working by vehicle tires forces the asphalt and some fines to the surface. However, bituminous pavements can be constructed which have and retain high coefficients of friction. This requires that nonpolishing aggregates be used and that proportioning be carefully controlled.

Roughness of the pavement surface also can substantially reduce the coefficient of friction by producing wheel slip and vertical wheel vibration. In a laboratory situation, losses of 30% were recorded.[33]

Coefficients of friction on earth or gravel roads vary widely. For example, braking performance on a smooth dry clay is excellent, but very poor if the surface is wet. Again, tightly bonded exposed coarse aggregate has a high coeffi-

[32]See, for example, J. G. Rose and B. M. Gallaway, *Transportation Engineering Journal of ASCE*, July 1977.
[33]See J. C. Wambold et al., *HRB Record 471*.

cient of friction, but a loose float rolls under the tires so that little braking force develops.

The preceding discussion has suggested the complexity of the problem of skid resistance. As would be expected, coefficients of friction vary widely; reported extremes have ranged from 1.0 or more down to 0.15. Muddy surfaces develop coefficients in the range of 0.10 and icy ones as low as 0.05. These latter values are far below recommendations for design as given in Table 8–8.

Studded tires were developed in Scandinavia about 1961 to alleviate the skidding problem and improve vehicle control under ice and snow conditions. By the mid-1960s all but a few states had enacted legislation legalizing them, at least during the winter months. Thirty-eight still had it in 1979. Serious pavement wear has been reported, in some cases 0.2 in./yr in the wheel track, bringing rainwater puddling and an accompanying splash and threat of hydroplaning. For these reasons, highway agencies have been faced with extremely high repair costs, so that today studded tires are viewed with disfavor by many highway officials. Although research is under way to develop materials and designs for both studded tires and pavements that will make them acceptable, their continued use is an unsettled question. In addition, the effectiveness of studded tires in reducing accidents is questionable. A review of accident records after studded tires had been permitted and then banned indicated a 0.6% increase in accidents in the December–March period of their use. One possible explanation is that drivers became less cautious while using them.[34]

One of the confusing factors affecting both research and field measurement of skid resistance is the multiplicity of devices employed. Among them are locked-wheel trailers, nonlocked-wheel devices, automobile methods, and portable field and laboratory testers. Often they give substantially different results.[35]

Even if high coefficients of friction under all conditions could be assured by improvements in tires and pavements, other factors prevent their use for design purposes. Tests conducted by the Bureau of Standards indicate that the maximum acceptable deceleration rate is 16.1 ft/s (or ½ g). This is produced by a coefficient of friction of 0.5 and indicates a force equal to one-half of the riders' weight trying to slide them from the vehicle seat or to rotate their upper bodies around lap seat belts if they are being worn. At values exceeding this level, injury to vehicle occupants could occur unless they were tightly restrained. Similarly, rapid deceleration would bring serious problems of cargo damage in trucks. Actually, only in emergencies do drivers employ high-friction factors. *NCHRP Report 154* indicates that on intersection approaches, the median driver develops a braking force of 0.2 of gravity.

A variety of issues makes decisions regarding investments to improve skid resistance difficult. First, it is often a wet or icy pavement phenomenon and 80% of fatal and 69% of all accidents occur on dry pavements. Second, there is the problem of legal liability of agencies and their officials, as discussed in Chapter

[34]See *NCHRP Report 61, 176,* and *183,* and *NCHRP Synthesis 32* for an evaluation of the effectiveness and damage-producing characteristics of studded tires.
[35]See *NCHRP Report 151* for added detail.

7. This liability may exist whenever a skidding accident occurs. However, if the agency inventories its system and finds coefficients of friction lower than some arbitrarily set level, this may establish a far stronger base for claims. Again, it may be that research recommendations will open new legal challenges. *NCHRP Report 37* recommends a minimum skid number of 37 for main rural highways and *NCHRP Report 154* proposes that, as intersection approach speeds rise, increasingly high specific skid numbers be provided with which agencies must comply, along with longer approach lengths. Do these then become standards? Another issue is that of noise; resurfacing to give high skid resistance has in some cases produced levels 10 to 15 dBA (See Chapter 13) higher and has brought a storm of complaints. Yet another quandary is the economic one of using an agency's limited funds to provide high skid resistance on stretches of low-volume roads and streets between intersections where braking is seldom required, thus disregarding other pressing needs.

Passing Sight Distance

On two-lane highways, opportunity to pass slow-moving vehicles must be provided at intervals. Otherwise capacity decreases and accidents may occur as impatient drivers risk head-on collisions by passing when it is unsafe to do so.[36] The minimum distance ahead that must be clear to permit safe passing is called the passing sight distance.

In deciding whether or not to pass another vehicle, drivers must weigh the clear distance available to them against the distance required to carry out the sequence of events that make up the passing maneuver. Among the factors that will influence their decisions are the degree of caution that they exercise and the accelerating ability of their vehicles. Because humans differ markedly, passing practices, which depend largely on human judgment and behavior rather than on the laws of mechanics, vary considerably among drivers. To establish design values for passing sight distances, engineers observed the passing practices of many drivers. Behavior of an appreciable percentage of them, but not of the average, was reconstructed.

The basic studies underlying passing sight distance standards were made during the period 1938 to 1941. Additional observations made in 1957 indicated that passing practices were somewhat more conservative even though vehicles had greater ability to accelerate. Assumed operating conditions were as follows:

1. The overtaken vehicle travels at a uniform speed.
2. The passing vehicle has reduced speed and trails the overtaken one as it enters the passing section.
3. When the passing section is reached, the driver requires a short period of time to perceive the clear passing section and to react and start acceleration.
4. Passing is accomplished under a delayed start and early return. The passing vehicle continuously accelerates; its average speed during occupancy of the left lane is 10 mph higher than that of the overtaken vehicle.

[36]Fatalities involving passing on two-lane roads total about 5000 per year.

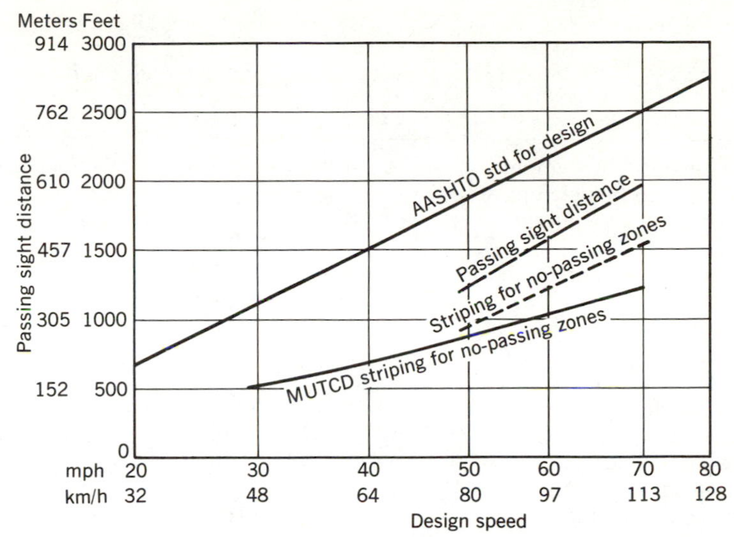

Fig. 8–18. Standard and recommended passing sight distances.

5. When the passing vehicle returns to its lane there is a suitable clearance length between it and an oncoming vehicle in the other lane.

Using these assumptions and observed behavior, the four distances that in sum make up passing sight distances were computed as shown in Table 8–9. These findings, adjusted slightly, represent the current AASHTO standards shown in Table 8–10 and on Fig. 8–18.

Research continues to develop passing sight distance requirements that more nearly conform to both theoretical considerations and driver behavior. One recent study proposed values based on $\frac{2}{3}d_2 + d_3 + d_4$.[37] Its recommendation is plotted as a dashed line on Fig. 8–18.

Requirements for design such as those for passing sight distance may not always be followed in other situations. This is shown by the plot on Fig. 8–18 of striping requirements for no-passing zones as given in the *Manual on Uniform Traffic Control Devices* which is based on the "short zone" concept which permits return to the lane over the solid stripe. It can be seen that the length of striping is roughly half the minimum passing sight distance called for by AASHTO standards. Another striping proposal by Harwood and Glennon (see references) is plotted as the dotted line on Fig. 8–18.

Decision Sight Distance

There are situations where drivers may be called on to make unexpected maneuvers which introduce a likelihood for errors in information reception, decision making, or control actions. Among the locations where this can happen are

[37]See D. W. Harwood and J. C. Glennon, *TRB Record 601*. See also W. G. Weber, *ITE Journal*, Sept. 1978 and D. L. Woods and G. D. Weaver, *TRB Record 737*.

TABLE 8–9. Elements of Safe Passing Sight Distance—Two-Lane Highways

Element	*Speed Group (mph)*			
	30–40	*40–50*	*50–60*	*60–70*
Average passing speed, V (mph)	34.9	43.8	52.6	62.0
Difference in speed between passing and passed vehicles, m (mph)	10	10	10	10
Initial maneuver:				
a = average acceleration (mphps)*	1.40	1.43	1.47	1.50
t_1 = time (s)*	3.6	4.0	4.3	4.5
d_1 = distance traveled (ft) = $1.47t_1 \left((V-m)+\frac{at_1}{2}\right)$	145	215	290	370
Occupation of left lane:				
t_2 = time (s)*	9.3	10.0	10.7	11.3
d_2 = distance traveled (ft) = $1.47Vt_2$	475	640	825	1,030
Clearance length:				
d_3 = distance traveled (ft)*	100	180	250	300
Opposing vehicle:				
d_4 = distance traveled (ft) = $\frac{2}{3}d_2$	315	425	550	680
Total distance, $d_1+d_2+d_3+d_4$ (ft)	1,035	1,460	1,915	2,380

*For consistent speed relation, observed values adjusted slightly. Metric conversion units: 1 mph = 1.6km/h.

TABLE 8–10. AASHTO Recommendations for Minimum Passing Sight Distance for Design of Two-Lane Highways

	Assumed speeds		*Minimum Passing Sight Distance (ft)*	
*Design Speed (mph)**	*Passed Vehicle (mph)*	*Passing Vehicle (mph)*	*Computed*	*Rounded*
20	20	30	810	800
30	26	36	1,090	1,100
40	34	44	1,480	1,500
50	41	51	1,840	1,800
60	47	57	2,140	2,100
65	50	60	2,310	2,300
70	54	64	2,490	2,500
75	56	66	2,600	2,600
85	59	69	2,740	2,700

*Metric conversion units: 1 mph = 1.6 km/h; 1 ft. = 0.3/m.

approaches to intersections, interchanges, toll plazas, and lane drops where information provided the driver may be incomplete. In these instances, A *Policy on Geometric Design* calls for the provision of "Decision Sight Distances." They are based on assigned time values for each speed and are, for 30 and 80 mph, respectively; detection and recognition, 1.5 to 3 s; decision and response initiation, 4.2 to 7.0 s; and maneuver or lane change, 4.5 to 4.0 s. The resulting decision sight distances range as follows: for 30 mph, 450 to 625 ft; for 50 mph, 750 to 1025 ft; and for 80 mph, 1250 to 1650 ft.

CURVATURE, SUPERELEVATION, AND SIDE FRICTION

The alignment of a road as shown on the plan view is a series of straight lines called tangents connected by circular curves. In modern practice it is common to interpose transition (spiral) curves between tangents and circular curves (see Fig. 6–4). The subsequent paragraphs present the laws of mechanics and the physical and human factors that control alignment design. Applications of them to highway practice appear in Chapter 9.[38]

As a vehicle travels the curved section of the road it is subjected to centrifugal force. If the surface is flat, the vehicle is held in the curved path by side friction between tires and pavement. The total of these friction forces (F_L and F_R of Fig. 8–19a) equals the centrifugal force which, in traditional units, is Wv^2/gR. Expressed in terms of the coefficient of friction f and the normal forces between pavement and tires, this relationship is

$$\frac{Wv^2}{gR} = (N_L + N_R)f = Wf$$

When velocity is stated in miles per hour, the radius in feet, and the equation is reduced, the relationships of coefficient of friction, velocity, and radius are

$$f = \frac{V^2}{15R} \qquad (8\text{–}4)$$

Centrifugal force acts above the roadway surface through the center of gravity of the vehicle (see Fig. 8–19*a*) and creates an overturning moment about the points of contact between the outer wheels and the pavement (point *B*). Opposing overturn is the stabilizing moment created by the weight *W* acting downward through the center of gravity. Overturn can occur only when the overturning moment exceeds the stabilizing moment. Modern passenger cars have low centers of gravity and, consequently, the overturning moment is relatively small. As a result, they generally slide sidewise rather than overturn. Many trucks have high centers of gravity so that relatively large overturning moments can be created. They may overturn before they slide.

Curved sections of modern highways are almost always superelevated—that is, the roadway surface is sloped upward toward the outside of the curve. By

[38]Vehicle performance on curves is dependent on many complicating factors far beyond the scope of this book. But these are so variable that they are ignored in highway design. For example, the sharpness of the curve that the vehicle travels varies substantially from that of the highway (see J. C. Glennon and G. D. Weaver, *HRB Record 390*). Also, downgrades combined with curvature and superelevation can make a maneuvering vehicle unstable (see W. Zuk, ibid.).

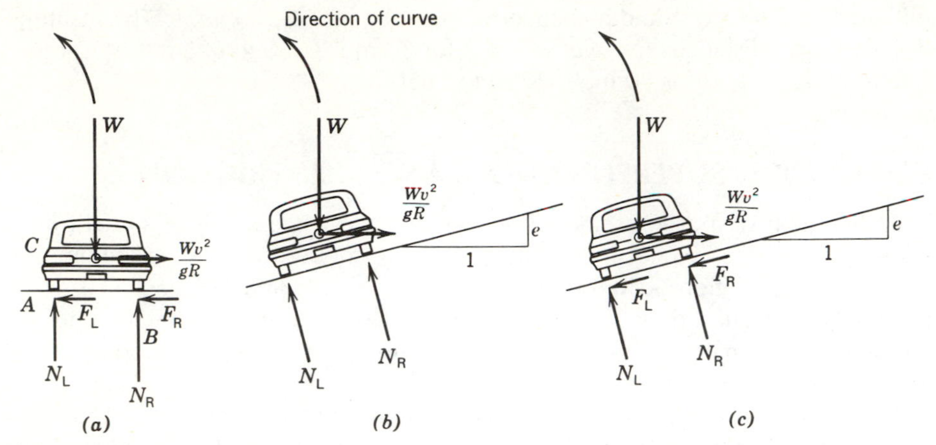

Fig. 8–19. Forces acting on motor vehicles in traveling in curved paths.

this means, the tendency of vehicles to slide outward or to overturn can be partially or entirely offset. For each combination of radius of curve and speed of travel, there is a specific superelevation that exactly balances centrifugal force. In this idealized situation, the reactions between tires and pavement act in a direction normal to the pavement surface. Forces acting on the vehicle are illustrated in Fig. 8–19*b*. The relationship among superelevation, velocity, and radius of curvature, expressed in units of feet and second, is

$$\frac{e}{1} = \frac{Wv^2/gR}{W}$$

where e is the superelevation stated in feet of rise per horizontal foot across the roadway. With velocity expressed in miles per hour this equation reduces to

$$e = \frac{V^2}{15R} \qquad (8\text{–}5)$$

When a vehicle travels at speeds greater than those at which the superelevation balances all centrifugal force, side friction again is needed to keep it in the curved path. Forces acting on the vehicle are shown in Fig. 8–19*c*. The coefficient of friction developed as the vehicle follows the curved path is

$$f = \frac{V^2}{15R} - e \qquad (8\text{–}6A)$$

In SI units of meters per second (m/s) and radius in meters (R_m), this equation becomes

$$f = \frac{(m/s)^2}{9.8R_m} - e \qquad (8\text{–}6B)$$

Maximum coefficients of side friction on dry pavements as determined by circle or curve tests normally range between 0.4 and 0.5. Somewhat lower coefficients are developed on wet pavements. Coefficients as low as 0.20 have been recorded at normal speeds on wet and glazed or bleeding bituminous surfaces, at 0.20 on snow, and 0.05 on glare ice or mud. Values with normal pavements and smooth tires have been reported as about 0.35 at 45 mph. At times, the statement appears that tire squeal on curves can be used as a measure of the coefficient of side friction and thus as a measure of maximum safe speeds on curves. Such is not the case.[39]

Full advantage of commonly realizable coefficients of side friction cannot be taken in highway design. Centrifugal force causes drivers to develop a feeling of discomfort and they instinctively slow their vehicles. Thus, unless the pavement surface is icy or muddy, the maximum coefficients of side friction recommended by AASHTO (see Table 8–11), which are based on driver reaction, are far below the values at which slipping occurs. Incidentally, a skillful highway designer takes advantage of this instinctive driver reaction and "builds in" speed control at critical locations. With reduced superelevation, side friction increases (see Equation 8–6) and a driver becomes uncomfortable and slows down.

TABLE 8–11. AASHTO Recommendations for Maximum Coefficients of Side Friction

Design speed (mph) Design speed (km/h)	20 32	30 48	40 64	50 80	60 97	70 113	80 129
Coefficient	0.17	0.16	0.15	0.14	0.12	0.10	0.08

The centrifugal forces that act on slow-moving vehicles are relatively small. When these vehicles travel around superelevated curves, side-friction forces acting outward from the center of the curve must be developed between tires and pavement; otherwise the vehicles will slide inward. For standing vehicles there is no centrifugal force, so that in this instance the coefficient of side friction must equal the superelevation. Since highways are used throughout the year, maximum superelevation must never exceed the minimum coefficient of friction that will develop under the most adverse weather conditions. Maximum superelevation recommended by AASHTO for any condition is 0.10. Where ice and snow conditions prevail, this maximum is reduced to 0.08. For urban situations, 0.06 or even 0.04 is the recommended maximum.

Application of the concepts developed here appear in Chapters 9 and 10, which follow.

PROBLEMS

8–1*A*. Compute the detection, recognition, decision, and response initiation time for the average driver approaching a blind intersection which has possible conflicts on the other three legs. At 40 mph, what is the distance traveled during this interval?

[39]*NCHRP Report 37* cites observations on a curve connecting two main rural highways showing actual friction numbers (coefficient of friction × 100) for cornering alone of 14 and for cornering and braking and for cornering and accelerating of 36.

8–1*B*. Work Problem 8–1A for an approach speed of 65 km/h. State the distance in meters.

8–2. In a written statement, clearly define and explain the significance of the following: (a) highway capacity; (b) level of service; (c) peak-hour factor; (d) thirtieth hour volume; (e) passenger-car equivalent.

8–3. Assuming that Fig. 8–3 is typical, determine the 8 AM, 11 AM, 2 PM, and 5 PM percentage of trucks and buses in the traffic stream in the direction of heavier flow.

8–4. The average daily traffic on a two-lane rural highway designed for 70 mph, stated in passenger-car-equivalents, is 8000.

- *a.* For an average traffic distribution, what is the thirtieth-hour volume?
- *b.* What are the levels of service for capacity and speed for this volume?
- *c.* Within what limits would you expect this thirtieth-hour volume to fall?

8–5. Assume that a city street carries an average daily traffic volume, stated in passenger-car equivalents, of 10,000 vehicles. Within what range might the thirtieth-hour volume fall?

8–6. An eight-lane freeway has a design speed of 70 mph. If a lane carries 1500 vehicles per hour, stated in passenger-car equivalents:

- *a.* What is the anticipated average speed under normal flow? Under forced flow?
- *b.* What are its capacity and speed levels of service?

8–7. For a six-lane divided freeway, find its capacity in passenger cars per hour in one direction:

- *a.* If it has a 70-mph design speed, 12-ft lanes, and shoulders 6 ft wide on each side.
- *b.* If it has 11-ft lanes, 4-ft shoulders, and an obstruction on the left side at the shoulder edge?
- *c.* If it has 10-ft lanes, 4-ft shoulders, and obstruction on each side at the shoulder edge?

8–8. For a two-lane highway with 12-ft lanes designed for speeds of 70 mph, find its capacity in the following situations:

- *a.* If the roadway has no features that limit capacity.
- *b.* If bridge railings are located 2 ft from the pavement edge on both sides.
- *c.* If the road is straight, has 4-ft shoulders, has a 5% grade 1 mi long, and the traffic stream contains 10% trucks.

8–9. An old two-lane highway in mountainous country has lanes 10 ft wide, 2-ft shoulders, and high shrubs trimmed vertically at the shoulder's edge. Trucks constitute 10% of the traffic. There are no steep grades. At capacity, how many passenger cars and how many trucks will this road carry?

8–10. What length and how many lanes are required for a weaving section to meet the following conditions: four approach freeway lanes; through freeway traffic, 5000 passenger vehicles per hour; weaving on, 800 passenger vehicles per hour; weaving off, 1200 passenger vehicles per hour; entering on ramp and leaving off ramp, 200 passenger vehicles per hour. Solve this problem assuming:

a. That the freeway to freeway curve of Fig. 8–11 applies.

b. That the collector–distributor curve of Fig. 8–11 is appropriate.

c. That Fig. 8–12 applies at a level of service at the boundary between *B* and *C*.

8–11. Determine for the cited operating conditions, the maximum approach volumes per clock hour for a 36-ft-wide street in an outlying business district in a city of 250,000. Load factor is 0.7, peak-hour factor 0.80, and percentage of green 50. Assume that the other conditions of Figs. 8–13 and 8–14 apply.

a. Operated as a two-way street with parking permitted.

b. Operated as a two-way street, with all parking prohibited.

c. Operated as a one-way street 36 ft wide with parking permitted.

d. Operated as a one-way street 36 ft wide with all parking prohibited.

8–12. Answer the questions of Problem 8–10 for a 60-ft-wide street in a fringe area of a city of 1,000,000 population. Load factor is 0.6, peak-hour factor of 0.70, and percentage of green 60.

8–13. From the traffic engineer of the community where your college is located or as a class project, determine the peak-hour traffic movements at a busy intersection that has separate left-turn lanes. Analyze it for level of service, assuming a two-phase signal, by the methods outlined in Fig. 8–15 and the text explanation.

8–14. Procure a copy of *TRB Circular 212* and, using it as a guide, do Problem 8–13 for a multiphase traffic signal.

8–15*A*. A vehicle is traveling at 50 mph. Assuming that the driver's detection, recognition, decision, and response-initiation time is 2 s, what will be the total stopping distance on a dry pavement for which f is 0.60? What will it be if attention is not on the road so that detection, recognition, decision, and response-initiation time is 4 s? What will it be if detection through response-initiation time is 2.5 s, f is 0.45, and there is a 5% downgrade?

8–15*B*. Work Problem 8–15*A* for a speed of 80 km/h. State the answer in meters.

8–16. One vehicle is following another on a two-lane two-way highway at night according to the safe-driving "rule-of-thumb" of one car length spacing for each 10 mph of speed. If both vehicles are traveling at 55 mph and the lead car crashes at that speed into the rear of an unlighted parked truck, at what speed will the following vehicle hit the wreckage? Assume a car length is 20 ft, reaction time is 0.5 s, and a coefficient of friction is 0.65.

8–17*A*. Vehicles often travel city streets adjacent to parking lanes at 35 mph (or faster). At this speed and setting detection through response-initiation time for an alert driver at 1 s, and f at 0.50, how far must the driver be away from a suddenly opened car door to avoid striking it? What will this distance be if detection through response-initiation time is 3 s?

8–17*B*. Work Problem 8–17*A* for a speed of 55 km/h. State the results in meters.

8–18*A*. A vehicle is traveling at 40 mph on a wet pavement coated with oil droppings ($f = 0.15$). The driver is alert, so that detection through response-initiation time is 1 s. What is the safe stopping sight distance for this condition? What would it be at a speed of 55 mph?

8–18*B*. Work Problem 8–18*A* for speeds of 55 and 90 km/h. State the results in meters.

8–19. The driver of a car traveling 45 mph requires 125 ft to stop after the brakes have been applied. What average coefficient of friction was developed between tires and pavement?

8–20. Investigate the program for skid-resistance testing and/or improvement of the local highway or street agency.

8–21. Vehicle performance is being tested on a large flat paved area. For this situation:

a. What coefficient of side friction must be developed to hold a car going 60 mph on a 1000-ft radius curve?

b. What is the minimum-radius curve that a vehicle traveling 60 mph can travel if the coefficient of side friction is 0.40?

8–22*A*. For a superelevation of 0.08 ft/ft, compute the minimum permissible radii of curves for design speeds of 30, 40, 50, 60, and 70 mph.

8–22*B*. Work Problem 8–22*A* for speeds of 50, 60, 70, 80, 90, and 100 km/h. State the answers in meters.

9 HIGHWAY DESIGN

Given the basic characteristics of drivers, vehicles, traffic and road, as presented in Chapter 8, the designer is in the position to develop the geometric details for a project, whether it is new or a reconstruction. First a design speed and ruling grade must be determined after weighing such factors as the road's importance, the estimated amount and character of traffic, the terrain, and the availability of funds. Design speed and ruling grade in turn provide the basis for setting minimum standards for vertical and horizontal alignment. Then the designer, to a large degree by trial and error, fits these or higher standards to the terrain as shown on aerial photographs, maps, and other exhibits to produce a plan and profile for the main roadway or roadways (see Fig. 6–4). Along with decisions concerning the continuous roadway, the designer also develops details of the geometry of intersections or interchanges, service roads, and similar features. Finally, particulars of signing, striping, traffic signals, if any, and other traffic-control devices must be worked out, as detailed in Chapter 10.

Possibly the most important single rule in highway design is consistency. Only by making every element conform to the driver's expectations by providing positive guidance through a variety of cues (see Chapter 8) and by avoiding abrupt changes in standards can a smooth-flowing, accident-free facility be produced. In addition, by careful attention to blending horizontal and vertical alignment and fitting structures to the landscape, the road's visual qualities can be greatly enhanced. For example, a flowing alignment is far better than a series of straight sections coupled by short, choppy horizontal and vertical curves.[1]

Experienced highway designers recommend that the signing for the highway be planned as an integral part of the preliminary layout studies. If directions to drivers can be planned to convey one simple message at a time, and if these directions can be followed smoothly, easily, and without undue haste or changes in speed, then the plans for the facility will be satisfactory.

An earlier series of "policy" publications of AASHTO, now being updated as

[1]With interactive graphics it is possible to display visually a "driver's eye" view of the road ahead. For more on this topic see Chapter 6 and B. L. Smith, *Transportation Engineering Journal of ASCE*, Aug. 1975. J. H. Looper, ibid., May 1976 demonstrates design approaches to preserve scenic values in narrow, confined canyons. Approaches employed in Glenwood Canyon, Colorado are described by N. J. Pointer II in *The ITE Journal,* Jan. 1979. The controversy and resolution surrounding the geometrics of the Interstate highway through Franconia Notch, N. H., is covered in articles by P. Hofman, C. Kapala, and B. Joslin, ibid. *TRB Record 717* describes the Interstate highway through Vail Pass in a high mountainous area of Colorado.

A Policy on Geometric Design of Highways and Streets, includes *A Policy on Geometric Design for Rural Highways (Blue Book), A Policy on Design of Urban Highways and Arterial Streets (Red Book), Standards for the Interstate System, Geometric Design Standards for Highways Other Than Freeways, Geometric Design Guides for Local Roads and Streets,* and *Highway Design and Operational Practices Related to Safety (Yellow Book).* It details accepted practice in the United States. Much of the material presented in this chapter is drawn from this source. The *Manual on Uniform Traffic Control Devices,* published by the Federal Highway Administration, is authoritative as to signs, signals, and pavements markings. *Compendium 1,* titled *Geometric Design Standards for Low-Volume Roads,* produced by TRB under an AID contract for Transportation Technology Support for Developing Nations, offers solid data on its subject. Among the chief advantages of these standards and guides are the pooling of collective knowledge and experience and the nationwide uniformity and consistency that result.[2]

HIGHWAY STANDARDS

No single set of geometric standards apply to all highways. To illustrate, it seems appropriate to design a mountain road that is primarily for passenger cars numbering 50 or fewer a day to the following standards: minimum speed, 20 mph; maximum grade, 16%; and total width of surfacing, 16 ft. At the other extreme, consider a major rural freeway on a primary network. Appropriate standards in this case could include: design speed, 70 mph; maximum grade, 3%; total width of roadway and shoulders including a median, 160 ft; and right-of-way width, including service roads, ranging from 250 to 400 ft. Standards for other facilities would fall between these extremes. For each highway segment, decisions regarding appropriate controls for every one of the many details must be reached. Furthermore, there are few "right" answers; for example, what constitutes "heavy" truck traffic which limits the maximum grade to 3%, and why should this limit be 3% rather than 2½ or 3½%? Again, why 4 to 1 side slopes rather than 3 to 1 or 6 to 1? In some instances, the commonly accepted standards are supported by research; in others they represent a pooling of the judgment of many engineers. Seldom, however, can they be considered as exact or beyond debate.

DESIGN SPEEDS[3]

Design speed is defined in *AASHTO Highway Definitions* as "a speed determined for design and correlation of the physical features of a highway that influence vehicle operation. It is the maximum safe speed that can be maintained over a specified section of highway when conditions are so favorable that the

[2]For a history of highway design, see a series of eight articles by F. W. Crow in *Public Roads* from Dec. 1974 through Dec. 1976.

[3]The safety aspects of speed are discussed in Chapter 12.

design features of the highway govern." AASHTO recommends that design speeds be set to the greatest degree possible to satisfy the needs of nearly all drivers both today and throughout the road's anticipated life. This is difficult at best, as indicated by Fig. 9–1. First, in the past average speeds increased over the years at roughly 1 mph/yr until about 1973. This trend may resume unless restricted by regulations such as the controversial 55 mph speed limit. Second, at a given time, individual preferences and judgments of suitable speed show considerable variation, as indicated by the form of the individual curves. Notice on Fig. 9–1 that, with low volumes and no speed restrictions, the 1973 curve shows 15% of the drivers exceeding 73 mph; they are sometimes referred to as the "reckless fringe" whose behavior may be disregarded in design.

Selection of the proper design speed is probably the single most-important decision, since this choice sets the limits for curvature, sight distance, and other geometric features. Since available funds are often limited, there is the temptation to reduce design speeds in order to save money. However, to do so may be unwise. The roadway section can always be improved and widened and paving strengthened at a future date. But alignment, grade, and sight distance, when once molded into the landscape and tied down with paving and rights of way are very difficult and expensive to correct.

AASHTO practice is to classify highways first as rural or urban, then as freeways, arterials, collectors, and local (see Chapter 8). Rural collector and local

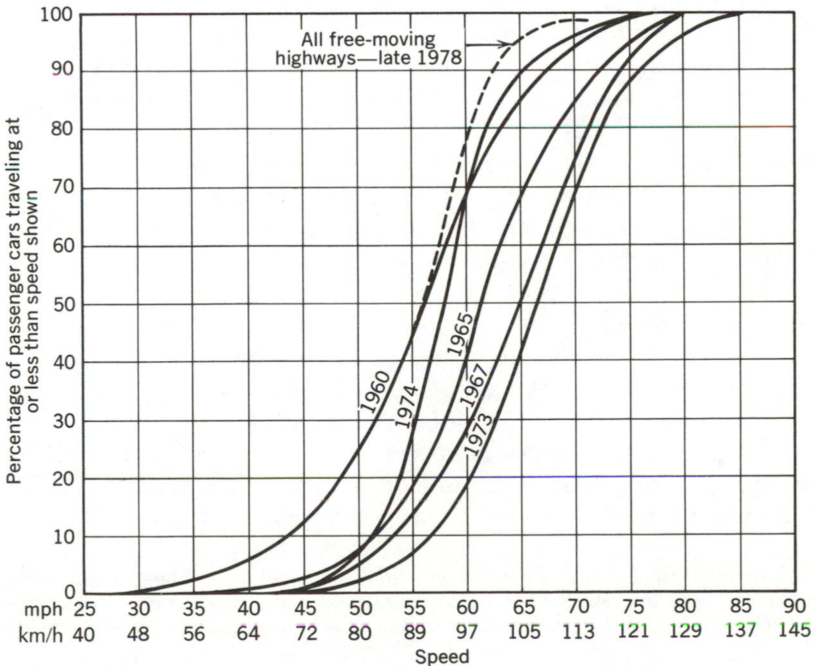

Fig. 9–1. Distribution of representative passenger-car speeds on rural Interstate highways.

facilities are classed as flat (or level), rolling, and mountainous. Minimum recommended design speeds are given below or in Table 9–1.

	Design Speed			
	Urban		*Rural*	
Facility	*mph*	*km/h*	*mph*	*km/h*
Freeways	50–60 preferred	80–97 preferred	70–60 in mountains	113–97 in mountains
Arterials	40–60 but 30 in built-up areas	64–97 but 48 in built-up areas	50–70	80–113
Collectors	30	48	See Table 9–1	See Table 9–1
Local	20–30	32–48	See Table 9–1	See Table 9–1

Although the highest presently-used design speed is 70 mph, many engineers argue that, wherever possible, design speeds up to 100 mph should be used. They contend that, even though driver and vehicle limitations,[4] safety, or fuel conservation measures[5] appear to preclude operating most present-day vehicles at such speeds, the higher standard of design provides some guarantee against future obsolescence as well as an increased margin of operating safety. In addition, it is well within the realm of possibility that a new generation of high-speed vehicles or entirely different transport modes may operate on many of these alignments.

The relatively high design speeds stipulated for freeways are in keeping with the concept that, regardless of cost, they are to form a nationwide high-volume, high-speed network. Design speeds for other highways are more greatly influenced by cost. For arterials, two choices are given. For rural collectors and local roads, design speed is related to volume (see Table 9–1).

Sudden changes in design speed along any highway should be avoided, particularly on high-speed roads. AASHTO recommends that this change be made over several miles in 10-mph increments at design speeds below 60 mph and in 5-mph increments at higher speeds.[6]

It is difficult to apply the design speed concept as defined above to low-volume roads in mountainous country. For economic reasons, the geometric features of certain sections are designed for speeds of 20 or 30 mph. And yet motorists drive far faster on straight or less sharply curved sections. In such instances, the highway designer must use controls such as reduced superelevation combined with easement curves, delineators, signs, and striping and rumble strips (if there is pavement) to alert motorists that they are approaching road segments having sharp curves or restricted sight distances.

[4]At speeds above 80 mph factors such as the gyroscopic action of the flywheel, motor, and wheels tend to take control of the vehicle from the driver. Also, reaction time becomes critical; on curves requiring maximum superelevation, responses are too slow to permit control of a vehicle in a 12-ft lane.

[5]See G. B. Pilkington II, *TRB Record 601*.

[6]See J. E. and J. P. Leisch, *TRB Record 631* for methods for implementing this policy.

TABLE 9–1. AASHTO Recommendations for Minimum Design Speeds in mph for Rural Collectors and Local Roads—Based on Current ADT*

Class	*Terrain*	*Average Daily Traffic*				
Collector		0–400	400–750	750–2000	2000–4000	Over 4000
	Level	40	50	50	50	60
	Rolling	30	40	40	50	50
	Mountainous	20	30	30	40	40
Local		0–50	50–250	250–400	Over 400	
	Level	30	30	40	50	
	Rolling	20	30	30	40	
	Mountainous	20	20	20	30	

*mph = 1.6 km/h.

DESIGN OF THE CROSS SECTION

Cross sections of typical highways of modern design are shown in Fig. 9–2. Dimensions or other requirements for the various elements are seldom stated absolutely by AASHTO; rather there is latitude to consider such variables as the volume, character, and speed of traffic and of the characteristics of motor vehicles and their operators. Factors affecting each element of the cross section are discussed in the sections that follow.

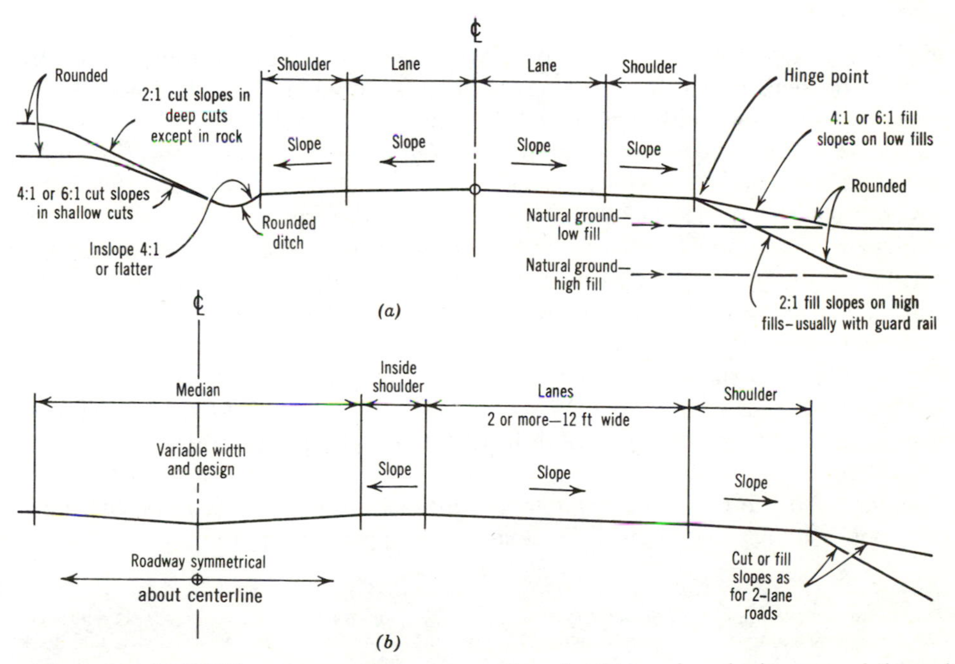

Fig. 9–2. AASHTO recommended cross sections for *(a)* two-lane highways and *(b)* multilane highways and freeways (half-section).

It is common practice in designing new highways to adopt a given cross section and employ it from end to end of the improvement. Seldom is this approach challenged on high-volume facilities. However, for low-volume facilities or for reconstructing old highways, it may be appropriate to modify features such as shoulder width in rough country or on longer bridges to reduce costs. Such proposals often meet a storm of protest on safety grounds, as happened when lower standards were proposed for the 3R Federal-Aid Program.

Given that a particular cross section offers the stipulated level of service, safety becomes a principal consideration. Unfortunately, knowledge is incomplete as to how the various elements and their interactions affect accidents, except for lane and shoulder width on two-lane rural roads. For them, a recent study concluded that, with 9 to 10 ft shoulders, 24 ft wide pavements had 18% fewer accidents than pavements narrower than 18 ft and 4% fewer than for 20 ft pavements. However there was no difference between 22 and 24 ft wide pavements. For 20, 22, and 24 ft pavements, 9 to 10 ft shoulders decrease accidents by about 35% compared to 0 to 2 ft shoulders, and 22% compared with 3 to 4 ft widths.[7] It must be recognized, however, that these differences in accident potential may not justify higher standards, particularly when traffic volumes are low. For, even if accident potential were the sole criterion for highway improvement, a variety of other expenditures might be far more cost effective than high standards (see Chapter 12).

Lane Widths

In meeting oncoming vehicles or passing slower ones, the position selected by a driver depends primarily on the paved or surfaced width of the highway. Originally this surfaced width was only 15 ft which was ample for horse-drawn vehicles. With the increase in motor-vehicle traffic the width increased first to 16 ft, then to 18 ft. Later two 10-ft lanes became a standard width for first-class paved highways. Today, 12-ft lanes are standard for freeways and other major traffic arterials although 14-ft widths have been recommended by some.[8] For two-lane rural highways, a 24-ft-wide surface is required for clearance between commercial vehicles and is recommended for main highways. For collectors, surfacing widths of 20 ft are considered adequate only for low volumes and few trucks. Minimum surfacing widths for local rural roads can be as low as 16 ft for 20 mph design speed and ADT less than 50, but range up to 22 ft. For urban streets, minimum design lane width is 12 ft; 11, 10, or even 9 ft are permitted where space is limited.

Recent research has brought another argument for wide lanes when there are frequent meetings or overtakings between passenger cars and large trucks. This

[7]Data from *NCHRP Report 197,* based on a review of many often-conflicting reports. This reference also offers procedures for comparing accident and construction costs and an extensive literature review.

[8]In Germany, lanes 12.3 ft (3.75 m) were standard for freeways. In 1976 this was lowered to 11.4 ft (3.6 m) on those carrying fewer than 5000 trucks per day.

is that, with strong cross winds, air disturbances can cause vehicles to swerve substantially within and even out of their lanes.[9]

Number of Lanes

In almost all situations, the number of lanes in a new segment of highway is set by bringing together estimates of traffic for the design year and of highway, street, or lane capacity at the desired level of service (see Chapters 3 and 8). Four lanes in one direction in a single roadway has been the accepted maximum. However, AASHTO policy recognizes dual-divided roadway 16 lanes wide consisting of 4 lanes in each direction for an inner freeway with 4 more freeway lanes in each direction on the outside. In some instances reversible lanes have been installed in the center of freeways having heavily unbalanced traffic flows between morning and night. Also, exclusive bus lanes are at times provided.[10] For mountainous areas the need for and location of climbing lanes for slow-moving vehicles can be explored on the basis of data such as are presented in Chapter 8. Changes in the number of lanes should not be made at interchanges or intersections.

Shoulders

The shoulder or verge (see Fig. 9–2) is that portion of the roadway between the edge of the traffic lane and the edge of the ditch, gutter, curb, or side slope. AASHTO sets its usable width as that of the pavement or other surface having strength to support vehicles. Shoulders provide a place for vehicles to stop when disabled or to stand for other reasons. If designs omit shoulders or if they are narrow, roadway capacity decreases and accident opportunity increases.

Shoulders on rural highways were originally 2, 3, or 4 ft wide and usually unpaved. Sometimes they were surfaced with gravel or similar material to provide hard standing at all times, but often they were earth and unusable in wet weather. Today, shoulders on major highways are usually paved. AASHTO recommends that when lane and shoulder are of bituminous construction, they be differentiated by color or texture. In the east, south, or midwest where rainfall is sufficient and frequent enough to support grass, turf shoulders so constructed as to provide firm support for vehicles are sometimes used (see Chapter 13).[11]

American practice calls for shoulders continuous along the full length of the roadway on almost all roads. In certain European countries occasional turnouts (called laybys in Great Britain) may be used instead on all but major roads.

One argument for wide, continuous shoulders is that they add structural strength to the pavement (see Chapters 19 and 20). Others are that outside shoulders increase horizontal sight distance on curves and provide for snow

[9]See D. H. Weir et al., *TRB Record 520*.

[10]On a short section of Interstate 95 just south of Washington, D. C., there are 27 lanes. These provide inner and outer freeways, connectors from and distribution to other facilities, plus four reversible lanes for buses and carpools.

[11]For a recent review of shoulder design practices in the United States, see *TRB Record 594* and *757* and *NCHRP Synthesis 63*.

storage during and after storms. Finally, they may reduce accident potential when vehicles stop for emergencies or other reasons.

Wide shoulders provide vehicles a place to stop off or partially off the traveled lanes. Of these stops, those for emergencies are primarily related to traffic volume and flow conditions. But leisure stops vary with other factors such as trip length, rural versus urban, scenic versus nonscenic, and whether or not other places to stop are available.[12] In any event, the argument for wide shoulders on all roads is inconclusive. For example, data from one study have indicated that on a mile of rural road carrying 400 vehicles per day, the chance of an accident involving parking maneuvers or standing or slow-moving vehicles is less than 1 in 100 per year in a given mile.[13] Again, it can be argued that, at least at lower traffic volumes, shoulders on long bridges or elevated structures are not justified on either accident-reduction or economic grounds.[14]

It is now common practice to paint a continuous narrow white stripe to mark the line between roadway and shoulder as a guide for drivers during adverse weather and poor visibility conditions. *The 1978 Manual on Uniform Traffic Control Devices* states that, "Edge lines shall be provided on all Interstate highways and may be used on other classes of roads." The evidence is that when such a stripe is present, drivers tend to stay in the traffic lane and fewer of them infringe on the shoulder. Experience in Kansas on 450 m of highways has shown a 14% reduction in accidents and 25% reduction in deaths with edge lines. However, the evidence to date is not conclusive. Those opposed to striping argue that on two-lane highways, keeping vehicles in the traffic lane positions them closer to opposing traffic and makes head-on collisions more likely.[15]

For all freeways, *A Policy on Geometric Design* recommends that the outside shoulder be paved at least 10 ft wide, with 12 ft called for if truck volume is more than 250 in the design hour. Recommended width for the left (median) shoulder is 4 to 8 ft, with at least 4 ft paved. With six or more lanes the median shoulder should be 10 ft wide, or 12 ft if truck volume in the design hour exceeds 250.

For rural arterials at ADTs less than 400, usable shoulder width is set at 4 ft minimum, with 8 ft suggested. It ranges to 8 ft minimum and 12 ft suggested when design hour volume exceeds 400. For urban arterials, where possible, similar shoulder widths without curbs, unless needed for drainage, are proposed. However, it is recognized that in many instances all available space is required for traffic so that shoulders must be omitted. The width of median

[12]On California freeways, one stop for every 18,000 vehicle-miles, including one mechanical stop every 26,000 miles, have been reported. On motorways in Great Britain there was one disabled vehicle every 20,000 m (see R. J. Salter and K. S. R. Jadaan, *TRB Record 536).* For Interstate freeways in South Dakota, M. R. Cheeseman and W. T. Voss, *HRB Record 162,* indicate, for passenger cars, one involuntary stop every 11,600 and voluntary ones every 2,600 vehicle-miles. For trucks and buses the values are, respectively, one for every 4160 and 1000 vehicle-miles. *NCHRP Report* 6 states that in the Holland Tunnel into New York City there is one stop each 6,800 mi.
[13]See *NCHRP Report 63.*
[14]See *NCHRP Report 197* for a review of the many studies on shoulders and accidents.
[15]See A. Taragin, *Public Roads,* Aug. 1957 and Aug. 1958, and articles by J. B. Musick and R. W. Williston, *HRB Bulletin 266.*

shoulders on four-lane divided arterials is set at 3 ft minimum; for six or more lanes, 8 to 10 ft widths are recommended.

For rural collectors, a graded shoulder 2 ft wide for ADTs less than 400 to 8 ft at ADTs over 2000 is called for. In this case, width is defined as extending from the edge of the surfacing to the point where shoulder slope intersects side slope. Urban collectors usually do not have shoulders; rather, parking lanes 8 ft, or preferably 10 ft, wide and gutters are prescribed.

For local rural roads, a graded shoulder extending outside the surfacing to an intersection with the side slope is stipulated. Its width is 2 ft for ADTs less than 400, 4 ft for higher volumes.

Cross Slopes

Cross slope is introduced in all tangent sections of roadway. Except where superelevation of curves directs all water toward the inside, slopes usually fall in both directions from the centerline of two-lane highways. Each half of a divided roadway is sloped individually, usually with the outside edges lower than the inside ones (see Fig. 9–2). For high-type pavements, this cross slope (or crown) is often 1 to 2%.[16] Having cross slope in one direction on multilane highways makes driving easier; however where rainfall is very heavy, it increases the depth of water in the roadway. On paved shoulders cross slopes are usually greater, in the range of 3 to 6%, with 4% most common. For gravel and turf, greater slopes are needed for satisfactory drainage, with 4 to 6% recommended for gravel and 8% for turf.

The cross sections of city street surfaces often are laid out with the pavement surface conforming to a parabola. This procedure makes the inside lanes that accommodate high-speed traffic flatter than the outer lanes. Also it places the steepest slopes adjacent to the gutters, which narrows the width devoted to carrying surface water. On very wide streets a parabolic crown makes the middle lanes almost flat unless gutters are unusually deep. Then some combination of uniform slope with a parabolic curve is used in place of the parabolic section.

Medians for Multilane Highways

Positive separation between opposing streams of traffic has proved to be an effective means for reducing headlight glare, conflicts, and accidents on multilane highways. Today medians in some form are an absolute requirement for all freeways. At intersections where roads or streets cross expressways or major city streets, medians, if wide enough, provide further advantages. They offer a refuge between opposing traffic streams so that cross traffic and pedestrians can traverse each stream as a separate maneuver. They also make space available for "left-turn" lanes; this clears the through lanes of turning vehicles and makes for smoother, safer operation and increased capacity.

[16]Steeper slopes are increasingly favored since they cause rainwater to flow away more rapidly and thereby reduce its thickness. This can be an important factor in reducing hydroplaning (see Chapter 8).

Wide medians are preferred wherever space and cost considerations permit. For rural sections of freeways, 60 to 90 ft widths are common. *A Policy on Geometric Design* states, however, that 10 to 30 ft widths may be appropriate in suburban or mountainous situations. Medians up to several hundred feet have been provided in some instances, thereby completely isolating one roadway from the noise, confusion, and headlight glare of the other.

For rural and urban arterials, medians 60 ft wide or wider are recommended, since they allow the use of independent profiles and reduce crossover accidents. Medians 22 to 60 ft wide permit drivers to cross each roadway separately; widths of 14 to 22 ft provide protection at intersections for turning vehicles. Under very restricted conditions, curbed medians 4 to 6 ft wide may be employed as a means of separating opposing traffic, protecting pedestrians, and providing locations for traffic-control devices.

As shown in Fig. 9–2, wide medians generally are depressed below the level of the roadway, with the inside shoulder sloped toward the median for drainage. At times, however, medians, particularly narrow ones, may be curbed and crowned, with drainage across the traveled way. Even wide medians are at times graded up into a wide mound. Some of these alternatives are shown in *A Policy on Geometric Design*.

Very serious and spectacular accidents result when vehicles traveling at high speed cross the centerline or median and collide head-on with or sideswipe those from the other direction. Concern over such occurrences has led to extensive research regarding the effectiveness of wide but traversable medians, and of various nontraversable barriers for positive separation.

With traversable medians, the width should be great enough to prevent most of the out-of-control vehicles from reaching the opposing traffic lanes. Fifty to 80 ft between lane edges has been suggested, but a specific value has not been stipulated. Cross slopes should not be greater than 6 to 1, with 10 to 1 preferred.

An alternative solution to wide medians combines a fairly wide median with some form of energy-absorbing device which will slow the vehicle without injury to its occupants. For example, it has been shown that dense, thick planting, such as multiflora rose hedges, can be used safely as crash barriers.[17]

For narrower medians, three means for reducing cross median accidents are being used. These are "deterring devices," "nontraversable energy-absorbing barriers," and "nontraversable rigid barriers."

Deterring medians incorporate such devices as two sets of double stripes painted on the existing pavement (called flush medians), raised diagonal bars, low curbings, and shallow ditches. They warn drivers and sometimes divert vehicles that are not too badly out of control back into the roadway. Sometimes they are designed to serve as refuges for left-turning or disabled vehicles. On the other hand, some fraction of vehicles cross into the opposing traffic lanes where the probability of a serious accident is high.

[17]See R. R. Skelton, *HRB Bulletin 185*.

Nontraversable, energy-absorbing barriers are devices such as a chain-link fence about 3 ft high, supported on light steel posts and augmented by cables at bottom and midheight. When struck by vehicles, they prevent intrusion into opposing traffic but yield sufficiently to minimize the tendency to bounce vehicles back into the traveled way completely out of control. Sometimes vehicles become entangled in the fence and are brought to a halt.[18] Although these barriers are effective, replacing damaged sections has been found to be costly as well as disruptive to traffic, since a lane must be closed while repairs are made.

Nontraversable rigid barriers are so designed that seldom does a vehicle get across this barrier into the opposing traffic lanes. On the other hand, it may be bounced back onto the roadway completely out of control and become involved in or be the cause of accidents there. Also, the upstream ends of some conformations of nonmountable curbs or railings provide a point of impact for vehicles, or turn them upside down. Until recently most of these barriers were very similar to a metal guard rail. The continuous horizontal element was of corrugated or box-beam cross section mounted on sturdy posts. Where medians were narrow, it was common to place barrier curbs along the inside shoulder and a guard-rail-type barrier with horizontal elements on both faces inside these curbs. Today, however, a high, nonmountable, sloped face concrete barrier called the New Jersey or G. M. barrier is favored. One conformation of it is shown later in Fig. 9–13. This can be cast or extruded in place or precast in sections and set into position with a crane.[19]

With narrow medians, headlight glare from vehicles traveling in one direction may be dangerous as well as annoying to motorists going the opposite way. Mounting a light barrier of lattice or expanded-metal on top of the vehicle barrier offers one solution to this problem.[20]

In some instances median widths for major streets and expressways at grade are variable; wide at intersections and narrow between.

Openings through medians on freeways to permit U-turns, although they would be useful to a few motorists, create a serious accident hazard. *A Policy on Geometric Design* recommends against them, except in rural areas where the spacing between interchanges is more than 5 mi. They may also be needed as emergency crossovers on elevated or depressed urban freeways carrying heavy volumes of traffic. In either instance, use, unless under police direction, is to be restricted to emergency vehicles.

[18]See articles by K. Moskowitz and W. F. Schaefer, and J. N. Beaton and R. N. Field, *HRB Bulletin 266;* and R. T. Johnson, *HRB Record 105.*

[19]The AASHTO publication titled *Guide for Selecting, Locating, and Designing Traffic Barriers* is authoritative. For an interesting report on the development of medians and median barriers see J. W. Hutchinson et al., *HRB Record 105.* Intensive research on median barriers and guard and bridge railing systems began in the late 1950s and usually involved crash testing. The literature is extensive and includes recent articles in *TRB Record 460, 488, 566, 586, 594, 631, 679* and *769; NCHRP Reports 54, 115, 118, 149,* and *153; Traffic Engineering,* Mar. and Sept., 1975 and H. E. Ross Jr. and D. G. Smith, *Transportation Engineering Journal of ASCE,* Jan. 1981. Recent tests of the New Jersey barrier are reported by J. G. Viner, *Public Roads,* June 1980. Indications are that redesign to fit minivehicles may be in order. For details of their design and construction, see E. C. Lokken, *Transportation Engineering Journal of ASCE,* Feb. 1974.

[20]See *NCHRP Synthesis* 66 for added detail.

Roadside Ditches

In cuts, a roadside ditch at the outer edge of the outer shoulder is provided. As indicated on Fig. 9–2, a 4 to 1 or flatter slope and rounding of the ditch bottom is recommended so that if a vehicle wheel strays off the shoulder, control will not be lost. Width and depth of the ditch is dictated by hydraulic considerations (see Chapter 11).

On lightly traveled roads, and particularly in areas of heavy rainfall, drainage requirements may call for a large ditch. Often this involves steep slopes both inside and outside the ditch; otherwise the roadway will become excessively wide. Very little accident data are available from which to measure the economic and other trade-offs between costs and land takings on the one hand and accidents on the other.

Cut or Fill Slopes

Earth fills of usual height stand safely on slopes of 1½ to 1.[21] Slopes of cuts through ordinary undisturbed earth remain in place with slopes of 1 to 1. Rock cuts as steep as ½ to 1 and sometimes ¼ to 1 are usually stable.[22]

In the past, these slopes were standard for many highway agencies because they involved a minimum of earthwork. In recent years, however, slopes generally have been flattened (see Fig. 9–2) to provide for safer operation, to facilitate plant growth and reduce erosion (see Chapter 13), where stability is a problem (see Chapter 14), and to decrease maintenance costs.[23]

Steep slopes on fills may create a serious accident hazard. If one wheel of a vehicle goes over the edge, the driver loses control and the vehicle may overturn. With flat side or back slopes, possibly 3 to 1 or flatter, and rounding at the hinge point (see Fig. 9–2) the car often can be directed back into the road, come to a stop, or continue down the slope without overturning. Flat fill slopes have the added advantage that they are visible from the vehicle for their full extent so that the road takes on a safer appearance. Drivers are assured that if necessary they may direct their vehicles onto or down the slope. As a result, vehicle positioning on roads with visible side slopes is closer to the edge and, on two-lane facilities, farther from opposing traffic.

A Policy on Geometric Design recommends for flat or rolling country, that 6 to 1 slopes be used on embankments less than 4 ft high and 4 to 1 slopes on higher fills. Only at heights greater than 20 ft would 2 to 1 slopes be permitted. Standards are somewhat less severe in moderately steep or steep terrain. Cut

[21]When slopes are expressed as 1½ to 1, 2 to 1 or the like, the first figure represents horizontal measurement, the second vertical measurement. Many prefer to express slopes by ratios stated as 1 on 2 or 1 on 4. In this instance the first figure represents the vertical and the second the horizontal dimension.

[22]*A Policy on Geometric Design* recommends stability analysis on slopes steeper than 2 to 1.

[23]Some soils, such as loess, stand almost vertically if undisturbed. With them, there may be less erosion with vertical cut slopes than with flat ones where cost or climate rule out grass or other plantings to hold the slopes.

slopes are never steeper than 2 to 1 except in solid rock or special soils. AASHTO policies, as they apply to other major highways and to urban freeways, are less stringent and permit steeper inclinations. In all instances it is stipulated that where cut or fill slopes intersect the original ground surface, the cross section is to be rounded to blend the slope into the natural ground surface (see Fig. 9–2).[24]

Sometimes in steep country the side slopes of the original ground approach 1½ to 1 (about 34° from the horizontal), and fills do not "catch" unless unduly extended down the mountain. This is objectionable, as it results in an unsightly scar and increased soil erosion. These long fill slopes may be eliminated by shifting the road into the mountain until the full cross section is "benched" into the hillside, but this requires increased excavation and end hauling and leaves extensive scar and increased opportunity for slides and erosion above the roadway. Often it is preferable to contain the embankment with a suitable retaining wall.[25] These may be handplaced stone, cement rubble masonry, concrete block, conventional reinforced-concrete T or counterforted designs, cribs assembled from timber, precast concrete or metal elements, or tied-back piling. Another technique is to reinforce the earth with metal or plastic bands (see Chapter 14). Figure 9–3 presents details of stone retaining walls sometimes used where hand labor is readily available or a natural appearance is desired. Figure 9–4 shows a crib wall of corrosion-resistant metal being installed. Face members of such installations often are of a steel which weathers to a brown color. In urban areas and at grade separation and interchange structures, similar retaining measures are often taken where narrow rights of way, small clearances, or appearance require that the fill slopes be contained.

Marginal Areas

Clear, flat, "forgiving" recovery areas alongside the traveled way are highly desirable, and whenever possible, new designs provide them. For rural areas, *A Policy on Geometric Design* proposes widths as follows: freeways, 80 to 150 ft; arterials, 30 ft;[26] and collectors, 10 to 20 ft. No specific values are proposed for low-volume facilities. For urban areas, widths, if specified at all, are narrower, recognizing restrictions on available space. Where clear, flat areas cannot be provided, a guard rail (see below), set back 2 ft from the shoulder, may be specified.

[24]For a discussion of the safety aspects of slopes and roadside ditches, see *NCHRP Report 158* and E. L. Marquis and G. D. Weaver, *Transportation Engineering Journal of ASCE,* Feb. 1976.

[25]J. W. Stiles, *HRB Record 302,* presents a brief summary of a Bureau of Public Road's publication of September 1967 titled *Typical Plans for Retaining Walls.* It gives detailed designs for reinforced-concrete T section and counterforted walls. Articles in *TRB Special Report 160* and *TRB Record 749* describe several other solutions.

[26]Based on observations which showed that few vehicles that left the roadway encroached further than 30 ft on marginal areas (see, for example, *TRB Record 601*) proposals were made in Congress to require the removal of all obstacles from a 30-ft recovery area on all roads traveled by the public. This effort failed when the enormity of the problem was recognized. Even so, some courts have accepted a "30-ft rule" in allowing negligence claims against highway agencies and officials (see Chapter 7).

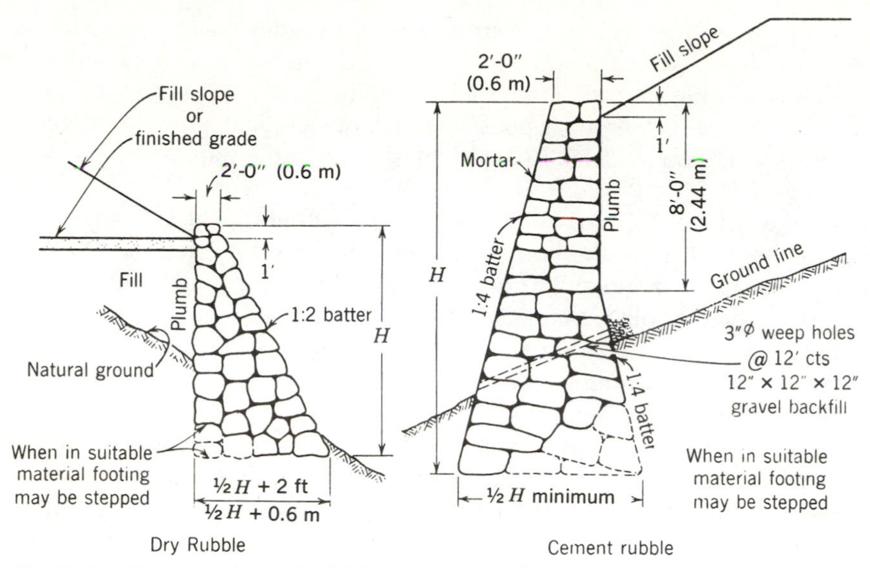

Fig. 9–3. Cross sections of rubble retaining walls.

Fig. 9–4. Installing a metal bin-type retaining wall. (Courtesy: Metal Products Division, Armco Steel Corp.)

On conventional highways and streets, unyielding objects near the traveled way abound. For example, a study of non-federal-aid roads found 71 unguarded obstacles per mile within 30 ft of the roadway edge. It has been estimated that in the United States there are 140 million poles supporting utility wires on highway rights of way. Collisions with them account for 5% of the highway fatalities.[27] In addition, there are unnumbered trees, sign and lighting supports, fire hydrants, bridge and guard rail ends, and other obstructions that can damage vehicles and injure their occupants.

Removing these objects from most roadsides is financially infeasible. With utility poles, there can also be legal constraints. Trees are desirable from environmental and aesthetic points of view, so there is a trade-off between them and safety.

In some instances, the effects of these roadside hazards may be mitigated with guard rail or impact attenuators. Ways are being sought to develop breakaway designs or retrofitting for utility poles and sign supports so that collisions may be less damaging to vehicles.[28] But no overall solution to the roadside hazard problem is in sight.[29]

Another roadside-related problem is that truck brakes sometimes fail on steep, sustained grades and the vehicle runs away. A method of stopping trucks without injury to driver or others or damage to the truck is to provide an escape ramp alongside the road. At times, this ramp can be directed uphill. Another approach is to provide a deep layer of loose rock in which the truck bogs down.[30]

Guard Rail and Bridge Rails

It is common practice to install a longitudinal barrier commonly called guard rail along the roadway on the outside of sharper curves or where side slopes are relatively high and steeper than 4 or 6 to 1. Its purpose is to keep the vehicle from leaving the roadbed, so that it does not overturn or strike a roadside object.[31] Similar installations often are placed in front of hazards near the traveled way such as bridge piers and abutments, sign or lighting supports, trees, utility poles, and in the gore area where ramps leave the main roadway. As with median barriers, a variety of designs and materials have been employed. The earliest were of wire mesh or horizontally strung cables. Until recently, steel and aluminum in corrugated plate, box, or pipe conformations were the most common materials. Today the New Jersey barrier is increasingly popular. As with designs for and research on median barriers discussed earlier, the aim is to se-

[27]See N. L. Graf et al., *TRB Record 571* or *Public Roads*, June 1975, C. P. Brinkman and K. Perchonok, ibid., June 1979, H. W. Miller, Jr. *ITE Journal* Nov. 1980, and D. H. Jones, *TRB Record 769*. H. E. Ross Jr. et al, ibid., report on hazards created by mailboxes.

[28]See, for example, G. K. Wolfe et al., *TRB Record 488* and H. E. Ross Jr. et al., *TRB Record 736*.

[29]See P. H. Wright and L. S. Robertson, *Transportation Engineering Journal of ASCE*, Nov. 1979.

[30]See *A Policy on Geometric Design* and *TRB Record 736* for added detail.

[31]One set of criteria to determine when guard rail should be installed appears in *HRB Special Report 81*.

lect materials, conformations, and placements that will do the least damage to vehicles and their occupants. That attention to these highway appurtenances is important is supported by the finding that, on Interstate freeways, 31% of the accidents involved striking guard rails and 18% some element of a bridge or overpass.[32]

The approaches to bridges, particularly where the road narrows, are the sites of many accidents. One study showed that on certain freeways 73% of the collisions with fixed objects involved either the approach guard rail or the end of the bridge itself. Among the recommendations for making such accidents less severe is to provide a smooth and structurally sound transition between guard rail and bridge rail.[33]

Bridge rails (see also Chapter 11) serve the same purpose on bridges as guard rail on the remainder of the highway. Where formerly designs were for appearance, increasing attention is now being given to safety. In some instances, open railings are giving way to solid ones patterned after the New Jersey barrier.[34]

Impact Attenuators

Impact attenuators are devices designed to lessen the destructive forces developed when a vehicle strikes a fixed object such as a bridge pier or abutment, a sign support or pedestal in the gore area, or some substantial fixed object at the roadside such as a large tree. Ideally, attenuators would be unnecessary, but often they must be employed. A variety of materials and designs have been tested and adopted, including metal drums, plastic containers or tubes, or hollow cylinders of lightweight concrete. These may be empty or filled with water or sand.[35]

Right-of-Way Widths

In the past, originally acquired rights of way have usually been too narrow.[36] Thus when it becomes necessary to add lanes or otherwise widen a road or street, developed property along one or both sides often had to be purchased at high cost. Sometimes several takings have been made at different times from the same property, each at considerable expense. Because of such experience in the past, engineers now consider it good practice to acquire rights of way wide enough to accommodate the ultimate expected development.

[32]See *NCHRP Report 149.*

[33] See *NCHRP Report 203* and *NCHRP Report 230.*

[34]For references on guard and bridge rail design and testing, see those given earlier for median barriers.

[35]See F. J. Tamanini, *ITE Journal*, Dec. 1978, *NCHRP Report 157*, and *TRB Record 488* and *566* for added detail and references.

[36]In some eastern states, original right-of-way widths were 5 chains (33 ft). In the Middle West and West, rural property was divided by the public lands surveys into sections 1 mi on a side. Around the boundaries of each section a strip 33 ft wide was dedicated to the public for land access, a total of 66 ft. Widths such as these became accepted as standard for almost all rural roads until the 1930s. They are still found on many lightly traveled rural roads.

Recommended right-of-way widths for rural and urban freeways at grade are given in Table 9–2. For elevated freeways on fills or for depressed facilities added widths over those shown in Table 9-2 will be required to accommodate the fill or cut slopes. With elevated freeways on structures space must be provided between the roadway and the property line. For an eight-lane elevated freeway with opposing roadways side by side, right-of-way width is 170 ft; this distance is 110 ft if one roadway is above the other. None of the widths shown includes provision for ramps.

For highways other than freeways and for streets, no specific right-of-way widths are indicated. It is recommended, however, that the acquisition be wide enough to accommodate all necessary elements, including flat side slopes; sight distance at intersections and on horizontal curves for appurtenances such as ramps, walls, and border areas; and potentially viable transit, bicycle, or pedestrian facilities.

With increasing attention to screening major arteries from adjacent property for noise control or aesthetic reasons (see Chapter 13) added width may be required for planting or sound barriers.

Fencing

Fencing serves to prevent unwanted intrusion of animals, people, vehicles, or machines onto the highway. In rural areas, its most important function is to control livestock, although on occasion it may serve to prevent access or crossings by persons, vehicles, or machines, except at designated points. A relatively low fence of barbed, smooth, or woven wire is generally effective. Depending on legal jurisdiction and circumstance, it may be built and maintained by the highway agency or by the landholder.

For high-speed, limited access facilities in urban and suburban areas, positive means must be taken to insure that pedestrians cross only at grade separations or other arranged and protected places. Fences for this purpose are really to protect people from their own folly, for many prefer to face the hazard of crossing

TABLE 9–2. Minimum Right-of-Way Widths for Rural and Urban Freeways at Grade in feet*†

	Rural		*Urban*		
Number of Lanes	*With Frontage Roads*	*No Frontage Roads*	*Restricted—No Frontage Roads‡*	*Normal—with Frontage Roads*	*Normal—No Frontage Roads*
4	225	175	135	—	—
6	250	200	170	295	175
8	275	225	195	320	200

*Source: *A Policy on Geometric Design.*
†*1 ft = 0.3048 m.*
‡*Includes a nontraversable median barrier.*

several lanes of fast-moving traffic rather than walk a short distance or climb a ramp or stairway.

For pedestrian control, chain-link fences 4 to 6 ft high are usually used, although a thick hedge has sometimes served. They are commonly placed in the marginal area outside the through lanes. At times they may be needed in the median as well. Fences are required along both sides of all access and egress connections to the through lanes. In this way these entries and exits do not provide a shortcut for pedestrians.

Serious accidents sometimes result when pranksters throw objects from overpasses or pedestrian overcrossings. In many instances highway agencies have placed chain link fence partially or entirely around the pedestrian area to prevent such occurrences.

Frontage Roads

For freeways and expressways to operate successfully, it is essential that direct access to them be denied adjoining property and some local roads or streets. This means that if local traffic and land use are to be served, it must be by means of frontage or service roads planned as a part of the main facility. The decision as to whether frontage roads are provided rests on a careful study of the cost and the political and legal aspects of each individual situation.

Standards for design will, depending on traffic volume and location (rural, suburban, or urban), be the same as for collectors or local classifications.

In most instances one-way rather than two-way frontage roads are recommended because they bring fewer conflicts at intersections with freeway crossings. However, two-way roads may be required to serve a highly developed area where the local street pattern is complex or where freeway crossings are widely spaced. To reduce the required width, frontage roads may, for most of their length, lie just outside the border area of the main artery. It is recommended, however, that frontage roads enter connecting cross streets at distances of at least 300 ft for rural and 150 ft for urban situations.

DESIGNING THE GRADE LINE

The grade line is shown on a profile taken along the road centerline and is a series of straight lines connected by parabolic vertical curves to which the straight grades are tangent (see Fig. 6–4). In laying this grade line, the designer must secure economy by keeping earthwork quantities to the minimum consistent with meeting sight distance and other design requirements. In mountainous country the grade may be set to balance excavation against embankment with the aim of getting the least overall cost. In flat or prairie country it is approximately parallel to the ground surface but sufficiently above it to allow surface drainage, to provide sufficient cover over troublesome native soils or those subject to frost heave, and, where necessary, to permit the wind to clear drifting snow. Where the road approaches or follows along streams, the height of the grade line may be dictated by the expected level of flood waters. As discussed earlier in this chapter, smooth, flowing grade lines are preferable to choppy ones of many short straight sections connected with short vertical curves.

With divided highways, separate grade lines in the two directions have many advantages, but alignment and grades must be coordinated to prevent headlights from vehicles going in one direction from blinding the drivers of vehicles going in the other.

Maximum and Minimum Grades

Passenger car speeds are unaffected by upgrades less than 3% and only those with low horsepower/weight ratios at grades up to 6 or 7%. Speeds increase somewhat on downgrades. On the other hand, truck speeds are reduced on relatively flat up or down grades (see Fig. 8–9). This indicates that short or "momentum" grades may be steeper than long or "sustained" ones. Furthermore, grades acceptable where traffic volume is light and passing is possible are not satisfactory where high volumes reduce passing opportunity. Also, steeper grades are more tolerable in mountainous terrain than in flatter country. In this connection it is to be noted that within reasonable limits nothing is gained by lengthening a road to reduce its steepness.

Maximum permissible grades for the several highway classes as set by *A Policy on Geometric Design* are given in Table 9–3. As would be expected, they are much more restrictive on high-speed, high-volume facilities than on low-speed, low-volume ones. In actual practice, steeper grades have been used; some greater than 20% will be found on city streets. Such values exceed the coefficient of friction between tires and muddy or icy pavements; with such designs the driver's appraisal of roadway conditions becomes a controlling factor in safe operation.

Standards setting minimum grades are of importance only when surface drainage is a problem, as when water must be carried away in a gutter or roadside ditch. In such instances AASHTO suggests a minimum of 0.5%, but this may be reduced to 0.35% or even lower under special conditions.

Vertical Curves over Crests[37]

All vertical curves should be as long as conditions permit, and under no circumstance should they be shorter than certain established minimums. Over crests, these minimums are usually dictated by sight distance requirements, as detailed

[37]No detailed explanation of the methods for computing elevations of tangents or points on parabolic vertical curves is included in this book. Complete descriptions will be found in all standard textbooks in route surveying, and many highway agencies do these as well as many other routine computations by computer. Briefly, the vertical distance from the intersection of the straight grade lines to the curve is equal to one-eighth the product of the algebraic difference in grades and the length of the curve in stations. This is called the maximum correction. The rate at which the curve departs vertically from both tangent grade lines is proportional to the square of the horizontal distance from the end of the curve. The correction at any intermediate point, then, is obtained by (1) multiplying the maximum correction by the square of the horizontal distance between the near end of the curve and the point and (2) dividing this product by the square of one-half the length of the curve.

TABLE 9–3. Maximum Permissible Grades for Highways (%)*

Design Speed		*Freeways†*			*Arterials*						*Collectors—Rural‡*			*Local—Rural†§*		
					Rural			*Urban*								
mph	*km/h*	*Flat*	*Rolling*	*Moun-tainous*	*Flat*	*Rolling*	*Moun-tainous*	*Flat*	*Rolling*	*Hilly*	*Flat*	*Rolling*	*Moun-tainous*	*Flat*	*Rolling*	*Moun-tainous*
20	32	—	—	—	—	—	—	—	—	—	7	10	12	8	11	16
30	48	—	—	—	—	—	—	8	9	11	7	9	10	7	10	14
40	64	—	—	—	—	—	—	7	8	10	7	8	10	7	9	12
50	80	4	5	6	4	5	7	6	7	9	6	7	9	6	8	10
60	97	3	4	6	3	4	6	5	6	8	5	6	—	5	6	—
65	105	—	—	—	3	4	6	—	—	—	—	—	—	—	—	—
70	113	3	4	—	3	4	5	—	—	—	—	—	—	—	—	—

*Source: *A Policy on Geometric Design*.
†Grades 1% steeper than values shown may be used for extreme cases in urban areas where development precludes the use of flatter grades and for one-way grades in mountainous terrain.
‡No prescribed values for urban collectors. Should be consistent with surrounding terrain.
§No prescribed values for local streets. Maximum is 15%.

subsequently. On occasion ease of riding or appearance may demand longer curves than does sight distance. Some engineers prefer that no vertical curve be shorter than 1000 ft. Minimum curve lengths suggested by AASHTO vary with design speed, with the length in feet equal to three times the velocity in miles per hour.

STOPPING (NONPASSING) SIGHT DISTANCE OVER CRESTS. Methods for determining minimum stopping distances are outlined and their numerical values for different design speeds are given in Table 8–8. Figure 9–5 shows the approved method for measuring the stopping sight distance over crests. It is the longest distance that a driver whose eye is 3.5 ft above the pavement[38] can see the top of an object 6 in. high on the road. On Fig. 9–6 are given the equations that express minimum length of curve in terms of algebraic difference in grades and stopping distance. Solutions to these equations and design standards based on these and other controls are given in the chart, from which can be found the minimum length of vertical curve to match a given design speed for algebraic differences in grades up to 16%. Curve lengths are given for both low (solid lines) and high (dotted lines) sight distances. Low values assume drivers slow down on wet pavement; high values assume they do not.

Headlights on some vehicles are mounted some 2.0 ft above the road surface. This is considerably below driver eye height, so that actual night sight distances over crests are shorter than in daylight. However, for design purposes, the daylight situation is assumed to control on the theory that drivers go more slowly at night.

There is sometimes difficulty in obtaining satisfactory gutter drainage at the crest of long, flat, vertical curves. The drainage-maximum curve on Fig. 9–6 indicates the curve length at which a slope of 0.35% is reached 50 ft from the crest, which is satisfactory by AASHTO standards.

PASSING SIGHT DISTANCE OVER CRESTS. Methods for determining passing sight distances are outlined in Chapter 8, and their numerical values are given in Table 8–10. Figure 9–5B shows the approved method for measuring passing sight distance over crests. It is the longest distance that a driver whose eye is 3.50 ft above the pavement can see the top of an oncoming vehicle that is 4.25 ft high.[39] Equations that express the relationships among passing sight distance, algebraic difference in grades, and length of vertical curve are as follows.

$$L = \frac{AS^2}{3093} \quad \text{where } S < L, \text{ lengths in feet} \qquad (9\text{–}1A)$$

$$L = \frac{AS^2}{943} \quad \text{where } S < L, \text{ lengths in meters} \qquad (9\text{–}1B)$$

[38]Until recently, the AASHTO design eye height was 3.75 ft and before that 4.5 ft. Vehicle designs have been proposed with driver eye height 3.2 ft.
[39]Earlier AASHTO standards set these heights at 3.75 and 4.5 ft.

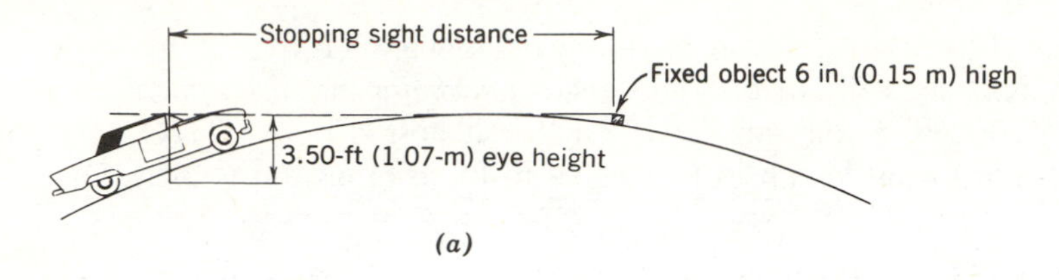

(a)

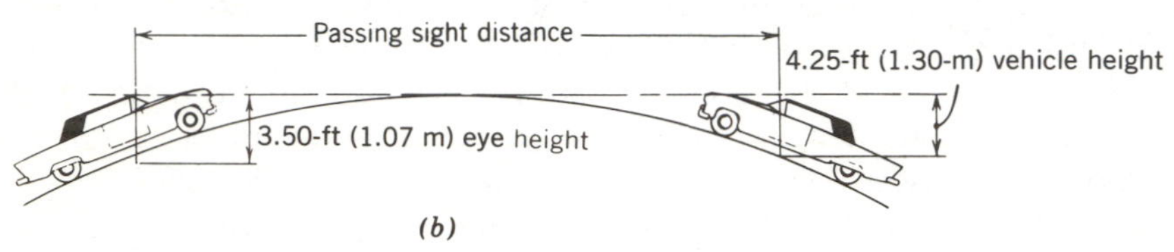

(b)

Fig. 9–5. Recommended procedures for measuring stopping and passing sight distances over crests.

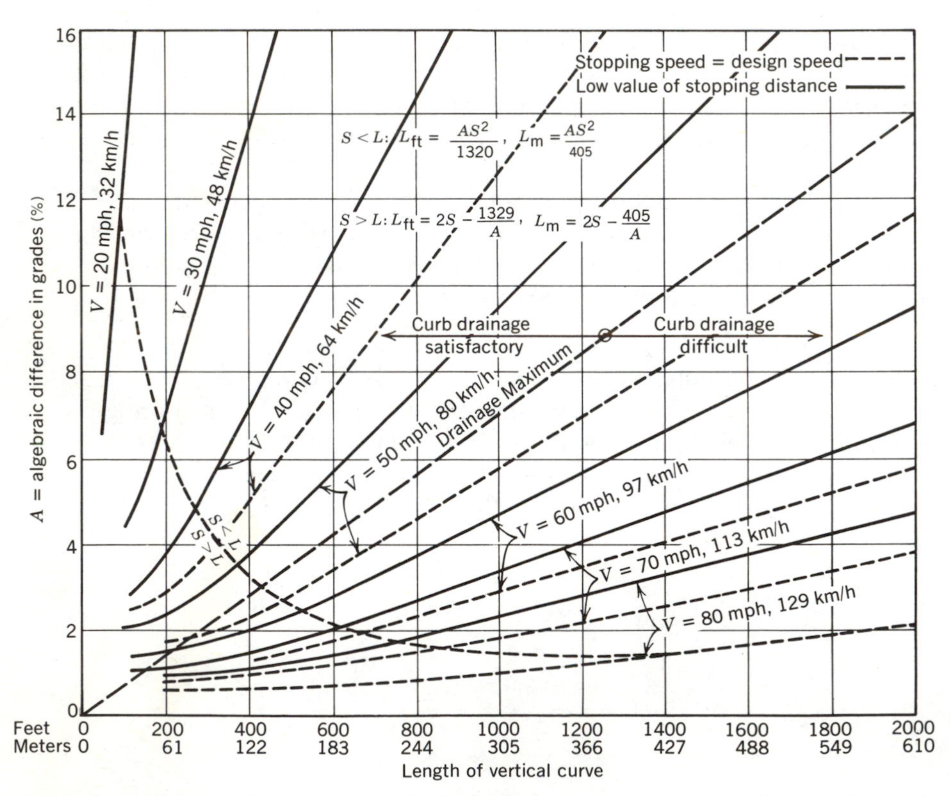

Fig. 9–6. Minimum lengths of vertical curves over crests to provide stopping sight distance.

$$L = 2S - \frac{3093}{A} \quad \text{where } S > L, \quad \text{lengths in feet} \tag{9–2A}$$

$$L = 2S - \frac{943}{A} \quad \text{where } S > L, \quad \text{lengths in meters} \tag{9–2B}$$

In these equations, L is the required length of vertical curve, S is the stipulated sight distance, and A is the algebraic difference in grades, expressed as a percentage.

Only where grades are flat is it generally practical to design two-lane highways for passing over crests, since much longer vertical curves are required to provide passing rather than stopping sight distance. Consider, for example, a crest formed by a 2% grade upward followed by a 2% grade downward. On a two-lane highway, for a design speed of 50 mph and accompanying passing sight distance of 1800 ft, a 4200-ft vertical curve must be used to provide passing sight distance, whereas a 700-ft vertical curve will meet the highest requirement for stopping sight distance. The vertical distances downward from the intersection of the tangent grades to the vertical curves are 21.0 and 3.5 ft, respectively. Or, stated differently, over the crest the road must be cut 17.5 ft lower to provide passing sight distance than if stopping sight distance only is provided. This example illustrates why, in rough country, it is extremely expensive to provide continuous passing sight distance on two-lane roads. It also illustrates why, if continuous passing opportunity is desired, a four-lane design over crests may be cheaper than a two-lane one, with the saving in excavation more than offsetting the cost of the extra lanes.

COMFORT CRITERION FOR VERTICAL CURVES, OVER CRESTS. If grade changes over crests are made too abruptly, the resulting changes in vertical forces acting on vehicle occupants cause discomfort. In such situations, a comfort criterion rather than sight distance sets vertical curve length. An example is that of a street rising on a steep grade that must change to nearly level at a cross street. One criterion for minimum vertical curve length for comfort is given by Equations 9–3A and 9–3B. They are:

$$L = \frac{AS^2}{46.5} \quad \text{for } L \text{ in feet and } S \text{ in miles per hour}^{40} \tag{9–3A}$$

$$L = \frac{AS^2}{400} \quad \text{for } L \text{ in meters and } S \text{ in kilometers per hour (km/h)}^{40} \tag{9–3B}$$

Vertical Curves in Sags

At least four criteria should be recognized in setting minimum lengths of vertical curves in sags. These are headlight illumination as controlling sight distance, rider comfort, drainage control, and general appearance from the standpoint

[40]At these lengths, vertical acceleration is 10% of gravity which means a 10% reduction in weight for car and occupants.

that short curves give a choppy rather than a flowing impression. The AASHTO manual recognizes headlight sight distance as usually controlling vertical curve length in sags. Since 1961, headlight positioning has been assumed as 2.0 ft above the road surface with a 1° upward divergence of the beam. Graphs showing the relationships among headlight sight distance, algebraic difference in grades, and low (solid lines) and high (dotted lines) vertical curve lengths are given in Fig. 9–7. This figure also shows maximum curve lengths for proper drainage when the road is curbed.

Comfort because of centrifugal force is more greatly affected on sag than on crest vertical curves, but the effects are hard to assess. According to *A Policy on Geometric Design* it is not a design control away from intersections since the curve length to prevent discomfort is only about 75% of that dictated by headlight reach. For short sag vertical curves approaching intersections, lengths given by Equations 9–3A or 9–3B are appropriate.

Where a highway passes under an overhead structure, such as an overpass, the driver's view along the highway may be blocked by the structure. In this instance also stopping sight distance sets the minimum length of sag vertical curve, but under a different set of controls. AASHTO recommends that eye height be set at 6 ft (that of a truck driver) and object height at 1.5 ft (that of the tail light of a vehicle). Recommended minimum structure vertical clearance is 15 ft; desirable is 17 ft. In all instances, it should be at least 1 ft greater than the legal limit to permit resurfacing.

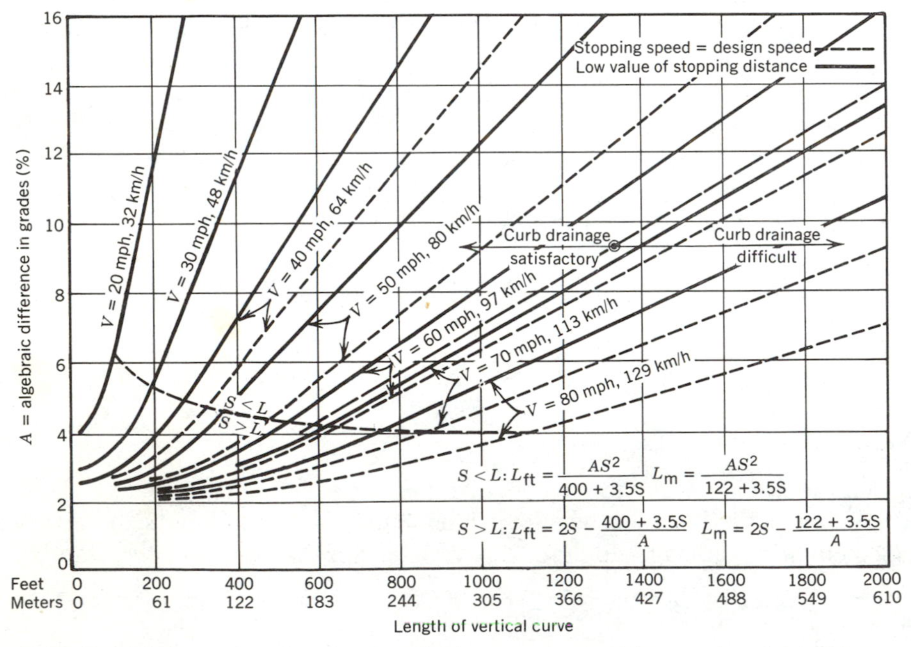

Fig. 9–7. Minimum lengths of sag vertical curves to provide stopping sight distance.

ALIGNMENT DESIGN

The alignment of a road is shown on the plan view and is a series of straight lines called tangents connected by circular curves. In modern practice it is common to interpose transition or spiral curves between tangents and circular curves (see Fig. 6–4).

Alignment must be consistent. Sudden changes from flat to sharp curves and long tangents followed by sharp curves are to be avoided; otherwise accident hazards are created. Likewise, placing circular curves of different radii end to end (compound curves) or having a short tangent between two curves is poor practice unless suitable transitions between them are provided. Long, flat curves are preferable at all times, as they are pleasing in appearance and decrease the possibility of future obsolescence. However, alignment without tangents is undesirable on two-lane roads because some drivers hesitate to pass on curves. Long, flat curves should be used for small changes in direction, as short curves appear as "kinks." Also, as indicated above, horizontal and vertical alignment must be considered together, not separately. For example, a sharp horizontal curve beginning near a crest can create a serious accident hazard. Neither should the substantially increased vehicle running costs associated with sharp curves, as discussed in Chapter 4, be overlooked.

This book presents the methods currently accepted for designing vertical and horizontal alignment for highways. These seem to be adequate for present and expected vehicle speeds. However, in designing test tracks for very high-speed test-car operation, engineers for motor manufacturers have measured the effects of human perception of motion on rider behavior and comfort. Designs based on these findings are different and interesting.[41]

Circular Curves

As has already been demonstrated in Chapter 8, a vehicle traveling in a curved path is subject to centrifugal force. This is balanced by an equal and opposite force developed through superelevation and side friction. From a highway design standpoint, neither superelevation nor side friction can exceed certain maximums, and these controls place limits on the sharpness of curves that can be used with a prescribed design speed.

Usually the sharpness of a given circular curve is indicated by its radius. However, for alignment design, sharpness is commonly expressed in terms of *degree of curve,* which is the central angle subtended by a 100-ft length of curve. Degree of curve is inversely proportional to the radius, a relationship expressed by the formulas

$$D = \frac{5729.58}{\text{radius}} \tag{9–4}$$

$$\text{radius} = \frac{5729.58}{D} \tag{9–5}$$

[41]See papers by K. A. Stonex and W. A. McConnell, *HRB Bulletin 149,* for added detail.

where D = degree of curve and the radius is in feet.[42]

Solutions to these equations, for typical curves, give the following results.

Degree of Curve	*Radius*		*Degree of Curve*	*Radius*	
	ft	*m*		*ft*	*m*
0°30′	11,459.16	3,491.75	6°00′	954.93	291.06
1°00′	5,729.58	1,746.38	10°00′	572.96	174.63
2°00′	2,864.79	873.19	20°00′	286.48	87.32

More precisely, degree of curve must be expressed under either the *arc* definition or the *chord* definition. For the arc definition, the degree of curve is the central angle subtended by a 100-ft *arc* of the curve. Radii for various degrees of curve, arc definition, are inversely proportional to the degree of curve and are given by Formula 9–5.

Under the chord definition, the degree of curve is the central angle subtended by a 100-ft *chord*. From this definition it follows that the precise relationship between radius and degree of curve is

$$\text{radius} = \frac{50}{\sin \frac{1}{2} D} \tag{9–6}$$

For flat curves of any stated degree, the radii obtained by arc and chord definitions are almost equal. However, as the curves become sharper, the results diverge. This divergence is not enough to affect geometric design standards, but must be recognized in computing and surveying the curves.[43]

A few agencies use "even-radius" curves, for which the radius is expressed in even figures such as 1000, 1500, or 2000 ft. When degree of curve for even radius curves is computed by Formula 9–4, the result does not come out in whole numbers. For example, D for a curve of 2000-ft radius is 2°51.88′.

Standards for Curvature

Equation 8–6A gives the theoretical relationships among velocity, circular-curve radius, coefficient of side friction, and superelevation. For design purposes this equation is more useful stated as

$$D_{max} = \frac{85{,}950(e+f)}{V^2} \tag{9–7}$$

[42]According to *Compendium 1,* Transportation Technology Support for Developing Countries, TRB, degree of curve in SI units is "the deflection angle subtended by a 20-m chord." A rough relationship based on this definition is

$$(\text{degree of curve})_{SI} = 0.328D$$

For sharp curves this relationship is further complicated by the "arc versus chord" definition problem. For these reasons, references to curves in SI units are always stated here in terms of radius.

[43]Space limitations preclude the inclusion here of the methods for computing and surveying circular and easement curves. These are found in any standard textbook in route surveying.

where

D_{max} = sharpest permissible degree of curve
V = design speed in miles per hour
e = design maximum superelevation in feet per foot
f = design maximum coefficient of friction (see Tables 8–11 or 9–4)

Equation 9–7, solved using recommended maximum superelevation limits leads directly to the listing of sharpest permissible curves given in Table 9–4. Note that SI units for speed and radius appear in the table. Until recently, the maximum permissible superelevation rate was 0.12, but it now is 0.10, except on low-volume gravel roads.

For curves flatter than those specified by Table 9–4 for a given design speed, the engineer must choose whether, in a given situation, it is better to reduce side friction, superelevation, or both. Suppose, for example, there is good reason to discourage travel faster than design speed. Then, on flatter curves along the road, superelevation could be set to develop side friction to its maximum permissible value. On the other hand, it might be desirable to minimize side friction by using full superelevation. The relationships between degree of curve and side friction to achieve these extremes, for a design speed of 30 mph, are shown by the solid lines with hatching on Fig. 9–8. Note that all combinations of curvature and superelevation falling between these limits represent acceptable designs.

Figure 9–8 shows for a maximum superelevations of 0.06, (dotted lines) and 0.10 (solid lines), the AASHTO recommended superelevation practice for design speeds of 30 to 80 mph. Similar curves or tables for other values of maximum superelevation appear in the design manual. They represent a compromise between (1) making the superelevation rate directly proportional to degree of curve and (2) counteracting all centrifugal force with superelevation until maximum superelevation is attained. One among many other possibilities would be to counterbalance completely the centrifugal force of a vehicle traveling at three-fourths of the design speed of the road.

Even for high-speed designs, vehicles traveling around flat curves develop relatively little centrifugal force. This can be compensated by a small amount of superelevation. On the other hand, if normal cross slope is retained, this is in effect a reverse superelevation on curves to the left. Thus there is the question of whether or not to superelevate flat curves. *A Policy on Geometric Design* recommends that, for design speeds of 30 mph, normal cross slope be maintained with 1°21′ or flatter curves. For 80 mph, only 0°15′ or flatter curves are without superelevation. In each of these instances the total side friction factor is less than 0.04 for an adverse cross slope of 0.02. This side friction is less than a third of that at which discomfort begins.

It must be recognized that side friction high enough to cause discomfort does not cause all drivers to slacken speed. For example, early studies by engineers of the Bureau of Public Roads found that, on curves sharper than 15°, 10% of

TABLE 9–4. Sharpest Permissible Horizontal Curves for Given Design Speeds and Superelevations*

Superelevation	Condition for Use		Maximum Permissible Degree of Curve, D, or Minimum Radius, R_{ft} or R_m						
			Design Speed						
		mph	20	30	40	50	60	70	80
		km/h	32	48	64	80	97	113	127
			Coefficient of Side Friction						
			0.17	0.16	0.15	0.14	0.13	0.12	0.11
0.04	Desirable for downtown arterials	D	45.0	19.0	10.0	6.0	3.75	—	—
		R_{ft}	127	302	573	955	1528	—	—
		R_m	39	92	175	291	466	—	—
0.06	Desirable for rural highways with snow and ice and for urban highways. Maximum for downtown arterials	D	49.0	21.0	11.25	6.75	4.25	2.75	1.75
		R_{ft}	117	273	509	849	1348	2083	3274
		R_m	36	83	155	259	411	635	998
0.08	Maximum for rural highways with snow and ice and for urban arterials	D	53.5	23.0	12.25	7.5	4.75	3.0	2.0
		R_{ft}	107	249	468	764	1206	1910	2865
		R_m	33	76	143	233	368	582	873
0.10	Maximum for rural highways and suburban freeways	D	58.0	25.0	13.25	8.25	5.25	3.5	2.25
		R_{ft}	99	229	432	649	1091	1637	2546
		R_m	30	70	132	198	333	499	776

*Source: *A Policy on Geometric Design*.

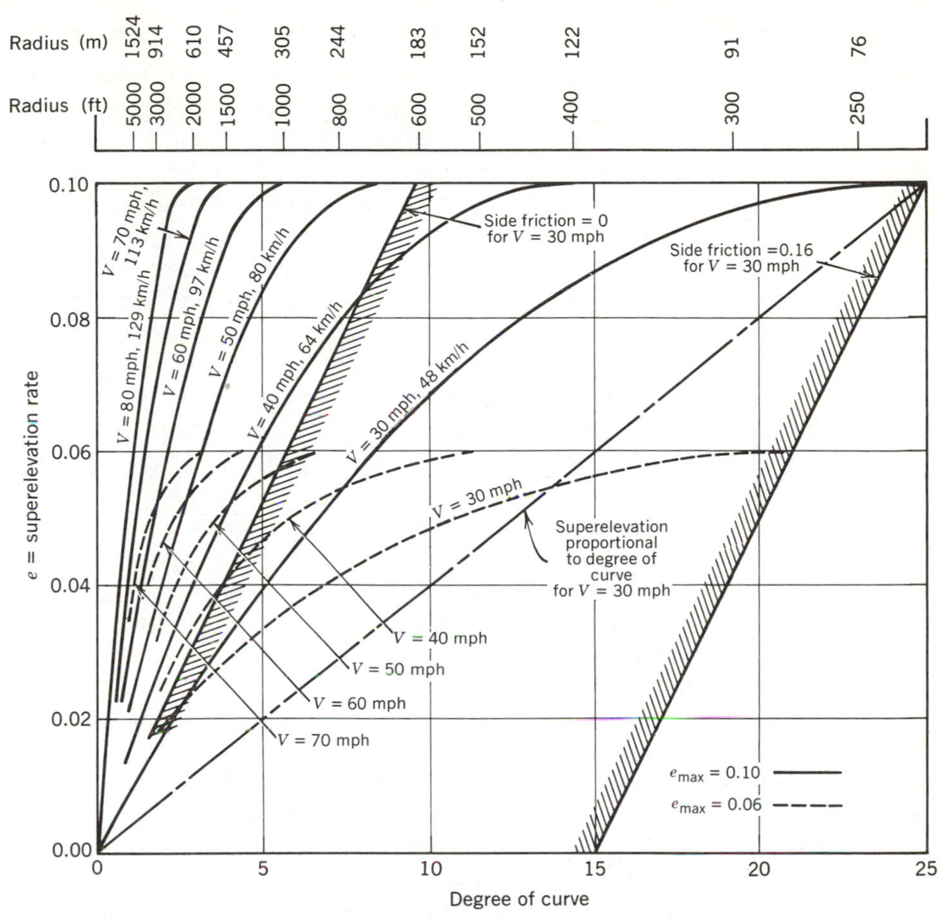

Fig. 9–8. Recommended superelevation rates for horizontal curves for e_{max} of 0.06 and 0.10. (Source: *A Policy on Geometric Design*.)

the drivers developed coefficients greater than 0.30; however a coefficient of 0.16 was seldom exceeded on curves of 6° or less.[44]

Accidents sometimes occur when vehicles traveling downgrade go out of control while entering curves. In this instance, the normal forces between individual tires and pavement are redistributed. If, then, the driver applies the brakes and changes the position of the steering wheel, "out of control" skidding may result. *NCHRP Report 184* indicates that this occurs only when downgrades, slippery surfaces, poor tires, and severe maneuvers are combined. One among the possible remedies is to provide pavements with high coefficients of friction at such locations.

[44]See A. Taragin, *Public Roads* (June 1954), or *HRB Proceedings 1954* for a detailed study of driver performance on horizontal curves.

Superelevation Runoff

Tangent sections of highways carry normal cross slope; curved sections are superelevated. Provision must be made for gradual change from one to the other. A common method is to maintain the centerline of each individual roadway at profile grade while raising the outer edge and lowering the inner edge to produce the desired superelevation (see Fig. 9–9). This involves first raising the outside edge of the pavement with relation to the centerline until the outer half of the cross section is flat (point *B* in Fig. 9–9); next the outer edge is raised further until the cross section is straight (point *C*); then the entire cross section is rotated as a unit until full superelevation is reached (point *E*). For smoother riding, *A Policy on Geometric Design* recommends that short vertical curves having a length in feet equal to the design speed in miles per hour be introduced into the edge profiles at their break points. These are not shown in Fig. 9–9. Also presented are three other methods for superelevation runoff and for the treatment of medians. These cannot be given here.

AASHTO recommendations are that runoff lengths vary both with the superelevation rate and design speed, but with minimums set for appearance and comfort. These minimums approximate the distance a vehicle travels in 2 s. Suggested values of runoff lengths for undivided highways when rotation of the cross section is around the centerline are given in Table 9–5. Multipliers for certain other situations are noted in the table.

Where the roadway lies in a cut and the grade line is nearly flat, lowering its inner edge may create a sag from which surface water will not drain. Then it may be advisable to accomplish all superelevation by raising the outer edge. This, of course, means that this edge must be elevated twice the usual distance. Also, with divided highways with narrow medians, it sometimes is better to rotate each roadway about the edge adjoining the median strip rather than its individual centerline. Otherwise the two sides of the median may be at considerably different elevations; this creates numerous problems. A discussion of these complexities appears in *A Policy on Geometric Design*.

Where the alignment consists of tangents connected by circular curves, introduction of superelevation usually is begun on tangent before the curve is reached and full superelevation is attained some distance beyond the point of curve. It is recommended that 60 to 80% of the runoff be on tangent. These percentages set the position of the cross slope at point *D* on Fig. 9–9.

Where the alignment includes easement curves (see below), superelevation is applied entirely on the easement curve, except for bringing the outer pavement edge to a level position (see Fig. 9–9). It follows that, at times, the superelevation application rate sets the minimum length of the easement curve.

Easement (Spiral) Curves

If a vehicle travels at high speed on a carefully restricted path made up of tangents connected by sharp circular curves, riding is extremely uncomfortable. As the car approaches a curve, superelevation begins and the vehicle is tilted in-

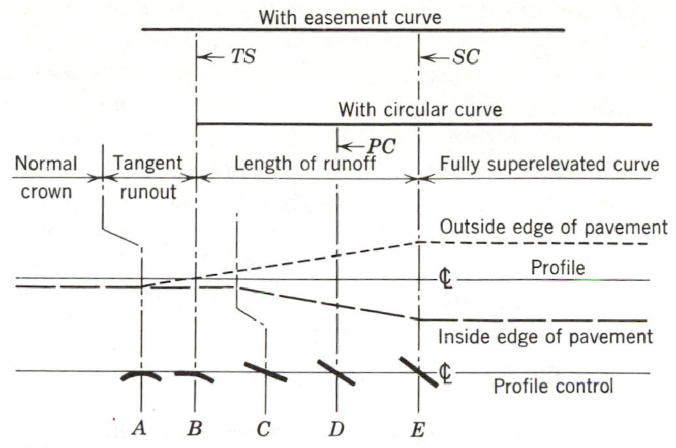

Fig. 9–9. Diagram of recommended method for attaining superelevation at the ends of horizontal curves when the pavement is rotated about its centerline. Vertical curves at break points in grade lines are not shown.

ward, but the passengers must remain vertical since there is no centrifugal force requiring compensation. When the vehicle reaches the curve, full centrifugal force develops at once, and pulls the riders outward from their vertical positions. To achieve a position of equilibrium, the riders must force their bodies far inward. As the remaining superelevation takes effect, further adjustments in position are required. This process is repeated in reverse order as the vehicle leaves the curve. When easement curves are employed, the change in radius from infinity on the tangent to that of the circular curve is effected gradually so that centrifugal force also develops gradually. By introducing superelevation along the spiral, a smooth and gradual application of centrifugal force can be

TABLE 9–5. Suggested Minimum Length of Superelevation Runoff for Two-Lane Pavements with 12-ft Lanes*

Super elevation Rate		*Length of Runoff for Design Speed of*‡						
	mph	*20*	*30*	*40*	*50*	*60*	*70*	*80*
	km/h	*32*	*48*	*64*	*80*	*97*	*113*	*129*
0.02		30	35	40	50	55	60	65
0.04		60	70	85	95	110	120	130
0.06		95	110	125	145	160	180	200
0.08		125	145	170	190	215	240	265
0.10		160	180	210	240	270	300	330

*Source: *A Policy on Geometric Design.*
†For wider roadways, lengths shown in table should be increased as follows:
four lanes, undivided, 50%; six lanes, undivided, 100%; six-lane highway, wide median, 1.2. See reference for other situations.
‡To obtain length of runoff in meters, multiply tabular values by 0.3048.

had and the roughness avoided. The recommended procedure for applying superelevation along the easement curve is shown on Fig. 9–9.

Easement curves have been used by the railroads for many years, but their adoption by highway agencies came much later. Many agencies do not use them today. This is understandable. Railroad trains must follow the precise alignment of the tracks, and the discomfort described here can be avoided only by adopting easement curves. On the other hand, motor-vehicle operators are free to alter their lateral positions on the road and can provide their own easement curves by steering into circular curves gradually. However, this weaving within a traffic lane (sometimes into other lanes) is dangerous. Properly designed easement curves make weaving unnecessary. Safety, then, is an argument favoring them. Another is that they give alignments a smoother, more flowing appearance.

For the same radius circular curve, the addition of easement curves at the ends changes the location of the curve with relation to its tangents; hence the decision regarding their use should be made before the final location survey. This difference in position is illustrated by Fig. 9–10. Changes made in providing an easement curve are as follows:

1. The entire circular curve (dashed curve on Fig. 9–10) is shifted inward on a line connecting the tangent intersection *PI* with the radius point of the circular curve O_c. This shift is designated as the "throw distance."
2. The ends of the shifted circular curve are removed (dotted curves in Fig. 9–10) and appropriate easement curves substituted.

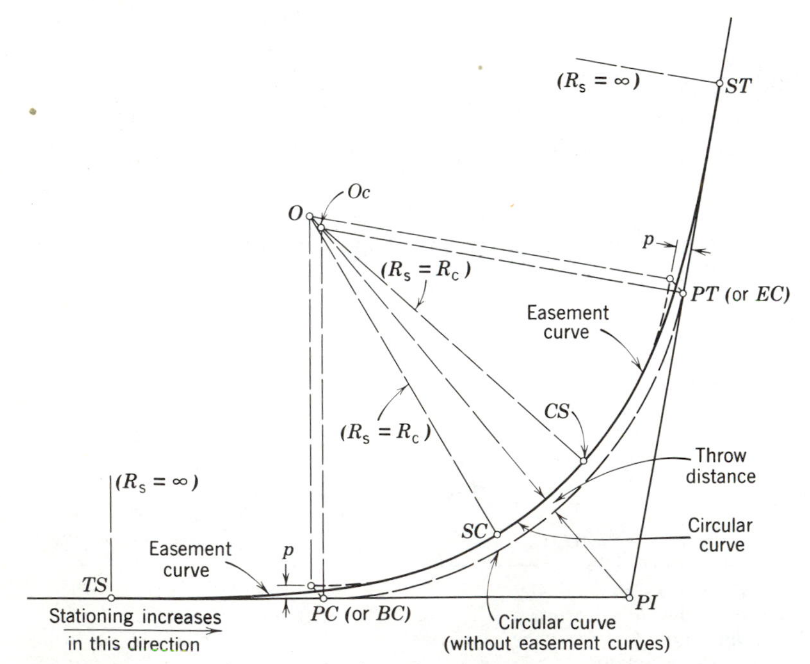

Fig. 9–10. A circular curve with and without easement curves.

The point of beginning of an ordinary circular curve is usually labeled the *PC* (point of curve) or *BC* (beginning of curve). Its end is marked the *PT* (point of tangency) or *EC* (end of curve). For curves that include easements, the common notation is, as stationing increases, *TS* (tangent to spiral), *SC* (spiral to circular curve), *CS* (circular curve to spiral), and *ST* (spiral to tangent).

The sharpness of the commonly used easement curves, measured in terms of degree of curvature, increases uniformly from their beginnings. If, for instance, easement curves 400 ft long are selected to connect each end of a 4° circular curve to its tangents, the sharpness of the easement curves will increase by 1° each 100 ft. At the *TS* or *ST*, where the curve begins, the degree of curve is zero and the radius infinite. At 100 ft along the curve, the spiral has the same radius as a 1° curve; at 200 ft, its radius equals that of a 2° curve; at 400 ft, where the easement curve ends (*SC* or *CS*), both easement and circular curves have the sharpness of a 4° curve and a common radius point. If length is added to this identical spiral, it will also fit sharper circular curves. Thus, if it is extended to 1000 ft, it will fit a 10° curve. A particular sharpness of easement curve, then, may be designated by its increase in degree of curve per 100 ft station, which is denoted by some writers with the letter k. For the curve described here, k equals 1°.

The sharpness or k value of an easement curve, and consequently its length L_s, depends on design speed. As mentioned earlier, motor-vehicle operators control the rate of application of centripetal force by steering into curves gradually. If drivers are to follow easement curves closely and retain their proper position on the roadway, this change in force must not be excessive. Based on this line of reasoning, the minimum length and k values of easement curves would be set by the formulas

$$(L_s)_{min} = \frac{0.00055V^3D}{C} \tag{9–8}$$

$$k_{max} = \frac{173{,}000C}{V^3} \tag{9–9}$$

where D is the degree of curve of the circular curve to which the spiral is joined and C is the rate of increase in centripetal acceleration. Values commonly proposed for C are 1 for railroads and 2 to 3 for highways.[45] Solutions to Equations 9–8 and 9–9 with a C value of two are given in Table 9–6.[46]

In the past, it has been argued that an easement curve had little value when the radial offset distance or "throw" (dimension p on Fig. 9–10) was less than 1 ft. Current practice indicates that easement curves should be used with radial offsets on the order of 0.3 ft.

Computations and field surveys for combined circular and easement curves

[45]See *A Policy on Geometric Design* for a far more detailed discussion.
[46]For recent proposed changes in easement curve selection see J. Craus and A. Polus, *TRB Record 631* and J. Greenstein, *Transportation Engineering Journal ASCE*, Nov. 1976.

TABLE 9–6. Minimum Lengths of Easement Curves for Various Design Speeds and Sharpness of Circular Curves

Design Speed		k, the Increase in Sharpness of Spiral per 100 ft of Curve Length (degrees)	Minimum Length of Easement Curve (ft)*									
			D—Sharpness of Circular Curve (degrees)†									
mph	km/h		1°	2°	3°	4°	5°	6°	8°	10°	12°	14°
40	64	5	20	40	60	80	100	120	160	200	240	280
50	80	2½	40	80	120	160	200	240	320			
60	97	1⅔	60	120	180	240	300	360				
70	113	1	100	200	300	400						
80	129	⅔	150	300	450							
90	145	½	200	400								
100	161	⅓	300	600								

*To obtain curve length in meters, multiply tabular values by 0.3048..

†Radius of circular curve in ft, $5730/D$, in meters, $1746/D$.

are somewhat more difficult and time consuming than those for circular curves alone. This explains, at least in part, why some American highway agencies did not adopt them. Detailed explanations, formulas, and tables of useful functions and dimensions for easement curves will be found in all standard textbooks in route surveying and in the standard drawings of many highway agencies. Because of space limitations, such information is omitted from this book.

As with circular curves, easement curves may be computed by either the arc or the chord definition. Arc-definition easement curves are used with arc-definition circular curves and chord-definition ones with chord-definition curves.

The American Railway Engineering Association (AREA) spiral, used by most railraods, is based on the chord definition. It is known as the ten-chord spiral because, regardless of its total length, it is laid out in ten equal segments. Points on other types of spirals are often staked out at convenient intervals along the roadway, such as 10, 25, or 50 ft, regardless of the number of segments which results.

Reverse Curves—Compound Curves

At times, particularly in rough country, a curve in one direction is followed by one turning the opposite way. In the past the *reverse curve* often was used, with the end point of one circular curve (*PT*) also the beginning point (*PC*) of the next. At this "point of reverse curve" (*PRC*), the roadway was made flat or with normal crown, and all superelevation was developed on the respective curves. For the low-speed operation of early days there was little objection to this practice, but with the higher speeds of modern vehicles reverse curves produced a very "rough-riding" roadway. Also, high-speed drivers cannot closely follow the intended path, which results in an accident hazard. Reverse circular curves are seldom used on modern highways. However, reverse curves that incorporate proper length easements between them are perfectly acceptable. If no easement curves are provided, curves in opposite directions should be separated by a tangent several hundred feet in length. Some roads have *compound curves* that change abruptly from one sharpness to another at a "point of compound curve" (*PCC*). Where the radii of the compounded curves are markedly different, the sudden change confuses and deceives drivers so that they shift position within the lane, and sometimes veer out of it. For this reason, compounding circular curves of greatly different radii is considered poor practice. AASHTO recommends, for rural highways, that the radius of the flatter curve never be more than 50% greater than that of the sharper one. For urban intersections, this ratio is 2 to 1 or less. Compound curves may be used, however, if a connecting easement curve is provided which introduces the change in radius gradually. Superelevation is changed along this transition section. Rules for determining curve length correspond to those stated earlier for spirals connecting tangents with circular curves. Computations and field layout are somewhat difficult. Explanations are found in most writings on easement curves.

Horizontal Sight Distance

As a vehicle travels around a horizontal curve, obstructions such as buildings, signs, or cut banks located inside the curve can block the driver's view ahead. Any specific combinations of sharpness of curve with position of obstruction establishes a horizontal sight distance, which is the greatest distance at which a driver can see an object lying in the roadway ahead. If the design is to provide for safe operation, this horizontal sight distance measured along the curve must equal or exceed the stopping distance for each design speed.

The accepted method for measuring horizontal sight distance places a 6 in. object and the driver's eye, which is 3.5 ft above the pavement, in the center of the inside lane (see Fig. 9–11). In this instance the inside lane is defined as the one closest to the sight obstruction. Figure 9–11 also shows the relationships

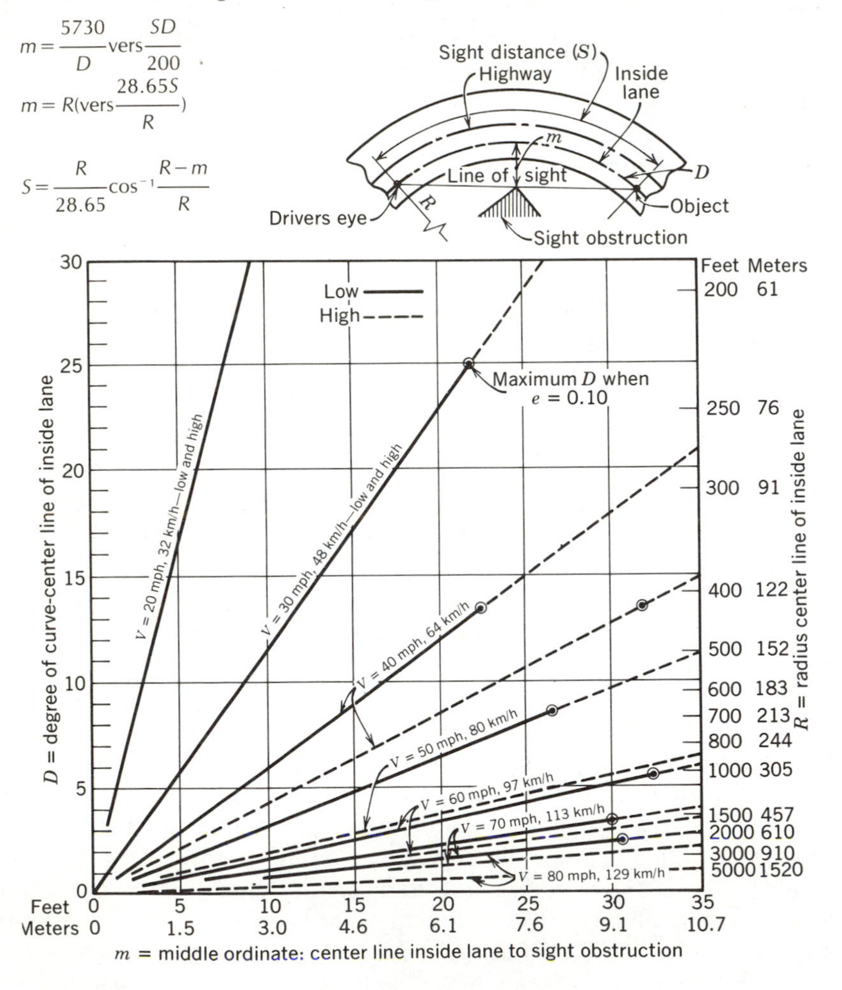

Fig. 9–11. Chart for establishing stopping sight distances on horizontal curves for open-road conditions. (Based on *A Policy on Geometric Design*.)

among obstruction position, degree or radius of curve, design speed, and low and high stopping distances. The figure does not apply if required sight distance exceeds curve length, at the ends of circular curves, or along easement curves. For these and like situations, the offset distances may be obtained easily by scaling from the plans. When computing the distance that a cut slope must be set back, it should be recognized that the line of vision passes the critical point of the bank at a level higher than the roadway shoulder. The required bank setback should be provided at this height. For straight grade lines this distance is 2.0 ft.

To provide a stipulated horizontal sight distance, it is at times necessary to set back obstructions or to move the cut slope outward along the inside of a curve beyond its location in the normal tangent cross section. If the design speed is high, this setback is large for sharp curves, and the consequent right-of-way or excavation costs are high. This is one reason why design speeds on mountain roads are often low. Where high design speeds are to be maintained regardless of cost, flat curves may be provided at little added cost since much of the right-of-way or excavation must be provided anyway to meet the horizontal sight distance requirement.

Night driving around sharp curves introduces an added problem related to horizontal sight distance. Motor-vehicle headlights are pointed directly toward the front and do not provide as much illumination in oblique directions. Even if adequate horizontal sight distance is provided, it has little useful purpose at night because the headlights are directed along a tangent to the curve and the roadway itself is not properly illuminated.

Horizontal sight distance requirements based on the same assumptions also must be applied to curved roadways at intersections and on curved bridges where the handrail blocks the view ahead. Details are not given here, but will be found in the AASHTO policy manual.

Studies of driver behavior indicate that restrictions on horizontal sight distance do not necessarily influence driver speed. For example, it was found that although few drivers exceeded safe speeds when horizontal sight distances were greater than 400 ft, with shorter sight distances *most* drivers stayed within a speed from which they could stop only if no allowance were made for detection, recognition, decision, and response-initiation times.[47] Findings such as these indicate that provision of adequate horizontal sight distance appears necessary to protect drivers from their own unsafe practices.

Curve Widening

On two-lane pavements provision of a wider roadway is advisable on sharp curves. This will allow for such factors as (1) the tendency for drivers to shy away from the pavement edge, (2) increased effective transverse vehicle width because the front and rear wheels do not track, and (3) added width because of the slanted position of the front of the vehicle to the roadway centerline. For

[47]See A. Taragin, *Public Roads*, June 1954, or *HRB Proceedings* 1954.

24-ft roadways on open highways, an added width of 1 ft is recommended for, for example, a 5° curve and 50-mph design speed or a 10° curve and 30-mph speed. For narrower pavement and sharp curves, widening assumes importance; with a 20-ft pavement, 250-ft radius, and 30-mph design speed, widening is 5 ft. Detailed recommendations appear in *A Policy on Geometric Design*.

Where vehicles must make sharp turns as is common at intersections or on turning roadways, both pavement widening and inside curve radii become important design features. This topic is discussed briefly in a following section on intersection design.

Combined Curvature and Grade

Horizontal curvature increases operating resistance. Also, as superelevation is added and taken off a road of constant grade, some sections along the actual vehicle path become steeper and others flatter than the profile grade. Where grades are relatively flat, no recognition is made of these factors in roadway design. In mountainous country where the combination of sharp curves with steep grades may occur at low design speeds, the effect of curvature on speed deserves attention. The practice of some agencies is to reduce the centerline grade for the length of the curve to compensate for the increased uphill resistance caused by curvature.

CHANNELIZATION

Purposes of Channelization

In modern highway practice, color and surface-texture differences in pavements, pavement markings (striping), raised bars, curbings, raised islands, guard rails, and fences are widely used to direct or control vehicle and pedestrian movements. When installed in accordance with principles such as those enumerated below, they have increased intersection capacity, improved traffic-flow conditions, and decreased vehicular and pedestrian accidents. Appropriate channelizing devices are included as an integral part of any new intersection or interchange. In addition, they offer an extremely useful and relatively inexpensive tool for correcting some of the deficiencies of existing roads and streets. Among the more important purposes of channelization are the following:[48]

1. By channelization, vehicles can be confined to definite paths. When drivers or pedestrians have free choice of routes through large all-paved intersections, their actions cannot be predicted by others. This creates confusion and congestion and often leads to accidents. In addition, accidents are more likely, since each vehicle is exposed for longer distances to others making conflicting movements.
2. By channelization, the angle between intersecting streams of traffic can be made

[48] *HRB Special Report 74*, titled "*Channelization*" is a classic. This publication offers a detailed discussion of channelizing principles and 80 typical examples. Numerous added examples appear in *A Policy on Geometric Design* and the *Manual on Uniform Traffic Control Devices*.

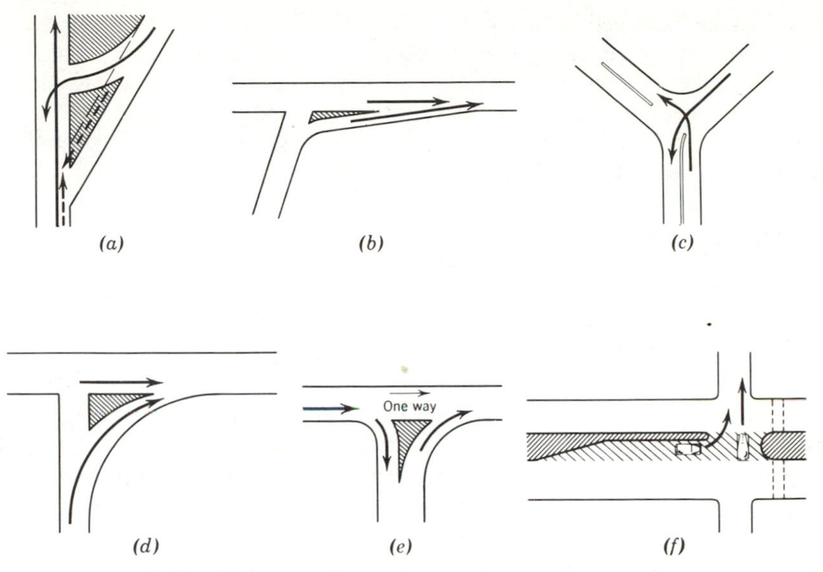

Fig. 9–12. Channelization techniques.

more favorable. When cross traffic meets at flat angles (see the dashed arrows of Fig. 9–12*a*), the accidents that occur usually are serious ones, just as head-on collisions are. The relative speed between the two vehicles is high, and practically all the kinetic energy of both is dissipated by damaging the vehicles or their occupants. As the direction of meeting swings away from head-on, accidents generally become less severe. It has now been established that, all things considered, the intersection of traffic streams at about right angles (75–105°) is most favorable (see Fig. 9–12*a*). Besides decreasing accident severity, meeting at this angle reduces the distance and time during which opposing vehicles can be in conflict, and thus reduces accident opportunity. The right-angle crossing also provides drivers with the most favorable condition for judging the relative position and speed of approaching vehicles.

3. By channelization, drivers can be forced to merge into moving traffic streams at flat angles and proper speeds (see Fig. 9–12*b*). Entry in this manner causes less disruption to traffic and has a smaller impact on the capacity of the main thoroughfare. Where traffic enters at larger angles, accident hazard is greater and fewer vehicles are accommodated, since the gap in the moving stream must be longer before entry can be made. Access to major arteries other than by merging usually should be subject to stop-sign or signal control.
4. By channelization, speed control can be established over vehicles entering an intersection. One method of accomplishing this is to bend the traffic stream (Fig 9–12*c*). Another is to funnel the vehicles into a narrowing opening (Fig. 9–12*d*). This causes drivers to feel hemmed in, and they react by reducing their speed: Funneling also is effective in preventing overtaking and passing in a conflict area. In cases where speed control is the aim, carefully planned superelevation is an important adjunct to channelizing.
5. By channelization, prohibited turns may be prevented (Fig. 9–12*e*).
6. By channelization, refuge may be provided for turning or crossing vehicles and for pedestrians. This is illustrated by the provision of turning lanes and of protected areas

for cross traffic and pedestrians at the center of the street (see Fig. 9–12*f*).

7. By channelization, points of conflict may be separated in such a way that the driver faces only one decision at a time. This reduces confusion and accidents because the driver can reach the proper decision in a shorter period of time.
8. Channelizing devices provide protected locations for essential traffic-control devices such as signs and signals and refuges for pedestrians.

Channelizing Devices

The various channelizing devices exert different degrees of control over driver and vehicle. On the one hand, pavement markings and changes in roughness or color of the pavement surface merely suggest the appropriate path and speed.[49] On the other hand, nonmountable curbs, guard rail, fences, and bumper blocks positively prohibit encroachment by the vehicle under usual circumstances.[50] Between the extremes fall such devices as raised bars made of concrete or bituminous mixtures or mountable curbs over which a vehicle may pass at low speed if the driver so desires. In selecting the device that will work best in a given situation, the designer must weigh such factors as the space available for the installation, its cost, traffic volumes and speeds, the seriousness of the accidents that may result from vehicles entering the prohibited area, and the influence of police supervision on driver behavior. For example, all these conditions differ greatly between rural and urban locations, and designs appropriate for one may not fit the other.

Curbing and raised islands bounded by curbs are particularly important channelizing devices. Figure 9–13 shows typical curb and median barrier cross sections.[51] The given dimensions are intended merely as guides, as curb details are not standardized among highway agencies. In setting curb heights it is important to note that skirt and fender edges or other low-hanging parts of passenger automobiles will catch curbs higher than 6 to 7 in.

Curbs of the configurations shown in Fig. 9–13 commonly are of concrete, although asphalt concrete is sometimes employed.[52] At times, they may be of granite or other quarried stone. Sometimes the "single curb" is constructed first and the pavement is placed against it (see Fig. 9–13*a*). Generally, where water is to flow along the gutter, combined curb and gutter (see Fig. 13*d*) is preferred. This conformation provides an excellent means for establishing proper pavement grades.

Special machines have been developed that form the curb by extruding concrete or asphaltic materials as the machine passes by. In other instances precast

[49]See Chapter 10 for further discussion of pavement markings.

[50]*NCHRP Report 150* investigated the effect of geometry and placement of certain curbs on vehicle behavior.

[51]The standard drawings of each highway agency include details of many curbing types, each of which is assigned a code number or letter. The roadway designer selects the type that meets the particular situation and designates it on the plans by giving its number.

[52]For details, *Construction Specifications for Asphalt Curbs and Gutters* is available from the Asphalt Institute.

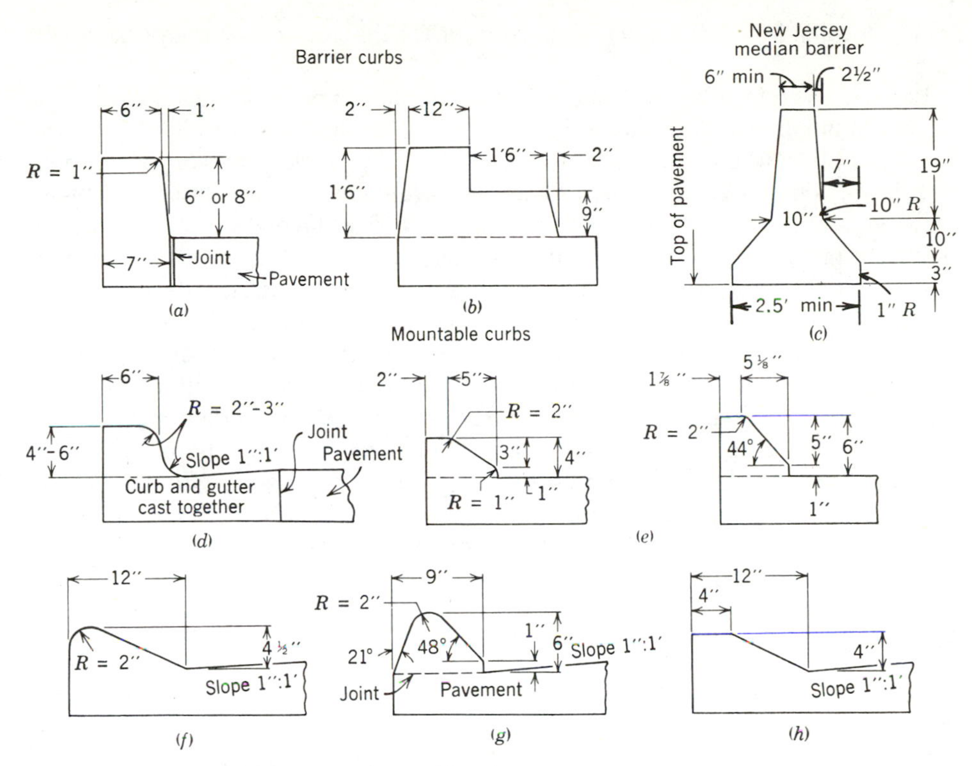

Fig. 9–13. Typical highway curbs and barriers.

curbing sections, stuck to the paving with epoxy resins, have been used to advantage.

In many situations, high visibility is an important requirement for curbings. In these cases it is common practice to use white portland cement in the concrete or to paint the curbing. Many agencies further increase visibility by providing indentations in the curb face which will reflect headlight rays back to the driver. Installing special reflectors in the curb face offers another method of increasing visibility.

Guard rail or New Jersey or G. M. barriers, discussed earlier, provide a positive means of confining vehicles to designated roadways and on occasion are used as a channelizing device. But vehicles striking them may suffer damage so that installations, if made, should be some distance removed from the traveled way.

DESIGN OF CONVENTIONAL HIGHWAYS, ARTERIALS, AND STREETS BETWEEN INTERSECTIONS

Conventional highways, arterials, and streets not only have intersections at grade (see below) but must be designed to accommodate traffic entering or leaving adjoining property. These movements interrupt the smooth flow of through traffic and lead to the conflicts that account for the commonly higher accident rate.

Ideally, these facilities will have 12-ft traffic lanes and shoulders 8 ft wide or wider. Multilane facilities should have a positive center divider to limit left turns through opposing traffic to suitable locations and provide a refuge for entering or turning vehicles clear of the through lanes. Often, however, limited right-of-way widths or other cost constraints result in less favorable conditions. As pointed out in Chapter 7, access from adjoining property for existing and new developments generally cannot be prohibited. It follows that the designer must attempt to limit it to favorable locations that provide adequate sight distance. This can be done because highway agencies have the legal authority to designate entry and exit locations and other conditions such as preventing left turns, limiting driveway length, or combining access to several properties (see Chapter 7) on the basis of reasonable traffic control and safety considerations.[53]

On major facilities it is common to provide one- or two-way frontage roads separate from through traffic for property access. Often existing facilities are reconstructed or otherwise modified to improve traffic flow and reduce accidents. This topic is discussed in Chapter 10.

SPECIAL DESIGN PROBLEMS

With the increased popularity and use of bicycles, designs to accommodate them on or parallel to highways are often required. Ideally, these would be entirely separated from vehicular traffic. Criteria must be set, such as design speed (15 to 20 mph) for relatively flat sections, width (6.5-ft minimum for two-way travel), and grades (5% maximum on short sections).[54] On existing facilities the shoulder or outer lane of the traveled way may accommodate bicycles. In suburban and some downtown areas, bicycles may be directed by signs to use sidewalks, in which case driveways and curbings must be so designed that the rider does not have to dismount and that wheelchairs and other transport vehicles for the handicapped can be accommodated.

At activity centers such as college campuses, shopping centers, and malls, travel is by a mixture of motor vehicles, bicycles, and pedestrians. Confusion and accidents result from the incompatibility of speed, weight, and controllability among the three. Discussion of this problem is beyond the scope of this book.[55]

[53]See *NCHRP Report 93*, E. A. Ziering, *TRB Record 747*, and *Guidelines for Driveway Design* published ITE. W. W. McGuirk and G. T. Satterly, Jr. *TRB Record 601*, discuss driveway accidents.

[54]An excellent reference on facilities for bicycles, pedestrians, and the handicapped is *Planning, Design, and Implementation of Bicycle and Pedestrian Facilities*, ASCE, 1975. See also *TRB Record 508, 538, 540, 570, 605, 629, 739,* and *743*, and the AASHTO publication, *Guide for Bicycle Routes*.

[55]For references, see the preceding list. Also, see *Transportation Planning for Colleges and Universities, Guidelines for Planning and Designing Access Systems for Shopping Centers,* and *Traffic Planning and Other Considerations for Pedestrian Malls*, published by ITE; and *TRB Record 498, 540, 634,* and *683*, and *TRB Special Report 153*.

DESIGN OF INTERSECTIONS AT GRADE

Except for freeways, all highways have intersections at grade, so that the intersection area is a part of every connecting road or street. In this area must occur all crossing and turning movements. Figure 9–14 schematically portrays typical intersections at grade, ranging from simple to complex.

The unchannelized intersections, shown down the left side of the figure, are

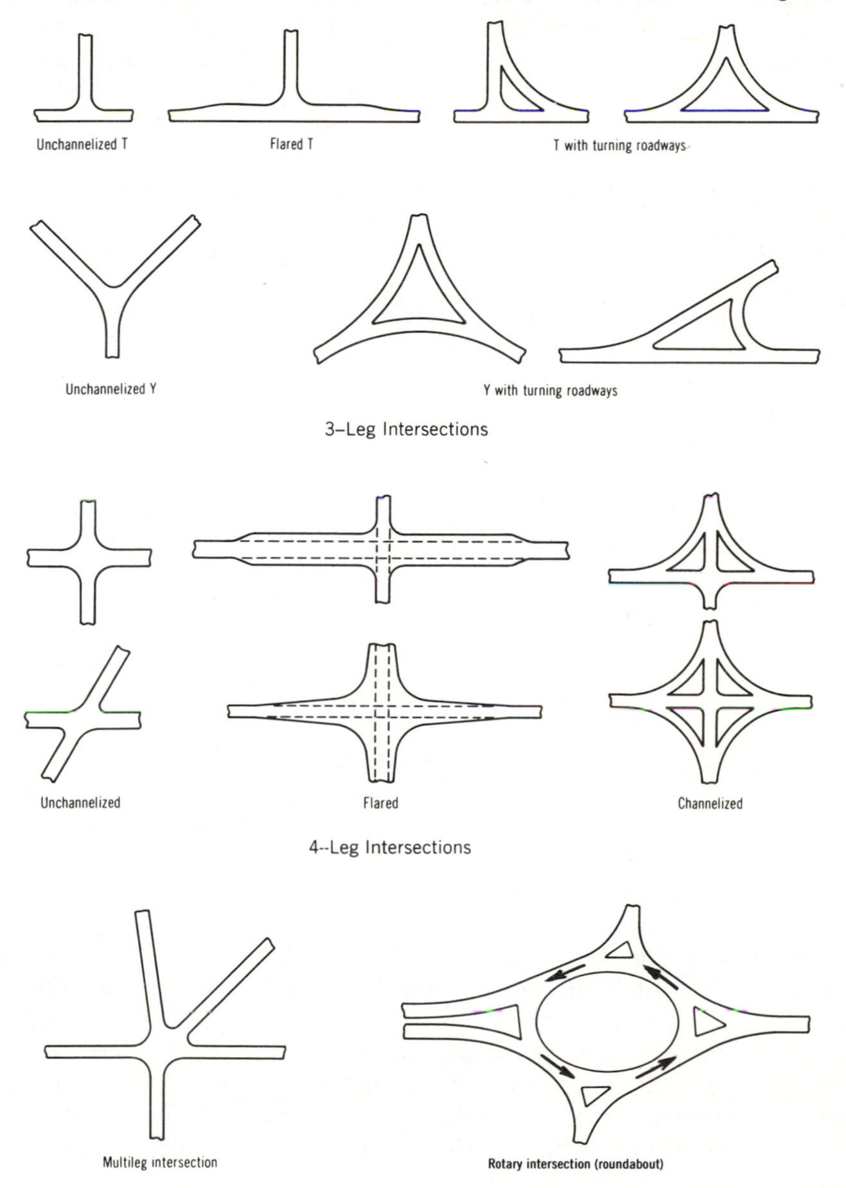

Fig. 9–14. General types of at-grade intersections.

cheapest and least elaborate. With them, the intersecting roadways have been joined by circular arcs in order to provide pavement under vehicles turning to the right. For right-angle intersections of roads or streets carrying little traffic, no further treatment except signs may be deemed necessary, with the possible exception of curbings to keep vehicles on the pavement or to channel surface drainage. Y intersections or other conformations where vehicles meet at unfavorable angles may demand channelization. Flared designs involve (1) widening the entering traffic lanes to permit deceleration clear of through traffic and (2) widening the leaving lanes to provide for acceleration and merging. The channelized designs shown on the right side of the figure are intended to direct approaching drivers to the particular paths by employing the principles of channelization enumerated above. However, some of the arrangements shown in Fig. 9–14 are considered to be obsolete, since they actually violate those principles.

Figure 9–15 shows several schematic layouts for situations where cross streets intersect divided expressways with frontage roads. If the frontage roads were omitted in Fig. 9–15*d*, it would also be typical. Note the acceleration and deceleration lanes to clear the through lanes of both right-turn and left-turn vehicles, channelizing islands, and pavement markings. A possible bus routing is suggested on Fig. 9–15*d*.

A careful traffic count and estimate of future changes for all movements, including turns, must precede the design of important intersections. Only in this way can those movements that are heavy be favored. This, coupled with knowledge of lane capacities leads to decisions regarding the number of lanes to be supplied. The speeds at which vehicles approach and move through the intersection and vehicle type govern many dimensions, particularly minimum sight distances in various directions, the radii of curves, and the lengths of the various turning and storage lanes. Likewise, a decision regarding the need for traffic signals currently or in the future affects certain features of the design. This topic is discussed further in Chapter 10.

In laying out intersections, characteristics of driver and vehicle and the possibility of accidents and their frequency and severity must be always kept in mind. As stated earlier, drivers should be confronted with one decision at a time. Furthermore, they should be guided into the proper channels for their intended routes and prevented from doing wrong or unpredictable things. It is important that spacious areas which permit "open-field running" be eliminated by the provision of directional islands that leave little choice of route. Pavement and islands that supply direction to converging streams of traffic should lead to blending at very small angles. It is preferable to use single large islands rather than several small ones, as single large ones are less confusing to drivers.

Adequate sight distances for vehicles entering intersections at anticipated speeds are required for safe operation. These encompass, first, seeing the roadway as the driver approaches, passes through, and travels beyond the intersection and, second, a clear view of vehicles that may be approaching on other intersection legs. For the latter situations, graphical *minimum sight triangles*

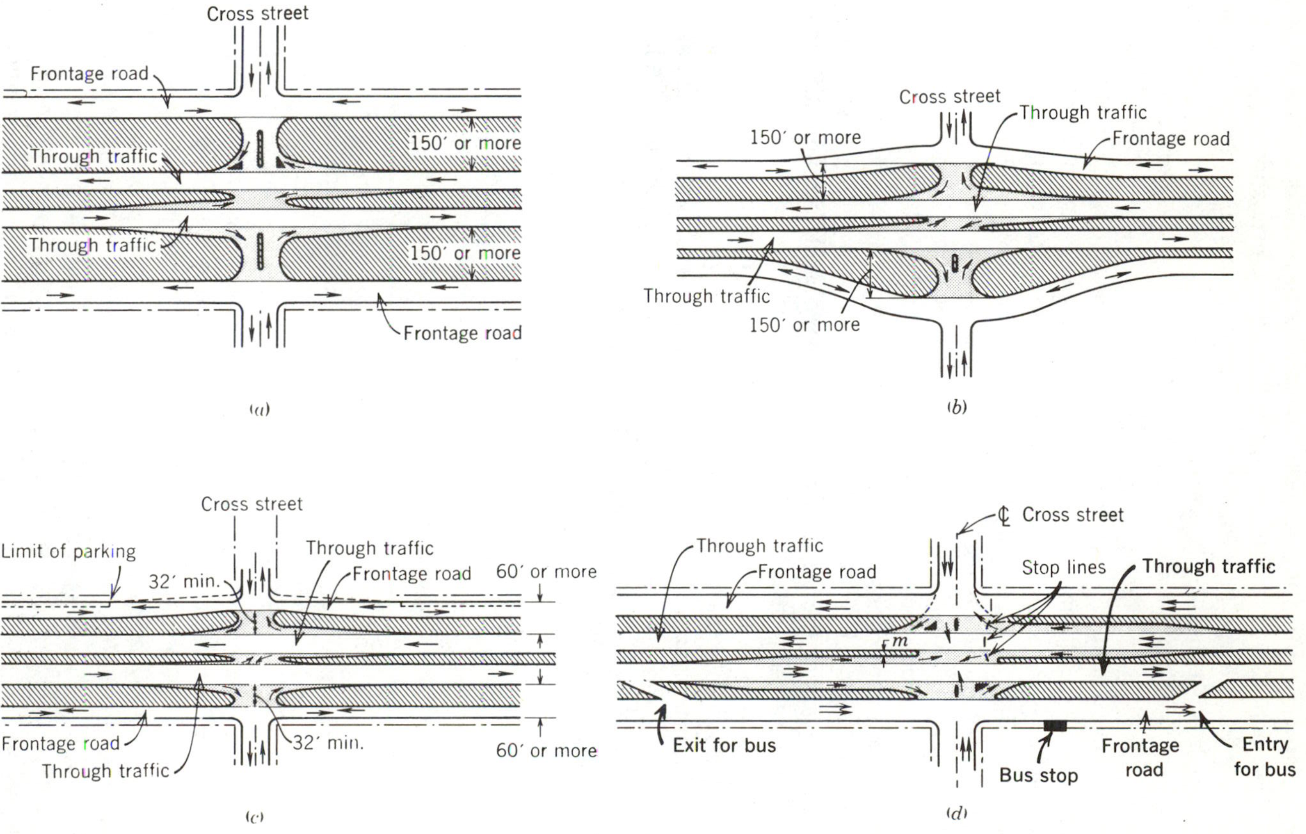

Fig. 9–15. Expressway intersections with frontage roads: *(a)* two-way frontage roads, wide outer separation; *(b)* two-way frontage roads, bulbed separations; *(c)* two-way frontage roads, minimum design; and *(d)* one-way frontage roads, minimum design.

drawn on a plan view of the intersection can be employed to determine available or needed sight distance. These would differ between situations where vehicles proceed through the intersection without stopping and where entry on some or all the legs is controlled by stop or yield signs or traffic signals. Detailed procedures for making such analyses are offered in *A Policy on Geometric Design*.

Buildings, signs, plantings, or other developments on adjoining private property which impair sight distance are often a problem. These cannot be removed by agency personnel, as that would infringe on private property rights. In many jurisdictions, ordinances limiting the height of plantings or stipulating setbacks are enacted, but if owners refuse to comply, court action may be necessary. Violators also may be subject to suits for negligence by injured motorists if they can prove the obstruction was a major contributor to an accident.[56]

Most important intersections must accommodate large trucks and the radii of all curves made long enough for them. As an illustration, for the inside edge of 90° turning roadways at low-speed intersections, AASHTO recommends, as a minimum, three centered compound curves with radii successively 180, 65, and 180 ft, with the radius point of the 65-ft curve 71 ft in from the tangent. Because vehicles do not track and their fronts overhang, minimum lane width near the center of the curve is 20 ft. In addition, a raised island having an area of about 125 ft is interposed between the turning roadway and the through lane that it is joining. *A Policy on Geometric Design* gives details such as these for typical situations.

In settled areas, street and intersection designs must consider the needs of pedestrians and bicyclists. Walking rates for pedestrians range from 2.5 to 6.0 ft/s with older people moving at the lower speeds. For design purposes in timing traffic signals, 4.0 ft/s is a common value but 3.0 ft/s is recommended where a substantial number of older people are involved. Widths of crosswalks commonly are 4.0 ft in residential areas and 6.0 ft in commercial areas. If bicyclists traveling in two directions are to be accommodated, a minimum width of 6.5 ft is recommended. With large pedestrian volumes, greater crosswalk widths than those listed may be needed.[57]

Sometimes a study of traffic-flow data indicates that relatively few vehicles perform a particular turning movement at a given intersection. For example, this could be expected where the streets intersect at oblique angles. If this movement complicates the design or increases congestion, it should be eliminated completely and provision made for it in some other way. To illustrate, vehicles desiring to make left turns often are directed to the right completely around an adjacent block; then they pass through the intersection with normal cross traffic.

Where left-turn movements between certain legs of an intersection are extremely heavy, two lanes rather than one may be provided in or adjacent to the

[56]See W. L. Moore Jr. and J. B. Humphreys, *TRB Record 541*.

[57]For further detail on intersection design to accommodate pedestrians, see Chapter 8, *A Policy on Geometric Design; TRB Record 540;* ITE committee recommendations in *Traffic Engineering*, May 1976; H. D. Robertson, ibid, Feb. 1976; and R. M. Cameron, ibid., Jan. 1977.

median on the approach leg. In such cases, traffic signal timing (see Chapter 10) is such that this path is free of opposing traffic.[58]

Certain design features of complicated intersections often require testing by actual use. In such cases it is common to place temporary barricades or channelizing islands of sandbags which can be easily shifted around. Sand sprinkled over the roadways indicates vehicle paths. After the design has been proved, the permanent installation is made.

The preceding paragraphs have merely suggested the complexities of proper intersection design. For added detail, including dimensional requirements to fit almost all situations and for examples of good designs, the reader is referred to *A Policy on Geometric Design*. It gives numerous examples of special intersection conformations to fit unusual situations.

ROTARY INTERSECTIONS

A rotary intersection, called a "roundabout" in England, is sometimes placed at the confluence of three or more intersection legs. Basically, it is a one-way road around a central island (see Fig. 9–14). It operates as a series of curved weaving sections placed end to end. At low traffic volumes, rotaries may minimize delay by substituting weaving for direct crossings of vehicle paths. However, at higher volumes or where traffic signals are required to reduce congestion, it has been found that capacity only approximates that of a single lane. Regarding them, *A Policy on Design on Urban Highways and Arterial Streets* states: "In general, rotaries are not recommended for use on urban arterial highways. A properly designed channelized intersection will usually operate better and result in considerable savings in cost and right-of-way areas." *TRB Research News,* Sept.–Oct. 1978, reaffirms this view. British experience with rotaries is more favorable.[59]

DESIGN OF FREEWAYS

Earlier discussions in Chapter 8 and this one provide the basis for sizing, dimensioning, and aligning stretches of freeway and weaving sections and ramps. Other decisions, unique in each case, must also be made, such as at-grade versus depressed versus elevated on fill or structure, whether or not there will be separate roadways in each direction, and the need to provide service roads. From these will flow the arrangements for crossings by or closings of intersecting streets that will not be connected to the freeway. A detailed discussion of these is beyond the scope of this book but appears in *A Policy on Geometric Design*.

In settled areas, freeway crossings solely for pedestrians and bicycles often are needed. Except for the hazards associated with mixing pedestrians and bicycles, these offer no serious problem if the freeway is either elevated or de-

[58]For ITE recommendations on double-lane left turns see *ITE Journal,* Feb. 1981. Z. A. Nemeth, *TRB Record 681* reports excellent accident experience for them.
[59]See K. Todd et al., *TRB Record 737* and *ITE Journal*, Jul. 1979.

pressed. But if the freeway is near ground level the solution is less satisfactory. Since the minimum vertical clearance above a roadway is about 16 ft, users of an overcrossing must climb a stairway or ramp. With undercrossings, vertical clearances total only about 8 ft, but policing becomes a serious problem, since the walkways are not visible until the approach ramp or stairway has been descended. A compromise solution that raises the freeway grade some 4 ft at the undercrossing has sometimes proved best. The pedestrian must still descend a short stairway or ramp, but the entire length of the walkway is visible from the approach sidewalks.

Modern freeways often carry tremendous traffic volumes. For example, the Hollywood Freeway near the downtown section of Los Angeles daily passes about 300,000 persons and 50,000 tons of freight, the equivalent of 800 railroad cars. The San Bernardino Freeway carries fewer persons, 250,000, but 110,000 tons of freight which is equivalent to 1800 box cars. At the same time, the costs of freeways are high. An extreme example is a single freeway-interchange complex in the Bronx, New York, which, at the time of construction, cost over $60 million; reconstruction of a major interchange and 3.5 mi of freeway in San Antonio, Texas was $63 million, and a 2.5-mi stretch of Interstate 95 in Virginia, south of Washington, D.C., which involves as many as 27 lanes, a complex of interchanges, and 4 lanes for buses, cost $88 million. But these extreme examples are not necessarily typical. Urban Interstate 94 between Minneapolis and St. Paul cost $56 million for 11.2 mi and the rural and urban Interstate Systems in their entirety will average $1.8 million per mile.

Design of Freeway Entrances and Exits

From Chapter 8 and the subsequent discussion of freeway operation in Chapter 10 it can be seen that the flow characteristics of and driver behavior at and near on and off ramps and weaving sections largely determine the overall effectiveness of the individual freeway or freeway system. For this reason, ramp design

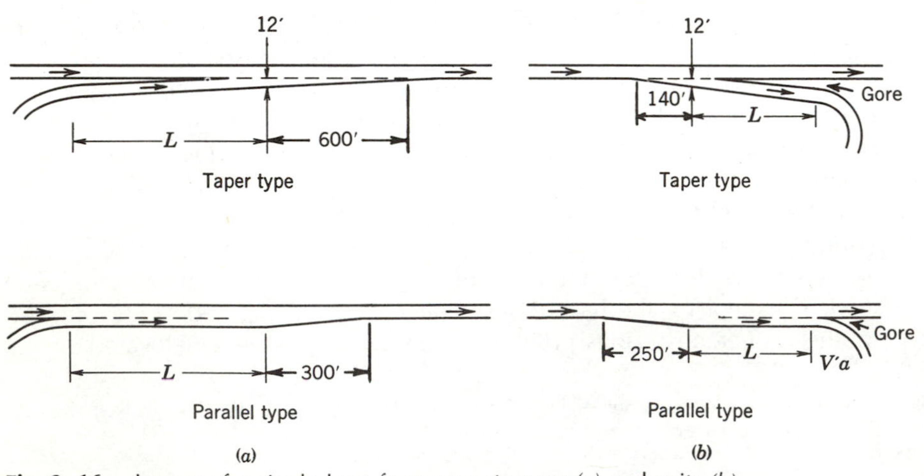

Fig. 9–16. Layouts for single-lane freeway entrances *(a)* and exits *(b)*.

and dimensioning have been the subject of intensive study since freeways were first constructed. Single-lane on ramps feed their vehicles into the outer continuing freeway lane. They may be designed either as a taper, blending into the through lane, or as an auxiliary lane parallel to the through one (see Fig. 9–16a)). In either case, sufficient length must be provided to permit the vehicle to accelerate and merge into the on-going stream at a speed near that of the freeway. Table 9–7 shows a sampling of minimum lengths of on and off ramps for various freeway and entry speeds. As shown on Fig. 9–16a measurement begins where the vehicle enters a path nearly parallel to the outer lane and ends at the spot at which the acceleration lane narrows below 12 ft. If the ramp is metered and entering vehicles must stop, entry speed is zero.[60] Also, as pointed out in Chapter 8, if ramp and outer lane volumes are too great to be accommodated by a single lane, a two-lane ramp is employed. In this case, a lane must be added to the freeway proper.

Vehicles leaving the freeway from the outer lane need a distance to decelerate clear of the stream of on-going traffic. As shown on Fig. 9–16*b*, either a taper or auxiliary lane should be provided. A few minimum deceleration-lane lengths *L* are given in Table 9–7. A two-lane exit is called for in situations where one lane does not meet the demand. In such cases, it is recommended that an auxiliary lane be added to the freeway at least 2500 ft before the point of exit. It has also been found that, even with one-lane exits, adding such an auxiliary lane for some distance before the point of exit reduces confusion and congestion.

It was common practice in early freeway designs to drop the outer freeway lane abruptly at off ramps. With this arrangement, drivers in the right-hand lane would be confronted with the decision to leave the freeway or to swing sharply

TABLE 9–7 A Sampling of Minimum Lengths for Freeway on and off Ramps*

Freeway Speed (mph)†	*On Ramp or off Ramp Design Speed (mph)†*				
	Stop	*20*	*30*	*40*	*50.*
	L = Length of ramp, ft‡§				
On ramps					
50	760	630	500	160	—
60	1170	1070	910	590	170
70	1590	1500	1330	1010	580
Off ramps					
50	435	385	315	225	—
60	530	490	430	340	240
70	615	570	510	430	340

*Source: *A Policy on Geometric Design.*
†Multiply by 1.6093 to obtain km/h.
‡Multiply by 0.3048 to obtain length in meters.
§Measurement of length as shown on Fig. 9–16.

[60]See R. A. Olsen and R. S. Hostetter, *TRB Record 605* for a study of merging behavior at entrance ramps.

into the adjoining lane. They might even run into the "gore" area separating the continuing freeway from the off ramp. Any of these choices creates confusion and accident potential such as striking other vehicles or colliding with abutments, piers, or sign supports in the gore. Good practice places lane drops downstream of the off ramp in either the inner or outer lane, depending on traffic and geometry. Visibility of the lane drop area, advance signing, striping, and pavement markings are imperative.[61]

Off ramps from the fast, left-hand lane have sometimes been employed. These are confusing to drivers as evidenced by a more-than-double accident rate and are not used today.[62]

Interchange Types and Characteristics

Geometric design of interchanges involves selecting the conformation best suited to a particular situation, considering such factors as topography of the site, traffic projections and character, land availability, impacts on the surrounding area and overall environment, economic viability, and financial constraints. It is a difficult task.

The functions of interchanges are (1) to provide grade separation between two or more traffic arteries and (2) to make possible the easy transfer of vehicles from one artery to the other or between local streets and the freeway. A cursory examination of the many existing interchange layouts would indicate that there was little rhyme or reason underlying the process. However, there is a basic order in the apparent chaos. For example, for the common situations where two continuing arteries cross at something close to right angles, the choice generally falls among the diamond, the partial or full cloverleaf, or direct connections for one or more of the left-turn movements. With all of these schemes, through movements on both arteries continue entirely or largely without interruption, but turning movements, particularly to the left, are handled in a variety of ways. Line drawing for typical examples of these interchange types are given in Fig. 9–17*a*, 9–17*b*, 9–17*c*, and 9–17*d*. *A Policy on Geometric Design* and *NCHRP Synthesis 35* show many additional conformations.

The simplest and generally least costly form of interchange is the *diamond*, shown in Fig. 9–17*a*. It is particularly adapted to situations where a freeway crosses a nonfreeway arterial. Flow on the freeway is uninterrupted, except for problems that may develop at points where ramp traffic enters or leaves. But traffic patterns on the arterial are complex, since the roadway must carry two through movements and accommodate four left turns, two of which must use the inside lanes of the arterial or separate turning lanes. When volumes are large, traffic signals generally are required. A modified three-level diamond de-

[61]See *Traffic Engineering*, Nov. 1976, for a survey of state practices on lane drops. H. L. Lunefeld and G. J. Alexander, *TRB Record 600*, discuss signing and K. M. Roberts and A. G. Klipple, *Public Roads*, Jun. 1976, driver expectations for them.

[62]*A Policy on Geometric Design*, *NCHRP Report 175*, and *Synthesis 35* are excellent references on ramp design, operational problems, and accidents.

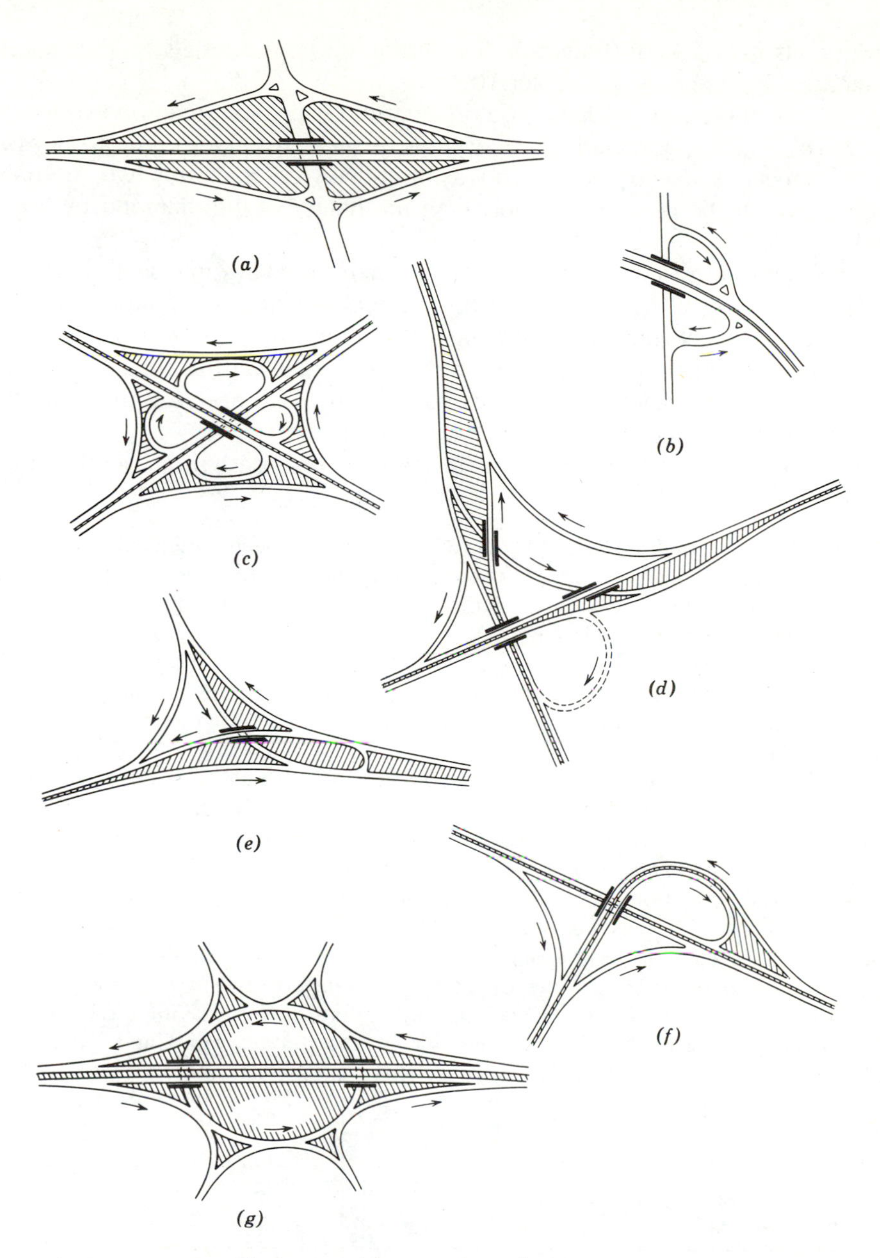

Fig. 9–17. General classes of freeway interchanges.

(a) Diamond. *(b)* Partial cloverleaf.

(c) Cloverleaf. *(d)* Directional.

(e) Y. *(f)* T or trumpet. *(g)* Through freeway with rotary (flyover with round-about).

sign permits free flow of straight-through traffic on both arterials.[63] An example of signal timing appears in Chapter 10.

A useful variation to the diamond configuration is the split-diamond shown in Fig. 9–18. This is particularly effective where the cross flow is on a one-way pair of streets. But even with two-way traffic on each of the two arterials, congestions can be reduced by separating the halves of the diamond by a city block.

Probably the most common interchange where freeways intersect arterials is the *cloverleaf* (Fig. 9–17*c*). With it, the intersecting arteries are separated, and in addition all eight turning movements are accomplished free of intersections where vehicle paths must cross. Turning vehicles peel off the right side of the roadway on which they enter the interchange and blend from the right into the roadway being entered.

There are several serious objections to the cloverleaf design, among them the following:

1. Cloverleaf interchanges take large areas of land. This has been particularly true where relatively high design speeds were used in proportioning the loops. In actuality the space required need not be excessive; it can be demonstrated that design speeds higher than 25 mph and loop radii longer than 150 ft offer little advantage. At higher speeds more time is required to traverse the longer loops. On the other hand, very short radii do not reduce area requirements much, since length equivalent to that for a 150 ft radius is needed to develop the necessary elevation differences between the ends of the loops.
2. Vehicles desiring to make a left turn must execute a 270° right turn and travel a substantially greater distance. This travel is unpleasant, if not hazardous, since it is on sharp curves and relatively steep grades.
3. Vehicles leaving the curved loop in one quadrant must weave through those entering the adjacent loop from the through roadways. With the usual cloverleaf pattern shown in Fig. 9–17*c*, this weaving must take place in or immediately adjacent to the through lanes carrying fast-moving traffic. Where weaving volumes are large, the weaving section must be long (see Chapter 8). This distorts the shape of and increases the land area required for the cloverleaf. Furthermore, drivers leaving the loops must simultaneously weave through vehicles leaving the through lanes and merge with those who remain in them so that they face two complex decisions at once.

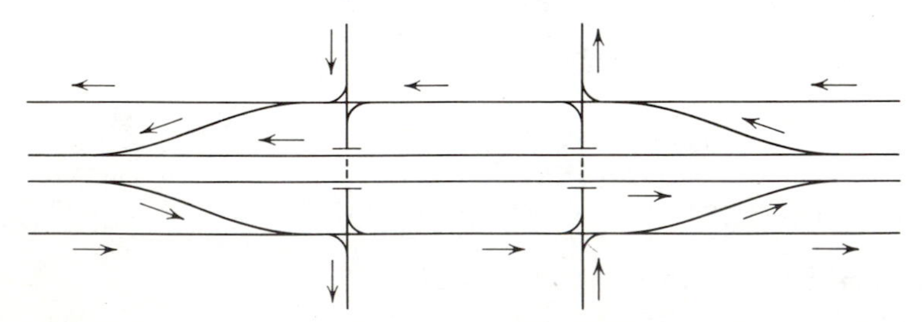

Fig. 9–18. Traffic flow through a split-diamond interchange where cross streets are one way.

[63]For a discussion of the effects or the dimensions of a diamond interchange on its operation see J. M. Turner and C. J. Messer, *TRB Record 682*.

Figure 9–19 diagrams an alternative cloverleaf design with *collector–distributor* roads. These can be applied to one or both through roadways if the costs of added land, paving, and structures can be justified. With this design, weaving and merging movements are separated. Also, the collector–distributor road provides an opportunity for speed adjustment clear of the freeway.

The partial cloverleaf (see Fig. 9–17*b*) offers connections by merging to the major freeway but calls for left turns through opposing traffic on the minor artery. It can be developed in many forms with the loops in different quadrants to fit topography and traffic patterns.

Figure 9–17*d* diagrams a form of directional interchange suitable at some freeway to freeway interchanges. It clearly demonstrates the basic concept that heavy left-turning movements can be accommodated by a 90° turn to the left. This is shown by the upper left to upper right connection. In contrast, the cloverleaf would call for a 270° turn to the right and the diamond for a left turn through opposing traffic. Figure 9–17*d* also indicates that one quadrant from a cloverleaf can be used to carry a supposedly light lower left to upper left turn. The conformation in Fig. 9–17*d* has a potential weakness. Accepted design practice requires that minor movements leave from and enter into the principal artery in the right-hand lane. The directional layout shown in Fig. 9–17*d* violates this rule and probably is appropriate only when the principal continuing route and major traffic movements are from upper left to upper right, and vice versa.

There are many interchange conformations for special purposes. Three of these are shown in Fig. 9–17. Figure 9–17*e* portrays a layout for a Y intersection. Here only one grade separation is required to eliminate all crossings at grade. It should be noted, however, that provision for vehicles traveling from

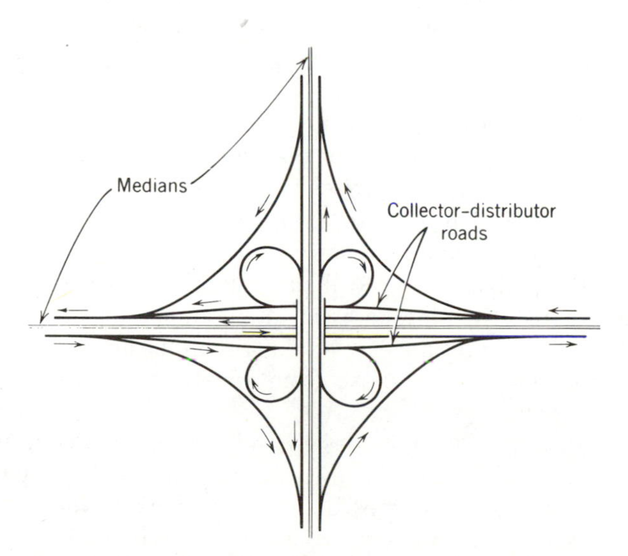

Fig. 9–19. Cloverleaf interchange with collector–distributor roads.

lower to upper left is circuitous. Also it requires two weaving motions and leaving one roadway and entering another on the left. It can be argued that this movement should be prohibited entirely. Figure 9–17*f* shows a T or trumpet interchange pattern suitable for three-way intersections. Note that traffic from lower to upper left must traverse a 270° turn but that all other turning movements are accomplished with curvatures not much greater than 90 °. A variation of the trumpet replaces the 270° loop with a directional roadway. Figure 9–17*g* diagrams a rotary intersection combined with an overcrossing (or undercrossing) for the freeway. They are effective only where a relatively low traffic volume is to be picked up and disbursed from several streets. The at-grade traffic circle has the disadvantages of rotary intersections discussed above.

Where freeways meet freeways and all traffic movements are heavy, interchanges often are designed with directional left turns in all four quadrants. A few of the possible conformations are shown in Fig. 9–20. The layout in Fig. 9–20*a*, which requires roadways at four levels, is often appropriate where land is expensive. Those in Fig. 9–20*b* and 9–20*c* are somewhat more open; commonly three levels of roadway are required to meet appropriate controls on grades and sight distance.

A whole spectrum of interchange layouts ranging from the relatively simple ones in Fig. 9–17 to the complex varieties of Fig. 9–20 can be employed. This is illustrated by Fig. 9–21 which portrays an interchange near San Francisco. With it, three left turns are directional; a fourth and minor one (upper left to upper right) is served by a cloverleaf loop which takes off from a service road. This appears near the bottom-middle of the photograph. Ramps to serve a cross road drop to ground level near the upper middle of the picture. All three levels of this interchange lie at or above ground level. This is necessary, since the ground surface is only slightly higher than the water level in nearby San Francisco Bay. Also, the openness of the design may be seen, with single columns replacing the more traditional piers running from one side of the structure to the other.

Signing for interchanges and their approaches is important, since proper direction to drivers is essential to smooth, accident-free operation. For example, advance warnings regarding the proper lane to occupy should begin a mile or more ahead of the interchange. Also, an overall layout can best be proved out in its early stages by deciding whether or not drivers can be alerted to and directed through the paths leading to their destinations. This problem is accentuated by the motorists' inability to assimilate complex messages at high speeds and by the short viewing distances, particularly on turning roadways. A parallel problem is that of directing motorists from adjoining or intersecting roads and streets to freeway entrances.

A particular concern in freeway design and signing is to keep motorists from entering exit ramps which will bring them head on to fast moving traffic. Careful design to make wrong-way entries difficult along with warning signs on red backgrounds and red-reflecting traffic markers are all helpful.[64]

[64]See N. K. Vaswani, *TRB Record 514, 628,* and *644.*

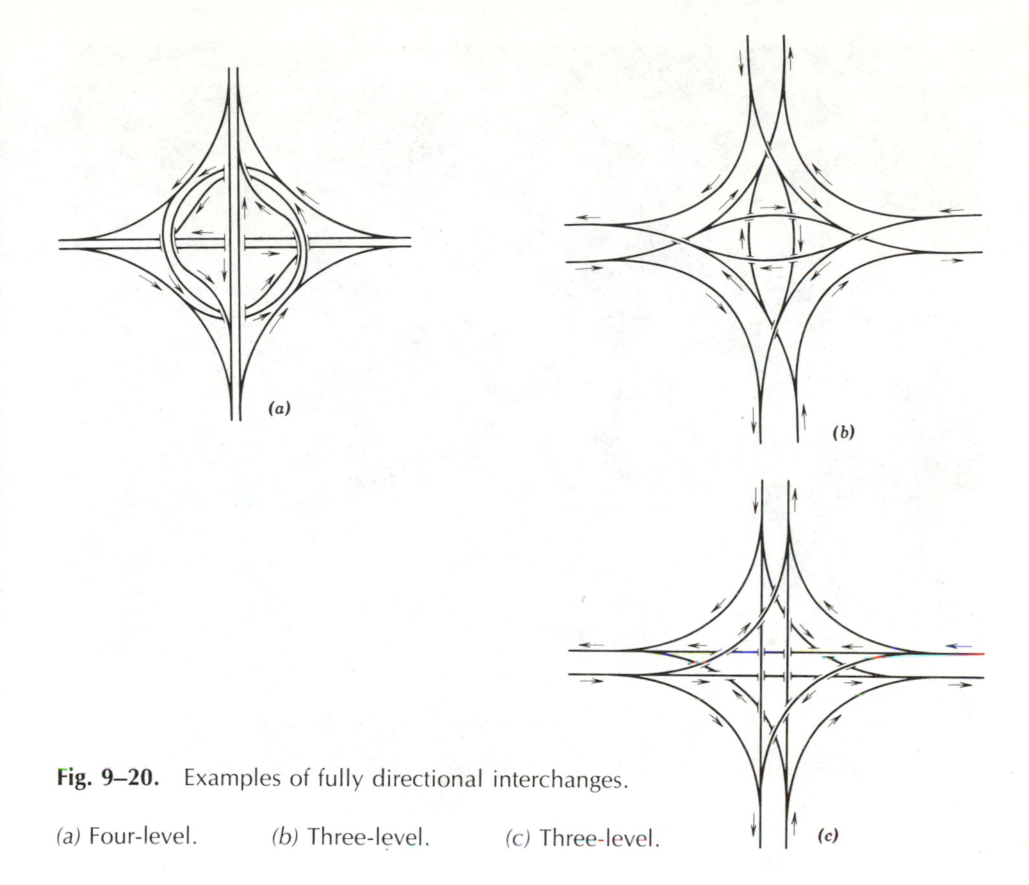

Fig. 9–20. Examples of fully directional interchanges.

(a) Four-level. *(b)* Three-level. *(c)* Three-level.

Interchanges offer serious challenges in layout and geometric and structural design. Further detail will be found in *A Policy on Geometric Design*. Diagrams showing clearance and bridge loading requirements are given in the section of Chapter 11 on bridges.

RAILROAD-HIGHWAY GRADE SEPARATIONS

Grade separation between major traffic arteries and main-line railroads is desirable. For example, AASHTO standards for Interstate highways require that "railroad grade crossings shall be eliminated for all through traffic lanes." Special appropriations have been made at both federal and state levels for this purpose. However, there is serious doubt that convincing economic arguments can be developed to support large investments in them unless vehicular volume is high or train passages are numerous.

The major design question for rural rail-highway separations is usually whether the highway shall go over or under the railroad. Where the railroad track lies in fairly deep cut or on a high fill, the answer is usually apparent, but

Fig. 9–21. Model of a freeway to freeway interchange with directional left turns in three of four connections. (Courtesy: California Department of Transportation.)

when it is near ground level the decision becomes more difficult. If the highway goes over the railroad, the structure itself may be lighter, as highway loads are much smaller than railroad loads. However, clearance must be greater—the usual minimum vertical distance from top of rail to underside of structure being about 23 ft, as against the 16-ft clear distance to be supplied above a highway. Then, too, the roadway passes over a crest, which requires a long, sweeping vertical curve to meet sight distance requirements. These two design controls make an overcrossing and its approaches much longer than an undercrossing at the same location. In settled areas, and particularly in business and residential districts, overcrossings are objectionable from an aesthetic and environmental point of view, and their construction may lead to objections and damage suits by property owners. If the highway goes under the railroad, other problems arise. Special provision must be made for removal of rainwater that falls within the crossing area. If the groundwater table is high in the area of the crossing, the water level must be lowered or the roadway must be sealed against leakage and made heavy enough that it will not float. At undercrossings, provision for uninterrupted train service during construction is both expensive and troublesome; often many underground utilities must be relocated and support provided for adjacent buildings. In all such cases, the problem requires careful study before a decision concerning the type of crossing is reached. Detailed design then follows accepted standards.[65]

PROBLEMS

9–1*A*. A proposed two-lane arterial in rolling country has an estimated design-hour volume of 400 automobiles, 50 trucks, and 10 buses. Design speed is 70 mph and sight distance will exceed 1500 ft for its entire length. For this situation, using data from Chapters 8 and 9, determine:

a. The expected speed and the capacity and speed levels of service in the design hour.
b. Minimum and advisable lane width.
c. Minimum and advisable shoulder width.
d. Steepest back slopes in low and high earth cuts.
e. Steepest side slopes on low and high fills.
f. Pavement and shoulder cross slope.
g. Desirable right-of-way width (traveled way plus marginal area).
h. Design stopping sight distance.
i. Design passing sight distance.
j. Maximum desirable grade.
k. Degree and radius for sharpest permissible horizontal curve for (1) normal and (2) snow and ice conditions.
l. Minimum length of superelevation runoff for (1) normal and (2) snow and ice conditions.
m. Minimum length of easement curve for sharpest horizontal curve.

9–1*B*. State the results of Problem 9–1*A* in SI units.

9–2*A*. Develop the answers to parts *b, c, d, e, f, g, h, j, k, l,* and *m* of Problem 9–1*A* for a rural road in mountainous country which is to carry 50 passenger cars, 30 trucks, and 20 buses per day.

9–2*B*. State the answers to Problem 9–2*A* in SI units.

9–3*A*. Projected design-hour traffic for each direction of a section of urban freeway in level country is 3000 passenger cars, 500 trucks, and 50 buses. Design speed is 50 mph and level of service is to be C or better. For this situation, using data from Chapters 8 and 9, determine:

a. The number of lanes in each direction required in the basic section.
b. The other design details called for in parts *b* through *m* of Problem 9–1*A*.
c. Develop designs for the inside shoulder and median of this facility.

9–3*B*. State the results of Problem 9–3*A* in SI units.

9–4*A*. A cement-rubble wall 15 ft (4.5 m) high from base to top is to retain

[65]See *A Policy on Geometric Design* for an extensive treatment of grade separations.

the side slope of a fill on a parkway project. Based on Fig. 9–3, draw a cross section of the wall. Show all dimensions.

9–4*B*. Dimension the wall called for in Problem 9–4*A* in SI units.

9–5*A*. Determine minimum vertical curve lengths to provide *(a)* design-speed stopping and *(b)* passing sight distances in each of the following situations.

	Grade Lines (%)	
Design Speed (mph)	*Approaching Intersection Point (+Grades Are Uphill)*	*Leaving Intersection Point (+Grades are Uphill)*
30	+6	−5
40	+5	−2
50	+2	−3
60	+5	+1
70	+2	0
70	+4	−3

9–5*B*. State the design speeds and results of Problem 9–5*A* in SI units.

9–6*A*. Determine design-speed minimum lengths of sag vertical curves for the conditions of Problem 9–5*A*, except that all signs are changed; + to − or − to +.

9–6*B*. State the design speeds and results of Problem 9–5*A* in SI units.

9–7. In an effort to discourage drivers from exceeding the design speed of a road, superelevations sometimes are set at values that develop the maximum recommended coefficient of side friction (see Table 8–11). Determine the superelevations to employ in the following situations:

a. Design speed 70 mph, (1) 2° curve, (2) 3° curve.

b. Design speed 50 mph, (1) 4° curve, (2) 7° curve.

9–8. For each of the combinations indicated below, develop profiles and profile controls similar to Fig. 9–9. The pavement is two-lane; each lane is 12 ft wide. Show stationing along the roadway by assuming that the *TS* is at station 20+00. For vertical dimensions, show differences in elevation between centerline grade and each edge of pavement at each point where the profile breaks. Determine superelevation for the circular curves from AASHTO recommendations on Fig. 9–8 for e_{max} of 0.10. Select easement curve lengths from Table 9–6. Tangent runout length is 50 ft.

a. Design speed 40 mph, 8° circular curve.

b. Design speed 50 mph, 6° circular curve.

c. Design speed 70 mph, 3° circular curve.

9–9. For each of the combinations indicated below, develop profiles and profile controls similar to Fig. 9–9 for the outside edge of the outer lanes of a six-lane freeway as it approaches or leaves a curve. The inner edges of the two opposing roadways maintain identical straight grade lines. Tangents run directly into circular curves; 60% of the run-off is on tangent. Show stationing along the edge of the roadway by assuming that the *PC* is at sta. 42 + 50. For vertical dimensions, show differences in elevation between the grades of the inner edges of the roadways and the outer pavement edges. Determine superelevation for the circular curves from AASHTO recommendations on Fig. 9–8 for e_{max} of 0.06.

a. Design speed 40 mph, 9° circular curve.
b. Design speed 60 mph, 4° circular curve.

9–10. A corner of an existing building is 30 ft from center line on a curved portion of a two-lane highway. Considering horizontal sight distance, what are the approximate low and high safe operating speeds if the curve is *(a)* 12°? *(b)* 5°? Lane width is 12 ft.

9–11. Cuts on the back-slopes of a highway lie on a 2 (horizontal) to 1 (vertical) slope. The grade line is straight. Superelevation is constant across pavement, shoulder, and to the toe of the cut slope. For a two-lane highway and 70-mph design speed, what distance is required between the centerline of the highway and the toe of the cut slope to satisfy low and high horizontal sight distance-requirements for *(a)* a 2½° circular curve ($e = 0.08$) and *(b)* for a 4° circular curve ($e = 0.10$)?

9–12. As a class project, plan the channelization for a troublesome intersection in the vicinity of your campus. This problem may involve traffic counts at peak hours and field measurements. Refer to *A Policy on Geometric Design* for the necessary detail.

9–13. Draw and show dimensions for taper and parallel single-lane freeway entrances for a freeway design speed of 60 mph and for entry speeds of 0 and 40 mph.

9–14. Do Problem 9–13 for freeway exits, substituting exit speeds for entry ones.

9–15. As a class project, develop a design to improve a hazardous location in your community.

9–16. As a class project, develop a plan layout and cross section for improving the safety and/or operational characteristics of an interchange in your community.

10 HIGHWAY OPERATIONS—TRAFFIC ENGINEERING

Each highway or transportation agency administers a given system involving some or all among freeways, expressways, rural or urban arterials, and local roads and streets. It is charged not only with maintaining that system in a physical sense (see Chapter 21) but also with operating and possibly modifying it to optimize traffic flow and safety. This broad obligation is defined here as "highway operations." Those in the agency responsible for this operating and optimizing effort often are designated as traffic or transportation engineers. Their responsibilities may also extend to the design and effective use of a whole spectrum of traffic guidance and control devices such as pavement markings, signs, traffic signals, motorist assisting arrangements, and lighting. Often, parking problems also fall in their domain. These functions, then, are the subjects treated in this chapter.[1]

TRAFFIC-CONTROL DEVICES

Pavement Markings

Among the earliest pavement marking was the longitudinal center stripe which appeared first in Wayne County, Michigan, in 1911. Today pavement markings of paint, epoxies, thermoplastics, inlays, and raised reflectors are among the most helpful instruments for traffic direction and control. Markings are used to delineate roadway centerlines, lane boundaries, no-passing zones, pavement edges, roadway transitions, turning patterns, the approach to obstructions, light-

[1]The Institute of Transportation Engineers (ITE) is the professional society of these engineers (see Chapter 2 for a brief description and reference to its publications). The Eno Foundation, Westport, Conn., publishes special reports and excellent articles in its *Traffic Quarterly*. Many associations such as the Highway Users Federation for Safety and Mobility and the American Trucking Association and manufacturers of paints, signs, traffic signals, roadway lighting, and computers also are active in a variety of ways. Foreign journals in English include *Traffic Engineering and Control*, London.

rail or bus clearances, stop lines, crosswalks, railroad crossings, and parking-space limits. Symbols, words, or numbers convey pertinent information such as that about speed limits and the nearness of schools. The quantity of paint consumed for these purposes is fantastic, about 30 million gal annually.

In the past, patterns and colors for pavement markings differed among agencies, but there is now substantial uniformity. Recommended pavement markings for all usual situations are described and illustrated in the *Manual on Uniform Traffic Control Devices,* published by the Federal Highway Administration, hereafter referred to as the *Manual.*[2] It has been prepared and periodically revised by a joint committee representing federal, state, county, and city highway officials, transportation associations, police, and manufacturers.

The *Manual* states that pavement markings shall be yellow, white, or red.[3] Black is permitted in combination with other colors when the pavement does not provide color contrast. In general, white is used in circumstances where vehicles may cross the markings, for example: lane lines, pavement-edge lines, channelizing lines, turn markings, stop lines, crosswalks, parking-space limit lines, and for words and symbols. In contrast, yellow lines delineate the separation of traffic flows, for example: centerlines of two-lane highways, double centerlines for multilane pavements, no-passing barriers or no-passing zones of two- and three-lane pavements and pavement-width transitions, channelizing a center lane for two-way left turns on undivided multilane highways,[4] or where obstructions must be passed to the right. Usually, directional lines such as the center stripes on two-lane highways and lane lines are dashed, with the standard ratio of stripe to gap at 1 to 3. Stripes of 10 ft and gaps of 30 ft are recommended for rural locations; shorter lengths may be appropriate in urban areas.[5] In contrast, the yellow barrier lines and white guidelines where crossing is to be discouraged are generally continuous.[6] Recommended stripe widths are 4 to 6 in.; but for emphasis a stripe double the usual width is recommended.

The ideal pavement marking installation would be cheap to install, permanent, readily visible in daylight and darkness, rain and snow, and with water standing on the pavement surface, and would clearly transmit the intended message. Also, it would not be destroyed by snow-removal equipment. Current approaches represent compromises among these factors, but intensive research to develop less costly and more effective solutions is continuous.[7]

Paints currently favored for striping are of modified alkyd resin with titanium dioxide (white) or lead chromate (yellow) pigments (AASHTO Designation M248). They can be supplied to dry at various rates, ranging from less than 60

[2]Available for purchase from the Superintendent of Documents, Government Printing Office, Washington, D. C. 20402.

[3]Studies of white and yellow centerline stripes showed white better at dawn and dusk and during nighttime wet conditions. Yellow was better when dry. See R. C. Johns and J. S. Matthias, *TRB Record 643*.

[4]See, for example, D. W. Harwood and J. C. Glennon, *TRB Record 681*.

[5]A 12–48 ft stripe-gap ratio with a raised reflective marker in the center of the gap has also been found to be effective.

[6]The *Manual* gives many illustrations of these and other marking patterns.

[7]See, for example, E. T. Harrigan, *Public Roads,* Dec. 1977 and H. J. Gillis, *TRB Record 762*.

s to more than 3 min. They are usually purchased under performance specifications which require that suppliers submit samples that are subjected to field wear tests by the highway agency.[8] Paint film thickness is in the range of 0.015 in., with about 60% solids. This calls for about 16 gal of paint per mile of continuous 4-in. stripe. In addition, 4 to 6 lb of glass beads with an average diameter of 0.02 in. (see AASHTO Designation M247) per gallon are added to the paint, usually by the "drop-in-during-application" method. The surfaces of some "flotation" beads may be chemically treated so that they are exposed for about half their diameter. At times the beads may be incorporated into the paint itself.

Striping paint is applied in a variety of ways. At one extreme it can be done by hand methods using a spray gun or a small self-propelled machine; at the other are trucks which travel as fast as 20 mph and place three parallel stripes of heated paint at once. Restriping frequency depends on a variety of factors, but has been reported to be as frequent as three times per year where traffic is heavy.

Pavement marking materials, also covered by AASHTO specifications, which reportedly last six or more times as long but have far higher first cost, also are in use. Among them are *hot thermoplastics* which are sprayed or extruded onto the surface at temperatures in the range of 425°F. A binder-sealer may at times be applied first. Glass beads can be incorporated. Minimum thickness is about 3/32 in. These also are subjected to performance tests.[9] Expensive machinery and skilled crews are required, so that use is restricted to situations where requirements are substantial. There are also cold, preformed plastic pavement marking materials. These consist of strips of plastic, coated on one side with a pressure-sensitive adhesive. To install them, a primer is applied to the pavement, the paper protecting the adhesive is removed, and the stripe is rolled or tamped onto the surface.

Stripe visibility is better with open graded pavements than with smoother surfaces. But even the most visible stripes with glass beads incorporated are difficult to see at night in inclement weather or when there is layer of water, slush, or snow on the surface. Many agencies place round, flat plastic disks along the lane lines. Today, there is an increasing use of raised reflectors less than 1 in. in height which are illuminated by the headlights. Colors are white for lane markings, blue to outline bicycle lanes, and red facing vehicles traveling in the wrong direction on freeways or entrance or exit ramps. Because they produce a rumble in the vehicle when it crosses the lane line or strays onto it, either the disks or the reflectors serve to alert inattentive drivers. Commonly these devices are fastened in place with epoxy. Unfortunately, earlier types often were torn out during snow removal, which made their use infeasible. A newer design seems to have overcome this difficulty.[10]

[8]See *Traffic Engineering*, Jan. 1976.

[9]See *ITE Journal*, Sept. 1980.

[10]See E. T. Harrigan, op. cit.; M. V. Jaganneth and A. W. Roberts, *TRB Record 651*; and C. W. Niessner, *Public Roads*, Sept. 1979.

There are situations where, for aesthetic or economic reasons, pavement markings consist of stone, brick, or concrete of a color contrasting with the pavement. Installation of this form of marker is particularly easy to accomplish with the brick or block pavements found in the older portions of some cities and towns.

At one time, large mushroom buttons were employed as pavement markings; today, however, their use is frowned upon.

It is usual practice to place letters, words, or arrows on the pavement at appropriate locations to indicate to drivers that they are approaching hazards such as railroad or school crossings or to mark straight through or turning lanes. Materials are the same as for other pavement markings. Details as to the dimensions for letters, arrows, and other symbols are available from FHWA. It should be noted that, because these markings are viewed by the driver from a very flat angle, they generally are greatly elongated, to 8 ft in rural locations.

As indicated in Table 12–3, pavement markings are considered to be very effective in accident reduction. However, much is yet to be learned of driver response to them.[11]

Raised Bars and Rumble Strips

Raised bars, called "jiggle bars," several inches high of concrete or asphalt often are employed as channelizing devices at the nose of traffic islands or to keep vehicles out of certain paved areas. They are not as high as curbs (see Chapter 9) and do not damage vehicles. They often are set diagonally to the vehicle path and oriented to deflect it back onto the right course. Sometimes they are painted or reflectorized.

Rumble strips are placed across roadways to alert motorists as they approach dangerous situations such as stop signs or abrupt changes in the oncoming alignmemt, grade, or profile. They are designed to produce a roar in the vehicle. Commonly they consist of several narrow transverse bands of asphalt-bound coarse aggregate. Rows of round plastic disks such as those used for lane marking are also effective.[12]

Delineators

Delineators are light-reflecting devices mounted on posts at the side of the roadway in series to guide the driver along the proper alignment. Requirements are that they be placed at a constant distance of 2 to 6 ft from the edge of the roadway, that the top of the reflecting head be 4 ft above the edge of the roadway, and that under normal atmospheric conditions they be visible for 1000 ft under the upper headlight beam. Individual reflective units are to have a mimimum dimension of 3 in., but elongated units may be substituted where two reflecting heads are specified. Color is the same as for edge lines.

[11]See, for example, W. J. Stimpson et al., *TRB Record 630,* and R. N. Schwab and D. G. Capelle, *ITE Journal,* May, 1980.
[12]See R. D. Owens, *HRB Record 170.*

Delineators are specified for the right side of expressways and freeways unless there is fixed lighting, and on at least one side of ramps. Red reflectors may be placed on the back side of ramp delineators to indicate that the driver is going the wrong way. It also is recommended that they be used on two-lane, two-way roads. Suggested spacings are between 200 to 528 ft on tangents; they should become progressively smaller on approaches to curves. Suggested maximum spacings on curves range from 20 ft with a 50-ft radius to 90 ft with a 1000-ft radius. Single delineators may be placed on the left side where the road curves to the right.

On freeways, two yellow delineators at 100-ft centers are stipulated along acceleration lanes. Roadway narrowings should be marked for their full length. Where curbs project into the traveled way their ends shall carry yellow reflectors if traffic moves to the right; white if it may pass on either side.[13]

Mileage and Object Markers

Distance markers every mile (thousand paces) were placed at the roadside during Roman times. These were usually large, upright, cylindrical stone monuments showing the distance to Rome or some other principal city. Many early American roads also had mileposts, often of stone. Today mileage markers or mileposts are installed to assist drivers in estimating their progress, for pinpointing accident locations, for distinguishing the positions of features such as bridges and culverts, to identify road sections for cost accounting purposes, and to designate where maintenance is to be done.

Object markers are installed at obstructions within or adjacent to the roadway or at the end of a road. The *Manual* proposes three choices as follows: Type 1, a cluster of nine yellow reflectors each about 3 in. in diameter, mounted on an 18-in. yellow and black panel; Type 2, a cluster of three of these yellow reflectors; or Type 3, a 1 by 3 ft vertical rectangle striped diagonally with black and reflectorized yellow or white. Types 1 and 3 are to mark objects within the roadway and types 2 and 3 for those adjacent to it. Markers on bicycle facilities resemble those described above.

Beacons

Beacons are standard circular traffic signal heads that flash 50 to 60 times per minute. Those for hazard identification are yellow and are employed to call attention to dangerous curves, obstructions, and approaching intersections, schools, midblock crosswalks, or other hazardous spots.

Speed limit beacons are yellow; they accompany speed limit signs. Intersection-control beacons are generally suspended on cables over the middle of the intersection. They flash yellow or red on appropriate faces; where visibility is poor, they may be preceded by hazard identification beacons. Stop sign beacons are mounted just above the stop sign.

[13]*NCHRP Report 130* offers an exhaustive discussion of delineators and object markers and an extensive bibliography. See also *TRB Record 681* and R. N. Schwab and D. G. Capelle, *Public Roads*, Dec. 1979.

Studies have shown that flashing beacons are an important adjunct to signs in affecting driver behavior. For example, signs indicating slippery pavement brought little speed reduction unless flashers were also used.[14] They also brought speed reductions at school zones, but only at high-speed locations. However, crossing guards or enforcement were even more effective.[15] Again, with flashers, driver behavior improved at approaches to intersections.[16]

Regardless of their purpose, beacons are only a supplement to and not a replacement for regulatory signs.

Color or Texture Differentiated Pavement

It is common practice to distinguish between lane and shoulder by employing contrasting color or texture in the surfacing. This already exists with concrete-asphalt combinations. With asphalt-asphalt situations, seal coats (see Chapter 19) can provide the contrast. Also, some installations with color or texture differences have been made in order to distinguish through lanes from ramps and turnoffs or for similar purposes. With portland-cement-concrete pavements, white cement or cement with iron oxide added can produce shades of white, red, yellow, black, brown, or buff. With bituminous pavements, a clear synthetic resin is available which has the properties of paving asphalts. Color can be added to this binder as desired.

Fixed-Message Signs

The most common device for warning, regulating, and informing drivers is the fixed-message sign. Although signs are not needed to confirm the driver's knowledge of the recognized rules of the road, they are essential wherever special regulations apply or where directions or notice of approaching hazards must be communicated to the driver.

The *Manual* treats signs under the following separate headings: regulatory; warning; guide for conventional roads, expressways, and freeways; and civil defense. In addition, signing is among the topics in sections on construction and maintenance, school areas, grade crossings, and bicycles.

There is wide variation in the number and nature of signs that may be installed. At one extreme are those for freeways, on which motorists are supplied detailed directions. They include, for example, warnings regarding destinations and lanes to occupy as far as 2 m from exits (see Fig. 10–1). There also may be variable message signs to notify motorists of changes in conditions. On the other hand, some low-volume rural roads have only directional and mileage signs at intersections and, possibly, warning signs for hazardous conditions. These may be widely spaced and very general, for example, "No curve warnings next 10 miles." There are situations where almost no signs are installed, since they will be vandalized or carried away as building material.[17]

[14]See F. R. Hanscom, *TRB Record 600*.
[15]See C. V. Zegeer et al., *TRB Record 597*.
[16]See R. B. Goldblatt, *TRB Record 644*.
[17]See *FHWA Report RD-77–39*.

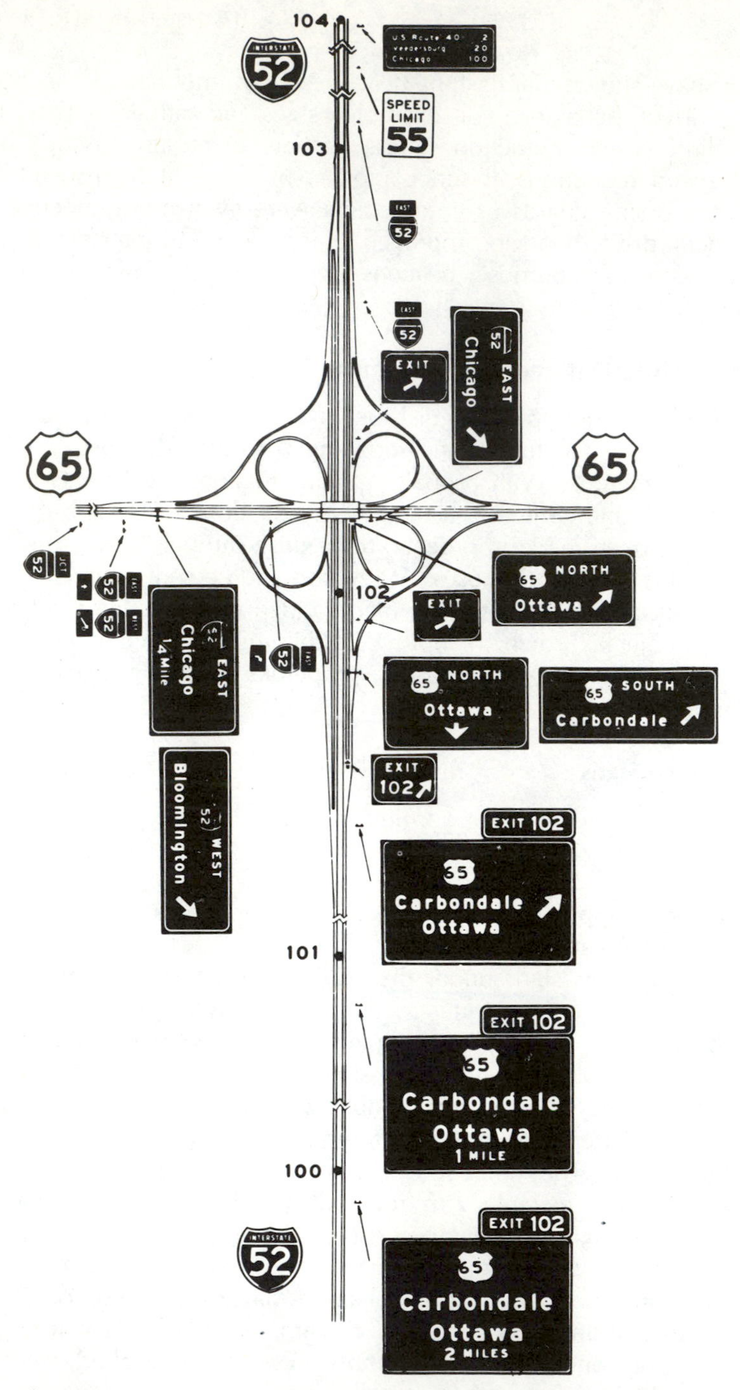

Fig. 10–1. Signing for a full cloverleaf interchange with collector–distributor roads. All signs are white letters on a green background except Interstate shields which are white or red on blue. (Source: *Manual on Uniform Traffic Control Devices*.)

Authority to install signs is usually vested by law or delegated to highway or police officials, although local elected officials may sometimes retain this power. The courts usually construe decisions on signing to be discretionary so that their absence does not constitute a right to sue (see Chapter 7). Placing signs on the right of way by other persons or organizations is seldom permitted in the United States, since these installations reduce the effectiveness of authorized signs. Under no circumstances should any sign or its support carry commercial advertising or messages other than for traffic control. Regulatory and warning signs should be installed sparingly, for excessive use decreases their authority.

MATERIALS FOR SIGNS. Signs and their appurtenances are intended to be permanent and therefore must be resistant to the elements and other corrosive influences in the harsh environment of the roadside. Insofar as possible, damage by vehicles and vandals should be minimized. Commonly, they are of sheet metal, protected by a rust-resisting coating, although waterproof plywood is used at times. Bolts, screws, fittings, and support structures generally are galvanized or otherwise treated to prevent corrosion and discoloration. Reflectorization, where needed, is by glass buttons or beaded sheeting. Materials and the signs themselves are available from manufacturers, although many agencies have their own sign shops. Extensive research to improve existing materials and develop new ones has been carried out, but is not reported here.

SUPPORTS FOR SIGNS. In the past, many signs at the roadside were mounted on steel posts. Today, however, these mountings are commonly of wood or breakaway design where the location can lead to impact by vehicles. Concrete bases for their supports should not project above ground level. Supports for street signs protected by curbings still are predominantly of metal.

Supports for the very large overhead or roadside signs on freeways and expressways are structures, and must be designed as such. Wind loads are of particular concern.[18] Designs that permit the support to break away on impact as a means of preventing injuries and vehicle damages are in common use not only for signs but also for luminaires and traffic signals. This topic has been intensively investigated.[19] Unfortunately, current designs may not be appropriate for the smaller, lighter automobiles now coming into use.

PURPOSES, SHAPES, COLORS, AND ILLUMINATION. Table 10–1 states the purpose and gives shape, color combination, and minimum dimensions for typical signs. Notice that words, shape, and color provide redundancies to emphasize the message. Warning signs are diamond shaped, except for the round railroad-crossing sign. Stop signs are octagonal. Informational and regulatory signs are rectangular. Warning signs are in black on a yellow background. Stop, yield, and "do not enter" or "wrong way" signs use great expanses of red. In-

[18]See *Structural Supports for Highway Signs, Luminaires, and Traffic Signals*, AASHTO.
[19]Recent research reports appear in *HRB Record 343, 346*, and *460; TRB Record 488, 594*, and *681*, and *Public Roads*, Dec. 1976.

TABLE 10–1. Partial Data on Recommended Signing Practice*†

Purpose of Sign	*Shape*	*Colors*	*Minimum Dimensions (in.)*
Warning of hazard	Diamond	Black on yellow	30 × 30
No passing zone	Triangular	Black on yellow	36 (vert.) × 48 × 48
Railroad crossing	Disk	Black on yellow	36 (diam.)
Regulatory	Vertical rectangle	Black on white	24 or 30 wide, height variable
Stop	Octagon	White on red	Primary 30 × 30 Secondary 24 × 24
Yield	Triangular	Red on white—red border	36 × 36 × 36
Parking—urban	Vertical rectangle	Red on white or green on white	12 × 18
Parking—rural	Vertical rectangle	Red on white	24 × 30
Informational (Guide)	Horizontal rectangle	Black on white, white on black, or white on green	
Scenic rest	Horizontal rectangle	White on green	
Recreational	Trapezoidal	White on brown	
Route markers	Shield or special	Black on white	
Route markers—Interstate	Shield	White on red and blue	
Construction and maintenance		Black on orange	
Pedestrian control	Vertical rectangle	Black on white	9 × 12 or 12 × 18

*Based on *Manual on Uniform Traffic Control Devices.*
†For design details see *Standard Highway Signs* available from The Superintendent of Documents, Washington, D.C. 20402.

formational signs are black on white or, for certain oversized signs on expressways and freeways, white on green with a white border. The sign dimensions given in Table 10–1 represent minimums. Larger sizes are desirable on expressways or where speed, hazard, accident experience, or competition from lights or other signs is extreme. Excessive use of large signs is to be avoided because it makes the regular sizes less effective.

Nonreflectorized signs are appropriate for most school-zone markings, for parking control effective only in the daytime or illuminated by street lighting, and for "Men Working" and other temporary warnings. Where speed limits differ between day and night, the *Manual* suggests that two signs, one black on white, unreflectorized, and the other white on black, reflectorized, be mounted on the same standard.

Signs having significance at night should be illuminated or reflectorized. Illumination, if other than by headlight, may be by lights in or behind the sign or by independently mounted floodlights. Reflectors, if for the legend and border, include reflector buttons and paint containing glass beads. If the background is reflectorized, it usually is by means of a paint or coating containing glass beads.

Overhead signs on freeways and expressways usually are illuminated since they are not lighted by the lower beam of headlights. But mountings for such signs are expensive, since there must be a maintenance walkway. In addition, lighting costs are in the range of $100 per year. It has been proposed that on tangents where vehicles are using high beams, reflectorized sheeting without illumination is sufficient. Another possibility is to use encapsulated reflective sheeting.[20] However, on curves, illumination is necessary.

Numerous research projects have been conducted to determine the effectiveness of various letter sizes and shapes and of sign positioning, layout, and symbols in getting the sign message to drivers during both daylight and darkness.[21] Conclusions from such studies have been incorporated into the standard pattern of letters used in the *Manual*.[22]

As indicated in Table 10–1, colors and color combinations for each signing purpose have been stipulated. Their selection represents compromises among factors such as legibility and ability to attract attention. For example, red ranks first from the standpoint of recognition and ability to attract attention if used as a background for white letters. On the other hand, black, which is highly visible, is almost entirely passive in attracting attention and providing stimulation or emotion. For example, it was found that white letters on a red background were considerably more effective than black on white in getting driver response to "wrong-way" signs at freeway ramp exits.

SIGNING PRACTICES. It is essential that the overall signing scheme for an urban area have only one interpretation, have continuity from sign to sign (see *NCHRP Synthesis 71*), give sufficient advance notice for drivers to change lanes or make appropriate decisions, and be relatable to road maps and other guides available to the motorist. As mentioned earlier, an excellent way to check the geometric design of a facility is to develop a plan for signing it.

Figure 10–1 is the scheme for signing one of several conformations of interchanges. Others are shown in the *Manual*. Among other details, Fig. 10–1 shows that important directional information is repeated several times. Figure 10–1 does not show identical signs outside the shoulder and in the median, a practice of some agencies.

[20]See R. N. Robertson and J. D. Shelor, *TRB Record 628*.

[21]See T. W. Forbes et al., in *HRB Record 70, 164, 216,* and *440* and *TRB Record 562* and *611* for a summary and bibliography covering much of this research. Other recent reports on sign effectiveness appear in *TRB Record 562, 600,* and *643*, and *Public Roads*, Sept. 1976, Sept. 1977, Jun. and Dec. 1979, and Mar. 1980. Studies of the brightness and legibility of signing materials are reported in *TRB Record 562* and *611*. *NCHRP Report 145*, titled *Improving Traffic Operations and Safety at Exit Gore Areas* reports in detail on experiments with signing at this high-hazard location.

[22]For example, better legibility is obtained by using a relatively wide spacing between letters rather than wider and taller letters with cramped space. Again, for conventional highway signs, upper case letters are more effective except for place names where lower-case following an initial upper-case letter is preferable. Also, it is important that sign messages be very simple. This stems from the finding that the driver, in the short time available for viewing, cannot observe and assimilate large amounts of information. Thus, on expressways or freeways, usually two messages, and no more than three, regarding direction or destination appear on a single sign (see Fig. 10–1). Other important "place" information is given on separate signs.

At turnoffs from limited-access facilities, either of two approachs to directional signing may be used. One is merely to indicate the immediate maneuver, without attempting to detail subsequent ones. The second is to diagram these later movements on earlier signs. For example, progressing upward on Fig. 10–1, the first "exit" sign after reaching the collector–distributor roads would have an arrow curving 90° to the right in place of the right-inclined arrow. At the turnoff to the loop, the right-inclined arrow would be replaced by one diagraming the 270° turn. Studies have shown this scheme reduces the number of stops and backups and unusual gore maneuvers.[23]

Signing for new construction, for reconstruction of existing roads, and while maintenance is under way has taken on added importance as drivers increasingly expect to drive through such projects with little delay or reduction in speed. An entire section of the *Manual* is devoted to this subject.

POSITIONING SIGNS. Positioning for individual signs depends on the purpose of the sign and the circumstances peculiar to each location. Whenever possible, locations should be standardized, since positioning provides another set of cues to motorists. Recommended positions for all usual situations are given in the *Manual*. For example, warning, regulatory, and advisory signs for rural roads should be placed 6 to 12 ft from the pavement edge with the bottom of the sign 5 ft or more above the roadway. On high-speed roads, warnings should be posted as much as 1500 ft ahead of the hazard. On rural roads, warning distance should be 750 ft; in urban areas 250 ft is usually appropriate. Where vehicles may park along the curb, a horizontal clearance not less than 2 ft between the curb face and near edge of the sign, and a minimum 7-ft height above the curb are prescribed.

The usual sign positions fall in the drivers' normal field of view, so that they do not have to look away from the road. And, unless signs are illuminated, they must show in the headlights if they are to be effective at night. Unfortunately, certain peculiarities of human vision compound some sign-location problems. For example, as drivers enter intersections, their vision focuses alternatively ahead and to the other approaches. During this shift between viewing points, drivers blink involuntarily and sight is blocked out. Thus, signs placed midway between these points of focus may be of little value.

SYMBOLIC SIGNS. With travel among countries becoming more common, the need for worldwide uniformity in signs and signing to replace some six distinct concepts has become increasingly important. Even before World War II, efforts were under way to develop a common system, and they are now well advanced. The 1978 edition of the *Manual* carries numerous symbols as substitutes for words. For example, the "No U Turn" symbol consists of a black arrow bent through 180° with a red diagonal line through it to indicate that the maneuver is prohibited.

The first step in introducing symbolic signs is to include both symbols and

[23]See A. W. Roberts et al., *TRB Record 531*.

words on separate signs mounted on the same post. Then the "educational" sign carrying the words can be removed when motorists are thoroughly familiar with the symbols. The *Manual* permits omission of the educational signs in situations where the symbolic sign is recognized by the public.

Changeable-Message Signs

Traditionally, signs have carried a single, fixed message which applied 24 hr a day. An exception has been to omit reflectorization on signs that apply only in daylight. Today, particularly on high-volume, high-speed facilities, it is often important to transmit appropriate messages to drivers but only when they are needed. A simple application is that of notifying the motorist of fog or gusty winds. These may be turned on by a police officer or automatically when reduced visibility, as measured by a backscatter unit, infrared or ultraviolet light, or lasers, causes the sign to turn on.[24] More complex arrangements also are in use. They carry messages such as ALL LANES OPEN or LEFT LANES BLOCKED AHEAD—USE EXIT 6. Installations following this concept are an essential part of any scheme to monitor and direct traffic on freeway systems, as discussed later.

An application of changeable message signs to smooth traffic flow is the *traffic pacer system*. With it, changeable message signs inform drivers as to the speed at which they can pass through the next traffic signal without stopping. An experimental installation in Warren, Michigan, although it did not decrease travel time, did substantially reduce stopping and starting which in turn reduced fuel consumption and contributions to air pollution. To date, this concept has had little application.[25]

A number of mechanical, electrical, or combined devices for displaying different messages have been developed under such names as flat scroll, drum, inert gas (neon), fiber optics, light-bulb matrix, electro-mechanical, flap matrix, electrostatic, vane matrix, and electronic disk matrix.[26] Among the variables that would affect selection of one over the other are reliability and number and complexity of messages.[27]

Except for cost, the possibilities with changeable message signs are limitless, as demonstrated by the elaborate scoreboards at sport stadiums and schedule boards in airports.

Effectiveness of Signs

As indicated earlier, too many signs make all of them less effective. Furthermore, large signs overwhelm any accompanying smaller, standard ones. And, too often, signs fail to accomplish the intended purpose. A common example is using stop signs to control speed in residential areas. First of all, only a few driv-

[24]See *NCHRP Reports 95* and *171*, and J. A. Wachtel, *Public Roads*, Dec. 1977.
[25]See, for example, C. E. Dare, *HRB Record 286*.
[26]See *NCHRP Synthesis 61 and FHWA Report RD 77-98* for added detail.
[27]See, for example, J. W. Hall and L. V. Dickinson Jr., *TRB Record 531* and C. L. Dudek et al. and S. H. Richards et al., *TRB Record 682*.

ers stop completely. A majority make roll stops, and a few ignore the signs.[28] In addition, speed reduction begins less than 200 ft from the sign and vehicles are at full speed after 200 ft or less. Not only are such installations ineffective, but they encourage motorists to ignore other signs and substantially increase fuel consumption, noise, and air pollution. Other studies have found that signs such as "Slow, Children at Play" had no effect on speeds.

Selection among sign messages also is important since drivers react differently to them. For example, response was shown in 63% of the cases to the personal risk message of "police control ahead" and in 55% of the cases where "breaks in the road surface" were indicated. Response to general warnings was far less, 18%; and to "pedestrian crossing" 17%. Apparently signs warning of slippery pavement were not threatening; they were ineffective unless accompanied by flashers.[29]

Changeable message signs that highlight personal risk also have been effective. For example, one reading "Bridge deck icy" brought reduced speed. Also those notifying operators that they were driving too fast or following another vehicle too closely caused them to slow down.

Another message problem results if signs carry improper or incorrect information. For example, if a "safe speed" indicator on a curve warning sign suggests a speed slower than drivers find comfortable, they will ignore subsequent messages. Again, if signs report that there is construction or maintenance work ahead when none is actually in progress, drivers will ignore all such signs.[30]

TRAFFIC SIGNALS[31]

In the United States alone, some 250,000 intersections have traffic signals, which are defined as all power-operated traffic-control devices except flashers, signs, and markings for directing or warning motorists, cyclists, or pedestrians. Each of these installations serves one or more of the following functions:

1. Provide for orderly movement of traffic.
2. Increase the traffic handling capacity of intersections.
3. Reduce the frequency of certain types of accidents.
4. Coordinate traffic under conditions of favorable signal spacing, so that it flows nearly continuously and at definite speeds.
5. Interrupt heavy traffic to permit crossings by other vehicles or pedestrians.
6. Control traffic lane use.
7. Provide ramp control at freeway entrances.
8. Interrupt traffic for emergency vehicles and at movable bridges.

[28]See, for example, R. F. Beaubien, *Traffic Engineering*, Nov. 1976.

[29]See F. R. Hanscom, *TRB Record 531* and *Traffic Engineering*, Sept. 1975.

[30]See W. M. Seymour et al., *TRB Record 484*.

[31]Much of the data presented here are from the *Manual*. A glossary of signal definitions appears in *Traffic Engineering*, Feb. 1976. Detailed specifications for signal components such as controllers, detectors, lenses, lamps, and standards are prepared by ITE committees made up of representatives of manufacturers and users.

Many laymen and elected officials believe that traffic signals provide a solution to all traffic problems at intersections and at other danger spots. This has led to many unsuccessful installations. Among the unfortunate results have been:

1. Excessive delays to motorists or pedestrians.
2. Disobedience to signal indications generally as well as at the particular installation.
3. Diversions of traffic to less advantageous routes.
4. Increased accident frequency, particularly involving rear end vehicle collisons and pedestrians.

As with signs, the directions given by signals are enforceable only if they are erected under legal authority, and the intention of each is stated in detail in laws or ordinances. Model legislation is found in the *Uniform Vehicle Code* and in the *Model Traffic Ordinance*, both prepared by the Committee on Uniform Traffic Laws and Ordinances.

Signals for Intersection Control

Signals for vehicular, bicycle, and pedestrian control are classed as "pretimed" where specific times intervals are allocated to the various traffic movements and as "traffic-actuated" where time intervals are controlled in whole or in part by traffic demands. These paragraphs discuss first the characteristics common to all classes of signals, followed by a brief description of the individual types.

PHYSICAL FEATURES OF SIGNALS. Modern signals for intersection control operate by electricity. Individual units must have separate red, yellow, and green lenses, 8 or 12 in. in diameter, each illuminated by its own light source.[32] Often lenses directing separate movements to the left or right and "walk—don't walk" indicators for pedestrian control are added. Lens assemblies with as many as five units may be arranged vertically with red at the top, horizontally with red to the left, or with one red lens at the top and the others in pairs below. They are mounted on pedestals or brackets outside the roadway limits, or are suspended above the intersection by cables, mast arms, or other supports. The recommended height for post-mounted signals, measured to the bottom of the housing, is 8 to 15 ft above the sidewalk or above the crown of the pavement where there is no sidewalk. Median mountings must be 4½ ft or more in height. Signals suspended over the roadway shall have minimum and maximum vertical clearances of 15 and 19 ft.

Two or more signal displays or "faces" visible to approaching traffic are prescribed for each through movement so that drivers of vehicles following immediately behind trucks or buses will almost always be able to see the signal indication as they approach the intersection.[33] They must be visible to approaching vehicles for distances ranging from 100 ft for an 85 percentile speed of 20 mph

[32]Twelve-inch lenses are required for intersections where the 85 percentile speed exceeds 40 mph, for all arrow applications, and for problem locations or situations.

[33]For a study of truck blockage of lines of sight see G. F. King et al., *TRB Record 597*.

to 700 ft for 60 mph. A warning sign is stipulated where a continuous view of at least two signal faces is blocked by physical conditions. A hazard indentification flashing beacon may also be installed ahead of the intersection. Only one signal face is specified for exclusive turning lanes, although a second is often added. On expressways and other facilities where signals must interrupt fast-moving traffic, common practice is to install three and sometimes four signal faces on each approach.

Special rectangular pedestrian WALK and DON'T WALK signals with the message either in words or symbols are called for under special conditions such as heavy pedestrian movement, school crossings, midblock crossings, where the vehicle control signals would confuse pedestrians, or where there is a separate time allocation for them. The optional flashing "walk" signal indicates possible pedestrian-vehicle conflict. The flashing "don't walk" message is to give time for pedestrians to clear the intersection or reach a median. In isolated locations, pedestrian signals commonly operate only when activated by a push button. Two signal faces for vehicles are required at nonintersection pedestrian crossings, and these must be augmented by signs, pavement marking, and parking restrictions.

The *Manual* assigns positive meaning to each color and its use. The steady circular green alone should be given only when traffic is permitted to proceed in any direction which is lawful and practical. When certain turns are permitted and others prohibited, the regular circular red lens facing traffic should be illuminated together with a separate green arrow for each permitted movement. Solid red alone means to stop and wait for a green indication, while flashing red has the same meaning as an arterial stop sign. Flashing yellow means proceed with caution. The use of steady yellow for caution is prohibited, except as a warning between green and red indications. The *Manual* also gives other permissible and prohibited color combinations.

SIGNAL LOCATIONS. For typical right-angle intersections, far-side, mast-arm, or cable-suspended locations 40 to 120 ft beyond the stop line are stipulated. If both signals are post mounted, they shall be placed on the far side, one on the right and the other on the left or median. The angle between a driver's normal direction of sight and either signal face shall not exceed 20°. The *Uniform Vehicle Code* prohibits any illuminated advertising sign that interferes with the effectiveness of traffic control devices, and the *Manual* recommends that local legal authority be established to prohibit them.

CONTROLLERS. Traffic-signal controllers activate the lights in individual signal heads. The simplest are timers for single intersections which provide one to several sequences of signal indications, possibly including flashing reds and yellows. Recently, solid-state miniprocessors for intersection control have become available. These receive inputs from detectors, interpret their meanings, and select the appropriate set of signal indications from those in the processor's memory bank. These devices can accept 96,000 instructions per second and respond very quickly to changes in traffic.[34] This same wide range of controller sophis-

[34]See J. F. Hahn and G. F. Eustis, Jr., *Traffic Engineering,* Apr. 1977.

tication is also available for progressive signals along a street and for area-wide systems.[35]

The rapid advances, miniaturization, and lowering costs of electronic devices now make it possible to combine the advantages of pretiming, coordinated movement, and traffic-actuation at single intersections or area-wide to reduce costs and delays substantially. Technology is advancing so rapidly that only by continued study is it possible to keep abreast of this field.

DETECTORS. The early traffic-actuated detectors were activated by electrical contacts on plates imbedded in the roadway. These were expensive to install and to keep in operating condition. There were and are "above the surface" detectors involving interrupted light beams or infrared, radar, or ultrasonic devices. Today, most detectors consist of wire loops under the surface of the pavement which register the presence of vehicles through changes in their magnetic fields. They may be placed along with the pavement or installed in slots sawed into the pavements later.

A variety of detector loop arrangements are employed. If only one vehicle is to be detected, only a single loop is called for. Longer loops or a more sensitive series of loops record the number of vehicles waiting to make a maneuver. Transverse ends of loops often are placed diagonally, in part to detect the presence of motorcycles.[36]

Placement of detectors or detector loops with relation to the stop line depends on the traffic-control plan. For example, a vehicle entering from the minor leg of an isolated intersection is expected to stop; here the detector is placed close to the stop line. On the other hand, vehicles on the main road approach the intersection at high speed and expect to continue through; in this situation a detector must be placed at some distance back from the intersection[37] and possibly another close in, to prevent entrapment at the stop line or to extend the green time.[38] In other circumstances, special attention to detector placement or selection of those with limited fields may be called for to prevent undesired activation. From this brief discussion, it should be clear that designing an effective detector system is not easy.

PRETIMED TRAFFIC SIGNALS. "Pretimed" traffic signals are set to repeat regularly a given sequence of signal indications for stipulated time intervals through the 24-hr day. They have the advantages of having controllors of lower first cost and that they can be interconnected and coordinated to permit vehicles to move through a series of intersections with a minimum of stops and other delays. Also, their operation is unaffected by conditions brought on by unusual vehicle behavior such as forced stops, which, with some traffic-actuated signal instal-

[35]See J. Barker et al., *Traffic Engineering,* Apr. 1977.

[36]*Traffic Engineering,* Feb. 1974, gives details on small-area loops, and ibid., June 1976 on those for large areas. See also *TRB Record 737*.

[37]Recommended distances range from 175 ft for 30 mph to 450 ft for 60 mph.

[38]See, for example, C. F. Zegeer and R. C. Deen, *ITE Journal,* Nov. 1978 and P. S. Parsonson, *TRB Record 681*.

lations, may bring a traffic jam. Their disadvantage is that they cannot adjust to short-time variations in traffic flow and often hold vehicles from one direction when there is no traffic in the other. This results in inconvenience, delays, and sometimes a decrease in capacity.

"Cycle length," the time required for a complete sequence of indications, ordinarily falls between 30 and 120 s. Short cycle lengths are to be preferred, as the delay to standing vehicles is reduced. With short cycles, however, a relatively high percentage of the total time is consumed in clearing the intersection and starting each succeeding movement. As cycle length increases, the percentage of time lost from these causes decreases. With high volumes of traffic, it may be necessary to increase the cycle length to gain added capacity.

Each traffic lane of a normal signalized intersection can pass roughly one vehicle each 2.1 s of green light.[39] The yellow (caution) interval following each green period is usually between 3 and 6 s, depending on street width, the needs of pedestrians, and vehicle approach speed.[40] To determine an approximate cycle division, it is common practice to make short traffic counts during the peak period. Simple computations give the number of vehicles to be accommodated during each signal indication and the minimum green time required to pass them. With modern control equipment, it is possible to change the cycle length and division several times a day, or go to flashing indications to fit the traffic pattern better.[41]

At many intersections, signals must be timed to accommodate pedestrian movements. The *Manual* recommends that the minimum total time allowed be an initial interval of 4 to 7 s for pedestrians to start plus walking time computed at 4 ft/s.[42] With separate pedestrian indicators, the WALK indication (lunar white) covers the first of these intervals, and flashing DON'T WALK (Portland orange) the remainder. The WALK signal flashes when there are possible conflicts with vehicles and is steady when there are none. Steady DON'T WALK tells the pedestrian not to proceed.

If pedestrian control is solely by the vehicle signals, problems develop if the intersection is wide, since the yellow clearance interval will have to be considerably longer than the 3 to 5 needed by vehicles. This will reduce intersection capacity and may call for a longer cycle time. On wide streets having a median at least 6 ft wide, pedestrians may be stopped there. A separate pedestrian sig-

[39]See Chapter 8 for a more detailed discussion.

[40]A. D. May, Jr., *HRB Record 221*, presents an extensive survey on laws relating to motorists' use of the amber period. He also reports on driver behavior with amber periods of various lengths. W. L. Williams, *TRB Record 644*, reported that, for the intersections studied, 85% of the drivers chose to stop if over 100 ft away when the amber indication came on; 85% continued if within 40 ft. Careful attention must be given to determining this "dilemma" and the appropriate amber time that is assigned. For added discussion see W. A. Stimpson et al., *ITE Journal,* Nov. 1980 and H. H. Bissell and D. L. Warren, *ibid,* Feb. 1981.

[41]See, for example, C. M. Abrams and S. A. Smith, *TRB Record 629*, and Feng-Bor Lin, *TRB Record 644*.

[42]Since walking times vary from 2.5 to 6 ft/s, this assumption should be evaluated in light of the pedestrian population.

nal activator must be placed on this median if pedestrian push buttons are incorporated into the overall control system.

In downtown locations where pedestrian movements are unusually heavy the *scramble system* of traffic control is sometimes used. With it, a separate interval is provided solely for pedestrians during which they may cross either street or proceed diagonally across both. Numerous reports indicate that with "scramble," vehicular capacity is lowered and pedestrian delays are longer; the principal argument for it is that "pedestrians and merchants like it." Alternatively, timing to favor pedestrians by setting their movements to lead or lag vehicles can reduce conflicts.

Signals are today being timed to speed buses through intersections by extending the green indication to favor them. With separate bus lanes, this can be done with a detector in that lane. Another scheme involves a driver-operated radio or strobe light which communicates bus position to the controller. Substantial reductions in bus travel times have been reported; however other motorists probably will be delayed.[43]

COORDINATED MOVEMENT. Fixed-time traffic signals along a street or within an area usually are coordinated to permit compact groups of vehicles called "platoons" to move along together without stopping. Under normal traffic volumes, properly coordinated signals at intervals variously estimated from 2500 ft to more than a mile are very effective in producing a smooth flow of traffic. On the other hand, when a street is loaded to capacity, coordination of signals is generally ineffective in producing smooth traffic flow.

Four systems of coordination—simultaneous, alternate, limited progressive, and flexible progressive—have developed over time. The *simultaneous system* made all color indications on a given street alike at the same time. It produced high vehicle speeds between stops but low overall speed. Because of this and other faults, it is seldom used today.

The *alternate system* has all signals change their indication at the same time, but adjacent signals or adjacent groups of signals on a given street show opposite colors. The alternate system works fairly well on a single street that has approximately equal block spacings. It also has been effective for controlling traffic in business districts several blocks on a side, but only when block lengths are approximately equal in both directions. With an area-wide alternate system, green and red indications must be of approximately equal length. This cycle division is satisfactory where two major streets intersect but gives too much green time to minor streets crossing major arteries. Other criticisms are that at heavy traffic volumes the later section of the platoon of vehicles is forced to make additional stops, and that adjustments to changing traffic conditions are difficult.

The *simple progressive system* retains a common cycle length but provides "go" indications separately at each intersection to match traffic progression. This permits continuous or nearly continuous flow of vehicle groups at a

[43]See S. R. Seward and R. N. Taube, *TRB Record 630*, and A. J. Richardson and K. W. Ogden, *TRB Record 718* for assessments and bibliographies.

planned speed in at least one direction and discourages speeding between signals. Flashing lights may be substituted for normal signal indications when traffic becomes light.

The *flexible progressive system* has a master controller mechanism that directs the controllers for the individual signals. This arrangement not only gives positive coordination between signals, but also makes predetermined changes in cycle length, cycle split, and offsets at intervals during the day. For example, the cycle length of the entire system can be lengthened at peak hours to increase capacity and shortened at other times to decrease delays. Flashing indications can be substituted when normal signal control is not needed. Also the offsets in the timing of successive signals can be adjusted to favor heavy traffic movements, such as inbound in the morning and outbound in the evening. Again, changes in cycle division at particular intersections can be made. The *traffic responsive system* is an advanced flexible progressive system with the capacity to adjust signal settings to measured traffic volumes (for added detail, see the discussion of Area-Wide Operation later in this chapter).

Where traffic on heavy-volume or high-speed arteries must be interrupted for relatively light cross traffic, semi-traffic-actuated signals are sometimes used. For them, detectors are placed only on the minor street. The signal indication normally is green on the main road and red on the cross street. On actuation, the indications are reversed for an appropriate interval after which they return to the original colors.

In settled areas push buttons providing pedestrian-actuated control along with advance warning signs and possibly flashers are included in such installations.

Semiactuated signals at minor intersections can be easily incorporated into a progressive system. In this instance, the green indication for the minor street is delayed until it fits into the progressive scheme. If there is no demand from the minor street, the signal indications do not change at all.

TIMING PROGRESSIVE TRAFFIC SIGNALS. Traditionally, the timing of progressive signals on a given street has been done using a time-space diagram such as Fig. 10–2. Today, this approach may be augmented or supplanted by computer analysis or simulation. Counts at appropriate times, both for the street in question and for important cross streets, represent the first step. From these counts appropriate cycle lengths for the entire system and cycle divisions for each intersection by time period can be worked out. Also, the most desirable travel speed along the street must be found. Stop-watch studies, a "floating car" that moves with the traffic stream, radar, or time lapse motion pictures all can be employed for gauging actual speeds. Starting with these basic data, the "offsets" in timing between the signal indications of the different streets can then be planned.

The graphical solution and the approaches employed in computer analyses all are based on some variation of the space–time diagram that has distance along the street as its ordinate and time in seconds as the abscissa (see Fig. 10–2). On this graph the slope of any straight line represents velocity, since the units are distance and time. To illustrate, the slope of the straight lines running

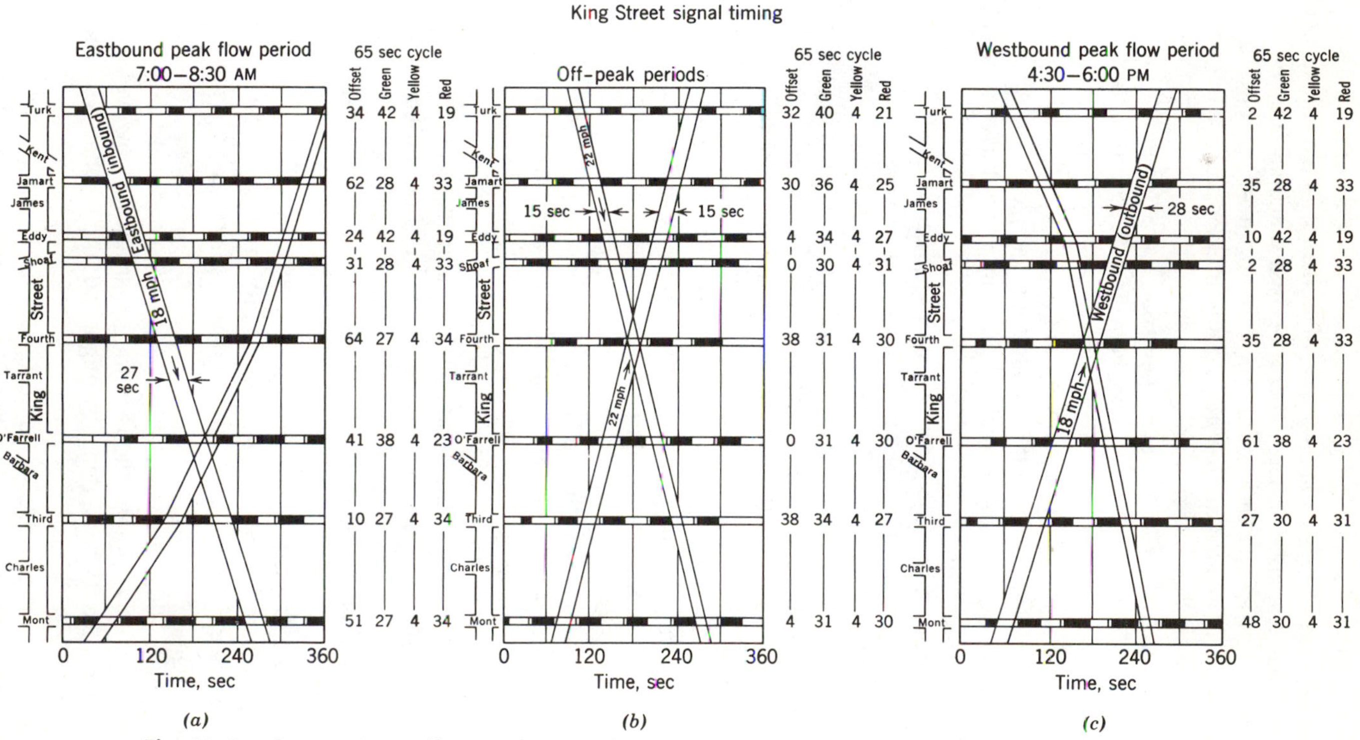

Fig. 10–2. Progressive traffic-signal timing for a typical street. (From *Traffic Engineering Handbook*.)

from upper left to lower right in Fig. 10–2*a* indicates the speed of 18 mph. The lines themselves show the location–time relationship for vehicles traveling at given speeds. For example, in Fig. 10–2*a* the two diagonal straight lines each represent the progress with time of an automobile traveling from Turk toward Mont at 18 mph. The heavy dashed horizontal lines opposite each street show the signal timing at that intersection. The wide, open spaces in these dashed lines represent "go" time, the narrow, open spaces are "caution" intervals, and the solid black segments are "stop" time. On a working model of the diagram these signal indication bands would be drawn separately on narrow strips of tracing paper or tracing cloth. By shifting the strips laterally, all possible combinations of signal offsets can be investigated on a single chart. Also the sloped lines representing speed might be marked out with black threads for easy shifting about on the diagram.

Figure 10–2 shows three combinations of offsets, each selected to fit a peculiar traffic-flow pattern. Timing for the morning peak hours (Fig. 10–2*a*) favors the heavy flow of inbound traffic by giving it a through band of 27 s, which is the maximum possible, given the selected total cycle time and demands for time for important cross streets. Signal offsets are timed for the most favorable speed of 18 mph. On the other hand, the light flow of outbound traffic receives a narrow through band, which means that some of the vehicles will be delayed at O'Farrell. Figure 10–2*c* shows the timing for the afternoon peak period, when heavy outbound traffic is favored. Timing for "offpeak" periods appears in Fig. 10–2*b*. In this case traffic flows with equal ease in either direction at the higher speed of 22 mph. However, the through bands are only 15 s wide, which limits the number of vehicles that can travel in a given squad.

The experienced traffic engineer can add many refinements to the procedure outlined here. As an example, by lagging or leading the signals at particular intersections, the main artery can be cleared in both directions to permit vehicles and pedestrians to cross at intermediate unsignalized locations. In the same way, main arteries can be cleared in one direction to permit easy left turns by vehicles traveling the other way. Often left turns can be reduced or prohibited at important intersections by providing for them at adjacent unsignalized locations, or by routing the vehicles through the intersection, to the right around the next block, and straight through the intersection again. Other techniques are based on the adjustment of cycle division at less important intersections. To illustrate, vehicles can be encouraged to move in platoons by limiting the green time at these locations.

The procedures illustrated by Fig. 10–2 are based on the assumption of a fixed cycle length at all intersections in the progression. However, there is the possibility that mixed cycle lengths might be more effective under some circumstances.[44]

Timing patterns for progressive traffic signals are often severely criticized by the uninformed on the grounds that they do not provide wide through bands in

[44]See, for example, J. B. Kreer, *Traffic Engineering,* Mar. 1977.

both directions. In most cases, this is physically impossible. Only in rare instances is the proper combination of block spacing and vehicle speeds to be found. Provision of through bands in both directions is particularly difficult where block spacings are irregular. This can be convincingly demonstrated by diagrams similar to Fig. 10–2.

Three-dimensional models of the space–time diagram have been employed to study signal timing over street grids.[45] Also, computer methods, some based on simulation, are available.[46]

TRAFFIC- AND PEDESTRIAN-ACTUATED SIGNALS. Traffic-actuated signals respond in a predetermined manner to the approach of vehicles from one or more legs of an intersection. They have detectors located on each approach lane and assign the right of way to the various traffic movements on the basis of demand. With modern detector and control equipment, signal operations can be adapted to many situations. As one example, consider an intersection between major and minor arteries. Here the green indication normally remains with the main street. On demand from the cross street, or for a left turn from the major artery, the signal indication changes at once if the main street has no traffic. With continued flow on the main street, the green indication remains until all vehicles have been cleared or until some predetermined period has elapsed. The minimum green period for the cross street or turn permits the passage of one vehicle. However, this interval is extended up to a set maximum by continued traffic.

Full traffic-actuated controls having several signal phases are common. With them, indications are offered to the various movements in rotation, except that those with no traffic demand are skipped. In some, traffic actuation may be combined with fixed-time coordinate movement at peak hours.[47]

PHASING DIAGRAMS. Phasing diagrams offer an excellent means for thinking through the timing of either traffic-actuated or fixed-time signals. Figure 10–3 is typical. It is for a four-leg intersection with separate left-turn lanes. With the phasing shown, all movements are free of conflicts. Phase 1 provides for left turns from both horizontal legs; phase 2 is for both through horizontal movements; phase 3 has left turns from the vertical legs; phase 4 is straight through in the vertical direction. Notice that left turns "lead" the through movements. With fixed-time signals, each of the four phases is given an allocated amount of time; with traffic-actuation, a phase is terminated when its traffic demand ceases or its allocated time expires. Also, if demand ends quickly on a particular left turn, the opposing straight through movement is permitted to start ahead of its pair.

There are alternative phasings to those shown in Fig. 10–3. A common one

[45]See, for example, *NCHRP Report 113,* Appendix D.

[46]See, for example, H. R. Leuthardt, *TRB Record 531* and T. K. Datta et al., *ITE Journal,* June 1979.

[47]At lower traffic volumes, traffic-actuated signals substantially decrease vehicle delay but have little effect as volume approaches capacity. See K. G. Courage and P. Papapanou, *TRB Record 630.* For methods of measuring intersection delays see W. R. Reilly and C. C. Gardner, *TRB Record 644.*

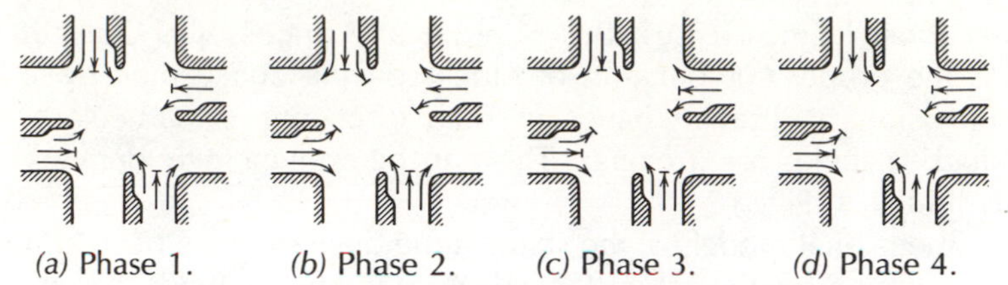

(a) Phase 1. (b) Phase 2. (c) Phase 3. (d) Phase 4.

Fig. 10–3. A typical signal phasing diagram for a four-leg intersection with provisions for separate left turns.

is to let vehicles entering from each leg proceed separately. This would offer a means of favoring a very heavy left-turn and through flow on one leg. Again, where straight-through movements are heavy or pedestrians are numerous, the free right turns of Fig. 10–3 could be limited.

The four-phase operation in Fig. 10–3 is for two major arterials. If, for example, the street shown in the vertical has low volumes, phases 3 and 4 often are combined. This introduces conflicts between left turns and opposing straight through traffic but reduces the amber time consumed in clearing the intersection after each phase.[48]

Diamond interchanges usually operate smoothly when through and turning movements along the cross street are low. But they become very congested at higher volumes. To ease this situation without major reconstruction, it is common practice to install traffic signals where the ramps connect to the cross street. Figure 10–4*a* shows the movements to be accommodated. Figure 10–4*b* demonstrates four phasing alternatives for the two sets of signals. "Lead" indicates that vehicles entering the cross street from the ramps of the grade-separated freeway (phase B) move before those leaving the cross street for the freeway (phase C). Figure 10–4*c* provides a more detailed examination of the lead-lead alternative and indicates the timing offset between the two sets of signals. (Note that time increases upward on the diagram.) In most instances, this timing offset increases as the distance between the left-side and right-side intersections increases. Congestion becomes more likely when this distance is very short. In this case, storage space for vehicles turning left from the cross street to the freeway is limited, and the queue they form may seriously encumber straight through movements.

Selecting among phasing alternatives and determining cycle length and split for a diamond is a difficult process. With fixed-time controls, this process begins with traffic counts to select the first settings, followed by adjustments based on field observation. Traffic-actuated signals also require careful study and observation before final timing is adopted. A near optimum arrangement may combine traffic actuation at off-peak with fixed time operation during peak periods. Several conventional and simulation approaches to the timing of such systems have been developed and compared. One of these involves control by micro-

[48] For simulation of the operation of an isolated intersection see K. G. Courage et al., *TRB Record 503* and T. W. Rioux and C. E. Lee, and C. J. Messer et al., *TRB Record 644*.

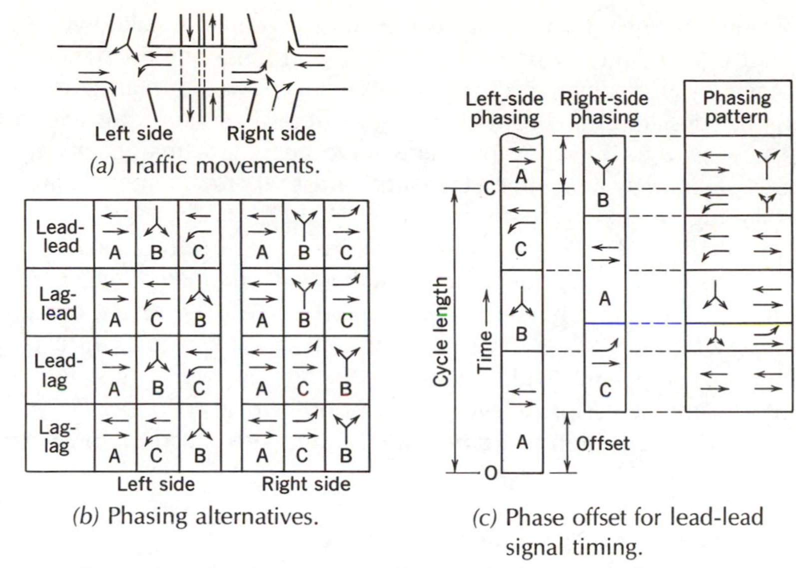

(a) Traffic movements.

(b) Phasing alternatives.

(c) Phase offset for lead-lead signal timing.

Fig. 10–4. Diagram illustrating traffic signal timing and phasing at a diamond interchange.

processor.[49] The results of one study found that timing based on simulation reduced delays and stops by 20 to 30% compared with conventional timing methods.

Another dimension to signal timing is to accommodate pedestrians and still not unduly delay motor vehicles. Details cannot be offered here.[50] Another special problem involves coordinating signals at railroad grade crossings to pass both vehicles and trains and still prevent accidents.[51]

FREEWAY AND EXPRESSWAY OPERATION

Freeways and, to a lesser degree, expressways are designed to be high-speed, free-flowing, low-accident facilities that operate unhindered by traffic controls. This is the usual situation, except at peak hours on urban and suburban facilities. Also (see Chapter 12) their accident rates are relatively low. One reason why freeways operate so well is their isolation from their surroundings, which comes from restricting access to widely spaced points and by barring pedestrians. But if congestion develops, this isolation makes correcting it difficult and is most frustrating to motorists who have no information whatsoever as to its

[49]For references, both on conventional methods and simulation, see *Control Strategies for Signalized Diamond Interchanges,* FHWA Technology Sharing Report TS-78-206. See also articles by C. J. Messer and D. J. Berry, *TRB Record 538;* C. H. Knapp et al., and C. J. Messer et al., *TRB Record 644;* and S. C. Tigor and J. W. Hess, *Public Roads,* Sept. 1975.

[50]See, however, F. L. Orcott Jr. and H. A. Walker Jr., *Transportation Engineering,* Jan. 1978 and S. A. Smith, *ITE Journal,* Nov. 1978.

[51]See *Traffic Engineering,* Dec. 1976, and the *Manual* for recommended practice.

cause or duration. Again, in case of accident or a vehicle breakdown, this isolation causes great difficulty both in calling for and rendering assistance.

Until recently, the attention of highway engineers was primarily focused on the planning, design, and construction of this freeway system; but with a large mileage now in use, operating problems have become a major concern. The situation is particularly acute in metropolitan areas where volumes on individual facilities are high and freeway use is great.

The operating ideal is that freeways flow freely at all times, possibly under full automation, with all accidents and breakdowns eliminated. Much research has been undertaken and numerous installations have been or are being made to serve freeway users better. The paragraphs that follow represent a progress report in this fast-moving field. To clarify the presentation it has been subdivided into (1) measures for relieving congestion at peak periods, (2) common means of incident detection and correction, and (3) advanced motorist advisory and control systems.[52]

Relieving Peak-Period Congestion on Freeways[53]

As shown in Chapter 8, freeway traffic in urban and suburban areas commonly peaks during morning and evening commuter hours. Volumes approach capacity day after day. As explained in Chapter 8, at this level of traffic, a few vehicles introduced into or departing from the traffic stream, or a brief slowing, may bring on a forced flow condition and severe congestion. The issue is to alleviate congestion without adding capacity.

The most common cause for breakdowns in the smooth flow of traffic is adding vehicles at on ramps. *Ramp control* is the method now being employed to prevent this flow disruption. With it, entry of vehicles at on ramps is restricted or prohibited when a breakdown of flow on the freeway is or appears to be threatened. It is argued that it is better to inconvenience or delay the few motorists who wish to enter from the ramp rather than the many traveling on the freeway. Also it has been found that the smoother operation that accompanies ramp metering reduces accidents.[54]

The manner in which flow in the traffic stream breaks down is complex, but it usually seems to occur some distance downstream from the point of entry. Apparently both the drivers in the outer lane of the freeway and those entering from the ramp tolerate very small headways in the merging area. However, after the merge has occurred, they find that they are following uncomfortably close behind the preceding vehicle and slow down to increase the following distance. This slowing in turn creates the disruption to smooth flow.

A simple solution to congestion at on ramps is to close the ramp completely before flow on the freeway reaches a critical level. This can be done either

[52]Ways of reducing traffic and the problems of accommodating transit and carpools on freeways are discussed in Chapter 3.
[53]For a review of the overall problem see *Freeway Traffic Management,* published by TRB.
[54]See B. T. Cima, *TRB Record 630*.

when the flow rate on the freeway reaches a predetermined level or by closing the ramp during peak periods. Another approach is to permit one to several vehicles to enter the ramp, get up to speed, and fit themselves into gaps in the flow in the outside freeway lane. More sophisticated computer-control systems also have been developed which pace entering vehicles by means of lights that flash in sequence or with a moving green band. This brings them into the gaps in the outer freeway lane at freeway speed. Unfortunately, this approach has not been satisfactory. With any ramp metering scheme, arrangements must be made to inform motorists of the operating arrangements and to designate alternate routes if ramp use is restricted or prohibited.

Congestion and accidents sometimes occur on freeways near off ramps. Here the problem involves vehicles attempting to weave into a heavily occupied outer lane. This situation often can be corrected by beginning a turnoff lane some distance upstream from the point of exit. Also, traffic may back up on the freeway if the exit ramp is blocked or there is insufficient capacity at its downstream end.

Where service roads or arterial streets parallel freeways it is sometimes possible to use the facilities together at peak hours to gain capacity. This involves coordination with local traffic administrators to see that signals and other traffic controls encourage transfer to the parallel streets. In addition, publicity and detailed directions by signing are imperative.[55]

In some instances, accepted geometric standards have been modified to reduce freeway congestion. For example, capacity was increased in the range of 16 to 22% by restriping three 12-ft lanes and a 10-ft shoulder to provide four 10½-ft lanes and a 4-ft shoulder.[56]

Traditional Means of Incident Detection and Correction

Traffic-related incidents other than congestion involve vehicle breakdown or accidents and loads spilled onto the roadbed. In addition, natural conditions such as rain, snow, or floods sometimes close or constrict sections of road, intersections, and other spot locations. In rural areas, highway agencies seldom make provision for systematically detecting such incidents. Rather, since these are unpredictable, response is made only after the trouble occurs. An exception might be emergency patrols instituted during storms or floods. Commonly, such incidents would be reported and aid provided by motorists, police, or maintenance personnel who happened by. Also, motorists often have access to nearby houses. On conventional streets in cities and towns, means of summoning aid are close at hand and response generally comes quickly.

Freeways and expressways are different. On them the motorists are effectively isolated from the surrounding community. Ramps at which they can leave are widely spaced; they are often several miles apart in rural areas and on toll fa-

[55]See, for example, C. J. Messer et al., *TRB Record 503;* S. Yagar, *TRB Record 562;* and A. D. May and A. J. Kruger, *TRB Record 630.*
[56]See W. R. McCasland, *TRB Record 666.*

cilities, and separated a mile or more in urban situations. Walking on freeway shoulders is hazardous; it is almost impossible on elevated structures that have neither shoulders nor walkways. Also, many motorists, particularly women, do not wish to leave their vehicles because of fear of robbery or attack. Thus, going for aid is not usually a viable solution. An added complication is that modern automobiles are complex and diverse in design. Except for changing tires, roadside repairs by the driver or a passing motorist are difficult; in fact, some 30% of the vehicles have to be towed away.

Highway patrols, particularly on toll roads, have recognized the need for responding to incidents by providing systematic surveillance. Some agencies have installed emergency radio devices or telephones at intervals along the roadside. In Ohio and Michigan, the aid of volunteers with citizen's band radios, particularly truckers, has been enlisted. In some instances, automobile clubs have developed distress signals for their members or even provided patrols. Some agencies hire professional observers. On some major facilities, such as long tunnels, toll bridges, or elevated highways, emergency vehicles and crews stand by on call and remove obstructing vehicles or debris as quickly as possible. Often this is made difficult because of inability to reach the affected location and, sometimes, the need to evacuate people.[57] In less difficult situations, police who reach the scene assume charge. They summon tow trucks or other equipment and an ambulance, if needed, by radio. Then they will keep traffic flowing as freely as possible until the road is cleared. A variety of proposals have been advanced to expedite such clearing operations, but they are not detailed here. All these approaches involve direct human intervention after the incident.

Advanced Motorist Advisory Systems

Freeway interruptions are costly. One estimate places the annual toll at 750 million vehicle-hours, 400 million gallons of fuel, and 41,000 accidents, this in addition to clearing and cleanup.[58] These costs can be reduced not only by relieving congestion but by advising and controlling motorists in a variety of ways.

As discussed earlier, striping and signs ahead of and adjacent to off ramps have been the usual devices for advising freeway users on paths to follow, destinations, and similar matters. These are commonly augmented by "after the incident" intervention by police or special teams that rush to the site. Today, however, some operating agencies are doing far more on a few freeway facilities. For example, they give directions or advice to motorists on such matters as route selection, changed itineraries to meet unexpected developments, and warnings of dangerous conditions. Also, remote controls can divert traffic from freeways, close ramps, or change traffic flow in other ways.

A simple way to direct or control traffic is through the vehicle's radio. For example, messages can be broadcast by commercial stations. Again, drivers can

[57]For a discussion of the operating problems of tunnels and long bridges, see *NCHRP Synthesis 31*.
[58]See S. C. Tigner, *Public Roads,* Sept. 1976.

be told by large signs to tune in on a stated frequency through which information is broadcast. Yet another approach is to transmit messages from a cable laid alongside the roadway.

Much more sophisticated approaches to providing for motorist control or advice are being installed and tested. These begin with a system for monitoring the individual freeway or the freeway system to determine traffic volumes and speeds and to locate incidents that may affect traffic flow. This has been done by observing closed-circuit television screens, with infrared devices that detect stopped or parked vehicles, and with a combination of detectors and on-line computers. An extension of this approach is to electronically track individual vehicles in and out of road sections and note cases where travel time is inconsistent. Output from such tracking-computer schemes could be vehicle volumes and speeds or exception reports that point out drastic volume or speed reductions. Personnel or possibly a computer at a central control station would review incoming information and, when necessary, alert police and others and give directions to motorists through appropriate changeable message signs.[59]

Systems with differing features have been installed on freeway networks in several U.S. cities. As an example, the Chicago scheme involves 196 directional miles on eight major expressways, with 1300 detector locations and 48 metered ramps. A real-time computer processes the information so that actions can be taken. In addition, results go to local radio and television stations by teleprinter. A Los Angeles system involves 84 m of freeway, with 52 ramps controlled, some of which have priority lanes for buses and carpools. The Los Angeles scheme also incorporates variable message signs.[60]

Research to develop complete motorist guidance systems and their components also has been undertaken. The overall concept is that, through a series of communications, advice would be given as to the most advantageous route to follow all the way from origin to destination, including parking. Messages delivered before motorists begin their journeys or while they are enroute indicate traffic conditions and direct them to new routes that avoid congested areas. Installation of a workable system of this sort lies some time in the future, but development of some of its components such as traffic monitoring systems, computers to analyze data, changeable message signs (see above), and radio communication to give oral directions or operate visual displays in the vehicle is well along. It is to be expected that soon these guidance and other emergency systems will be combined with the area-wide real-time control systems described below to make traffic flow more smoothly.

[59]The literature on these advanced systems is extensive. An excellent bibliography appears in *Transportation Engineering Journal of ASCE,* Nov. 1978. Other recent references include *NCHRP Report 169, TRB Special Report 128 and 153,* articles in *TRB Record 495, 503, 533, 536, 600, 601, 643,* and *737,* and *Report FHWA 77-47, 48, 49.*

[60]See S. C. Tignor, *Public Roads,* Sept. 1976 and, with H. J. Payne, ibid., June 1977. For a description of the system on the New Jersey Turnpike see P. M. Weckesser and K. W. Dodge, *Traffic Engineering,* Apr. 1976; and, for Baltimore, J. W. Erdman and C. I. Barnfield, ibid., Jul. 1976.

HIGHWAY AND STREET OPERATION IN URBAN AREAS

Traffic in urban areas flows on a combined network of freeways, expressways, arterials, collectors, and local streets. In operating the arterials and local streets, traffic engineers make use of pavement markings, signs, and signals. These signals are controlled in a variety of ways, ranging from fixed settings to computers. A summary of these approaches follows.

Arterial and Collector Streets

Before the coming of freeways and expressways, arterial routes were the principal carriers of through traffic. Even with a fully developed freeway and expressway system, arterials carry traffic to the nearest access points; often they offer the most advantageous routes for relatively long trips. Most arterials are existing highways or streets of considerable length along which cross traffic is regulated by signals or stop signs. In addition, arterials provide access to adjacent property, but often with restrictions on entry and exit locations, street parking, cross-centerline left turns,[61] and other conflict activities. With a properly selected network of arterials it is usually possible to increase the capacity of the entire street system, reduce delays to through and local traffic, and decrease accidents. Even so, arterial streets are considered a "make do" substitute for controlled access facilities when traffic volumes exceed about 20,000 vehicles per day.

Arterial routes and through streets are usually incorporated into the master plan for the community. Although there are no fixed rules regarding selection of routes, studies of traffic volumes, origins and destinations, and accident experience are basic tools for planning them. Suggested criteria include the following.

1. Arterials should be at least 48 ft wide. They must carry at least one lane of traffic in each direction with parking and two without. They should be at least ½ mi in length.
2. Arterials should skirt neighborhood areas rather than penetrate them.
3. In a grid system of streets, arterials should be spaced about 2000 to 3000 ft apart.[62]
4. Where accident hazard is not a factor, minimum volumes to justify arterials are 300 vehicles per average hour during the day and 450 vehicles hourly during peak periods.

In many instances, peak-hour traffic volumes on existing streets designated as

[61]Where roadway width permits, cross-centerline left turns, except at specified locations, can be prevented by installing a central, continuous raised island. If this island is wide enough, a protected left-turn lane can be provided in it. Pavement markings alone provide a less positive means to accomplish this end. Another alternative is to mark off a single center lane from which left turns in either direction are permitted (see the *Manual* and *TRB Record 737* for details). Often the parking space provided in the original design is eliminated to provide the width needed while still retaining all the traffic lanes.

[62]See *NCHRP Report 187* for a method for analyzing this question in depth.

arterials exceed the capacity of the conventional two-way streets; however, a variety of techniques may aid in increasing their capacity. Among these are[63]:

1. Prohibition of parking during peak hours on one or both sides of the street. Sometimes this parking ban extends throughout the day.
2. Provision of an extra lane adjacent to signalized intersections to bring intersection capacity nearer to street capacity between intersections. This is often accomplished by eliminating parking for several hundred feet on each side of the intersection.
3. Permitting right turns on red or making special provision for these turns outside the through lanes.[64]
4. Eliminating left turns at congested intersections.
5. Reversing the direction of traffic in the center lanes to provide more lanes in the direction of heavier flow.

In most modern installations traffic is directed into or away from the reversible lanes by means of overhead signal lights.[65] At times these may be replaced or augmented by vertical markers that are placed and removed by special crews. As an alternative to reversible lanes, it sometimes is feasible to make an entire street reversible.

Collector streets, if used, form a smaller mesh grid which picks up traffic from service streets and carries it to arterials. Standards and controls for them are less exacting than for arterials.

In residential areas built in a gridiron pattern without arterials, certain streets may be heavily traveled. Again, drivers sometimes elect to travel streets that parallel crowded arterials or collectors. At times stops signs have been installed at frequent intervals to slow traffic. As noted, these are not very effective. A better approach may often be to block these movements with physical barriers (see Chapter 12).

Large industrial plants, commercial enterprises, or amusement facilities such as drive-in theaters almost always front on arterial streets. These create sudden traffic peaks, and often present serious congestion and accident situations unless special provisions such as left-turn lanes or traffic lights are installed to serve them.

One-Way Streets

One-way streets are those on which vehicular traffic moves in only one direction. In many cities much of the downtown street grid is operated on a one-way

[63]For a more detailed listing see W. R. McShane and L. G. Pignataro, *TRB Record 597;* E. M. Hall et al., *TRB Special Report 153;* and *NCHRP Report 110, 113,* and *169.*

[64]All states permit right turns on red (RTOR) after a stop unless they are specifically signed against. H. W. McGee, *TRB Record 644,* reports that accidents with this maneuver were 3% of the totals for vehicles and for pedestrians. However, *Status Report* of the Insurance Institute of Highway Safety, Dec. 9, 1980, reports a 20% increase in crashes involving right turns at signalized intersections with RTOR and a 57% increase in pedestrians struck. See also articles by W. E. Baumgaertner and D. Galin, *ITE Journal,* Jan. 1981. For guidelines for RTOR see H. W. McGee, *Transportation Engineering,* Jan. 1978.

[65]See *Traffic Engineering,* Jan. 1977, and the *Manual* for details.

basis with opposing traffic using alternate streets. In numerous other locations, pairs of one-way streets serve as major traffic arteries.

The widespread adoption of the one-way traffic plan stems from a number of important advantages over two-way operation, some of which are:

1. *Greater capacity*. More vehicles can be accommodated by the same street system (see Figs. 8–13 and 8–14).
2. *Increased average speed and fewer stops*. Progressive signals can be timed to give a full-width through band on each one-way street, even where block spacing is irregular. There are fewer delays at intersections because the number of possible conflicts is greatly reduced. This is illustrated for two-lane streets by Fig. 10–5. At intermediate unsignalized intersections traffic can cross freely during the breaks in through traffic.[66]
3. *Improved pedestrian movement*. At signal-controlled intersections of two one-way streets, one crosswalk is completely free from turning vehicles during each phase of the signal. Turns across the opposite crosswalk are from one direction rather than from two (see Fig. 10–5). At unsignalized intersections and midblock locations, pedestrians can cross during the breaks in traffic.
4. *Reduction in accidents*. By the elimination of conflicts listed here, one-way operation reduces most types of accidents.
5. *Other advantages*. Among these are the elimination of headlight glare, greater ease of movement for emergency vehicles, and a possible reduction in police attention to traffic.

One-way operation often presents difficulties of at least temporary nature. Transit routings must be revised or a contra flow lane reserved for operation in the original direction. Travel distances to reach certain locations are often increased, a condition that may seriously affect particular businesses. For example, a food store may lose most of its patrons if the street on which it fronts is made one-way inbound so that potential customers pass it only in the morning. Sometimes accidents increase, particularly when traffic is sped up.

Proposals to install one-way streets often meet with opposition from business-

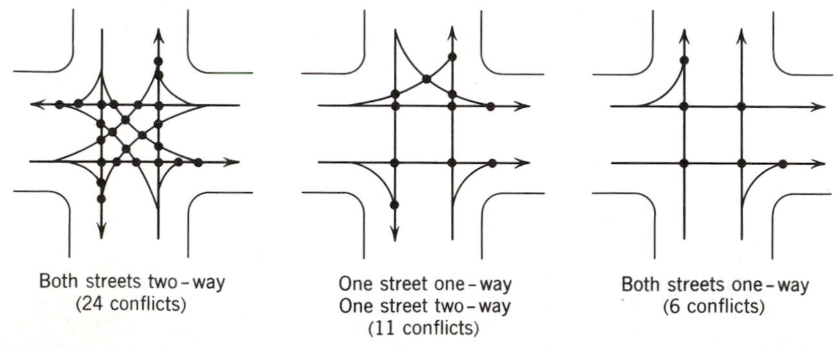

Fig. 10–5. Potential conflicts between vehicles at an intersection of two 2-lane streets; 2-way versus 1-way operation.

[66]See R. L. Vecellio, *Transportation Engineering Journal of ASCE*, Sept. 1977 for a study of movements on one-way streets.

people and others who fear that their interests will be adversely affected. Sometimes the initial step is to try a one-way operation on one pair of streets. Only after these have been accepted by the public can permission be obtained to expand the plan. There have been numerous instances where objections have prevented so much as a trial of a one-way street plan. In other cases public opposition has forced city officials to abolish the plan after a trial. In at least one large city one-way streets were installed, forced out by business interests, and finally restored after motorists had deluged city officials with protests.

"One-way preference" streets, in pairs, have also been used as a compromise measure. With them, more lanes of a street are given to the favored direction and traffic signals are timed to provide it with wide through bands.

Traffic Control

Stop and yield signs and traffic signals are the most important control devices for street operation. The *Manual* calls for two-way stops on the less important of two intersecting streets where reliance on the right-of-way rule, high speeds, or restricted sight distance indicates high accident hazard. It states that multiway (4-way) stop sign installations should be used only where volumes on the intersecting roads are approximately equal, and only when any one of the following "warrants" or qualifications is met.

1. As an interim measure when traffic signals are needed.
2. When an accident problem (as demonstrated by five or more accidents in a year) can be corrected by multiway stop signs.
3. When total vehicular volume of 500/hr enters the intersection for 8 hr an average day, combined vehicular and pedestrian volume of 200/hr from the minor highway, and with an average delay to minor street vehicles of 30 s. Volume warrants are reduced by 30% when the 85-percentile approach speed on the major highway or street exceeds 40 mph.

Often there is pressure from the public and its elected officials to install stop signs where the situation does not justify them. As already indicated, there is strong evidence that drivers tend to ignore such controls. In particular, four-way stop sign installations often have been used where they cannot be justified economically or on the basis of proven accident data (see Chapter 4).

The "yield" sign as a compromise between the full stop and no control is recommended by the *Manual* under certain conditions. Its merits include decreased operating costs, lower contributions to air pollution, passage time reductions of 2 to 6 s, and, in some instances, lower accident frequency.[67]

The *Manual* also provides minimum warrants for traffic signal installations based on vehicular or pedestrian volumes, accident experience, progressive movement, interruption to continuous traffic to permit cross traffic to move as a part of a network, or a combination of these. Those involving vehicular vol-

[67]See *NCHRP Report 41* and *ITE Journal*, Oct. 1978.

umes alone and for interruption of continuous traffic are given in Table 10–2.[68]

As with stop sign installations and unrealistically low speed limits, public pressures often force the installation of traffic signals. Often they lead to an increase in accidents.[69]

Area-Wide Operation of Urban Highways and Arterials

Today, systems for optimizing traffic movements on freeways and arterials are in operation or being installed in many urban areas, some 135 in the United States alone. There are almost no standards, and levels of sophistication vary widely. The simpler ones are pretimed; that is, they reset signal cycle times and splits for individual intersections by time of day. The most complex gather data on traffic volumes and speeds from detectors placed at strategic locations, transmit this data by leased telephone lines or radio, process the information with a computer operating on real time, and reset all the signals every few minutes to optimize traffic flow. Of intermediate complexity are computer "look-up" schemes which choose among a number of preprogrammed alternatives.

At the heart of the more complex systems are computers programmed to (1) assimilate the data on traffic flow rates and speeds, (2) apply a real-time simulation program to this information in order to develop optimum cycle times, phasing, and cycle splits for signals at individual intersections, and (3) direct the resetting of all the signals. Overrides to favor buses or other multiple-occupant vehicles can be included. These programs are far too complex to describe here. Among them are TRANSYT, from Great Britain, and SIGOP, developed in the United States.[70] Of these urban traffic-control systems, one of the most elaborate is that for New York. It will start with 800 intersections under unified control, but will have 8000 when fully implemented. Such systems seem to be well justified on economic, energy, and environmental grounds. An operating system in Toronto, Canada, which involves 864 intersections, showed savings as follows: delays, 20%; stops, 53%; accidents, 13%; and travel time, 44%.[71] However, it is essential that some system for incident detection and control, as discussed earlier, be included in the overall scheme.

[68]Extensive research leading to revised warrants has been sponsored by NCHRP. For a bibliography, see *NCHRP Research Results Digest 78*. K. R. Agent and R. C. Deen, *TRB Record 737,* offer warrants for left-turn phasing. C. F. King, *TRB Record 629,* proposes revised pedestrian warrants.

[69]D. Solomon, *Public Roads,* Oct. 1959, and G. F. King and R. B. Goldblatt, *TRB Record 540,* report a change in pattern with rear-end, head-on and sideswipe accidents increasing and angle and miscellaneous classes decreasing. Accidents decreased with signal installations on five- and six-leg intersections. T. T. Wiley, *Traffic Engineering,* May 1958, reports that in one large city accidents at school crossings were seven times as numerous with traffic signal controls as with other forms.

[70]For a few of the many descriptions of and comparisons between these systems and of traffic simulation models in general, see the entire issue of *Traffic Engineering,* Apr. 1975; E. L. Cleary, ibid., April 1977; T. P. Weldon and P. S. Parsonson in *Transportation Engineering,* Oct. 1977; D. G. Gibson and P. Ross, ibid., Dec. 1977; *Public Roads,* Sept. 1977, Mar. and June 1978, Sept. and Dec. 1979, and Mar. 1980; and *TRB Record 509, 531, 536, 538, 567, 596, 597,* and *682*.

[71]Articles describing some of these systems appear in *Traffic Engineering,* Apr. 1975, Jul. 1976, Apr. 1977, and *ITE Journal,* Jul. 1978 and Feb. 1979. See also *NCHRP Reports 29, 73,* and *124*. For details of the downtown Chicago system see R. Q. Pool, *Transportation Engineering Journal of ASCE,* Jul. 1979.

TABLE 10–2. Minimum Vehicular Volumes Warranting Traffic Signals*

Number of Lanes for Moving Traffic on Each Approach		*Minimum Vehicular Volume for Each of Any 8 hr/day*			
		Warrant Based Solely on Volume†		*Warrant for Interruption to Continuous Traffic†*	
Major Street	*Minor Street*	*On Major Street (Total Both Approaches)*	*On Minor Street (One Direction Only)*	*On Major Street (Total Both Approaches)*	*On Minor Street (One Direction Only)*
1	1	500	150	750	75
2 or more	1	600	150	900	75
2 or more	2 or more	600	200	900	100
1	2 or more	500	200	750	100

*Based on the *Manual on Uniform Traffic Control Devices.*

†When the eighty-fifth percentile speed of the major street exceeds 40 mph, or the intersection lies within a built up area of an isolated community having a population of 10,000 or less, the minimum vehicular volume warrant is 70% of the stated standards.

Regardless of the sophistication or lack of it in the traffic control system, networks in most urban areas will occasionally become oversaturated and traffic will bog down. *NCHRP Report 194* proposes both signal timing and nonsignal approaches to alleviate these situations.

RURAL HIGHWAY AND ROAD OPERATION

A limited mileage of rural highways carrying large traffic volumes have been or will be improved to freeway standards, and the preceding discussion on freeway and expressway operation generally applies to them. But, as indicated in Chapter 2, most of the rural mileages consists of roads carrying relatively small numbers of vehicles. Undoubtedly these roads will continue to operate with little surveillance. Activities of the individual highway agencies will consist primarily of physical maintenance, striping, signing, and the installation of traffic signals, flashers, or fixed lighting at important or dangerous intersections and other trouble spots.[72]

Standard practices for striping and signing rural roads are given in the *Manual*. For example, a dashed yellow stripe designates the centerline where passing is permitted. A solid yellow stripe on the near side indicates "no passing" for that direction; two solid yellow stripes indicate "no passing in either direction." Lengths of these zones for 85 percentile speeds are plotted on Fig. 8–18 of this book. It should be noted that the "short-zone" sight distance requirements are roughly one-half those employed in geometric design.[73]

There are many miles of rural road on which, because of restricted sight distance, passing is prohibited by striping or signing. Studies have been made to explore the feasibility of electronic motorist information systems which would make passing possible where sight distance is restricted. In one experiment, roadside detectors located approaching automobiles; the driver of the test vehicle was informed that the road was clear for 1500 ft ahead by a continuous tone on the car radio.[74] In some extreme cases, mirrors have been installed to show drivers opposing vehicles approaching over crests, around sharp curves, or at blind intersections.

Where rural roads intersect, it is common practice to install stop or yield signs on some or all approaches. In some cases, there are traffic signals. Warrants for stop signs and signals generally are the same as given in Table 10–2 for streets. An exception is that, in recognition that urban and rural environments differ, volume warrants for traffic signals may be reduced by 30% in isolated communities.

Enforcing speed limits and other traffic rules is difficult in rural areas since

[72]F. W. Walker and S. E. Roberts, *TRB Record 562,* cite one study that shows a 52% nighttime accident reduction with fixed illumination.

[73]For recent studies and bibliographies dealing with the marking of "no passing" zones see G. W. Van Valkenburg and H. L. Michael, *HRB Record 366,* and R. J. Waldorf, *Traffic Engineering,* Feb. 1977.

[74]See D. Niebur, *Public Roads,* Aug. 1968.

motorists know that the chances of detection and arrest are slight. For example, with four different sign conformations, motorists traveled 40 mph in 15-mph school zones. With flashing beacons and "speed violations" signs, speeds dropped only to 34 mph.[75] It seems clear that, in rural areas, unsafe behavior must be largely controlled by perceived threats to the motorist's personal safety or well being.

NIGHT VISIBILITY—HIGHWAY ILLUMINATION

Motor-vehicle headlights provide the only illumination for most of the rural highway mileage, but fixed lighting at important intersections and points of hazard is becoming more and more common. In some instances, high-volume rural freeways have continuous illumination, although this practice has been questioned. In urban areas, on the other hand, fixed lighting for main arteries and residential streets alike is widely found. Here, where the population density is much greater, added benefits, such as improved police protection and freer, safer pedestrian movement, offer further arguments favoring street lights.

Roughly 45% of the highway fatalities occur during daylight and 55% during darkness. On the basis of vehicle-miles driven, however, the accident rate in both urban and rural areas is several times as great at night as during the day. This is explained in part by the fact that a greater percentage of night drivers, and particularly those in the early morning hours, may have been drinking or are fatigued. On the other hand, some 30% of the increase can be attributable to poorer nighttime visibility.

Principles of Night Visibility

Nighttime seeing under headlights or fixed illumination involves many complexities. First of all, the level of brightness of road or objects may vary roughly from 0.003 to 8 foot-lamberts (fl) a multiple of 2000 to 3000. This range covers much of the adaptive power of the eye and involves reaction through both its cone and rod mechanisms. Again, the driver's vision must be rapid and gained at high speeds, with the pattern of viewing varying with the situation being observed. Furthermore, contrast in brightness between object and background assumes extreme importance at low levels of illumination. Then too, there are wide variations in seeing ability among individuals. For example, drinking drivers observe nighttime situations less quickly. Again, the eye adaptability to changes in brightness at low levels of illumination decreases substantially with increases in age and as the level of illumination changes.[76]

In night-driving situations, the manner in which seeing occurs varies with both the absolute level of brightness and the relative brightness of road surface

[75]See R. J. Rosenbaum et al., *TRB Record 541.*

[76]For greater detail and an extensive bibliography on the complexities of nighttime seeing, see the series of 12 literature summaries by O. W. Richards. The last three of these appear in *HRB Record 70, 164,* and *179.* Other discussions appear in *NCHRP Report 99, HRB Record 216* and *377, Public Roads,* June 1979, and *TRB Special Report 156.*

and object. When an object appears darker than its background, discernment is by *silhouette*. If the object is brighter than its immediate background, seeing is by *reverse silhouette*. When direct illumination of about 1 foot-candle (fc) intensity is provided on the side facing the driver, variations in brightness permit discernment by *surface detail,* without general contrast with the background.

Under headlight illumination, the upper portions of persons and vehicles appear in reverse silhouette. Here the reflective quality of the object being viewed assumes particular importance. The reflection factor of a white, diffusing surface is about 98%. For light gray objects however, the factor is about 14%, for medium gray about 7%, and for black only about 3%. Persons in white clothing against dark backgrounds can be seen twice as far away as those in dark clothing. On the other hand, the reflecting quality of the pavement has particular importance under headlight illumination, since the lower portions of persons and vehicles and most other hazards first appear as dark objects in silhouette against a brighter roadway. A light-colored, rough-textured pavement or seal coat that reflects light back to the driver is highly desirable. Surfaces that become mirror-like when wet are particularly to be avoided where illumination is by headlight, for almost no light reflects back from them toward the light source and seeing by silhouette is poor.

Data such as those cited above indicate that pedestrians and bicyclists can do much to protect themselves by wearing light-colored clothing or reflective devices. If dress is dark and nonreflective, the pedestrian can achieve visibility by displaying an unfolded white pocket handkerchief.

Effectiveness of Headlights

Motor-vehicle headlights have been improved in successive steps. In 1940 the sealed beam was introduced as a standard component of all cars. It was followed in 1955 by an improved sealed beam and in 1957 by the dual-headlamp system. Sealed beams for two-lamp cars were improved further in 1958 and 1970 and for dual-lamp systems in 1970. In 1978, permissible candlepower on high beams was increased from 75,000 to 150,000. These halogen lamp installations improve seeing by 20%, but are only available as replacements on four-lamp automobiles. Further improvements under discussion and test include (1) those to provide better aiming by more careful installation, better inspection, and required load-leveling devices; (2) adding a third beam with a sharp cutoff of its higher portion to reduce glare when vehicles meet on two-lane roads; and (3) introducing polarized headlighting. The first of these, better aiming, seems to offer primarily administrative and enforcement rather than technical problems. Adding a third beam means higher cost, vehicle redesign to provide space for a far more complex headlamp, and a control system considerably more elaborate than the simple button on the floor. Operating it might be so confusing to drivers that its purpose would be defeated.[77] The polarized light system is discussed below as a separate topic.

[77]See B. Adler and H. Lunenfeld, *TRB Record 502* and *Special Report 156*.

Even with today's improved headlamps, nighttime speeds on two-lane roads under headlight illumination are generally too fast for safety. This fact is illustrated by Fig. 10–6 which relates detection distances to cumulative percentages of drivers for two situations: (1) a pedestrian in light clothing (17% reflectance) positioned 2 ft from the edge of the driving lane and (2) a reflecting sign (96% reflectance) positioned 6 ft back.

From Fig. 10–6 it can be seen that one driver in five, while in the process of meeting another vehicle, would not detect a pedestrian at the roadside in time to stop from 30 mph, and over half could not stop in time from 40 mph. And these data assume 17% reflectance. As indicated above, if the pedestrian were in dark clothing or the hazard were a dark colored animal, this detection distance would be cut in half. Thus, unless the object is silhouetted in oncoming headlamps, the chances of colliding with it would be substantially greater than those stated above. Also, because glare recovery takes time, the effects on seeing which result from a meeting with an opposing vehicle apply for almost half a mile; they begin over 2000 ft before the meeting and continue possibly 600 ft beyond. Even worse, seeing distance may be reduced to as little as 100

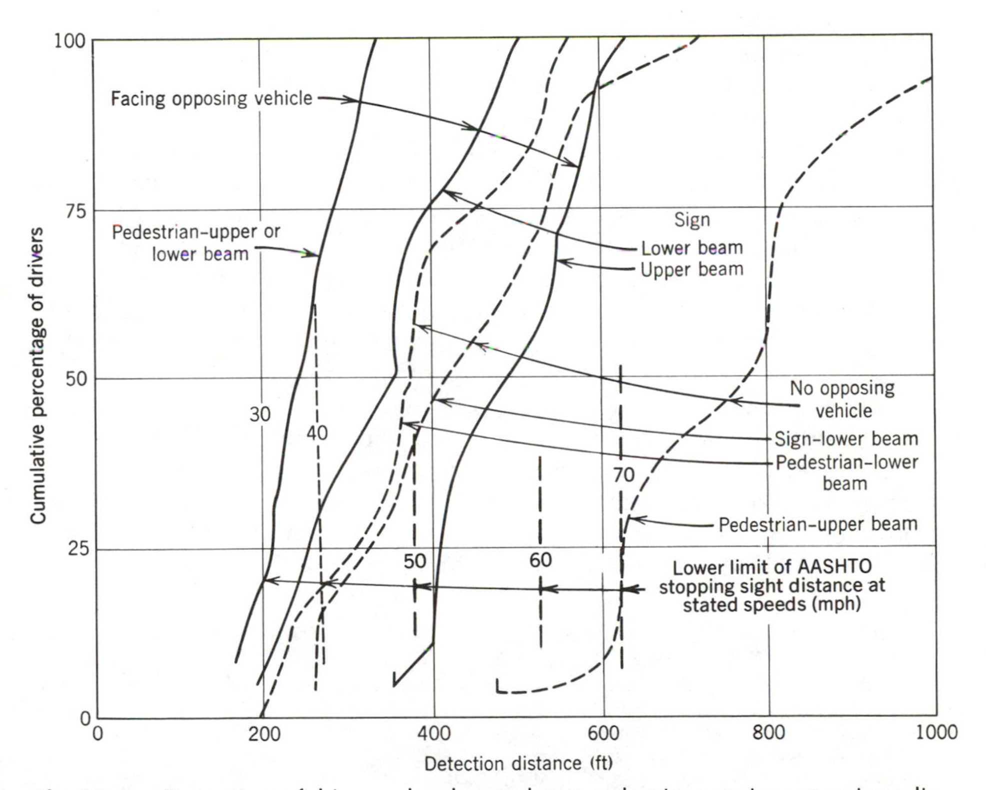

Fig. 10–6. Percentage of drivers who do not detect pedestrians or signs at various distances under headlight illumination. Headlamps of both vehicles on same beam. (Source: R. N. Schwab and R. H. Hemion, *HRB Record 327*.)

ft during the near approach of a vehicle with upper beam headlamps.[78] On divided highways, motorists can be screened from opposing headlamps with planting or a glare-reduction screen mounted on the median barrier. Unfortunately, there is no such remedy on two-lane roads.

It can also be concluded from Fig. 10–6 that, even where there are no opposing headlights, many motorists driving with lower beams may not detect dark objects in time to avoid striking them. Additional reductions in visibility such as from tinted, dirty, fogged or rain-wet windshields, and tinted contact lenses also reduce night seeing by various amounts.[79]

Research on driver eye movements under headlight illumination has found yet another contributing factor to nighttime accidents. It is that, in setting vehicle speeds, drivers focus primary attention on the roadside only 75 to 100 ft ahead of the vehicle. Thus, drivers do not take full advantage of headlight illumination, limited as it is.[80] There are other substantial differences in daytime and nighttime eye movements. Also, drinking greatly impairs seeing under headlamps.[81]

The *polarized headlight system* has in the past and again, after a lull, received particular attention as a means for reducing headlight glare from opposing vehicles. With it, the light beam would be polarized into 45 planes rising upward to the right. The driver would view the polarized light from opposing headlamps through a movable polarized visor, called an analyzer. It also would, when observed from the driver's side, polarize the light into planes slanting 45° upward to the right. The effect of these orientations is almost completely to block out light from opposing vehicles. On the other hand, polarized light reflected from objects illuminated by the driver's headlamps would not be blocked out.

Arguments favoring the polarized headlamp system include better night visibility which means higher safe driving speeds, greater driver comfort, and reduced fatigue. Arguments against it include higher costs, much greater (2½ to 3 times) energy requirement with the attendant problems of heat dissipation, a serious diminution of the light from tail lights of preceding vehicles, and the loss of seeing by silhouette in the lights of approaching vehicles.[82]

Light intensities in tail lights and turning and stop indicators on vehicles have been substantially increased over the years to improve visibility. However, their placement has been largely unchanged. Studies have been made to test operator reaction time for various geometric arrangements of stop and turning rear-end lights and combinations of red, amber, and green colors. Several were found to

[78]See *TRB Special Report 156,* R. L. Austin et al., *TRB Record 540,* and V. D. Bhise et al., *TRB Record 611* for details and extensive bibliographies. G. Helmers and K. Rumar, *TRB Record 502,* report findings on the relationship between obstacle visibility and the reflective quality of the roadway surface.

[79]For references on sign visibility under headlamps, see the section on signs.

[80]See *NCHRP Report 99* and *TRB Special Report 156.*

[81]See, for example, R. D. Hazlett and M. J. Allen, *HRB Record 216.*

[82]See R. N. Schwab and R. H. Hemion and discussions by M. J. Allen and R. W. Oyler, *HRB Record 377.*

be more effective than those commonly employed.[83] Again, taxicabs with brake lights mounted at one level on top of the trunk had 54% fewer rear-end accidents.[84] On the other hand, reflective license plates,[85] side-mounted turn indicators,[86] or accelerator position indicators[87] did not prove to be desirable.

Fixed Highway and Street Lighting[88]

It is accepted practice in the United States to provide lighting from fixed sources for almost all streets in urban areas. Also, many urban freeways and interchanges and some rural interchanges and intersections are illuminated. The 1966 Highway Safety Act required that the individual states develop uniform standards for highway lighting. The AASHTO publication *An Informational Guide for Roadway Lighting, NCHRP Report 152,* and the 1977 American National Standards represent current responses. However, pressures for energy conservation and from high energy costs are forcing many agencies to use lower than stipulated values.[89]

For freeways, the *Guide* does not recommend continuous fixed lighting in rural areas. Furthermore, it calls for full or partial lighting of rural interchanges only with heavy traffic volumes or where delineation might be needed in diverging and merging areas or at ramp terminals. It exempts from fixed lighting interchanges where there is ample room for signing and the layout and detail are typical. For urban freeways, the *Guide* presents a series of warrants too lengthy to present here. In sum, they call for lighting where the freeway or interchange is in a developed area that is lighted, where three or more interchanges occur in 1.5 mi, or where local government finds sufficient benefits to pay some or all of the costs. Fixed interchange lighting is considered to be warranted when the adjacent freeway or surrounding area is lighted or where total ramp traffic or that on the crossroad exceeds 10,000 vehicles daily for urban, 8000 for suburban, or 5000 for rural surroundings.

Highway and street lighting, except on major downtown arteries, generally is designed to illuminate the roadway and thus provide seeing by silhouette. For freeways and interchanges, the *Guide* recommends an average horizontal illumination level of 0.6 to 0.8 fc when the light source is at its lowest output because of age or dirt. This calls for about 1.0 fc at the time of installation. The

[83]See R. G. Mortimer, *HRB Record 275*.
[84]See Institute for Highway Safety, *Status Report,* Mar. 2, 1978.
[85]See C. B. Stoke and W. L. Sacks, *TRB Record 502*.
[86]See E. I. Farber et al., *TRB Record 600*.
[87]See R. G. Mortimer and S. P. Sturgis, *TRB Record 600*.
[88]The most valuable sources of current information on highway and street lighting are the *Roadway Lighting Handbook,* FHWA and the *Journal of the Illuminating Society of America*. The Oct. 1977 issues gives the current "American National Standard for Roadway Lighting" of the American National Standards Institute under the sponsorship of the Illuminating Engineering Society. See also *TRB Record 737*.
[89]For example, the state of Oregon has stopped illuminating freeways carrying fewer than 40,000 vehicles per day.

uniformity ratio (average illumination divided by lowest illumination) is to be 3:1 to 4:1.[90] The *Guide* also proposes "adaptation lighting" on the "leaving" end of continuously lighted freeways to provide an opportunity for eye adjustment to headlight illumination.

The justification of continuous freeway lighting in terms of economic and other social gains has been neither proved nor disproved. On two occasions, detailed studies of driver behavior under continuous lighting along sections of the Connecticut Turnpike have been undertaken to determine its cost and to assess the benefits in terms of greater comfort, improved traffic operation, increased capacity, and reduced accidents, but the results did not argue conclusively for or against lighting.[91]

Proposed illumination levels for highways other than freeways and for streets are based on the lowest level of lamp output. For expressways, the standards propose 1.0 fc in residential, 1.4 in intermediate, and 2.0 in downtown areas. In these situations, seeing is by surface detail. For sidewalks in commercial areas, recommended level is 1.0 fc, and for residential streets the value is 0.4. Requirements for collector facilities fall between those for arterials and for minor streets. As for freeways, the uniformity ratio is set at 3:1 or 4:1.

The *Guide* also recommends illumination levels for low mounted bridge railings and the walls of tunnels and underpasses. Some notion of the adaptive power of the human eye is given by the requirements for tunnels over 500 ft long. Multipliers of 1/10 to 1/15 progressively reduce illumination levels from possibly 10,000 fc to a minimum of 5.[92]

Highway Lighting Installations

LIGHT SOURCES. Earlier, fixed sources for street and highway lighting were of the mercury-vapor, incandescent (filament), or fluorescent types. Today new and more economical types include high- and low-pressure sodium and metallic halide (mercury with iodides added). The trend is toward high-pressure sodium.[93] Common wattages for all types range from about 175 to 1000. All of these produce about the same visibility for the same level of illumination, since color differences do not materially affect vision.

[90]Criteria for fixed lighting are stated in terms of a given level of *illuminance* on a horizontal surface. But, with seeing by silhouette, the *luminance* or light flux leaving the surface in the viewer's direction is the important measure. It has been stated that luminance can vary by a factor of three with such factors as pavement surface characteristics and wet versus dry. Illumination engineers are aware that the illuminance criteria are unsatisfactory, but to date have felt obliged to retain them for want of a better approach. See, for example, L. E. King et al., *HRB Special Report 134* and W. J. M. van Bommel, *Journal of the Illuminating Society of America,* Oct. 1978.

[91]See A. Taragin and B. M. Rudy, *Public Roads,* Aug. 1960, and *NCHRP Report 60* for reports of those studies. The second study found, among other things, that the differences between an illumination level of 0.2 and 0.6 fc were not readily discernible to the human eye. *FHWA Reports RD77-37* and *38* state that the effects of lighting on freeway accidents is not clear.

[92]For authoritative discussion of tunnel lighting, see J. A. Thompson and B. I. Fansler, *Public Roads,* Oct. 1968.

[93]See *FHWA Reports RD77-37* and *38,* and *TRB Circular 173.*

Fluorescent lamps often are mounted below eye level to provide continuous lighting along the sides of a roadway, on bridges, or on the walls of tunnels or underpasses. Because of the length and number of lamps, these fixtures are quite large.

LUMINAIRES. Luminaires distribute light from the sources into definite patterns that best suit particular situations. For example, beams can be concentrated in two directions along the street; a symmetrical pattern is used when the fixture is centered in the street or an asymmetrical form when mounting is near the curb. Four-way or uniform distributions may be used at intersections. Recommended practice places luminaires 40 ft or more above the roadway, although at present the predominant mounting height is between 25 and 35 ft. With high mountings, a more uniform illumination can be maintained even though individual units are widely spaced. High mounting also greatly reduces the blinding effect of direct glare. Of necessity, individual lamps are larger with wide spacings. For the higher mountings, recommended spacing may be in the 200-ft range, with fixtures placed on both sides of the road. On curves, more than half and sometimes all the units are placed on the outside. The usual practice is to suspend the luminaires over the roadway, sometimes on cables and again on mast arms extending outward from the roadside. Sidewalk illumination is gained by proper luminaire selection.

The relative economy and effects on accidents of various lighting installations has been the subject of numerous studies. Variables include type and brilliance of light source, luminaire height and spacing, and installation, maintenance, and, in some cases, accident costs.[94]

Increasing attention is being focused on accidents when motor vehicles strike luminaire supports. One argument favoring brighter lamps on wider spacings is that the chance of collision with their supports is decreased. Also, as discussed under sign supports, much attention has been focused on designing them with lightweight materials and breakaway bases.

For interchanges, there is a growing trend toward mounting a few luminaires on high poles, some as high as 150 ft. For example, in Houston, Texas, four 10,000-W luminaires on 150 ft supports replaced 96 units.[95]

Trees are a definite asset to a community, and their mutilation or removal to provide adequate street lighting is usually unwarranted. Much can be done by cooperation between the agencies charged with the two responsibilities. For example, luminaire position, height, and spacing can be fitted to tree-planting patterns. Again, new trees can be selected from the globeheaded or upright types that do not conflict so seriously with lighting installations.

[94]See, for example, articles in *HRB Record 377* and *416, FHWA Reports 77-37* and *38,* and M. S. Janoff et al., *Journal of the Illuminating Society,* Jan. 1978. P. C. Box, *Traffic Engineering,* Oct. 1976, found a benefit-cost ratio of 4:1 favoring full- versus half-on lighting on a major artery. F. W. Walker and S. E. Roberts, *TRB Record 562,* found that lighting reduced nighttime accidents at rural intersections by 51%; from 1.89 to 0.91 per million entering vehicles.

[95]For a discussion of such systems, see R. C. LeVere, *Civil Engineering,* May 1971.

PARKING[96]

Widespread automobile ownership has brought serious parking problems to all urban locations. In the older residential areas of many cities, every available space is often filled both day and night with parked vehicles. Residents have difficulty finding space, and vehicles left overnight make street cleaning and policing difficult. The older shopping areas which do not have adequate parking space for their customers have suffered substantial losses in patronage to suburban shopping centers. Street congestion has been particularly acute in and near central business areas of the larger cities as more and more people have chosen to drive downtown rather than to travel by public transportation. How this situation will change, if at all, with gasoline shortages and high fuel costs is an unanswered question.

Trucks also affect parking, but in different ways. Truck or rail-truck terminals for long-haul vehicles are being placed in outlying or industrial areas to avoid downtown congestion. Parking, if needed, is provided off the street.[97] Goods delivery vehicles offer different problems. For newer buildings and at shopping centers truck parking is provided off street. However, short-term parking for pickup and delivery during daytime, particularly to small businesses in downtown areas, can greatly complicate vehicle movements by blocking lanes and consuming parking space. One study found that delivery trucks were actually driven only an hour a day. However, they made about ten deliveries and averaged a half hour per stop. To accommodate these vehicles, curb space often has been reserved solely for deliveries. For example, one city devoted 6% of its curb space to delivery parking. It reported that only 1.5% of the trucks were illegally parked. In another city, however, almost half the trucks were parked illegally. To relieve such situations, efforts have been made to require deliveries at night. But this means providing added staff which is particularly difficult for small businesses. Among the most costly proposals to get trucks off downtown streets is to provide separate underground roadways and unloading sites.[98]

Among the reasons for today's serious condition is that parking has often been "everybody's business." Responsibility for providing and controlling it often is divided among property owners, merchants, private investors, planners, parking and transit authorities, and other governmental agencies. Recently, controls on parking as a mechanism for restricting automobile use to reduce air pollution have added air quality agencies to the scene. In addition, responsibility for assigning the available space among users, setting fees, and for policy on or enforcement of regulations often is assumed by, among others, city council, police, traffic engineer, and transit agency. Many of their decisions are subject to

[96]The role of parking in transportation planning and procedures for assessing parking demands and needs are discussed in Chapter 3. This section deals with the means for providing parking and some of the associated problems.
[97]See *TRB Record 496* and *511*.
[98]See *HRB Special Report 120; TRB Record 496, 591, 668, 758,* and *772,* and P. Ross, *Public Roads,* Dec. 1978.

review in the courts. Creation of agencies whose primary concern is parking has been one commonly suggested way to resolve this continuing dilemma.[99]

Parking Requirements as Related to Land Use

Land use in today's automobile-oriented society calls for a place to store vehicles. For new developments or for reconstruction, local governments commonly have employed zoning as a means of providing off-street parking in order to keep these vehicles from the streets. Zoning provisions, then, offer one measure of the apparent parking needs of various land uses. These would vary greatly among jurisdictions. For one moderate-sized western city, a few from the many requirements for off-street parking call for spaces as follows: residences, per unit, single family, 2 to 4 depending on location (one covered); studio apartments, 1.25; commercial developments per 1000 ft^2 of floor area, offices and banks, 6.7; retail, 6.7 for intensive to 2 for open lot; shopping centers, 3.6; manufacturing in light manufacturing area, 3.3; hospitals, per bed, 0.67. Some bicycle spaces also are stipulated. Exceptions in the cases of commercial and industrial developments may be permitted if adequate public transportation is available or where a group pools its needs into one or more facilities. Waiving the parking requirement on downtown office buildings is a strategem now being employed by some cities to encourage both downtown redevelopment and the use of public transit. Further adjustments can be anticipated by changes dictated by energy, pollution, and other considerations.

Zoning is not an effective mechanism for developing new parking in built-up areas, but other strategems, some of which are discussed below, can be employed.

Parking and Its Relation to Downtown Activity

The coming of the private automobile coupled with the rapid growth of suburban areas substantially altered the retail shopping patterns of many cities. Although overall business activity in central business districts increased, its relative share of the total metropolitan market declined as contrasted with suburban sales, particularly at shopping centers. Favoring the shopping centers are such factors as decreased travel time, availability and low cost of parking, decreased congestion, short walking distances, and the ease with which purchased goods can be brought to the vehicle. Favoring downtown may be greater selection of goods and possibly cheaper prices. Although parking is but one factor, studies have indicated that downtown areas with good, convenient, and cheap parking have been better able to maintain their positions. Other downtown demands for space such as for offices, seem to have been less affected by suburban competition as evidenced by high-rise construction in downtown areas. This problem

[99]A comprehensive reference that treats most of the aspects of the parking problem is *HRB Special Report 125*. Also, the periodical *Parking* and other publications of the National Parking Association provide much useful information. A. B. Rappaport, *TRB Record 644*, discusses parking management.

is far too complex to develop in this book; it is mentioned as one of the challenges facing the urban planner.

On-Street Parking

The space devoted to on-street parking in downtown areas and along major highways is being steadily reduced. There are many instances where parking on principal arteries is banned, at least during morning and evening rush hours, to increase street capacity. Tow-away zones with high charges to reclaim vehicles help to make these restrictions effective. Also, more and more space is being reserved for mass-transit and commercial loading zones.

On-street parking is seldom a good use of limited street space. In the first place, it substantially reduces capacity, as illustrated in Figs. 8–13 and 8–14. Congestion and confusion associated with parking also increase travel times and accidents.

Vehicle positioning for on-street parking is almost always parallel to and against the curb. The *Manual* recommends that individual parking stalls be dimensioned 8 ft wide and 22 to 26 ft long. Alternatively, "paired parking," with two stalls each 20 ft in length may be laid out with an 8-ft clear space between pairs for vehicle maneuvering. A clear space of at least 20 ft is stipulated adjacent to crosswalks and at approaches to intersections. At major cross streets this setback should be 50 ft or more. Although parallel parking accommodates fewer vehicles, it offers much less disruption to moving traffic and reduces accidents as compared with angle positioning.[100] For such reasons, angle parking is seldom recommended. For example, a common restriction limits 45° angle parking to streets at least 54 ft wide.

Legally, the right of the government to regulate on-street parking is firmly established. In 1805 in England, Lord Ellenborough, a famous British jurist, asserted that "the King's highway is not to be used as a stable yard." He established the principle that streets are primarily for the free passage of the public and anything that impedes that passage, except in an emergency, is a nuisance that may be abated. Parking, even in front of one's own property, was classed as a privilege subject to control and not as a right.[101] In the United States, authority to regulate parking stems from the *police power,* the right of government to legislate to protect health, safety, and morals. Under it, public officials are free, within reasonable limits, to establish rules to control on-street parking and to set penalties against violators.[102] And the police power can be exercised without compensation to property owners or others who may suffer loss as a result (see Chapter 7).

Parking meters have proved to be an effective means for regulating on-street parking. First installed in Oklahoma City in 1935, there are now roughly 2 million of them in use in some 4000 cities of the United States. Some cities, mostly

[100]See, for example, J. B. Humphreys et al., *TRB Record 722.*
[101]*Rex* v. *Russell,* 6 East 427, 102 Eng. Rep. 1350.
[102]Recommended curb-parking legislation will be found in the *Uniform Vehicle Code.*

under 50,000 population, have removed theirs, mainly because of complaints from merchants.

Parking meters offer an accurate time check on parkers, thus discouraging overtime and all-day users. Short-time parking as contrasted with longer space use is encouraged, with the turnover often two to three times as great with meters. Police time for parking enforcement is cut approximately in half. Often double parking is substantially decreased. On the other hand, many motorists resent the charge for parking, the nuisance of carrying small coins, the time constraint, and the threat of a fine. Some refuse to trade in areas where meters are used if meter-free parking is also available nearby.

On-street parking meters produce substantial revenues[103]; the average reported in 1970 was $108 per year per meter. *HRB Special Report 125* states that in cities over 100,000 population, the cost of meter maintenance averaged $15 per year and coin collection $10 per year. Most of the cities place meter revenues in the general fund and use them for ordinary city expenses. The remainder apply this income to traffic improvement and the development of off-street parking. A disadvantage of meters is that, with earnings so great, city officials will sacrifice traffic improvements in order to retain the revenues. Since authority to install parking meters stems from the police power, it apparently cannot be challenged as long as the charges reasonably approximate the cost of street space, including rights of way, construction and maintenance, and the expense of regulation.

Off-Street Parking[104]

Off-street parking is a substantial enterprise of both governmental and private sectors. By the early 1970s, city governments alone had over $5 billion invested, with revenues of $250 million annually. Investments in the private sector would probably be several times this amount if they were counted up through all purposes, including the home garage, factory and other work places, and facilities associated with businesses of all sorts.

GENERAL REQUIREMENTS. The gross space per U.S. standard size car in a parking lot or parking garage usually ranges between 250 and 350 ft² per vehicle. In general, less floor space is required where cars are positioned by attendants as contrasted with driver parking. Other variables such as lot or building dimensions and layout also have important bearing. First cost per parking space has a wide spread, from possibly $600 for lots where land is inexpensive to perhaps ten times that for elevated structures counting land costs, and roughly double that for underground facilities.

Provision for all possible demands for parking in downtown areas has never been economically feasible. As already indicated, approximately 250 ft² of floor

[103]Over $200 million in 1976 in cities with populations over 50,000.

[104]See *HRB Special Report 125* for added details on all aspects of off-street parking. R. I. Strickland, *ITE Journal,* Nov. 1980, discusses stall design for small cars.

space per vehicle is required. On the other hand, the average office worker occupies less than 100 ft^2. Thus, if all office workers drove downtown, two persons per car, parking space would exceed office space. Neither can downtown parking be furnished for all shoppers as is done in modern suburban shopping centers. There parking space often exceeds floor area by two or more times.

Location and layout of parking facilities must be fitted to traffic conditions on adjacent streets. For example, if parking lots or garages are designed to discharge onto heavily traveled, narrow streets, both street traffic and parkers are seriously inconvenienced. On the other hand, by connecting off and on ramps of freeways directly into parking installations, the adjoining streets can be completely freed of many vehicles. Again, the design must match the rate at which vehicles can enter and leave the facility to the intended clientele.

The willingness of users to walk from parking place to destination is of particular importance. Variables affecting walking distance include type of parking facility, urban area population, and trip purpose and its closely related variable, length of time parked. Results of comprehensive parking studies give a notion of walking distances. For example, combined figures where parking was free or pay, curb or off-street, and legal or illegal gave averages based on population ranging from 200 ft in cities under 25,000 through 280 ft in cities between 50,000 and 100,000, to 560 ft in cities over 500,000. In addition, such factors as the nearness of competing shopping areas or parking locations, the charge for parking, and the type of goods being bought may influence potential parkers equally as much as does walking distance.

Environmental and political considerations may carry heavy weight, particularly with public projects. For example, only an underground design, with the surface largely restored, would be acceptable at a downtown park site.

PARKING LOTS. The parking lot is the simplest off-street parking facility. Usually the area is subdivided with curbs or bumpers, surfaced with a bituminous or concrete pavement, and marked out into parking stalls and driveways. On private lots attendants often park and return vehicles and collect fees. In many cases public lots are equipped with parking meters and are operated in conjunction with on-street parking. An abridged set of recommended dimensions for parking lot layouts for standard sized American and imported (American subcompact) cars is given in Fig. 10–7. Those for standard cars are based on a 9-ft stall width, which is midway between the workable minimums and maximums of 8.5- and 9.5-ft; for imported cars stall width is 7.5-ft.

As indicated, the widespread swing to compact and subcompact cars means that a portion of the parking stalls can be smaller. A variety of layouts are possible. One would be to dimension some of the bays for compact cars only. Unless such combined facilities are policed, drivers of compacts may appropriate any better-located, larger stalls, thereby limiting the capacity for larger cars.

Advantageous sites for parking lots often are found in the interior of large blocks or facing on back streets where property values are low. At times, by careful planning, small or irregular-shaped lots can be developed to good advantage.

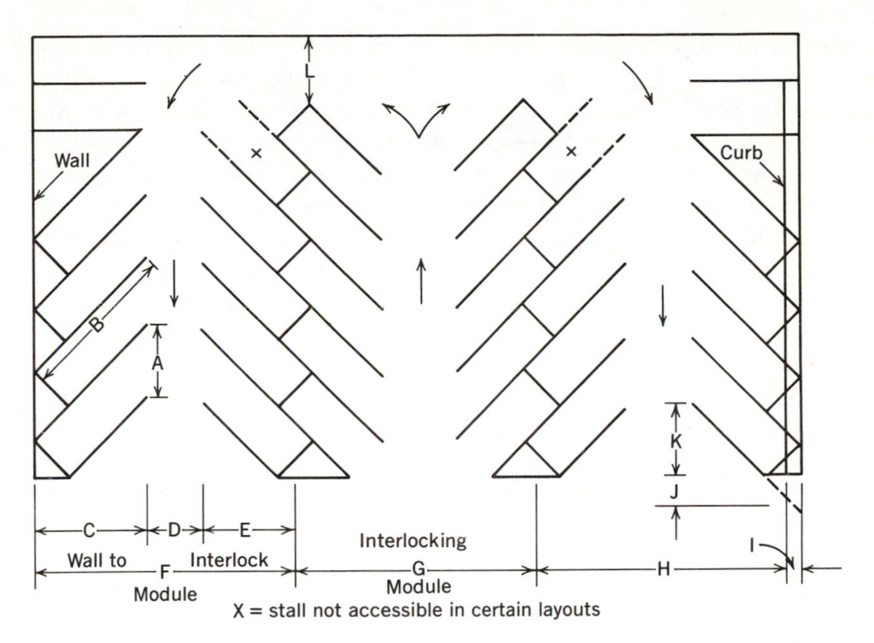

Dimensions (ft)	On Diagram	Standard U.S. Vehicles*				Foreign and Subcompact†			
		45°	60°	75°	90°	45°	60°	75°	90°
Stall width, parallel to aisle	A	12.7	10.4	9.3	9.0	10.5	8.7	7.8	7.5
Stall length of line	B	25.0	22.0	20.0	18.5	—	—	—	—
Stall depth to wall	C	17.5	19.0	19.5	18.5	—	—	—	—
Aisle width between stall lines	D	12.0	16.0	23.0	26.0	11.0	14.0	17.4	20.0
Stall depth, interlock	E	15.3	17.5	18.8	18.5	16.0	16.7	16.3	15.0
Module, wall to interlock	F	44.8	52.5	61.3	63.0	—	—	—	—
Module, interlocking	G	42.6	51.0	61.0	63.0	—	—	—	—
Module, interlock to curb face	H	42.8	50.2	58.8	60.5	—	—	—	—
Bumper overhang (typical)	I	2.0	2.3	2.5	2.5	—	—	—	—
Offset	J	6.3	2.7	0.5	0.0	—	—	—	—
Setback	K	11.0	8.3	5.0	0.0	—	—	—	—
Cross aisle, one-way	L	14.0	14.0	14.0	14.0	—	—	—	
Cross aisle, two-way	—	24.0	24.0	24.0	24.0	—	—	—	—
Module—wall to wall	—	—	—	—	—	43.0	47.4	50.0	50.0

*Stall width 9.0 ft
†Stall width, 7.5 ft.

Fig. 10–7. Recommended parking layout dimensions at various angles. (Source: *HRB Special Report 125*.)

Careful consideration should be given to the appearance of parking lots fronting on the streets. Often, for example, by setting the pavement a few feet back from the sidewalk, the lot can be screened with a border of shrubs or other planting.

Fringe (modal transfer) parking on lots or in terminals located outside the downtown area is an increasingly common approach to solving downtown congestion and parking difficulties. As noted in Chapter 5, federal assistance is available to pay for such facilities. With them, motorists travel downtown from the lots on fixed-guideway transportation vehicles such as trains or by special express buses operating on freeways or expressways, and sometimes on exclusive bus lanes. Some of the patrons may not park. Rather they are dropped off and picked up again by a family member (called "kiss and ride"). Others may ride bicycles and leave them in special storage facilities. A few may walk. However, a large majority (85% in one case) park an automobile in the lot. Modal transfer parking lots commonly are constructed and operated by the public or private transit agency that provides transportation in and out of the city center. Some are free; others have a daily or monthly charge. Effective designs, among other features, permit easy drop off and pickup of those who do not bring cars and provide for heavy peak loads at entry and exit.[105]

Where parking demands for different purposes do not conflict in time, the same spaces may be shared. For example, park and ride needs daytime space Monday through Friday, while shopping demand peaks Saturdays and possibly at night. Churches and sports facilities need parking primarily on Saturday, Sunday, or at night. Even relatively close-in special parking may be put to multiple use. For example, a shuttle bus system makes parking for Soldiers' Field stadium near downtown Chicago available to persons destined for the business district. In most instances, difficulties with such multiple uses are political or institutional rather than with the physical arrangements.

Parking lots and, in many instances, parking garages are a necessary adjunct of schools at all levels, shopping centers, stadiums and sports arenas, amusement parks, and airports, since many patrons of such installations arrive by private automobile. Each has its peculiar problems in terms of peak demands and other controls on sizing and design details. These are beyond the scope of this book.

MULTISTORY PARKING BUILDINGS.[106] Multistory parking buildings have been constructed in many urban and suburban locations where land values are high. At times, the valuable ground floor has been devoted to stores or other businesses and the remainder to parking. Access and egress to parking garages often is by fairly steep ramps which may be either straight or circular, depending on site conditions or design preference. In some, vehicles are parked by attendants; in others, customers park their own cars. Often gasoline and lubrication sales and washing, greasing, and mechanic services add supplementary income.

Parking buildings are relatively cheap. Often the walls are open, the ceilings

[105]See, for example, articles in *TRB Record 505, 557,* and *590; TRB Special Report 153;* and A. J. Pacelli, *Transportation Engineering Journal of ASCE,* Jul. 1980. D. I. Riley and J. W. MacIsaac, ibid., Jul. 1978, report on a park and ride scheme for Seattle.

[106]For added detail on multistory parking facilities, references include *HRB Special Report 125,* the *Transportation and Traffic Engineering Handbook,* and R. E. Weant, *Parking Garage Planning and Operation,* Eno Foundation, 1978.

low, and heating, ventilating, and certain other refinements omitted. Maintenance costs also are low.

Designs of parking garages are fitted to customer requirements. For service to all-day parkers, provision must be made to receive and discharge almost all the vehicles in short periods of time. Different designs would be more suitable for shopper parking, where loading and unloading peaks are not so sharp. In a few cases where site use was to be temporary, modular designs of steel and/or precast concrete which could be disassembled and moved have been employed.

UNDERGROUND PARKING GARAGES. In downtown areas where property values are extremely high, underground facilities built under parks or plazas may offer the only feasible way to provide off-street parking. The first large garage of this kind, with a capacity of 1500 cars, was completed under Union Square in San Francisco in 1942. It has been a marked success financially. A garage under Grant Park, Chicago, holding 2350 cars went into service in 1954. By 1967, Pittsburgh had 7800 parking spaces, some above and some below ground. One underground facility was designed to accommodate a 13-story building above it. Another, under Mellon Square in the downtown heart, adds a most attractive park while accommodating 1040 spaces on six levels. Many other cities also have similar installations. Often these are an integral part of a large redevelopment scheme such as the Renaissance Center in Detroit.

MECHANICAL PARKING GARAGES. Since World War II, private investors have built a number of mechanical devices that park and retrieve cars. They are usually placed on small lots in crowded downtown areas. Difficulties include congestion at the point of entry and discharge and possible breakdown.

TRUCK AND BUS TERMINALS. As indicated above, in many large cities congestion in commercial and industrial areas is relieved by providing special off-street terminals for trucks. Large and unwieldy over-the-road trucks drive to destinations located out of the congested areas but on or near the main highways into the city. Here loads are sorted for distribution in the city and placed in lighter, more maneuverable vehicles. Possibly the largest of these is located on Manhattan Island close to the Holland Tunnel. Often such terminals serve the dual functions of market for perishables such as vegetables and transfer point to small delivery trucks.

Off-street bus terminals conveniently situated near entering highways, the downtown area, and local mass transportation are in widespread use. One of the most elaborate was constructed adjacent to the Lincoln Tunnel in New York City by the Port Authority. It serves both intercity and commuter buses. The combination of priorities for buses through the tunnel, ready access in the terminal to the city's subway system, and difficult and costly parking for private automobiles all make use of this facility attractive.

DEVELOPING OFF-STREET PARKING. There are a host of legal and financial knots that must be untied to develop public off-street parking. As a means of solving these problems, state legislators have, to different extents, delegated powers to municipalities or transit authorities, including permission to create separate agencies to deal specifically with parking. As examples of these delegations, authority may be granted to (1) levy taxes or create special assessment districts; (2) acquire land by eminent domain; (3) finance by revenue or general-obligation bonds; (4) arrange for commercial uses in public parking facilities or for private enterprise to build and operate the facilities; (5) construct facilities; and (6) operate them on completion. Often on-street and off-street facilities will be integrated functionally and financially with parking-meter revenues reserved to alleviate parking difficulties or to pay off the financial obligations. The choice of method for providing off-street parking accommodations will vary with the size and economic characteristics of the city, its political and business mores, its tax and debt structure, the magnitude of its parking needs, and a host of related factors. Certainly there is no "best" method.[107]

Private interests have developed and will continue to develop many off-street parking facilities as an investment or to provide for employees or customers. But alone or as part of an overall scheme, private capital is interested only when investments in parking promise an attractive return. Again, privately owned parking facilities, once constructed, may be withdrawn for other uses that promise greater income. To illustrate, it may be more attractive financially to convert a private parking lot into a building site. Another complaint is that rates are set to gain maximum profit rather than most effective use. For example, greatest revenues and fewer operating problems may come from all-day parkers or from renting spaces by the month. But most effective use in a broader public view may demand rates that encourage shoppers and other short-time customers and discourage all-day parkers. On the other hand, owners of private parking facilities sometimes complain that public agencies, free from taxes and the pressure to make money, offer unfair competition to private garages.

Without question, private investors and the public both have a stake in privately owned parking facilities. One extreme suggestion for protecting the public interest is that governmental agencies take over all private facilities. Another less drastic approach is to regulate private parking facilities used by the public as public utilities.

PROBLEMS

10–1. Examine the marking and signing practices on a major traffic artery near your campus and point out any differences between them and the *Manual on Uniform Traffic Control Devices*.

10–2. By a visit to a local highway or street agency, find out how much and what types of paints and other pavement-marking materials are used. Also determine the period of time between remarkings and the procedures and equipment employed for major projects.

10–3. Make rough traffic counts or estimates for a troublesome intersection

[107]For details of legislation in various states, see G. A. Culp, *HRB Record 168*. See also, *HRB Special Report 125*.

near your campus. Then determine whether it can best be regulated by *(a)* placing stop signs on the minor street, *(b)* employing four-way stop signs, *(c)* installing fixed-time traffic signals, or *(d)* installing traffic-actuated signals. Use the warrants cited in Chapters 8 and 10 to support your conclusion.

10–4. Extend the study called for in Problem 10–3 to include an economic analysis based on data and procedures given in Chapter 4.

10–5*A*. Traffic signals with a master controller for flexible progressive operation have been installed at eight intersections on the main street of a small town. Intersection spacing is uniformly 300 ft center to center. For this situation, develop progressive timing for the signals. Proceed step by step as follows:

a. On a full-size sheet of drafting paper, plot street spacing as the ordinate versus time in seconds as the abscissa (see Fig. 10–2). It will be more convenient to plot street spacing as 0.682 times the distance in feet, for then slopes of lines on the graph are in miles per hour.

b. Mark out signal timing for each intersection on separate narrow strips of paper. For this simplified problem, assume that total cycle time is 60 s, divided at all intersections into 30 s green, 6 s yellow, and 24 s red. Also, procure black string or thread for laying out the through bands.

c. Determine signal offsets to give equal-width through bands in both directions (see Fig. 10–2*b*). Observations indicate that 14 mph is the prevailing vehicle speed. Use a trial and error procedure, shifting the signal-interval strips and thread as required.

d. How wide are the through bands? Also determine the number of cars that can go through uninterrupted on each band.

e. Determine signal offsets to give the maximum-width through band to vehicles traveling in one direction.

10–5*B*. Follow through the steps outlined in Problem 10–5*A*, but work in SI units. Assume that intersection spacing is 200 m center to center. For part *c*, set prevailing speed at 25 km/h.

10–6. Using the chart and strips developed for parts *a* and *b* of Problem 10–5*A* determine answers for parts *c, d,* and *e* of that problem for a prevailing vehicle speed of 20 mph.

10–7. Using the instructions of Problem 10–5*A* as a guide, review the timing of a system of progressive signals located near your college.

10–8. By observation or an interview with the engineer in charge, develop a phase diagram similar to Fig. 10–3 for the traffic signals at an intersection near your campus. On the diagram indicate, if it is fixed time, the intervals given to each phase. If it is traffic-actuated, note the maximum times allocated to each phase. Then consider and explain how cycle length and phasing might be changed to reduce overall delays to traffic.

10–9. Do Problem 10–8 for a diamond interchange, using Fig. 10–4 as a reference.

10–10. If ramp metering is used to reduce congestion on a freeway near your college, investigate its effectiveness. If it is not, explore the feasibility of installing it at a nearby freeway on-ramp.

10–11. Outline a plan for a system of one-way streets for a town or city near your college.

10–12. Determine the maximum nighttime speed of travel from which the average driver can stop before striking a large, dark boulder that has rolled down a cut slope into a roadway in front of him if

a. There is no opposing traffic and headlamps are on lower beam.

b. The driver is facing opposing traffic. Assume that the pedestrian curves of Fig. 10–6 apply.

10–13. Same as Problem 10–12 but for the driver whose visibility lies at the 25 percentile level.

10–14*A*. For an actual or hypothetical parcel of vacant land designated by the instructor, make a tentative layout for a parking lot having the maximum number of stalls but still meeting the standards for stall and lane dimensions. Assume 45% diagonal vehicle positioning and 50% compact cars.

10–14*B*. Dimension the layout called for in Problem 10–14*A* in SI units.

11 DRAINAGE

This chapter discusses the means for collecting, transporting, and disposing of surface water originating on or near the right of way or flowing in streams crossing or bordering that right of way. On an average, one highway construction dollar in four is spent for culverts, bridges, and other drainage structures. Substantial added expenditures are demanded on rural roads for ditches, dikes, channels, and erosion-control installations. In urban and suburban locations, major capital investment goes into storm drains and their appurtenances. In addition, routine cleaning and repair of drainage facilities coupled with the expense of rebuilding after heavy storms takes a substantial share of maintenance funds.

The attack on surface-drainage and erosion-control problems must begin with the location survey. Ideal locations from a drainage standpoint would lie along the divides between large drainage areas. Then all streams flow away from the right of way and the drainage problem is reduced to caring for the water that falls on roadway and back slopes. In contrast, locations paralleling large streams are far less desirable as they cross every tributary where it is largest. Again, ideal locations avoid steep grades and heavy cuts and fills, both of which raise difficult problems in erosion control. Admittedly, surface drainage is only one among many considerations in location, but it warrants careful attention.[1]

Once the location is established, analysis of surface-drainage problems follows three basic steps: (a) hydrology—estimating the peak rates of runoff to be handled, (b) hydraulic design—selecting the kinds and sizes of drainage facilities to most economically accommodate the estimated flows, and (c) making certain the design does not create erosion or other environmentally unacceptable conditions.

ECONOMY IN DRAINAGE

True economy in drainage design means finding the solution that is cheapest in the long run. The first step is to estimate the first cost, maintenance outlay, and anticipated loss and damage for each reasonable solution. These are then compared by one of the methods outlined in Chapter 4. In these computations, the

[1]For added discussion see *Highway Drainage Guidelines—Hydraulic Considerations in Highway Planning and Location*, AASHTO.

annual charge for possible flood damage or economic loss equals the estimated losses from floods of various magnitudes, multiplied by the probability that these floods will occur in any one year. For example, if losses from any flood exceeding the design flow total $50,000, and hydrologic studies indicate that such a flood will occur once in 25 yr, the annual charge for possible flood damage equals $50,000 x 1/25, or $2000. This approach can be extended to recognize different costs of damage for floods of various magnitudes.[2]

Economy problems in drainage are diverse, and do not lend themselves to any single set of assumptions or rules. Commonly, the answer sought is the flood frequency on which the design should be based. For a major highway carrying large volumes of heavy traffic, losses to motorists and to the economy in general may be tremendous if the road is closed frequently by floods and washouts. On the other hand, on a lightly traveled rural road, economic losses alone will not justify large capital expenditures to prevent occasional interruptions to traffic. For the major highway, studies might prove that drainage facilities should accommodate a 50-yr flood, whereas designs based on a 5-yr flood would be more reasonable for the low-volume rural road. In another instance, capacity to accommodate infrequent floods would be appropriate if smaller waterways resulted in flooding of valuable property. However, it might be cheaper in the long run to inundate bottom lands every few years rather than to provide channels and structures large enough to protect them. As a third example, damage to embankments overtopped by flood waters will generally be much more severe in arid regions where the slopes are unprotected than in humid regions where they are covered with grass or other growth. Thus, other things being equal, more generous designs to prevent overtopping would be appropriate in arid regions than in humid ones.

Where drainage problems of any magnitude have alternative solutions, economy studies based on reasonable estimates of costs and possible future damage represent the best approach. It is true that flood frequencies and some of the costs must be roughly approximated, and that catastrophic occurrences warp our perspective.[3]

LEGAL ASPECTS OF DRAINAGE

Agencies have legal responsibility for damage to private property resulting from changes they make in natural drainage patterns within the limits that "water must flow." Thus, damage claims against an agency could be established if erosion, silting, or flooding of private property resulted when the flow of several small streams was concentrated into a single structure or channel. Again, a cause

[2]For further discussion of the "cost of risks," see the chapter on probability in E. L. Grant, W. G. Ireson, and R. S. Leavenworth, *Principles of Engineering Economy,* J. Wiley, New York, 1982. Solution of a typical drainage economy problem involving risk appears in Robley Winfrey, *Economic Analysis for Highways,* International Textbook Co., Scranton, Pa., 1969.

[3]See, for example, *HRB Record 479,* which reports on several events in 1972, a year of devastating floods.

of action might be created if water backed up against a highway structure or embankment and inundated lands or property or resulted in injury or death because of poor design or inadequate maintenance. However, liability supposedly is restricted to those damages that are a direct consequence of the improvement. It follows that the agency normally would not be liable if an unprecedented storm caused a stream to overflow a newly constructed channel, provided this channel had a capacity equal to the natural one. This follows the rule that engineering decisions, if based on accepted practice, do not provide a cause for action. In each case, responsibility of the highway agency would be determined by negotiation or court proceedings. The important point is that improper or inadequate drainage designs may bring not only criticism and charges of incompetence, but damage suits as well.[4]

Designs for drainage also must satisfy a variety of environmental laws and regulations. These are discussed briefly in Chapter 13.

HYDROLOGY

Hydrology is that branch of physical geography dealing with the waters of the earth. The branch of hydrology of particular concern to highway engineers deals with the frequency and intensity of precipitation and the frequency with which this precipitation brings peaks of runoff that equal or exceed certain critical values. Of importance also is the distribution of precipitation through the seasons insofar as it influences the moisture underlying highway surfacings (see Chapter 14) and the growth of grasses, shrubs, and other plants useful for erosion control or roadside improvement (see Chapter 13).

It should be understood at the outset that predictions regarding future rainfall or runoff from accumulated records, whether by statistical approaches, formulas, or simulation, rest on the laws of probability, in other words, the chance that a given event will or will not take place. To illustrate, consider the statement that a culvert is designed to carry a "50 yr" flood. This means that, if past experience is repeated, the chances are 1 in 50 that the structure will flow full or be overtaxed once in a particular year. It does not mean that the design flood or a larger one will occur exactly one time in 50 yr; in fact, the chances are only 64 in 100 that a flood of this magnitude will occur in a given 50 yr period. On the other hand, several floods of this or greater magnitude could occur in successive years or in a single year, but the chance for either combination is extremely small.

Precipitation occurs as rain or in frozen form, particularly as snow. It results when warm, moisture-laden air is cooled as it flows over a mountain barrier (orographic lifting), is forced upwards by a cold-air mass (frontal lifting), or rises through cooler air (convective lifting), as in summer thunderstorms. The duration and intensity of precipitation produced by these storm types is markedly

[4]For detailed discussions of legal responsibilities in general, see Chapter 7. For those more specifically related to drainage see *Highway Drainage Guidelines—Legal Aspects of Highway Drainage*, AASHTO, and *NCHRP Report 134*, Chapter 4.

different. Many other factors also have pronounced influence. Among them are geographical position in the general circulation of the earth's atmosphere, shifts in this circulation pattern, the moisture content of the air, wind velocity and direction, elevation, and the steepness and aspect of slopes. It follows that rainfall data collected at scattered gauges offers only an approximation of the precipitation in the immediate area of each gauge. Variations will be even greater along the length of a highway.

Of the moisture that falls, some is returned to the atmosphere by evaporation from land and water surfaces and by transpiration through plants. Another portion percolates through the ground, sometimes emerging as springs. This underground water travels slowly and generally adds little to extreme floods. The remaining precipitation flows overland in a thin sheet until it reaches streams or channels. These in turn deliver the water to larger streams that take it to an ocean or inland sea. Lakes, reservoirs, swamps, or diversions for human use often intervene in this progression.

As would be expected, the ratios of runoff to precipitation vary widely. Most of the rain that falls on rocky or bare, impervious slopes, roofs, or pavements runs off quickly while only a small percentage of that falling onto plowed land or heavy forest litter may run off at all, and that at a slow rate. Slope and surface moisture at the time of rainfall introduce added variables. The snow melt rarely causes major floods on small streams, as the water comes off in long sustained moderate flows. As drainage areas become larger, increasing amounts of water are devoted to swelling the streams to flood stage, and this "channel storage" slows the rush of water downstream. For these and other reasons, the determination of peak runoff from rainfall records is difficult.

The uncertainties inherent in predicting runoff from rainfall records would be of no concern if long-time, countrywide records of the flow from all or at least a large number of representative small drainage areas were available. Some 5000 watersheds in the United States are gauged and the data reported by the participating states to the U.S. Geological Survey under a program titled National Small Streams Data Inventory (NSSDI). However, this coverage is too meager, the data too incomplete, and record duration too short to provide a reliable method for designing individual drainage facilities.

NCHRP Report 136 (1972), offers an assessment of the procedures for, and the accuracy of, peak runoff estimates from small ungauged rural watersheds. From an historical viewpoint, the report indicates that, since 1852, more than 100 equations involving over 50 variables, as well as several other approaches, have been proposed. In a comparison of actual records of 493 watersheds with procedures used by state highway agencies, it found two-thirds of the predictions off by at least 25%; in one in five cases actual runoff was overestimated by a factor of 3. Its main conclusion was that

> presently used methods for estimating runoff on ungauged rural watersheds are unsatisfactory on a nationwide basis. Consequently, designers should make the best possible use of existing prediction methods, with full realization of the high probability of

error, and giving careful consideration to the increased cost of overdesign versus the possible consequences of an underestimation of peak flow.

The following paragraphs present three methods now used for estimating peak flows and discuss other approaches without offering solutions based on them. In using them, engineers must realize that predictions regarding future runoff are subject to the laws of chance. Again, they must remember that changes in land use may alter the historical runoff characteristics of the drainage basin. Finally, they must understand that, for small ungauged watersheds, today's "state of the art" is such that any method of prediction may be subject to substantial error.[5]

Runoff from Stream-Flow Records

There are gauging stations on many large and occasionally on small streams. Where a runoff record has accumulated for a considerable time, these data can be reduced to a graph that predicts the frequency of recurrence of floods equal to or exceeding given magnitudes.[6]

A method has been advanced by FHWA to extend stream-flow data to ungauged watersheds as large as 50 and possibly 100 mi^2. Its first step is to determine the probable maximum peak runoff for the area. The equation and a plot of it appear on Fig. 11–1. It represents an envelope incorporating all "period of record" instantaneous maximum flood peaks for over 1000 gauging stations. Designs based on it would involve virtually no risk so that no further analysis would be needed for any drainage structure which, for other reasons than for drainage, has this or larger capacity.

For situations other than the "no-risk" one, the FHWA method involves solving one among a set of three-, five-, or seven-parameter regression equations. These are offered for 24 hydrophysiographic zones covering the United States, including Alaska, Hawaii, and Puerto Rico. Each gives the estimated 10-yr runoff peak. Values for the parameters are provided in the reference, but cannot be given here.

Also offered is a three-parameter, all-zone equation. It is:

$$q_{10} = 1.28015A^{0.56172}R^{0.94356}DH^{0.16887} \tag{11–1}$$

where

q_{10} = estimated 10-yr runoff peak in cubic feet per second

[5]The discussion that follows is based primarily on *Reports FHWA R. D. 77-158* and *159* and *NCHRP Report 136*. See also *Highway Drainage Guidelines,* AASHTO, R. K. Linsley et al., *Engineering Hydrology,* 3rd ed. McGraw-Hill, New York, 1982, G. Fleming and D. D. Franz, *Journal of the Hydraulic Division of ASCE,* Sept. 1971, V. L. Gupa and S. A. Moin, ibid., Oct. 1974, and S. L. Chiang, ibid., Jul 1975. For recommended procedures for the state of New York see L. H. Irwin and J. L. Nieber, *TRB Special Report 160.* A federal interagency group called the Water Resources Council is working to establish a single approach for all federal agencies.

[6]Several methods for reducing stream-flow records into graphs of discharge versus recurrence interval are in use. Each gives somewhat different results. For details of these methods see R. K. Linsley, et al., op. cit., R. K. Linsley and J. B. Franzini, *Water Resources Engineering,* 3rd ed. 1979, or other standard textbooks or handbooks in hydrology and hydraulic engineering.

A = area of watershed in square miles

R = an iso-erodent factor, defined as the annual mean rainfall kinetic energy times the maximum 30-min annual maximum rainfall intensity. This parameter is influenced by such factors as position on the earth, air-circulation patterns, and elevation. Values for it for a few representative locations are: Fairbanks, Alaska, 10; Phoenix, Arizona, 40; Denver, Colorado, 70; Portland, Maine, 90; Washington D.C., 150; St. Louis, Missouri, 185; New Orleans, Louisiana, 700; and two high-rainfall locations on Maui, Hawaii, 3200.

DH = difference in elevation in feet in the main channel from its most distant point to the drainage structure site

An examination of Equation 11–1 shows clearly that the runoff peak is most strongly influenced by R, the iso-erodent factor, considerably by area A, and far less by DH, the elevation difference. It is to be emphasized that regardless of the values assigned to R or DH, runoff will always fall below the limit line on Fig. 11–1. As an example, if A is 10 mi², R is 100, and DH, 250 ft, q_{10} is 900 ft³/s, or 1/60 of the probable maximum shown on Fig. 11–1.

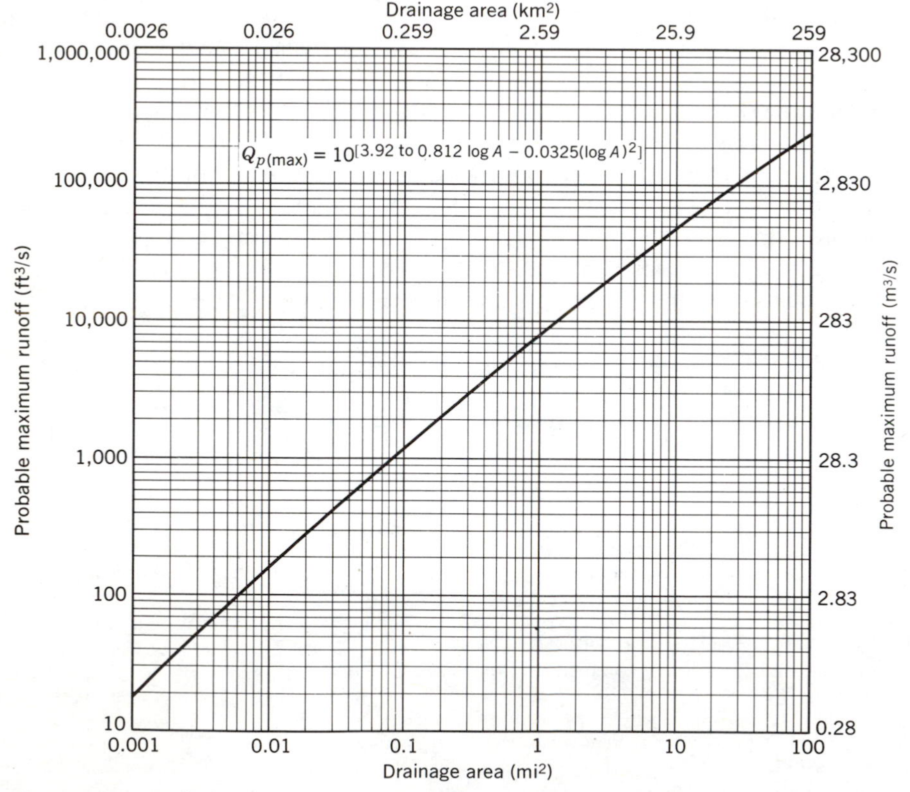

Fig. 11–1. Probable maximum runoff peak curve for small watersheds in the United States and Puerto Rico. (Source: *Report FHWA RD 77-158 or 159.*)

The FHWA method also gives multipliers for converting 10-yr to other return periods. Values for these are: mean annual, $0.46921q_{10}^{1.00243}$; 50 yr, $1.45962q_{10}^{1.02342}$; and 100 yr, $1.64380q_{10}^{1.02918}$. With relationships such as these, it is possible to select and determine the cost of providing drainage structures for various return periods. These values can be combined with damage estimates to select the most economical alternative.

The FHWA method also gives procedures for computing the effects of ponds, marshes, or other water storages on peak flows.

It cannot be emphasized too strongly that predictions of peak flows can be subject to substantial error. The method described above gives most probable values. But the standard error for the three-parameter all-zone equation is 119% and is about the same for the five- and seven-parameter solutions.

Runoff Prediction by the Rational Method[7]

The rational method which was first proposed in Ireland in 1851, may be reduced to the formula:

$$Q = ciA_d \tag{11–2}$$

where

Q = runoff in cubic feet per second

c = a "runoff" coefficient expressing the ratio of rate of runoff to rate of rainfall

i = intensity of rainfall, inches per hour for a duration equal to the time of concentration

A_d = drainage area in acres

This expression, as written, is not dimensionally correct. It gives numerically correct results, however, because 1 in./hr/acre and 1 ft^3/s represent the same amount of water per unit time, within 0.8%.

Although the rational formula has been employed to estimate flows from large drainage areas, some researchers have recommended that the limit be 200 acres; others have proposed a 500-acre maximum.

Suggested coefficients c for the rational formula are given in Table 11–1. Where ground cover is dissimilar, the drainage area is sometimes subdivided and a composite coefficient obtained by weighting the coefficients for each section according to area. Rainfall intensity i is obtained from records of nearby stations of the U.S. Weather Bureau. These records are reduced to a graph showing rainfall intensity versus duration for various recurrence intervals (see Fig. 11–2). Actual selection of the value for rainfall intensity rests on estimates of the acceptable frequency of occurrence of the design flood and on the time of concentration for the area. The latter is the time period required for water to reach the outlet from the most remote point in the basin. For paved highway surfaces, the recommended time is 5 min; where the water is from grass, 10 min is proposed. For larger areas, this time can be considerably longer, since

[7]This brief discussion draws heavily on *Design of Urban Highway Drainage*, FHWA.

TABLE 11–1. Suggested Values of Coefficients of Runoff, c, for Use in the Rational Formula.

Type of Drainage Area	*Coefficients of Runoff, c*
Concrete or bituminous pavements	0.8–0.9
Gravel roadways, open	0.4–0.6
Bare earth (higher values for steep slopes)	0.2–0.8
Turf meadows	0.1–0.4
Cultivated fields	0.2–0.4
Forested areas	0.1–0.2

periods for overland flow and that along the roadway gutter or other channel must be summed. The drainage area A_d is determined from topographic maps, aerial photographs, or rough field surveys comparable in accuracy to compass and pacing. Greater precision is not justified.

The rational method rests on a number of assumptions. One of these is that the basin is in equilibrium: Outflow equals rainfall less all retention in the basin, and this retention is stated solely in terms of surface characteristics. Factors such as surface conditions when rainfall begins and the retarding effect of overland flow and channel storage are neglected. Another assumption is that rainfall is of uniform intensity throughout the area during the time of concentration. This does not recognize that elevation, steepness and aspect of slopes, and other features that affect rainfall may be quite different over the basin. If the time of concentration is set at a low value, this substantially increases the assumed rainfall intensity and predicted peak runoff. The assumptions just listed and others not mentioned become particularly susceptible to error as the size of drainage area increases. With correct values for rainfall and for the runoff coefficient, the rational formula always overstates the runoff, with the error increasing as the size of basin increases. Thus, as indicated, its application should be confined to relatively small areas.

Another assumption under the rational method is that large floods occur with the same frequency as excessive rainfall. For storm drains and other installations in urban areas or in other cases where the surface is nearly impervious, this assumption is probably valid. There are many situations, however, where a large flood follows excessive rainfall only if the surface is already saturated by previous rain. In this case, the maximum runoff will occur only with the combination of excessive rainfall and previous rain. Then, if the design is made for a one-in-10-yr rainfall and the ground is wet before such a storm but once in five times, the probability of a stated flood is one-tenth times one-fifth or one-fiftieth, or once in 50 yr.

In spite of its limitations, the rational method is widely used in estimating runoff from small basins in both rural and urban areas.

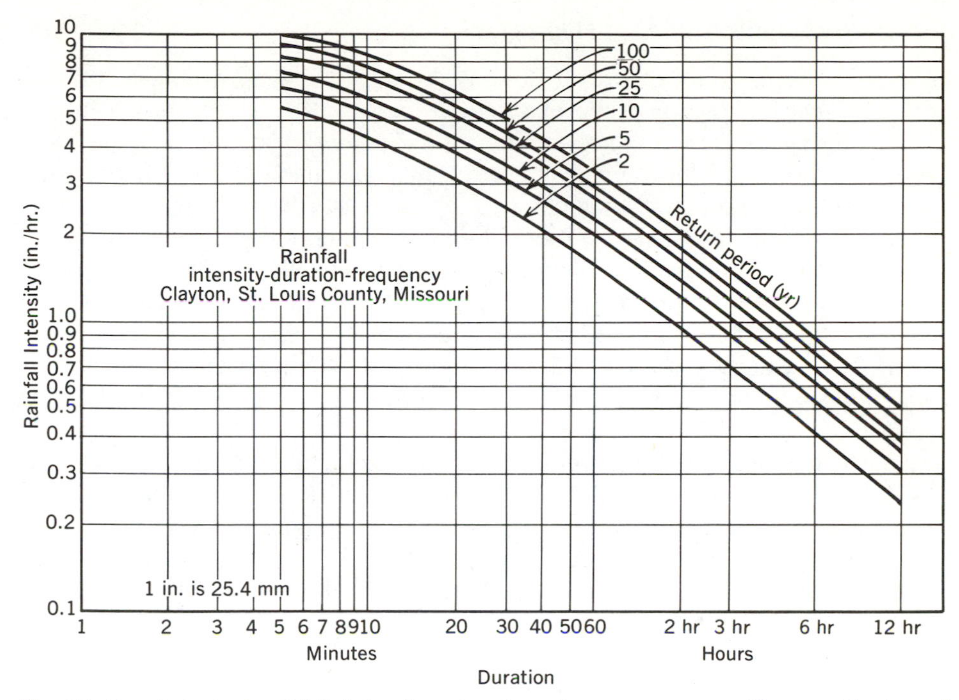

Fig. 11–2. A typical rainfall intensity, duration, frequency curve. (Source: *Design of Urban Highway Drainage*, FHWA.)

Runoff Prediction by Empirical Formula

Formulas expressing the waterway opening to accommodate extreme floods have been widely employed in the past. The most common of these is that of Talbot:

$$a = cA_d^{3/4} \tag{11–3}$$

where

a = waterway opening in square feet

c = a coefficient, ranging from 1.0 for steep, rocky ground through 0.6 for hilly country of moderate slopes, to 0.2 for level terrain not affected by snow

A_d = drainage area in acres

The Talbot formula was first proposed in 1887 when practically nothing was known regarding hydrology or hydraulic design. Limited investigation has established that, for certain portions of the midwestern United States, results from the formula approximate the combination of a 10-yr flood and a velocity of 10 ft/s through the culvert. Otherwise the Talbot formula has little scientific verification. Its widespread adoption in the highway field probably can be attributed to its simplicity and the lack of something better.

Runoff Predictions from Unit Hydrographs

A *hydrograph* is a plot of stream flow as the ordinate against time as the abscissa. A *unit* hydrograph is the hydrograph produced by a 1-in. runoff from a given drainage area for a typical or specified rainfall distribution of some specified duration. With the rainfall characteristics specified, the unit hydrograph presents clearly the runoff characteristics of the basin. It has been shown that, within reasonable limits, similar (but not equal) rainfall of the same duration and distribution will produce unit hydrographs of substantially similar shape. It can then be assumed that, for a specified type of storm, the time scale of the unit hydrograph for a basin is constant. The ordinates will be approximately proportional to the volumes of runoff. Thus the unit hydrograph provides a means for relating the runoffs to be expected from storms of like type but of differing intensity.

Unit hydrographs can be extremely useful in predicting peak flows at various flood-recurrence intervals, which are the highway engineer's principal concern. For example, short-term runoff records of a basin can be extended to cover the full period for which rainfall records are available. Again, runoff predictions can sometimes be made for local ungauged areas from the records of similar areas for which the runoff has been measured.[8]

Runoff Predictions by Statistical Approaches

A number of statistical approaches aimed at reliably predicting peak flow have been explored. Two of these were simple linear and stepwise multiple regression analyses. The aim in all cases was to develop equations that give runoff in terms of appropriate topographic, climatological, and hydrologic factors. The factors to include in the equation along with appropriate coefficients and constants are determined by finding the combination that best fits the observed data. The FHWA approach discussed earlier under the topic Runoff from Stream-Flow Records represents one application of such statistical analyses.

NCHRP Report 136 gives the results of 24 "hydrologic/statistical experiments" based on data from 493 watersheds. Predictors employed totalled 223. The best three of 81 equations were determined; each of them gave peak flow for a stated return period directly or in logarithmic form. Of these three, a single rational equation yielded the best results. Pertinent variables included length of main stream and tributaries, 6-hr precipitation with a 10-yr return period, the mean annual flood, a watershed slope factor, two regional factors, and mean July temperature. When compared with state design practices, the predictions of the equations were better in seven cases, worse in eight, and inconclusive in 17. This explains the recommendation cited earlier that the states continue to use their present approaches.

[8]For added discussion of this and the next two topics, see R. K. Linsley et al. and R. K. Linsley and J. B. Franzini, op. cit.

Runoff Predictions by Simulation

A computer-based mathematical simulation is based on a moisture-accounting procedure that recognizes precipitation, evaporation, interception and depression storage, soil moisture, ground water, subsurface flow, interflow, surface runoff, and stream flow. Output is a continuous hydrograph of the stream's flow. The difficulty at present is that to calibrate the model requires rainfall and stream flow records for 3 yr or more, plus data on soil permeability. Such records are available for only a few small watersheds. It may well be that in time a master data bank for relatively large areas can be developed which will make the synthesis approach a workable tool for highway agencies.

HYDRAULIC DESIGN PRINCIPLES

Hydraulic design for highways includes the basic principles of fluid flow, particularly those relating to open channels and closed conduits.[9] In most cases, solutions are aimed at the single problem of accommodating occasional large volumes of storm water. The methods of analysis are the same as for other branches of hydraulic engineering.

Of necessity, highway construction disrupts established drainage patterns. To illustrate, roadway cuts intercept water that earlier had moved overland across the right of way. Again, the flow from several small streams may be collected and passed under the highway at a single location. Often, streams are diverted to channels or culverts that differ from their predecessors in length, cross section, and flow characteristics. Construction operations may remove ground cover, loosen soils, or create dust so that subsequent rain will bring erosion or muddy streams. These examples show that almost every drainage installation upsets balances established by countless earlier storms. In particular, there is the possibility of higher velocity of flow or an abrupt dissipation of energy which may bring unsightly and costly erosion. Erosion in turn creates debris that is transported downstream and deposited at points where the velocity slackens. A cardinal rule of drainage design could well be that existing drainage patterns and soil cover be disrupted as little as possible. Necessary changes must not at any point bring velocities that will create new erosion problems. Disregard for this simple rule has created many serious maintenance problems and, even worse, has made the surroundings of many highways very unsightly and brought down the wrath of conservationists on all involved.

Water standing in pools along the highway sometimes creates a health hazard by providing a breeding place for mosquitoes. Furthermore, water passing through joints and cracks in the pavement or drawn from roadside ditches by capillary action may soften the subgrade and contribute to pavement failure.

[9]For the basic concepts and formulas see J. K. Vennard and R. L Street, *Elementary Fluid Mechanics*, 6th ed., Wiley, New York (traditional and SI versions), 1981; R. L. Daugherty and J. B. Franzini, *Fluid Mechanics with Engineering Applications*, 7th ed. McGraw–Hill, New York, 1977, or other standard textbooks or handbooks in these fields.

Another rule of drainage, then, might be that all water be drained away after every storm. There will be numerous exceptions, but the consequences should be weighed carefully for each of them.

Water draining from highways carries with it soil and dust that accumulate during dry weather. Discharging it directly into streams, lakes, or urban storm drains may often be objectionable. Among the possible solutions to this problem in rural areas is that of providing a storage basin or pool which slows or ponds the water, thereby permitting the sediments to settle out. Deicing salts create other problems.[10]

DRAINING THE ROADWAY AND ROADSIDE

Water that falls on the roadway flows laterally or obliquely from it under the influence of cross slope or superelevation in the pavement and shoulders. Figure 11–3 shows typical cross slopes for rural two-lane highways. (Figure 9–2 is representative of multilane highways.) When the roadway lies on fill, the flow may be permitted to continue off the shoulder and down the side slope to the natural ground (see Fig. 11–3*a*). Little erosion results if the slopes are protected by turf or if the water flows across the roadway and down the slope as a sheet. Where slopes are unprotected, they may wash badly if irregularities in pavement or shoulder concentrate the water into small streams. A likely place for such concentrations is the low points of sag vertical curves. One technique for preventing washing of side slopes is to retain the water at the outer edge of the shoulder as indicated in Fig. 11–3*a*.[11]

When roadways are in cut, water from the traveled way and backslopes is collected in a roadside channel commonly of trapezoidal, rounded, or triangular cross section (see Fig. 11–3*b*). Dimensions, slope, and other characteristics of this channel are determined by the flow to be accommodated and safety considerations. Particular care must be exercised where roadside channels discharge to prevent erosion of the toe of the fill slope.

Figure 11–3*b* also indicates that an intercepting channel, sometimes called a crown ditch, may be used at the top of the cut slope. It prevents erosion of the cut slope by surface runoff from the hillside above. Crown ditches often are constructed with a motor grader that excavates a small channel and forms the spoil into a dike on the downhill side. In cases where the natural ground is already protected by grass or ground cover the surface should not be disturbed. Rather, the channel should be formed with the natural ground as its bed and a small dike of topsoil or other imported material as the bank.

It is common practice to build crown ditches as single, continuous channels placed a short distance uphill from the top of the backslope. This procedure must be used with greatest caution; otherwise unsightly erosion is bound to result. Evidence of this is abundant along many existing roads. Methods for controlling scour where the entire flow must be retained in the ditch are discussed

[10]See J. Jodie, *TRB Record 556* for a report on the quality of urban storm water.
[11]For a brief discussion of slope-protection methods see Chapter 13.

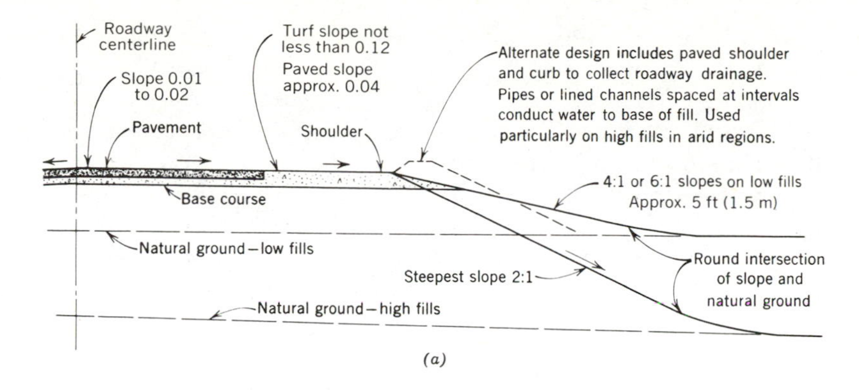

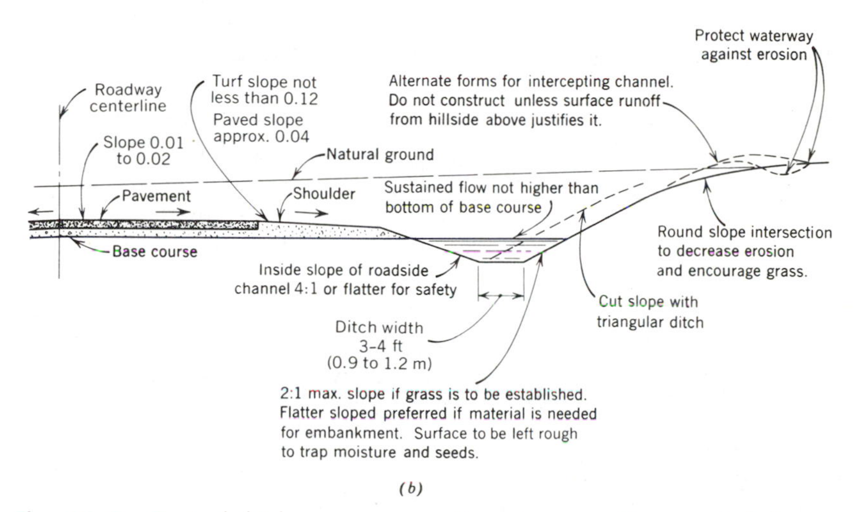

Fig. 11–3. Typical highway cross sections incorporating good drainage features.

in the section on channel design. As an alternative, it is sometimes possible to limit the flow to a safe amount by diverting it at intervals. One technique is to drop the water down the slope to the roadside ditch by means of specially designed channels or conduits. In open country there are often situations where diversions can be made at intervals to ditches leading away from the road along contour lines.

ROADWAY DRAINAGE IN URBAN AREAS

Water falling on or near city streets or sidewalks and on urban freeways and expressways generally is directed by the cross slope to the gutters and along them to curb or gutter inlets and from them into underground storm drains. Installations of this kind are very expensive compared to those suitable for rural areas. On the other hand, these expenditures have greater justification, as urban facilities generally carry larger volumes of vehicular traffic and often serve bicycles and pedestrians as well. Furthermore, property damage from flooding may well be extremely high in urban areas.

Gutters and inlets on urban streets and highways are generally designed to limit the spread of water over the traveled lanes to some rather arbitrary maximum. AASHTO recommendations are that, for at-grade arterials, the water not encroach on the outside lane by more than 6 ft for a 10-yr flood. A 50-yr return period is proposed for depressed facilities. It also recommends that designs for storm-water inlets and connections for gutters and depressed medians be based on the same return periods. Main storm drains for freeways are to accommodate 50–100 yr storms; for arterials the return interval is 20–50 yr. Estimates of flows are to be based on the rational formula using values of c, the runoff coefficient, between 0.8 and 0.9 for pavements, 0.4 and 0.6 for gravel, and 0.1 and 0.7 for grass.

The design of gutter and median inlets for freeways and urban streets deserves particular attention. They must pass the design flood without clogging with debris. At the same time, they must not be a hazard for vehicles, which means that the inlet must be flush with the surface of gutters or median. Where bicycles may be involved, openings in grating must be so dimensioned and directed that the narrow tire will not be trapped.[12]

Details of design for drainage facilities for urban highways is outside the scope of this book. The reader is directed particularly to *Design of Urban Highway Drainage*, FHWA and the *Journal of the Hydraulic Division of ASCE*.

CHANNELS[13]

Designs for roadside and crown ditches, gutters, stream channels, and culverts flowing partially full are all based on the principles of flow in open channels. For uniform flow, the basic relationships are commonly expressed in the Manning formula:

$$V = \frac{1.49}{n} R^{2/3} S^{1/2} \tag{11–4}$$

where

V = mean velocity in feet per second
n = Manning's roughness coefficient (see Table 11–2)

[12]FHWA *Hydraulic Engineering Circular No. 13, NCHRP Synthesis of Highway Practice No. 3,* Dah-Chen Woo, *Public Roads,* June 1978, C. F. Izzard, *TRB Record 631,* and A. W. Brune et al., *Journal of the Hydraulic Division of ASCE,* Dec. 1975, offer designs for efficient and safe gutters, medians, and inlets.

TABLE 11–2. Representative Values of Roughness Coefficient, *n*, Various Channel Linings *(Manning Formula)*

Type of Lining	*Value of* n
Ordinary earth, smooth graded	0.02
Jagged rock or rough rubble	0.04
Rough concrete	0.02
Bituminous lining, likely to be wavy	0.02
Smooth rubble	0.02
Well-maintained grass—depth of flow over 6 in.	0.04
Well-maintained grass—depth of flow under 6 in.	0.06
Heavy grass	0.10

R = hydraulic radius in feet; this is the area of the flow cross section divided by the wetted perimeter
S = slope of the channel

Also,

$$Q = VA = \frac{1.49}{n} AR^{2/3}S^{1/2} \tag{11–5A}$$

where

Q = discharge in feet per second
A = area of the flow cross section in square feet

In SI units, this equation is

$$Q = VA = \frac{1}{n} AR^{2/3}S^{1/2} \tag{11–5B}$$

In this equation, the definitions for the individual terms are as given above, except that meters replace feet. At times, these equations are modified to state the dimensions of particular channels or conduits more directly, but in all cases the same fundamental concepts apply.

Water progressing down mild slopes in open channels is in "subcritical flow," but that traveling steep slopes is in "supercritical flow" (see Fig. 11–4). The proper solution to many channel problems rests on this distinction. Subcritical flow exists when the depth of water in the channel is greater than the "critical depth," and supercritical flow when the depth is less than critical. Physically, critical depth is illustrated by the depth at which water flows over a weir. Mathematically, critical depth occurs when the velocity head ($V^2/2g$) is half the "mean depth." Mean depth, in turn, is the area of the flow cross section divided by its width at the liquid surface. It follows that critical depth is independent of channel slope and roughness and has a fixed value as long as the discharge and channel dimensions remain constant. It should be noted, however, that channel roughness does enter the calculation for "critical slope," at which uniform flow at critical depth occurs.

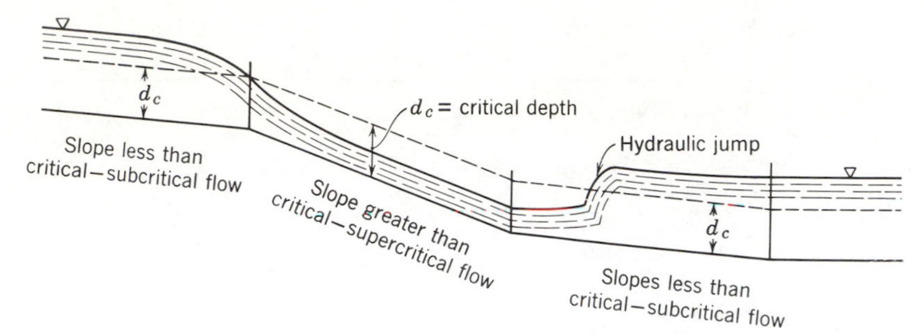

Fig. 11–4. Diagram illustrating the effect of critical depth on flow in open prismatic channels and partially full culverts.

The effects of critical depth on flow characteristics are illustrated by Fig. 11–4. At the crest of the steep slope, the velocity increase between subcritical and supercritical flow occurs smoothly over some distance. Lowering of the water surface begins upstream from the change in slope. Where the stream grade flattens again, transition from supercritical to subcritical flow takes place abruptly in a "hydraulic jump," in which turbulence absorbs some of the energy from the flowing stream. Some distance downstream from the hydraulic jump the channel grade flattens even more. This causes a further decrease in velocity and an increase in stream depth that begins some distance upstream from the grade change. Figure 11–4 demonstrates that, with subcritical flow, the depth of water at a given point may be altered by downstream conditions, in other words, "control is downstream." Thus, with subcritical flow, the effects of grade change, channel constriction, junction with another stream, and other modifications extend upstream in a "backwater curve." Methods for computing these are offered in the references cited earlier. On the other hand, unless supercritical flow is submerged to a depth greater than critical, it is unaffected by downstream conditions. Thus, with supercritical flow, "control is upstream."[14]

The aim in channel design is to find the cross section that will be cheapest to construct and maintain. Side slopes 2:1 or flatter are essential except in rock or other hard materials, or where the channel is lined. For unlined channels, it can be generalized that the best cross section requires the least total excavation. However, this rule applies only when construction can be accomplished with conventional equipment. In particular, designs that have numerous changes in dimensions or that must be executed by hand-labor methods will prove costly unless labor rates are very low.

Open-channel designs can be accomplished by solving the Manning equation numerically. As this procedure is tedious and time-consuming, charts have been

[14]When velocities become substantially greater than critical, designs for expansions, contractions, bends, side-stream entries, channels, bridge piers, and culvert entrances demand extreme care. See C. E. Behlke and H. D. Pritchett, *HRB Record 123,* for an example and references.

developed to solve the more common problems. One appears[15] as Fig. 11–5. Explanations and two examples appear in the legend. Notice that critical depth plays an important part in the analysis.

Channel design is not complete until the possibility of erosion is eliminated, within practical limits. The first step is to check actual velocity against maximum safe values for unprotected earth (See Table 11–3). Where channel scour is indicated, means for reducing velocity to safe levels or for protecting the channel should be adopted.[16]

An effective way to reduce velocity is to reduce the flow by diversion. In some instances this can be done, as has been pointed out in the discussion of crown ditches. In many cases, however, diversion is not possible. A cursory examination of the Manning equation might indicate that substantial velocity reduction could be gained by making channels wider, but unfortunately this is not so. In some instances velocity can be reduced somewhat by increasing channel length and thus decreasing the slope. However, velocity varies as the one-half power of the slope, so that large slope changes are required before velocity is changed appreciably. Furthermore, this solution usually is limited to open country where rights of way or easements for drainage cost little so that ditches can be laid out at an angle to the highway centerline. It is also possible to reduce the slope by introducing baffles, checks, or drops into the channel. These devices have been constructed of many materials, including sod, treated timber, corrugated metal, bales of hay, gabions (wire baskets filled with stone), and concrete.[17] However, many drops have washed out during the first severe storm. Those with adequate notch capacity and with suitable provision for energy dissipation on the downstream apron may be more costly in the long run than a continuous smooth lining. Furthermore, drops in roadside channels are a hazard to traffic and, in humid regions, interfere with mowing operations.

In humid areas, grass offers effective and attractive protection for roadside ditches and other small channels subject to intermittent flow. Grass roughens the channel and raises the safe velocity as compared to unprotected earth (see Tables 11–2 and 11–3). Many agencies sow grass on slopes and line the channels with sod promptly after grading operations. The effectiveness of a covering of loose stones laid over the native material (rip-rap[18]), and of artificial liners of various fibers also have been investigated.[19] A rubble lining of large stones

[15]This is one of a group of charts offered by FHWA as Nos. 3 and 4 of its Hydraulic Design Series under the titles *Design Charts for Open Channel Flow* and *Design of Roadside Drainage Channels,* respectively. These reports are reprinted in *Compendium No. 5* of the TRB series on Transportation Technology Support for Developing Countries.

[16]An excellent basic reference is Text 5 of *Compendium 9* in the TRB Series Transportation Technology Support for Developing Countries.

[17]See, for example, D. J. Poche and W. C. Sherwood, *TRB Record 594,* and M. A. Burroughs, *Civil Engineering,* Jan. 1979.

[18]For a study of the stability of rip rap see M. A. Stephens et al., *Journal of Hydraulic Division of ASCE,* May 1976.

[19]See FHWA *Hydraulic Engineering Circular No. 15.*

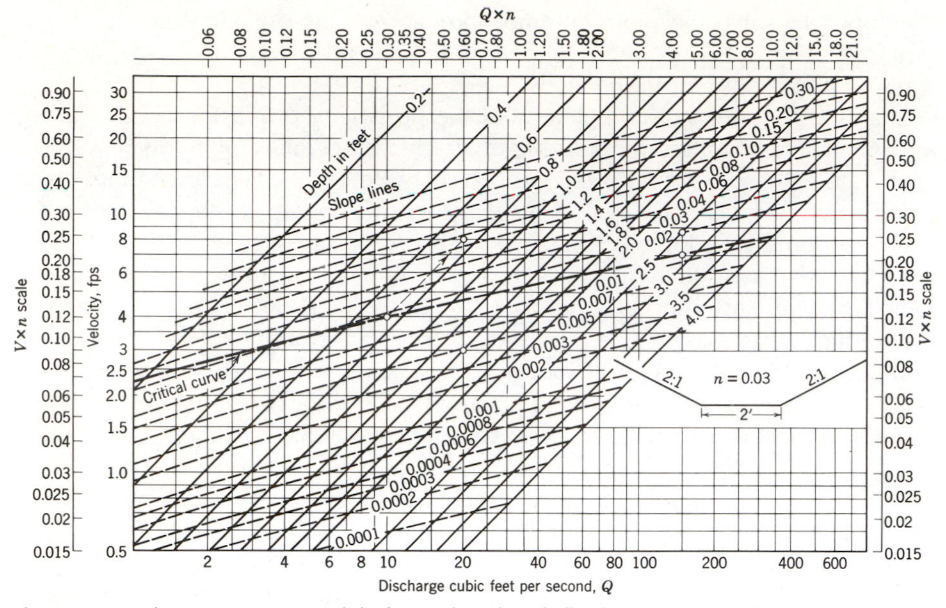

Fig. 11–5. Flow in a trapezoidal channel with 2-ft flat bottom and sides on 2:1 slopes. Critical depth is independent of channel roughness and must be read only from Q scale; critical slope is dependent on channel roughness and is read at intersection of critical depth and vertical line $Q \cdot n$. (Source: *FHWA Hydraulic Design Series No. 3*.)

Example 1. Given: Discharge. $Q = 150$ ft³/s; roughness, $n = 0.03$; slope, $S = 0.02$. Read upward from $Q = 150$ to intersection with $S = 0.02$, and find normal depth, $d_n = 2.5$ ft and normal velocity, $V_n = 8.5$ ft/s. Also, for $Q = 150$, critical depth, $d_c = 2.8$ ft; critical velocity; $V_c = 7.0$ ft/s, and critical slope, $S_c = 0.013$ or 1.3%. Thus flow is supercritical and control is upstream.

Example 2. Given: Discharge, $Q = 10$ ft³/s, roughness, $n = 0.06$; slope, $S = 0.5\%$ or 0.005. Read down from $Q \times n = 10 \times 0.06$ or 0.6 to slope, $S = 0.005$, and find normal depth, $d_n = 1.4$ ft and $V \cdot n = 0.09$. Compute $V = 0.09/0.06 = 1.5$ ft/s. To find critical depth and critical velocity for 10 ft³/s discharge, read upward from $Q = 10$ to critical depth, $d_c = 0.72$ ft, and critical velocity, $V_c = 4.0$ ft/s. Critical slope for roughness, $n = 0.06$ is read at intersection of diagonal line $d_c = 0.72$ and $Q \cdot n = 0.6$ and equals 0.07 or 7%

bound with cement mortar is another effective means of channel protection, particularly for roadside ditches. It is rough and has high erosion resistance but may be costly, as the individual stones are laid by hand. Paving with concrete or dense-graded bituminous mixtures is often done on small and large channels alike. As these linings can be placed successfully on steep side slopes and have relatively smooth surfaces, designs can be hydraulically efficient and thus less costly. Where the design results in high velocities, provision for energy dissi-

TABLE 11–3. Maximum Safe Velocities When Channel Erosion Is To Be Prevented.

Type of Lining	*Allowable Velocity (ft/s)*
Well-established grass on any good soil	6
Meadow type of grass with short, pliant blades, heavy stand, such as bluegrass	5
Bunch grasses, exposed soil between plants	2–4
Grains, stiff-stemmed grasses that do not bend over under shallow flow	2–3
Earth without vegetation:	
Fine sand or silt, little or no clay	1–2
Ordinary firm loam	2–3
Stiff clay, highly colloidal	4
Clay and gravel	4
Coarse gravel	4
Soft shale	5

pation must be made at the lower end of the channel unless discharge is into a rocky stream bed or into a pool deep enough to force a hydraulic jump.[20]

Streams carrying sediment bring particularly difficult channel-design problems. It has been established that the ability of water to transport solids along the stream bed varies as a power of the velocity greater than the square. It follows that any sizable velocity decrease will cause the stream to drop part of its load. This debris may clog the channel or fill culverts, causing troublesome and sometimes costly overflow and heavy maintenance expense. The most acceptable solution is to design channels and structures so that the upstream velocity is equaled or exceeded throughout. Then the sediments will be transported on past the highway as before. In situations where the initial velocity cannot be maintained, sediments may be trapped in strategically placed debris basins designed for easy cleaning with mechanical equipment.

At times highway improvements require that existing streams be transferred to new channels. For example, stream alignment may be straightened as in Fig. 6–3. Again the road may intersect a stream twice as it cuts directly across a meander or looping bend. Here the first cost is often far less to carry the stream in a channel paralleling the road than to construct two bridges or culverts. Changes like these increase the slope of the channel and often improve its hydraulic properties. Both measures bring increased velocity unless the excess energy is absorbed by means of drops or other devices. Without such controls, unstable conditions are set up in streams having erodible beds. During succeeding storms, the current will cut and fill until a new balance is established. Often these adjustments in grade and alignment are sizable in amount and extend far

[20]See *NCHRP Report 108* and *FHWA Hydraulic Engineering Circular No. 14*.

up and down stream. It follows that channel changes on streams with erodible beds must be planned with extreme care.[21]

DIKES

Dikes are earth embankments constructed to contain or divert stream flow. Where all construction is to be above the level of the existing ground, dikes are used alone. Often, however, dike–channel combinations represent the most economical solution, as the dike can be made of spoil from the channel. In the past, dikes were often built by casting or dumping the materials loosely into place; but modern practice demands that construction be in compacted layers as for roadway embankments (see Chapter 15).

Unprotected dike faces exposed to swiftly moving water are subject to erosion. Defensive measures include those previously mentioned for channels. Also, bank-protection installations such as rock blankets, fabrics, gabions, sacks filled with concrete, or precast concrete elements have proved effective. Special care must be taken to extend the protection out into the channel or a distance below the stream bed so that the water cannot undercut it. A wide variety of solutions appropriate on rivers or ocean or lake fronts have been devised but cannot be offered here.

Where highways cross wide streams or flood plains, a portion of the crossing is generally on a bridge, the remainder on approach embankments. "Spur dikes," generally parallel to the direction of the stream, often are used to direct flow through the bridge. Few quantitative measures of the relative effectiveness of various shapes, locations, and conformations of such spur dikes have been available in the past. The results of recent model studies, too detailed for inclusion here, have led to tentative criteria for spur dikes that can be invaluable aid to designers.[22]

Where dikes are constructed to turn the flow of a stream, the need for face protection sometimes can be eliminated by placing the dike at a distance back from the channel proper. Then an area of dead water cushions the bank and prevents scour.

CULVERTS

The term "culvert" encompasses practically all closed conduits used for drainage with the exception of storm drains. Culverts must be classed as stock products, in that standard designs are used repeatedly. This is in direct contrast to the situation for bridges that span larger streams, for which special designs are made in almost every case. Culverts are far more numerous than bridges, and more money is spent on them. In fact, about one-sixth of the highway construction dollar goes for these smaller drainage structures. Clearly, culvert design warrants serious attention.

[21]For added discussion see H. W. Shen et al., *TRB Record 736*.
[22]FHWA *Hydraulic Engineering Circular No. 11* offers details of many bank-protection devices.

Culvert Types

The more common culvert types and the materials of which they are made are shown in Fig. 11–6. For smaller openings, pipe in stock sizes is generally chosen, with the pipe arch as a substitute where headroom is limited. For openings of moderate size, pipe and box culverts compete for favor. For larger openings, single- or multiple-span box culverts are generally used, although one or more large-diameter pipes of reinforced concrete or bolted metal plates sometimes are preferred. Bridge culverts replace box culverts when the foundation is nonerodible and a paved floor is unnecessary. Arch culverts may be economical under high fills where loading is heavy.

Under normal circumstances, selections of culvert type and material are

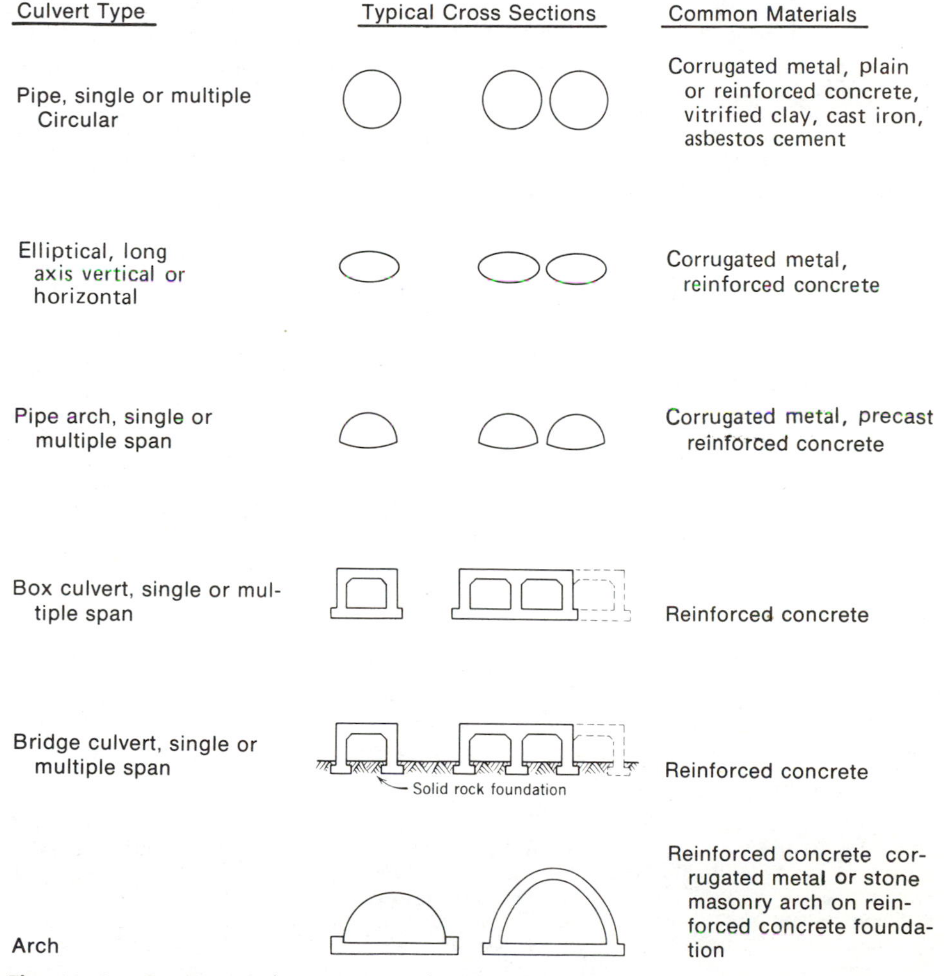

Fig. 11–6. Common culvert types and materials. Metals include galvanized iron and steel and aluminum alloy.

based on comparative costs. At times, however, other factors may control. For example, the presence of corrosive agents in the soil may bar certain materials unless a means of protection can be devised. Again, if the structure location is remote, the portability and ease of erection of light, prefabricated metal sections may make them particularly desirable. At times, factors such as the availability of skilled labor or time limitations may govern. In any event, the decision must be based on careful study of all pertinent factors.

Prefabricated Metal Culverts[23]

Accepted materials for prefabricated metal culverts are cast iron, corrugated and zinc-coated (galvanized) copper-bearing pure iron or copper-steel, and corrugated aluminum alloy. All have relatively high resistance to corrosion. In forming the metal pipe, individual flat sheets of stipulated thickness and wide enough to provide 2- to 5-ft culvert lengths are bent to the selected cross section. Joints are lapped and fastened either by spot welding or with cold-driven rivets of the base metal. During forming, circumferential corrugations are pressed into the metal. These vary with pipe diameter. For example, for iron and steel pipe with diameters 12 to 36 in., corrugations are ½ in. deep on 2⅔-in. centers; for larger diameters these dimensions may be, respectively, 1 in. and 3 in. Individual sections are assembled at the shop into lengths convenient for transportation and field handling. Field connections between these lengths are made with corrugated metal bands pulled tight with corrosion-resistant bolts.

Large diameter pipes, arches, and arch culverts of corrugated metal with some spans exceeding 50 ft are made up into segments of manageable size that can be assembled and bolted into a unit on the site. Sometimes the specifications require that pipe sections be coated with bituminous material to provide added protection against corrosion. For situations where the stream carries sand, gravel, or other abrasives, the invert may be paved with bituminous mastic.[24]

Nonmetallic Culverts[25]

Culvert pipe of plain or reinforced concrete, cast iron, asbestos cement, or vitrified clay is commonly made at the plant in standard lengths. Jointing between individual sections, using specified materials and methods, follows bedding of the pipe. Some installations are constructed of plain concrete in place using a slip form or inflated rubber tube to maintain the inside opening.

[23]See AASHTO specifications *M36, M167, M190, M196, M218,* and *M219* in its publication *Transportation Materials* for details on materials and dimensions.

[24]*NCHRP Synthesis 50* gives an excellent summary and extensive bibliography on corrosion resistance and other durability problems of culverts. *AASHTO Designation M190* gives the requirements for bituminous pipe coatings and linings. See also *TRB Record 713*, and *762*.

[25]See *AASHTO Designations M64, M170, M206,* and *M217* for specifications, dimensions, reinforcing, and other details of these types of pipe. See the standard specifications and drawings of the state highway departments for details of jointing practice.

Culvert Loads and Stresses

Culverts are loaded vertically by vehicle wheels and superimposed fill and horizontally by passive or active earth pressure. Particularly for pipe culverts, the magnitudes of these loads are uncertain. Major factors influencing them are depth of cover; nature and density of the overlying and adjacent soils; trench width and depth (if the pipe is set in a trench rather than on a flat surface); deformation of the pipe under load; and field-construction procedures.

Stress analysis for pipe culverts is complex because they are indeterminate structures subject to complex loadings. Also, stress calculations are based on assumptions regarding the method of support under the pipe, soil support at its sides, and load distribution over it. For example, results are far different between point support in a flat-bottom trench and distributed support when the pipe is carefully bedded or is cradled in concrete. It must be concluded that theoretical analysis can define the range in which loads and stresses lie, but that the results may not be quantitatively correct for particular situations.[26]

Corrugated metal pipe is more flexible than pipe of most other materials and can tolerate considerably greater deformations. Under vertical loads, the sides of corrugated metal pipe tend to deform laterally against the adjacent backfill. The resulting horizontal earth pressure substantially increases the load-carrying ability of the pipe. Many agencies require that large-diameter pipe of corrugated metal be distorted vertically by temporary shores before backfilling is begun. Removal of the shores after backfill results in large horizontal pressures against the pipe and creates a more favorable stress distribution.

Some 40 years ago, Spangler of Iowa State University developed an "imperfect ditch" approach which involved imposing a layer of straw in the fill over the culvert. This transfers a substantial portion of the weight of the overlying fill from the pipe to the adjoining embankment.[27]

Because of the difficulties and uncertainties outlined here, highway agencies seldom make structural designs for pipe culverts. Instead, using research findings and past experience as a guide, they have developed standard plans to fit all usual situations. Thus, for corrugated-metal pipe, highway agencies prescribe plate thickness for various pipe sizes and fill heights. For concrete, vitrified clay, and cast-iron pipe, they specify the particular strengths or classes of pipe to be used in each situation. Strong reliance is placed on the recommendations of the various manufacturers.

Design and construction of box, bridge, and arch culverts of reinforced concrete follow the fundamental rules developed in courses in structural engineering and will not be discussed here. It should be noted, however, that each highway agency has standard drawings covering culvert designs appropriate for the more common heights and widths of openings, fill heights, and skew angles.

[26]The literature on loads and stress in culverts is too extensive to detail here. *FHWA Reports 77–5, 6,* and *7* present CANDE, a computer-based approach to analysis. Recent articles also appear in *HRB Record 413* and *443* and *TRB Record 510, 517, 518, 548, 556, 640, 678,* and *785*.
[27]For a recent application of this method and a bibliography see A. C. Scheer and G. A. Willett, Jr., *HRB Record 262*.

Culvert Installation

Standard procedures for installing culverts are directly related to loads and stresses as discussed above. They are spelled out in detail in the specifications of each agency. Particular attention is devoted to bedding and to backfill in order to protect the culvert and to prevent subsequent settlement in the roadway surface. Many agencies specify that backfill materials be brought to the proper moisture content, placed in small lifts, and compacted with power-driven tampers. Some permit the substitution of clean sand or gravel for the regular backfill material, in which case careful jetting or puddling usually produces satisfactory compaction. The principles of soil compaction discussed in detail in Chapters 14 and 15 apply equally to trench backfill. Of necessity, equipment for trench backfill is smaller since the operation must be conducted in close quarters.

Culverts through embankments demand particular attention to protect them from damage by construction equipment and to secure proper soil compaction around them. Some agencies require that the embankment first be constructed above the level of the culvert crown, after which a trench is dug for the culvert.

Culvert Location

Culverts usually are installed in the original stream bed with their grades and flow lines conforming to those of the natural channel. In this way, disturbance to stream flow and the erosion or deposition problems it creates are held to a minimum. In rolling and mountainous country in particular, marked departures from channel alignment either upstream or downstream may direct the current to one side of the channel, causing erosion there and deposition on the opposite side. On the other hand, culverts on substantial skews are longer and more costly than those at right angles or on small skews. Often the best solution involves reducing large skews somewhat and providing a channel change and erosion protection at one or both ends of the structure.

In rough country, culverts can sometimes be advantageously located on a bench on the side of the canyon rather than in the channel. Under high fills, positioning in the channel is costly as culverts must be very long and carry heavy loads. Use of the side hill location reduces both length and load. On the other hand, the erosion threat at the outlet may require expensive control measures. Also, objections to ponded water because of health or safety hazards and threats to stability of the fill must be overcome. Where the stream bed is steep, a compromise solution with the culvert entrance lowered to the stream bed level and the culvert on the sidehill may be most satisfactory. Curvature or breaks of grade to make the culvert conform to the channel should be used if the design is cheaper and is hydraulically and structurally sound.

Inverted siphons (depressed culverts) should be avoided whenever the water carries sediment or debris. Even though the velocity at peak flow may keep the barrel clear, deposits may collect as the discharge decreases. Also, stagnant water trapped in the sag may be objectionable.

Headwalls and Endwalls

Most culverts begin upstream with headwalls and terminate downstream with endwalls. Headwalls direct the flow into the culvert proper, while endwalls provide a transition from the culvert back to the regular channel. Hydraulically, then, they function differently. Both retain the embankment and protect it from washing by flood waters. Common types are diagramed in Fig. 11–7. Most headwalls and endwalls are cast in place of reinforced concrete, although rubble masonry and timber have been used at times. Units prefabricated of corrugated metal or precast of concrete are sometimes installed with pipe of the same materials. In all cases, cutoff walls extending below the level of expected scour should be incorporated in the design. Often a paved apron or energy dissipator extending beyond the cutoff wall is a wise addition.

Straight headwalls and endwalls are selected mainly with smaller pipe culverts (see Fig. 11–7*a*). They are hydraulically inefficient as entrances. In recent years, some agencies have been omitting endwalls and, sometimes, headwalls from small pipe culverts. Instead, the pipe is extended beyond the toe of the embankment. As pointed out below, from a hydraulic point of view this design is inefficient, since the entrance loss for a projecting thin-walled pipe flowing full is about 0.8 velocity head.

The L type headwalls (Fig. 11–7*b*) direct the flow from roadside ditches into culverts under the road. They create a serious accident hazard, and many agencies are replacing them with gutter inlets covered with grates. For large culverts, wing type walls (Fig. 11–7*c*) are most widely used. Entrance losses with them are about 0.15 of a velocity head as contrasted with a loss of 0.05 with hydraulically designed entries (see below). Flared, U, and warped walls (Figs. 11–7*d*–*g*) have special applications.

Hydraulic Design of Culverts

The aim in hydraulic design is to find the type and size of culvert that will, in the long run, most economically accommodate the flow of the stream. In almost

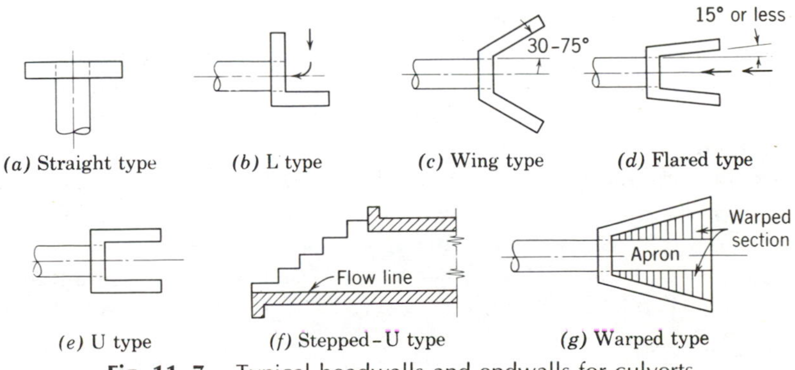

Fig. 11–7. Typical headwalls and endwalls for culverts.

all cases, the primary control on culvert size and inlet conformation is the permissible level of the headwater pool upstream from the structure. Under certain circumstances, high headwater may have serious consequences, and cannot be tolerated. For example, in settled areas high headwater may inundate valuable property. Again, a relatively low fill may be overtopped, bringing lengthy interruptions to traffic and serious damage to pavement and embankment. Sometimes high velocities through the culvert may produce erosion problems downstream or threaten damage to the culvert or its appurtenances. On the other hand, there are many situations where high headwater will cause little damage or inconvenience. In these cases the extra head can be recognized in the design, with an accompanying reduction in culvert size and cost. In effect, then, culvert design becomes an economy problem in which structure costs for various headwater elevations are balanced against the estimated costs of possible damage or inconvenience.[28]

Some agencies have established arbitrary design criteria for culverts in terms of the height of the headwater pool. This approach might require, for example, that "a culvert just pass a 10-yr flood without static head at the crown of the culvert at entrance, and that design of culvert appurtenances be balanced to avoid serious damage from head and velocity obtaining in a 100-yr flood." Headwalls would be proportioned for zero freeboard against (possibly) a 100-yr flood.

Flow through conventional culverts may occur in numerous fashions, some of which are illustrated in Fig. 11–8. The diagrams under case I demonstrate three flow patterns with low headwater. The most significant characteristic is the free water surface at the entrance and through the culvert barrel. Notice that this free water surface exists even when the headwater pool is somewhat above the crown of the culvert. Flow under cases I*a* and I*b* are similar to flow over a weir. Case II depicts three common situations where the culvert entrance is submerged. In case II*a* the headwater surface is above the crown of the culvert by something more than one-fifth the culvert height. Flow is comparable to that through an orifice. The barrel does not flow full and will not do so, except for *HW* at least nine times culvert height. In case II*b* the culvert flows full because the barrel slope is too flat to overcome friction losses. In case II*c* tailwater submerges the outlet, which causes the barrel to flow full at the outlet, and, under most circumstances, to flow full for its entire length. Cases II*b* and II*c* are practically identical with problems of flow in closed conduits of limited length as treated in fluid mechanics.

For designs combining low headwater, low tailwater, and steep culvert slopes (Fig. 11–8, case I*a*), flow through the culvert will be supercritical. Unless downstream channel flow is also supercritical, a hydraulic jump will occur near the culvert outlet. For this flow pattern, control is at the culvert entrance, which functions much like a weir. The relationship between discharge and headwater

[28]See *Highway Drainage Guidelines—Hydraulic Design of Culverts,* AASHTO, and *L. W. Mays, Journal of the Hydraulic Division of ASCE,* May, 1979.

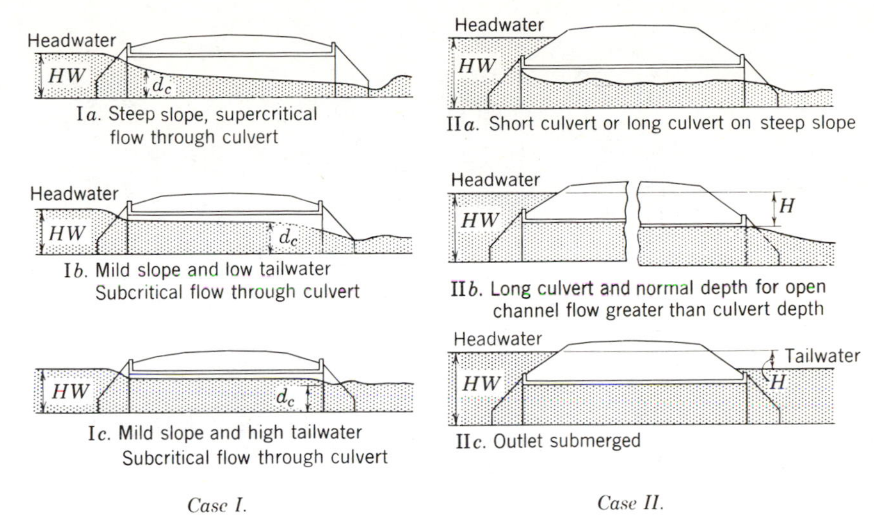

Fig. 11–8. Typical flow patterns for culverts with square-edge entrances.

elevation can be determined for box culverts from Fig. 11–9, which is representative of a series of nomographs published by FHWA as *Hydraulic Engineering Circulars 5* and *10*. These cover the usual sizes of box culverts and also pipe culverts of concrete and corrugated metal for common conditions of flow. Figure 11–9 indicates that wing type headwalls are the most effective hydraulically of the three shown; straight or flared types are poorer, and the U type is least effective of all.

Where low headwater, low tailwater, and culvert slopes less than critical are combined (case I*b* of Fig. 11–8), flow in the barrel is subcritical but passes through critical depth near the outlet, which becomes the control section. A backwater curve and headwater elevation can be computed as a problem in nonuniform flow by methods outlined in textbooks in fluid mechanics. The combination of low headwater, high tailwater without outlet submergence, and mild barrel slope (case Ic of Fig. 11–8) is seldom found in practice. If it is encountered and critical depth is not submerged, the solution is essentially the same as outlined for case I*b*.

Where culverts are to operate with submerged entrances, the first design step is to determine whether the culvert in question will flow full. If there is a free outlet (cases IIa and II*b* of Fig. 11–8), this depends on culvert length, slope, and roughness. If the outlet is submerged (case II*c*, Fig. 11–8), the culvert will always flow full for its entire length.

For culverts flowing full, head difference between down and upstream water surfaces is computed by applying the Benoulli equation

$$H = \left[1 + k_e + \frac{29n^2L}{R^{4/3}} \right] \frac{V^2}{2g} \qquad (11\text{–}6A)$$

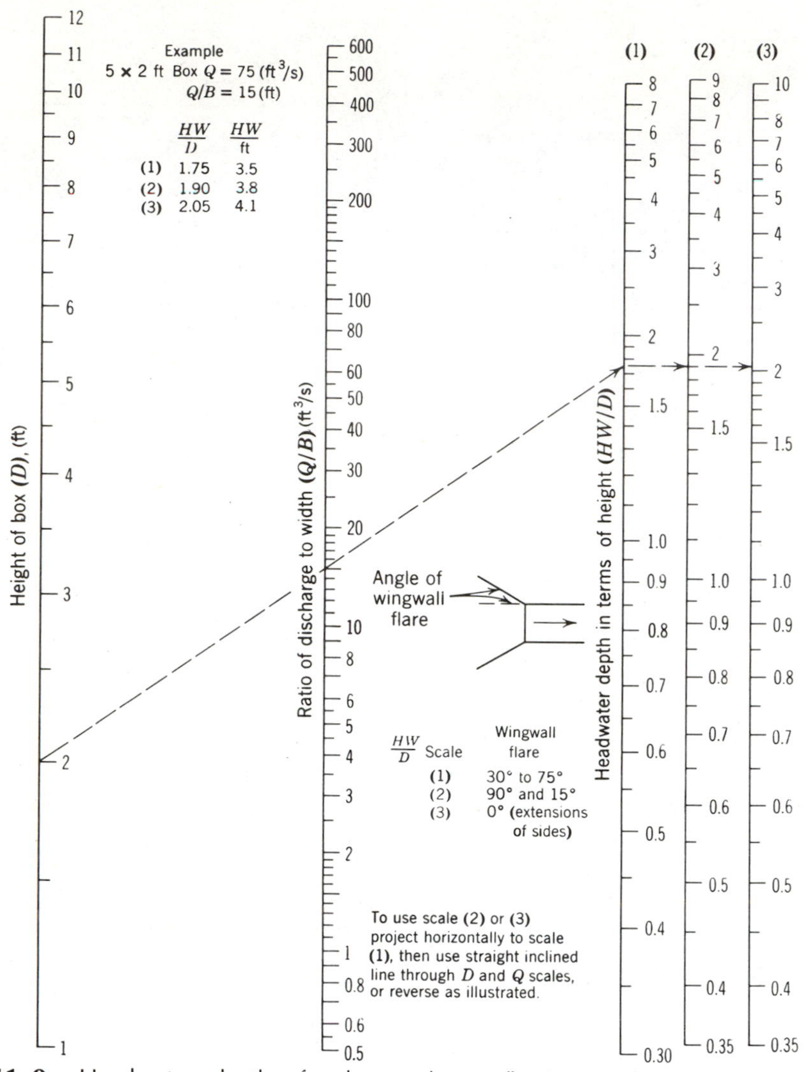

Fig. 11–9. Headwater depths for box culverts flowing under entrance control. (Source: *FHWA Hydraulic Engineering Circular No. 5.*) 1 ft = 0.3048m, 1 ft^2 = 0.0929m^2, 1 ft^3/s = 0.02832m^3/s

where

H = difference in elevation between the surfaces of the headwater and tailwater pools (Fig. 11–8, case IIc), or between the surface of the headwater pool and the crown of the culvert outlet (Fig. 11–8, case II*b*), in feet

$V^2/2g$ = velocity head in feet

k_e = coefficient of entrance loss, ranging from 0.9 for corrugated metal projecting from the fill through 0.7 for the least effective entrances to 0.2 or less for carefully rounded entrances

n = Manning's roughness coefficient (see Table 11–4)
L = length of the culvert, in feet
R = hydraulic radius in feet

In SI units this equation is

$$H = \left[1 + k_e + \frac{2gn^2L}{R^{4/3}}\right]\frac{V^2}{2g} \tag{11–6B}$$

In this equation also, the definitions for the individual terms are as given above, except that meters replace feet. In metric units, 2g equals 19.6.

Figure 11–10 is a nomograph which solves the Bernoulli equation for case IIc situations. It applies to box culverts of the usual sizes and with a range of entrance conditions. By using both Fig. 11–9 and 11–10 and accepting the higher value for the headwater elevation that is found, these two graphs will solve many box culvert flow problems. Figure 11–11 is a sample of the information in the nomographs available in the two FHWA publications for concrete or corrugated-metal pipe culverts of various conformations.

Culvert Inlet Design

In hydraulic engineering terms, the entrances of highway culverts of standard design would be classed as square-edge, as contrasted with rounded, tapered, or bell-mouth. It is true that the upstream end of corrugated metal pipe is flared slightly, that concrete pipe generally is laid with the bell end of the bell and spigot joint upstream, and that the corners of concrete box culvert entrances have been chamfered. However, none of these constitutes rounding in the hydraulic sense. It has been demonstrated in carefully controlled laboratory experiments that by generously rounding the entrances of circular pipes or tapering the inlets of rectangular or square cross sections, the flow accommodated at

TABLE 11–4. Suggested Values of Roughness Coefficient, n, for Various Culvert Materials

Kind of Culvert	*Value of Manning's* n
Concrete pipe with rough joints	0.013
Concrete pipe, ordinary joints, reasonably smooth	0.012
Concrete pipe, excellent joints, steel forms	0.011
Concrete box culverts, plywood forms, smooth	0.012
Concrete box culverts, ordinary formwork	0.013
Concrete box culverts, rough, with sediment deposits	0.016
Vitrified clay pipe	0.013–0.014
Corrugated metal pipe, riveted, ½-in corrugations	0.024
Corrugated metal pipe, paved invert, ½-in corrugations	0.019–0.021
Corrugated metal pipe, 2-in. corrugations	0.030
Corrugated metal structural plate pipes and arches	0.0302–0.0328

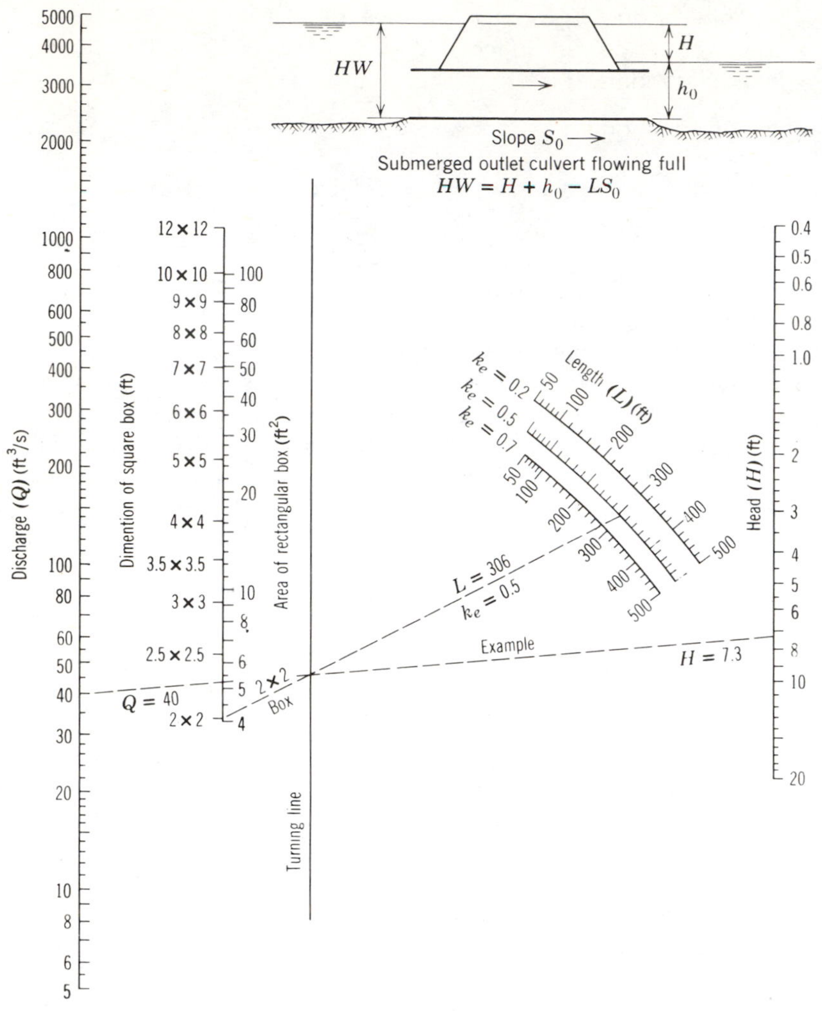

Fig. 11–10. Relationships between head, quantity of flow, and dimensions for concrete box culverts flowing full, $n = 0.12$. (Source: *FHWA Hydraulic Engineering Circular No. 5*.) 1 ft = 0.3048m, 1 ft² = 0.0929m², 1 ft³/s = 0.02832m³/s

a given head can be substantially increased or, conversely, at a given flow the headwater surface can be lowered. This comes about in long pipes on steep slopes because the conduits with special entrances flow full for their entire lengths, while those with square-type entrances do not. For the rounded entrances the full difference in elevation between downstream and upstream water surfaces is effective in forcing water through the conduits.[29] To provide this augmented capacity, however, pressure less than atmospheric must be developed and maintained at the throat of the culvert entrance. Unfortunately, later exper-

[29]For experiments with circular pipe, see L. G. Straub et al., *HRB Research Report 15–B;* for those on tapered inlets for box sections, see R. H. Shoemaker, Jr. and L. A. Clayton, ibid.

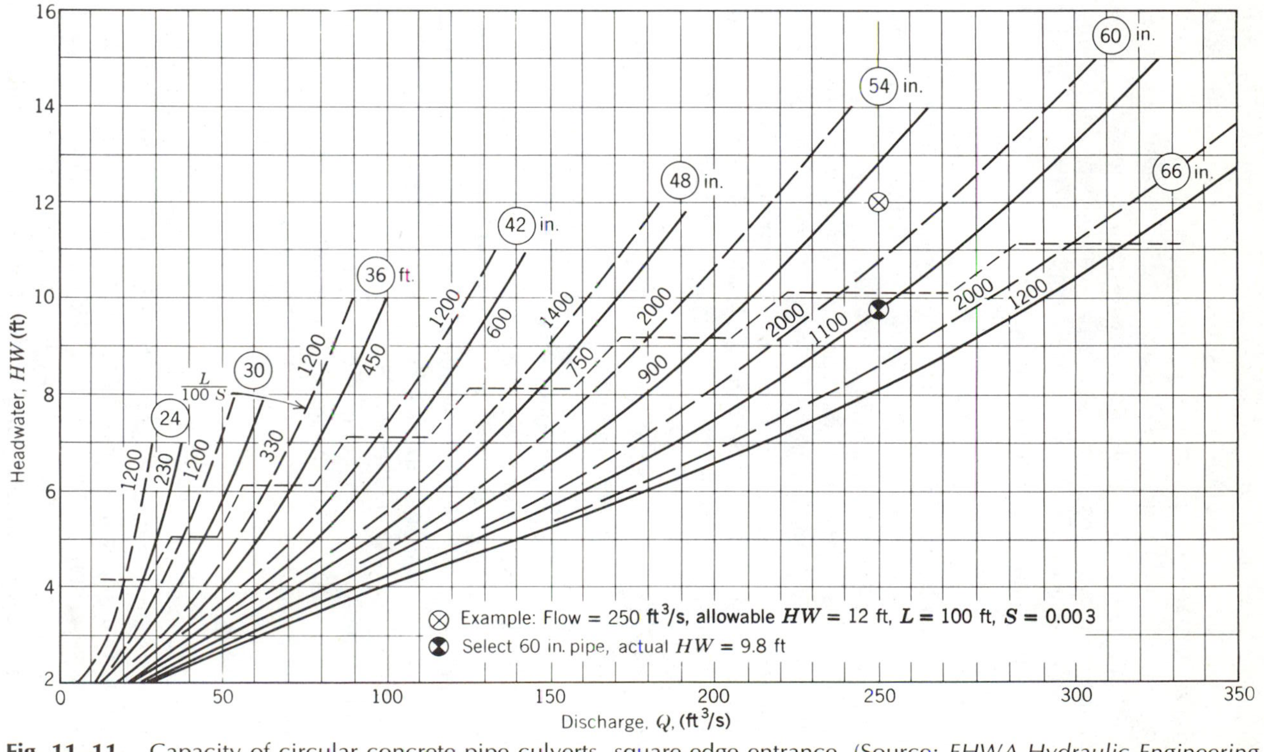

Fig. 11–11. Capacity of circular concrete pipe culverts, square-edge entrance. (Source: *FHWA Hydraulic Engineering Circular No. 10,* with data from two charts combined.) 1 ft = 0.3048m, 1 ft^2 = 0.0929m^2, 1 ft^3/s = 0.02832m^3/s

iments indicate that such favorable behavior cannot be counted on in actual installations.[30] Slight channel turbulence upstream, vortex formation at the entrance, and air entrained in the water or penetrating from the downstream end of the culvert all conspire to break the vacuum and bring atmospheric pressure at the culvert entrance. Experiments aimed at reshaping the entrance or using auxiliary devices to prevent loss of the vacuum have not been successful as yet. It would seem then that long culverts on steep slopes must be designed assuming entrance control, and that no practical method now exists for utilizing the fall through the barrel.

Even though advantage cannot always be taken of the total head difference through long, steep culverts, modified entrance design can often be extremely effective toward improving culvert flow. Figure 11–12 shows a rough summary of some of the findings at the Bureau of Standards. A generously rounded entrance (curve *H*) is most effective, while a protruding thin-walled pipe (curve *A*) is poorest. The capacity of the square-end pipe with headwall (curve *D*), which represents common design practice, falls between these extremes. From these curves it can be generalized that by careful attention to inlet design, marked reductions in culvert sizes can be obtained without increasing the elevation of the headwater pool.

For larger culverts, beveled, tapered, or tapered and sloped entries to box culverts offer substantial improvements in capacity over standard designs; or, conversely, design flows and headwater controls can be accommodated with significantly smaller barrel cross sections. A beveled entry results when the upstream end of the culvert roof is given a 1 to 1 batter for $\frac{1}{25}$ of the culvert depth. Tapered and slope-tapered inlets are diagramed in Fig. 11–13. For a flow of 1000 ft^3/s, and headwater of 10 ft, required cross section areas in square feet for the different c.asses of entries are as follows: square edge, 98; beveled edge, 84; side-tapered, 66; and slope-tapered, 42. It is recognized that such special inlets cost substantially more to build than traditional ones. But, particularly where the culvert barrel is long, this cost difference in the entry can be offset several times over by reduced cost of the straight barrel section.[31]

Culvert Outlet Design

Culverts on steep slopes or carrying large flows often discharge at such high velocities that they create serious erosion problems in unprotected channels. Some velocity reduction can be gained by roughening the floor and walls of the culvert or by taking advantage of the greater flow resistance of corrugated metal pipes. Usually, however, the excess energy is dissipated at the discharge end of the culvert by creating a hydraulic jump if the flow is supercritical, or by generating turbulence in some other manner. One effective means is to direct the flow into a basin lined with rock, either of one size or graded from sand through

[30]See *National Bureau of Standards Reports to the Bureau of Public Roads No. 4444*, 1955; *No. 4911*, 1956; and *No. 5306*, 1957.
[31]See *FHWA Hydraulic Engineering Circular No. 13*.

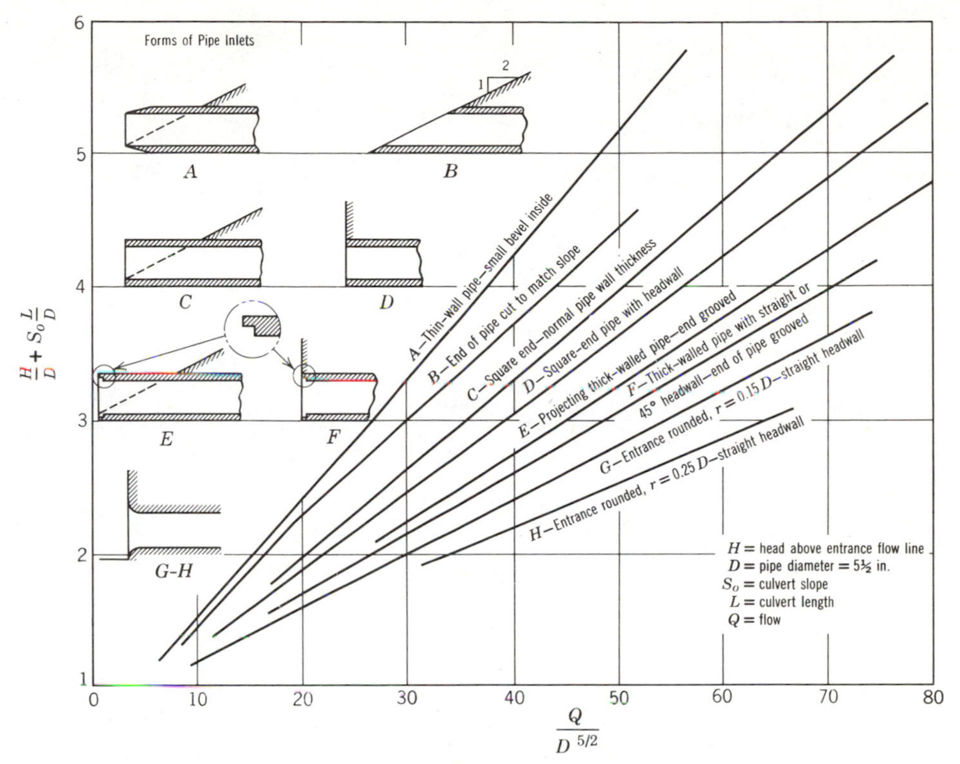

Fig. 11–12. Flow characteristics of several circular pipe inlet models. Entrance submerged, entrance control.

large boulders. At times a downstream sill may be provided. Other devices have also been developed. Drops in the upstream channel or drop inlets to the culvert proper, if designed to produce free fall in the stream, sometimes offer an economical means for velocity control.[32]

Debris Control

Streams in flood often carry brush and occasionally transport large branches, whole trees, or other sizable objects. There are many instances where this floating debris has clogged culvert entrances and raised the headwater elevation till the road was overtopped or adjacent property damaged by flooding. Where possible, culverts should be designed to pass expected debris. For example, where a stream may carry large floating objects, a single large-span box culvert is preferable to a multispan structure of the same total opening. As a possible alternative, the curtain wall separating the barrels of the multispan culvert might be extended upstream, with its top slanting downward. Debris will ride up on this wall, or at least be turned to pass more easily through the opening. In many cases, upstream debris racks of wire, timber, steel rails, piling, or other mate-

[32]See *FHWA Hydraulic Engineering Circular No. 14* and *TRB Record 785.*

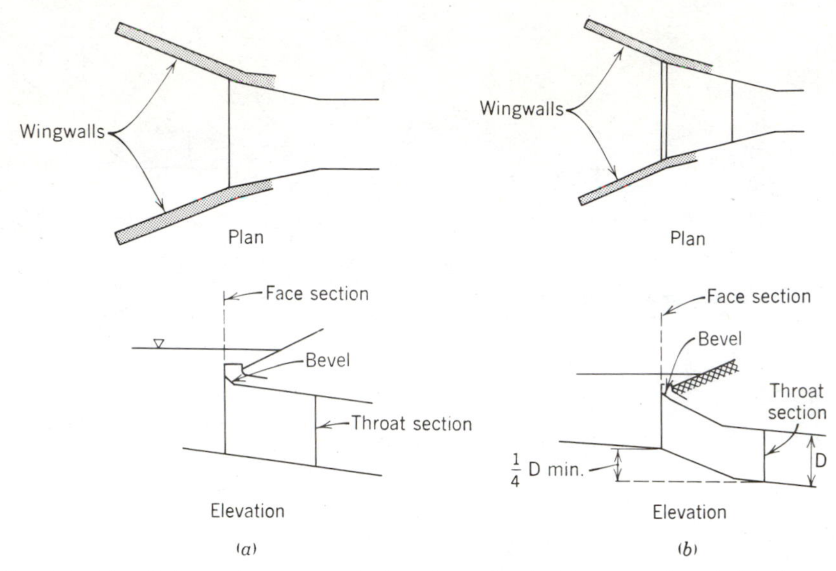

Fig. 11–13. Schematic plan and elevation of *(a)* side-tapered and *(b)* slope-tapered inlets for box culverts.

rials may offer the most reasonable solution to the problem. With such installations, maintenance crews must remove the debris following each flood.

At times, it may be desirable to trap sand or gravel carried by the stream rather than to pass it through the structure. Again, in mountainous areas, large rocks and boulders may be tumbled along the stream bed. Here also provision must be made either to pass this detritus through the structure or to trap it upstream where it will do no harm.

Design of debris-control devices depends on the form of debris or detritus to be handled, the volume of flood water, and individual site conditions. Experience in similar situations is a most useful guide.[33]

Dips and Other Low-Cost Water Crossings

A dip is formed by lowering the roadway grade to the level of the stream bed from bank to bank of the stream. Vertical curves at each end form transitions back to the regular grade line. Washing of the roadway surface is prevented by a curtain wall of concrete, timber, gabions, or cement-rubble masonry along the downstream edge of the traveled way.

In arid regions where streams flow at infrequent intervals, dips often provide an economical substitute for culverts on roads carrying low volumes of traffic. They offer a large waterway area at little cost. If properly designed, they are little damaged by flood water and are self-cleaning, so that maintenance costs

[33]See *FHWA Hydraulic Engineering Circular No. 9,* S. A. Gilje, *Public Roads,* Mar. 1979. and *NCHRP Synthesis 70.*

are low. With long transitions at the ends, they ride smoothly. Principal disadvantages are interruptions and hazard to traffic when water is flowing.

A dip–culvert combination has sometimes been chosen to good advantage. Here the grade line is only partially lowered. Pipe culverts under the road surface at stream bed level carry small flows without inconvenience to traffic. The larger waterway capacity of the dip comes into play during major floods. Dip–culvert combinations must have carefully streamlined profiles, and the downstream side must be protected by pavement. Provision for energy dissipation at the downstream toe is also required.[34]

BRIDGES[35]

Highway bridges fall into two classes; those that carry vehicular traffic and pedestrians over larger streams and other bodies of water and those that separate traffic movements at interchanges and street or pedestrian over- or undercrossings. Although bridges are relatively few in number, each structure presents unique problems and involves a large expenditure. In American highway practice, bridge design is considered a distinct function and usually is carried out in a separate department of the highway agency. At times, private engineers are called in to advise on particularly difficult problems and some highway agencies assign the complete design to such consultants.

Highway bridge engineers utilize the same analytical tools as other structural engineers—principal differences are in clearance and loading requirements. Students who are aiming toward highway bridge design should devote their major attention to structural engineering. Among other subjects, they must master the analysis of indeterminate structures and the applications of computers to structural design.[36]

Hydraulic Problems

Stream-flow records, particularly those developed by the U.S. Geological Survey, provide the usual method for estimating discharge under bridges. This and other methods were discussed earlier in this chapter. An analysis of the channel to determine the relationships among peak flow, waterway opening, water-surface elevation at the structure and upstream from it, and flow velocity must then be made. A major factor is the degree of contraction of the flowing water

[34]See *Compendium 4* of Transportation Technology Support for Developing Countries, TRB.

[35]See Chapter 6 for a discussion of bridge location.

[36]The basic AASHTO document from which highway bridge designers work is *Standard Specifications for Highway Bridges*. Other AASHTO publications concerned with bridges are the *Manual of Foundation Investigations* and *Construction Manual for Highway Bridges and Incidental Structures*. Other references on the structural design of highway bridges are far too numerous to give here. Those interested are referred to the various publications of TRB such as *TRB Record 664* and *665* and the *Structural Engineering Journal of ASCE*. In addition the Federal Highway Administration has issued *Standard Plans for Highway Bridges, Vol. 1, Concrete Superstructures; Vol. 2, Structural Steel Superstructures; Vol. 3, Timber Bridges; Vol. 4 and 4a, Typical Continuous Bridges;* and *Vol. 5, Pedestrian Bridges*.

in the approach channel. Final determination of structure proportions and required channel modifications result from this study.[37] Closely allied to the hydraulic aspects of bridge design is the effect of bridge openings and approach fills on the possible flooding of low lying areas in flood plains crossed by the highway. Here economic, legal, and social implications must be considered, and cooperative planning with all affected groups and agencies is required.[38]

Where bridge piers must be set into streams with erodible beds, possible undermining by scour becomes a primary consideration. To date, knowledge regarding the depths to which stream beds will be disturbed by flood waters is fragmentary. The dilemma for the designer is that if the estimate is oversafe, the foundation is more costly than necessary; but, if scour is underestimated, the foundation may be undermined, bringing destruction to the entire bridge. Among the research findings to date are that the least scour occurs when the pier offers the least resistance to flow, which means that piers aligned with the flow and of the smallest possible cross section are superior where scour is a problem. It is also known that scour increases with depth of flow, and thus is a greater problem in streams having high ratios between flood stage and normal stage. Studies of the effect on scour of flexible mats surrounding bridge piers have also been conducted.[39]

Clearances for Highway Vehicles

Recommendations of AASHTO for horizontal and vertical clearances for tunnels, bridges, and underpasses are summarized in Fig. 11–14. For bridges, the AASHTO recommendation is that roadway width be that of the approach roadway, including full shoulders. However, it stipulates that for low traffic speeds and volumes bridge roadway width may be reduced with the absolute minimum 8 ft wider than the approach traveled way.

Waterway Clearances

Where highways cross navigable waterways, the question of suitable vertical clearance becomes important, since higher bridges with longer spans almost always are more expensive. A corollary question concerning the bridge structure is whether it should be high enough to meet the clearance requirements so that highway and water traffic are undisturbed or whether a movable span or spans should be installed. In the latter case highway or water traffic would suffer certain delays each time a ship passed.

[37]*FHWA Hydraulic Design Series No. 1* and *Texts 10* and *11* of *Compendium 9* of Transportation Technology Support for Developing Countries, TRB, are authoritative.

[38]This complex subject is beyond the scope of this book and the reader is referred to the water resources literature. For a beginning, see A. J. Knepp et al., *Public Roads,* Sept. 1976.

[39]*NCHRP Synthesis 5* offers an excellent analysis of this complex problem, summarizes present knowledge and solutions, and gives an extensive bibliography. See also P. C. Klingeman, *Journal of the Hydraulic Division of ASCE,* Dec. 1973, C. J. Posey, ibid., Dec. 1974, and K.N. Derucher and C.P. Heins, *TRB Record 785*. A. G. Anderson, *HRB Record 123,* gives a case history of a spectacular bridge failure resulting from scour and of subsequent model studies.

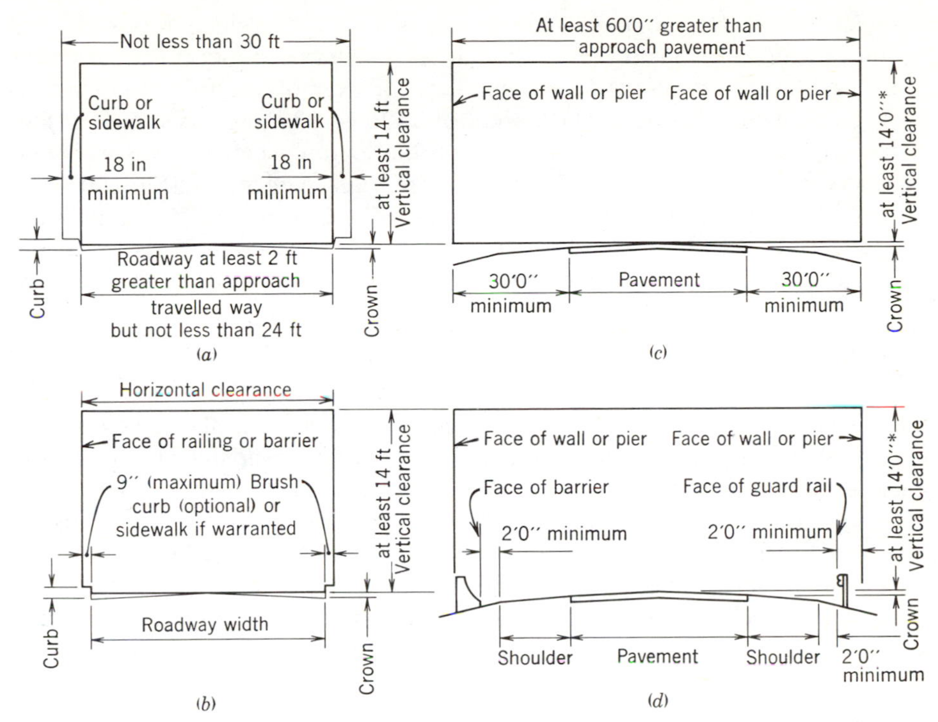

Fig. 11–14. Recommended horizontal and vertical clearance requirements for *(a)* tunnels, two-lane roadways; *(b)* bridges; *(c)* underpasses, general condition; *(d)* underpasses, limited condition. (Note: *Minimum vertical clearance 16 ft on rural Interstate and trunk highways.) (1 in. = 2.54 cm, 1 ft = 0.3048 m)

Since 1899, authority to approve or disapprove clearances over navigable waters has rested with the Corps of Engineers, Department of the Army. Highway officials have at times claimed that the cost of providing the stipulated clearances far exceeded their economic worth. This issue is certain to be a point of argument in the years ahead.[40]

Bridge Loadings

Highway bridges are designed to resist loads produced by the weight of the structure (dead load), the weight and dynamic effect of moving loads (live load and impact), and wind loads. Structures on curves must resist centrifugal forces developed by moving vehicles. Under certain circumstances, stresses resulting from temperature change, earth pressure, buoyancy, shrinkage, rib shortening, erection, ice and current pressure, and earthquakes must also be considered.

Trucks and other heavy vehicles that produce the larger live loads have a wide variety of total weights, axle loads, and axle spacings. For design pur-

[40]For a presentation of the various viewpoints and an extensive bibliography, see E. W. Weber, W. Kurylo, W. E. Cleary, N. L. Caruthers, and E. E. Dittbrenner, *Transactions, ASCE,* 1958.

poses, AASHTO has adopted standard vehicles that produce representative loadings. Figure 11–15 shows two alternative vehicles used for bridge design for trunk highways. The HS 20-44 designation, for example, is for a truck–semitrailer combination having a total weight of 36 tons distributed as shown in Fig. 11–15. The 44 indicates the year in which the loading standard was adopted. For minor highways, the standard vehicles are trucks weighing 20 tons (H-20), 15 tons (H-15) and 10 tons (H-10). Distribution of these total weights among the four wheels of the truck is as indicated in Fig. 11–15. Design loads for longer spans are a combination of uniform lane load and a single-moving concentrated load.

Bridge Railings

Bridge railings are designed as structures to keep vehicles in the traveled way, or at least to prevent them from plunging through or over the railing. Earlier practice focused attention also on making railings visually pleasing. However, the frequency of accidents involving them (22% of fixed-object collisions on freeways) has brought concentration on designs or retrofittings that minimize damage to vehicles and injury to their occupants.[41] Research to this end parallels that outlined in Chapter 9 for guard rail and median barriers. References appear there.

Bridge Types

In the simplest terms, bridges consist of substructures of abutments and piers which support superstructures that carry the roadway between these supports. Types include, among others, slab, girder, truss, arch, cable-stayed, and suspension bridges, each with a distinctive form of superstructure. Rigid or continuous frame structures are bridges in which the substructure and superstructure are rigidly joined. A further distinction is made in terms of materials, the most common of which are reinforced concrete, structural steel, and timber. Aluminum has also been used.

The suitability of the various bridge types is governed primarily by the length of individual spans. Short-span structures, ranging up to about 60 ft, are generally either *(a)* reinforced-concrete rigid frames with slab decks (similar in cross section to the bridge culverts shown in Fig. 11–6); *(b)* T beams or box girders of reinforced concrete; or *(c)* steel or prestressed concrete I beams with reinforced-concrete decks. Precasting and prestressing of the reinforced-concrete portions of these structures is now common practice.[42] A combination of timber stringers with timber or reinforced-concrete deck is sometimes used for spans of less than about 20 ft. Bridges of somewhat longer spans are often *(a)* girder type rigid frames of reinforced concrete or steel; *(b)* T beams or box girders of reinforced

[41]See, for example, *FHWA Report RD77-40*.
[42]See *NCHRP Synthesis 53* for examples and a bibliography.

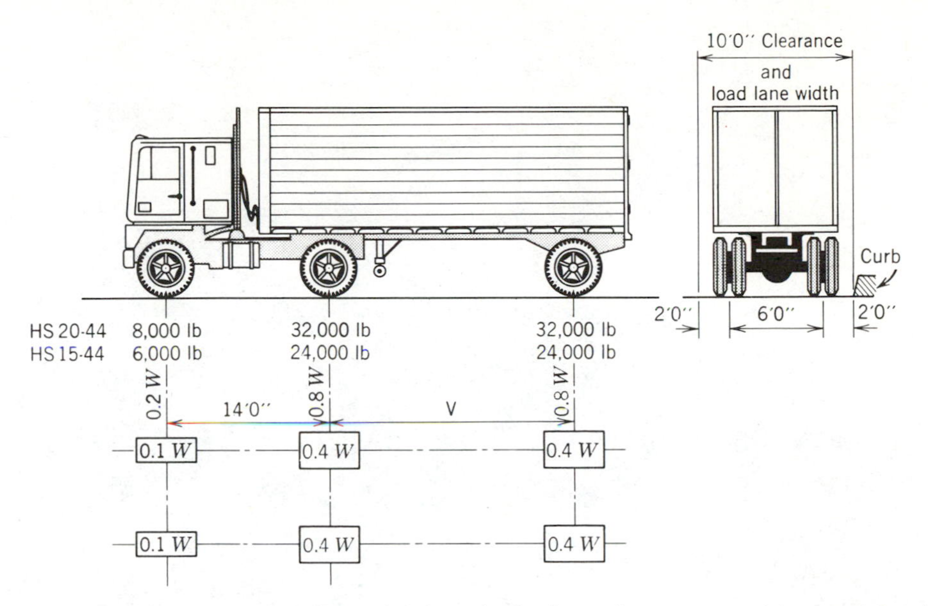

Fig. 11–15. Recommended design loadings for bridges where W = combined weight of first two axles which is the same as for corresponding H (m) truck and *V* variable spacing, 14 to 30 ft inclusive. Spacing to be used is that which produces maximum stresses. 1 ft = 0.3048 m; 1 lb = 0.453 kg.

concrete; or *(c)* steel plate girders with reinforced-concrete decks. When spans greatly exceed 300 ft, steel trusses or arches of steel or reinforced concrete are usually favored. Spans greatly in excess of 500 ft are generally steel trusses, cable-stayed, or suspension bridges. Where provision is made to pass ships through rather than under the roadway level, the channel span generally is selected from the vertical-lift, swing, or bascule types.

Combination of several bridge types in single structures is illustrated by the two parallel Chesapeake Bay bridges pictured in Fig. 11–16. The two-lane structure shown on the left was opened to traffic in 1952. Its three-lane twin, positioned 450 ft to the north, was completed in 1973. It can be operated with reversible lanes to accommodate unbalances in traffic demands. Beginning in the far distance of the photograph, the new bridge has a series of 60 ft spans; following these are 100 and 200 ft spans; and next are nine plate-girder units each 302 ft long. The adjoining cantilever structure covers 1350 ft in three spans. The suspension bridge across the main ship channel has a center span of 1600 ft and end spans each of 675 ft. Behind the suspension bridge and out of the picture in the foreground is a series of nine deck cantilever trusses, a through cantilever structure over the secondary ship channel, and shorter spans leading to the shore. The new structure incorporates many design refinements developed in the last two decades. The photograph shows two of these: the

Fig. 11–16. Old and new bridges over Chesapeake Bay. (Photograph courtesy United States Steel Corp.)

change in conformation of the piers and the positioning of the anchorage for the suspension cables.

One of the important design principles illustrated by the Chesapeake Bay bridge is that, for economy, increases in pier height are accompanied by increases in the lengths of individual spans. Thus, other things being equal, low bridges have short spans and high bridges, long spans. Another rule is that, where clearance above the stream bed or water surface is not a factor, deck type structures with the roadway above the upper chord of truss or girder are superior to through structures. With deck type designs, lateral bracing is simpler and cheaper and the view is not obstructed by structural members. This difference is shown in the position of the stiffening trusses on the suspension-bridge sections in the foreground of Fig. 11–16.

Bridges should harmonize with their surroundings. For example, exposed steel girders may be entirely appropriate on a structure crossing an industrial artery, but a rigid frame with exposed surfaces of concrete or stone may be more suitable on a landscaped parkway. Again, in mountainous country, large bridges often are visible from the approach roadway or from roadside areas and other vantage points. Here, graceful arches of steel or reinforced concrete may well be most appropriate from an aesthetic if not from a cost point of view.

Obsolescence and Deterioration of Bridges—Inspection of Bridges

Many bridges built before 1930 had lane widths less than 10 ft. Design live loads sometimes were 5 tons. Until 1941, the AASHTO specifications permitted 10-ft lane widths and 10-ton truck-loadings on certain state highways. Many bridges of these and other inadequate designs are still in service. In addition, many bridges constructed later are geometrically substandard and structurally inadequate for today's traffic. Physical deterioration through lack of maintenance is also widespread. For example, on the federal-aid system alone, of the 248,000 bridges recently inventoried, about 9000 were structurally deficient or closed and 31,000 functionally obsolete. It can be supposed that the situation of off-system bridges is considerably worse. Furthermore, many bridges that meet some acceptable load or width standard may be subject to damage. Frequently drivers unwittingly or willfully impose overloads beyond posted limits. And through truss structures, particularly on tortuous alignments, are vulnerable since a vehicle striking a main truss member will bring complete collapse of the span.

The December 15, 1967 collapse of the suspension bridge over the Ohio River at Point Pleasant, West Virginia, which took 43 lives, brought the problem of bridge obsolescence and structural weakness into sharp focus. In this instance, stress corrosion and fatigue in the lower limb of the eye of an eye bar brought the failure. In 1927, when the bridge was designed, stress corrosion was an unknown phenomenon; furthermore, the break was so located that discovery before failure was not possible under normal inspection procedures. In one way, the failure served a valuable purpose, since it forced attention to bridge obsolescence and deterioration. Federal legislation now requires that attention be given to bridge inventories and that standard procedures and training programs for inspections be developed. In addition, some federal matching funds on an 80–20 ratio have been appropriated for use on both federal-aid and off-system roads.

Bringing even a fraction of U.S. bridges up to modern load and geometric standards is financially infeasible in the near future. Rather, overloads must be controlled by setting and enforcing limits. For bridges that are too narrow, strategies such as making them one lane or directing traffic with striping, signing, or installing guard rail at approaches will be needed.[43]

Bridge deterioration presents another set of difficulties. Rust damage to old steel structures, spalling or cracking of concrete, and rotting of timber are continuing problems. A recent difficulty is the spalling of concrete decks under the combined effects of frost and deicing salts (see Chapters 20 and 21). It is clear that these bridge problems, particularly on secondary highways and local roads, will plague highway engineers in the years ahead.[44]

[43]See D. L. Woods et al., *Traffic Engineering,* Mar. 1976.
[44]For a recent analysis of this problem see *NCHRP Report 222.*

PROBLEMS

11–1. A rural stream near St. Louis, Missouri drains 20 mi^2. Fall in the main channel is 250 ft. For this situation:
a. From Fig. 11–1, determine the maximum rate of runoff that can be expected.
b. Using Equation 11–1, determine the estimated 10-yr peak rate of runoff.
c. For this situation, determine the mean annual and 50-yr peaks.

11–2. Work Problem 11–1 for a drainage basin of 25 mi^2 and a channel fall of 20 ft.

11–3. Provide the answers called for in Problem 11–1 for a drainage basin of 40 mi^2 near New Orleans, Louisiana. Channel fall is 20 ft.

11–4. A parking lot has an area of 2.5 acres. Based on the rational method and assuming that Fig. 11–2 and Table 11–1 apply, what peak rate of runoff can be expected once in 5 yr? Once in 50 yr?

11–5. Work Problem 11–4 for a grass playing field.

11–6. A channel of roughly graded clay and gravel ($n=0.03$), with a 2 ft flat bottom and 2:1 side-slopes (see Fig. 11–5) is laid out on a 6% grade. Design flow is 120 ft^3/s.
a. How deep must the channel be to provide 6 in. of vertical freeboard against overtopping?
b. Would erosion be a serious problem? Explain.

11–7. Assume that, at a downstream point, the grade of the channel described in Problem 11–6 flattens to 0.2%:
a. How deep must the channel then be to provide 6 in. of freeboard?
b. Would you expect a hydraulic jump near the transition point? Explain.
c. Would the flatter channel require lining for erosion-control purposes?

11–8. A channel with dimensions as shown in Fig. 11–5 lies on a 3% grade and is lined with heavy grass. Design flow is 20 ft^3/s.
a. How deep is the water in the channel?
b. Does the grass offer suitable protection against erosion? Explain.
c. Is the flow in the channel supercritical or subcritical?

11–9. A concrete box culvert is 8 ft by 8 ft in cross-section and 400 ft long ($n=0.012$). It lies on a 0.1% grade and will flow full with the outlet unsubmerged at the design discharge of 500 ft^3/s (case II*b*, Fig. 11–8). How high will the headwater rise above the crown of the culvert if:
a. The entrance is square-edged ($k_e=0.7$)?
b. The entrance is rounded ($k_e=0.2$)?

11–10. Work Problem 11–9 assuming that the culvert is rectangular in cross section and is 10 ft wide and 6 ft high.

11–11. A culvert under a high fill is 300 ft long and is laid out on a 3% grade. Design flow is 200 ft^3/s. Headwater can rise 10 ft above the flow line at the inlet without causing serious damage. The channel is rocky, so that erosion upstream or downstream is not a problem. Tailwater elevation is below the culvert crown. Select a concrete box culvert of square cross section and standard size (4×4, 5×5, 6×6, etc.) for these conditions assuming:

a. That the culvert entrance has a 90° flare ($k_e = 0.7$ if the culvert flows full).

b. That the culvert entrance flares 45° ($k_e = 0.2$ if the culvert flows full).

11–12. A 48 in. concrete pipe 600 ft long with a square-edge entrance and on a 1.5% grade flows part full at a discharge of 125 ft³/s (case II*a* of Fig. 11–8).

a. What is the headwater elevation if the invert elevation is 100.0 ft?

b. Is some arrangement for energy dissipation necessary to prevent erosion if the downstream channel is lined with well-established grass? (*Note:* For a given rate of flow, velocity in a part-full pipe must always be greater than in a full pipe.)

11–13. A 36-in. concrete pipe ($n = 0.012$) 450 ft long is on a 1% grade. Design flow is 100 ft³/s. How high, in feet, will the headwater rise above the invert?

11–14. A 24-in. pipe 450 ft long has a uniform fall of 2 ft in its full length. It will flow full and the outlet will be submerged by 2 ft (case II*c*, Fig. 11–8). What flow will the pipe carry when the inlet is submerged 3 ft above the crown of the pipe:

a. For concrete pipe ($n = 0.012$)?

b. For corrugated metal pipe ($n = 0.024$)?

11–15. Work Problem 11–11 for a circular concrete pipe culvert using $n =$ 0.012. Culvert sizes are in 6-in. increments.

12 HIGHWAY ACCIDENT PREVENTION

Motor-vehicle accidents claim an appalling toll throughout the entire world. In the United States alone, 1978 deaths totaled 51,500. This is four times as many as at work places. In addition, 2,000,000 were injured, of whom 150,000 were permanently disabled. In all there were 18.3 million individual accidents in that year, involving 31.5 million vehicles. Of these, 45,500 involved fatalities, 1,400,000 resulted in nonfatal injuries, and 16,900,000 brought only property damage.[1] Annual economic waste from motor-vehicle accidents (see Chapter 4) is over $34 billion. Yet, in spite of great public clamor and strong legislative efforts to reduce them such as the 1966 and subsequent Highway Safety Acts, the National Motor Vehicle Traffic and Safety Acts, federally mandated state programs, and a variety of state and local actions, motor-vehicle accidents remain one of our most critical unsolved problems. Much must be done to disprove the jaundiced statement that 50,000 lives per year is the human sacrifice that must be made to automotive transportation.

As will be pointed out in more detail below, efforts to reduce motor-vehicle accidents are proceeding on many fronts.[2] A notion of the scope of these efforts is given by Table 12–1, which lists 37 programs and estimates of their potential for fatality and injury reduction over a 10-yr period. Table 12–1 also gives the estimated expenditure required to prevent one fatality and ranks the programs on this basis as well as by potential to reduce both accident classes.[3]

Table 12–1, extensive as it is, does not reflect all the efforts to reduce highway accidents or their severity. For example, item 1, mandatory safety belt usage, is but one approach to reducing severity. An alternative that is being actively pursued is passive restraint by air bags or by other devices which do not

[1]These statistics and many others which follow are from *Accident Facts*, published annually by the National Safety Council. Other quoted sources include the publications of the Insurance Institute for Highway Safety and the Highway Users Federation for Safety and Mobility.

[2]There is a vast literature on all aspects of highway safety, ranging through the professional and trade journals of medicine, public health, engineering, education, trucking, and insurance, among others.

[3]In an attempt to focus attention on specific accident prevention measures, NHTSA in 1979 announced priorities for the next 5 yr as follows: 55 mph speed limit; occupant restraint systems; alcohol and drug abuse; pedestrians, bicyclists and school buses; motorcyclist and moped accidents; driver licensing; and the high accident rates for 16 to 24-yr-old drivers.

require driver attention. There are also many other vehicle-related measures such as crash-minimizing bumpers and nonlocking brakes that are being developed and tested by manufacturers and others, sometimes in response to pressures from governmental agencies. The problem is to sort out the best measures among the many possible. One approach is that of cost effectiveness, as illustrated in Table 12–1. The difficulty is to have reliable estimates of either costs or benefits. And some people object to applying cost-effectiveness as a measure at all.[4] Federal concern is reflected by the creation and funding of the National Highway Traffic Safety Administration which reports directly to the Secretary of Transportation. A number of other safety activities are carried out through the Federal Highway Administration. State agencies also are active.[5] Some organizations, such as the Insurance Institute for Highway Safety, focus solely on highway safety problems, and others, such as the National Safety Council, place highway safety among its principal concerns.

Because there are so many approaches to the highway accident problem, it is helpful to subdivide it into a nine-element matrix. The ordinate is time-related, segregated into precrash, crash, and postcrash; the abscissa is broken into three categories: human, vehicle and equipment, and environment.

To those working in highway safety, it is disheartening to observe the apparent unconcern of highway users for their own safety. For example, seat-belts were used in 1979 by only 11% of the drivers, down from 18.5% in 1976. Driver explanations for lack of use were primarily related to comfort and convenience. Also only about 7% of infants and small children are properly constrained when riding in motor vehicles. This is in spite of widespread publicity pointing out that such restraints reduce by one-half the possibility of being killed. Again, accident reports show improper driving as a specific item in the chain of events leading to 73% of the fatal and 83% of all accidents.

The commonly advanced reason for driver unconcern about accidents is that they are rare events. For example, the probability that, in a given year, an average driver will be involved in an accident is 1 in 8. Based on population, the annual probability of injury is 1 in 110 and of being killed 1 in 4300. Looking at fatalities in a different way, there are about 30 chances of dying from other causes to 1 involving an automobile accident. The appalling toll results, then, not from the high probability of accidents to individuals, but because well over 140 million drivers and their passengers are exposed to these low levels of risk which do not register strongly in their minds. Furthermore, professionals studying behavior as a factor in accidents point out that there are drivers who refuse to even entertain the thought that accidents can happen to them. As a consequence, they see no reason to follow any of the rules of safe driving.

[4]For an objective and thought-provoking examination of the issues in highway accident prevention see R. F. Baker, *The Highway Risk Problem,* Wiley, New York, 1971. Others include *NCHRP Report 162* and R. H. Binder, *Traffic Engineering,* Dec. 1976. *TRB Special Report 178* is a 1977 assessment of federal highway safety programs. For a mutlifaceted analysis see K. R. Agent and R. K. Deen, *TRB Record 541*.

[5]See *TRB Circular 197* and *TRB Record 603*.

TABLE 12–1. Estimated 10-Yr Reduction in Motor-Vehicle Fatalities and Injuries for a Variety of Countermeasures.*

	Fatalities Only				*Fatalities + Injuries*	
Countermeasure	*Rank†*	*Number Forestalled*	*Cost ($) Millions*	*$ per Fatality Forestalled*	*Rank‡*	*Number Forestalled*
Mandatory safety belt usage	1	89,000	45.0	506	1	3,220,000
Highway construction and maintenance practices	2	459	9.2	20,000	33	18,000
Upgrade bicycle and pedestrian safety curriculum offerings	3	649	13.2	20,400	26	11,200
Nationwide 55-mph speed limit	4	31,900	676.0	21,200	2	415,000
Driver improvement schools	5	2,470	53.0	21,400	15	113,000
Regulatory and warning signs	6	3,670	125.0	34,000	10	143,000
Guardrail	7	3,160	108.0	34,100	13	52,800
Pedestrian safety information and education	8	490	18.0	36,800	31	19,200
Skid resistance	9	3,740	158.0	42,200	9	195,000
Bridge rails and parapets	10	1,520	69.8	46,000	17	15,300
Wrong-way enter avoidance techniques	11	779	38.5	49,400	23	3,290
Driver improvement schools for young offenders	12	692	36.3	52,500	25	27,000
Motorcycle rider safety helmets	13	1,150	61.2	53,300	21	14,400
Motorcycle lights-on practice	14	65	5.2	80,600	37	1,680
Impact-absorbing roadside safety-devices	15	6,780	735.0	108,000	6	158,000
Breakaway signs and lighting supports	16	3,250	379.0	116,000	12	127,000
Selective traffic enforcement	17	7,560	1,010.0	133,000	5	296,000
Combined alcohol safety action countermeasures	18	13,000	2,130.0	164,000	3	153,000
Citizen assistance of crash victims	19	3,750	784.0	209,000	8	0
Median barriers	20	529	121.0	228,000	20	2,740
Pedestrian and bicycle visibility enhancement	21	1,440	332.0	230,000	18	24,200

Tire and braking system safety critical inspection—selective	22	4,591	1,150.0	251,000	7	180,000
Warning letters to problem drivers	23	192	50.5	263,000	36	3,760
Clear roadside recovery area	24	533	151.0	284,000	29	20,700
Upgrade education and training for beginning drivers	25	3,050	1,170.0	385,000	14	131,000
Intersection sight distance	26	468	196.0	420,000	32	18,300
Combined emergency medical countermeasures	27	8,000	4,300.0	538,000	4	148,000
Upgrade traffic signals and systems	28	3,400	2,080.0	610,000	11	133,000
Roadway lighting	29	759	710.0	936,000	24	29,600
Traffic channelization	30	645	1,080.0	1,680,000	27	31,500
Periodic motor vehicle inspection—current practice	31	1,840	3,890.0	2,120,000	16	71,900
Pavement markings and delineators	32	237	639.0	2,700,000	35	9,210
Selective access control for safety	33	1,300	3,780.0	2,910,000	20	50,300
Bridge widening	34	1,330	4,600.0	3,460,000	19	51,000
Railroad-highway grade crossing protection (automatic gates excluded)	35	276	974.0	3,530,000	34	1,080
Paved or stabilized shoulders	36	928	5,380.0	5,800,000	22	35,800
Roadway alignment and gradient	37	590	4,530.0	7,680,000	28	23,000

*Source: *National Transportation Trends and Choices,* DOT, 1977.
†Ranking based on cost per fatality forestalled.
‡Ranking based on cost of fatality plus injury accidents forestalled.

ACCIDENT STATISTICS

Compared with other forms of transportation, travel by private automobile is hazardous. Passenger deaths in the United States per 100 million passenger-miles for the 1974–1976 period averaged as follows: passenger automobiles, 1.4; buses, 0.18; intercity buses, 0.03; railroad passenger trains, 0.07; and scheduled air transport, 0.06.

Automobile travel is relatively safe in the United States as compared with other countries. Deaths per registered vehicle in England have been 1.4 times the American rate; the multipliers for West Germany, France, Spain, and Greece are, respectively, 2.2, 2.5, 2.5, and 8.

With the exception of a small increase in 1965, the death rate per vehicle-mile in the United States decreased until 1977. One percent increases were recorded in 1978 and 1979. However, the overall death rates were only 20% of those in 1925 and 40% of those in 1943. But this decrease seems to come from "forgiving" highway designs, better passenger cushioning and constraint in the vehicle, and improvements in medical and surgical techniques, medication, and hospital care rather than because of better driving.

Figure 12–1 diagrams nationwide accident experience for 1978. It can be seen that situations are quite different between rural and urban locations. Although roughly 55% of the vehicle miles are run up in urban areas, only 36% of the fatalities occur there. On the other hand, about 72% of all accidents occur in urban areas. The high fatality rate in rural areas and in predominantly rural states (rates in New Mexico 2.5 times those in Connecticut) unquestionably is related to higher speeds while the high urban rate for all accidents indicates greater accident opportunity but with speeds generally lower. About 32% of the urban fatalities were pedestrians struck by motor vehicles; in contrast only 11% of those killed in rural areas were walking. Daylight and dark also have

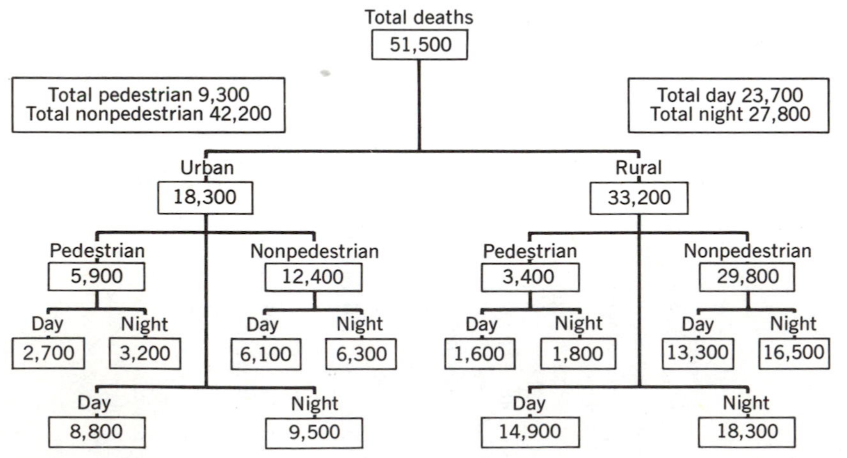

Fig. 12–1. Principal classes of motor-vehicle deaths in the United States in 1978. (Source: *Accident Facts*. 1979 edition.)

markedly different accident records; although total daytime mileages are roughly triple nighttime mileages, 54% of the nationwide deaths occurred at night. The early morning hours after midnight are particularly hazardous with an accident rate six to eight times as great as daytime.

Because of widespread publicity, there is the common impression that holiday driving is particularly hazardous. The number of deaths on holidays is 24% greater, but vehicle-miles are 4% greater, so that the fatality rate per vehicle-mile is 20% higher. This may be explained by such factors as driver fatigue and unfamiliar routes. With more occupants per vehicle, injuries and deaths per accident also are higher.

The relative frequency of highway accidents associated with different vehicle movements or maneuvers varies substantially with vehicle and highway type. This is reflected in Table 12–2, which shows the percentage of fatalities in several movement categories for the rural and urban Interstate Systems and for all rural and urban facilities, with the latter predominantly built to lower standards. The variability in the percentages can be readily explained in terms of the differences in design as outlined in Chapter 9 and discussed in detail later in this chapter.

Table 12–2 also permits comparisons for all highways and streets among rural and urban situations and between the percentages and rates for fatalities and all accidents. Particularly striking are the differences between rural and urban rates.

TABLE 12–2. Percentage of Accidents Associated with Various Vehicle Movements

	*Interstate Fatalities**		*All Highways and Streets†*			
			Rural		*Urban*	
Maneuver	*Rural*	*Urban*	*Fatalities*	*All*	*Fatalities*	*All*
Single vehicle	62	51	48	41	29	11
Multiple vehicle						
Angle	‡	‡	15	23	19	39
Head on			14	18	5	24
Wrong way	5	5	§	§	§	§
Out of control	7	8	§	§	§	§
Rear end	16	15	5	6	4	9
Broadside, side-swipe and other	5	9	8	12	7	16
Pedestrian	5	12	10‖	0‖	36‖	1‖
Totals	100	100	100	100	100	100
Rate per 100 million vehicle miles	2.8	1.8 (1.6 in 1973)	5.3	800	2.4	1700

*Source: *Public Roads* June 1972.
†Source: *Accident Facts, 1978 ed.*
‡Angle accidents do not occur on freeways.
§Wrong-way and out-of control accidents are included in angle and head-on above.
‖Includes non-traffic accidents.

These clearly reflect the higher rural speeds which lead to more fatalities, and, on the other hand, the urban problem of far more but less-severe accidents because of the greater number of intersections and other conditions which lead to conflicts at lower speeds. Also reflected is the extreme severity of pedestrian accidents.

The difference in accident exposure under various conditions is well illustrated by statistics for buses, as follows, stated in accidents per million vehicle-miles. They are: city, 66; suburban, 34; intercity 5, and school 16.[6]

From an overall point of view, Tables 12–1 and 12–2 and the statistics cited above highlight the difficulties in selecting and evaluating the value of the various approaches to accident reduction. Also, it can be seen that far different programs will be undertaken and appraisals made if the focus is primarily on fatalities as contrasted with schemes that somehow weigh a variety of consequences such as fatalities, permanent disabilities, other injuries, vehicle damage, and the many economic consequences.

THE MULTIPLE APPROACHES TO HIGHWAY ACCIDENTS

Motor-vehicle accidents, as all other accidents, are events that take place without one's foresight or expectation. Ordinarily they happen very rapidly. Furthermore they are the culmination of a chain of unfortunate events. If, by any means, any link of this chain can be broken, a potential accident will be prevented. Among the most glaring weaknesses in the present efforts to reduce highway accidents is that too often those concerned follow a "single focus" approach. Highway engineers may think only of improvements in roads, manufacturers and regulatory agencies of safer vehicles, educators of training for drivers and pedestrians, medics of better and more prompt care of the injured, and law-enforcement and licensing officials of control by strict enforcement and keeping the unfit from driving. Actually, all have important and overlapping roles. Some of the efforts currently under way by this diverse group are discussed below.

ACCIDENT REPORTING AND ANALYSIS

The first step in highway accident prevention is to have accurate and detailed information of all circumstances surrounding past accidents. This has been a difficult task for a number of reasons. First of all, a large quantity and variety of information is needed for the diverse interests of those using the data. Often reporting is difficult and fragmentary, given that first priority goes to caring for the injured and keeping traffic moving. Sometimes weather and lighting make assessment difficult. If a reporter does not reach the scene quickly, as is often the case, many details may be omitted from the report. And biases can often distort or oversimplify reporting; for example, attention may be focused on law viola-

[6]See G. L. Fleischer, *TRB Record 706,* for an analysis of commercial vehicle accident factors.

tion and exclude all other factors. Finally, many less-serious accidents go unreported.

Among the data that are needed are exact accident location and time; vehicle speeds, paths, and damage; light, weather, and road situations; driver and other occupant data; vehicle condition; law violations; and witnesses, if any. In the United States, this reporting is done on forms provided at either the state level or by an organization such as the National Safety Council. Federal legislation requires that all accident reports be compatible with state and federal systems so that certain data will be provided on all accident reports. In addition, almost all the states have laws that require vehicle operators to report injury or property-damage accidents, but in most instances, property damage less than a stated amount (such as $100 to $200) need not be reported.

To facilitate reporting and assure that certain statistical information is gathered, these forms begin with a checklist. However, because this is not sufficient, space is also provided for sketches and a narrative description. These are particularly important for followup studies that can lead to corrective measures.

Accident reports, to be most useful, must be unbiased. It is all too easy for the engineer to emphasize vehicle or road conditions while a patrol officer may concentrate on law violations. In particular, reporters must resist the temptation to assign the accident to a single cause or to fix legal responsibility. Otherwise they overlook some hazard that, if uncorrected, will contribute to a similar accident at a later time.

Accident reports are collected at designated central locations such as the state highway patrol or state motor vehicle department. Sometimes duplicate copies are provided for traffic engineers; however, their use is usually restricted because of their confidential nature. Parties involved in the accidents may be permitted to see them. Attorneys may accompany clients but otherwise generally are denied access.[7]

As indicated, it is essential that reporting and computer or other processing systems pinpoint the exact location of accidents. In rural areas this usually is in relation to mileage on the particular road. Another procedure for rural locations is to tie the accident site to the state plane coordinate system. Another uses public-land-survey coordinates. In urban areas, locations are commonly identified by street along with position in or distance from an intersecting street.[8]

After the accident data are assembled, the next step is to separate out locations where accidents of various sorts most frequently occur and at the same time develop other pertinent data such as geometric features and traffic volumes so that priorities for attention can be set. Often this is done on a strip map or, for intersections, on an accident spot map with accident locations marked by round-headed pins or some other identifying symbol. It is common to distin-

[7]Space limitations precluded the inclusion of accident reporting forms here. Details of local procedure, including reporting forms, can usually be obtained from a highway agency, highway patrol, or local police. See also the *Transportation and Traffic Engineering Handbook*.
[8]See *TRB Record 543, 643,* and *706* for examples.

guish among fatal, injury, and property damage accidents with different colored pins. At times, these maps may be displayed prominently to keep attention focused on the accident problem. Then a cluster of markers at a particular location is a signal for further investigation. Large agencies have more elaborate procedures for identifying road segments and spot locations where accidents are particularly prevalent.[9]

As indicated earlier, accident statistics are collected and reported in a variety of ways. One of these is in terms of fatal, injury, property damage, or total occurrences. At times, injury accidents may be segregated by severity (see Table 4–3). Again, as shown in Table 12–2, reporting may be by single vehicle, collision, and pedestrian or by maneuver or object struck. The most widely employed base for such statistics is by vehicle-mile, for example, fatalities per 100 million vehicle-miles. Vehicle registration or population also are common denominators. For locations such as intersections, number of passages by vehicles or pedestrians may be a more appropriate measure. Another approach is to equate accidents to the number of potential conflicts between vehicles or the observed number of near misses, evasive actions, or brake applications.[10] Others have suggested that the rate base for single-car accidents should be the number of vehicles using the system, the base for two-car accidents the square of the number of vehicles, and the base for vehicle-pedestrian accidents in terms of the product of vehicle number times pedestrian numbers.

Accident statistics are reported through many outlets such as the Federal Highway Administration, the National Highway Traffic Safety Administration, the National Safety Council, and state transportation agencies, to name a few.

Accident statistics must be used with caution for they are averages, and, as such, may obscure important variables. As an example, accidents are commonly reported and often used on a vehicle-mile basis, without recognizing such variables as design standards or traffic volume, both of which affect accident rates.[11]

A concerted effort will be required if detailed, accurate, rather than unreliable, fragmented accident data are to be available in the future. The FHWA report, *Safety-Related Information Needs* (1979), lists 769 information items that are pertinent. These cover motor-carrier safety, 103 items; highway design

[9]For other procedures for identifying high accident situations, see *NCHRP Report 162* and *197,* and articles by J. A. Deacon et al., and G. D. Weaver et al., *TRB Record 543,* T. K. Datta et al., *TRB Record 643,* and J. W. Sparks, *Traffic Engineering,* Jan. 1977. Typical recent accident experience data with which to make comparisons by direct or statistical methods include: for intersections in rural municipalities with signal, stop, and yield controls, J. T. Hanna et al., *TRB Record 601;* for bridges, K. R. Agent and R. C. Deen, ibid.; and for pedestrians, R. L. Knoblauch, *TRB Record 629.* P. H. Wright and L. S. Robertson, *Traffic Engineering,* Aug. 1976, found curvature often associated with rural accidents. In 80% of their cases, there was a curve within 500 ft of the accident site. Curves sharper than 6° were particularly troublesome, as were steep down grades.

[10]For details of this procedure see *NCHRP Report 219* and articles in *TRB Record 486, 562, 630, 667,* and *709.* Since conflict analysis is time consuming, locations for study must be selected in some other manner.

[11]For example, B. V. Chatfield, *HRB Record 469,* found the fatality rate substantially higher on Interstate freeways of lower volume. See also *NCHRP Report 197.*

standards, 473 items; and other highway programs, 193 items. Of these, 232 were classified as high priority. The report also suggests suitable approaches, including sampling and proxy measures and appropriate agencies for data gathering. Estimated cost of such a program is $20 million.

FACTORS IN HIGHWAY ACCIDENTS AND THEIR REDUCTION

The Driver

Accident reports indicate that improper driving, often accompanied by law violations, is in the chain of events leading to 73% of the fatal and 83% of all highway accidents. But it is not enough merely to criticize drivers, for their task is complex. It has been estimated that on the order of 75 separate decisions are taken for every mile driven. The question, then, is how to prevent the very few wrong decisions that lead to accidents. Four aspects of this extremely complicated subject are discussed here.[12]

PHYSICAL CHARACTERISTICS OF DRIVERS. As has been mentioned earlier, the 140 million licensed drivers in the United States cover almost the entire spectrum of human abilities. A few simple measures of these abilities for the majority of drivers have been utilized in setting highway design standards. But are there those who lack these or other abilities or have deficiencies such that they should be denied driving privileges in order to reduce accidents?

One question is that of age.[13] With increasing age, driver reflexes slow and certain physical abilities diminish and it might appear that older people would have more accidents. However, based on the fatality accident experience per licensed driver, the 10% "over 65" have only 65% the involvement rate of all drivers, while the involvement rates of the 10% "under 20" and 12% in the "20 to 24" group are, respectively, 180 and 170% of the average. It has been reported that oldsters drive about 50% less, which makes their vehicle-mile rate 1.3 times the overall average. Apparently they are more cautious and make fewer wrong decisions in the sense of narrow safety margins[14] and aggressive behavior than do young drivers. In sum, more could be done to reduce accidents by barring those 24 and younger than by eliminating the "over-65" group.

Persons identified as suffering from epilepsy, heart disease, diabetes, and mental illness have an accident rate roughly twice that of the public in general. But medical records are not complete enough to establish the relative contribution of such conditions to accidents. One study indicated that only 0.6% of drivers fell into this group. In the same vein, it has also been proposed that

[12]Two excellent references are *Human Factors in Highway Traffic Safety Research,* T. W. Forbes, ed., Wiley, New York, 1972 and *Road User Behavior and Traffic Accidents* by R.Naeaetaenen and H. Summala, American Elsevier, New York, 1976.

[13]For a classic analysis of the problem of aging and driving, see B. W. Marsh, *Traffic Engineering,* Nov. 1960.

[14]For example, a smaller percentage than average of young drivers use seat belts. See *TRB Record 739*.

some of the high-speed single-car accidents are in reality a means of self-destruction adopted by the mentally ill or emotionally disturbed.

Physical defects in seeing, hearing, and from other causes are not major contributors to accidents. Drivers with these defects were involved in only 1.3% of the fatalities and 0.6% of all accidents, and, of the pedestrians killed, 5% had these disabilities. It must be concluded that the physical attributes that are weighed so heavily in the usual driver-licensing procedures (see below) are of doubtful importance in the overall accident picture.[15]

DRIVER EDUCATION. Driver improvement by education on first thought seems to offer great promise as a means of accident reduction. However, a study sponsored by the Pure Oil Company and the American Trucking Association raised serious doubts. It was found that 90% of all drivers and 100% of those with records of traffic violation rated themselves above average in driving skills and in obeying traffic laws. Thus, any effective educational program must first overcome this superiority complex. The survey also found that the typical slogan-type program based on "speed kills" or "slow down and live" is probably doomed at the outset, since, to most drivers, it applies to others but not to them.

Since World War II, formal driver education has been given increasing attention. Many of the nation's high schools have courses; in addition, many schools offer parallel courses for adults. There are also a large number of commercial driving schools, and many industrial concerns give training to employees who will operate motor vehicles or trucks.[16]

Authoritative persons have stated that there is no clear evidence to justify high-school driver education. For example, a recent and controversial study by the Insurance Institute for Highway Safety found no substantial decrease in fatal crash involvement with increased percentages of 16 and 17-yr-old licensed drivers who had taken high school driver education. In fact, with driver education leading to a higher percentage of 16 and 17-yr-old licensees, fatal crash involvement rose substantially. However, several other studies have demonstrated that formally trained drivers have roughly half to four-fifths the accidents of the untrained group. Also, it has been claimed that those who take training voluntarily would have had better accident records anyway because they have stronger motivations and better attitudes. In sum, the effectiveness of driver education as an accident reduction measure, at least at the high-school level, is uncertain.

[15]K. H. Antia, *HRB Highway Research News,* Autumn, 1969, reports that less than 1% of drivers have vision so bad that, after correction, they should not drive. See also B. L. Hills, *TRB Record 681*. A. Burg, *HRB Record 216,* recommends frequent eye testing for more elderly drivers since their ability to see may change rapidly. R. S. Coppin and R. C. Peck, *HRB RECORD 79,* report that totally deaf males, but not females, have an accident-involvement rate 1.8 times that of their nondeaf counterparts. Again, eliminating all drivers with such physical handicaps would not substantially reduce the number of accidents.

[16]There is an extensive literature on driver and other traffic-safety education. See, for example, the periodical *Traffic Safety* published by the National Safety Council. Many articles appear in the trade journals of the motor-carrier industry.

New approaches to education for both new drivers and corrective purposes have been undertaken, including simulated rather than actual driving.[17] Also, the effectiveness of traditional course content and sequencing has been investigated.[18]

Experimental public educational programs through television, radio, and newspapers have been undertaken on such subjects as safe driving practices, the use of safety belts, and drinking and driving. In some instances they have been effective; in others not.[19]

DRIVER FRAME OF MIND. Evidence is substantial that safe, lawful, and courteous driving is closely associated with emotional makeup, social adequacy, and attitudes toward risk taking. It has been found that accident repeaters are apt to be aggressive and intolerant of others; they tend to resent authority; they are inclined to have an exaggerated opinion of their importance and abilities; and they are likely to be lacking in responsibility and to act impulsively and on the spur of the moment. In one study it was found that 25% of the accident repeaters stated that people "made fun of them"; in the accident-free group only 2½% had this impression. It was also found that accident repeaters had many times the number of unfavorable reports from courts, credit bureaus, and social service agencies than did the accident-free group. Stated differently, as a group, accident- and violation-free drivers tend to be more mature, conservative, and intellectual in their interests and tastes, have a higher aspiration level, and are products of happier family backgrounds.

Although it is known, then, that "a man drives as he lives," testing procedures to detect drivers of high accident potential usually have low validity. To eliminate even a small proportion of accident-likely drivers, a large number of safe drivers would also have to be rejected. Unfortunately, also, neither personnel nor proven techniques in psychotherapy are available to correct the underlying causes of accident likeliness.[20] It should be recognized, on the other hand, that if the aim is to select a small number of potentially safer drivers from a large population, behavioral and physical testing may be effective.

A particular problem in driver attitude concerns the teenage and under-25 drivers. In a Minnesota study, the under-25 group represented some 20% of the driving population but accounted for 30% of the arrests for drinking and 50% for those for speeding. As noted above, nationwide in 1977 they represented 22% of the drivers but were involved in 38% of fatal accidents. A California

[17]See, for example, G. R. Hatterick and R. F. Pain, *TRB Record 629* and J. C. Prothero and T. A. Seals, *TRB Record 672*.

[18]See, for example, articles by R. M. Weiers and M. H. Jones, *TRB Record 629* and K. L. Schmitt, *TRB Record 672*.

[19]J. W. Hutchinson et al., *HRB Record 292,* reported that an 18-month series of 2 to 3 min short films shown on television brought a substantial reduction in accidents. On the other hand, a nine-month saturation television campaign on safety belt use "had no effect whatsoever." (See the various issues of *Status Report,* published by the Insurance Institute for Highway Safety.)

[20]See K. H. Antia, *op. cit.* Also see papers by D. E. Cleveland, R. V. Rainey, et al., and E. D. Heath, *HRB Bulletin 212* and *HRB Bulletin 285*.

study showed teenage males and females with double the accident and violation rates of adults of the same sex. In both age groups rates for males were double those for females.[21] These facts readily explain the high insurance premiums assessed against the automobiles they drive. The reasons for this unfavorable accident experience are complex. Apparently they include overconfidence and a greater willingness to accept narrower margins of safety.[22]

TEMPORARY DRIVER CONDITIONS. *Exhaustion and drowsiness* decrease a driver's ability to drive safely. Among the research findings on this subject are the following:

1. In driver simulation tests, work decrement occurs within the first 2 hr after driving begins. However, a refreshment pause postpones the onset of fatigue.
2. In actual situations, the abilities of drivers who had been previously deprived of sleep for a considerable period were found to be seriously impaired.
3. Efficiency may be lowered by sustained vehicle operation without the driver being aware of it. This has sometimes been referred to as "trip hypnosis."

Conclusions such as these may seem obvious; however, both educational and enforcement measures may be necessary if accidents from driver fatigue are to be reduced in number.[23]

Drinking drivers and pedestrians offer the most serious of all highway accident problems. The magnitude can be demonstrated statistically by facts such as: *Half of all highway fatalities stem from crashes involving the abusive use of alcohol;* or, in the same vein, *two-thirds of the drivers killed had measurable alcohol in their blood*. Of these, 85% *had 0.10% or higher; almost one-half had 0.20% or higher*. Again, *alcohol was detected on more than 70% of the drivers killed in single-vehicle accidents and of the pedestrians killed, some 40% showed blood alcohol levels higher than 0.10%*. Since the National Standard on Alcohol and Highway Safety defines the 0.10% blood alcohol level as presumptive evidence of intoxication, it becomes clear that alcohol is the predominant contributor to highway accidents.

Accidents which involve alcohol are usually far more serious than those which do not. As indicated above, there was presumptive evidence that two-thirds of the drivers in fatal accidents had measurable alcohol in their blood; on the other hand only 15% of those in all accidents showed this condition.[24]

The effects of alcohol are complex and highly variable among individuals. For the majority, blood alcohol levels less than 0.05 produce some sedation and tranquillity. Above that level, there is an inability to coordinate visual scanning

[21]See G. S. Ferdun et al., *HRB Record 163*.

[22]For a procedure for evaluating willingness to take risks see H. T. Zwahlen, *TRB Record 464*.

[23]See, for example, B. D. Greenshields, *HRB Record 122,* and R. R. Safford and T. H. Rockwell, *HRB Record 163*.

[24]Statistics cited here are from a few of the many reports on the topics of alcohol and accidents. An early and authoritative one is *Alcohol and Highway Safety,* a report to Congress by the Secretary of Transportation, 1968. A more recent one is T. J. Smith, *TRB Record 520*. See also J. A. Waller, *TRB Special Report 151*. M. Burns and H. Moskowitz, *TRB Record 739* discuss tests for drinking drivers.

(sensing) and vehicle control (psychomotor abilities). Simple reaction time does not increase markedly, so that the difficulty arises because of an inability to translate information into proper action. Also, alcohol affects vision; it has been found that two cocktails are like wearing dark glasses. Again, the inability to distinguish between light and dark is seriously impaired. At the same time alcohol, which acts as a depressant and first affects the part of the brain that produces self-restraint, causes behavior changes which increase drivers' confidence in their abilities, so that they take chances they would ordinarily not take. For example, drinking drivers make less use of seat belts than others. Thus both physically and mentally the drinking driver is accident-susceptible. Unfortunately also, the common belief that coffee or other stimulants offset the effects of alcohol has been proved to be false. Rather the reduction in the effects of alcohol comes only as the body burns it up, which is at a relatively fixed rate of about 0.010 to 0.015%/hr. Thus, if one ounce of 80 proof whiskey (a jigger in a highball) produces 0.02% alcohol in the blood, well over an hour is needed to burn it off.

Human beings show a wide variation in individual susceptibility to alcohol; in a controlled experiment involved 1000 persons, 105 were "intoxicated" (by objective tests) when the blood-alcohol level was 0.05%, which is but one-half the recommended legal measure of drunkenness. Sixty-seven of the group were classed as "sober" when the alcohol level was eight times as great. Effects are also influenced by the amount of food or liquid in the stomach and by the amount of fatty tissue in the body.

Looking at the driving population as a whole, drinking even moderately greatly increases the chance of accidents. One study showed that drivers with alcohol blood levels of 0.05 to 0.10% (two to four cocktails per hour) have twice as great a chance of accident involvement as those whose blood is alcohol free or below the 0.05% level. At blood alcohol levels of 0.10 to 0.15% the chance is roughly seven times greater. Another researcher developed risk factors for blood alcohol percentages as follows: negative to 0.04, 1; 0.05 to 0.09, 3; 0.10 to 0.14, 12; 0.15 or greater, 27.

As would be expected, alcohol was involved in 50% more accidents in the early morning hours than otherwise; however, weekend nights were no worse than weekday ones.[25]

It has been reported that of the licensed drivers in the United States, two out of three drink; of total drivers, 45% are classified as low-volume and 16% as high-volume social drinkers; another 4% are "escape" drinkers; and 3% are alcoholics. It is with these last two or possibly three groups that much of the problem lies, since they may be driving at blood alcohol levels where performance is seriously impaired. But with such a large percentage of the population using alcohol, generalized programs aimed at "If you drink, don't drive" have little popular support.

One approach to the drinking driver problem is "get tough." For example, in

[25]See T. J. Smith, *TRB Record 520*. See also a series of FHWA publications on drinking and driving.

Sweden and Finland the law stipulates mandatory roadblocks and chemical tests. For the violator, without exception, there will be several months in jail, license revocation, and cancelled insurance. Early reports on such programs were glowing; for example, it was claimed that only 10 to 12% of the fatally injured drivers had been drinking; this in contrast to the 50% figure in the United States. More recent evaluations do not support these claims.[26] Neither does the recent experience in Great Britain. There the death rate fell dramatically on enactment of punitive laws; it soon went back to nearly the old rate as potential offenders realized the remote chance of being apprehended.

Currently, a combination of "get tough" and "behavior modification" is being tried to mitigate the "drinking driver" problem. In New York State, first offenders are handled through an administrative-judicial approach stressing education. In California, motorists suspected of "driving under the influence" are taken into custody (possibly handcuffed) and delivered to a custodial agency. Their vehicles are impounded and they must pay charges for towing and storage. They are then given one of the accepted tests for alcohol and, if the measurement is unfavorable, they may be jailed until released on bond or on their own recognizance. They must appear in court for arraignment where there is a choice between pleading guilty or being remanded for trial before judge or jury. Conviction on a first offense carries a substantial fine, typically $350. A second conviction leads to incarceration for at least 48 hr, a far heavier fine, loss of driving privileges for 5 yr, and informal probation for a year. On recommendation that rehabilitation is likely, the sentence may be reduced, but offenders are required at their own expense to attend a 1-yr class which combines viewing "horror" movies of automobile accidents, discussion of the arrest circumstances of group members, and group therapy. Programs such as these are still being evaluated but, as of today, the problem of the drinking driver remains largely unsolved.[27]

Drugs such as marijuana, opiates, amphetamines, antihistamines, aspirin, and barbiturates affect driver behavior and probably increase highway accidents.[28] It can be presumed that, since they dull or accentuate the driver's senses, reponses in emergencies will be different than usual. A review of knowledge on this subject is beyond the scope of this book.

Exhaust fumes and *smoking* may contribute to accidents and to deaths within automobiles, although factual evidence to this claim is far from conclusive. One estimate has placed deaths inside vehicles from carbon monoxide intoxication, which were not classified as highway accidents, at 500 per year. Those involved would presumably be smokers who already had a high carbon monoxide level in their blood. Then, in a traffic jam, the carbon monoxide in the exhaust

[26]See M. H. Wagner, *TRB Record 609*.
[27]See for example, N. Rosenberg et al. and W. S. Moore, *TRB Record 609*.
[28]One study found users of Valium and other tranquilizers five times more likely to be involved in accidents.

of the preceding vehicle was drawn into the victim's car and produced a lethal level that brought unconsciousness followed by death.[29]

Speed

"Speed too fast for the conditions" has been given as a contributing factor in some 37% of the fatal, 17% of injury, and 13% of the total motor-vehicle accidents, and in 52% of motorcycle fatalities. The psychological reasons that drivers go too fast are related to overall behavior patterns. For example, maladjusted persons as a group drive faster than others. Again, speed often is associated with the thrill that accompanies risk taking and with showing off and braggadocio. However, because it has certain unique aspects not associated with behavior, speed is treated as a separate topic.

From the statistics just cited two facets of the speed-accident relationship appear. The first is "speed too fast for the conditions." Sixty miles per hour may be conservative on a rural highway or on a freeway where conflicts are few and where long sight distances give ample opportunity to take the simple measures necessary to avoid an accident. On the other hand, where accident opportunity is high and sight distances are short, as on a congested street with parking adjacent to the through lanes, 30 mph may be too fast. There the driver must make complex decisions under difficult circumstances. Thus, safe speed can be measured in terms of the driver's ability to recognize and cope with situations that have accident potential.

The second aspect of speed and accidents is this: If an accident does occur, how does speed affect its severity? Statistics show that where excessive speed is a factor, fatalities are 2.3 times and injuries 1.3 times as likely as in the average accident. Another study found the probability of an injury resulting in death to be 1 in 47 for all injury accidents, but 1 in 9 at travel speeds over 60 mph. In terms of absolute speed, the laws of mechanics offer a basic explanation of the fact that higher speeds increase accident severity. A moving vehicle and *the bodies of its occupants* have kinetic energy; in amount it increases as the square of the velocity. When an accident occurs, all this kinetic energy must be dissipated, much of it in the form of damage to vehicle and occupants. If the energy to be dissipated at 30 mph is set at 1, amounts for other speeds are as follows: 60 mph, 4; 90 mph, 9; 120 mph, 16. For the jet airplane traveling 600 mph, the figure is 400. It should be apparent, then, that speed is a vital factor in accident severity.

There has been a dramatic decrease in vehicle speeds and highway accidents after the imposition of the 55 mph speed limit in 1974. Before then, 50% of the vehicle speeds on the Interstate System exceeded 65 mph; in late 1977, only 10% did. Estimates are that there are 4000 fewer deaths and 81,000 fewer injuries annually. This has been cited as a clear demonstration of the relationship

[29]See T. H. Rockwell and F. W. Weir, *TRB Record 520.*

between speed and accidents. Enforcement plays an important part; one study showed average speed reductions on the Interstate System of 5 mph without enforcement and 10 mph with it. Also, accident severity decreased; on the Interstate System in one state the fatality/injury ratio fell from 1 in 9.41 to 1 in 11.71.[30] As indicated, there has been a small but continuing increase in speeds since 1974. Since the rate of increase of fatalities has substantially exceeded that in vehicle-miles, this may be a further demonstration of the speed-accident relationship.

High-speed differences in the traffic stream also contribute to accidents. Vehicles traveling at the average speed have the fewest involvements; as speeds, either higher or lower, depart from this average, accidents increase. This argues for setting realistic maximum and minimum speed limits.[31]

In some states, posted speeds for trucks may at times be lower than for passenger cars. One study found that truck drivers seldom observed the lower limits, which probably explains why such intended speed restrictions had little effect on accidents.[32]

Many low-volume roads were originally built with narrow roadways, sharp curves, and limited sight distances. Surfacing was commonly earth or gravel, which discouraged driving at high speeds. Several studies have shown that when such roads were paved without improving the geometrics, speeds increased and accident frequency went up, 73% in one instance. However, a recent analysis (see *Public Roads,* Mar. 1981) found an average rate increase after resurfacing of only 2.2%. Moreover, data from individual sections varied so greatly that a generalized conclusion could not be supported.

A variety of methods for getting drivers to slow down to safe speeds have been explored. Some of these are discussed in Chapter 10.

Driver Licensing

Driver licensing is legally an exercise of the police power, in that it protects the public health and safety. But because automobile transportation is so vital to living and making a living, the right to drive is given legal recognition and protection. For this reason, arbitrary or over-zealous laws or regulations can be challenged in the courts. Also, testing procedures under which the right to drive could be denied may be subject to judicial review.[33]

To date, it has not been possible to establish strong correlations between the

[30]Numerous studies of the effects of the 55-mph speed limit have been made. See, for example, W. J. Kemper and S. R. Byington, *Public Roads,* Sept. 1977; H. S. Dawson Jr., ibid., Sept. 1979; articles in *TRB Record 567, 609,* and *643; Traffic Engineering,* Nov. 1975, Mar. 1976, Sept. 1976, and Feb. 1977; and *Transportation Engineering Journal of ASCE,* May 1980.

[31]See David Solomon, *Accidents on Main Rural Highways,* Bureau of Public Roads, July 1964.

[32]See J. W. Hall and L. V. Dickinson, Jr., *TRB Record 486.*

[33]For a detailed discussion of the legal aspects of driver licensing and of the practices of several states see J. H. Reese, *HRB Special Report 123.* Projected developments in driver licensing are discussed at length in *TRB Special Report 151.*

visual, written, or "behind the wheel" driving tests given once in, say, 3 yr and drivers' accident reports.[34] Their correlation with accidents is in the range of 0.10 to 0.20. Those with traffic-law violations range from 0.30 to 0.50.[35] These low correlations reflect such factors as the lack of reliability in present testing procedures, that accidents are rare events, and that only a fraction of the law violations are detected. However, this statement is not to be construed to mean that driver licensing should be abandoned. For, even though most of those with poor records cannot be denied licenses, some high-risk drivers may be eliminated and others who have difficulty with the tests may choose not to drive or will learn of their limitations.

As indicated before, drivers' psychological characteristics can be related to their driving habits and violation and accident records. But such tests have been reliable only for selecting a few potentially better drivers from large groups and not as a means of denying the right to drive, since their use would deny this right to many in order to eliminate a few.[36]

There appears to be a growing trend toward license suspension for those whose records show multiple arrests or accidents. As indicated above, in some states conviction for "driving under the influence" brings automatic revocation or at least restrictions such as limiting driving to work purposes. Again, a point system may be followed under which reported convictions or accidents are tallied by the licensing agency. When the driver's score reaches a stipulated level the license is suspended. Such schemes may have merit in controlling the most reckless drivers. Unfortunately, even if effective, they will not result in a substantial reduction in accidents. One analysis indicated that 70% of the drivers involved in crashes had no previously reported accidents or violations. Another study found that if all drivers whose records showed three or more law violations in 2 yr were barred from driving, 96% of the accidents would still occur.

Efforts are under way in some states to create better attitudes in drivers by individual or group therapy. The mechanism is usually to use licensing provisions or court orders to compel attendance at scheduled interviews or classes. Some of these procedures, particularly group education meetings, have proved to be effective, at least in the short run.[37]

[34]See R. L. Henderson and A. Burg, *HRB Special Report 130* for present and proposed approaches to vision testing. Burg, *HRB Record 216,* and Henderson and Burg, *HRB Special Report 134,* indicate that there is a small but significant relationship between certain visual characteristics and accidents. The best test is the one for binocular dynamic acuity. J. A. Waller and J. T. Goo, *HRB Record 225,* report studies relating the accident and violation experience of drivers with and without medical conditions to scores on driving tests. Some correlation was found, but not enough to provide a basis for denying licenses.

[35]See L. G. Goldstein, *TRB Record Special Report 151.*

[36]See, for example, J. E. Uhlander and A. J. Drucker, *HRB Record 84.*

[37]See, for example, W. C. Marsh, R. S. Coppin, and P. C. Peck, *HRB Record 163,* and, with A. Lew, *HRB Record 195.* Many more reports will be found in the psychological literature and that of professional organizations such as the American Association of Motor Vehicle Administrators.

Traffic-Law Enforcement

It is generally considered that fear of arrest and punishment causes drivers to conform better to traffic laws and regulations and thus reduces accidents. For example, driver behavior at intersections was observed to be better when a policeman was present.[38] On the other hand, there is evidence to suggest that enforcement crusades have little lasting effect. For example, in several experiments, vehicle speeds and driver behavior were recorded before and after an intensive enforcement effort. No significant changes were found either in speeds or in the number of law violations. Other studies have shown that many drivers ignore speed limit and speed zone signs that do not conform with their usual driving habits or that affect their personal well-being.[39]

Many obstacles have made consistent law enforcement difficult. Among these are the archaic laws in force in many jurisdictions. For example, a few states prohibit or limit arrests for speeding based on evidence from electronic speed-measuring devices, and their accuracy has been challenged in the courts. Again, police cars must be conspicuously marked, apparently to prevent drivers from being caught unawares. The fact that many law violators are not apprehended and those who are often escape with light penalties or none at all brings disregard for laws and for law-enforcement officials. But public attitudes and those of the courts may be changing with the increasing attention to the highway accident toll.

The development of state highway patrols as highly effective professional organizations, the upgrading of the local police, and quicker, more efficient and rational handling of traffic violations by the courts or special tribunals all indicate more effective law enforcement. It can be anticipated that this, in turn, will mean better disciplined drivers and a reduction in accidents.

Among the difficulties faced by law enforcement officers is that arresting traffic-law violators is but one of many duties. Other overriding responsibilities or legal requirements such as keeping traffic moving on major arteries, aiding or securing aid for crash victims, filing accident reports, and court appearances usually must take precedence over law enforcement. Among the procedures adopted by police agencies for stretching their very thin law enforcement manpower is to keep accurate accident records and to concentrate their efforts on high-accident locations.

Traditionally, police officers have driven vehicles capable of the high speeds needed to pursue and overtake violators of traffic and other laws. Often, either pursued or pursuing have been involved in serious accidents. Arguments too many to give here have been offered advocating or proposing abandonment of this "hot pursuit."

Among other organizations, the Committee on Uniform Traffic Laws and Ordinances is doing much to suggest more workable and consistent approaches

[38]See P. J. Cooper, *TRB Record 540*.
[39]See, for example, the discussion on signs in Chapter 10.

through publications such as the *Model Traffic Ordinance* and *Uniform Vehicle Code*. However, final authority rests with the legislatures of the individual states.

The Vehicle

Intensive efforts are under way by automobile manufacturers, governmental agencies, and private researchers to produce a safer vehicle and one that will be less severely damaged in crashes. Both the precrash and during-crash aspects of the problem are involved in these efforts.

Even though much in the public eye, defects in vehicle and tire design and in-service behavior are not among the major contributors to the precrash chain leading to highway accidents. Various statistics have been cited ranging from 11 down to 2% involvement in fatal accidents. The 11% figure was for turnpikes, with 10% attributed to tire failures, where speeds were high and many other types of accidents were made unlikely by the highway's design. This percentage would then agree with a 5% figure cited for all highways. A recent study of tire failures on toll roads in Illinois indicated that at least 0.9% but no more than 2.4% of all accidents involved tire disablement.[40]

The effects of horsepower or horsepower-weight ratio of vehicles on accidents has also been studied. The finding most frequently cited was by David Solomon.[41] It was summarized as: "The highest involvement rate both day and night occurred at the lowest horsepower of 110 or less. Among the higher horsepower groups there is little difference in involvement rate in relation to horsepower." Also, certain special vehicles may have characteristics that may make them more prone to certain kinds of accidents; for example, jeeps are susceptible to rollover. It must be recognized that these statistics not only reflect the character of the vehicles but of roadway conditions and drivers as well.

The "during crash" aspects of vehicle safety mainly revolve around constraining or cushioning the vehicle occupant during and after the crash. Experimental vehicles have been designed with occupant protection the main concern. There have also been many modifications in all vehicles and their interiors. Among them are seat belts, padded dashboards, collapsible steering columns, the removal of all projecting knobs and levers, and doors that are stiffened to resist impact from the side and that remain closed during a crash. It has also been proposed that vehicle designs be based on the "fail safe" idea used, for example, on railroad locomotives. With it, if the driver becomes incapacitated or inattentive, the vehicle stops. Present arrangements for gas pedals, for example, do not follow this principle, since an incapacitated or careless driver may actually cause the vehicle to accelerate. The practical difficulty with this scheme is, of course, the reorientation of 140 million drivers!

The National Highway Traffic Safety Administration (NHTSA) has been em-

[40]See J. S. Baker and G. D. McIlraith, *HRB Record 272*.
[41]See "Accidents on Main Rural Highways," Bureau of Public Roads, 1964.

powered by the National Safety Standards Act to set and enforce safety standards to be followed by all motor-vehicle manufacturers. The list of requirements already issued is far too space consuming to be given here and many additional ones will be forthcoming. Many of these have been and will be controversial. For example, since 1973, arguments have raged over the requirement that a passenger restraint system consisting of air bags that inflate when the vehicle is in a collision be installed as standard equipment. Automobile manufacturers have argued that the air bags are effective only in front-end crashes and are a hazard in themselves. It also is pointed out that seat belts, if employed, are almost equally effective. Spokesmen arguing for the air bags countered that it has been clearly demonstrated that the public will not use seat belts.

As of 1979, experimental installations of air bags or self-positioning restraining systems have been made with encouraging results. NHTSA has now ruled that air bags or some other passive restraint system that operates when the car door is closed must be phased in over 3 yr, beginning with 1982 models. Directives such as this can, however, be reversed by Congress or the courts.

Ways of improving the crashworthiness of vehicles by designs that provide safe fuel storage in crashes, strengthen them structurally, and fill cavities with foam have also been studied.[42] Other concerns include vehicle-passenger interactions such as (1) the effects, for various vehicles, of injury severity in relation to the object struck and the angle of impact[43] and (2) the ways in which humans are injured in various vehicles and the tolerance of their bodies for impacts.[44]

For trucks, NHTSA plans to focus on brakes, visibility, rear underride protection (rear bumpers to prevent cars going under them), cab ride quality and comfort, multipiece wheels which may blow apart, and occupational protection for drivers to diminish kidney damage and other ailments.

Another focus of attention is the high cost of vehicle repair after accidents. Design changes that will substantially reduce repair costs have been initiated. For example, shock-absorbing bumpers that cushion impacts and prevent front and rear vehicle damage under a 5 mph-crash have been stipulated. This requirement has been challenged on the basis that it increases vehicle weight and cost, but proponents claim it is cost effective. All in all, many changes to make vehicles safer can be anticipated in the years ahead, with much of the argument centering around the stipulation in the National Safety Standard Act that requirements must be reasonable, practicable, and appropriate.

The 1966 National Traffic and Motor Vehicle Safety Act directed vehicle and supply manufacturers to notify purchasers of specific deficiencies which might make vehicles unsafe. Later acts required callbacks and repair at no cost to owners. There have been numerous callbacks for a variety of defects. Two of the notable ones concerned the gasoline tank mountings in the Ford Pinto and the Firestone 500 steel-belted radial tire. Each of these deficiencies not only led

[42]See, for example, D. Friedman and R. Tanner, *TRB Record 586*.
[43]See J. W. Garrett, *TRB Record 586*.
[44]See J. W. Melvin et al., *TRB Record 586*.

to recall and correction or replacement, but also brought multimillion dollar claim settlements against the manufacturers from accident victims or their survivors.

Unfortunately, legislation arising from other national priorities may bring vehicles that are less safe. For example, energy-conservation measures are forcing manufacturers to produce a fleet that includes many more lightweight passenger cars. Recent studies have found that, in two-car accidents, injuries are substantially higher to occupants of these cars. This applies not only when the collisions are with a heavier vehicle, but also when both are lightweight. Insurance claims are also higher for light vehicles, particularly for imported subcompacts.

Compulsory vehicle inspections to correct safety deficiencies have been advocated as a means of accident reduction. It has been shown that 20% of the vehicles failed such tests although such failures have not been directly linked to accidents. Currently, some states have compulsory vehicle inspection laws. In others, the highway patrol or other agencies sometimes carry out inspections and may issue citations or otherwise require that deficiencies such as badly worn tires be corrected. As of 1980, no compulsory vehicle inspection program had been adopted nationally, although it is called for in the 1966 law. A principal difficulty is to develop an acceptable inspection plan.

Motorcycles

Accidents involving motorcycles and motor scooters account for an increasing number of deaths and injuries on our highways. Registrations nationwide increased from 600,000 in 1960 to 5.0 million in 1978. Possibly an equal number of machines are unregistered. During the same period deaths increased from 750 to 5100, or from 2% to 10% of total motor-vehicle fatalities.

The hazards of motorcycle riding are well documented. For example, a study in Great Britain indicated a motorcycle fatality rate per vehicle-mile 20 times that for automobiles and an injury rate three times as great. Data for the United States for 1975 found deaths per vehicle-mile 5.6 times as great for motorcycles and injury accidents roughly double the automobile rate.[45]

The high fatality rate for motorcyclists is understandable. First of all, as a group, they probably are willing to take more risks. Secondly, in contrast to automobiles, the rider is not cushioned in and protected by a vehicle of considerable mass. Finally, riders, on impact, continue their forward motion at precollision velocity, usually head first, until they strike a vehicle or fixed object or slide to a halt. Any of these can bring injury or death.

Helmets have been proved to be effective, dramatically reducing the often fatal injuries to motorcyclists. In one state, fatal accidents fell from 10 to 7 per 10,000 registered cycles when a helmet law was enacted. Again, it was found that 95% of unhelmeted riders died of head and neck injuries as contrasted with one in three helmeted persons. Yet, after Congress in 1976 repealed provisions withholding certain funds if a state had no helmet law, 25 states had by 1978

[45]See S. F. Polanis, *Traffic Quarterly*, Jan. 1979 and *Status Report, op. cit.*, June 21, 1979.

repealed previously enacted laws requiring helmets. The rationale has been that government should not deprive individuals of their right to make choices concerning their personal safety. The counter argument is that government should protect individuals from their own folly and, furthermore, that safety is of public concern since often society must carry the costs of caring for those who do not take care of themselves.

The motorcycle accident problem remains largely unsolved, and may be a continuing one since studies have shown that younger, novice riders and not the experienced "leather jacket" crowd are those most frequently involved.

Bicycles

There are 100 million bicycles available and 50 million now in fairly constant use in the United States. Deaths totaled 1000 in 1978 from collisions with motor vehicles. As with motorcyclists, the bicyclist often suffers severe injury or death in such collisions, since there is little cushioning. Even striking a standing or slowly moving vehicle can have serious consequences. Designs that separate vehicles from bicycles or restrict potential conflicts to locations where vehicles are moving very slowly (see Chapter 9) can reduce accidents significantly.[46]

Collisions with motor vehicles are not the only contributor to bicycle accidents. Vehicles are involved in only 26% of the total number of serious injuries for dedicated riders. Falls account for 36%; other bicycles, 13%; and animals for 10%.[47]

As indicated earlier for motorists, many bicycle riders seem to have little concern for their own safety. Law violation such as failure to heed stop signs, travel in hazardous patterns through traffic, and ignoring bicycle paths and sidewalks in favor of the roadway are common. Only a few riders wear head protection, and many travel at night without lights or reflectors. As indicated earlier, education of bicyclists in safe practices is a high priority item for the years ahead.[48]

Pedestrians

As indicated in Fig. 12–1, pedestrian highway deaths were 36% of the urban and 10% of the rural total. But their percentage in the "all accident" figures is low. This is to be expected, because pedestrians are so highly vulnerable to serious injury or death if struck by motor vehicles. The aim of design is wherever possible to keep vehicles and pedestrians apart.

An analysis of urban accidents that involved primarily pedestrian behavior found that, in 35% of the cases, pedestrians darted or dashed out into the street away from intersections, 17% dashed out at intersections, 7% were hit by turn-

[46]D. F. and D. Y. Lott, *TRB Record 605,* indicate that bikeways reduce accidents by 51%. They also give other bicycle accident statistics.

[47]See J. A. Kaplan, *Transportation Engineering,* Jul. 1977 and K. R. Agent et al., *Transportation Engineering Journal of ASCE,* May 1980.

[48]See *TRB Record 683.*

ing vehicles, 5% walked into vehicles, and 4% were struck while outside the roadway. Another breakdown of the same data shows that variables in the accident chain include one or more among the following: appearing suddenly, 44%; running, 39%; walking or running into vehicles, 17%; and under the influence of alcohol or drugs, 6%.[49] Again, then, the demonstrated unwillingness of people to recognize hazards and their propensity to move impulsively or to take actions that put them in jeopardy make the pedestrian safety problem difficult.[50]

Accident prevention for school children has been a focus of particular attention.[51] Various associations such as automobile and service clubs and the police are deeply involved. In addition, schools often provide crossing guards, and pedestrian safety is part of almost all curricula. In such efforts, attention to the very young is particularly important, since the rate for 5-yr-olds is three times that for age 11 and older children for whom the rate is constant.[52]

Postaccident Emergency Services

It has been claimed by various authorities that 25% of the permanently disabled would not be crippled and that 5500 to 20,000 lives would be saved annually if prompt, effective medical service were made available at accident sites and if rapid evacuation to hospitals were assured. Advances in providing such services have been made over the years, including radio communication to bring quick attention, better-equipped ambulances, and more highly trained paramedics operating them. Another improvement is on-board instruments which develop and transmit by radio information on the patient's condition and vital signs to a doctor. Emergency treatment to be administered in the ambulance on the way to the hospital can then be prescribed. Today, also, most hospitals have excellent emergency facilities to give treatment.

Even though there have been many advances in emergency services, there are still unsolved problems. Among them are high costs and delays in reaching and getting away from the accident scene over traffic-clogged roadways. One solution attempted on a trial basis has been to summon by radio helicopters specially equipped with first-aid equipment and staffed with medically trained personnel. To date, however, such special emergency services are only experimental.[53]

[49]See R. L. Knoblauch, *TRB Record 629*. For a different approach see W. G. Berger and H. D. Robertson, *TRB Record 615*.

[50]For a summary of engineering approaches to pedestrian safety see *TRB Record 743* and *Public Roads*, Dec. 1978.

[51]For an excellent reference on safety planning for schools, see *ITE Journal*, Aug. 1978. See also *A Program for School Crossing Protection*, available from ITE.

[52]See M. L. Reiss, *Transportation Engineering*, Oct. 1977.

[53]For further discussion of these experiments see E. M. Wilson and J. S. Matthias, *TRB Record 498*, and J. M. Waters, *TRB Special Report 153*.

ENGINEERING APPROACHES TO HIGHWAY ACCIDENTS

Analyzing Situations for Accident Reduction

Procedures for reporting and pinpointing the location of individual accidents, for presenting the details about them, and for focusing attention on high-accident road segments or spots were described earlier in this chapter. The next step is to analyze these situations and, if possible, reduce the hazards.

To eliminate or reduce accident opportunity, new construction or reconstruction of existing facilities to standards that eliminate most of the hazards is the ideal solution. However, because money is limited, attention must be focused on high-accident locations. Once they are selected, assembling roadway or intersection plans and accident reports, along with site visits, is the traditional next step. Today, however, many agencies have developed "photolog" data for their systems (see Chapter 3) so that certain details not on either plans or accident reports are available without going to the site. For many situations it is next appropriate to develop a combined collision and condition diagram (see Fig. 12–2). The collision portion shows schematically the detail of each accident that has occurred within the study period. On many such diagrams the time of occurrence is printed along each arrow. The condition part depicts a plan view of the location. More elaborate studies might include profiles or sight-distance data.[54]

Given the pertinent information on the site and the nature of the accidents, the engineer then develops an improvement scheme using such tools as striping, signing, signals, curbs, channelizing islands, barriers, lighting, or other devices, and even reconstruction, which have been effective in similar situations. Figure 12–2 shows a typical "after improvement" layout and the change in accident experience that resulted. Figure 12–3 shows a more elaborate solution involving reconstruction from a traffic circle to an intersection controlled by traffic signals.[55]

Highway Design in Relation to Accidents

As of today, the most advanced highway designs are found in freeways. They offer free flow, high capacity, and reduced accident opportunity as illustrated in Tables 4–6 and 12–2. In looking toward accident reduction, then, it is useful to start with the freeway as a model, and, to the greatest extent possible, incorporate its features in other facilities, but at the same time to recognize that it also has limitations.

Among the reasons for fewer accidents on freeways are the following:

1. Fewer pedestrian accidents because pedestrians and vehicles are separated.

[54]See D. M. Litvin and T. K. Datta, *TRB Record 706,* for a description of automated collision diagrams.
[55]See C. W. Dale, *TRB Record 528* and *Traffic Engineering,* Aug. 1976, for added examples.

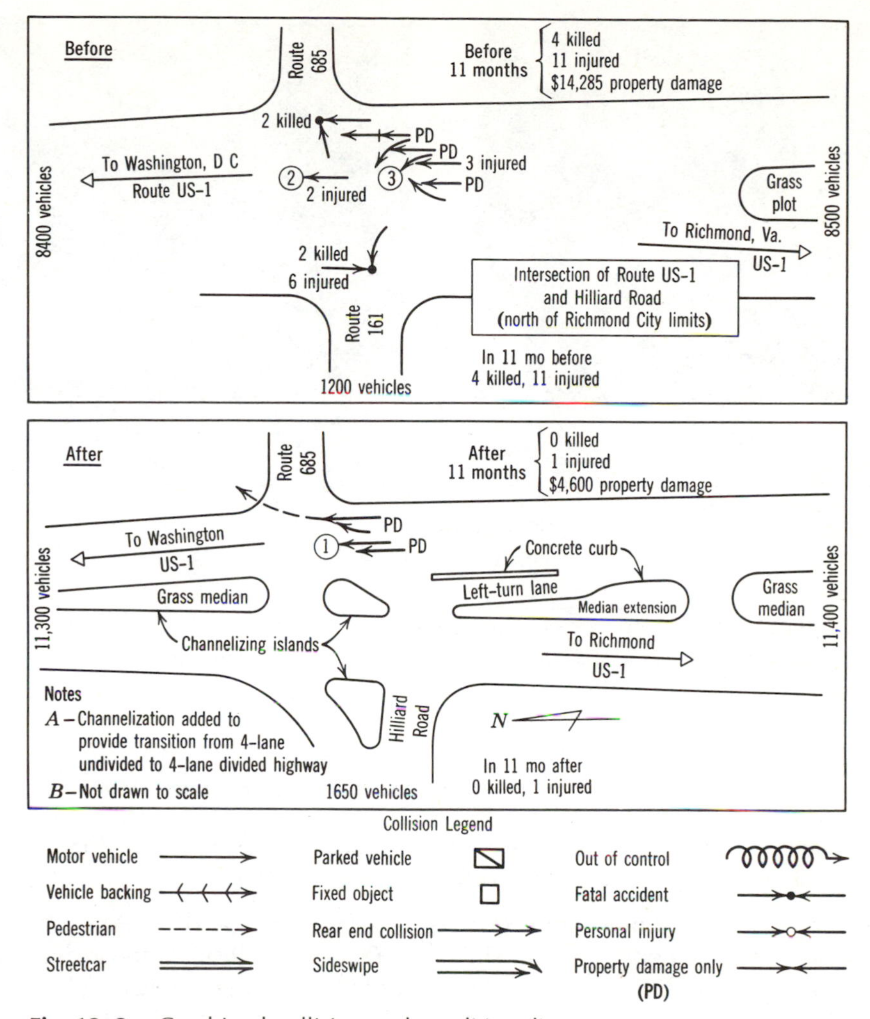

Fig. 12–2. Combined collision and condition diagram.

2. Prevention of head-on, or head-on sideswipe collisions by eliminating intersections and by positive center separation.
3. Reducing marginal accidents by eliminating parking, except by disabled vehicles, and providing wide shoulders for them.
4. Reducing entry and exit incidents by limiting access and departure to a few carefully selected and dimensioned points and using appropriately designed weaving sections where such maneuvers are necessary.
5. Minimizing the effects of single-car and other accidents involving vehicles out of control with features such as (see Chapter 9)
   a. Providing wide shoulders, traversable ditches, and flat side slopes which permit recovery or prevent overturning.
   b. Eliminating almost all cross-centerline accidents with wide medians or effective barriers.

Before (2 yr)		After (22 months)
5.0	Accident rate*	0.6
176	Total accidents	21
27	Injury accidents	7
31	Persons Injured	9

*Accident rate is per million vehicle passages

Fig. 12–3. Reconstruction of a traffic circle to a signal-controlled intersection to reduce accidents. (Courtesy: Federal Highway Administration.)

c. Suspending signs and lights or using breakaway supports.
d. Setting bridge abutments, walls, and poles back or isolating them with guard rail.
e. Cushioning other fixed objects with impact attenuators.

6. Providing illumination or reflectors at danger points to alert drivers and reduce the hazards associated with poor night visibility.

At the same time it must be recognized that freeway-type designs do not preclude all accident opportunities. There are rear-end collisions and sideswipes which bring one-fourth of the fatalities. These result when drivers follow too closely or fail to adjust their speeds to safe levels where visibility is limited by rain, fog, or snow.[56] Neither can freeway design always be forgiving when out-of-control vehicles leave the traveled way between exits or fail to negotiate exit ramps. As indicated above, it can mitigate them. Another difficulty, as indicated earlier, is that the higher velocities that accompany free flow often make accidents more severe.

Expressways, the next highest design category, offer more opportunities for accident. They have only partial control of access and commonly have some

[56]Many drivers feel secure when trailing another vehicle at a safe following distance which the National Safety Council suggests is one car length (20 ft) for each 10 mph (100 ft clear distance at 50 mph). This is far short of the safe nonpassing sight distance, which at 50 mph is at least 350 ft. If, then, there is a stopped vehicle or other obstruction ahead, it can be avoided only by swinging around it. There have been instances where, with visibility limited, more than 100 vehicles have been involved in a succession of rear-end pileups which began when a single vehicle became disabled on the roadway. Following close behind trucks is even more dangerous, for the driver has little view of the road ahead. This is discussed by M. L. Reiss and H. Lunenfeld, *TRB Record 562*.

intersections at grade. Then, as illustrated by Table 4–6, other reductions in design standards for conventional highways and streets produce added accident opportunities and higher accident rates. Choosing among strategies and compromises to reduce accidents is difficult.

Accidents are among the primary concerns in all phases of highway planning, economy, finance, design, operation, construction, and maintenance. Specific approaches to accident reduction, with references, appear at appropriate places throughout this book.

Making Existing Highways and Streets Safer[57]

Techniques for determining high-accident highway or street segments or spot locations were discussed earlier. There then comes the problem of selecting which among the many possible projects to undertake. A recent survey indicated that in 75% of the cases selection was on a "common sense" basis, rather than by theoretical analysis. This can mean that choices are based on "rules of thumb" such as number of accidents (probably fatalities) in a given time period or pressures from citizens or politicians. Once the selection is made, many of the design options listed above offer possible solutions.

For spot locations, a careful analysis of past accidents, possibly using collision and condition diagrams, will suggest the most likely approaches. However, when an overall accident reduction program is contemplated, past experience can be a useful guide. For example, Table 12–3 is a ranking, based on a nationwide survey, of 23 different strategies for accident reduction. It shows their estimated "before and after" accident experience and costs, project cost, and benefit-cost ratios. Averages such as these are suspect as they may not reflect local situations, but they can serve as a starting point for a more detailed selection procedure. In addition it must be recognized that the table lists only safety benefits based on savings in accident costs. But final project selection must also weigh economic, financial, energy, and environmental concerns as well.

The approaches given above to improving safety apply principally to rural highways or major traffic arteries. Residential and other low-traffic streets also deserve attention. For new subdivisions, safer layouts result with limited access, curved streets, staggered intersections, and few connections to the major traffic arteries on the periphery. Followups show only one-eighth as many accidents as with the traditional gridiron pattern. By modest expenditures, it is sometimes possible to break up existing gridiron systems by introducing diagonal curbs or traffic circles at intersections or closing streets at strategic points to create cul-de-sacs or "play" streets. Reducing street width is sometimes helpful.[58] Unfortunately, less positive methods for speed control such as stop signs (see Chapter 10) are often ineffective.

[57]Valuable references include *NCHRP Report 162*, the AASHTO *"Yellow Book" titled Highway Design and Operational Practices Related to Highway Safety*, the FHWA *Handbook of Highway Safety Design and Operating Practices, TRB Record 488, 709, 753,* and *773*. K. R. Agent et al., *Transportation Engineering Journal of ASCE*, May 1976, and *Implementing Highway Safety Improvements*, published by ASCE. For low-volume roads see *NCHRP Report 214, NCHRP Report 231*, and F. J. Tamanini, *Transportation Engineering Journal of ASCE*, Jul. 1981, discuss safety appurtenances.

TABLE 12–3. Safety Benefits of Certain Highway Improvement Programs*†

Rank	Code	Improvement	Service Life	Number of Projects	Average Annual Accident Experience: Before Accidents	Before Injuries	Before Fatalities	After Accidents	After Injuries	After Fatalities	Annual Percent Accident Reduction Expected: Accidents	Injuries	Fatalities	Annual Benefit All Projects	First Cost Project $	Benefit Cost Ratio $
1	23	Shoulder widening or improvement	20	46	1,917	585	35	1,353	465	21	29	20	41	4,754,088	36,200	28.83
2	64	Installation of striping and/or delineators	4	2,000	5,849	3,351	113	5,060	2,599	60	13	20	48	17,493,150	1,094	26.49
3	25	Skid treatment/grooving	20	96	1,117	395	27	580	275	7	48	30	74	6,372,399	32,395	20.12
4	60	Installation or upgrading of traffic signs	4	775	3,727	1,538	80	2,879	1,038	23	23	33	27	8,031,275	2,278	15.03
5	57	Signing and/or marking	10	3,046	191	142	34	192	82	22	0	42	35	3,638,462	536	14.94
6	63	Installation or improvement of median barrier	10	23	962	479	48	994	449	4	3	6	91	12,712,351	270,070	13.73
7	66	Localized lighting installation	10	115	1,119	546	20	1,022	499	6	9	9	73	4,393,558	19,363	13.24
8	62	Installation or improvement of road edge guardrail	10	1,651	2,077	844	69	1,716	719	29	13	15	56	12,273,743	4,546	10.97
9	50	Flashing lights replacing signs only—railroad crossing	10	56	36	17	7	2	1	.06	94	93	99	2,014,882	25,065	9.41
10	60/64	Signs/striping combination	4	465	5,858	2,844	90	4,464	2,108	65	24	26	27	9,982,024	8,270	8.80
11	61	Breakaway signs or lighting supports	4	527	195	41	2	127	23	0	35	44	100	658,538	599	7.25
12	11	Traffic signals, installed or improved	10	609	9,408	4,181	87	7,698	2,840	43	18	32	49	17,688,205	26,660	6.36
13	26	Skid treatment/overlay	20	126	3,071	1,627	37	2,552	1,194	26	17	27	30	4,747,692	80,796	6.09
14	55	Automatic gates replacing signs only	10	101	43	17	11	0.2	0.03	0.2	99	99	100	3,100,704	37,872	5.44
15	10	Channelization, including left turn bays	10	612	5,815	2,618	83	4,481	1,860	30	23	29	65	17,982,781	50,091	3.94
16	20	Pavement widening, no lanes added	20	241	951	489	26	715	301	4	25	38	87	7,238,024	60,188	3.88
17	13	Sight distance improved	10	142	338	205	6	234	127	4	31	38	36	859,826	13,896	2.97
18	12	Channelization and signals	10	36	887	333	1	609	215	0.5	31	35	50	620,449	64,846	1.78
19	58	Automatic gates replacing active devices	10	166	28	11	3	5	3	0.1	81	75	96	948,528	33,921	1.13
20	42	Changes in horiz. and vert. alignment	20	69	423	219	6	332	150	2	21	32	69	1,350,900	211,055	0.91
21	31	Replacement of bridge or other major structures	30	163	113	84	5	63	33	0	44	60	47	1,548,858	118,475	0.90
22	21	Lanes added, without new median	20	96	1,482	595	7	1,224	531	5	17	11	31	900,544	114,987	0.80
23	30	Widening existing bridge or other major structures	20	354	565	291	3	198	76	2	65	74	33	1,103,632	75,440	0.41

*Source: *An Evaluation of Highway Safety Programs,* DOT, July 1977.
†Computed at 8% interest rate.

Accidents at Railroad Crossings

Protection or elimination of the 220,000 highway–railroad crossings at grade to reduce accidents has been a focus of major attention for many years. In 1978, 1100 fatalities, or 2% of the total, occurred at such locations. But these accidents receive great attention because of their spectacular nature, with vehicles badly mangled and with one fatality for each 12 injury accidents, as contrasted with one to 39 for all highways. In part for this reason, both federal and state governments have programs of grade-crossing elimination.

Grade separation, the most positive solution to railroad crossing accidents, has been challenged on the grounds that other investments in highways and in safety could show a greater return, and that in most cases far less costly investments in warning or crossing-protection devices would be more suitable. Furthermore, funds are available for a relatively few projects, so that grade separation cannot be the whole solution.

Some 50,000 of the nonseparated crossings at grade are currently equipped with active, train-actuated devices. The remainder are marked with the X-shaped crossbuck sign. Studies have indicated that flashing lights, wigwags, or crossing gates substantially reduce accidents. Gates are particularly effective; in California, accidents were reduced by 70% and deaths in the remaining accidents by 64% when they were installed.[59] Numerous other studies have been made dealing with predicting accidents, the economic consequences of various warning devices and grade separation, and the effectiveness of various signs. Selecting which of many crossings to protect and in what manner is difficult, with many variables including rural versus urban, daily number of trains and vehicles, and crossing visibility and geometry to consider.[60]

Transporting Hazardous Materials

Trucks transport flammables, explosives, munitions, nuclear materials, and caustic liquids or solids. Some of these can turn into dangerous gases. These present hazards both to motorists and to people along the highway. Regulating their handling and the design and operation of the transporting vehicles and the necessary highway appurtenances are another function of state governments. This problem, although of concern to highway engineers, is outside the scope of this book.[61]

[58]See, for example, R. H. Borowski, *TRB Record 722* and L. C. Orlob, *Traffic Engineering*, July 1975. A summary of approaches used in London appears in *Transportation Research News*, TRB, May–June 1979.

[59]See W. R. Schulte, *TRB Record 611*.

[60]See *NCHRP Report 50* and its addendum, J. H. Sanders, *TRB Record 562*, J. B. Hopkins and F. R. Holstrom, ibid., J. Coleman et al., *Public Roads*, Mar. 1977 and Mar. 1979, R. J. Ruden and J. Coleman, *ITE Journal*, Mar. 1979. L. W. Margler and M. B. Rogozen, *Transportation Engineering Journal of ASCE*, May 1980, and the *Railroad Crossing Handbook*, FHWA.

[61]See *A Guide for Control and Cleanup of Hazardous Materials*, AASHTO. Also *TRB Record 554* and 693 and *TRB Circular 219*.

SAFETY IN HIGHWAY CONSTRUCTION AND MAINTENANCE

During highway construction and maintenance it is often necessary to alter the usual traffic patterns and to have workers and equipment engage in activities alongside the highway or to cross lanes of moving traffic. Even with posted warnings and other notices of these activities, it is difficult to get drivers to reduce speed, since they generally do not associate them with their personal safety. All too often, also, warning devices have been left in place when no activities are in progress, which causes drivers to ignore all such messages.

The *Manual on Uniform Traffic Control Devices* devotes an entire section to traffic control for construction and maintenance. It recommends, first of all, that plans and specifications spell out in detail how such operations are to be carried out and that personnel be trained for it. It then gives details on signs (black letters on a brown background) and standards for striping, pavement markings or plastic cones, barricades, and flashers or other warning or directional devices. Spacings and other dimensions are specified. Also, instructions for flagging, including details of flags, clothing for flaggers, and procedures for directing motorists and operating pilot cars are spelled out.

Even though careful attention has been given to making construction and maintenance sites safe for motorists and workers, their accident experience is substantially worse than for other sections of highway.[62] Such accidents become a fruitful source of lawsuits for negligence against contractors, agencies, and individuals (see Chapter 7).

Increasingly, research to improve safety on construction and maintenance sites is being undertaken. One study involved substituting raised pavement markers for temporary striping with paint.[63] They are more visible and can be easily installed and removed, while paints require machines to apply and must be sandblasted for removal. Studies of collisions with barriers made of timber have led to their abandonment except as delineators in urban areas. Instead, short sections of New Jersey style concrete barriers may be set in series as needed.[64]

Safety problems in construction and maintenance will assume increasing importance in the years ahead as more attention is given to the rehabilitation of existing highways and less to construction on new alignments.

Maintenance operations in particular often must interrupt or interfere with normal traffic flow. The *Manual* recommends the same practices for maintenance as for construction. Often, however, maintenance is for far shorter periods and temporary markers such as traffic cones that can be set out and removed easily are better than striping and signs. There is a tendency in maintenance not to follow the standards in detail, since to do so is very time

[62]See, for example, *ITE Journal,* Apr. 1979 and R. J. Paulsen et al., and J. L. Graham et al., *TRB Record 693.*
[63]See C. W. Niessner, *Public Roads,* Sept 1979.
[64]See articles in *TRB Record 693, 703,* and *769* for details of this and other studies. *NCHRP Synthesis 1* is an earlier reference.

consuming and costly. For example, to meet warning standards requires two signs, one 1500 ft ahead, and a flagman when a lane on a two-lane highway is temporarily blocked. However, failure to conform to accepted methods not only breeds accidents but creates the opportunity for lawsuits.

Protecting construction and maintenance workers from traffic is difficult and requires special attention. Reflective orange-colored vests and hard hats are musts and training and on-the-job instructions on self-protection should be required. In some instances, energy-absorbing devices have been installed on service vehicles.[65]

PROBLEMS

12–1. From the latest edition of *Accident Facts,* published annually by the National Safety Council:

a. Fill in the boxes in Fig. 12–1 for the reported year. Explain any marked differences between the new data and that in the figure.

b. Determine the number of deaths, the mileage death rate, and the population death rate for the state in which your college is located. Compare the record of your state with that of the United States as a whole and explain any marked discrepancies.

12–2. Investigate and prepare a brief report on the motor-vehicle accident-prevention program of the community where your college is located.

12–3. Investigate and report the penalties accompanying arrest for driving under the influence of alcohol or drugs in the state where your college is located.

12–4. Investigate and report on any unsafe practices of bicyclists or pedestrians on your campus.

12–5. On the basis of accident reports in the file of your local police department, plot a collision diagram for the accidents in the last calendar year or several years at one or more dangerous intersections. Then prepare recommendations for improvements to the intersection. The discussion of channelization and intersection design in Chapter 9 and of pavement markings, signing, and signals in Chapter 10 should be helpful.

12–6. A two-lane rural road 0.35 mi in length has had 2 fatal, 5 injury, and 20 property damage accidents over a 5-yr period. If the ADT totals 10,000, compute the fatal, injury, property-damage, and overall accident rates. Compare these rates with those given in Table 4–6. From these results, rate this road section as relatively safe, average, unsafe, or very dangerous.

[65]See, for example, J. F. Carney III and R. J. Sazinski, *Transportation Engineering Journal of ASCE,* Jul. 1978 and E. L. Marquis et al., *HRB Record 460.*

13 TRANSPORTATION AND THE ENVIRONMENT

The construction and operation of any transportation facility or system inevitably affects the environment. As a result, they become a major focus of attention and conflict. Some of these concerns, such as pollution or depletion of resources, might be considered as worldwide. At the other end of the scale are matters at the community or neighborhood level such as the effects of air pollution, noise, and paving over the landscape. Particularly in the developed, more affluent countries such as the United States, environmental legislation and court actions based on it have delayed or even blocked many public and private projects. Among those affected are dams, power plants, pipelines, ports, industrial and residential projects, and, of concern here, highways and facilities for public transportation. Without question, the years ahead will see many more legislative and judicial battles pitting those attempting to carry through such projects against conservationists, preservationists, and others who see their interests or those of the public adversely affected.

Highway and public transit officials, in carrying out their mandates to plan, construct, operate, and maintain their systems, must recognize and cope with the issues and constraints imposed by laws and court rulings concerning environmental issues. Furthermore, they must consider on the one hand, making facilities aesthetically pleasing and "good neighbors" in the community and, on the other, having them efficient and safe for users.

This chapter first presents the legislative, judicial, and administrative framework for decisions regarding environmental matters; it then discusses the implications and design considerations of highways and transit as neighbors; finally, aesthetic and environmental aspects of roadside development are presented.[1]

[1] *HRB Special Report 138* offfers a broad-gauge approach to this topic and is an excellent beginning point for further study.

LEGISLATIVE, JUDICIAL, AND ADMINISTRATIVE ASPECTS OF TRANSPORTATION AND THE ENVIRONMENT

Concerns for the environment as it affects motorists or transit users and those who view transportation facilities is of long standing. Early on, expenditures for erosion control and roadside improvement were authorized as an integral part of construction and maintenance. Controversy, where it occurred, was between those who advocated expenditures to make highways aesthetically pleasing and those who felt the money could be better spent in building more miles of road or providing better alignments and surfacings.

There are, however, newer highway and transit related aesthetic and environmental issues that have been and are the subject of continuing legislative and judicial controversy. These include billboard, junkyard and litter control, air quality, noise, and neighborhood intrusion.[2]

Highway Beautification Legislation

"Beautification" is a broad term which, under federal legislation, encompasses such activities as (1) landscaping and roadside development within the right of way, (2) acquisition of interests in and improvement of strips of land adjacent to the highway for the restoration, preservation, and enhancement of natural beauty, and (3) the acquisition and development of publicly owned and controlled rest and recreation areas and sanitary and other facilities. Great impetus was given to such activities by the Highway Beautification Act of 1965. It should also be recognized that many other highway and transit expenditures relate directly to beautification. Good location, design, construction, and maintenance practices as discussed elsewhere in this book can all contribute to "beautification" as defined above. One estimate is that when all efforts toward environmental protection and enhancement are totalled, they amount to about 12% of all Federal-aid funds.

OUTDOOR ADVERTISING. Control and removal of outdoor advertising has been for many years a controversial element in highway beautification. From a legal standpoint, the federal position is clear. The U.S. Supreme Court, in 1954, in *Berman* v. *Parker,* confirmed the public right to control billboards, stating that "The concept of public welfare, for the purpose of which a legislature may exercise the police power, is broad and inclusive, and the value it represents are spiritual and aesthetic as well as physical and monetary." In the main, state courts agree with this view.

In the 1958 Federal Highway Act, Congress addressed the problem of unsightly billboards when it stipulated that it is "in the public interest to encourage and assist the states to control the use of and to improve the areas adjacent to

[2]See *Traffic Engineering* Mar. and Sept. 1976 for a listing.

the Interstate System by controlling the erection and maintenance of outdoor advertising signs, displays, or devices within 660 ft of the edge of the right of way and visible from the main traveled way." The 1959 act and subsequent legislation has exempted urban areas zoned commercial or industrial from control, but included the federal-aid primary system in the legislative ban.

Although compensation for billboard removal would not be required when the police power is exercised, Congress provided for compensation to property owners by grants from the general fund (see Chapter 5). Current legislation stipulates that states without acceptable legislation be penalized by reducing their federal aid by 10%.

The federal legislation outlines in detail which signs are and are not acceptable, and provides a grace period for removal of unacceptable ones.[3]

JUNKYARDS. Junkyards have been another target of legislation. The Highway Beautification Act of 1965 called for their screening from view from the Interstate and federal-aid primary systems or removal by at least 1000 ft. It also provides 75% federal aid for financing screening or removal and a 10% reduction in federal aid for states that do not enact appropriate legislation.[4]

LITTER CONTROL. Collecting litter from the roadside is a costly operation. Legislation at the state and local level attempting to control litterers is almost universal. Commonly it provides fines against those who throw refuse from vehicles or dump it along the right of way. In addition, some states have considered and a few enacted legislation that requires a refundable deposit on bottles and cans that otherwise might be thrown along the roadside. These "bottle" laws are highly controversial and pit environmentalists against bottlers and merchants. The problem of litter control is complex and has several aspects, but further discussion is beyond the scope of this book.

Environmental Legislation

The National Environment Policy Act of 1969 and the Federal-Aid Highway and Environmental Quality Improvement Acts of 1970 are without question the key legislative enactments. Their provisions set forth a broad national policy which affects all sorts of activities and projects, both public and private.

LEGISLATIVE INTENTS. Title I of the 1969 act states its broad purpose. It provides in part as follows:

(a) The Congress, recognizing the profound impact of man's activity on the interrelations of all components of the natural environment, particularly the profound influences of population growth, high-density urbanization, industrial expansion, resource exploitation, and new and expanding technological advances and recognizing further the critical importance of restoring and maintaining environmental quality to the overall

[3]See *NCHRP Report 119* and R. D. Netherton, *TRB Circular 164,* for a detailed presentation on federal and state billboard legislation and the constitutional questions that surround it.
[4]See *NCHRP Report 112* for added detail.

welfare and development of man, declares that it is the continuing policy of the federal government, in cooperation with state and local governments and other concerned public and private organizations, to use all practicable means and measures, including financial and technical assistance, in a manner calculated to foster and promote the general welfare, to create and maintain conditions under which man and nature can exist in productive harmony, and fulfill the social, economic, and other requirements of present and future generations of Americans.

(b) In order to carry out the policy set forth in this chapter, it is the continuing responsibility of the federal government to use all practicable means, consistent with other essential considerations of national policy, to improve and coordinate federal plans, functions, programs, and resources to the end that the nation may—
   (1) fulfill the responsibilities of each generation as trustee of the environment for succeeding generations;
   (2) assure for all Americans safe, healthful, productive, and aesthetically and culturally pleasing surroundings;
   (3) attain the widest range of beneficial uses of the environment without degradation, risk to health or safety, or other undesirable and unintended consequences;
   (4) preserve important historic, cultural, and natural aspects of our national heritage, and maintain, wherever possible, an environment which supports diversity and variety of individual choices;
   (5) achieve a balance between population and resource use which will permit high standards of living and wide sharing of life's amenities; and
   (6) enhance the quality of renewable resources and approach the maximum attainable recycling of depletable resources.

(c) The Congress recognizes that each person has a responsibility to contribute to the preservation and enhancement of the environment.

Title II of the 1969 act establishes a Council on Environmental Quality in the Office of the President and the 1970 act provides staffing to support it. Among its functions is the coordination of all federal environmental activity. Also, as advisor to the president, it will provide a final review of policies and programs developed by the many federal agencies.

The 1970 act makes clear, however, that the primary responsibility on environmental matters rests on state and local governments. But, in reality, because there are many direct federal activities and because federal grants support many others, the federal role is nevertheless predominant.

By 1977, it became clear that environmental actions involve extremely complex political processes calling for joint actions of a variety of agencies and others operating at several levels of government. Attempts were made to correct the shortcomings of earlier legislation by amendments enacted in that year.

ENVIRONMENTAL IMPACT STATEMENTS (EIS). Provisions of the 1969 act require that "environmental impact statements" be submitted for many projects. They state that:

Included in every recommendation or report on proposals for legislation and other major federal actions [including expenditures of federal funds] significantly affecting the quality of the human environment [must be] a detailed statement by the responsible official on

(i) the environmental impact of the proposed action,
(ii) any adverse environmental effects which cannot be avoided should the proposal be implemented,
(iii) alternatives to the proposed action,
(iv) the relationship between local short-term uses of man's environment and the maintenance and enhancement of long-term productivity, and
(v) any irreversible and irretrievable commitments of resources which would be involved if the proposed action should be implemented.

Thus, the act makes clear that federal approval of any major federal-aid highway or public transportation project would be contingent on preparing and submitting an environmental assessment of it.

To implement the provisions of environmental legislation, the Federal-Aid Highway Act of 1970 stipulated certain duties of the Federal Highway Administration. Possibly the most important was to submit a report to Congress by June 30, 1972, which would outline a procedure for assessing environmental impacts. That report indicated that the following factors should be covered in the statements:

1. Need for fast, safe, and efficient transportation.
2. Public services.
3. Costs of eliminating or minimizing adverse effects.

Among the adverse effects to be considered were:

1. Air, noise, and water pollution.
2. Destruction or disruption of manmade or natural resources, aesthetic values, community cohesion, and the availability of public facilities and services.
3. Adverse employment effects, and tax and property value losses.
4. Injurious displacement of people, businesses, and farms.
5. Disruption of desirable community and regional growth.

These recommendations have been put into effect by a series of regulations. More recently they have been extended to include such additional concerns as:

1. Energy impact analysis.
2. Cost effectiveness analysis, including the alternative of improvement to existing public transportation.
3. Consideration of effects on neighborhoods and minorities.
4. Consideration of the effects on the economic viability and development plans for central cities.

All of these requirements apply equally to public transportation projects involving federal money.

Another important recommendation of the 1972 report was that the states and not the federal government should prepare the impact statements and that the Federal Highway Administration should not attempt to develop detailed technical guidelines for such studies, since no single approach would be valid nationwide and among projects. Rather, each state was to develop a process for making impact studies and create competent, interdisciplinary capabilities to carry them out "in house" or with specialists from other governmental agencies or consultants. Today, capabilities to produce EIS studies are readily available.[5]

The discussion just concluded deals only with federal laws and procedures. The states have paralleled the federal pattern. It should be clear, then, that in the years ahead all transportation agency activities will be heavily influenced by environmental laws and the requirements stemming from them.[6]

THE COURTS AND ENVIRONMENTAL LAWS. Traditionally, the courts have denied aggrieved parties the right to sue to overturn administrative decisions as long as those decisions conformed to the legislative delegation of power, were free of fraud, and were not arbitrary, discriminatory, or unreasonable. However, in recent years this precedent, which is founded on the doctrine of *nonreviewability,* has been substantially weakened by the courts, particularly for environmental cases. This was demonstrated by the Overton Park decision. In this case, the location of a segment of Interstate 40 took 26 acres from a downtown park. The Federal Highway Statute stated that "the Secretary of Transportation may not approve any transportation project which requires the use of parkland unless he finds that (1) there is no feasible and prudent alternative and (2) all possible planning to minimize harm to the park is included in the project." The secretary approved the project. The U.S. Supreme Court overruled that approval on the basis that parks could be invaded only if the disruptions from alternative routes reached "extraordinary magnitude." Furthermore, the court stated that the nonreviewability doctrine does not shield an administrator's action from a "thorough, probing, in-depth review." Some lawyers feel that this ruling will permit courts to substitute their judgments for those of administrators. On the other hand, a federal appeals court has taken an opposite view. In *Pennsylvania Environmental Council* v. *Bartlett* it held that a finding by the city council that a trout stream in a park was not of "significance" for recreational purposes freed the project from environmental requirements.

Another aspect of court interpretation of federal law is *timeliness* as illustrated by the *La Raza Unida* v. *Volpe* case in California. In this instance, the state argued that federal environmental and relocation laws did not apply to the project because the state had not yet applied for federal aid. The court ruled that federal law did apply because a project is treated as a federal highway as long as the state preserved the option of obtaining federal aid.

An interesting by-product of court decisions such as those just described ap-

[5]See Chapter 3 and *NCHRP Synthesis 40*.

[6]For further detail and a review of current approaches see J. G. Rau and D. C. Wooten, *Environmental Impact Analysis Handbook,* McGraw Hill, New York, 1980, *TRB Record 561, 583,* and *603;* and *TRB Special Report 138*. E. C. Muse et al., *TRB Record 716,* describe an EIS for rapid transit in Miami, Florida.

pears in the Federal Highway Act of 1973. Construction of U.S. 281, the McAllister Freeway north of San Antonio, Texas, a federal-aid project, first proposed in 1956, was blocked by a court action because it transgressed on certain park lands. A specific provision in the 1973 act now states that, with repayment of federal-aid money and "notwithstanding any other provision of the federal law or any court decision to the contrary, . . . the expressway shall cease to be a federal-aid project." The state was then able to complete the highway with its own funds, but not until 1978.[7] Congress, in effect, circumvented both the "reviewability" and "timeliness" situations described above without changing the basic law. It chose the same course in freeing the controversial Alaska pipeline from court-imposed delays based on environmental laws.

Another series of cases revolves around the question of *standing to sue*. Usually, only those parties damaged by a project are entitled to their "day in court." However, in several recent environmental cases, the courts have permitted bona fide conservation organizations to bring suit even though no economic loss or other harm flows to them from the proposed project.

Stipulations that governmental agencies pay attorneys' fees may have a strong effect on the number of lawsuits brought on environmental issues. Without such payments, only affluent individuals or organizations can afford to go to court. In the *Alyeska* v. *Wilderness Society* case the U.S. Supreme Court apparently halted a trend toward allowing attorneys' fees by stipulating that "it would be inappropriate for the judiciary, without legislative guidance, to reallocate the burdens of litigation . . ." Thus the decision now largely rests with legislative bodies, although there may be exceptions.[8]

This discussion of the courts and environmental law is not intended to be either authoritative or complete. Rather, its aim is to suggest some of the issues that will be before legislative bodies and the courts in the years ahead. All will have substantial impact on agency procedures as well as on many other public and private enterprises.[9]

ENVIRONMENTAL IMPACTS OF HIGHWAYS AS NEIGHBORS

From any listing of adverse environmental effects of transportation, air and noise pollution are the focus of the most community and neighborhood concern and complaint (see below). Other factors include traffic-induced vibration from rough pavements or vehicles,[10] accidents (see Chapter 12), and annoyances such as water pollution, dust, dirt, and litter (see Chapters 11 and 21). Without

[7]See *Civil Engineering,* June 1980.
[8]For more on this subject see *NCHRP Research Results Digest 103*.
[9]See *HRB Circular 135* and *136,* or *Selected Studies in Highway Law, Vol. 3,* published by TRB, for discussion of these issues.
[10]See, for example, F. Chilton et al., *TRB Record 541*.

question, all of these affect the quality of communities and neighborhoods and possibly the physical and psychological well-being of residents.[11]

Air Pollution and Its Control

Air pollution from both fixed and mobile sources has become a major concern in most intermediate- and large-sized urban areas in the United States and the rest of the developed world. It results from discharges into the air of nonreactive pollutants including carbon monoxide (CO), sulfur dioxide (SO_2), sulfates (SO_4), particulate matter such as dust, smoke, and lead, with its long-range effects, and reactive pollutants including hydrocarbons (HC), carbon dioxide (CO_2), nitric oxide (NO), nitrogen dioxide (NO_2), and ozone (O_3), which involve atmospheric transformation processes.

The highway air-pollution problem has two dimensions. The first deals with the area-wide effects of primarily reactive pollutants; the second with high concentrations of largely nonreactive pollutants at points or corridors along or near highways. The motor vehicle is a primary contributor to both forms, accounting for an estimated 70% of the CO, 50% of the HC, and 30% of the NO_X. Other transportation sources account for roughly 20% more of each.

Area-wide conditions become particularly bad when temperature inversions trap pollutants near the ground surface and there is little or no wind, so that concentrations become extremely high. For some individuals, eyes burn and breathing is difficult. It is charged that lives generally are shortened and some deaths actually result from these exposures. Also, certain kinds of vegetation are killed, stunted, or the foliage burned. An emergency approach is to call a "smog alert" when predicted atmospheric and air circulation patterns indicate high levels of pollutants. With them, warnings are issued of the hazard, industry and power plants might be required to burn natural gas rather than oil, attempts are made to curtail motor vehicle use, and even outdoor activities at schools may be restricted.

For the longer run, federal and state laws and regulations based on them are the mechanisms being employed to reduce area-wide air pollution. Urban areas have been delineated and limits set for the various contaminants. Where these are exceeded in an area, corrective measures are required; even for those meeting the standards, there may be prohibitions against increasing present levels.

A variety of middle- and long-term approaches for reducing area-wide pollution have been employed or suggested. One is land use control; for example, new industrial, commercial, or residential developments might be prohibited where their construction and use would add pollutants. An alternative might be to lower the emissions from present sources to compensate for the inputs from new projects. Modifications to existing fixed installations can be called for: an

[11]It has been proposed that, by interviews to determine public attitudes, measures of environmental capacity for highways and streets be developed. These would provide a measure paralleling vehicular capacity. See C. P. Sharpe and R. J. Maxman, *HRB Record 394*.

example would be scrubbing the exhaust gases of power plants or substituting low-sulfur for high-sulfur coal to reduce releases of SO_2.

Lowering pollution from motor vehicles is already the subject of intensive efforts. These fall into four general categories. These are: (1) reducing the output of pollutants by individual vehicles by making them more fuel-efficient (see Chapter 4) and controlling pollutant output[12]; (2) limiting vehicle travel by reducing the need or desire through fuel rationing or pricing, or possibly appeals to patriotism; (3) shifting travelers to other modes (see Chapter 3); and (4) cutting fuel consumption and pollution output for the remaining vehicles by improving traffic flow conditions (see TSM, Chapters 3 and 10).

As stated above, nonreactive pollutants create the more serious difficulties at specific locations or along highway corridors. It follows that before and after analyses of proposed individual projects would take account of them. No attempt is made here to quantify the amounts of the various pollutants generated by specific operating conditions. This situation is changing too rapidly. However, Fig. 13-1 is given to illustrate the complexities. It shows how the outputs of CO, a nonreactive pollutant, and HC and NO_x, reactive ones, vary with average operating speed and therefore would be affected by improvements in traffic flow. However, it does not distinguish between the effects of constant slow speeds and congestion induced by slowing and accelerating. Neither are data nor procedures offered that go the next step and provide a time frame for dispersal over an area at various distances from the roadside. Such analyses are extremely complex and usually rely on computer modeling.[13]

The longer-range implications of stringent air-quality controls reach into many facets of future urban and possibly rural growth and development. As indicated, maximum levels for individual pollutants for stated return periods have been legislated. Also, standards to be met by new vehicles and, in some states, inspection requirements for others, have been set. If these stated requirements are stringently enforced, plans for industrial and other land use development may be affected. In some instances, industries may be forced to move or cease operation. Those planning highways and other transportation schemes will be forced to consider area-wide and local air quality during both construction and use. The impacts on economic and other activity can be great and are probably unpredictable.

The laws and regulations regarding air pollution have forced highway and other transportation agencies to develop or hire expertise in air pollution and its control. Furthermore, of necessity if not by choice, they have worked closely with air pollution control and other agencies at all levels of government. Some projects have been substantially delayed or blocked by administrative or legal action. Whether the past zeal for a less-polluted environment will continue is

[12]A substantial reduction has begun and will continue with turnover in the vehicle fleet because the replacements must meet stringent pollution-control requirements. In addition, there may be added reductions if vehicle inspection and correction is mandated and enforced by the states to avoid losing federal-aid funds.

[13]Recent references include *NCHRP Report 200* and articles in *TRB Record 648* and *714*.

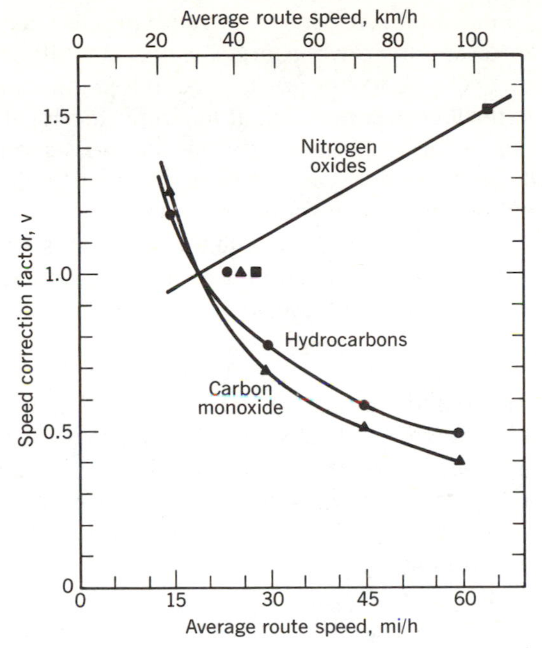

Fig. 13–1. Average speed adjustment factors for pollutants produced by gasoline engine vehicles. (After E. C. Sullivan.)

an uncertain question, given that many of the costs and inconveniences to the public are immediate and the benefits diffuse and long-range.[14]

Noise and Noise Control

NOISE CHARACTERISTICS. Noise is defined as unwanted sound. One of the unfortunate consequences of urbanization and mechanization on our way of life has been a substantial increase in noise levels. Today, serious efforts are well along on many fronts to reduce them. Highway and public transportation agencies are deeply involved.[15]

Noise is measured in decibels, with the common unit being the dBA. This single unit combines the sound intensities from all frequencies, but gives greater weight to frequencies above 1000 per second because human beings react more

[14]Space limitations permit only this brief discussion of air pollution and its effects on transportation. Among the many recent references are articles in *TRB Record 580, 599, 648, 670, 714, 722, 731,* and *789,* and E. C. Sullivan, *Transportation Engineering Journal of ASCE,* Nov. 1975. See also A. C. Stern, *Air Pollution,* (in 5 volumes), Academic Press, Inc. and E. Obert, *Internal Combustion Engines and Air Pollution,* Intext Educational Publishers.

[15]Control of noise levels in work places and on construction sites to improve working conditions and to prevent hearing loss is among the main purposes of the Federal Occupational Safety and Health Act of 1970 and of the regulations issued by the Occupational Safety and Health Administration of the Department of Labor.

strongly to them. Sounds at a level of 1 dBA can barely be detected by the human ear; but it can discriminate when pressures are 10 billion times as great. To apply a measuring scheme to this great range in hearing capacity, the measure of intensity, the decibel, has been set at ten times the logarithm of pressure to the base 10. Thus, an increase of 10 on the decibel scale means a tenfold increase in intensity. For example, an increase in the decibel level from 60 to 70 dBA means a tenfold increase in sound intensity.

The decrease in sound intensity with distance from the source is influenced by several factors. Measurements taken near highways show doubling the distance results in a lowering of 3 dBA over clean, level ground and 4.8 dBA over lush growth.

Noise in settled areas is a combination of a relatively continuous ambient background level and sporadic or instantaneous peaks. It comes to the ear from a variety of sources at different distances and directions. Also, as indicated, it is recorded on a logarithmic scale. Thus, noise level relationships are very complex and the theoretical aspects will not be discussed here. Rather, attention will be directed to noise as a transportation problem.

As indicated, some sustained (ambient) noise is always present. In a quiet residential neighborhood at night it is in the 32 to 43 dBA range; the urban residential daytime limits are about 41 to 53 dBA. In industrial areas the range is 48 to 66 dBA; in downtown commercial locations with heavy traffic it is 62 to 73 dBA. A notion of what these sound levels mean in terms of communication comes from such facts as the following: at 65 dBA a conversation is difficult at 3 ft; at 75 dBA, you must plug a finger in the nonlistening ear to hear a telephone conversation.

Human reactions to transportation-produced noise are subjective and very complex. In sum, studies have shown that complaints generally focus on interference with speech, TV viewing, or sleep and are related to increases above ambient noise levels. Increases up to 9dBA bring only sporadic protests; they range from sporadic to widespread with increases in the 9 to 16 dBA range. At increases greater than 16 dBA, complaints will be widespread and there may be community action. Other studies have shown that where highways are near residences, peak noise levels of about 70 dBA bring a few complaints; at 75 dBA complaints are likely; and at 80 dBA there will be letters or petitions of protest. There seems to be little correlation between protests and social measures such as income level or length of residence in the affected area.

A first approach to the control of highway noise is to limit or reduce noise emissions of the vehicles themselves by design and by enforcement of noise legislation. These vehicle noises have several sources: engine, drive train, fan, tires, air inlet and exhaust, and, at times, horns. For automobiles, total noise and tire noise, which is a major contributor, are generated at or near the road surface and increase as the third power of velocity. With diesel trucks, noise levels are almost constant with speed because engine speed is almost constant. Exhaust noise is the principal component and offers complexities because it is generated about 8 ft above the pavement surface. To illustrate, exhaust noise may be low near the roadside at ground level but produce protests from occupants of the upper stories of adjacent buildings. Also, barrier walls must be higher to be effective against it. Motorcycles and motor scooters have become major noise sources; many of them are far noisier than automobiles and equally as noisy as diesel trucks.

Controlling vehicle noise at the vehicle is difficult. Given that vehicle and muffler are designed to meet reasonable standards, operators may fail to maintain them; they may even alter them. Recently many enforcement agencies have instituted noise control patrols which monitor noise at the roadside and issue citations to violators.

The other general approach to noise control is employed by transportation agencies. It is to design the facility and its appurtenances and, sometimes, the surroundings, so that the noise reaching human ears does not exceed acceptable levels.[16] Strategems include:

1. Incorporating noise level control into the design or redesign of the facility. Among the alternatives are to depress and sometimes cover the roadway or to install sound barriers of earth or masonry. Unfortunately, these may trap air pollutants.
2. Keeping developments at appropriate distances from noise sources through zoning or building permit controls. An alternative is to require that new buildings or other activity centers be oriented and designed to control the noise level.
3. By redesigning or insulating existing buildings.

The first step in designing for noise control is to establish acceptable noise levels. Those proposed by the Federal Highway Administration are shown in Table 13–1. It is to be noted that some of these standards are set for building exteriors and some for building interiors. Table 13–2 shows anticipated noise reductions in going from the exterior to the interior of typical structures. It can be seen that structural material, window position (open versus closed) and window glazing all affect noise attenuation.

The acceptable noise levels cited in Table 13–1 are in terms of L_{10}, a level not to be exceeded more than 10% of the time. Other percentages are sometimes set. Another approach is to specify the peak noise, such as is generated by the passing of a diesel truck.[17] Neither procedure covers all situations. For example, the federal standard based on the "percentage time" concept does not control a short, loud burst of sound. On the other hand, the "peak noise" approach does not directly reflect the relationship between traffic volume and sustained noise levels.

PREDICTING HIGHWAY NOISE LEVELS. A variety of procedures for predicting noise levels at various distances from highways, with and without barriers, and for given numbers of vehicles of different types have been developed. The

[16]The discussion that follows is based primarily on *NCHRP Reports 173* and *174*. Other recent references include *TRB Record 580, 685, 740,* and *789, TRB Special Report 152,* and *Guide on Evaluation and Attenuation of Traffic Noise,* AASHTO.
[17]Without shielding, this is on the order of 90 dBA at 50 ft.

Table 13–1. Design Noise-Level and Land-Use Relationships*

Land-Use Category	*Design Noise Level, L_{10}†*	*Description of Land-Use Category*
A	60dBA (exterior)	Tracts of lands in which serenity and quiet are of extraordinary significance and serve an important public need, and where the preservation of those qualities is essential if the area is to continue to serve its intended purpose. Such areas could include amphitheaters, particular parks or portions of parks, or open spaces which are dedicated or recognized by appropriate local officials for activities requiring special qualities of serenity and quiet.
B	70 dBA (exterior)	Residences, motels, hotels, public meeting rooms, schools, churches, libraries, hospitals, picnic areas, recreation areas, playgrounds, active sports areas, and parks.
C	75 dBA (exterior)	Developed lands, properties or activities not included in categories A and B above.
D	—	Undeveloped lands. Requirements to be established in cooperation with local officials.
E	55 dBA (interior)	Residences, motels, hotels, public meeting rooms, schools, churches, libraries, hospitals, and auditoriums.

* Based on requirements of the Federal Highway Administration.
†L_{10} is noise level that will be exceeded 10% of the time.

one proposed in *NCHRP Report 174* for preliminary investigation is illustrated here by Figs. 13–2 and 13–3. These nomographs consolidate the findings of many analyses and observations. In-depth prediction methods are far more complex.

Figure 13–2 is a nomograph for predicting approximate L_{10} noise levels at various distances when there is a direct line of sight between the vehicle stream and the observer. Example 13–1, which follows, illustrates its use.

Table 13–2. Noise Reductions as Related to Building Type and Window Condition*

Building Type	*Window Condition*	*Noise reduction due to the Structure (dBA)*
All	Open	10
Light frame	Ordinary sash	
	closed	20
	with storm windows	25
Masonry	Single glazed	25
Masonry	Doubled glazed	35

* Data from Federal Highway Administration. For more detail see *NCHRP Reports 173* and *174*.

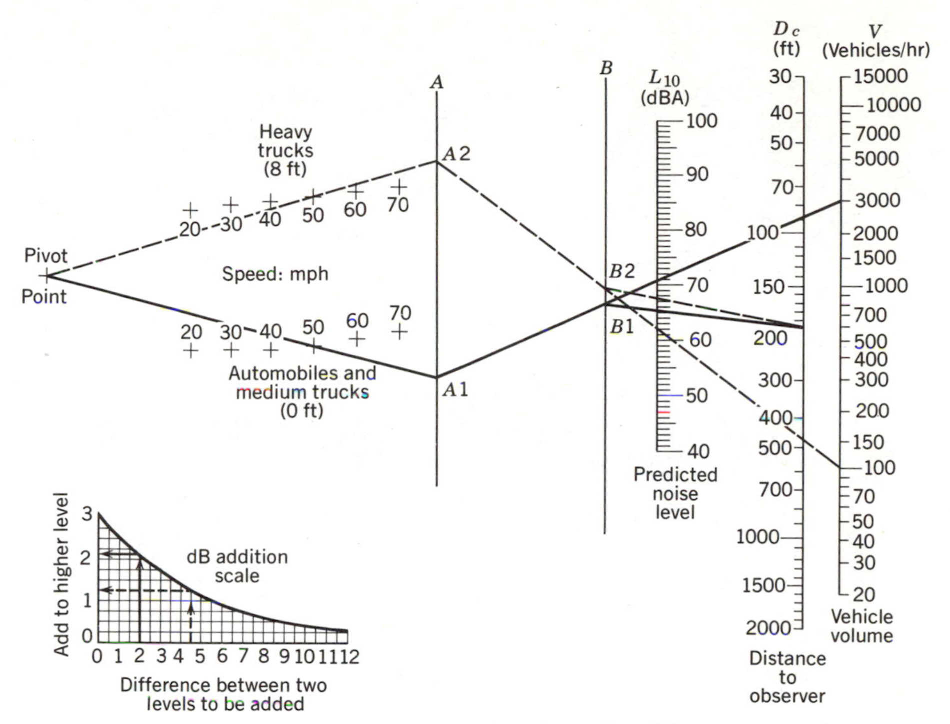

Fig. 13–2. Nomograph for predicting noise levels produced by a stream of motor vehicles when the receiver is in the direct line of sight with the noise sources. (From *NCHRP Report 174* somewhat modified.) 1 ft = 0.30m; 1 mph = 1.6 km/h

Example 13–1

Given:

traffic: automobiles, 2000 vehicles/hr; medium trucks, each counted as 10 automobiles, 100; heavy diesel trucks, 100

vehicle speed average: 50 mph

distance from roadway to observer: *L/S* = 200 ft

The solid and dashed lines on Fig. 13–2 show the steps in the solution.

For automobiles and medium trucks, with equivalent total volume of 3000 vehicles/hr.

Step 1. Draw a straight line from the pivot point through 50 mph speed to pivot point *A*1 on turning line *A*.

Step 2. Draw a straight line from point *A*1 to 3000 vehicles/hr (2000 + 10 × 100) on the *V* line. This establishes point *B*1 on turning line *B*.

Step 3. Draw a straight line from point *B*1 to 200 ft on the D_c line, the distance from road to observer. Read 66 dBA on the L_{10} line. This is the noise level if there were only automobiles and medium trucks on the highway.

Step 4. Follow the same procedure (shown by dotted lines on Fig. 13–2) to determine that 100 diesel trucks alone produce a dBA level of 68.

Step 5. Using the chart at the lower left of Fig. 13-2 and the difference of 2 between the individual noise levels of 68 and 66, determine that the combined noise level is greater than 68 by slightly more than 2. The final level, then, is 70 dBA.

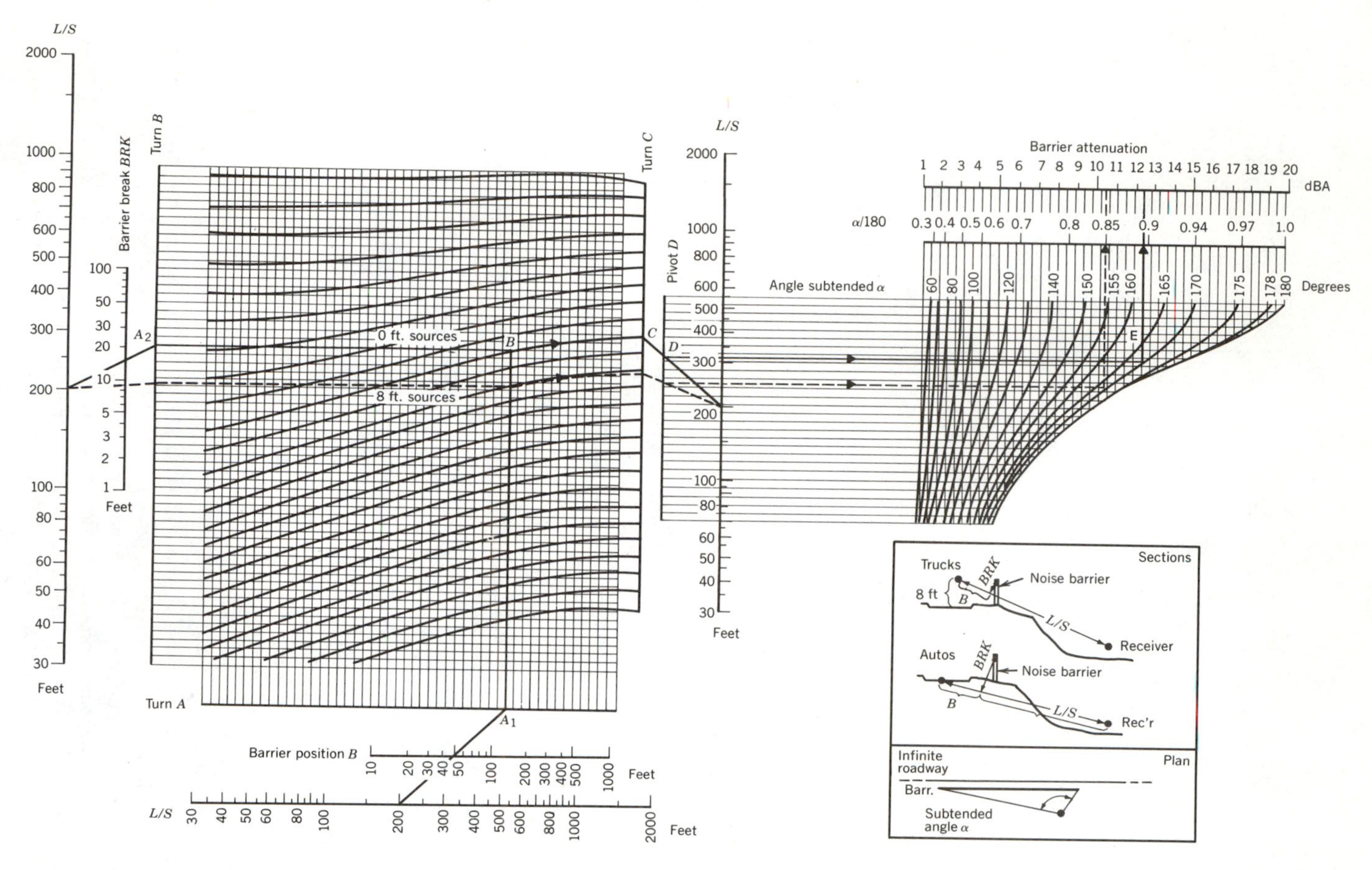

Fig. 13–3. Noise attenuation resulting from barriers which break the line of sight between vehicle noise sources and receiver. (From *NCHRP Report 174*, somewhat modified.) 1 ft = 0.30m; 1 mph = 1.6 km/h.

Figure 13–3 gives the reduction in dBA when an effective sound barrier such as an earth bank of solid, heavy barrier wall is interposed between source and observer. Example 13–2 extends Example 13–1 to show its use.

Example 13–2

Step 6. From Example 13–1, steps 3 and 4, bring forward the sound levels of 66 and 68 dBA respectively, for automobiles and medium trucks and for diesel trucks.

Step 7. From a cross section for the site (see diagrams at lower right of Fig. 13–3):

a. Determine the distance *B* from the noise source to the barrier, in this case 50 ft.

b. Determine *BRK* for automobiles and medium trucks. This is the perpendicular distance between the top of the barrier and the line of sight connecting the noise source (pavement surface level) and the observer. In this instance, *BRK* is 15 ft.

c. Determine *BRK* for diesel trucks, for which the noise source is 8 ft above the pavement surface. Here *BRK* is 9 ft.

Step 8. From a plan view of the site, determine, for the location of the observer, the angle (α) subtended between the ends of the sound barrier. For a continuous barrier, this angle is 180°. In this example, α is 170°.

Step 9. For the passenger car, medium truck situation, determine the noise attenuation provided by the barrier. The steps, shown by the solid lines, are:

a. Enter the left-hand nomograph at the bottom with *L/S* of 200 ft, draw a straight line through barrier position *B* at 50 ft, and intersect turning line *A* at A_1; draw a line vertically from this point.

b. Enter the same nomograph from the left with *L/S* of 200 ft, draw a straight line through barrier height *BRK* of 15 ft, and intersect turning line *B* at A_2. Draw a line horizontally through A_2 until it intersects the vertical line at point *B*.

c. Follow the gently curved path from point *B* to point *C* on turning line *C*.

d. Connect point *C* with 200 ft on line *L/S* in the right-hand nomograph. Establish point *D* where the connecting line intersects pivot line *D*.

e. Draw a horizontal line from point *D* to the curve for a subtended angle of 170° (point *E*).

f. Project a vertical line upward from point *E* and read the sound attenuation offered by the barrier. In this instance it is 12.5 dBA.

Step 10. Determine the sound attenuation for diesel trucks. This follows the procedures outlined in step 9 and is shown by dotted lines on Fig. 13–3. The reduction is 10 dBA.

Step 11. Compute the lowered dBA levels. These are:

a. For automobiles and medium trucks, 66 – 12.5 or 53.5 dBA.

b. For diesel trucks, $68 - 10 = 58$ dBA.

Step 12. Compute the lowered combined dBA level. This is done as in step 6, entering the lower left nomogram on Fig. 13–2 with $58 - 53.5$ or 4.5 and reading a combined increase of 1.2 dBA. The combined dBA is then $58 + 1.2 = 59.2$ or 59. It follows that the barrier reduces the noise level by 11 dBA.

It must be recognized that the information presented here merely introduces a very complex subject. Among the many other problems are those associated with noise reflected from buildings, walls, barriers, and the undersides of elevated structures. To deal with the noise problem, agencies and consultants have developed both equipment and procedures for measuring noise levels along existing highways and expertise at predicting them for designing new or correcting deficiencies in existing facilities.

MATERIAL FOR NOISE BARRIERS. Where space permits, earth barriers alone or in conjunction with a depressed roadway are excellent noise barriers.[18] Vertical walls, to be effective, must be free of openings and have substantial mass. For example, *NCHRP Report 174* recommends that they weigh at least 4 lb/ft^2 of surface. Commonly these heavier walls are constructed of cast-in-place or precast concrete, concrete block, or brick.[19] Often decorative designs are used to improve their appearance. Lighter construction of metal, wood or plaster is less effective, since sound is transmitted by or through the wall. Planting brings little reduction in intensity, but may sometimes be psychologically effective.

Compensation for Transportation-Induced Impairment to the Quality of the Environment

Air pollution, smoke, dust, fumes, and noise all can impair the desirability and value of property adjacent to or near transportation facilities. However, except in special cases, as noted below, compensation is seldom awarded for such damage. In some instances, the courts have ruled that because the damage is suffered in common by the general public it is *damnum absque injuria* (injury without legal damage) and constitutes neither a "taking" nor "damage in the constitutional sense." Other courts have founded noncompensability on "necessity," that is, that if the rule were otherwise, transportation facilities could not be constructed or operated.

Where no property is taken, exceptions to the denial of compensation generally result under "inverse condemnation" proceedings when it is shown that the damage is special to the affected property and not common to all. As an example, the suit would charge that traffic noise is far greater for the affected property than for its neighbors. Where there is a property taking, some courts have held that environmental damage can be included in setting severance damage (see Chapter 7) to the property that remains after the taking. Other courts have barred consideration of environmental effects in setting severance damages unless the damage is special to the property in question. The situation to date, then, seems to be that although individual claims for environmental damage to property may be allowed, sweeping claims for them by communities or neighborhoods will not.[20]

[18] See, for example, K. C. Sinha and N. R. Wienser, *TRB Record 685*.
[19] See, for example, E. C. Lokken, *Transportation Engineering Journal of ASCE*, Nov. 1976.
[20] See *NCHRP Research Results Digest 99* for added detail, particularly as it applies to noise.

PRESERVING SPECIAL LANDS AND SITES

Since 1966, federal legislation has declared it to be national policy to preserve natural beauty, park and recreational lands, wildlife and waterfowl refuges, wetlands, and historical sites. As indicated earlier, this can lead to difficulties and delays in locating and building transportation facilities, since the agency must prove that no feasible alternative to using the site exists and that all possible planning has been done to minimize harm. Other federal legislation, first enacted in 1906, calls for archaeological and paleontological salvage. Again, locations and schedules are often affected.

Numerous governmental agencies and private interests have cooperated in preservation and salvage activities. Many of them have specialists on their staffs. Expenditures are substantial; DOT alone committed an average of $65 million in each of several recent years to historic preservation. However, a discussion of the problems associated with preservation and salvage are beyond the scope of this book.[21]

ENVIRONMENTAL AND AESTHETIC ASPECTS OF THE ROADSIDE[22]

Roadsides

Roadsides are the entire right of way, except for the traveled way. Also of concern are the abutting lands that affect the appearance or utility of the road. Roadside development is sometimes thought of as roadside beautification, to be accomplished after the road itself is completed, if at all. It has become apparent, however, that early attention to aesthetics and to "naturalizing" the roadside not only provides a more pleasing environment for travelers but also results in lower maintenance costs and safer highways. It also has been claimed that, by improving the visual environment, capacity can be increased.[23] Today, these factors are given careful consideration in every phase of the highway program.

It would be a mistake to assume that attention to the roadside is a recent phenomenon. There was a Highway Research Board Committee on roadside development as early as 1933. The chapter on "Highway Landscape" in L. I. Hewes, *American Highway Practice* (Vol. 1, Wiley, New York, 1942) outlined early developments and devoted particular attention to the aesthetics of roads in the national parks. Also, the 1944 report *Interregional Highways,* which was the fore-

[21]Among many references on site preservation are *HRB Special Report 138* and *TRB Circular 206. A Design Guide for Wildlife Protection,* AASHTO, and D. L. Smith, *TRB Record 647* deal with that topic and *NCHRP Report 218 A* and *B* and P. W. Shudiner and D. F. Cope, *TRB Record 736* with wetlands.

[22]The aesthetic aspects of location practices and design of the traveled way are discussed in Chapters 6 and 9, respectively.

[23]See W. L. Smith and J. E. Faulconer, *HRB Record 377*.

runner of the Interstate System, emphasized the importance of attention to the roadside.

Control of roadside development on the highway right of way proper rests with the administering agency. The degree to which planting for appearance and screening can be used is defined by legislative stipulations, if any, coupled with administrative decisions as to whether or not expenditures are warranted. In some instances, the decision may be to limit improvements to the simplest erosion control measures; in others an all out effort may be decided upon. At times, the authority to improve or not to improve the roadside may serve to force an aesthetically pleasing development of adjoining property. For example, in some states roadside improvement in excess of simple erosion control is carried out only in those political subdivisions that have acceptable ordinances prohibiting billboards along the affected highways.[24]

Designing the Roadside Environment

Early highways were patterned after the railroads of the same period. The principal concern was to fashion the cheapest roadway that would carry traffic under all conditions of weather. Little attention was given to appearance. Horizontal and vertical alignment consisted of long straight sections connected by short curves. Shoulders were narrow or nonexistent, and side slopes were as steep as soil or rock would stand. Rights of way were narrow. Drainage ditches, channels, and structures were designed to provide protection for the roadbed without regard for erosion outside the roadway limits. These economies in first cost inevitably brought unsightly conditions and high maintenance costs. With the passage of time, design standards have been gradually modified, and now are far different.

As pointed out in Chapter 6, location practices have marked effects on the appearance of the finished highway. Often a pleasing vista or a view point can be developed with little sacrifice in cost or distance or scenic easements to enhance driving pleasure can be acquired (see Chapter 7). Long sweeping horizontal curves are preferable to short curves connected by long tangents. Choppy or broken-backed grade lines should be flattened and smoothed even though they meet the demands of geometric design. In rough country, cut depths and fill heights should be as small as possible to reduce scar and slope erosion to a minimum. Sometimes retaining walls (see Chapter 9) are helpful. In many other ways also the locator can by careful attention improve the attractiveness of the road. Computer graphics (see Chapter 6) which will enable designers to "see" their designs will provide added insights.

Modern highway design provides wide roadbeds, shallow, wide gutter ditches, and flat back slopes in cuts and flat side slopes on fills. Tops of cut banks and toes of fills are rounded to blend into the original ground (see Fig.

[24]The legal aspects of acquiring rights of way and controlling adjacent areas are discussed in Chapt. 7. Billboard and junkyard control is treated earlier in this chapter.

13–4 and Fig. 11–3). These features provide a safer roadbed and one of more pleasing appearance. Furthermore, erosion occurs more slowly or can be more easily prevented on the flatter slopes. One result is decreased expenditures for cleaning gutters and ditches. In addition, ditches and slopes can be dressed or mowed with power equipment, and this further reduces maintenance costs.

Wide rights of way are extremely important in roadside development. In rural areas, they permit blending the road into the natural landscape, provide space to plant screening in front of objectionable signboards and other unsightly objects, isolate the highway visually from its surroundings, and free the roadside

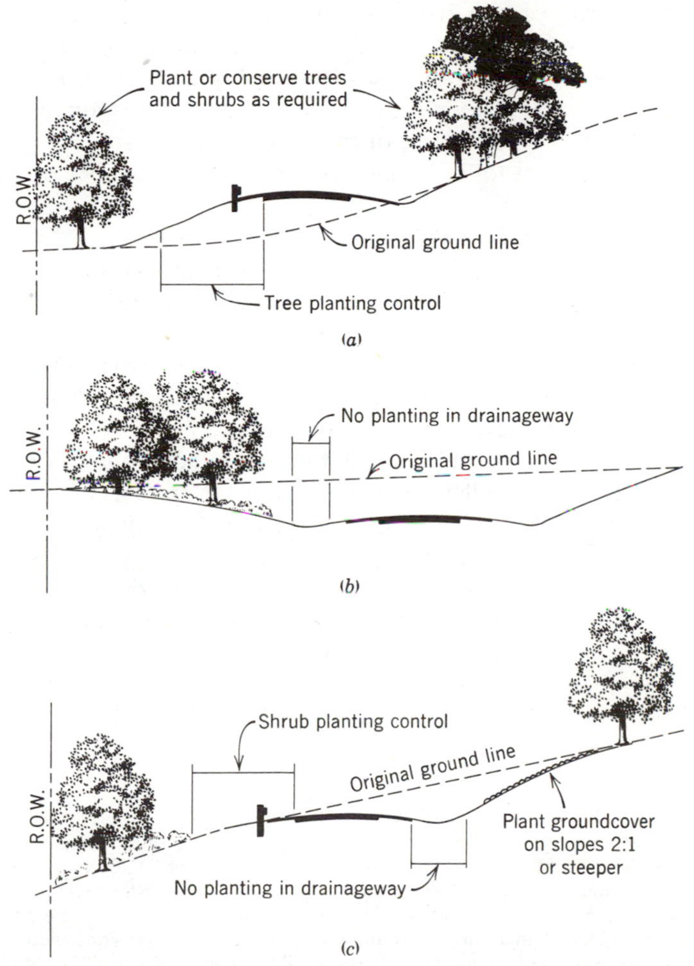

Fig. 13–4. Cross sections of rural highways employing good aesthetic practices; *(a)* roadway on fill; *(b)* roadway in cut; *(c)* roadway near ground level. (Source: *A Guide for Highway Landscape and Environmental Design*, AASHTO, slightly modified.)

of hazards. Also, access to necessary service facilities can be more easily controlled when they are some distance from the traveled way. In urban areas, wide rights of way on freeways and like facilities make room for planting or noise barriers to screen adjoining property from the sounds of heavy traffic or, in other cases, separate the road from unsightly industrial developments.

There is the same need to make detailed designs for roadside development as for any other highway feature. By a study of aerial photographs and other maps, coupled with on-the-ground inspection, it is possible to determine which shrubs and trees in the roadside are to be retained, which are to be removed or trimmed, and where new planting is required. One of many gains from such an approach is that travel can be made less monotonous.

Control of soil erosion during and after construction is one of the most important roadside problems. In the first place, badly eroded slopes and ditches are unsightly. Then the materials that are washed away muddy and pollute streams, disrupt drainage by filling ditches and culverts, and are deposited in unacceptable locations. It follows that modern designs must not permit erosion to occur.[25]

From an aesthetic point of view, natural flowing lines rather than plane surfaces are highly desirable on all cut and fill slopes. Transitions between cuts and fills deserve special treatment. For example, forming cut and fill slopes in "bell mouths" offers a pleasing transition. The end of each cut slope conforms to a cone with its apex located a suitable distance above the roadway at the intersection of the slope and the ground surface. For fills the cone is inverted. Contours of these transitions have a bell-mouthed appearance. "Daylighting" small cuts on the downhill side of the roadway to provide better views or small roadside parking areas is an excellent practice. Structures such as overcrossings can also be made more aesthetically pleasing. Bridges sweeping across the roadway in one span or with slanted and tapered piers may seem less stark. Again, breaking up the harsh straight lines of piers and bridge decks by generous rounding or with stone facings can make them blend with their surroundings.[26]

In planning roadside development, the designer must always be conscious of the cost of maintenance, including the expense of mowing, trimming, and snow removal. These problems are discussed further in Chapter 21.

[25]Design of ditches and channels to control erosion is discussed in Chapter 11. There is a voluminous literature on erosion control both in books and in publications of the U.S. Soil Conservation Service. *Compendium 9* of Transportation Technology Support for Developing Countries, TRB, *NCHRP Reports 220* and *221, HRB Special Report 135, NCHRP Synthesis of Highway Practice No. 18,* and *Vol. XII* of the National Association of County Engineers *Action Series* are excellent beginning points for further study, since they cover the important aspects of erosion control in a highway context and offer extensive bibliographies. Articles in *TRB Record 594* and *705* describe straw and fabrics as barriers for sediment control.

[26]*A Guide for Highway Landscape and Environmental Design,* AASHTO, is an authoritative reference as, for urban situations, is *The Freeway in the City,* FHWA. See also *Practical Highway Esthetics,* ASCE, and *Visual Values for the Highway User,* FHWA, C. L. Benson, *Civil Engineering,* Sept. 1976, and G. Dallaire, ibid., Oct. 1976. W. N. Melhorn and E. A. Keller, *HRB Record 452,* propose a numerical scheme for landscape appraisal.

Construction Practices and the Roadside

Careful examination of construction practices is an important aid in roadside development. Where possible, taking soil for fills from along the right of way (called side borrow) should be avoided in favor of sources out of sight of the roadway. Grading operations must be so planned that suitable topsoil is salvaged for future planting. Boulders or rock fragments resulting from blasting should be buried in the fills or disposed of otherwise. Suitable trees standing a proper distance from the traveled way should be protected from construction machinery and from damage during blasting. Objects of interest, such as rock outcrops in cut slopes or particularly attractive trees lying near the toes of fills, often can be preserved to add variety.

Control of dust by watering or with dust palliatives during construction in settled areas is a "must." Also, procedures for excavation and for placing embankments must minimize the danger for slope erosion. Today's specifications may require that grass or other permanent slope protection such as sprayed-on asphalt or mixed-in lime or cement be placed immediately after grading is completed.[27] It may be required that the tops of uncompleted fills be shaped and compacted if heavy rainfall is anticipated. There are instances where construction forces have had to plan culvert placement and earthwork to keep a stream flowing and clear for spawning fish. Also, as pointed out in Chapters 19 and 20, batch and mixing plants for paving operations must meet stringent dust control standards.[28]

In the past, environmental controls over construction were left primarily to the contractor and the administering agency. Today, they involve requirements and monitoring from federal, state, and local officials and ad-hoc citizen groups concerned with air, water, dust, noise, traffic, and protection of sensitive nearby areas such as wetlands. One agency reports that simply to designate the location of a borrow pit before operations can begin requires permission of ten governmental agencies.

Final cleanup after construction demands particular attention. Ragged slopes in borrow areas and along the roadside must be dressed to encourage the return of native plants and shrubs. However, the common practices of applying a "sandpaper" finish to cut and fill slopes is to be avoided, for with the first heavy rain they will erode badly. The best practice is to leave a series of small benches that will trap seeds and water and offer a foothold for plant growth. Often the slopes are protected with a fabric.

Vegetation and the Roadside

Over the continental United States, soil, topography, temperature, and amount and character of rainfall are very different. Plants that grow well in one area

[27]See, for example, *NCHRP Report 220* and *221, NCHRP Synthesis 18,* and *TRB Record 641* and *642.*

[28]See, for example, *HRB Special Report 135,* P. L. Cole and S. Steinborn, *Journal of the Construction Division of ASCE,* Mar. 1974, and J. R. Gordon, *TRB Record 551.*

may not be suitable in another. In addition, planting and maintaining vegetation is costly; in arid areas irrigation may be required; in snow country plantings may be killed by de-icing salts (see Chapter 21). Yet the use of turf, ground cover, shrubs, or trees in roadside development is appropriate throughout the country. Each of these is discussed below briefly and in general terms.

TURF. That portion of the United States roughly east of San Antonio, Texas (longitude 97 West), and the coastal areas of Oregon and Washington are classed as "humid." In humid regions, turf or other ground cover can be rapidly established and relied upon to prevent erosion of slopes 3 to 1 or flatter (or 2 to 1 where there has been liberal rounding at crest and toe). Recommended practice is to sow grasses immediately after grading. In the past this was not possible because the planting season was limited. However, varieties of grass and other ground cover are now available for sowing through the spring and summer and to fit most conditions of exposure and soil type.[29]

Turf is most commonly developed by *seeding,* farm equipment often being used. In some instances seeding has been done by spraying seed and fertilizer with compressed air; again, some seeding has been done from aircraft. *Sodding,* solidly or in strips, with grass already growing at another location, is another common method. Because of high cost, sodding is done sparingly. A typical application would be the lining of a drainage channel where severe erosion could occur before seeding or sprigged grass could become established.

Topsoiling, or covering the surface with soil containing dormant seed, soil micro-organisms, and plant food, has been widely used to provide favorable conditions for plant growth. Topsoiling is expensive, and the trend is toward conditioning the existing soils by mulching (see below) and with fertilizers. Before topsoil is placed on steep slopes, the underlying material must be roughened; otherwise the thin soil layer will slump down when softened by rain. On inclines steeper than 1½:1, it is common practice to hold topsoil in place by means of wooden frames or with longitudinal boards staked normal to the surface. Where rains are heavy, these measures may prove unsatisfactory unless the topsoil is covered with straw or other vegetable matter. This, in turn, is held down by wire mesh staked to the slope.

Mulching, with or without topsoil, is a highly recommended practice. Straw, hay, roadside cuttings, or local materials such as pine branches, leaf litter, moss, sawdust, tobacco stems, cottonseed hulls, or threshed soybean plants are spread uniformly over the surface, often by passing them through a special blower. The spread material is then worked into the previously loosened surface by means of disks, soil pulverizers, or with sheepsfoot or similar tamping rollers. On banks too steep for ordinary equipment, rollers are often propelled up and down the slope by winches mounted on trucks or tractors. Mulching retards washing of the soil, adds organic material, and holds moisture between rains. It is extremely effective for erosion control as it binds the soil together and pro-

[29]See articles in *HRB Record 411* and *TRB Record 506, 551,* and *674.*

vides favorable conditions for the growth of native plants. On high, steep slopes in rough country where rainfall is intense, mats of brush embedded in the slopes are sometimes used to supplement mulching. Certain resins and emulsions have been effective in preventing soil erosion and in holding seeds and fertilizers against the action of wind and water.[30] Again, mats or meshes, sometimes combined with muslin holding seed, fertilizer, and mulch, have been used instead of the conventional mulching process.

TURF SHOULDERS. In the past, but less frequently today, turf shoulders have been common throughout the East, Middle West, and South. They are pleasing in appearance, furnish effective erosion and dust control, and have been reported to be cheaper than other permanent shoulder surfacings in first cost and maintenance.[31] Stated disadvantages include a tendency in some cases to build up higher than the adjacent pavement and in others to subside below it. Turf will not resist continued traffic and, at turnouts or where pavements are narrow, a pavement of some kind must be substituted.

Another disadvantage of turf shoulders has been that they softened badly during rainy weather. Extensive research on this problem carried out in the 1950s indicates that satisfactory turf shoulders can be constructed. Recommended material is a compacted, stabilized surface course meeting AASHTO requirements (see Chapter 16) with organic materials, fertilizers, and ground limestone included to the extent permitted by stability. Mowing should be no lower than 3 in.[32]

GROUND COVER. "Ground cover" has been defined as "low-growing herbaceous or woody plants not more than 3 ft high at maturity." Both low shrubs and vines are included. Ground cover is an alternative to grass in controlling erosion by wind and water. It also serves as insulation that reduces sloughing caused by freezing and thawing. For protecting slopes and other roadside areas the best ground covers are thicket or mat-forming plants, those that root from decumbent branches, and those that spread from suckers and shoots. Bushy, dense-foliaged plants and those producing litter with great water-holding capacity afford the greatest protection against erosion. Rapid-growing species with inconspicuous flowers to prevent distraction and vandalism are to be preferred. Fire resistance or the ability to sprout after burning is important. Plants that are subject to disease and insect damage, that crowd out more desirable species, that are poisonous or irritating to the skin, or that may become agricultural pests are not acceptable.[33]

For median strips and islands, erosion control is but one of the functions of ground cover. Properly selected plants prevent headlight glare without affecting

[30]See *HRB Special Report 135,* W. T. Hottenstein, *Public Roads,* June 1970, and *HRB Record 206* for further details and bibliography.

[31]Untreated gravel and oiled shoulders are cheaper, at least in first cost, but are not classed as permanent.

[32]See *HRB Special Report 19* for a summary of accepted practice and an extensive bibliography.

[33]See *HRB Special Report 135, HRB Record 425,* and *TRB Record 551.*

sight distance and provide a contrasting background that implements traffic direction.

TREES AND SHRUBS. Trees and shrubs offer an effective means for providing interest, variety, and beauty to the roadside. Those native to the area generally are more desirable than imported varieties. For rural roads, the objective is the preservation or, where necessary, the recreation of a natural foreground in harmony with the distant view. Existing, well-placed trees should be preserved, while unpleasing or view-obstructing growth should be removed. Replanting should be considered only where irregular introduction of trees and shrubs serve to highlight the natural beauty or where it is particularly desirable to screen unsightly or distracting objects or activities. Trees should be placed back a distance from the traveled way to provide a recovery area for vehicles that run off the road. *A Guide for Highway Landscape and Environmental Design* sets a usual minimum of 30 ft from lane edge. Row planting along rural roads is not considered good practice as it spoils distant views and is monotonous. It is better to create occasional points of interest or to call attention to intersections, bridges, or other points of hazard by carefully planned group planting.

Formal arrangement of trees and shrubs is more appropriate on urban freeways and expressways than in rural areas. Often continuous planting is desirable to screen unsightly roadside conditions or to insulate residential areas from the road. Serious effort should be made, however, to avoid monotony and sameness over long stretches of the route. One effective means is to group flowering trees and vines at appropriate locations.

Parking Turnouts and Rest Areas

Parking turnouts to permit stopping off the traveled way by school buses, mail carriers, and others are an important adjunct to major rural highways. As these roads approach urban areas and on urban expressways and streets, off-road stops for public-transportation vehicles also become necessary. Provision for parking clear of the road at points of scenic or historical interest serves to prevent road obstruction and accidents.

A number of states have developed statewide systems of wayside rest areas. Since 1938, this activity has been encouraged by the fact that federal aid can be used to develop them, even including sanitary facilities. Important elements in site selection are: natural features that make the area attractive, easy accessibility at safe locations, sufficient area (usually 1–3 acres), and existing shade. Locations where public use will create a fire hazard or otherwise affect adjoining property should be avoided. Experience to date indicates that locations close by cities and towns are not satisfactory as they are monopolized by townspeople. In developing sites, adequate driveways and parking space separated from the traveled way, bumper rails or curbs to confine vehicles, and, possibly, fencing should be provided. Benches, water supply, and comfort facilities are highly desirable. Often they are a center for providing information to motorists. Without question, the provision of highway rest areas is worthwhile. However, many highway officials resist their establishment because users are untidy and sometimes vandalic, which makes maintenance difficult and expensive.[34]

PROBLEMS

13–1. Obtain and analyze an environmental impact statement prepared by or for the transportation agency that wishes to build a freeway or some other transportation facility.

13–2. Review the recent trends in federal and state legislation, administrative actions, and court rulings and determine if there have been any marked changes in direction or attitudes toward the control of vehicle emissions.

13–3. As indicated in the textbook, federal environmental law requires that each state prepare an implementation plan for air-pollution control. By contacting the appropriate air-pollution agency, find out the current status of the plan for the "control region" in which your college is located.

13–4. A parkway with a wide median contains an earth noise barrier which blocks cross-median noise. It carries 1000 automobiles per hour in one direction at 40 mph. Based on Fig. 13–2 and Table 13–1:

a. What would the L_{10} noise level 100 ft from the edge of the roadway?

b. At what distance from the roadside would an acceptable L_{10} noise level be reached for (1) a tranquil area and (2) a playground?

13–5. A two-lane highway at grade is expected to carry a total of 600 passenger cars and 80 medium and 30 heavy-duty trucks at 50 mph in the design hour. Assuming that noise is generated at the roadway centerline and that Fig. 13–2 and Tables 13–1 and 13–2 apply:

a. Find the outside L_{10} noise level at 200 ft from the roadway centerline.

b. Determine the L_{10} decibel level inside a light-frame building 200 ft from the roadway centerline with (1) windows open and (2) with ordinary windows closed.

c. Would this condition be acceptable for a residence with (1) windows open and (2) windows closed?

13–6. By studying the construction specifications or by interviews with local officials, determine the procedures that are employed to:

a. Prevent slope erosion during and after construction.

b. Control dust on construction sites.

c. Control the noise created by construction operations to protect nearby residents or businesses.

13–7. Determine from its annual or other reports, the amount and percent of total construction and maintenance costs that the highway agency of your state spends for

a. Installing and maintaining grass, plants, shrubs, and trees.

b. Picking up litter.

[34]*A Guide on Safety Rest Areas for the National System of Interstate and Defense Highways*, AASHTO, and *NCHRP Synthesis of Highway Practice No. 20* provide authoritative sources for the planning and layout of rest areas. See also J. M. Tyler and C. V. DeVere, *TRB Record 498*, and B. N. Lord, *TRB Record 536*.

14 SOILS AND MATERIALS

An understanding of the basement soil (or subgrade) and other materials used in the construction of pavement structures for highways and other transportation facilities is particularly important. The subgrade is normally defined as the supporting structure on which the pavement surface and its special undercourses rest. In cut sections, it is the original soil lying below the layers designated as base and subbase material. In fills, the subgrade consists of imported material from nearby roadway cuts or from borrow pits. Figure 14–1 shows two of the many combinations of pavement surface, base, and subbase which are employed by transportation agencies.

Before 1920, attention was focused largely on the pavement surface, and little notice was given to the subgrade and base materials or to the manner in which they were placed or compacted. Later, increased vehicle speeds brought demands for higher design standards which resulted in deeper cuts and higher fills. In many instances, subsidence or even total failure of the roadway resulted. Study of these failures indicated that the fault lay in the subgrade and not in the pavement. This led to the investigation of the properties of subgrade soils and their performance under service conditions. Now, most transportation agencies have established detailed procedures for the investigation of subgrade materials.

Similarly, most agencies have developed procedures to evaluate other materials used in the construction of pavement structures. These materials include aggregates for subbase, base, and surface layers and binders such as portland and asphalt cements. In recent years, many agencies have been utilizing geotextiles (filter fabrics) to aid in construction over soft ground or as filters in drains.

This chapter summarizes present knowledge of the behavior of soil as a subgrade material, of aggregates used as subbase, base, or surface materials, of portland and asphalt cements available as binders, and of geotextiles available for soil reinforcement and drainage applications. Although explicit answers to all the complex problems discussed are not available, much progress has been made. Research is still needed to improve the present state of practice.

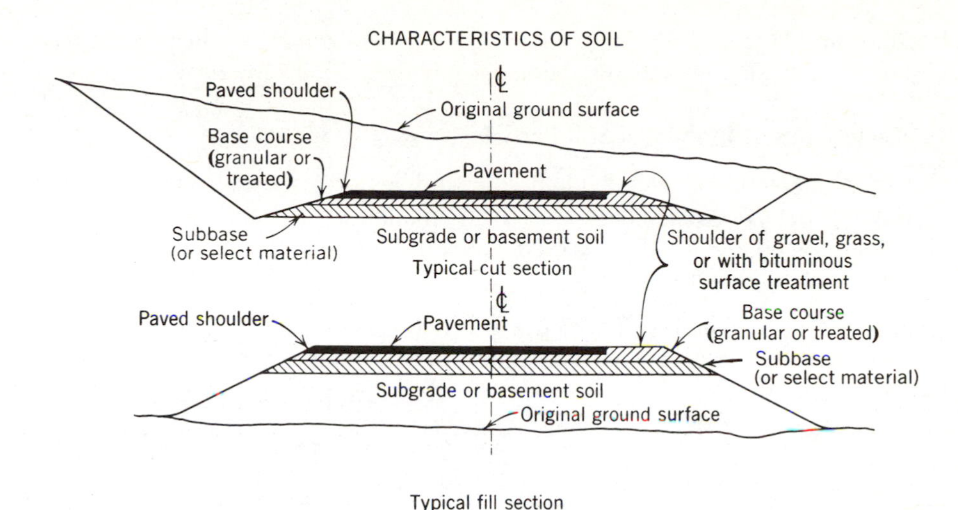

Fig. 14–1. Typical roadway cross sections (note alternative shoulder treatments).

CHARACTERISTICS OF SOIL

Constituents of Soil

Soils consist largely of mineral matter formed by the disintegration or decomposition of rocks. This disintegration into soil may be caused by the action of water, ice, frost, or temperature changes, or by plant or animal life. Soils near the surface may contain humus and organic acids resulting from the decay of vegetation. Almost all soils contain water in varying amounts and in free or adsorbed form.

In most instances, soils are blends or mixtures of particles of many sizes, shapes, and parent materials. Considerable variation in these characteristics are found in samples of apparently like soils taken from almost adjacent locations. For this reason, soil behavior is far more difficult to predict by the principles of soil chemistry than would be the behavior of a steel for which the chemistry is known. Furthermore, complete changes in soil types at frequent intervals are the rule rather than the exception; it is not uncommon to find five to ten distinct soil types along a mile of road. One estimate indicates that in the continental United States alone, there are more than 15,000 separate and distinct soil types.[1]

Caution must be exercised in applying American experience and testing procedures as described here to other areas of the world. For example, soils found in arid or humid areas or in temperate, tropical, or arctic climates may have substantially different properties. A case in point is the recent experience with

[1]See, e.g., articles in *HRB Record 284 and 374.*

laterities and lateritic soils so prevalent in tropical regions. For these soils, techniques for assessing engineering behavior have not been fully developed.[2]

Characteristics of Individual Soil Particles

Certain characteristics of individual soil particles are especially useful in predicting the behavior of a soil mass. These include *grain size, grain shape, surface texture,* and *electrical surface charges* resulting from chemical composition and molecular structure.

GRAIN SIZE. The grains of which a soil is composed have been classified in terms of size by the American Association of State Highway and Transportation Officials as follows:

		U.S.A. Standard Series	
Class	*Particle Diameter (mm)*	*Passing*	*Retained*
Gravel	75–2.0	3 in.	No. 10
Coarse sand	2.0–0.425	No. 10	No. 40
Fine sand	0.425–0.075	No. 40	No. 200
Silt	0.075–0.002	No. 200	—
Clay	0.002–0.001	—	—
Colloidal clay	Smaller than 0.001	—	—

Certain other classification schemes substitute 0.050 mm (No. 270 sieve) for 0.075 mm (No. 200 sieve) as the break between fine sand and silt. Regardless of the differences in the various grain-size classifications, all have a common aim; to establish a basis for relating particle size to soil behavior. General characteristics of the various particle-size groupings are as follows:

Gravel consists of rock fragments usually more or less rounded by water action or abrasion. Quartz, the hardest of the common rock-forming minerals, usually is the principal constituent. Well-rounded pebbles and boulders that have undergone long wear are almost entirely quartz. Gravels that are only slightly worn and therefore rough and angular commonly include other minerals or rocks such as granite, schist, basalt, or limestone.

Coarse sand frequently is rounded like the gravel with which it is found and generally contains the same minerals.

Fine sand particles commonly are more angular than coarse sand particles because the film of water that usually surrounds the finer particles has served as a buffer to protect them from abrasion.

Silt grains usually are similar to fine sand and have the same mineral composition. Often they are found as rock flour in glacial moraines. They may also be produced by chemical decay. Occasionally, silts contain pumice, loess, or other materials foreign to the associated sand. The presence of silt in fine soils

[2]See, e.g., articles in *TRB Records 497, 612, and 733,* and *Compendium No. 7,* Transportation Technology Support for Developing Countries, TRB.

may be detected by its grittiness if a tiny amount is placed in the mouth and bitten between the teeth.

Clays result almost entirely from chemical weathering and are often plate-like, scale-like, or rod-like in shape. Because of their small size, their performance is strongly influenced by moisture and surface chemistry.

Colloidal clays are finer clay particles that remain suspended in water and do not settle under the force of gravity.

Another characteristic of soil particles directly dependent on grain size is the relationship between surface area and volume or weight. Surface area is a second-power function of diameter; volume and weight are third-power functions. Thus, as particles become smaller and smaller, properties related to surface area rather than volume become more and more important.

For most purposes, coarse-grained materials are far more satisfactory as construction materials. Silty soils cause difficulties in areas where the ground freezes or where the movement of moisture by capillary action is present. Soils containing any great percentage of clays can be extremely troublesome and often cause design and construction difficulties. They should never be placed close under the roadway surface. One helpful "rule of thumb" is that soils which yield abundant crops may be poor for road building.

GRAIN SHAPE. The shape of the larger soil particles as found in nature often indicates their strength and toughness. Rounded particles found in stream deposits have undergone considerable wear and are probably quite strong. On the other hand, flat and flaky particles probably have not been subject to such treatment and may be weak and friable and not suitable for many uses.

Granular soil mixtures such as those used for base courses contain little clay, and the properties of the larger soil particles have an important influence on their behavior. Here, an angular or roughly cubical particle shape produced by crushing strong, tough rock or gravel increases the resistance of the soil mass to deformation under load since the individual grains tend to lock together. On the other hand, rounder particles tend to roll over each other. Most specifications for base courses or ballast require that the aggregate contain a stated percentage of crushed rock particles.

Very fine-grained particles, in the clay and colloid sizes, generally have the shape of thin flat plates or rods (see Fig. 14–2). In the extreme case of bentonite, the ratio of length to width has been reported to be as high as 250:1. These elongated shapes, combined with the effects of surface charges and moisture, bring some of the troublesome behavior of clays which will be discussed in more detail later.[3] Also, the thickness of individual clay particles varies greatly. For example, a typical particle of kaolinite may be 100 times as thick as one of montmorillonite.

[3]See E. B. Kintner, A. M. Wintermyer, and Max Swerdlow, *Public Roads,* Dec. 1952, for more detail on the size, shape, and structure of clays as determined with the electron microscope. See also W. C. Ormsby and L. H. Bolz, *Public Roads,* June 1968. Studies of the structure of calcium montmorillonite clays using x-ray defraction by D. Senich, T. Domirel, and R. L. Handy appear in *HRB Record 209.*

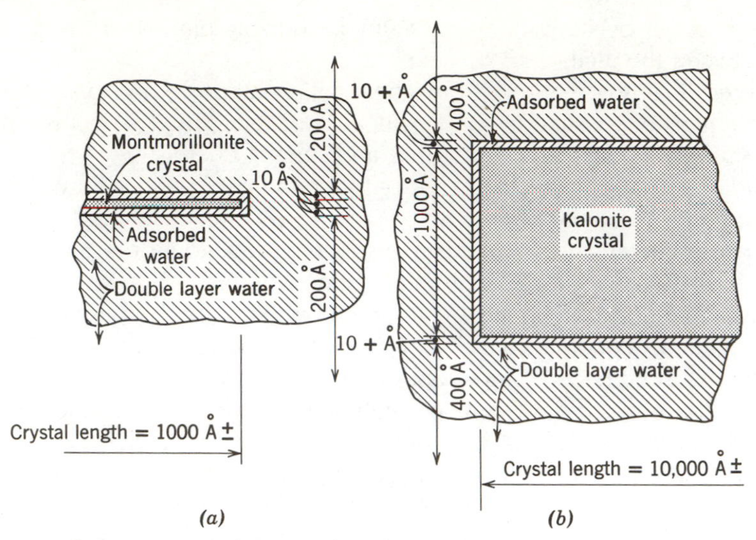

Fig. 14–2. Soil moisture on clay particles of typical dimensions (Å = Angstrom unit = $\frac{1}{10,000}\mu = \frac{1}{10,000,000}$mm). (*Note:* Figures are not to same scale.) (After T. W. Lambe.)

SURFACE TEXTURE. The surface texture of the larger soil particles greatly influences their performance in granular soil mixtures. For example, the grains of wind-blown and beach sands and of crushed quartz often have slick surfaces. The coefficient of friction developed between these surfaces is low. As a consequence, the particles can more easily slide over each other and the soil mixture that contains them has little resistance to deformation under load. Some authorities assert that the rough surface of freshly crushed rock is more important than its angular shape in developing strength in coarse-grained soils.

With soil grains of small size, the effect of surface texture is overshadowed to a large degree by other factors.

CHEMICAL COMPOSITION AND ELECTRICAL SURFACE CHARGES. As indicated earlier, soils are variable mixtures of many different materials in a variety of sizes and shapes. Largely for this reason, few routine tests based on soil science or colloidal behavior have been developed. On the other hand, these theories are extremely helpful in explaining soil behavior. Furthermore, many of the empirical tests and design procedures are successful to varying degrees in measuring these and other properties indirectly.

Chemical and surface-charge characteristics of the coarser soil particles apparently have little influence on the behavior of the soil mass because surface areas are small in relation to volume.[4] However, in soils containing appreciable amounts of clay, these properties are very important. As indicated, clay particles

[4]On the other hand, reactions between the surface charges of aggregates and various soil-stabilizing agents may determine their effectiveness. Similarly, incompatible charges on the aggregate surface and the asphalt account for pavement deterioration by "stripping" of the asphalt from the aggregate.

are plate-like or rod-like in shape. The sides of these particles generally have a negative charge, while the edges or ends are charged positively. Soil structure in the natural state usually is "flocculated" with adjacent grains oriented side to end or face to edge. However, particle reorientation into a parallel or "dispersed" structure by manipulation at high moisture contents can drastically change strength, permeability, and other properties of the soil.

Clay particles in which certain chemicals predominate, silicon for example, have larger electrical surface charges than those containing certain others, such as aluminum or iron. These more active soils exert greater influence on available water; it follows that they would attract thicker moisture films and tend to swell and shrink more with changes in moisture content than soils that are less active. In the case of the mineral montmorillonite, there is very little bonding force between the successive sheets which make up the individual clay particles. Water, when available, enters between these sheets, causing even more swelling.

These surface charges also seem to be at least partially responsible for some fine grain soils being *thixotropic*. *Thixotrophy* is defined as a change in strength over time. For example, a material will lose strength when manipulated, but it regains some strength after standing for a time. It has been proved that a clay which has some strength may become weak when manipulated and regain strength when it is undisturbed for a period. Research indicates that the strength loss comes when manipulation at a relatively high moisture content forces the particles into a parallel orientation; over time, however, the particles realign to produce compatibility in the electrical charges.[5]

Clay soils also contain cations attracted to the soil particles. These are of such elements as sodium, potassium, magnesium, or calcium. In general, those of lower valences, such as sodium, have the greatest attraction for water and, for this reason, react more vigorously to changes in moisture content. For example, it has been said that sodium favors very thick layers of water, probably in tens of molecular layers, and with no abrupt change in behavior between successive layers. On the other hand, calcium develops only a few molecular layers of well-oriented water, with a sharp break between the nonliquid and liquid phases.[6] The practice of adding lime to troublesome soils to produce an exchange of calcium ions for those that are more troublesome is common. (See Chapter 17 for more detail.)

Effects of Moisture on the Behavior of Soils

Soils engineers agree that the properties of a soil mixture containing significant fines are influenced more by moisture than by any other factor. Soils that have

[5]See, e.g., J. K. Mitchell, et al., *ASCE Journal,* Soil Mechanics and Foundation Division, SM 4, 1965, and C. L. Nalezny and M. C. Li, *HRB Record 209*.

[6]No attempt is made here to offer a comprehensive treatment of soil chemistry and electrical surface charges. Starting points for further study include *HRB Record 119* and *209; HRB Special Report 103;* publications of the Soil Mechanics Division, ASCE; the book by J. K. Mitchell, *Fundamentals of Soil Behavior,* Wiley 1976; and a variety of soil science and chemical publications. Other references also will be found in the various indexes of the Transportation Research Board.

ample strength and supporting power under one set of moisture conditions may be entirely unsatisfactory if the percentage of moisture changes.[7] One difficulty with soils as subgrades is that they are subject to such moisture changes. Means to prevent these changes are impractical, costly or both.

Soil grains are surrounded by films of water (see Fig. 14–2). This water is attracted by the molecular charge of the soil grains and has a higher boiling point, lower freezing point, and greater viscosity than ordinary water. The first few molecular layers, labeled adsorbed water in Fig. 14–2, are almost solid and are more nearly like ice. Then as the distance from the particle increases through the double-layer water, its properties change and finally become those of free or gravitational water.[8] The force required to pull the solidified water from the mineral surface is extreme, generally ranging from 100 atm for the outer layers to 10,000 atm for the closest molecules. Smaller forces are required to remove the double-layer water, with the force required to move it parallel to the particle surface much less than that to remove the water entirely. However, none of this water can be removed from the soil by pressures normally developed in construction.

The thickness and viscosity of the water film varies with changes in the amount of water available, with the electrolytic charge and crystal structure of the particle, the number and kind of attraction cations, the chemicals in the water, and with pressure and temperature.[9] However, for a particular soil being placed or in service in the subgrade, the amount of water is one of the two variables (the other being density) under the control of the design or construction engineer.

Figure 14–2 also serves to show the size and shape of typical clay particles and their relationship to the attracted water. An oven-dry soil will retain little more than the adsorbed water layer. For the air-dry condition, the moisture layer is somewhat thicker. The full thickness of film exists only when there are no interferences from other particles. However, the great degree to which surface moisture affects the volume of the clay–water combination that makes up a fine-grained soil can readily be visualized from this figure. In addition, the relative magnitudes of moisture influences between two types of clay can be seen by comparing the mineral-water dimensions for montmorillonite (Fig. 14–2*a*) and kaolin (Fig. 14–2*b*). Similar relationships also exist between the attached cations and their water jackets.

The long dimension of the kaolinite crystal represented in Fig. 14–2*b* is 0.001 mm, which is the division point in size between clays and colloids. The finest silt particle would have five times and the finest sand particle 74 times this length. Furthermore, they would probably be more nearly cubical in shape. It

[7]Work by E. J. Barenberg et al., *TRB Record 572,* indicates that an increase of 1% in the moisture content of granular layers increased pavement damage by a factor of 700.

[8]Terminology for describing the moisture layers is not consistent among writers. Some refer to the entire film as "adsorbed" water and label the inner layer as "solidified" water and the outer as "cohesive" water.

[9]*HRB Special Report 103* titled "Effects of Temperature and Heat on the Engineering Behavior of Soils" contains numerous papers and extensive bibliographies.

then becomes clear that moisture influences assume rapidly decreasing importance as particle sizes increase.

The moisture in partially saturated soils has another property that greatly influences soil performance—the ability to bind the particles together by tensile forces in the water film. These "water bonds," due to surface tension, tend to pull the soil grains together as the soil dries. This surface tension depends on water chemistry and temperature of soil.[10] However, water bonds are not the only source of cohesion in clays. Molecular forces, called van der Waals forces, are also present.

The forces exerted by surface tension decrease as the moisture content increases. When the soil becomes saturated, the forces exerted by it disappear entirely.

In summing up, it may be said that the properties of soils composed largely of coarse materials are primarily controlled by the characteristics of the particles, but for soils composed largely of clays and colloids the properties are primarily controlled by the chemistry and surface charges of the soil and its attached cations.

Effect of Density on the Behavior of Soils

The density of a soil is its mass per unit volume. It is sometimes expressed as "wet weight," or the total weight including water. It is more commonly expressed as "dry weight," which is the weight of the soil particles alone, excluding the weight of the contained water.[11] As a particular soil becomes more dense, it will contain a greater number of particles, and the (pore) volume remaining for air and water will be decreased.

For coarse-grained soils, increased density and, to a lesser degree, decreased moisture content improve the physical properties of soil which are of primary importance in construction of pavements. Strength is increased; consolidation under loading and the rate of water movement through the soil are decreased. High compaction of subgrades and bases of coarse-grained materials to obtain these advantages is accepted practice. On the other hand, trouble may result from overcompaction of clays that have a high affinity for water. Unless confined by superimposed loads such as the weight of an overlying fill, these materials, if overcompacted, later will take on water and expand. This results in an uneven road surface and possible failure of the roadway.

[10]A familiar demonstration of water bonds in coarse particles is provided by the stability of moist sand compared with dry sand. When moist sand is molded to a given shape, it will retain that shape until all surface moisture evaporates; then it will crumble and become free-flowing. This procedure is used to determine when sand is "surface-dry" *(AASHTO Designation T84)*.

[11]The density of a soil is usually expressed in pounds per cubic feet, but engineers should understand the meaning of the terms "void ratio" and "porosity." Void ratio is the ratio between the volumes of voids or pores in the soil (V_v) and the volume of the solid particles (V_s). This ratio may be greater or less than unity, depending on the degree of compaction. Porosity is the ratio of the volume of voids (V_v) to the total volume of the soil plus its voids ($V_v + V_s$), given as a percentage. If the void ratio $\frac{V_v}{V_s}$ ratio is represented by e the porosity *(n)* will equal $\frac{e}{1+e}$ 100.

Effect of Compaction Method on the Strength of Soils

Many roadway design methods are based on strength tests on laboratory-compacted samples. Some compaction methods, such as static load, produce little deformation or strain even in wet samples of fine-grained materials. On the other hand, at high foot pressures the mechanical or kneading foot compactor produces high strains in these wet, fine-grained soils. Impact hammers and vibratory devices produce strains between these extremes. Figure 14–3 shows the strength–moisture relationship at 5% strain of a fine-grained soil compacted by the three methods. Although moisture–density relationships are identical for all, the strength with compaction on the wet side of optimum varies markedly with moisture and with compaction method. An explanation of the strength differences is illustrated by the small circles attached to the strength curves. Below optimum moisture content, soils compacted by all methods retain their flocculated structure. On the wet side, this structure is partially maintained if compaction is at low strains. At high strains, however, the particles are reoriented into a parallel or dispersed structure, which greatly reduces strength. It follows that the choice of compaction device used in preparing specimens for strength tests can markedly affect the results of these tests. Similarly, field compactive devices such as sheepsfoot rollers that produce a kneading action can seriously weaken a wet, fine-grained soil.

Moisture Migration in Soils

Under usual present-day methods of construction, the moisture content of soils in roadway structures is carefully controlled. Since efforts to keep moisture out after construction have generally been unsuccessful, the original moisture in liquid or vapor form will move through the soil until a balance between the availability of water and the soil's ability to attract it is reached. This adjustment may require several years; furthermore, in some instances water will move in and out of the soil with changes in the weather or the seasons. If substantial increases in the quantity of water are involved, strength losses or expansion, particularly differential expansion, may lead to serious problems.[12]

The difficulties indicated here can be expected everywhere that transportation facilities are constructed. There are also situations where moisture migration creates special problems. These are discussed later under the topics of capillarity and frost action.

TESTS FOR SOILS

The preceding discussion shows that there is wide variation in the characteristics of different soils and that the performance of each individual soil is affected by its moisture content and density. A number of physical tests have been de-

[12]Many observations of the variations of moisture content over time have been made. A few reports appear in *HRB Record 203, 276,* and *301, TRB Record 612,* and *790.*

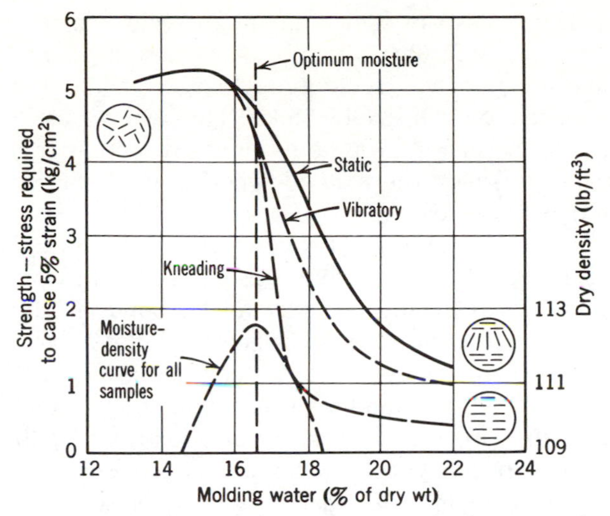

Fig. 14–3. Influence of compaction method on strength of a silty clay. (After H. B. Seed.)

veloped to measure soil performance. Standard procedures for making most of them appear in *Transportation Materials, Part II, Tests* or as *Interim Test Methods* issued by AASHTO. Many of the tests also appear in the publications of the American Society for Testing Materials (ASTM). Wherever possible, the appropriate test designations have been given here to make reference easier. A parallel designation scheme is employed for recommended specifications in *Transportation Materials, Part I, Specifications* or as *Interim Specifications*.

Physical Tests for Particle Size

Particle sizes for gravels and for coarse and fine sands are determined by "sieve analysis."[13] A sample of the material is dried and then shaken through a series of sieves ranging from coarse to fine, and the amount (percentage of sample dry weight) retained on each sieve is weighed and recorded.

The standard sieve sizes of the American Association of State Highway and Transportation Officials for soil aggregates and similar materials are as follows:

Sieve description									
in inches*	2	1½	1	¾	⅜				
by number†						4	10	40	200
Opening in millimeters	50.0	37.5	25.0	19.0	9.50	4.75	2.00	0.425	0.075

*Sieves in this group have square openings of the size indicated.
†Sieves in this group also have square openings. The sieve number designates the number of openings per lineal inch across the sieve.

[13]See *AASHTO Designation T27*. Standards for sieves are found in *AASHTO Designation M92*, and *ASTM Designation E11*.

Sieve analysis is not feasible for determining particle size of materials much finer than the No. 200 sieve (0.075 mm). Rather these are found by observing the rate at which the grains will settle through a gas or liquid. This settlement phenomenon is related to grain size by Stoke's law, which states that the rate of settlement of a solid through a given liquid or gas is proportional to the square of the diameter of the solid. The method specified by AASHTO is commonly called the hydrometer test (see *AASHTO Designation T88*).

In general terms the hydrometer test is conducted as follows: a sample of the material passing the No. 10 sieve is thoroughly mixed with water containing a dispersing agent which dissipates any electrolytic bonds in the material that might cause flocculation and accelerate settlement. After 12 hr the entire mixture is agitated until particles are in suspension in the water. The mixture is then placed in a graduated flask and the solids are permitted to settle under the pull of gravity. As the larger particles settle out, followed by those of smaller and smaller size, the specific gravity of the liquid decreases, and this change is recorded by a special hydrometer that is read at prescribed intervals. This change in specific gravity is then related to grain size of the material by Stoke's law. Precise control of temperature and of other possible variables is required if the results of this test are to be satisfactory.

Tests To Evaluate the Influence of Soil-Moisture Interactions

By increasing the moisture content of a soil, its consistency can be varied from semisolid to plastic to liquid. Experience has shown that the percentage of moisture at which these changes take place can be directly correlated with the behavior of the material in service. Tests on the portion of the sample which passes a No. 40 sieve (0.425 mm) determine the percentage of moisture (based on dry weight) at which each change in consistency takes place. They are called the Atterberg tests after the Swedish agronomist who suggested them. These were developed for engineering use by A. Casagrande.

LIQUID LIMIT (AASHTO DESIGNATION T89). The liquid limit (LL) signifies the percentage of moisture at which the sample changes, with a decrease in moisture, from a viscous or liquid state to a plastic one. If the sample is wetter than the liquid limit, a grooved sample of the soil in a standard cup will flow when lightly jarred 25 times (see Fig. 14–4). If the sample is drier than the liquid limit, the groove will not close when the sample is jarred. At the liquid limit the soil particles have been separated by the water just widely enough to deprive the soil mass of its shearing strength.

Many tests on each sample would be required to reach the moisture content at which the groove in the sample would close at exactly 25 blows. Several procedures for relating moisture content at other blow counts to that for 25 blows are detailed in the test procedure. Also, since the liquid limit is often a determining factor in the "go or no go" decision to accept or reject base course and other materials, a check or referee test is given.

PLASTIC LIMIT (AASHTO DESIGNATION T90). The plastic limit (PL) signifies the percentage of moisture at which the sample changes, with decreasing wet-

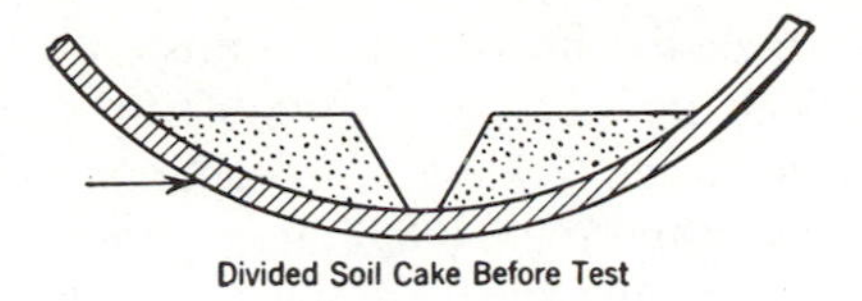

Divided Soil Cake Before Test

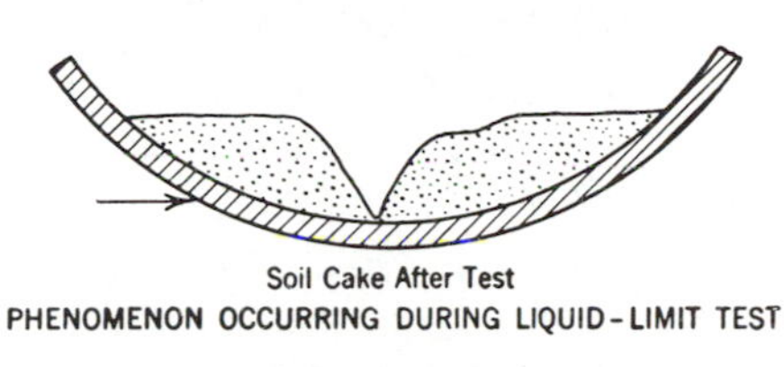

Soil Cake After Test

PHENOMENON OCCURRING DURING LIQUID-LIMIT TEST

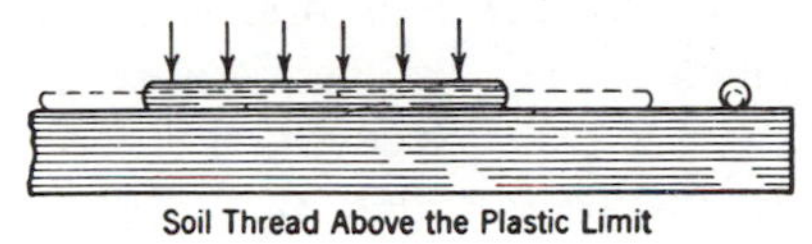

Soil Thread Above the Plastic Limit

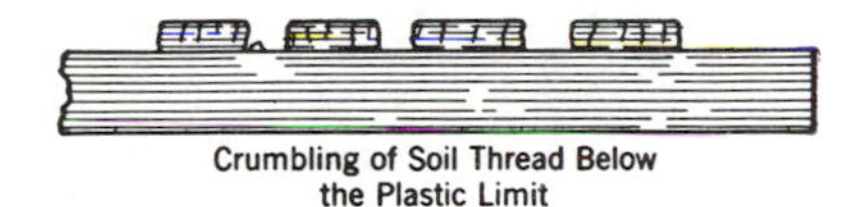

Crumbling of Soil Thread Below the Plastic Limit

PHENOMENON OCCURRING DURING PLASTIC-LIMIT TEST

Fig. 14–4. Diagram illustrating the Atterberg tests for fine soils.

ness, from a plastic to a semisolid state. In this condition the soil mortar begins to crumble when rolled into threads ⅛ in. in diameter (see Fig. 14–4). Additional moisture causes the soil to become plastic. Certain soils, such as clean sands, are nonplastic; that is a plastic limit cannot be determined.

PLASTICITY INDEX (AASHTO DESIGNATION T90). The plasticity index (PI) of the sample is defined as the numerical difference between its liquid limit and its plastic limit. It also is stated as a percentage of dry weight. It measures, in some combination, the fineness and shapes of the soil particles, the interplay of the attractive forces tending to hold the clay–mineral flakes together, the thickness and viscosity of the water film, and the quantity and electrical charges of the cations. For a coarse-grained soil, or for a fine-grained soil with few particles of clay of colloid size, a small increase in moisture above the plastic limit provides enough particle separation to destroy the attractive forces which provide shearing strength. This means that the difference in numerical value between the plastic limit and the liquid limit is small, so that the plasticity index also is small. On the other hand, for a soil high in clays or colloids, considerable water will be required before the attractive forces are overcome and the strength of the mass is destroyed. In this case, the numerical value of the plasticity index is high. Thus the plasticity index is an indirect method for measuring the amounts and moisture affinities of the clays, colloids, and cations in the soil.

Experience amassed over the last 40 yr demonstrates conclusively that soils with high plasticity indices are much less desirable for subgrade or base courses than those having lower indices. Many construction agencies use the PI as a primary control in selecting the materials that go close under the pavement. A common specification for base courses requires that the plasticity index shall not exceed 6; and some agencies insist that it be no greater than 3.

Experienced soils engineers have for years employed quick "hand-feel" tests to roughly predict the plasticity index of soils. These tests may include (a) thread toughness at a moisture content approximating the plastic limit, (b) air-dried strength, and (c) dilatancy, or the tendency of moisture to come to the surface when a moist ball of the material is jarred sharply. Soil mortars having high PIs will produce tough threads, have high dry strengths, and show almost no dilatancy. Soils having low PIs will react in an opposite manner. One study showed that a skilled soil technician could, on the basis of these tests, predict the PI with a standard error of about 1%.[14] Recently attempts have also been made to relate Atterberg limits to soil moisture tension.[15] These results show good correlation for the liquid limit and fair correlation for the plastic limit for the soils investigated.

NONPLASTIC SOILS (NP DESIGNATION). Fairly clean sands, some rock dusts, and certain other materials are classed as nonplastic since they cannot be rolled into threads as required to determine the plastic limit. As a rule, nonplastic soils make excellent road materials when properly confined under a wearing course. Some of them, well graded rock dust, for example, form hard durable surfaces when wetted and compacted. Others, like clean sand, displace easily under load and their use for base course or fill brings difficult construction problems, but they may have other desirable features such as facilitating subsurface drainage.

SHRINKAGE TESTS OF SOILS (AASHTO DESIGNATION T92). These tests measure the changes in volume and weight that occur as a pasty mixture of soil (minus No. 40 sieve) and water is dried from near the liquid limit to constant weight at 110°C. Test results are stated in terms of shrinkage limit, volumetric change, and lineal shrinkage.

Large values for the shrinkage factors indicate that the soil may be troublesome. However, the test for shrinkage as such has in the main been superseded by various other tests for volume change.

OTHER TESTS FOR VOLUME CHANGE. As already indicated, the volume of certain soils increases when they are given a chance to absorb water. Soils that make good subgrade and base courses expand very little while those that are poor swell more. To illustrate, volume change when the specimen is soaked is measured routinely as a step in the California Bearing Ratio design method described later. A common specification requirement limits the volume change of base-course materials to 1%.

[14]See J. W. Spencer, *HRB Record 284*.
[15]See A. Gadallah et al., *TRB Record 497*.

Again, a test for volume change is included as part of the Hveem stabilometer design method, also described later. Here the pressure exerted by a confined sample measures its tendency to expand during soaking.

SAND EQUIVALENT TEST (AASHTO DESIGNATION T-176). F. N. Hveem of the California Department of Transportation developed a "Sand-equivalent" test for quick field determination of the presence of undesirable quantities of clay-like materials in soil–aggregate mixtures on a volume rather than a weight basis. Essentially the test is performed by shaking vigorously a sample of fine aggregate (passing the No. 4 sieve) in a transparent cylinder containing a special aqueous solution that includes calcium chloride, glycerine, and formaldehyde as flocculating agents. The strength of the solution has been adjusted to exaggerate the volume of montmorillonite and other troublesome clays while not exaggerating the volume of kaolinite clays. After shaking, the mixture stands in the cylinder for exactly 20 min. Then the levels of the top of the sand and of the top of the clay suspension are noted. The "sand-equivalent" is the ratio between the height of the sand column and the combined heights of sand and expanded, saturated clay, expressed as a percentage. Thus, higher values of the sand-equivalent indicate superior materials. Permissible moving average values from the 1978 California specifications are 21 for subbases; 31 for aggregate bases; 45–50 for aggregates for various types of asphalt concrete; and 76 for concrete sand.[16]

Tests for Determining the Density of Soils

THEORY OF SOILS COMPACTION. In early days, the soil forming highway and similar embankments was piled up as it fell from the transporting vehicle. No attempt was made to provide uniform layer thickness or density. Rather, fills were left to settle or "season" for a period before the final surfacing was placed. Subsequent settlement was often large and nonuniform and support for the surfacing was poor.

Today, great care is taken to spread embankment materials in layers that are uniform in quality and thickness. After placement, each layer is brought to uniform moisture content and compacted to a high density. By these means, subsequent compressibility, resilience, and volume change are reduced and embankment strength is enhanced.

Soil density (weight per cubic foot) varies with the peculiarities of the soil itself, the moisture content, and the compactive device and method that are used. Thus a standard weight per cubic foot cannot be set, but must be determined in each instance. Principal variables in the soil proper are:

[16]For a discussion of the rationale of this test, see F. N. Hveem, *HRB Proceedings, 1953*. Comparisons of sand-equivalent test results with percent passing No. 200 versus plasticity index for some 2400 samples are given by W. G. O'Harra in *HRB Proceedings 1955*. See also R. A. Pettit, *Civil Engineering,* May 1971. Details of mechanical devices to remove operator-induced variables from the test will be found in the AASHTO test description.

1. Specific gravity of the soil particles themselves. It may vary from 2.0 to 3.3, but usually is between 2.5 and 2.8.
2. The particle-size distribution of the soil. A mass composed entirely of spheres of one size in the densest possible condition will contain 74% solids and 26% voids. If smaller spheres are introduced into the mass, the percentage of solids will increase. This idea extended to soils indicates that particle-size distribution may greatly affect density.
3. Grain shape of soil particles. Sharp, angular particles will resist shifting from a loose to a more compact state. Flaky particles in soil will cause a decrease in density as they cannot easily be compacted.

The influence of moisture content on the density of soil is illustrated by curve *A*, Fig. 14–5. This curve was obtained as follows. Designated percentages of water were added to dry samples of a particular soil. Each sample was then compacted by an identical procedure, and the weight of soil per cubic foot of compacted material was obtained. Curve *A* shows that the densest sample was obtained when 9% (by weight) of water was added to the dry material. It also shows that 129 lb of dry material would be needed to make a cubic foot of soil, under the stated conditions.

Although the curve in Fig. 14–5 is for a particular soil and for a designated compaction method, similar tests have proved beyond doubt that the general form of the moisture–density curve remains the same for other soils and for other compaction methods. For each compaction method, there is an *optimum moisture content* at which a given soil can be compacted to greatest density. For any other moisture content, the soil will be less dense.

For low percentages of moisture (say, below about 5% for the coarse-grained soil in Fig. 14–5), water attaches in the individual particles; air fills the remaining voids. Neither the water nor the air lubricates the soil mass. Hence, friction between the grains prevents further densification when the compactive effort is applied. But water in larger amounts serves to decrease surface tension and lubricate the soil somewhat. Then the compactive effort can rearrange and densify the grains. As the amount of water is increased, this lubricating action becomes more and more effective until at the optimum moisture content the greatest density is obtained. With fine-grained soils at low moisture content, the flocculated structure is not rearranged by the compactive effort; with more water a reduced degree of flocculation and a greater density can be produced.

Above the optimum moisture content the soil weight per cubic foot drops off again, but the reason for the decrease is entirely different. In Fig. 14–5, curve *B* shows the dry weight of soil (of specific gravity 2.65) in a cubic foot of a mixture of soil and water without air voids.[17] For this condition, if a certain moisture percentage is assigned, the volumes occupied by water and by soil are fixed. As the amount of water increases, the amount of soil must decrease accordingly, since both water and soil are incompressible. Thus curve *B* represents the max-

[17]The equation for this curve is

$$\text{dry weight of soil per cubic foot} = \frac{62.4}{\dfrac{1}{\text{specific gravity of soil particles}} + \dfrac{\%\ \text{moisture}}{100}}$$

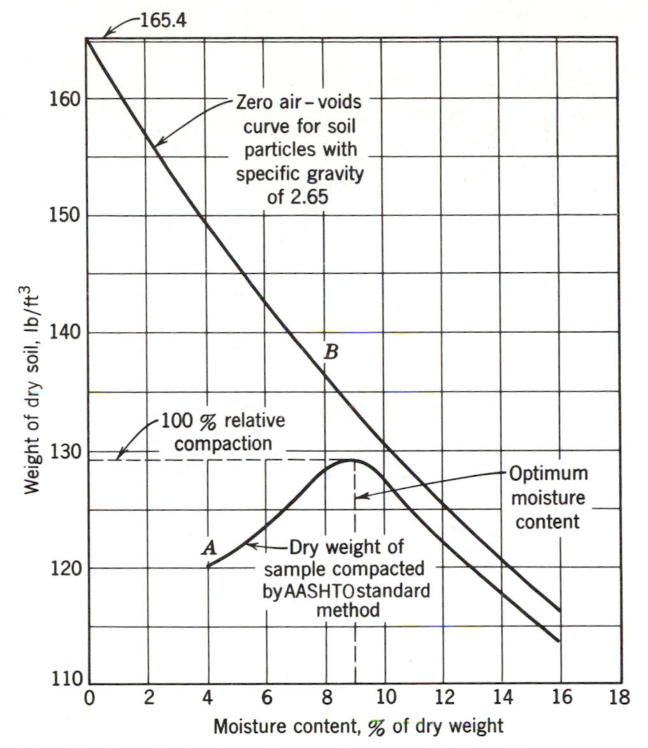

Fig. 14–5. Typical moisture–density relationships.

imum possible weight of soil that can be forced into a cubic foot of space along with a given percentage of water. Densities obtained in practice do not reach those indicated by curve *B,* as some air voids are always present in compacted soils. This explains why the actual dry weights shown by curve *A* are slightly less than the theoretical values at moisture percentages above the optimum. This difference in the ordinates of the two curves represents the small amount of air that always is present in soil–water mixtures.

As mentioned, different soils have different maximum densities and optimum moisture contents. Figure 14–6 shows the moisture–density curves obtained by compacting a variety of soils in the same standard manner.[18] The compactive method or amount of compactive effort expended changes both maximum density and optimum moisture content. This is illustrated by Fig. 14–7, which shows moisture–density curves for the same soil compacted in several different ways. The solid lines on Fig. 14–7 give the results obtained by the standard and modified AASHTO tests (a modification that employs 4.5 as much energy). For this particular soil, compaction at higher energy lowers the optimum moisture

[18]It must be recognized that the moisture–density relationships found in the laboratory may not be obtained in the field. For example, in the laboratory the sample is pulverized and an intimate mixture of soil and water obtained. At low percentages of moisture such an intimate mixture could not be obtained during construction with the heavy clay of curve 6 of Fig. 14–6 since the clay would be in clods.

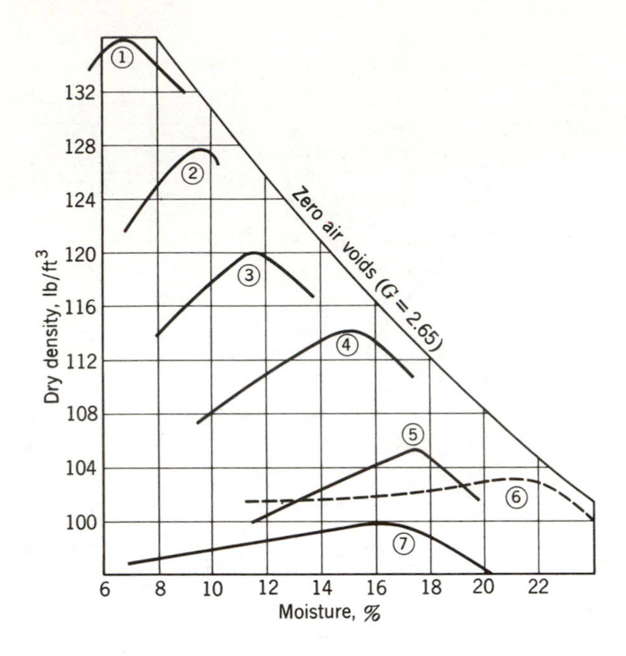

Number	Description	Sand (%)	Silt (%)	Clay (%)	LL	PI
		Soil Texture and Plasticity Data				
1	Well-graded loamy sand	88	10	2	16	NP
2	Well-graded sandy loam	72	15	13	16	0
3	Medium-graded sandy loam	73	9	18	22	4
4	Lean sandy silty clay	32	33	35	28	9
5	Loessial silt	5	85	10	26	2
6	Heavy clay	6	22	72	67	40
7	Very poorly graded sand	94	6		NP	NP

Fig. 14–6. Moisture–density relationships for seven soils, each compacted by the AASHTO standard method. (After A. W. Johnson.)

content by 5% and raises the density from 105 to 117 lb/ft³. The dotted lines show the densities obtained with certain pneumatic and sheepsfoot rollers.

Figure 14–7 can be used to demonstrate that by increasing the density of a soil it is possible to maintain its "strength" under service conditions. Assume that one sample of this soil, containing 17% moisture, is compacted by the AASHTO standard method and another by the modified AASHTO method. Dry densities of these samples are shown to be 103 and 112 lb/ft³, respectively. It may be that, as compacted, even the less dense of these will carry the superimposed loads without failure. Suppose, however, that each of these samples becomes saturated, as may well happen in time. Assuming no swell, the moisture content of the samples will be 24 and 19%, respectively, as indicated by the intersection of the horizontal lines *aa'* and *bb'* with the zero air-voids curve. For this soil, the moisture content of the wetter sample exceeds the plas-

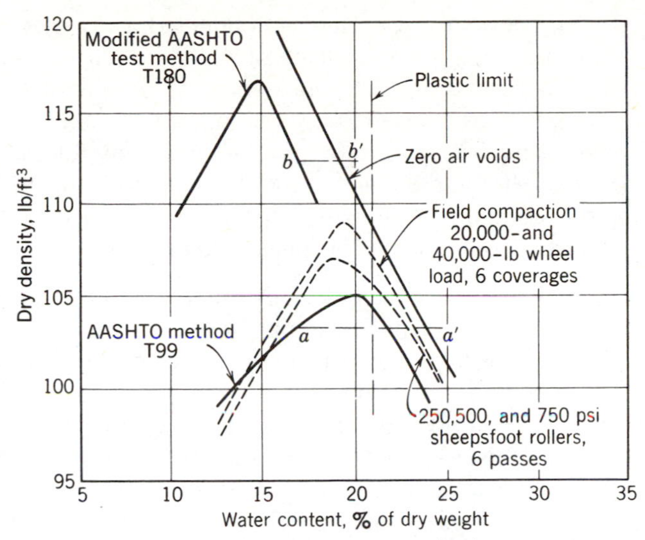

Fig. 14–7. Effect of compaction method on moisture–density relationships. Tests made on a silty clay soil having 10% sand, 63% silt, 27% clay; LL = 36, PI = 15, Sp. gr = 2.72 (modified after A. W. Johnson).

tic limit, which means that the material may lose much of its supporting power. The more highly compacted one will probably remain stable.

The foregoing discussion of compaction assumes that a soil, once compacted, will remain at almost constant volume in service. For the better soils this is true, or nearly so. However, many of the poorer soils, compacted at optimum moisture content, will swell if added moisture becomes available. They require special treatment. The swelling will not occur if the external pressure on the soil mass is great enough, which in turn means that such materials, if buried deeply enough, will not cause trouble. When confinement under a weight of fill is impractical, volume change can be reduced by compacting the soil at some higher moisture content at which the desire of the soil for water is more nearly satisfied.[19]

LABORATORY TESTS FOR SOIL DENSITY. Tests for density may be divided into two classes: laboratory tests to set a standard for density, and field tests to measure the density of a soil in place in the roadway structure. Laboratory tests may in turn be subdivided on the basis of compaction procedure, into "static," "dynamic" or "impact," and "tamping foot" or "kneading" methods.

Static Tests. Some agencies have used a static test to determine maximum den-

[19]There is a large literature on the theory of soil compaction. A suggested starting point for further study and references is A. W. Johnson and J. R. Sallberg, *HRB Bulletins 272 and 319*. More recent studies and extensive bibliographies appear in *HRB Records 22, 177, 235, 304, and 438*. Similarly, as to the swelling properties of clays, see H. B. Seed et al., *HRB Bulletin 313, TRB Record 790,* and J. K. Mitchell, op. cit.

sity of laboratory samples. One such test is conducted as follows. About 4000 g of soil containing a designated percentage of water are placed in a cylindrical mold 6 in. in diameter and 8 in. high. The sample is compressed under a load of 2000 lb/in.^{2}, applied at a speed of 0.05 in./min. When the full load is reached, it is held for a period of 1 min and then gradually released. Using the known dry weight of soil, mold diameter, and the measured height, the dry density of the sample is computed. Enough samples are processed to delineate the peak of the moisture–density curve. This peak value represents the standard.

Dynamic or Impact Tests. Most agencies determine optimum moisture content and maximum density with dynamic or impact tests. Samples of soil, each containing a designated percentage of water, are compacted in layers into molds of specified size. Compaction is obtained with a given number of blows from a free-falling hammer of prescribed dimension and weight that has a flat, circular face. The peak of the moisture–density curve represents standard density.

Details of three impact tests are given in Table 14–1. For the AASHTO tests, the volume of the compacted sample is held constant by shaving it to proper height. Density varies with dry weight of the trimmed sample. For the California impact test, sample weight is constant, usually 5 lb, and density varies with the height of the compacted sample. The AASHTO standard test is used by most transportation agencies, while the modified AASHTO test is employed by a few. The California impact test (nicknamed the "rathole" test) is used only by a few highway departments.

The AASHTO tests may be performed on that portion of the soil which will pass a No. 4 sieve or on all that passes the ¾ in. sieve (19 mm).[20] The California impact method requires that material retained on the ¾-in. sieve be removed from the sample but that it be replaced with an equal weight of the ¾ in. to No. 4 gravel from the same soil.

Tamping-Foot or Kneading-Compaction Tests. A group of West Coast engineers, working as a subcommittee of the American Society for Testing Materials, standardized a tamping-foot compactor originally developed by Hveem of the California Department of Transportation *(AASHTO Designation T190)*. It is used primarily to compact samples for the Hveem stabilometer design method. Material is fed into a rotating mold and is compacted by many repetitions of load applied through a tamping shoe shaped like a sector of a circle. About one-fourth of the specimen is covered in each application. Compaction comes through a kneading action, as contrasted with static pressure or impact. Proponents of this method have developed substantial evidence that kneading compaction, compared to other laboratory methods, provides better correlation with field particle orientation and densities obtained with tamping or pneumatic rollers.

[20]If larger particles are in the soil, their percentage of the total weight of the sample and their specific gravity is determined. Then an adjusted density with the coarse particles included is computed for comparison with field results. See *AASHTO Designation T224* for a graphical solution.

TABLE 14–1. Details of Impact Compaction Tests

	Name of Test		
Test Details	*AASHTO* Standard*	*Modified† AASHTO*	*California Impact*
Diameter of mold, in.	4 or 6	4 or 6	2.86
Height of sample, in.	5 cut to 4.58	5 cut to 4.58	10–12
Number of lifts	3	5	5
Blows per lift	25 or 56	25 or 56	20
Weight of hammer, lb	5.5	10	10
Diameter of compacting surface, in.	2	2	2
Free-fall distance, in.	12	18	18
Volume, net, ft^3	$\frac{1}{30}$ or $\frac{1}{13.33}$	$\frac{1}{30}$ or $\frac{1}{13.33}$	Varies

**AASHTO Designation T99*
†AASHTO Designation T180

FIELD TESTS FOR DENSITY OF SOILS IN PLACE. Field tests provide a means of comparing the densities obtained in the laboratory. Usually this comparison is made on the basis of *relative compaction,* which is defined as follows.

$$\text{relative compaction} = \frac{\left\{\begin{array}{l}\text{dry weight per cubic foot of soil in place in roadway} \\ \quad \text{structure}\end{array}\right\}}{\left\{\begin{array}{l}\text{dry weight per cubic foot for a soil sample at optimum} \\ \quad \text{moisture content compacted in a prescribed standard} \\ \quad \text{manner}\end{array}\right\}}$$

With many agencies, relative compaction is the sole measure by which the acceptability of a completed roadway structure is measured.

Field Density and Moisture Content by Sampling. The procedure for determining relative compaction by sampling is as follows:

1. Dig a small sample of the compacted material through the full depth of the layer to be tested.
2. Obtain the wet and dry weights of the sample. From these also determine the moisture content of the sample.
3. Determine the volume that the sample occupied in the fill by finding how many pounds of a material of known unit weight are required to fill this space. Sand or water poured or pumped into a flexible rubber liner have served this purpose.
4. From the dry weight of the sample and the known volume that it occupies in the fill, obtain the dry weight per cubic foot.
5. Determine the relative compaction of the soil in the fill by dividing its dry weight per cubic foot by the laboratory standard density.[21]

[21]A variety of devices for measuring the volume of the hole in the fill have been employed. These include sand cones *(AASHTO Designation T191),* rubber ballons *(AASHTO Designation T205),* and oil *(AASHTO Designation T214).*

Where the material is fine-grained, undisturbed samples from the fill may be taken in blocks, chunks, or cores. Volume of the sample can be determined in the laboratory. (See *AASHTO Designation T233*.)

Field density control by sampling presents many difficulties. In the first place digging out the test sample and measuring the volume of the hole are time consuming. Often they interfere with construction operations so that construction equipment must be delayed or diverted from the test area, which is costly. Secondly, because density testing is so time consuming, only a few tests will be taken, so that field control is spotty, at best.[22] Finally (as explained subsequently), the accuracy of test results is often subject to serious question.

Field Density and Moisture Content Measurement with Nuclear Devices. In recent years, the state transportation agencies and the larger local agencies have increasingly adopted nuclear devices for determining in-place densities and moisture contents.[23]

The principle of measurement by nuclear means is relatively simple. A source of nuclear energy provides both gamma rays and high-velocity neutrons. The gamma rays are reflected to a degree dependent of the density of the material through which they travel. The high-velocity neutrons are slowed down by any hydrogen atoms they contact, hence more neutrons are slowed down if a given soil has a higher moisture content. The intensity of the reflected gamma rays and slowed down neutrons are measured separately by Geiger-Muller counter tubes. Gauge readings are easily converted to density and percent moisture using calibration curves or microprocessors.

At present, the portable devices are of either the transmission or backscatter types (see Fig. 14–8). The former requires that a probe containing the energy source be placed in a tube in a small predriven hole in the soil layers. It measures density and moisture content for any depth of the probe, which can be up to about 10 in. The backscatter device is placed directly on the top of the soil

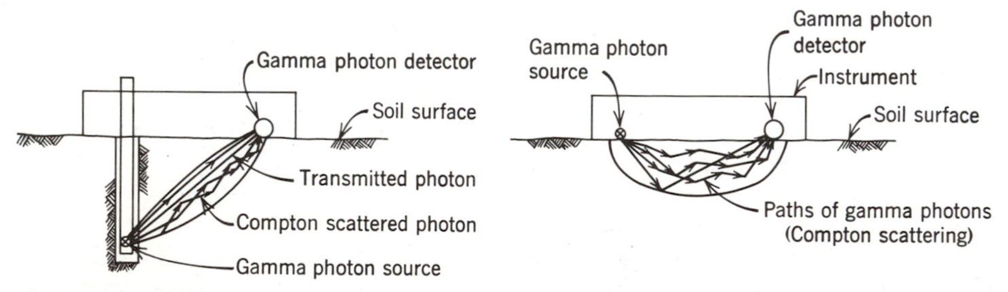

Fig. 14–8. Nuclear devices for measuring soil density: *(a)* transmission type, *(b)* backscatter type.

[22]New York DOT has developed a rapid test method for compaction control to eliminate this problem. For details refer to W.H. Peak, *TRB Record 593*.
[23]As of 1973 all state highway departments had engaged in nuclear testing and most use them for compaction control. (See *TRB Circular 166*.)

layer or is supported slightly above it to provide an air gap between instrument and soil surface. It measures density and moisture content to a depth of 3 to 4 in.

Nuclear devices overcome some of the disadvantages of the sampling method. Since determinations can be taken quickly and while construction equipment is operating, delays are minimized, and this can reduce construction costs. Also, more readings can be taken, which leads to greater confidence that prescribed densities have been obtained and also permits a statistical approach to compaction control. At the same time newer equipment has overcome, at least in part, some of the earlier problems such as the danger of operator exposure to radioactivity, the need for frequent recalibration, and the effects of soil chemistry.[24]

A self-contained unit called the Road Logger is commercially available. It travels over the fill at about 3 mph and continuously records density and moisture content.

Field Density and Moisture Content Measurement with Nuclear Devices (Subsurface). Several transportation agencies also use nuclear gauges to measure moisture and density at depths ranging from 1 to 20 ft or more. Problems with unreliable first generation equipment, questionable results, and lack of apparent need have slowed their spread.[25]

VARIABILITY IN RESULTS OF LABORATORY AND FIELD COMPACTION TESTS. Figures 14–6 and 14–7 indicate clearly that moisture–density relationships change with soil type and compactive effort. Another important question concerns the correlation between densities obtained in the laboratory and those that are suitable and attainable for the roadway as constructed. More specifically, the question might be asked, "Would a general specification that called for 95% relative compaction by one of the standard methods give proper results with all soils?"

Figure 14–9 shows the variation in unit weights and the nonuniform results among several standard laboratory compaction tests. It serves to indicate the complexity of the problem, even in the laboratory. Then there is the further question of how closely laboratory compaction of a small sample in a steel mold simulates the effect of actual compaction equipment operating over large areas and with far less accurate control of moisture. If, as some engineers claim, the tamping-foot compactor most closely approximates actual field compactors, the suitability of impact methods as a control for all soils must be questioned.

The reliability of field-density measurements introduces yet another variable

[24]Measurement of density and moisture and asphalt content with nuclear gauges has been the subject of intensive research. More recent studies, including bibliographies listing earlier work, appear in *HRB Record Nos. 66, 107, 177, 290, 301, 412,* and *438* and *NCHRP Reports 43* and *125*. For a description of a statistical approach to nuclear acceptance testing see W. G. Weber, Jr., and T. Smith, *HRB Record 177*. *NCHRP Report 138* appraises several methods for measuring moisture. For test methods refer to *AASHTO Designations T-238* and *T-239*.

[25]For details, see *TRB Circular 195*.

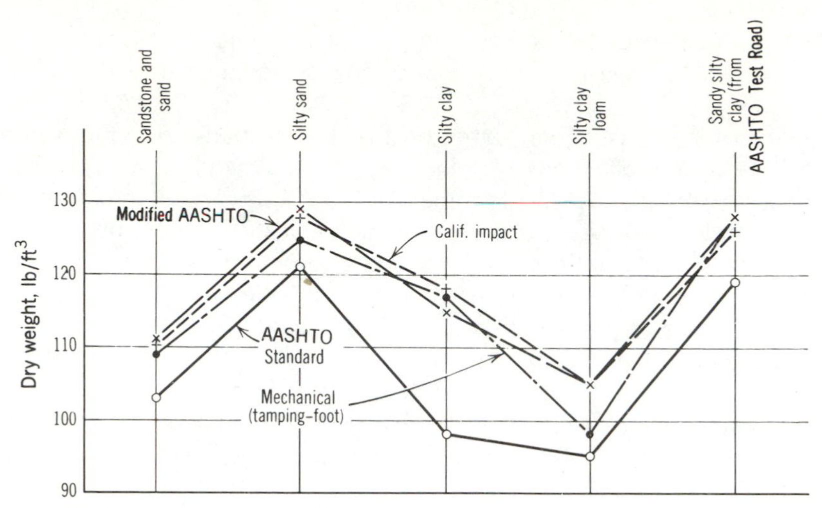

Fig. 14–9. Maximum densities of several soils as determined by different laboratory methods. (Data from F. N. Hveem, *HRB Bulletin 159*.)

into compaction control. Figure 14–10 shows the range obtained in volume measurements with several commonly used devices. Another investigator reported variations up to 10% in the unit weight of sand used for volume measurement if the jar holding the sand were struck sharply or if the fill were vibrated by construction equipment. There are conflicting reports on the accuracy of nuclear devices. Unpublished studies by the California Department of Transportation indicated that, at the 95% confidence level, the standard deviation was 2.5 lb/ft^3 with either the transmission gauge or the backscatter gauge corrected by the air-gap ratio. From data such as these it must be concluded that there are many uncertainties in determining and controlling the density of compacted embankments.[26]

In sum, four variables affect the accuracy of measurements of relative density. These are (a) changes in the soil itself, (b) the sampling method, (c) the accuracy of laboratory testing for standard density, and (d) the accuracy of testing for field density. Under such circumstances substantial variation in results is to be expected. This is borne out by test results such as those shown in Fig. 14–11.

Strength Tests for Soils

Soil tests for determining strength or supporting values of soils may be divided into two groups. One of these is used in foundation investigations and is concerned with the load carrying capacity and rate and amount of consolidation in soils that support foundations. Tests are conducted on the soils in place or on (almost) undisturbed samples. In transportation studies, the application of these methods is limited to bridge-foundation studies or other special problems. For

[26]However, the alternative of not checking density entails high risk which is generally considered to be unacceptable.

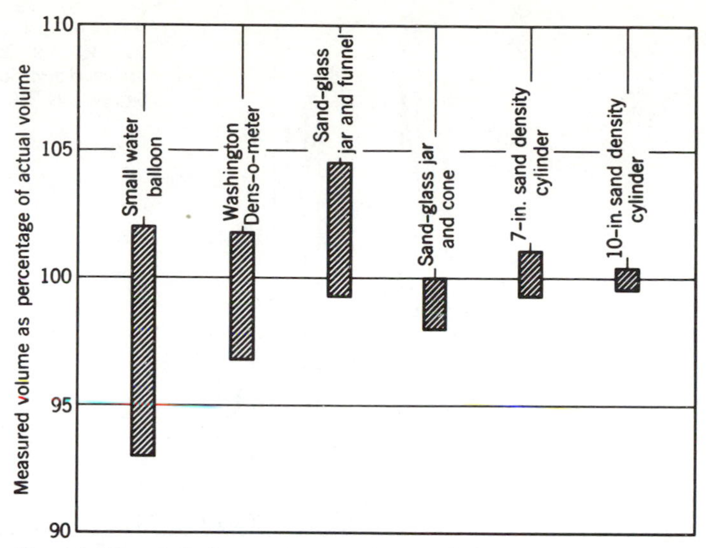

Fig. 14–10. Relative accuracy of several devices for measuring field densities; 90% of test values fall within ranges represented by bars. (After J. F. Redus, *HRB Bulletin 159.*)

discussion of them the reader is referred to the standard textbooks in soil mechanics.

The second group of strength tests is designed to measure the supporting power of disturbed soils as recompacted under standard procedures. They provide data for one of the several techniques used to set layer thicknesses for pavement systems.

CALIFORNIA BEARING-RATIO METHOD *(CBR).* This method, commonly referred to as the *CBR,* combines a load-deformation test performed in the laboratory with an empirical design chart to determine the thicknesses of pavement, base, and other layers. The test originated with the California Department of Transportation although this agency now uses the Hveem stabilometer method (see below). With modifications the original *CBR* test has now been standardized under *AASHTO Designation T193*. Each agency has developed its own design chart as has the Corps of Engineers, U.S. Army, for airports as well as highways.[27]

The California bearing-ratio test is conducted in general as follows (see Fig. 14–12):

Step 1: Disturbed samples of the soil, at different moisture contents, are compacted in three layers by static load or impact hammer into cylindrical steel molds 6 in. in diameter and 8 in. in height. The resulting specimen depth is about 5 in.

[27]For a discussion of the development of the *CBR* method and its application to airport design, see *Transactions,* ASCE, 1950, pp. 453–589. Details of current practice are given by D. N. Brown et al., *Transportation Engineering Journal,* ASCE, Nov. 1969.

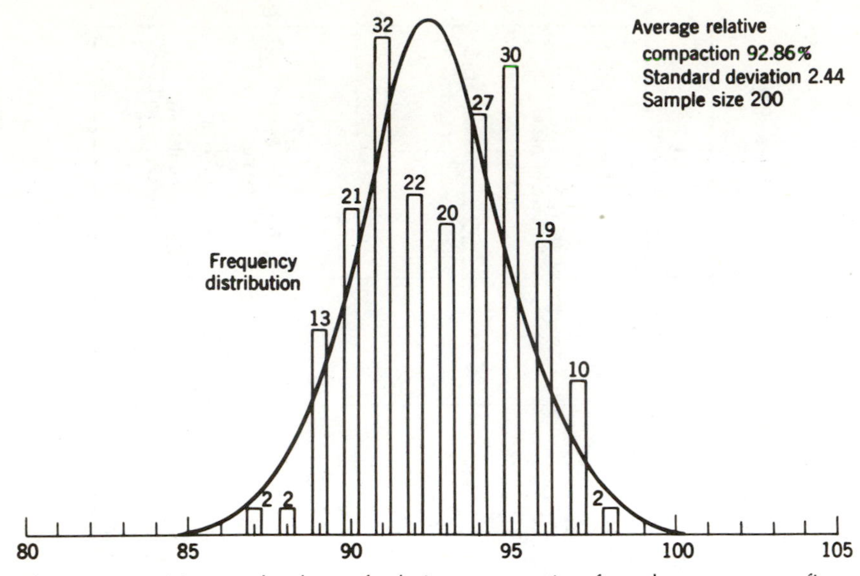

Fig. 14–11. Measured values of relative compaction for a homogeneous fine-grained soil where 90% was required. (Source: *Public Roads*, Apr. 1969.)

The 5.5 lb hammer for the AASHTO standard density test is used in the impact method.

Step 2: The moisture–density curve is plotted, and the sample with greatest dry density is selected.

Step 3: The specimen, still in the mold, is immersed in water and soaked for four days to simulate saturation that may occur in service. Expansion of the specimen from soaking is measured.

Step 4: A small cylindrical piston, 3 in.^{2} in end area, is forced into the still-confined test specimen. Load-deformation data are gathered as the specimen is penetrated. Usually the piston passes through a surchange ring weighing 10 lb or more which provides confinement for the material and simulates the effect of overlying pavement and base.

CBR load-deformation curves for a variety of soils are shown as Fig. 14–13. From such curves, or the load-deformation data itself, the California bearing ratio *(CBR)* is computed as follows:

$$\text{California bearing ratio} = \frac{\left\{\begin{matrix}\text{load carried by test specimen at 0.1 in.} \\ \text{piston penetration}\end{matrix}\right\}}{\left\{\begin{matrix}\text{load carried by standard crushed rock} \\ \text{base at 0.1 in. piston penetration}\end{matrix}\right\}} \times 100$$

Thus, the *CBR* states the quality of the material in terms of that of an excellent base course, which has a *CBR* of 100. As examples, the disintegrated granite subbase of Fig. 14–13 has a *CBR* of 70, whereas the *CBR* of adobe is only about 5. It should be recognized that the *CBR* testing procedure is not always conducted

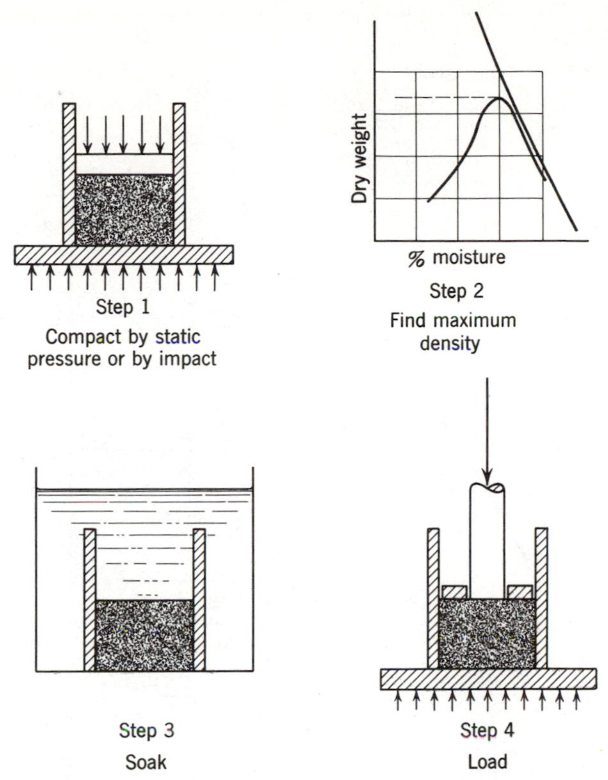

Fig. 14–12. Laboratory procedure for finding the *CBR* of a soil.

in the same manner. For example, some agencies measure relative bearing value at 0.2 in. rather than 0.1 in. penetration. This change could seriously alter the results for certain soils, as would be the situation with the disintegrated granite of Fig. 14–13. Other agencies adjust their designs to recognize differences in rainfall, groundwater, or frost conditions. It follows that direct comparison among the *CBR* designs of different agencies often is not possible.[28]

Proponents of other design methods criticize the *CBR* on the basis that the penetration of a piston into a confined specimen does not simulate the shearing forces that develop in the supporting materials under a flexible pavement. One cited example is that soil mixtures of rough-surfaced or angular coarse aggregate with small amounts of troublesome clays will rate extremely high under the *CBR* method because the course material keys together in the mold and resists penetration of the piston. In the roadway, on the other hand, this soil reportedly will perform badly because the clays destroy the shearing strength by lubricating the soil mass.

HVEEM STABILOMETER METHOD. The Hveem stabilometer measures the

[28]For a comparison of *CBR* results between static and dynamic compaction, see *HRB Record 466*.

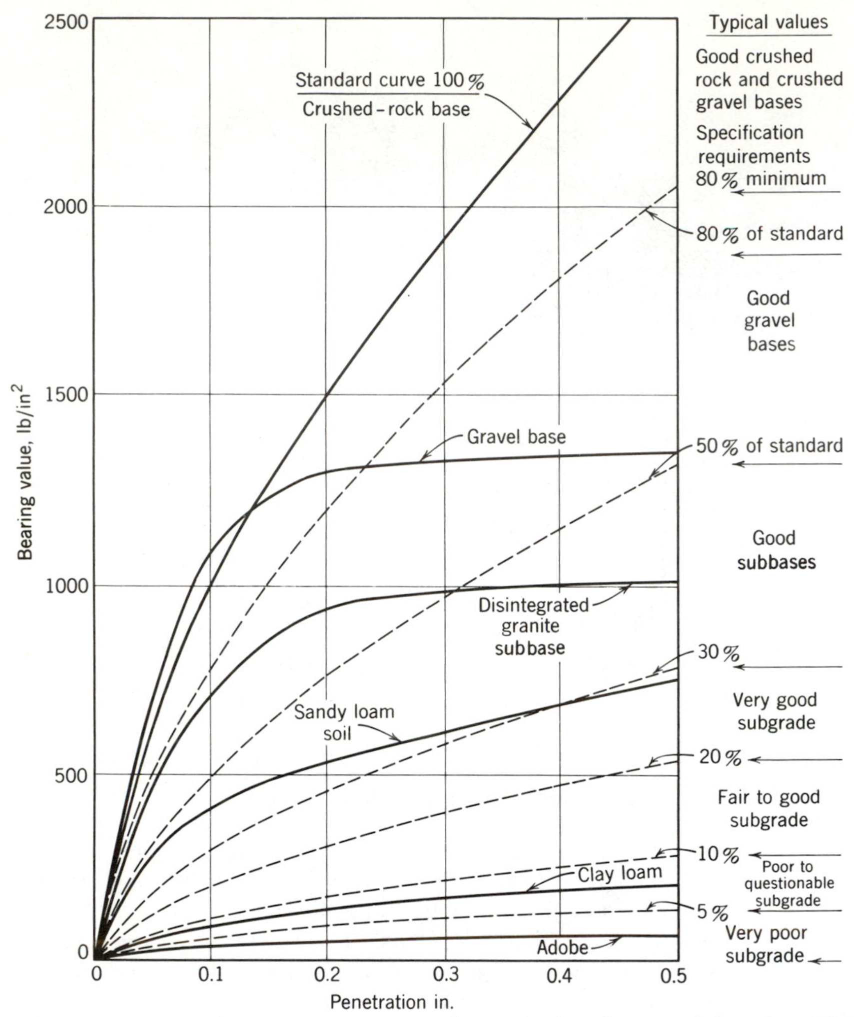

Fig. 14–13. Load-penetration curves for typical soils tested by the *CBR* method. (After O. J. Porter.)

horizontal pressure developed in a short cylindrical sample loaded vertically on its ends (see Fig. 14–14). This device was first conceived in 1930 to measure the stability of both field and laboratory samples of bituminous pavement and has subsequently been modified and improved. The test procedure also utilizes devices to measure the tendencies of subgrade soils to absorb water and expand.[29]

[29]See *AASHTO Designation T190*. Greater detail on the stabilometer method is given under Test Method Calif. 301 in the *Materials Manual* of the California Department of Transportation. Suggested modifications to the stabilometer method to provide closer correlation with the findings of the AASHO Road Test were offered by F. N. Hveem and G. B. Sherman in *HRB Record 13*. Minnesota approaches to design using the stabilometer are discussed in *HRB Record 329*.

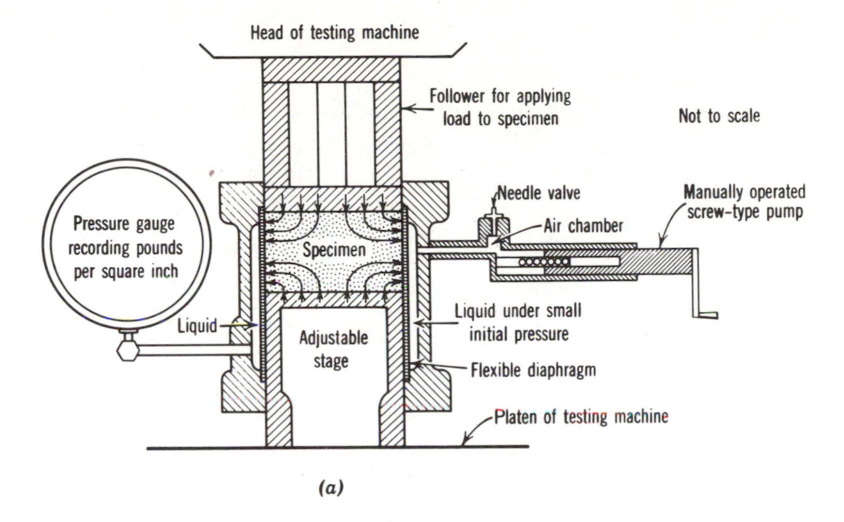

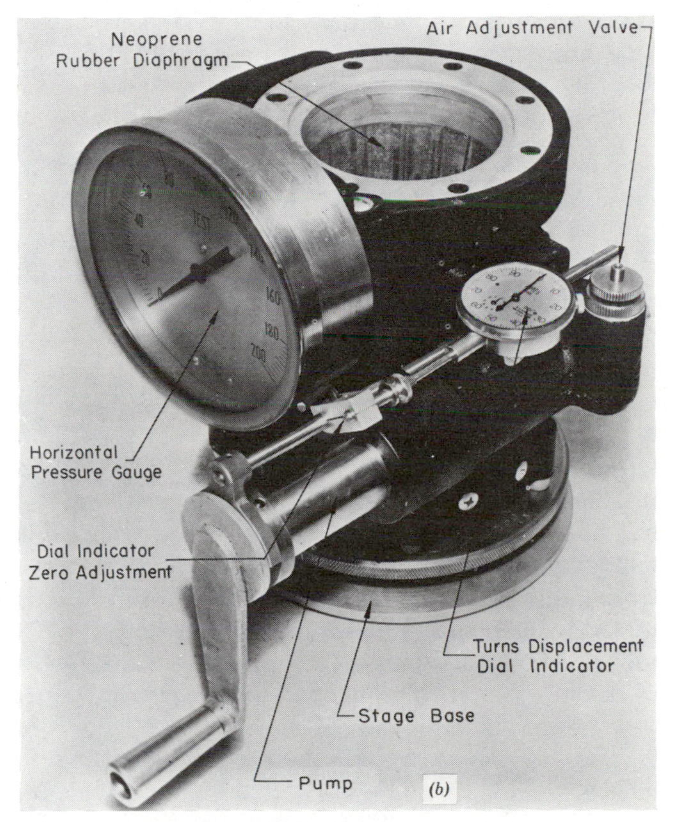

Fig. 14–14. *(a)* Diagram and *(b)* photograph of the Hveem stabilometer. (Courtesy F. N. Hveem.)

The stabilometer procedure of the California Department of Transportation is as follows:

Prepare and Compact Specimen. These are short cylinders about 4 in. in diameter and 2½ in. high. Normally four are prepared at different known moisture contents. Compaction is by the tamping-foot (kneading) compactor.

Test for Exudation Pressure. Each compacted specimen, still in the steel mold, is compressed until water exudes from it. This water completes several electric circuits wired in parallel through the base plate of the exudation measuring device. Exudation pressure is the load in psi at which five of six circuits are activated.

It has already been pointed out that, with time, moisture may move into the soil and cause a reduction in strength. The *CBR* procedure recognizes this situation by soaking specimens for four days before loading. California engineers found by many tests that soils underlying pavements exuded moisture under pressure of about 300 psi. They also found that soaking did not produce this condition in certain fine-grained materials. In effect, the exudation pressure test is a different means for determining the moisture content of the soil in the road. Moisture content in stabilometer test samples is set to bracket the 300 psi value on which design is based.

Test for Expansion Pressure. After the exudation pressure test is completed, samples still in the steel mold are subject to the test for expansion pressure. A perforated brass plate is placed on the sample, after which it is covered with water and allowed to stand for 16–20 hr. Expansion during this period is prevented, but the pressure to prevent it is measured.

As mentioned earlier, soils that expand when water becomes available to them should be placed under a sufficient weight of fill to prevent swelling. This test, then, provides a quantitative measure of the weight required to prevent expansion.

Stabilometer Test. After the expansion test is completed, the specimen is enclosed in a flexible sleeve and placed in the stabilometer (see Fig. 14–14). Vertical pressure is applied slowly, at a speed of 0.05 in./min, until it reaches 160 psi. The developed horizontal pressure is read immediately. Then half the load is removed from the specimen and the horizontal pressure is reduced to 5 psi using the displacement pump. Finally the turns of the displacement pump needed to bring the horizontal pressure to 100 psi are determined. This "displacement" procedure is primarily intended to measure the penetration of the flexible diaphragm into the interstices of the sample. Without this correction, surface roughness of the sample would distort the stabilometer results.

The resistance value R of the soil is computed by the formula

$$R = 100 - \frac{100}{\frac{2.5}{D}\left(\frac{P_v}{P_h} - 1\right) + 1} \qquad (13\text{–}2)$$

where

R = resistance value
P_v = vertical pressure (160 psi)
D = turns displacement reading (range approximately 2–5)
P_h = horizontal pressure in psi at P_v of 160 psi

Notice that the resistance value of a fluid where $P_h = P_v$ will be 0. The R value of an infinitely rigid solid ($P_h = 0$) will equal 100.

TRIAXIAL DESIGN METHODS. A few agencies use triaxial compression tests as their basic roadway-design tool. Important elements of the test device are shown in Fig. 14–15. The sample is encased in a flexible membrane and confined laterally by fluid enclosed in a lucite or other container. Then load is applied vertically (see *AASHTO Designation T234*).

For the "open" system triaxial test, lateral pressure is held constant by releasing fluid from the container as increased load causes the sample to expand laterally. Variables here are vertical load and vertical deformation, which, when plotted, give a load-deformation relationship comparable to that for steel, concrete, or other structural materials. Typical load-deformation curves are shown as Fig. 14–16. Test results are related to thickness of base and pavement by empirical formulas or design charts which recognize factors such as traffic volume and weight, groundwater conditions, rainfall, and freezing and thawing. The Kansas and Texas Highway Departments, among others, use "open" system triaxial testing for design purposes.

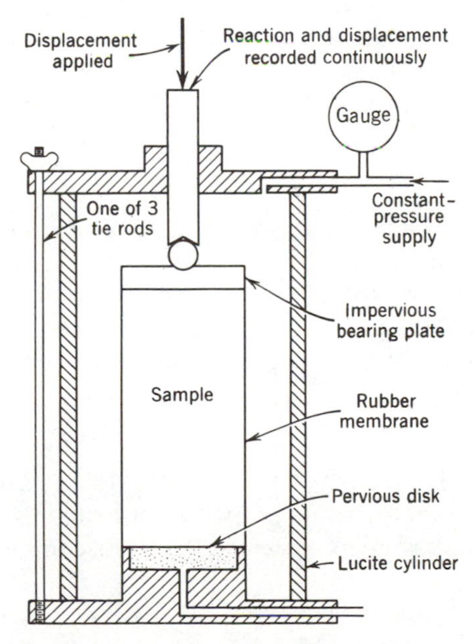

Fig. 14–15. Essential features of triaxial test (open system).

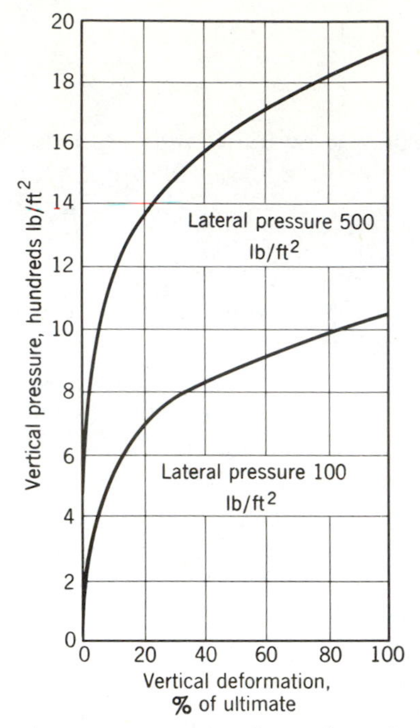

Fig. 14–16. Typical results of soil test using open triaxial device.

Some agencies use the "closed" system triaxial test. For this, all fluid is retained in the encircling container. As a vertical load is applied, the sample tends to deform laterally, which increases the pressure in the surrounding fluid. Variables are vertical load and fluid pressure. Actually, the Hveem stabilometer is a "closed" system triaxial device, although the connotation is usually applied only to tests on tall cylinders.

PLATE-BEARING METHODS. Large-scale plate-bearing tests are used for design purposes by certain highway and airport agencies. Some test large samples prepared in the laboratory; others load the undisturbed ground or compacted fill during construction. Loading is through rigid steel plates of various diameters, commonly ranging from 12 to 36 in. in diameter. As with other test methods, correlations have been developed between bearing test results and field performance.[30]

DYNAMIC MODULUS. During the past 10 to 15 yr, a number of agencies have shown interest in the dynamic properties of materials as a basis for evaluating their behavior in pavement systems. The most common techniques deter-

[30]See, e.g., A. C. Benkelman and S. Williams, *HRB Special Report 46;* G. R. Ingimarsson and W. S. Housel, *HRB Bulletin 289;* and W. H. Zimpfer, ibid.

mine the modulus of elasticity of each material type by repeated-load triaxial or diametral tests. These procedures are illustrated in Fig. 14–17.[31]

In each method, recompacted or undisturbed samples can be tested to determine the influence of such factors as temperature, degree of saturation, density, and age on the dynamic response of pavement materials.

It has generally been found that the modulus of asphalt concrete in psi ranges from 100,000 to 1,000,000, depending primarily on temperature; for cement treated bases from 500,000 to 3,000,000; for aggregate base and subbase from 10,000 to 50,000 or more; and for fine-grained soil from 1,500 to 50,000, depending primarily on water content.[32]

CLASSIFICATIONS FOR SOILS

Soil classifications, based on physical tests or other information, represent groupings into which all soils of like characteristics can be separated. Once a soil has been classified, its performance should be predictable from the known behavior of others in the same group. Many systems of classification have been proposed and have been very useful for their intended purposes. But no single grouping will fit the many diverse problems of soil science.[33] A classification suited to agricultural or geological uses does not satisfy the requirements of the civil engineer; neither is a classification best suited to the foundation engineer's problem entirely satisfactory for the highway engineer. Descriptions of some of the classifications developed for highway and airport purposes follow.

Textural Classification

The textural classification is based purely on grain-size distribution. It was developed about 1890 before the influence on soil behavior of grain shape, colloidal particles, and other variables was understood. It is simple and can be applied with little experience. Soils alike under this classification, however, may show widely different performance in service.

A textural classification into three groups—sand, silt, and clay—is often followed. (Note that the gravel portion is excluded.) A corresponding "triangle textural classification" diagram (see Fig. 14–18) is helpful. The sum of the perpendicular distances from any point within an equilateral triangle to the three sides is constant and may be arbitrarily set as 100%. The position of every point inside the triangle then represents the sieve analysis (or three-way textural classification) of a soil of particular grading. After the grain-size distribution of a sample has been determined, the chart is read as follows: spaces vertically upward,

[31]For detailed discussions of these methods, the reader is referred to *TRB Special Reports 140* and *162* and C. L. Monismith and F. N. Finn, *Transportation Engineering Journal of ASCE,* Jan. 1977.
[32]The elastic modulus is similar to Young's modulus, the applied dynamic stress divided by the recoverable or elastic strain.
[33]*TRB Record 642* deals with soil taxonomy which is not treated in this book.

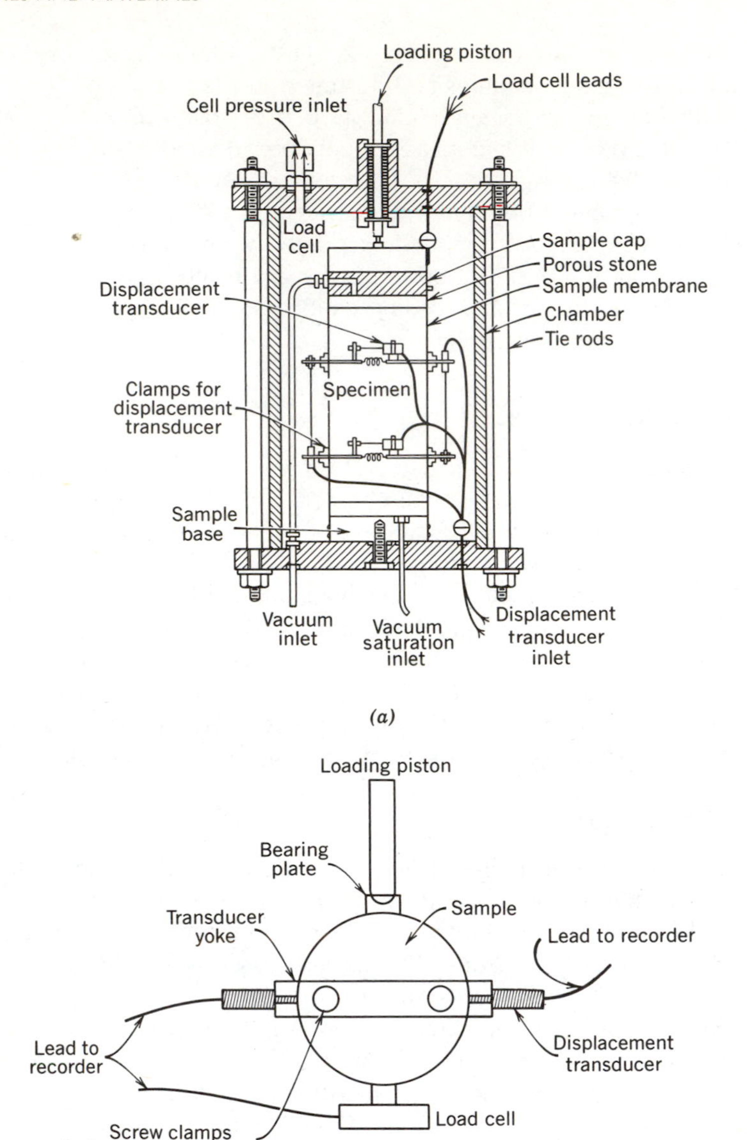

Fig. 14–17. Schematic of tests for dynamic modulus. *(a)* Repeated load triaxial device (not to scale). (After *TRB Special Report 162,* slightly modified.) *(b)* Diametral test device (not to scale.)

starting with zero at the bottom, represent clay percentages; spaces left to right, diagonally downward, starting with zero at the left, represent silt; spaces right to left, diagonally downward, starting with zero at the right, represent sand. For example, the three dotted lines in Fig. 14–18 represent clay 28%, silt 25%, and sand 47%. The soil described by these percentages (point *P*) is therefore classified as a clay-loam.

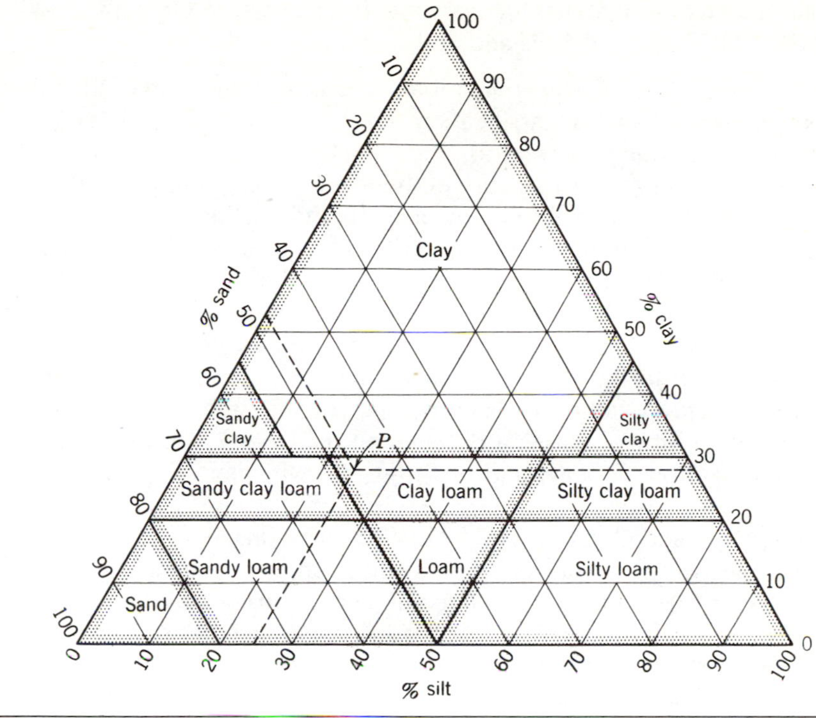

Grade	Size of Particles (mm)
Coarse sand	2.0–0.25
Fine sand	0.25–0.05
Silt	0.05–0.005
Clay	Smaller than 0.005

Fig. 14–18. Textural classification for subgrade soils of sand and smaller sizes.

Soil descriptions reflected by the triangular classification diagram are often used in engineering work. Soils composed almost entirely of sand and silt are called loams. Thus, with one exception, every soil containing less than 20% clay is a loam, with a prefix "sandy" or "silty" added to indicate which predominates. These names appear along the lower part of the chart. Soils containing 20–30% clay are also called loams, with the term "clay" added to indicate the higher clay content. These names appear above the "loam" classifications in the chart. All other soils are designated as "clays," with prefixes of "sandy" or "silty" added to indicate the remainder of the material. The textural classification points up the important principle that the behavior of soils containing roughly 30% or more clay depends solely on the characteristics of the clay, since the larger particles "swim" in the clay mass. With lower percentages of clay the soil has a "skeleton" of coarse particles which tends to give it greater strength and stability.

American Association of State Highway and Transportation Officials (AASHTO) Classification of Soils and Soil-Aggregate Mixture[34]

The AASHTO classification shown in Table 14–2 is an adaptation of the original Public Roads Administration classification which was developed about 1928 to classify soils as road surfaces or bases. The AASHTO scheme represents changes that seemed desirable after a 15-yr use of the older grouping. It retains the group designations and descriptions used previously, but differs in other ways from the Public Roads classification. In the AASHTO classification, the *A*-2 grouping merits particular attention. With the exception of the excellent *A*-1 materials and the *A*-3 sands, it encompasses all materials having a skeleton of coarse particles as measured by 35% maximum passing the No. 200 sieve.

Subgroups added to the *A*-2 classification indicate the properties of the soil mortar which have a strong influence on the suitability of the soil. The third digit which, for example, makes a designation read *A*-2-7, shows that the fine material is an *A*-7 plastic clay, and that the soil is therefore the least desirable material in the *A*-2 class.

The subgroups added to the *A*-7 classification indicate in more detail the properties of the different clay soils. Thus subgroup *A*-7-5 includes those materials that may be highly elastic as well as subject to considerable volume change under variations in moisture content. Subgroup *A*-7-6 includes soils that are subject to extremely high volume change.

As an example of the workings of the AASHTO classification, consider a soil which when tested shows:

Sieve analysis:	% passing No. 40	55
	% passing No. 200	30
Liquid limit		42
Plasticity index		9

The classification procedure (using Table 14–2) is as follows:

1. Under the heading "general classification" it is found that the soil is a granular material, as less than 35% passes the No. 200 sieve: thus the material falls into groups *A*-1, *A*-3, or *A*-2.
2. Under the heading "Sieve analysis—% passing No. 40" it fails to meet the requirements for group *A*-1 as more than 50% passes the No. 40 sieve, but does meet those for group *A*-3 or *A*-2.
3. Under the heading "Plasticity index" it falls outside group *A*-3, which is nonplastic. It must then be in group *A*-2, under which the PI of 9 meets the demands of subgroups 4 or 5.
4. Under the heading "Liquid limit" the value of 42 finally places the sample in classification *A*-2-5.

[34]This classification was originally presented in the 1945 *HRB Proceedings,* pp. 375–392. It was adopted by AASHO as recommended practice in 1949 and appears in *Specifications for Transportation Materials* as *AASHTO Designation M145.*

TABLE 14–2. American Association of State Highway and Transportation Officials Classification of Soils and Soil-Aggregate Mixtures, with Suggested Subgroups

AASHTO Designation M145											
General Classification	*Granular Materials (35% or Less Passing No. 200 Sieve)*							*Silt-Clay Materials (More Than 35% Passing No. 200 Sieve)*			
	A-1			*A-2*							*A-7*
*Group Classification**	*A-1-a*	*A-1-b*	*A-3*	*A-2-4*	*A-2-5*	*A-2-6*	*A-2-7*	*A-4*	*A-5*	*A-6*	*A-7-5, A-7-6*
Sieve analysis: % passing:											
No. 10 (2.00 mm)	50 max.										
No. 40 (0.425 mm)	30 max.	50 max.	51 min.								
No. 200 (0.075 mm)	15 max.	25 max.	10 max.	35 max.	35 max.	35 max.	35 max.	36 min.	36 min.	36 min.	36 min.
Characteristics of fraction passing No. 40 (0.425 mm)											
Liquid limit				40 max.	41 min.	40 max.	41 min.	40 max.	41 min.	40 max.	41 min.
Plasticity index	6 max.		NP	10 max.	10 max.	11 min.	11 min.	10 max.	10 max.	11 min.	11 min.†
Group index‡	0		0	0		4 max.		8 max.	12 max.	16 max.	20 max.
Usual types of significant constituent materials	Stone fragments, gravel and sand		Fine sand	Silty or clayey gravel and sand				Silty soils		Clayey soils	
General rating as subgrade	Excellent to good							Fair to poor			

**Classification procedure:* With required test data available, proceed from left to right on above chart, and correct group will be found by process of elimination. The first group from the left into which the test data will fit is the correct classification.

†Plasticity index of *A-7-5* subgroup is equal to or less than LL minus 30. Plasticity index of *A-7-6* subgroup is greater than LL minus 30.

‡See group index formula for method of calculation.

A general classification for the soil can also be obtained from Table 14–2. The material is a gravelly soil, as indicated by the *A*-2 description; the fines are of a silty nature, as indicated by the subgrouping 5. As subgrade material it is on the border between good and fair.

An important feature of the AASHTO classification is the "group index" which is based on the service performance of many soils. It permits a more precise prediction of soil behavior than is possible by soil classification alone and is used by some highway designers to guide in determining the combined thickness of pavement and base over a given soil. The "group index" is defined by the empirical equation,

$$\text{group index} = 0.2a + 0.005ac + 0.01bd$$

in which

a = that portion of percentage passing No. 200 sieve greater than 35% and not exceeding 75%, expressed as a positive whole number (1–40)

b = that portion of percentage passing No. 200 sieve greater than 15% and not exceeding 55%, expressed as a positive whole number (1–40)

c = that portion of the numerical liquid limit greater than 40 and not exceeding 60, expressed as a positive whole number (1–20)

d = that portion of the numerical plasticity index greater than 10 and not exceeding 30, expressed as a positive whole number (1–20)

Unified Soil Classification System[35]

The Corps of Engineers, U.S. Army, in 1942 tentatively adopted a new system of soil designation, referred to as the airfield classification. It was developed by Casagrande, and was designed so that soils could be classified by experienced engineers after visual and manual inspection. Later, after some modification, the system was adopted by both the Corps and the Bureau of Reclamation, and was renamed the Unified Soil Classification System.

The Unified System uses letters instead of numbers to designate the various groups. Mechanical analysis and liquid- and plastic-limit tests are the primary classification tools. Principal symbols and soil designations are as follows:

For coarse-grained soils (more than 50% retained on No. 200 sieve) :

1. Gravels or gravelly soils, symbol *G*.
2. Sands and sandy soils, symbol *S*.

Subdivisions of sands or gravels are:

a. Well-graded, fairly clean material, symbol *W*. Thus *GW* signifies well-graded gravel, and *SW* well-graded sand.

[35]For greater detail, see *U.S. Army Technical Memorandum 3-357* (revised 1960). A classic presentation that describes much of the reasoning underlying the system appears in *Transactions, ASCE,* 1948, pp. 901–930. T. K. Liu, *HTB Record 156* gives a detailed comparison of the AASHTO, Unified, and FAA classification systems and an extensive bibliography. See also the Asphalt Institute *Soils Manual,* MS-10.

b. Coarse materials with clay binder; symbol *C,* in combinations *GC* and *SC.*
c. Poorly graded, fairly clean material; symbol *P,* in combinations *GP* and *SP.*
d. Coarse materials containing silts or rock flour, symbol *M,* in combinations *GM* and *SM.*

For fine-grained soils (more than 50% passing No. 200 sieve):

1. The inorganic, silty, and very fine sandy soils; symbol *M* (from the Swedish terms *mo* and *mjala,* flour), used for fine grained, nonplastic or slightly plastic soils.
2. The inorganic clays; symbol *C.*
3. The organic silts and clays; symbol *O.*

Each of these fine-grained soils is grouped according to its liquid limit into:

a. Fine-grained soils having liquid limits less than 50, that is, of low to medium compressibility; symbol *L,* in combinations *ML, CL,* and *OL.*
b. Fine-grained soils having liquid limits greater than 50, that is, of high compressibility; symbol *H,* in combinations *MH, CH,* and *OH.*

Highly organic soils, usually fibrous, such as peat and swamp soils having very high compressibility are not subdivided but placed in one group; symbol *Pt.*

The Unified Classification suggests several ways for quick field identification of the fine-grained portions of a soil. For example, a sample may be mixed with sufficient water to give a putty-like consistency, formed into a pat, and dried completely. The higher the dry strength of the pat when broken in the fingers, the greater the plasticity. High dry strength is characteristic of clays and colloids; pats of silts or fine silty sands break easily. Fine sands will cause the pat to feel gritty on the fingers; silts will be smooth. Dilatancy or reaction to shaking is another aid to classifying the soil mortar. Enough soil to form a pat with a volume of about half a cubic inch is mixed with water to a soft but not sticky consistency. If the material is primarily fine sand, shaking and jolting the sample in the palm of the hand will bring water to the surface of the pat. Squeezing it between the fingers causes the moisture to disappear. Silts react less completely and clays not at all.

Federal Aviation Agency (FAA) Classification System

The Federal Aviation Agency has another system of soil classification. With it, there are 13 soil types ranging from *E*-1, the best, to *E*-13, the poorest. Details and references for it are given in the last two items in the preceding footnote.

Pedological Classification

Soil science (pedology) is the basis for the pedological classification. Its principle is that *like soils are developed on like slopes when like materials are weathered in like fashions*. As an illustration, weathering of limestone produces a reddish silty clay soil. Where slopes and climate are alike, the depth and nature of this soil mantle are similar, regardless of geographical location. Again, it should be possible to predict the behavior of loess, a windblown deposit of sand, silt, and clay particles, wherever it is found.[36]

Under the pedological classification, soils produced in like manner from the same parent rocks are grouped together, as it is assumed that they possess similar engineering properties and require the same engineering treatment. The number of samples subjected to physical tests may be reduced below that usually required, as the test results serve more as a check on pedological classification than as primary design tools. This procedure is in direct contrast to the other classification methods, which rely almost exclusively on physical tests and make little use of pedological information.

As mentioned, like soils occur under like conditions of materials, slopes, and weathering. In addition, similar landforms or soil patterns are developed. Thus, such elements as surface-drainage patterns, erosion characteristics, and soil color reflect the nature of the soil. As examples, the shapes of the gullies reveal soil texture or clay-pan development; color patterns often reflect ground-water conditions; ridges with parallel axes may indicate wind-blown sands and silts. To the trained observer a study of landforms, particularly on aerial photographs, can lead to a pedological classification and to accurate prediction of the soil conditions to be encountered in a given area. Study of stream-bed characteristics, glacial activity, and other signs may lead to the discovery of sources of construction materials.[37]

Many state transportation agencies use engineering soil survey reports and maps (see below) based on the pedological approach as part of their working tools. Some have developed their own engineering soil maps and manuals from this source.

SOIL SURVEYS

A preliminary soil investigation is an integral part of highway reconnaissance and preliminary location surveys. Soil conditions as well as directness of route, topography, right of way, neighborhood disruption, environmental considerations, and other factors must be weighed in fixing the position of the road. Normally, this preliminary soil study consists of an examination of soil maps and pedological classifications for the area when they are available and a visual examination coupled with a small amount of sampling and testing where the findings might influence the final decision. There will, of course, be situations for which comprehensive soil study will precede selection of the final route. For example, if one proposed location crosses a marsh while the alternative remains on stable ground, careful testing, preliminary design, and cost studies might precede the final decision.

[36]See, e.g., articles in *HRB Record 212, 374, 405,* and *426*.

[37]For detailed discussions of the techniques for such studies see *HRB Bulletins 213, 299 and 312; HRB Record 33;* Q. L. Robnett and M. R. Thompson, *HRB Record 315,* and *HRB Special Report 102*. For the use of color and infrared aerial photographs for soil-mapping purposes, see e.g., J. P. Minard and J. P. Owens, *HRB Bulletin 316;* J. R. Chaves, *Public Roads,* Apr. 1962; *HRB Record 63, 142, 156, 319;* and *HRB Special Report 102*.

Most agencies make a detailed study along with the final location survey. Recommended procedures appear in *Specifications for Transportation Materials, AASHTO.*[38] Such a survey is to provide pertinent information about soil and rock for a decision on one or more of the following subjects:

1. Location of the proposed construction, both vertically and horizontally.
2. Location and evaluation of suitable borrow and construction materials.
3. Need for and type of subgrade or embankment foundation treatment or drainage.
4. Need for special excavating and dewatering techniques.
5. Development of detailed subsurface investigations for specific structures.
6. Investigation of slope stability in both cuts and embankments.
7. Selection of roadway or area pavement type and section.

An early phase of the soil survey is the collection and examination of all existing information. This may include the identification of soil types from geologic and agricultural soil maps, aerial photographs, and other sources, investigation of groundwater conditions, and an examination of existing roadway cuts and other excavations. A review of the design, construction procedures, and present condition of roads that traverse the area often is extremely helpful.[39]

Soil exploration along the right of way usually is made by means of auger borings and test pits. There is no definite rule to follow except that sampling should be at frequent enough intervals to fix the boundaries of each significant soil type. Test holes should extend a significant distance below the subgrade elevation, with a recommended minimum of 5 ft. Deeper holes may be appropriate to locate bedrock, adverse ground such as peat or muck, and groundwater conditions that would influence design or construction. A complete and systematic record should be made for each hole. Its location, the nature of the ground, origin of the parent material, landform, and agricultural soil name, if known, should be recorded. Each soil layer should be described according to its thickness, texture, structure, organic content, relative moisture content, and degree of cementation. The depth of seepage zones or the free water table and of bedrock should be included when they are encountered. On the basis of these reports, many agencies plot a *soil profile,* which records the test data in visual form. It shows the location of each test hole, a soil profile along the roadway centerline, and the range of soil-profile characteristics for each distinct soil type.

The locations at which laboratory samples will be taken are found from a study of the auger borings or soil profile. Disturbed samples of earth may be obtained to a depth of roughly 6 ft with a hand-operated soil auger or to depths

[38]See *Investigating and Sampling Soils* and *Rock for Engineering Purposes, AASHTO Designation T86.* This reference describes the usual equipment and procedures for soil exploration. *Compendium No. 2* of Transportation Technology Support for Developing Countries, TRB is also an excellent reference on identifying and developing material resources. See also *NCHRP Synthesis 33* and *AASHTO Manual on Foundation Investigation,* 1978.

[39]Advances in remote sensing (aerial surveys) including color and infrared photography and radar, have made this a most useful tool for identifying soil types, locating likely borrow or material sites, detecting possible groundwater problems, and other investigation. See, e.g., *HRB Special Report 102;* H. T. Rib, *Public Roads,* June 1968; J. R. Chaves and R. I. Schuster, *HRB Record 63;* C. W. Mintzer and R. A. Struble, *HRB Record 156;* M. G. Tanguay and R. D. Miles, *HRB Record 319* and several articles in *HRB Record 421.*

up to about 30 ft if the auger is power-driven. Specimens in hard or boulder-filled soil may be gained by digging test pits. Many highway agencies have specially trained crews and elaborate equipment for making soil surveys. Power-driven augers are common. Wash-boring rigs, churn drills, and rotary drills are available for exploration to greater depths or in hard or rocky materials. Special devices have also been made for taking undisturbed samples from deep down in swamps and bogs.

Two geophysical methods are also in use for subsurface exploration. One of these, called the refraction seismic method, relies on the principle that the speed of travel of shock waves through the earth's surface is different for different materials. For example, waves travel through light, loose soils at approximately 600 ft/s, but through dense, solid rock at speeds approaching 20,000 ft/s. The test device records the time it takes for the shock from a disturbance charge to reach several points at increasing distances away (see Fig. 14–19). If the earth's crust is of uniform composition for some depth, these time intervals are directly proportional to the distance from the point of explosion. If, on the other hand, the surface layer is underlain by a harder, denser material, the time interval to more distant points is shortened because the shock wave travels downward into the denser material, along its upper margin, and up again to the recording device. By plotting the time of wave travel against shooting distance, the number and thicknesses of various underlying layers and the depth to bed-

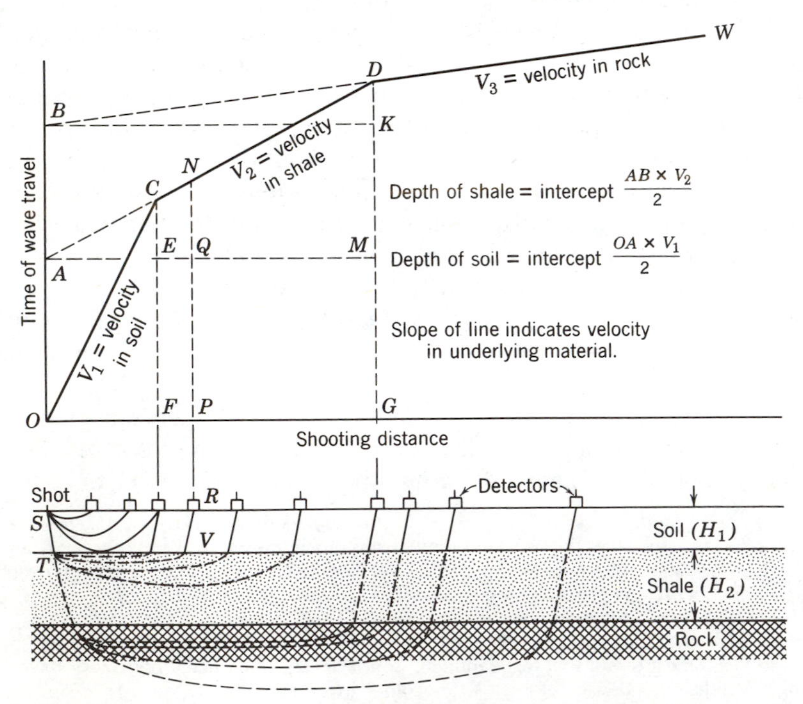

Fig. 14–19. Refraction seismic method of subsurface exploration.

rock can be predicted (see Fig. 14–19). This test has proved particularly useful in determining the depth to rock. It can be deceiving in badly tilted, stratified materials.

A simple adaptation of the refraction seismic method has been developed for exploring underground conditions in roadway cuts. The equipment is light and portable, weighing only 40 lb. The shock wave is produced by striking a 6-in.-square steel plate with an 8-lb sledge hammer.

The variation in the electrical resistivity between different subsurface materials offers another method of soil exploration. Ordinary moist soils containing moderate amounts of clay and silt and some electrolytic agent have comparatively low resistance. On the other hand, sand, loose dry soils, and solid rock have relatively high resistivity. In performing the test, direct current is made to flow through the soil between two electrodes. The drop in potential is then measured, not between these supply electrodes, but between two others placed intermediately at the third points. It has been established empirically that this voltage drop measures the resistance of the entire soil layer to a depth equal to the spacing of the intermediate electrodes. The change in resistivity with depth is obtained by recording the resistivity at various electrode spacings. By comparing a plot of these finds with those for a deposit of like material that has already been exposed, layer thicknesses can be found. This method is generally quite reliable, although there have been occasions when it has not furnished completely dependable information regarding the presence of solid rock. It is particularly useful in area exploration, as when it is desired to find localized sand and gravel deposits.[40]

ROADWAY DESIGN FOR UNUSUAL SOIL CONDITIONS

Among the problems in roadway design and construction are stability of fills and slopes, drainage, capillarity and frost heave, permafrost, elasticity, and rutting. They are discussed in the following paragraphs.

Stability of Fills and Slopes

At times, materials in cut banks will slip downward onto the road. Again, portions of high fills slide outward and downward, often carrying portions of the roadway or shoulder along. Such failures are often spectacular and gain considerable attention. Figure 14–20*a* illustrates a common failure, in which slipping occurs along a seam of wet or weak material. Figure 14–20*b* shows slumping of a fill. Notice that the shearing surface approaches the arc of a circle and that the entire disturbed mass has rotated around the center of the circle. This pattern of failure is common in fills or cut slopes of relatively homogenous nongranular materials.

Slides may occur during construction or at a later date after the road is in

[40]For more information of subsurface exploration by special methods, and a bibliography see *HRB Record 81* and *Compendium No. 2*, op. cit., Text 3.

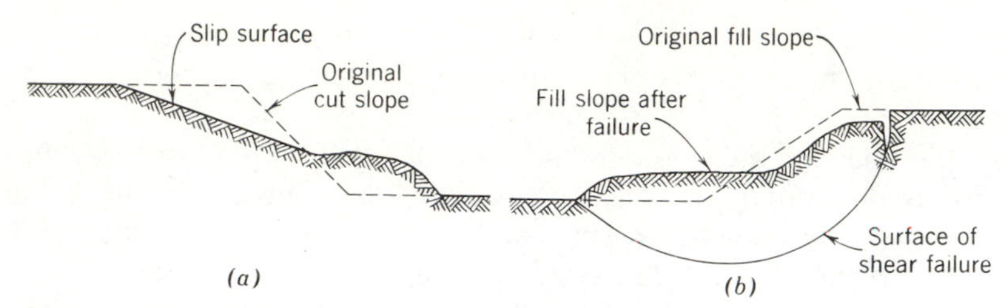

Fig. 14–20. Roadway failures caused by slipouts.

service. It may be better practice to risk some sliding or adjustment of large cut slopes than to flatten them all and thus increase the pay yardage. Areas that threaten persistent slides should be avoided if possible during location. Minor slope-adjustment slides and ledge falls that occur during construction must be removed and usually are used to widen fills or to flatten their slopes.

Slides may result from mud flows, slope adjustments, or movement caused by underground water or by undercutting a rock strata. Less frequently, coarser material such as rock talus, coarse sand or gravel, or weathered debris from above the road will threaten grading operations or the completed highway.

Many troublesome slides result when an underlying inclined surface of shale, soapstone, clay, or like material is lubricated by seepage water (see Fig. 14–20). During the rainy season or after hard rains or heavy snowfall, this surface may become lubricated, causing the entire mass above the slippage surface to move. Such slides can be controlled by removing all or a large portion of the material above the slippage surface or by some drainage device that keeps water from the surface of weakness. If the slip plane is lubricated by surface water from above the cut slope, the water may be intercepted by surface ditches farther up the slope or sometimes by oiling or otherwise waterproofing the area of penetration. If the water comes from a previous layer at a fairly shallow depth below the surface (say, 10 ft or less) underdrains may be installed to cut the flow off higher up. If the threatening mass is too deep for these procedures, flows sometimes may be intercepted by drainage tunnels of small cross section backfilled with coarse sand over open pipe. These drains would run on a downgrade from the apex zone to a gravity outlet. The water must always be intercepted before it can lubricate the critical slip surface. On most of these wet-slope readjustment slides, barriers such as lines of piling, rock windrows, or ordinary breast or retaining walls will fail.

The California Department of Transportation has developed a "hydrauger method" for stabilizing slopes which has proved both cheap and effective. A special machine drills holes into the cut or fill slope on a slight upward inclination. A perforated pipe which intercepts and carries off the flows is then driven into these holes. This agency has also developed other techniques for preventing fill slipouts or slope failures.[41]

Slides of dry material have been checked by sturdy masonry breast walls at

[41]See T. W. Smith, *HRB Record 57* and S. E. Lamb, *TRB Record 749*.

the foot of the slides. Some apparently threatening dry debris masses fail to move when cut into, probably because the cubical or angular-shaped rock fragments on such slopes tend to lock together. The dribbling of an occasional loose rock or boulder down the slope onto the road often can be prevented by covering the slope with wire mesh or by treating the surface with a thick cement grout, or it can be intercepted by a wall or heavy fence along the inside of the road. The cost of handling bads slides obviously must be compared with the cost of alternative designs using half-trestle or similar construction which avoids deep cuts and high fills.

In recent years reinforced earth, a material formed by combining earth and reinforcement, has been used extensively to stabilize fills and for retaining structures. Reinforced earth is a construction material composed of soil fill (silt, sand, gravel, and stone of all sizes) strengthened by the inclusion of rods, bars, fibers or nets which interact with the soil by means of frictional resistance. As with reinforced concrete, the benefits depend on a combination of the tensile strength of the reinforcing and the shear bond with the surrounding soil. The shear bond is a function of the size, geometry, and type of loading of the structure as well as the types of materials, drainage, and other factors. A detailed discussion of the theory, design, and performance of reinforced earth walls is beyond the scope of this book. For details, the reader is referred to numerous articles in ASCE and TRB publications.[42]

Any extensive discussion of slope stability is beyond the scope of this book, and for further information the student is referred to textbooks and literature in soil mechanics.[43]

Drainage

Permeability indicates the degree to which fluids pass through soils. Coarse-grained soils such as gravels and sands are highly permeable; clays and fine-grained soils are almost impermeable. In general, highly permeable soils make excellent subgrade material, as coarse-grained soils are stable whether dry or saturated. However, permeable soils may carry seepage water and create serious stability problems by furnishing a source of supply for capillary moisture, by increasing pressure on slippage surfaces or by producing seeps in cut slopes.

Figure 14–21 shows two situations in which the removal of water from previous soils may be advantageous. In Fig. 14–21*a*, the high water table would provide a source of capillary moisture which, on rising, would saturate the subgrade and base course and cause weakness or lead to frost heave. The installation of deep underdrains could lower the water table so that capillary

[42]See articles in *HRB Record 282* and *TRB Record 510,* and *749*. See also, papers by K. Lee, and J. Binquet and K. Lee in ASCE *Journal of Geotechnical Engineering,* Oct. 1973 and Dec. 1975, and by K. M. Romstad et al., ibid., May 1976.

[43]Discussions of this problem from the highway literature include *TRB Special Report 176* titled *Landslides: Analysis and Control; Compendium 2* Texts 9 and 13 and *Compendium 13,* Transportation Technology Support for Developing Countries, TRB; and articles in *HRB Record 223, 323, 457,* and *463,* and *TRB Record 749.*

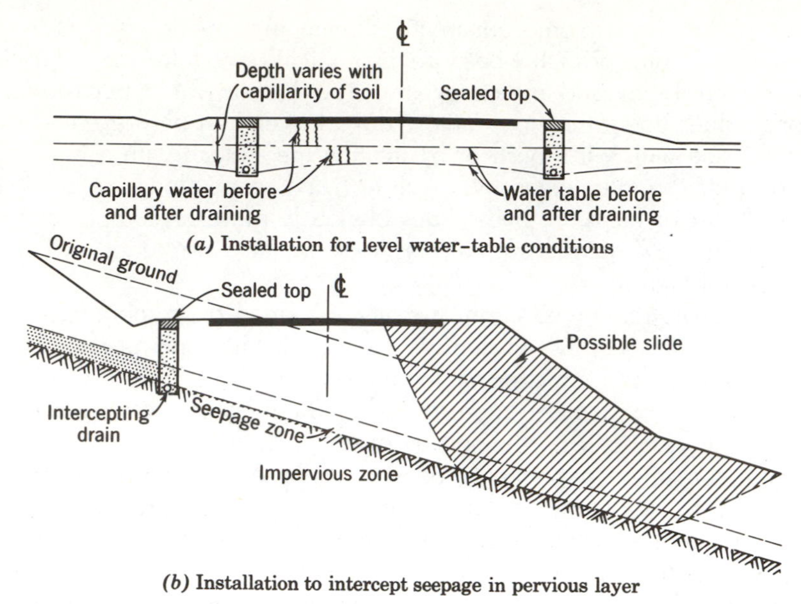

Fig. 14–21. Typical underdrain installations.

moisture would not rise high enough to be troublesome. Figure 14–21*b* shows the use of an underdrain to intercept seepage water which otherwise might cause a slippage at the surface between pervious and impervious layers and lead to a slipout. All subdrains need gravity outlets.

Subdrains may not be effective in impermeable soils through which water flows very slowly, such as those classified in the *A*-4 to *A*-7 ranges. Also, they will not remove much moisture already in such soils.

Engineering studies have greatly extended previous knowledge about the proper design of subdrains. Some resulting recommendations are:

1. Pipe should be used to carry away the collected water. Older "French drain" installations which contain no pipe are not as effective. Pipes of corrugated metal, concrete, bituminized fiber, or vitrified clay, perforated to permit entry of water, are satisfactory. Also, unperforated concrete, fiber, and vitrified clay pipe, laid with open joints, can be used. The number, size, and location of openings into the pipe must be studied carefully to prevent excessive entrance velocity which will cause silting of the pipe.
2. Filter material with which the trench is backfilled, or the openings in the filter cloth, should be fine enough that the adjacent soil is not washed (piped) into the drain.[44] For example, if the drain is installed in a silty soil, the backfill material should be similar in grading to concrete sand. At times, a coarser material may be placed toward the trench center. An open-graded asphalt concrete has sometimes been employed for this purpose.
3. The top of subdrains should be sealed with impervious soil to prevent the entrance of surface water. Earlier practice which combined subdrains with drains for surface

[44]For criteria for graded filters refer to H. R. Cedergren, *Drainage of Highway and Airfield Pavements,* John Wiley and Sons, 1974. For fabrics refer to *FHWA Report RD-80-02,* "Evaluation of Test Methods and Use Criteria for Geotechnical Fabrics in Highway Applications."

water by extending the filter material to the surface has proved unsatisfactory.

4. Intercepting drains (see Fig. 14–21*b*) to be effective must extend downward into the impervious zone.
5. Pipe must be laid with the flow line at least 48 in. below the finished grade and be carefully bedded in gravel or filter material.[45]

Capillarity

Capillarity permits the movement of free water through the pores and fine channels of the soil. In coarse-grained materials such as sand and gravel, the behavior of free water is governed almost entirely by gravity, and the water will tend to seek a level as if in an open channel. As grain sizes become smaller and channel diameters decrease, soil–water interactions influence the behavior of the water. For example, if the pores of a soil are not full of water and if free water is available, capillary forces will tend to pull free water through the soil until all voids are full. Such movement can take place in any direction, but upward movement usually creates the most serious problems. Capillary action is most pronounced in "dirty gravels" or soils composed mainly of fine sands, silts, or clayey silts as they are fairly permeable and have channels through which moisture can pass. Yet the diameters of the openings are small enough so that the capillary forces are high. Clays and colloidal soils are practically impermeable and are little subject to capillary flow.

There is evidence that large amounts of water can be raised considerable distances by soils subject to capillarity.[46] If the surface of the soil is open and this moisture evaporates as fast as it rises, no damage may result. But, if evaporation is slow or the surface is sealed by pavement or some other impervious blanket, this capillary water accumulates and saturates the subsurface layers. Many surface failures have resulted. This problem assumes particular importance where a seal coat or bituminous blanket is applied to roads with gravel or similar surfaces. Also, as discussed below, capillarity coupled with ground freezing creates a host of problems.

Failures Resulting from Frost Action

Failures variously described as frost heaving, frost boils, and spring breakup often occur when soils subject to capillary action freeze and thaw. Unfortunately, this combination occurs frequently in northern regions where glacial action has produced capillary soils and winters are severe.[47]

[45]For added detail on subsurface drainage for highways see, e.g., H. R. Cedergren, *Seepage, Drainage, and Flow Nets,* 2nd ed., Wiley, New York, 1977; U.S. Waterways Experiment Station, Vicksburg, Miss., *Technical Memorandum 183-1; HRB Record 68, 203, 215, 360,* and *373;* TRB Record 733; G. Ring, *Public Roads,* Dec. 1978, and H. R. Cedergren *Drainage of Highway and Airfield Pavements,* op. cit.

[46]Test data show a capillary rise of 45 in. in 24 hr in coarse silt with a 78–120 in. maximum rise. A capillary flow of 2.7 ft^3/ft^2 of area in 24 hr at 18 in. above groundwater also is on record.

[47]An excellent summary on roadway design considerations in frost areas is given in *NCHRP Synthesis 26*. See A. F. DiMillio and D. G. Fohs, *Public Roads,* Mar. 1980 for an overview of the frost action problem and possible solutions.

A simplified explanation of frost heaving is as follows:

1. When the soil temperature decreases below the freezing point, water in the larger voids freezes, but that in the capillary tubes does not, because its freezing point is lower (see Fig. 14–22).
2. This capillary water is drawn to the frozen particles with tremendous force. Tensions as high as 8000 cm of water have been reported.
3. When this water comes in contact with the ice particles, it freezes and increases the thickness of the ice layer, thus raising the overlying material. Pressures of about 200 lb/in.^{2} can be developed, which far exceed the usual superimposed load.
4. Water drawn from the soil pores is replaced with moisture supplied by capillary action from the groundwater below, and that in turn is added to the growing ice lens.
5. Growth of a lens is halted and another below it begins to form when the temperature between them becomes low enough to freeze the capillary moisture. The rate at which cooling takes place governs the number and thickness of lenses.

In Scandinavia, ground freezing occurs to a depth as great as 6–10 ft. But even in the northern United States, where the ground seldom freezes more than 2–3 ft below the surface, heaving or raising of the road surface by 6 in. or more is not uncommon. In one case heaving of 2 ft has been reported. Very often heaving is not uniform but occurs only in stratified layers or pockets of fine sand, silts, or clayey silts which so often occur in regions that have been subject to glacial action. Thus a small section of the road surface will rise and create a serious accident hazard. Differential heaving where the grade line passes from cut to fill is common. At times, large rocks may be thrust upward and protrude through the surface.

At least two added problems related to frost heave occur with wide roadbeds. First, heaving to the sides as well as vertically may cause wide longitudinal cracks in the pavement. Second, snow that is plowed from the traveled lanes onto the shoulder provides an insulating blanket. Shoulder freezing depth is decreased and differential heaving between the traveled lanes and the shoulder may develop.

Heaving accompanies freezing. But with spring thaws, the ice lenses melt, and free water concentrates to form soft spots in the subgrade called "frost boils." Harmful settlement of the road surface follows. When water remains under the surface until vehicles break through, the action is describes as "spring breakup." Frost action also brings bank slides and fill slope instability.

Methods for identifying frost-susceptible materials have been extensively studied. One rule proposed by Casagrande in 1948 is that soils having more than 3% smaller than 0.02 mm would be subject to frost heave. Other measures based on standard tests would classify as frost susceptible all soils with more than 35% (others say 15%) passing the No. 200 sieve or having sand equivalent values lower than 60.[48] Laboratory procedures for frost-susceptibility testing have been developed.[49]

[48]*HRB Record 33* offers an excellent and detailed summary of present-day practices.
[49]See articles by C. W. Kaplar and R. M. Leary et al., *HRB Record 215,* J. H. Zoller, *Public Roads,* Sept. 1973, and C. A. Pagan and V. K. Khosla, *TRB Record 497 and 532.*

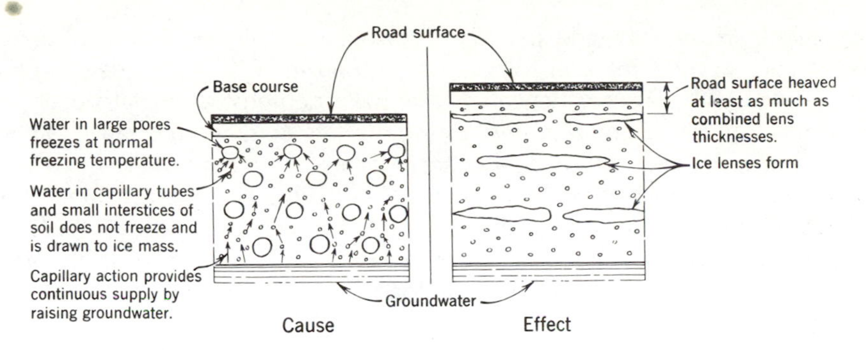

Fig. 14–22. Mechanics of frost heave.

Predicting the depth of freezing is difficult, since it is dependent on many meteorological and physical factors. Among them are air temperatures, variations in soil moisture, and roadway configurations. The temperature component sometimes is recognized by a "freezing index," which is defined as the degree days below 32°F. Predictions of freezing depth can be made theoretically; more commonly however, they are based on field observations.[50]

Several remedies for frost heave have been used. Among them are:

1 Remove the soil that is subject to capillary action for a considerable depth and replace it with a granular noncapillary material. The practice varies. Some agencies remove susceptible materials to the full depth of frost penetration; others stop at one-half this depth.
2. Install subdrains to lower the groundwater below the reach of extensive capillary action (usually about 10 ft).
3. Excavate the soil to the frost line and place an impervious seal such as an asphaltic membrane or a layer of granular, noncapillary material at this level to cut off the flow of capillary water. Then put back the original material.
4. Eliminate or reduce the depth of frost penetration by providing an insulating layer. For example, layers of extruded polystyrene foam have been effective in small experimental installations.[51]
5. Increase the effective grain size of frost-susceptible materials with chemicals. This solution may become economical when supplies of frost-free materials become exhausted.

Even where freezing does not bring frost heaving, a marked strength loss re-

[50]The movement of water through soils under temperature gradients and its relationship to frost heaving has been studied intensively. An early and classic experiment by S. Tabor is described in *Public Roads,* Aug. 1930. A. R. Jumikis, Rutgers Univ., has conducted many theoretical and experimental studies (see *HRB Record 276* for a recent example and extensive bibliography). See also articles in *HRB Record 33, 215, 276, 301, 304, 360, 393, and 431* and *TRB Record 497, 532 and 572*. Discussions of the freezing index by A. L. Straub and F. G. Wegmann and L. K. Moulton and J. H. Schaub appear in *HRB Record 68,* and *Transportation Engineering Journal,* ASCE, Nov. 1969, respectively. See also G. McCormick, *Transportation Engineering Journal,* ASCE, Aug. 1971.
[51]See E. Penner et al., *HRB Record 128,* and A. L. Straub and W. G. Williams, *HRB Record 181.*The effect of surface color is reported by A. L. Straub et al., in *HRB Record 276*. J. A. Horton et al., and H. R. J. Walsh, *HRB Record 429,* describe a test installation in Indiana. See also *TRB Record 675* for recent work on use in concrete pavements.

sults when a frozen subgrade and base course thaw. This has been attributed, at least in part, to a decrease in soil density and an increase in the degree of saturation and pore pressure. Repeated tests have demonstrated that frozen soils match the supporting power found during the late summer and fall. However, as thawing from the surface downward progresses, strength falls rapidly and may decrease as much as 50% by the time all frost is out of the ground. These findings offer strong support for the idea of the "regional factor" in the design method stemming from the AASHO Test Road. They also help to justify the action of many northern states where axle loads are severely limited during spring months.[52]

Permafrost

In northern districts of Canada and in Alaska, Scandinavia, and Siberia, a frozen-soil condition known as "permafrost" presents serious problems. The ground is frozen permanently and often to great depths. During the summer the top few feet thaw from the top down; this leaves an impassible quagmire if the soil is finegrained and has a high moisture content. In the winter the ground freezes solid again, with freezing progressing downward from the surface.

Permafrost creates a variety of problems in road building. To illustrate: When the first sections of road were built through permafrost areas on the Alaska Highway, conventional procedures were followed. The thick moss and other plant growth and the thawed soil below them were removed, and the fill was constructed on the frozen material. Sections so built failed almost immediately, as the blanket of relatively warm fill material furnished heat which melted the underlying ice and left the road without support. On subsequent construction, care was taken to prevent thawing of the ground. Vegetation was left in place, and often brush and other material were added to provide insulation.

In general, designs for permafrost conditions aim to interfere as little as possible with the permafrost regime. For example, structures are built free of the ground to permit freezing and thawing to occur as usual.[53]

Elasticity and Rutting

Elasticity means that soils are springy and compressible under load; usually they return to shape or "rebound" when the load is removed, although some residual deformation may occur after repeated load repetitions. Elasticity is common in soils whose fines consist mainly of flat, flaky particles such as mica. It can be

[52] *HRB Bulletin 207* is a report covering extensive studies of the effects of freezing on load-carrying capacity carried out from 1949 to 1959 by 11 state highway departments. A variety of procedures for evaluating this strength loss have been explored.

[53] The literature on permafrost is extensive. See, e.g., *HRB Special Report 58,* which reports on Russian practices. See also *Geotechnical Engineering for Cold Regions,* O. B. Andersland and D. M. Anderson ed., McGraw-Hill, New York, 1978 and *Proceedings of International Conferences on Permafrost,* 1963, 1973, published by National Academy of Sciences, and 1978, published by Canadian National Research Council. *TRB Record 755* contains a discussion on pavement design considerations.

detected on construction work by a characteristic "rubbery" action of compression and rebound under heavy wheel loads. Under repeated manipulation and reworking of the soil with the consequent particle reorientation, elasticity usually becomes worse until the soil loses much of its strength.

As indicated earlier, highly elastic soils should never be placed close to the roadway surface if there are many repetitions of heavy loads, since their deformation is sufficient to cause early cracking in the pavement. This problem has become more and more acute as the number of heavy wheel loads on major highways has been increased.

One set of observations has shown that when a heavy wheel load passes, measurable compression and rebound occurs at least 20 ft under the surface; but more than 80% is concentrated in the top 3 ft. From a design standpoint, then, the aim is to sum up the total expected compression and rebound in the pavement structure to be sure it does not produce fatigue failure in the pavement. One such procedure was developed by the California Department of Transportation.[54] Measurements conducted in a device called the resiliometer, which applies and removes fluid pressure from Hveem stabilometer specimens, show close correlation to field observations made with the Benkelman beam (see below). California now sets maximum permissible deflections which vary with pavement type and traffic to ensure good performance and to design overlays.[55] Studies in Virginia, which correlated strength coefficients with Benkelman beam field measurements in typical pavement structures have led to a procedure that recognizes resilience as a design control.[56]

Extensive field observations of surface deflections under load were first made on the WASHO Test Road and intensive laboratory and field activities to appraise resilience have continued since that time. The most common measuring device prior to 1970 was the Benkelman beam, a long, slender bar which slips into the space between the dual tires on a truck wheel. It is instrumented to measure deflections of the pavement surface as the wheel slowly rolls past the reference point. Other schemes for applying stress to the pavement include static plate loads and impact or vibratory devices.[57] In all instances, the aim is to correlate deflection, rate of travel of a shock wave, or some other measure of strength with actual pavement performance.[58]

Field measurements of deflection or similar characteristics serve three purposes. First, as indicated above, they introduce an important additional variable into pavement design. Second, they provide a means for detecting road segments where failures can be expected so that remedial measures such as adding overlays to provide added strength may be considered. Finally, as indicated in

[54]See E. Zube and R. Forsyth, *HRB Record 189*.

[55]See R. W. Bushey et al., *TRB Record 572*.

[56]See N. K. Vaswani, *HRB Record 239*.

[57]For a description of several nondestructive procedures to measure deflection, including the Benkelman beam, see *AASHTO Designation T256* and *TRB Circular 189*.

[58]Recent publications dealing with deflections, resilience, the effects of frost on load-carrying capacity, and the design of overlays, include *NCHRP Reports 21, 35,* and *76, HRB Record 190, 291, 300, 327,* and *345* and *TRB Record 471, 497, 510, 572,* 666, and *700*.

the section on frost action, they provide a rational basis for controlling loads during the critical spring breakup period.

On heavily traveled highways surfaced with bituminous pavements, high densities in all layers must be obtained to prevent later formation of "ruts" in the wheel tracks. The usual testing procedures or specifications for construction do not explicitly measure this variable. They may well do so in the future as the results of research now under way are applied.[59]

MINERAL AGGREGATES

In 1978, mineral aggregate consumption in transportation facilities including base courses, pavements and structures, was over 1000 million tons. The most common materials are broken stone or slag, crushed or uncrushed gravel, and sand. Waste glass, lightweight fired clay, and other reclaimed materials as well as marginal low quality aggregate also have been employed.[60] Because of the large quantities involved, the cost of transporting mineral aggregates long distances is prohibitive, and they are usually obtained from relatively nearby sources. Over the country, commonly used aggregates differ widely in parent material, strength, toughness, surface roughness and chemical charges, porosity, resistance to polishing, and other characteristics; of necessity, criteria for acceptance and design and construction practices have been adjusted to meet local situations. In all cases, however, control over certain characteristics of the aggregates is imperative for satisfactory pavements; for this purpose specifications and tests have been developed. Even so, many unresolved problems involving aggregates still exist.

Mineral Aggregates for Soil Surfaces and Base Courses

Mineral aggregates are the sole constituent of soil surfaces and untreated base courses. For treated bases, the desirable properties of these native materials have been enhanced by adding bitumen, cement, lime, or salts. To keep their costs low, specifications for them are not as stringent as for aggregates for bituminous or portland-cement-concrete pavements. They usually combine controls on grain-size distribution and plasticity of the fines. Testing for these characteristics has been discussed earlier in this chapter. Specific requirements are presented in Chapter 16 for soil surfacings and Chapter 17 for bases.

[59]See for example, C. L. Monismith et al. and S. F. Brown, *TRB Record 537,* and K. Majidzadeh, *TRB Record 671. TRB Record 616* is an excellent overview of the wheel track rutting problem.
[60]See, e.g., *HRB Record 430* and *NCHRP Report 135.* For a report on waste glass see W. R. Malisch, et al., *HRB Record 307.* Laboratory and field studies of lightweight aggregate by B. M. Gallaway and W. J. Harper are described in *HRB Record 150* and *236,* by Gallaway and E. R. Hargett in *HRB Record 273* and by D. W. Bonifay et al., in *HRB Record 353. NCHRP Report 166* and *TRB Record 734* provide detailed discussions on use of waste materials as substitutes for aggregates. Properties of power plant aggregate (clinker and fuel ash) are given in *TRB Record 595.* Use of low quality aggregates is discussed in *NCHRP Report 207* and *TRB Record 741 and 762.*

Aggregates for Bituminous Pavements

GENERAL REQUIREMENTS. In bituminous pavements, aggregates constitute 88 to 96% by weight or something more than 75% by volume. Certain general requirements should be met by all mineral aggregates for pavements. Those given by AASHTO for Dense-Graded Bituminous Road and Plant-Mix Surface Course (*old AASHTO Designation M62*) are still representative.

> Aggregates shall be of uniform quality, crushed to size as necessary, and shall be composed of sound, tough, durable pebbles or fragments of rock or slag with or without sand or other inert finely divided mineral aggregate. All material shall be free from clay balls, vegetable matter, and other deleterious substances, and an excess of flat or elongated pieces. Slag shall be air-cooled blast-furnace slag of reasonably uniform density and quality. Excess of fine material shall be wasted before crushing.

TESTS FOR PARTICLE SIZE. For "dense-graded" pavements, particle sizes of the aggregate range from coarse to dust; for "open" pavements, one or more layers of coarse rock of uniform size are used; and, for sheet asphalt, the mineral aggregate is carefully graded sand and mineral dust. Although the particle sizes and size distribution vary between pavement types, control is necessary for every type and is obtained by screening the material through standard sieves. Particle sizes for the common pavement types will be discussed in later pages.

In the case of aggregates, inherent variances in the material, errors in sampling and testing, and segregation both in the samples and in handling aggregates in large quantities can mean that the sieve analysis may not be truly representative of the material actually incorporated. Techniques for minimizing such discrepancies have been developed.[61] Some of these are discussed under the topic of "quality assurance" in Chapter 15.

Grain size as measured by sieve analysis *(AASHTO Designation T27)* is only a partial gauge of the "size" parameters of mineral aggregate particles. Others include shape, angularity, and surface texture. It can be argued that these or other descriptors should be added to the specifications. One proposal is that "packing volume," the actual space that a particle preempts, including voids, is a suitable measure, at least for aggregates in a single size range.[62]

TEST FOR STRENGTH AND TOUGHNESS. As already mentioned in the general requirements, aggregates should be sound, tough and durable. The common criterion is the Los Angeles Rattler Test *(AASHTO Designation T96)*. It has largely supplanted the older Deval Test for Abrasion *(AASHTO Designations T3 and T4)*.

The rattler test is conducted in a hollow steel cylinder closed at both ends. The cylinder is 28 in. inside diameter, 20 in. long, and contains a steel shelf which projects radially inward 3½ in. It is mounted with its axis horizontal on stub shafts attached at the ends. A sample of coarse aggregate, all retained on

[61]See particularly *NCHRP Reports 34* and *69* for added detail.
[62]See E. Tons and W. H. Goetz, *HRB Record 236*.

the No. 8 (2.36 mm) sieve, is charged into the cylinder along with a prescribed number and size of steel spheres, and the cylinder then is rotated for 500 revolutions at a speed of 30–33 rpm. After testing, the sample is sieved on a No. 12 (1.70 mm) sieve, and that portion which passes through is discarded. The loss (percentage of wear) is the difference between the original and final weights of the test sample expressed as a percentage of the original weight.

FHWA recommendations (1979) are that coarse aggregates for dense-graded mixes have a coefficient of wear of 50 or less; for open-graded mixes or for coarse aggregates for bituminous concrete where requirements are more exacting or construction or service requirements are more severe, the requirement is 40 or less. There are situations where, because strong, tough aggregates are extremely expensive, an agency may accept higher percentages of wear; in cases where excellent aggregates are readily available, lower percentages may be set.

TESTS FOR SOUNDNESS. Soundness is the resistance of aggregates to deterioration from the effects of actions such as freezing and thawing. The traditional and more common test for soundness is with sodium or magnesium sulfate *(AASHTO Designation T104)*. This test subjects dried and sized samples of fine or coarse aggregate to immersion in a saturated solution of sodium or magnesium sulfate, followed by draining and oven drying. The liquid penetrates the interstices of the individual particles and, on drying, creates pressures which cause splitting, crumbling, cracking, or flaking of the surface. Usually, five immersion-drying cycles are run. After washing and drying, the sample is examined visually and also sieved again to determine the change in particle size. Results are reported as "percentage loss," which is measured as the percentage by weight which passes a sieve on which the particles were originally retained.

An alternative test for soundness involves freezing and thawing *(AASHTO Designation T103)*. This test has the same purpose as the sodium sulfate test and is conducted in a comparable way. Samples are first immersed in water, next frozen for 2 hr, then thawed for ½ hr. The number of cycles is not specified in the standard test, but ranges from 16 to 50. An alternative procedure first subjects the sample to a vacuum. Results are reported as the findings of visual examination and as "percentage loss" as defined previously.

Although most agencies generally employ the sodium or magnesium sulfate test, *AASHTO Designation T103* speaks favorably of the freezing and thawing test. It states that "the results of the method are considered more reliable for determining the quality of aggregates than those obtained by other methods of soundness testing on discrete particles of aggregates."

Some agencies do not specify soundness tests for aggregates for bituminous pavements. In general, they are located in areas where the temperature of the pavement seldom falls below freezing. However, *AASHTO Specification M29* for Fine Aggregates for Bituminous Paving Mixtures calls for a weight loss of less than 15% on the sodium sulfate test. Similarly, the now discontinued *AASHTO Specification M79* for Crushed Stone and Crushed Slag Surface Courses prescribed a weight loss of less than 12% with sodium sulfate.

Recently, the sodium or magnesium sulfate and freezing and thawing tests have been challenged as not duplicating natural conditions nor correlating with the field performance of aggregates in portland-cement concrete. Some of these criticisms probably would also apply to aggregates for bituminous mixtures. Extensive research to improve these approaches and to develop new ones has been carried out.[63]

DEGRADATION TESTS. Some aggregates degrade in the presence of water. The California Department of Transportation developed a test, now AASHTO T–210, to measure their degradation when subjected to mechanical agitation in water. For coarse aggregate, 2500g, all retained on the No. 4 (4.75 mm) sieve, is prepared to a standard gradation, washed, and then agitated for 10 min. A sedimentation test paralleling the sand-equivalent test for soils is performed on the passing No. 200 (0.075 mm) sieve material to obtain a measure of the types and amounts of fines generated. The resulting durability index (D_C) can range from 0 to 100, FHWA specifications set a minimum value at 35. For fine aggregate, all of which passes the No. 4 (4.75 mm) sieve, the test is similar, except that sample size is 500g. At the end of the 10-min period a sand-equivalent test is performed to establish a durability index, D_F. FHWA specifications normally set this index at 35.

Other agencies, including the Oregon and Washington Departments of Transportation, have developed similar degradation tests. Descriptions will be found in their materials testing manuals.

TESTS FOR AGGREGATE–BITUMEN AFFINITY AND FOR SWELL. If a pavement is to be strong and durable, the binder must adhere firmly to the aggregate particles. If the binder separates or "strips" from the aggregate, the pavement may disintegrate under traffic. Often, also, the pavement surface becomes pitted as aggregates are pulled loose by vehicle tires. Furthermore, if the paving mixture swells, interlock and internal friction are destroyed and stability is lost.

Aggregates with greater affinity for water than for asphalt often are described as *hydrophilic*, meaning that they love water as contrasted to *hydrophobic* which, interpreted literally, means "fear of water." If an aggregate is hydrophilic, that is, if the chemical bonds between aggregate and water are stronger than those between aggregate and asphalt, pinholes in the asphalt may develop through which water will reach the aggregate. Such aggregate–water interfaces also may occur at the sharp edges of crushed particles. In time, the water will break the aggregate–asphalt bond. On occasion, the stripped asphalt may flush to the surface making it slick. Also, after stripping, fines that are susceptible to swelling when wet will be in contact with water and will expand and disrupt the pavement structure.

There are antistripping agents that may be added to asphalt cements; also emulsions now are furnished in anionic and cationic forms. In addition, tests to control stripping are employed by all state transportation departments.

[63]See *HRB Special Report 80* and *NCHRP Reports 12, 15, 65* and 66 for detailed discussions and extensive bibliographies.

The most common test for stripping is given in *AASHTO Designation T182.* For it, either wet or dry aggregates in the ⅜–¼ in. size range are mixed with the selected binder and immersed in water for 16–18 hr. Stripping is evaluated by visually observing whether the surface area of aggregate remaining coated is greater than or less than 95%. The California Department of Transportation has a parallel test for aggregates for seal coats and open-graded mixes. With it, the coated sample, covered with water, is agitated for 15 min before observation.

The immersion-compression test (*AASHTO Designation T165*) also is widely used. It indirectly measures the tendency of aggregates to strip or swell under the effects of moisture. Duplicate sets of molded compression cylinders of bituminous-surfacing mixture are tested, one set dry and the other after immersion in water. Differences in strength serve to measure the effect.[64]

The stripping problem is of long standing, and many agencies know from experience which of the commonly used aggregates will or will not be affected. But existing and improved tests for stripping will remain important, since new aggregate sources and modified binders will most certainly come onto the scene.

TESTS FOR AGGREGATE SHAPE AND SURFACE TEXTURE. Rounded, relatively smooth aggregate particles are desirable in portland-cement concrete because mixes made with them are more workable. In contrast, angular and cubical shapes and rough surface texture have proved to be better in asphalt pavements because those with aggregates having such characteristics have shown greater stability. Thin or elongated pieces or dirt are, of course, always undesirable.

The particles of coarse and fine aggregate made by crushing suitable quarry stone or slag have angular shapes and rough surfaces. Commonly oversized material in gravel deposits also is crushed, so that a sizable proportion of the particles in such aggregates also has these qualities. None of the AASHTO specifications for aggregates for bituminous pavements make specific stipulations to control shape or surface texture. However, individual agencies sometimes do so. For example, for its Type A coarse aggregate for asphalt concrete, the California Department of Transportation requires 90% by weight of particles with at least one face produced by fracturing. For fine aggregate passing the No. 4 sieve, 70% by weight of the particles passing the No. 8 sieve must meet this requirement. Testing is by visually examining individual particles.

As indicated earlier, the end being sought by using crushed aggregate is to produce a stable pavement. This may also be achieved by testing bituminous mixtures for stability. This approach is discussed in Chapter 19.

[64]For a survey of practices in testing for stripping see *HRB Circular 67*. Discussions of the nature of adhesion and stripping and an extensive bibliography appear in *HRB Special Report 98*. See also A. T. DiBenedetto, *HRB Record 340*. Stripping problems in several western states are discussed by W. L. Eager, *HRB Record 51*. For a listing of earlier studies see *HRB Bibliographies 23, 25, and 29*. Recent studies on stripping or moisture damage are reported in *TRB Record 515* and *712* and in *NCHRP Report 192*.

TEST FOR RESISTANCE TO POLISHING. High coefficients of friction between tires and pavement are an important criterion in pavement design. With properly designed bituminous as well as portland-cement-concrete pavements, the rubber tire is in contact with the aggregate, not the binder. If initially or in time the aggregate surface becomes polished and slick, the coefficient of friction will become dangerously low.

Skid resistance, in which aggregate polishing is a major factor, has been a topic for research for many years (see Chapter 8). Currently almost all large highway agencies include skid resistance measurements on existing pavements as a part of their road inventory procedures. In addition, numerous highway test sections have been constructed to evaluate aggregate frictional reaction under traffic. Variables include aggregate class, size, shape, and mix ingredients and proportions. Also, a variety of laboratory devices have been developed to permit pretesting before actual construction is carried out.[65] Petrographic examination also provides a valuable means for identifying polishing or nonpolishing particles.[66]

It has been charged that aggregates produced from limestone are particularly susceptible to polishing. On the other hand, the claim is made that if the parent rocks are sandstones or from the fine-grained igneous types, polishing is not so severe; rather, tire wear pulls out individual grains and a pitted, rough surface remains. Actually, more careful analysis is required. For example, limestone coarse aggregates containing relatively large amounts of sands that are insoluble in diluted hydrochloric acid are very resistant to polishing.[67]

Most researchers assert that in dense-graded mixes, the polishing resistance of the coarse aggregate is of primary concern if high coefficients of friction are to be maintained. However, it has been demonstrated that friction factors can be substantially improved by incorporating silica sand into the mix.[68]

SPECIFICATIONS AND TESTS FOR MINERAL FILLER. Often it is desirable to add dust to dense-graded mixtures, since this can reduce void content, decrease permeability, and increase the tensile strength. Mineral filler is the term used to describe such additives. Common materials include finely powdered limestone or slag, hydrated lime, portland cement, trap-rock dust, and flyash. Some agencies accept diatomaceous earth; others exclude it.

AASHTO Specification M-17 stipulates percentages passing by weight as follows: No. 30 sieve (0.600 mm), 100; No. 50 sieve (0.300 mm), 95–100; No.

[65] *Recent publications, with extensive bibliographies, include HRB Record 120, 236* and *341,* and *HRB Special Report 101.* W. A. Goodwin, *HRB Special Report 101,* describes and evaluates several laboratory devices. California's laboratory procedures for rating screenings for seal coats are described by E. Zube and J. Skog in *HRB Record 236.* See also S.H. Dahir et al., in *TRB Record 584* and several articles in *TRB Record 788.*

[66] See J. E. Gray and F. A. Renninger, *HRB Record 120* and S. H. Dahir and W. E. Meyer, *TRB Record 602.*

[67] For reference, see the two previous footnotes and W. C. Sherwood and D. C. Mahone, *HRB Record 341,* and the discussion by F. P. Nichols, Jr. and G. Balmer in its addendum.

[68] See, e.g., *HRB Bulletin 219,* and W. C. Burnett et al., *HRB Record 236.*

200 (0.075 mm), 70–100. It also stipulates that for all materials other than hydrated lime or portland cement the PI be 4 or less.[69]

Aggregates for Portland Cement Concrete Pavements

GENERAL REQUIREMENTS. Mineral aggregates form about 75% of the volume or roughly 80% of the weight of normal paving concrete. It follows that, if a paving is to be strong, sound, and durable, the aggregates must have like properties. This is of particular importance in areas where winter brings freezing temperature, snow, and sleet. Under these conditions the concrete is subject to freezing and thawing, the pounding and wear from tire chains and studded tires (if permitted), as well as the action of calcium chloride, salt, or other deicing or antiskid agents. To make this high quality certain, agencies have almost without exception prescribed that aggregates pass appropriate tests for strength, soundness, wear, or combinations of these three.[70]

SPECIFICATIONS FOR FINE AGGREGATES. The specifications for fine aggregates of most state highway agencies follow fairly closely those proposed by AASHTO *(AASHTO Designation M6)*. A general requirement is that fine aggregates shall consist of natural sand or, subject to approval, other inert materials with similar characteristics, or combinations thereof, having strong, hard, durable particles. It is usual to reinforce this requirement by placing maximum limits on the amounts of deleterious substances. Recommended and maximum permissible limits proposed by AASHTO appear in Table 14–3.

TABLE 14–3. Maxium Amounts of Deleterious Materials Permitted in Fine Aggregates for Concrete

Substance	*Recommended Permissible Limits (% by wt)*	*Maximum Permissible Limits (% by wt)*
Friable particles	0.5	1.0
Coal and lignite	0.25	1.0
Materials passing No. 200 sieve		
In concrete subject to surface abrasion	2	4
In all other classes of concrete	3	5
Other deleterious substances such as shale, alkali, mica, coated grains, soft and flaky particles		To be specified*

*The Illinois Department of Transportation sets a maximum on all these at 3 to 5%, depending on aggregate class.

[69]For added discussion of mineral fillers and their effects see L. H. Csanyi et al., *HRB Record 51*, E. Zube and J. Cechetini, *HRB Record 104*, R. H. Gietz and D. R. Lamb, *HRB Record 256*, and F. N. Finn, *NCHRP Report 39*. For a discussion of infrared spectroscopy as a means for evaluating fillers, see B. Chaiken, et. al., *Public Roads*, June 1962.

[70]Aggregates for bituminous pavements and the tests applied to them were discussed previously. In general, the same selection and control procedures apply to aggregates for concrete, except that some requirements for concrete may be more exacting.

As has been indicated earlier, the polishing of certain fine aggregates has been a contributing factor to low coefficients of friction. With this in mind, some agencies limit the percentage of manufactured sands that can be used. Little uniformity in these requirements appears among the different road-building agencies.

The presence of organic impurities may prevent hardening of cement. The standard test *(AASHTO Designation T21)* requires that a sample of sand be mixed with sodium hydroxide solution which turns dark when organic material is present. Sands failing this test are considered to be satisfactory if 2 in. cubes made with cement and this sand develop 95% of the strength of cubes made with the same sand after it had been washed in a 3% solution hydroxide solution (see *AASHTO Designation T71*).

Fine-aggregate grading, by weight, as recommended by AASHTO is as follows:

Percent passing ⅜ in. sieve	100
Percent passing No. 4 sieve	95–100
Percent passing No. 16 sieve	45–80
Percent passing No. 50 sieve	10–30
Percent passing No. 100 sieve	2–10

Gradings generally like the above are specified by the various highway departments. Some permit a wider variation in the percentages; others have closed these limits somewhat and, in some cases, have interposed additional sieves into the group. Many engineers believe that a "percent passing" specification alone does not assure uniformity of the fine aggregates, since sands ranging from coarse to fine all will fall within the usual grading limits. To prevent these fluctuations the AASHTO specifications and those of many of the states also require that the "fineness modulus"[71] of the sand from a given source remain relatively constant. The usual permitted variation is 0.2 either way from the fineness modulus of an original representative sample.

The strength of fine aggregates is measured by compression tests of sand–cement mortars. AASHTO specifications do not prescribe this control, but many agencies do.

Soundness of fine aggregates is measured by their resistance to deterioration under the action of solutions of sodium or magnesium sulfate or under freezing and thawing. Specifications of AASHTO and of a number of highway departments favor a five-cycle sodium sulfate test. Several others substitute magnesium sulfate as the agent or permit the use of either. The maximum loss under the AASHTO specifications is 10%.

SPECIFICATIONS FOR COARSE AGGREGATES. The various state specifications for coarse aggregates follow the pattern proposed by AASHTO *(AASHTO Designation M80)*. The general requirement is that coarse aggregates consist of

[71]To determine the fineness modulus, add the total percentages by weight retained on the 3, 1½, ¾ and ⅜ in., Nos. 4, 8, 16, 30, 50, and 100 sieves, and divide this sum by 100.

crushed stone, gravel, blast-furnace slag, or other approved inert materials of similar characteristics, or combinations thereof, having hard, strong, durable pieces free from adherent coatings. Limits to the amounts of deleterious materials, based on AASHTO recommendations, appear in Table 14–4. Few of the individual specifications conform exactly to the limits shown. However, they do outline the pattern of current practice. Some states are forced to accept higher percentages of certain objectionable materials because they are present in all local deposits and it is not economical to import aggregates from distant sources.

TABLE 14–4. Maximum Amounts of Deleterious Materials Permitted in Coarse Aggregates for Concrete *(AASHTO M-80)*

	Maximum Allowable (% by wt.)	
	Pavements	*Bridge Decks*
Chert (< 2.40 specific gravity)	3.0	3.0
Coal and lignite	0.5	0.5
Clay lumps	3.0	2.0
Material passing No. 200 sieve	1.0	1.0
Sodium sulfate soundness	12*	12*

*18% if magnesium sulfate is used.

Coarse-aggregate gradings for concrete, as recommended by AASHTO, are shown in Table 14–5. Size numbers 467 (1½ in. to No. 4) and 357 (2 in. to No. 4) are representative gradings commonly selected for paving concrete. Gradings often are stated separately for two size ranges, coarse and fine. Aggregates of size No. 3 (2–1 in.) and No. 57 (1 in. to No. 4) would be combined to produce the desired grading of 2 in. to No. 4 aggregate. Size Nos. 4 and 67

TABLE 14–5. Selected Grading of Coarse Aggregates for Concrete*

Range in Size	*Size† Number*	*Percentage by Weight Passing Laboratory Sieves Having Square Openings*							
		2½	*2*	*1½*	*1*	*¾*	*½*	*⅜*	*No. 4*
½ in. to No. 4	7	—	—	—	—	100	90–100	40–70	0–15‡
¾ in. to No. 4	67	—	—	—	100	90–100	—	20–55	0–10‡
1 in. to No. 4	57	—	—	100	95–100	—	25–60	—	0–10‡
1½ in. to No. 4	467	—	100	95–100	—	35–70	—	10–30	0–5
2 in. to No. 4	357	100	95–100	—	35–70	—	10–30	—	0–5
1½–¾ in.	4	—	100	90–100	20–55	0–15	—	0–5	—
2–1 in.	3	100	90–100	35–70	0–15	—	0–5	—	—

*From *AASHTO Designation M80*.
†Based on Standard Sizes of Coarse Aggregate for Highway Construction *(AASHTO Designation M43)*.
‡Not more than 5% shall pass a No. 8 sieve.

likewise can be combined, with the separation between them on the ¾-in sieve. The trend is toward separating coarse aggregates into two sizes, and a majority of the states now follow that practice.

A test to measure the abrasion resistance of coarse aggregates for pavements is required by every state. The Los Angeles Rattler Test, prescribed by AASHTO specifications, is most common and represents the procedure of a large majority. AASHTO recommends a maximum percentage of wear of 40, which is typical of current practice.

Tests for soundness of coarse aggregates for concrete pavements are required by most of the states. Sodium sulfate is the most common test medium. Some use magnesium sulfate as the agent or permit it as a substitute for sodium sulfate. AASHTO specifications, which allow either agent, limit the loss after five test cycles to 12% when sodium sulfate is used. Several states substitute the freezing and thawing test *(AASHTO Designation T103)* for the sulfate test, or use it as a supplement. A typical limit for the freezing and thawing test is 15% loss after 50 cycles.

Extensive research has been conducted to develop tests for frost susceptibility of concrete aggregates. One approach is through petrography, the study of the structure, composition, texture, and other characteristics of the individual particles. The other is to test concrete made with suspected aggregates.[72]

ALKALI–AGGREGATE REACTION. Alkali–aggregate reactions in concrete sometimes result when cements containing alkali are combined with (a) aggregates containing either certain silicate or silica minerals, or (b) aggregates produced from certain carbonate rocks.

Alkali–silica reaction was first reported by Stanton, after an investigation of the failure of a year-old section of concrete pavement near Salinas, California in 1938. Since then deterioration of concrete from this cause has been found in numerous localities, including several western and plain states and certain areas in the South. With alkali–silica reaction, a gel slowly forms around the affected aggregate particles, and the concrete expands and loses strength. In concrete pavements, surface cracking and blowups result. In dams and similar structures, misalignment of gates and machinery is one of several serious consequences. As a practical means of preventing it, agencies using doubtful aggregates specify that all cement be of low-alkali content. The usual requirement is that the total sodium and potassium oxide (expressed as Na_2O) not exceed 0.6%.

Alkali-carbonate reactions can cause map cracking of concrete, sometimes in a few months and again after periods of 20 yr or more. If suspected aggregates are used, cements of low-alkali content should be specified.[73]

[72]See e.g., the articles by R. D. Walker and L. Dolar-Mantuani in *HRB Record 120,* several papers in *HRB Record 210,* J. W. Harman, Jr. et al., *HRB Record 328,* and *NCHRP Reports 15, 65,* and 66.
[73]A detailed discussion of alkali-aggregate reaction is beyond the scope of this book. However, the literature is abundant. One test *(ASTM Designation C227)* involves measuring the length change of mortar bars over a period of several months. Quicker tests for alkali–silicate and alkali–carbonate reaction are *ASTM Designations C289* and *C586,* respectively. An evaluation of testing procedures for alkali–silicate reaction by B. Chaiken and W. J. Halstead appears in *Public Roads,* June 1959, or *HRB Bulletin 239*. See also C. W. DePuy, *HRB Record 124,* and L. Dolar-Mantuani, *HRB Record 268*. Several valuable articles on alkali–carbonate reaction appear in *HRB Record 45*.

BITUMINOUS MATERIALS

Bituminous materials for road-building purposes are viscous liquids. Consistencies at normal temperatures range from something slightly thicker than water to hard and brittle materials that, when cold, will shatter under a hammer blow; but even the hardest of them will flow if subjected to long, continuous loading.

Regardless of the type of pavement in which they are used, bituminous binders must be in liquid form when combined with the aggregates. This fluid state may be produced either by using a liquid material or by making harder asphalts liquid by heating, by dissolving in solvents, or by emulsifying in water.

In completed pavements, the action of the binders depends greatly on the aggregates with which they are combined. If pavements are of the "open" type, consisting entirely of coarse particles held together by bituminous materials, these binders in themselves must resist the abrasive forces produced by vehicular traffic. The cohesive strength required to perform this function is gained by using a tenacious, heavy binder. On the other hand, if the aggregates contain fine particles, cohesion is developed by surface tension in the thin bituminous films surrounding these fines, just as water films develop cohesive forces in fine-grained soils. For mixes containing fines, then, less viscous bituminous liquids may be used successfully as binders.

Sources of Bituminous Materials

All bituminous materials are hydrocarbons, which are combinations of hydrogen and carbon. The lighter and more volatile members of this hydrocarbon family include natural or manufactured gas, gasoline, kerosene, and diesel oil. Heavier combinations provide lubricating oils and paving materials. Some of the hydrocarbons used in paving occur naturally, but most of them are by-products from the manufacture of gas, liquid fuels and lubricants, or coal gas and coke. Principal sources of bituminous materials for pavements are listed in the paragraphs that follow.

NATIVE ASPHALTS. Native asphalts came from among other sources, Trinidad Island and Bermudez off the north coast of Venezuela. Softened with viscous petroleum fluxes, they were once extensively used as binders for asphaltic surfaces. Trinidad Lake asphalt when ready for fluxing contained approximately 40% organic and inorganic insoluble matter, whereas that from Bermudez had about 6% insoluble matter. With the development of petroleum asphalts, native asphalts have become relatively unimportant. The *AASHTO Standard Specifications for Transportation Materials* no longer give requirements for them; however in Great Britain they are often combined with petroleum asphalts.

ROCK ASPHALTS. Rock asphalts are natural deposits of limestone or sandstone impregnated with bituminous material. They occur in various parts of the United States, especially in Alabama, Kentucky, Oklahoma, Texas, Utah, and California. They have generally made extremely durable and stable road surfaces, but high transportation costs have limited applications to the general areas of occurrence. The percentage of bitumen varies widely between deposits, 4.5 and 18% representing the extremes. Often the rock asphalt as mined or quarried must be processed by adding mineral aggregate, asphaltic binder, and a fluxing oil.[74]

PETROLEUM ASPHALTIC MATERIALS. Petroleum asphalts were first used in the United States for road treatment in 1894, when crude petroleum from the Summerland wells was sprinkled on earth roads in Santa Barbara County, California. Production of paving materials from California and Mexican crude petroleums followed. Asphaltic-paving materials now come from domestic crudes originating in Kentucky, Ohio, Michigan, Illinois, Mid-Continent, Gulf-Coastal, Rocky Mountain, California and Alaska fields. Foreign sources include Mexico, Venezuela, Colombia, and the middle east. Some 32 million tons were used in 1980.

A flow chart for a typical refinery producing fuels, lubricating oils, and paving materials is diagramed in Fig. 14–23. It is to be observed that motor fuels and lubricating oils are distilled off and only the remaining heavier hydrocarbons are processed into paving materials. The exceptions are the so-called "liquid" or "cutback" asphalts. For them, the asphalt cements are liquified to different degrees by cutting them back with one of the lighter fractions removed earlier in the production sequence. Some notion of the common uses of the various forms of asphaltic binders is given in Table 14–6. Brief descriptions, specifications, tests, and common problems with the various bituminous binders are outlined in the sections which follow.

Descriptions of and Specifications for Bituminous Binders

ASPHALT CEMENTS (AC). These are used as binders for almost all high-type bituminous pavements. They are semisolid hydrocarbons remaining after lubricating oils as well as fuel oils have been removed from petroleum (see Fig. 14–22). Prior to 1970, the consistencies of asphalt cements were normally given in terms of penetration (AASHTO T49), the distance that a standard needle penetrates a sample under known conditions of loading, time, and temperature. The softest grade commonly used for paving is 200–300 penetration, the hardest is 60–70 penetration. All asphalt cements are so viscous that both aggregate and binder must be heated before mixing and placing of pavements can be accomplished.

The most common procedure for grading asphalt cements is based on viscosity rather than penetration and appears as *AASHTO Designation M226* in the 1978 *AASHTO Specifications for Transportation Materials*. Three viscosity-based specifications are provided for: Two of these grade on the original asphalt (AC gradings); the other employs gradings measured on the residue from the Rolling Thin Film Oven Test *(AASHTO Designation T240)* which was adopted in 1974 by seven West-Coast States (AR gradings). These specifications reflect the

[74]Use of shale oil is discussed in *TRB Record 549* and *777*.

TABLE 14–6. Typical Uses of Asphalt (after Asphalt Institute)

Type of Construction	Asphalt Cements														
	Viscosity Graded —Original					*Viscosity Graded —Residue*					*Penetration Graded*				
	AC-40	AC-20	AC-10	AC-5	AC-2.5	AR-16000	AR-8000	AR-4000	AR-2000	AR-1000	40–50	60–70	85–100	120–150	200–300
ASPHALT–AGGREGATE MIXTURES															
ASPHALT CONCRETE AND HOT LAID PLANT MIX															
Pavement base and surfaces															
Highways		x	x	x	x[1]	x	x	x	x[A]	x[1]		x	x	x	x[1]
Airports		x	x	x			x	x				x	x		
Parking areas		x	x			x	x	x				x	x		
Driveways		x	x				x	x				x	x		
Curbs	x	x	x			x	x				x	x	x		
Industrial floors	x	x				x	x				x	x			
Blocks	x										x				
Groins	x	x									x	x			
Dam facings	x	x									x	x			
Canal and reservoir linings	x	x									x	x			
COLD-LAID PLANT MIX															
Pavement base and surfaces															
Open-graded aggregate															
Well-graded aggregate															
Patching, immediate use															
Patching, stockpile															
MIXED-IN-PLACE (ROAD MIX)															
Pavement base and surfaces															
Open-graded aggregate															
Well-graded aggregate															
Sand				x	x									x	x
Sandy soil				x	x									x	x
Patching, immediate use															
Patching, stockpile															

Type of Construction	Liquid Asphalts																									
	Rapid Curing (RC)				*Medium Curing (MC)*					*Slow Curing (SC)*				*Emulsified*												
														Anionic							*Cationic*					
	70	250	800	3000	30	70	250	800	3000	70	250	800	3000	RS-1	RS-2	MS-1	MS-2	MS-2h	SS-1	SS-1h	CRS-1	CRS-2	CMS-2	CMS-2h	CSS-1	CSS-1h
ASPHALT–AGGREGATE MIXTURES																										
ASPHALT CONCRETE AND HOT LAID PLANT MIX																										
Pavement base and surfaces																										
Highways																										
Airports																										
Parking areas																										
Driveways																										
Curbs																										
Industrial floors																										
Blocks																										
Groins																										
Dam facings																										
Canal and reservoir linings																										
COLD-LAID PLANT MIX																										
Pavement base and surfaces																										
Open-graded aggregate								x								x	x	x					x	x		
Well-graded aggregate		x					x	x	x		x	x	x						x	x			x		x	x
Patching, immediate use		x	x				x	x	x			x	x						x	x					x	x
Patching, stockpile							x	x			x	x														
MIXED-IN-PLACE (ROAD MIX)																										
Pavement base and surfaces																										
Open-graded aggregate		x	x	x				x	x		x	x				x	x	x					x	x		
Well-graded aggregate		x					x	x			x	x							x	x			x		x	x
Sand		x	x			x	x	x											x	x					x	x
Sandy soil	x	x	x				x	x											x	x			x		x	x
Patching, immediate use		x	x				x	x	x			x	x						x	x					x	x
Patching, stockpile							x	x			x	x														

Application																																								
ASPHALT-AGGREGATE APPLICATIONS																																								
SURFACE TREATMENTS																																								
Single surface treatment				X	X									X	X		X	X	X				X	X					X	X						X	X				
Multiple surface treatment				X	X									X	X		X	X	X					X					X	X						X	X				
Aggregate seal				X	X				X					X	X		X	X	X				X	X					X	X						X	X				
Sand seal																	X					X	X						X	X						X	X				
Slurry seal																																		X	X					X	X
PENETRATION MACADAM																																								
Pavement bases																																								
Large voids			X										X					X	X										X	X						X	X				
Small voids				X										X			X												X	X						X	X				
ASPHALT APPLICATIONS																																								
SURFACE TREATMENT																																								
Fog seal																																		X^2	X^2					X^2	X^2
Prime coat, open surfaces																X	X				X	X																		
Prime coat, tight surfaces																X				X	X													X						
Tack coat																X													X					X^2	X^2	X				X^2	X^2
Dust laying																					X													X^2	X^2					X^2	X^2
Mulch																																		X^2	X^2						
MEMBRANE																																								
Canal and reservoir linings	X										X																														
Embankment envelopes	X	X									X	X																													
CRACK FILLING																																								
Asphalt pavements																																		X^3	X^3					X^3	X^3
Portland-cement-concrete pavements	X^4										X^4																														

[1]For use in cold climates.
[A]For use in bases only in cold climates.
[2]Diluted with water.
[3]Slurry mix.
[4]Rubber asphalt compounds.

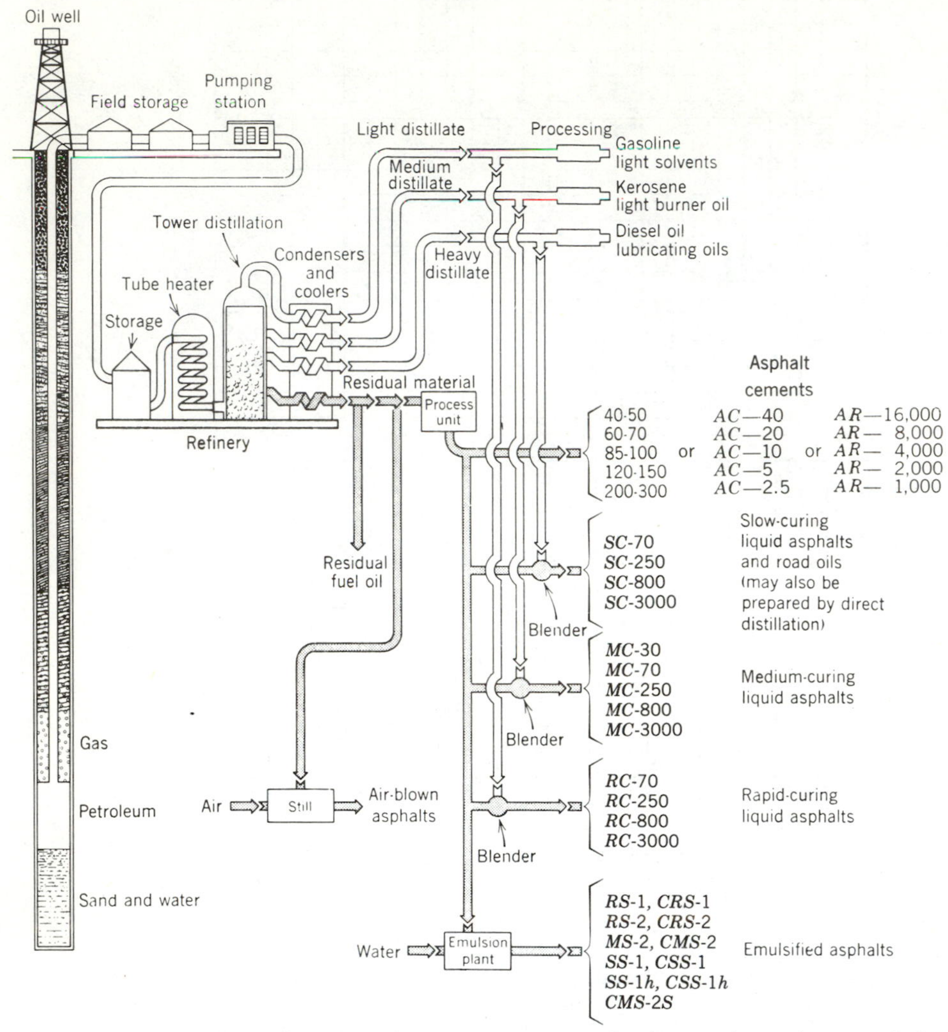

Fig. 14–23. Simplified flowchart showing recovery and refining of petroleum asphalts. (Asphalt Institute chart, somewhat modified.)

fact that there is no single "right" way to control the properties of all paving asphalts, since they come from a variety of sources and are processed in different ways. Table 14–7 shows an example of these specifications. Working temperatures for these and other bituminous materials are given in Table 14–8.

TABLE 14–7. Specifications for Asphalt Cement Graded by Viscosity at 60°C (140°F) *(AASHTO Designation 228)* (Grading based on original asphalt).

Test	Viscosity Grade				
	AC-2.5	*AC-5*	*AC-10*	*AC-20*	*AC-40*
Viscosity, 60°C (140°F), poises	250 ± 50	500 ± 100	1000 ± 200	2000 ± 400	4000 ± 800
Viscosity, 135°C (275°F), Cs-minimum	80	110	150	210	300
Penetration, 25°C (77°F), 100 g, 5 s-minimum	200	120	70	40	20
Flash Point, COC,°C (°F)-minimum	163(325)	177(350)	219(425)	232(450)	232(450)
Solubility in trichloroethylene, percent-minimum	99.0	99.0	99.0	99.0	99.0
Tests on residue from Thin-Film Oven Test:					
Viscosity, 60°C (140°F), poises-maximum	1000	2000	4000	8000	16000
Ductility, 25°C (77°F), 5 cm/min cm-minimum	100*	100	50	20	10
Spot test (when and as specified)† with:					
Standard naphtha solvent	Negative for all grades				
Naphtha-xylene-solvent, % xylene	Negative for all grades				
Heptane-xylene-solvent, % xylene	Negative for all grades				

*If ductility is less than 100, material will be accepted if ductility at 15.6°C (60°F) is 100 minimum.

†The use of the spot test is optional. When it is specified, the engineer shall indicate whether the standard naphtha solvent, the naphtha-xylene solvent, or the heptane-xylene solvent will be used in determining compliance with the requirement, and also, in the case of xylene solvents, the percentage of xylene to be used.

TABLE 14–8. Suggested Spraying and Mixing Temperatures (°C) for Bituminous Materials and Road Tars*†

Type and Grade of Asphalt	Pugmill Mixture Temperatures‡		Spraying Temperatures**	
	Dense-Graded Mixes	Open-Graded Mixes	Road Mixes	Surface Treatments
Asphalt cements				
AC-2.5	115-140	80-120	—	130+
AC-5	120-145	80-120	—	140+
AC-10	120-155	80-120	—	140+
AC-20	130-165	80-120	—	145+
AC-40	130-170	80-120	—	150+
AR-1000	105-135	80-120	—	135+
AR-2000	135-165	80-120	—	140+
AR-4000	135-165	80-120	—	145+
AR-8000	135-165	80-120	—	145+
AR-16000	150-175	80-120	—	—
200-300 pen.	115-150	80-120	—	130+
120-150 pen.	120-155	80-120	—	130+
85-100 pen.	120-165	80-120	—	140+
60-70 pen.	130-170	80-120	—	145+
40-50 pen.	130-175	80-120	—	150+
Cutback asphalts *(RC, MC, SC)*§				
30 (*MC* only)	—	—	—	30+
70	—	—	20+	50+
250	55-80‖	—	40+	75+
800	75-100‖	—	55+	95+
3000	80-115‖	—	—	110+
Emulsified Asphalts				
RS-1, *CRS*-1	—		—	20-60
RS-2, *CRS*-2	—		—	50-80
MS-1, *MS*-2, *CMS*-2	10-70¶		20-70	—
MS-2h, *CMS*-2h	10-70¶		20-70	—
SS-1, *CSS*-1	10-70¶		20-70	—
SS-1h, *CSS*-1h	10-70¶		20-70	—
Road Tars				
RT 1-2-3	—		15-55	15-55
RT 5-6	—		27-65	27-65
RT 7-8-9	65-105		65-105	65-105
RT 10-11-12	80-120		80-120	80-120

*Sources: Asphalt Institute for Asphaltic Materials; *AASHTO Guide Specifications* for Tars.
†°F = 9/5°C + 32.
‡Temperature of mixture immediately after discharge from the pugmill rather than temperature of asphalt cement or cutback asphalt.
§Application temperature may, in some cases, be above the flash point of the material. Caution must therefore be exercised to prevent fire or an explosion.
‖Rapid-curing *(RC)* grades are not recommended for hot pugmill mixing.
¶Temperature of the emulsified asphalt in the pugmill mixture.
**The maximum temperature (asphalt cement and cutback asphalt) shall be below that at which fogging occurs.

LIQUID ASPHALTS (OR CUTBACKS). These are liquid petroleum products that consist of asphalt cement cut back with a liquid distillate (gasoline, kerosene, or diesel). The amount of cutback determines the viscosity. The less viscous products contain up to 40% diluent, the more viscous as little as 15%. The type of cutback determines the rate of setting. In recent years, the use of cutbacks have been discouraged for two reasons. First, the cutback is usable fuel; and second, air-pollutants are given off. With the restrictions the Department of Transportation and the Environmental Protection Agency as well as state environmental agencies are placing on the use of these materials, it is anticipated that their availability will be limited in the near future.

Slow-Curing (SC) Road Soils. These products harden or set very slowly. They can be produced directly through distillation or can be manufactured by "cutting back" asphalt cement with a heavy distillate (diesel). They are more fluid than asphalt cement but are more viscous than the very light grades of lubricating oils. The lightest grade *SC*-70[75] has the consistency of light syrup at room temperatures. Under similar conditions, *SC*-250 resembles a light and *SC*-800 a heavy molasses, and *SC*-3000 will barely deform. In turn, *SC*-3000 is somewhat less viscous than a 200–300 penetration asphalt cement, the softest in its group. Specifications for the various grades of *SC* oils no longer appear in *AASHTO Specifications for Transportation Materials*.

Medium Curing (MC) Cutback Asphalts. These are asphalt cements fluxed or cut back with distillates in the the boiling range of kerosene (325–535°F). Thus stiffening is faster than for the slow curing liquid asphalts. Specifications for medium cure cutbacks appear in Table 14–9. Consistencies and application temperatures (Table 14–8) of the individual grades of *MC* asphalts parallel those for the *SC* series. However, one grade has been added, the *MC*-30. This is very fluid.

Fluidity of the various *MC* grades is controlled by the amount of solvent. *MC*-70 may contain as much as 45% diluent by volume and *MC*-3000 as little as 20%. Consistency of the binder after the solvent evaporates depends on the asphalt cement originally chosen, which is generally of 120–150 penetration.

Rapid-Curing (RC) Cutback Asphalts. These are asphalt cements cut back with a petroleum distillate such as gasoline or naphtha. The boiling range of these is 250–400°F; thus they evaporate rapidly. Rapid-curing products are used when a quick change from the liquid state of application back to the original asphalt cement is desired. Specifications for rapid-curing cutbacks are given in *AASHTO M81*.

Consistencies, application temperatures, percentages of solvent, and penetration of the original asphalt cement closely follow those for comparable designations of the *MC* series. Penetration of the residue after distillation approximates that of the original asphalt.

[75]70, 250, 800, 3000 indicate the viscosity in centistokes at 140°F (60°C).

TABLE 14–9. Specifications for Medium Curing Liquid Asphalt *(AASHTO Designation M82)*

	MC-30		*MC-70*		*MC-250*		*MC-800*		*MC-3000*	
	min.	*max.*	*min.*	*max.*	*min.*	*max.*	*min.*	*max.*	*min.*	*max.*
Kinematic Viscosity at 60°C (140°F) (See Note 1) centistokes	30	60	70	140	250	500	800	1600	3000	6000
Flash point (Tab. open-cup), °C (°F)	38 (100)	—	38 (100)	—	66 (150)	—	66 (150)	—	66 (150)	—
Water Percent	—	0.2	—	0.2	—	0.2	—	0.2	—	0.2
Distillation test:										
Distillate percentage by volume of total distillate to 360°C (680°F)										
to 225°C (437°F)	—	25	0	20	0	10	—	—	—	—
to 260°C (500°F)	40	70	20	60	15	55	0	35	0	15
to 315°C (600°F)	75	93	65	90	60	87	45	80	15	75
Residue from distillation to 360°C (680°F). Volume percentage of sample by difference	50	—	55	—	67	—	75	—	80	—
Tests on residue from distillation:										
Absolute viscosity at 60°C (140°F) (See Note 4) poises	300	1200	300	1200	300	1200	300	1200	300	1200
Ductility, 5 cm/cm, cm (See Note 2)	100	—	100	—	100	—	100	—	100	—
Solubility in trichloroethylene, percent	99.0	—	99.0	—	99.0	—	99.0	—	99.0	—
Spot test (See Note 3) with:										
Standard naphtha	Negative for all grades									
Naphtha-xylene solvent, -percent xylene	Negative for all grades									
Heptane-xylene solvent, -percent xylene	Negative for all grades									

Note 1. As an alternate, Saybolt Furol viscosities may be specified as follows:

Grade *MC-70*—Furol viscosity at 50°C (122°F)—60 to 120 s
Grade *MC-30*—Furol viscosity at 25°C (77°F)—75 to 150 s
Grade *MC-250*—Furol viscosity at 60°C (140°F)—125 to 250 s
Grade *MC-800*—Furol viscosity at 82.2°C (180°F)—100 to 200 s
Grade *MC-3000*—Furol viscosity at 82.2°C (180°F)—300 to 600 s

Note 2. If the ductility at 25°C (77°F) is less than 100, the material will be acceptable if its ductility at 15.5°C (60°F) is more than 100.

Note 3. The use of the spot test is optional. When specified, the engineer shall indicate whether the standard naphtha solvent, the naphtha xylene solvent, or the heptane xylene solvent will be used in determining compliance with the requirement, and also, in the case of the xylene solvents, the percentage of xylene to be used.

Note 4. In lieu of viscosity of the residue, the specifying agency, at its option, can specify penetration 100 g, 5 s at 25°C (77°F) of 120 to 250 for grades *MC-30*, *MC-70*, *MC-250*, *MC-800*, and *MC-3000*. However, in no case will both be required.

EMULSIFIED ASPHALTS. These are mixtures in which minute globules of asphalt are dispersed in water or in an aqueous solution by means of an emulsifier. Asphalt content is in the range of 55–70% by weight. Because these globules have like electrical charges, they do not coalesce until the emulsion breaks or the water evaporates.

Emulsions offer the asphalt in liquid form for application or mixing at normal temperatures. When the water evaporates, the paving asphalt remains. Products with fast *(RS)*, medium *(MS)*, and slow *(SS)* breaking times make emulsions suitable for a variety of purposes. The base stock is usually asphalt cement in the 100–200 penetration range; however some grades have a harder base (40–90 penetration). In many situations either emulsions or cutbacks will be satisfactory. Emulsions are superior with wet aggregates as the water medium carries the asphalt into intimate contact with the particle surfaces. Emulsions are now recommended as alternates to cutbacks for energy and environmental reasons.

Emulsifying agents include the soap of fatty and resinous acids, tallow derivatives, glue, and gelatin. Some for the cationic group (see below) also contain oil distillates. Emulsions are formed by forcing asphalt, water, and emulsifying agents together by passing them through a colloid mill or by violent agitation. Some of the processes are patented and the products may appear under trade names.

Before 1957, emulsions were all *anionic;* that is, the globules carry a negative charge. These emulsions are very effective in coating electropositive aggregates such as limestones. On the other hand, they tend to "strip" from aggregates high in silica which have strong electronegative surface charges. There are now *cationic* emulsions, in which the emulsified particles carry a positive charge. These are very effective with highly siliceous aggregates, but may strip from highly alkaline ones that carry strong positive surface charges. It is claimed that, over all, cationic emulsions fit a wider range of aggregates. In addition, they show better adherence to wet aggregates in that they positively displace water on the surface of the aggregates, break quickly on contact, and are less affected by humidity and lower air temperatures. Thus, with them, construction can be carried out over a longer working season and there will be less hazard from rain.[76]

When frozen, ordinary emulsions will break down so that they must be protected if extreme cold is expected. Special emulsions which are not damaged by freezing are available. There are also "inverted" emulsions; in these the asphalt is the continuous phase with minute globules of water suspended in it. Example specifications for cationic emulsions are given in Table 14–10. Similar specifications are also available for anionic emulsions (see *AASHTO Specification M140).*

[76]*AASHTO Designations M140* and *M208* cover anionic and cationic emulsions. See M. J. Borgfeldt and R. L. Ferm, *HRB Proceedings 1962,* and A. O. Bohn, *HRB Record 67,* for explanation of the surface chemistry of emulsions and the breaking reaction. Also see *NCHRP Synthesis 30* and the *Asphalt Institute Publication MS-19.*

REJUVENATING OR RECYCLING AGENTS. Under weathering and aging, the resins in some asphalts change to asphaltenes; this causes the binder to harden and crack. Emulsified petroleum resins, sprayed over the surface, penetrate into the pavement and soften and "rejuvenate" the binder. Care must be exercised in applying the resin, since too much will produce a slick pavement surface. Similar products are used as recycling agents.[77]

BLOWN OR OXIDIZED ASPHALTS. These result when air is blown through heated asphaltic materials. They have higher softening points than normally refined asphalts of comparable penetration, which makes them suitable for roofing and similar applications. Highway uses are limited largely to the waterproofing of structures and filling joints in concrete pavements.

ROAD TARS. Tars are a by-product of the destructive distillation of coal. In 1867 a tar-aggregate mixture was laid in Prospect Park, Brooklyn. Soon thereafter similar installations were made in other locations. However, the early mixtures required a month or more to harden, and in addition there were numerous failures. With modification of production methods, adequate specifications, and more efficient testing, these difficulties were overcome, and at present tars are employed in the East, Middle West, and South.

Tars are produced by the gashouse, coke-oven, or water–gas methods. The American Society for Testing Materials makes the following designations:

Gashouse coal tar. Coal tar produced in gashouse retorts in the manufacture of illuminating gas from bituminous coal.

Coke-oven tars. Coal tar produced in by-product coke ovens in the manufacture of coke from bituminous coal.

Water-gas tar. Tar produced by cracking oil vapors at high temperatures in the manufacture of carbureted water–gas.

Tars are affected by the character of the oil or coal and by the methods and temperatures involved in their production. As with petroleum asphalts, tars are supplied in a number of forms from light liquid to semisolid and as cutbacks, so that they fit a wide range of construction methods.

The AASHTO classification for road tars, used by many agencies, includes 14 grades: *RT*–1 to *RT*–12, and *RTCB*–5 and *RTCB*–6. *RT*–1 is a light oil suitable for application as a tack or prime coat at normal temperatures. As the number designations become larger the tars become more viscous until *RT*–12 can be processed only at elevated temperatures (see Table 14–8). Uses of the more viscous grades parallel those for the road oils and cutbacks shown in Table 14–6. *RTCB*–5 and *RTCB*–6 are tars cut back with a quick-evaporating solvent for low-temperature application and quick setting. Detailed specifications appear under

[77]See *1980 Proceedings of the Association of Asphalt Paving Technologists* and *NCHRP Synthesis 54 for Proposed Specification for Recycling Agents*. See also *Civil Engineering,* Dec. 1978, *NCHRP Report 224 and TRB Record 777*.

TABLE 14–10. Specifications for Cationic Emulsified Asphalt *(AASHTO Designation M208)*

Type	Rapid-Setting				Medium-Setting				Slow-Setting			
Grade	CRS-1		CRS-2		CMS-2		CMS-2h		CSS-1		CSS-1h	
	min.	max.	min.	max.	min.	max.	min.	max.	min.	max.	min.	max.
Test on emulsions:												
Viscosity, Saybolt Furol at 77°F (25°C), s									20	100	20	100
Viscosity, Saybolt Furol at 122°F (50°C), s	20	100	100	400	50	450	50	450				
Settlement,* 5-day, %		5		5		5		5		5		5
Storage stability test† 24-hr, %		1		1		1		1		1		1
Classification test‡	passes		passes									
or												
Demulsibility,§ 35 ml 0.8% sodium dioctylsulfosuccinate, %	40		40									
Coating, ability and water resistance:												
Coating, dry aggregate					good		good					
Coating, after spraying					fair		fair					
Coating, wet aggregate					fair		fair					
Coating, after spraying					fair		fair					
Particle charge test	positive		positive		positive		positive		positive‖		positive‖	
Sieve test, %		0.10		0.10		0.10		0.10		0.10		0.10
Cement mixing test, %										2.0		2.0
Distillation:												
Oil distillate, by volume of emulsion, %		3		3		12		12				
Residue, %	60		65		65		65		57		57	
Tests on residue from distillation test:												
Penetration, 77°F (25°C), 100 g, 5 s	100	250	100	250	100	250	40	90	100	250	40	90
Ductility, 77°F (25°C), 5 cm/min, cm	40		40		40		40		40		40	
Solubility in trichloroethylene, %	97.5		97.5		97.5		97.5		97.5		97.5	

*The test requirement for settlement may be waived when the emulsified asphalt is used in less than 5 days time, or the purchaser may require that the settlement test be run from the time the sample is received until the emulsified asphalt is used, if the elapsed time is less than 5 days.

†The 24-hr storage stability test may be used instead of the 5-day settlement test.

‡Material failing the classification test will be considered acceptable if it passes the demulsibility test.

§The demulsibility test shall be made within 30 days from date of shipment.

‖If the Particle Charge Test is inconclusive, material having a maximum pH value of 6.7 will be acceptable.

AASHTO Designation M52. *HRB Bulletin 350* is a symposium on coal-modified tars.

BITUMEN–RUBBER MIXTURES. An experimental pavement bound with a bitumen–rubber mixture was laid in Holland in 1929. The first use of this binder in the United States was made in 1947 when a section was laid in Akron, Ohio. The pavement was conventional, except that finely divided rubber amounting to 5–7½% of the bitumen by weight was included. Since that time experimental roads have been laid by (among others) the state transportation agencies of Virginia, Ohio, Texas, Massachusetts, California, Colorado, and Utah, the cities of New York and Baltimore, and in Great Britain. In addition, bituminous binders modified with rubber have been used for seal coats.

Certain advantages are attributed to rubber additives. For example, skid tests in Virginia, reported in 1950, showed very little improvement in coefficients of friction on newly laid pavements but considerable advantage after six months. Early tests by the Bureau of Public Roads indicated both favorable and unfavorable results. Rubber added in powdered form brought unfavorable consequences; when preblended with the asphalt, it improved the stability of some but not all laboratory specimens. Recently, greater elasticity, reduced temperature susceptibility and brittleness, and longer life in pavement have been claimed. In sum, however, the conclusions of a 1954 analysis by the Bureau of Public Roads still appears to be valid. It stated that an appraisal of the "real economic value of the addition of rubber to asphalt must wait on further observation of the behavior of experimental pavements under the influence of age, weather, and traffic."[78]

Since 1970 Arizona and others have employed recycled tires in asphalt rubber systems for seal coats and for asphalt pavement interlayers. They expect the use to increase considerably as the cost of conventional maintenance materials rises.[79]

OTHER ADDITIVES FOR BITUMINOUS BINDERS. Certain chemical additives have been used for some time to improve the coating effectiveness and bonding power of asphalts to hydrophilic aggregates and thereby to minimize the "stripping" problem. Often these agents are fatty or resin amines. Lime, either in powdered form or as a slurry with which the aggregate is coated before mixing, also has proved effective.

Earlier applications of "nonstripping" additives have been in cold-laid and cold-mixed construction. Recently, research has been conducted to determine their usefulness in hot asphalt mixes by evaluating the influence of heat on their effectiveness. Indication is that some of these additives are damaged by heat; but that even so, under certain circumstances some of them substantially im-

[78]For the Virginia experience, see T. E. Shelburne and R. L. Sheppe, *HRB Bulletin 27*. Other references include R. A. Crawford, *HRB Record 236*, P. D. Thompson, *HRB Record 273*, and M. I. Darter et al., and J. E. Fitzgerald and J. S. Lai, *HRB Record 313*.

[79]For more information, see G. R. Morris et al., *TRB Record 595*. See also *American Society for Testing and Materials, STP 724*.

prove hot pavement mixtures against deterioration from moisture effects. Asbestos fibers or fiberglass also have been investigated.[80] Additives have also been developed to give a colored rather than a dark surface to bituminous pavement.

EPOXY RESIN BINDERS. Epoxy resins for highway purposes are available in clear, dark, rigid, and flexible forms and thus can be applied over either concrete or bituminous pavements. Hardening is accomplished by mixing resin and hardener immediately before application; the resulting material is thermosetting—that is, it will not soften under heat or the action of solvents such as water or petroleum fuels. Because of their cost, the use of epoxy resins to date has been largely restricted to bridge resurfacing or other special nonskid seal coating operations, or to bonding new concrete to old. *AASHTO Designation M200* gives both specifications and recommended practice for installing coal-tar epoxy protective coatings.

SULFUR ASPHALT MIXES. With the rapid increase in asphalt prices since 1975, many agencies have looked for methods of extending the available supply. The Federal Highway Administration and others have studied the use of sulfur[81] for sand mixes, sulfur extended asphalt *(SEA)* and plasticized sulfur (complete replacement). Based on field and laboratory studies, it appears durable pavement surfaces can be constructed; however, there are some potential health hazards since hydrogen sulfide (H_2S) and sulfur dioxide (SO_2) are given off at temperatures above 300°F.

Tests for Bituminous Binders

The consistency, quality, and certain other properties of bituminous binders must be carefully controlled if successful pavements are to be constructed with them; specifications and tests furnish this control. Typical specifications for the more common types have been given earlier. Tests that assure that the binders conform to these specifications or those of other agencies are briefly described in the paragraphs that follow.[82]

TESTS FOR CONSISTENCY. As indicated, bituminous binders are temperature susceptible; that is they become less viscous as the temperature increases. Materials of different origin, which have the same consistency at room temperature, may have far different viscosities at a mixing temperature of 300°F. Furthermore, at room temperatures, some asphalt cements are non-Newtonian in behavior—that is, the rate of shear stress is not proportional to the strain. For these reasons, asphalt cements are commonly graded on the basis of viscosity (see Table 14–7), although some agencies still use the penetration grading.

[80]See J. H. Kietzman, *HRB Bulletin 270,* and W. A. Garrison, *HRB Record 117.*

[81]See papers by D. Saylak et al., and G. D. Love, *Transportation Engineering Journal of ASCE,* Feb. 1975 and Sept. 1979 and T. W. Kennedy et al., *TRB Record 659* and *741.*

[82]An excellent summary on the development of asphalt tests is given by W. J. Halstead and J. Y. Welborn, *Public Roads,* June 1975.

Because there are difficulties in conducting certain tests such as penetration and ductility at other than room temperatures, these have been standardized at 25°C (77°F). But tests for viscosity, which are based on rate of flow, can easily be carried out at higher temperatures. For example, 60°C (140°F), which represents summer temperature of the road surface, or 135°C (275°F), to approximate mixing conditions, both are employed.[83]

The tests for consistency of petroleum asphalts specified in Tables 14–7 through 14–10 are kinematic or absolute viscosity, Saybolt–Furol viscosity, penetration, and softening point. For tars, the Engler viscosity and float tests are used. All these tests are comparative; they merely assure the engineer that the binder being tested will have properties similar to those of another binder that has already served successfully. Careful control of testing temperatures is imperative, as the consistencies of bituminous materials change rapidly with variations in temperature. For liquid asphalts, a temperature decrease of about 40°F changes the consistency to that of the next higher grades of the series.

Kinematic viscosity, in stokes or centistokes (one one-hundredth stoke), is a scientific unit for expressing the viscosity of liquids. This unit has been adopted as a measure of viscosity for liquid asphalts, replacing Furol viscosity.

Several devices can be used to measure kinematic viscosity. All of them measure the time of flow of the sample between two points in a capillary tube under controlled temperature and accurately reproducible head. Several sizes of capillary tubes are available along with a standard oil for calibration purposes. (See *AASHTO Designation T201*).

For viscosity-graded asphalt cements the *absolute viscosity* test *(AASHTO Designation T202)* is employed; for it, a vacuum rather than constant head is used to cause the asphalt to flow. The test is run at 60°C (140°F); the units are poises.

The *Saybolt–Furol viscosity test (AASHTO Designation T72)* is the present control of consistency for emulsified asphalts and an alternate control for liquid asphalts. The Saybolt–Furol viscometer is a special cylindrical vessel approximately 1.2 in. in diameter and 5 in. high, enclosed in an oil bath. The outlet tube in the base is about 0.12 in. in diameter and ½ in. long. The viscosity is defined as the time in seconds required for 60 ml of the oil to flow by gravity from the completely filled cylinder. Testing temperature is carefully controlled by regulated heating of the oil bath and, in order to make testing times reasonable, is different for oils of different fluidity. Some measure of the differences in viscosities of liquid asphalts can be gained by noting in Table 14–9 that the lightest (30 grade) product at 77°F flows out in 75–150 s, while the heaviest (3000 grade) requires 300–600 s, at 180°F.

The *Engler specific viscosity test (AASHTO Designation T54)* is the control for consistency of liquid tar products, *RT*–1 to *RT*–6 and *RTCB*–5 and *RTCB*–6.

[83]Testing methods for the viscosity of bituminous materials have been the subject of intensive research; e.g., *Public Roads,* Aug. 1959, Jan. 1960, Feb. 1962, June 1966, and June 1975; *HRB Record 24, 67, 117, 134, 178,* and *231; HRB Bibliographies 35* and *40,* and the annual *Proceedings, Association of Asphalt Paving Technologists.*

The Engler viscometer is a shallow cylindrical vessel about 4 in. in diameter equipped with a slightly tapered outlet tube about 0.11 in. in diameter and 0.8 in. long. The vessel is calibrated at 25°C by filling it with water and measuring the time in seconds needed for the passage under gravity of 50 cm^3. Engler specific viscosity, a dimensionless number, is the quotient resulting when the flow time of the bituminous material at specified temperature is divided by the flow time of water at 25°C. The nearly constant ratio of 4:1 exists between Furol and Engler viscosities when the tests are conducted at the same temperatures, so that the Furol test would be satisfactory for testing tars.

The *float test (AASHTO Designation T50)* is for heavier tars, *RT–7* to *RT*–12. The specimen for the test is a small tapered plug of bitumen. This plug is molded into a brass collar which is threaded to fit into the bottom of a small aluminum dish. The assembly, chilled to 5°C, is placed in a water bath held at the specified test temperature until the water breaks through the plug and into the dish. Results are given as the time in seconds, measured from the placing of the assembly in the bath until the water breaks through. The time is shorter for soft materials than for hard ones.

The *penetration test (AASHTO Designation T49)* is to determine the consistency of asphalt cements as delivered and after heating, and of the residues after distillation of medium- and rapid-curing asphalt oils and of emulsions. A standard needle penetrates the sample vertically under known conditions of loading, time, and temperature. The needle must be hardened and highly polished, as small irregularities will greatly affect results. Load, time, and temperature of test are normally 100 g, 5 s, and 77°F, respectively. Units of penetration are in hundredths of a centimeter. Increased values for penetration indicate softer asphalts. In contrast, higher numerical results for the viscosity tests indicate harder asphalts.

The *softening-point test—ring and ball method (AASHTO Designation T53)* is the control for consistency of the residue of tars after distillation. A brass ring ⅝ in. in diameter and ¼ in. thick is filled with a melted sample of the material to be tested. Sample and solid steel ball ⅜ in. in diameter are placed in ethylene glycol at 41°F. The ball is then placed in the center of the upper surface of the bitumen in the ring, and heat is applied in such a manner that the temperature of the bath is raised 9°F each minute. The softening point is the temperature of the bath at the instant that the softened bituminous material, pushed downward by the ball, deforms 1 in.

TEST FOR DUCTILITY *(AASHTO Designation T51).* A ductile material is one that will elongate (be led out) when subjected to tension, as contrasted to a brittle material that will break rather than stretch. Minimum values for ductility apply directly to asphalt cements, to their residues after heating, and to residues of road oils, cutbacks and emulsions. The test is intended to provide assurance that the binder in the completed road will be ductile rather than brittle, so that the pavement surface will distort rather than crack and fail by fatigue under the effects of repeated loads.

The material to be tested is heated and poured into a special mold which produces a specimen with a thickness of about 0.4 in., a width varying from 0.8 in. at the ends to 0.4 in. at the center, and a length between grips of about 1.7 in. This specimen is placed in a special tension machine which pulls the ends apart horizontally at designated speed. Testing is conducted under water at 77°F; the usual test speed is 5 cm/min.[84] Resulting ductility is given as the distance in centimeters that the specimen stretches before breaking; a ductility of 100, which is common, indicates that the specimen stretched to something over 20 times its original length.

TESTS FOR SOLUBILITY OR FOR PERCENT BITUMEN *(AASHTO Designation T44).* Bitumen, the active cementing portion of asphaltic binders, is, by definition, soluble in trichloroethylene. Petroleum products are almost pure bitumen. By specifying some minimum percentage of solubility of the asphalt or residue after heating, the purchaser avoids accepting materials unduly adulterated with sand, dirt, or other impurities. For tars, the solvent is carbon disulfide. Tars contain some organic matter, commonly designated as "free carbon," the amount varying with the source of coal and production conditions. Specified limits on this free carbon range from 12 to 25%, depending on the grade.

Earlier solubility tests for asphalts used carbon disulfide which is flammable, or carbon tetrachloride which is toxic.

DISTILLATION TEST. Bituminous materials of the same consistency may have widely different characteristics. For example, a liquid slow-curing oil of a given grade has the same viscosity as a rapid-curing oil made up of asphalt cement cut back with gasoline. The difference can be determined, however, by ascertaining the percentages distilled off at each of several temperatures and by measuring the consistency of the residue (see *AASHTO Designation T78*). Reference to Table 14–9 for medium curing liquid asphalt shows the kerosene distilling off and a residue of paving-grade asphalt. For rapid-curing oils, parallel values exist except that the light gasoline solvent distills away at a lower temperature. A modification of the distillation test determines the percentage of water in a bituminous material.

THIN-FILM OVEN TEST.[85] For asphalt cements the "thin-film oven" test *(AASHTO Designation T179)* replaces the more complex distillation test. In it, a 50-m sample is distributed in a ⅛ in. layer over a pan 5.5 in. in diameter. It is heated in an oven for 5 hr at 325°F. Time and temperature were chosen to fit the observed hardening of asphalts mixed with aggregates in the laboratory. There also is evidence that the thin-film oven test approximates the hardening that occurs when asphalts are mixed with aggregates in commercial plants and the aging that occurs in the pavement.

An alternative to the thin-film oven test for measuring changes in the viscosity and other properties of asphalt cements during mixing in the hot plant is used

[84]Some agencies performed the test at 39.2°F. Also, a force ductility test has been used. See D. I. Anderson et al., *TRB Record 595*.

[85]This has largely supplanted the "loss-on-heating" test, for which the sample was ⅞ in. deep.

in several Western states. It is called the rolling thin-film oven test (see *AASHTO Designation T240*). With the oven at 325° F, a thin layer of asphalt is caused to flow around the periphery of a horizontal glass cylinder. Also, it is subjected to an intermittent blast of air. As with the thin-film oven test, excessive changes in properties are cause for rejection.

FLASH POINT. Bituminous materials are flammable, and the lighter vapors, when mixed with air, create explosive mixtures. If the processing temperature is so high that vapors are driven off, suitable precaution to eliminate fire hazards must be taken. The flash point is the temperature at which a bituminous material, during heating, will evolve vapors that will temporarily ignite or flash when a small flame is brought in contact with them. Comparison of working temperatures (see Table 14–8) with specified flash points for the various types of binders indicates that often fire or explosion is a real hazard. This is particularly true with the heavier grades of *RC* and *MC* cutbacks which have low flash points but require fairly high processing temperatures.

To make the test, the material is heated in an open cup, and at intervals a small flame is applied near its surface. For cutback products of flash points lower than 200°F, the Tag open-cup test is used *(AASHTO Designation T79);* for those of higher flash point, the Cleveland open-cup test is specified *(AASHTO Designation T48).*

Some agencies have substituted the Pensky–Martens closed tester *(AASHTO Designation T73)* for the open-cup devices. They have found that silicones added to asphalts to prevent foaming will blanket the surface of the sample. These suppress the evolution of vapors, thereby distorting the results of the flash-point test. However, the Pensky–Martens device continually stirs the sample and breaks up the silicone film.

TESTS FOR HOMOGENEITY OF PETROLEUM ASPHALTS *(AASHTO Designation T102).* Petroleum asphalts that have been overheated during refining, that are the residuals of the "cracking" process, or that have been air-blown at high temperatures contain small carbon flecks and are classed as "heterogeneous." Normal steam-refined or native asphalts and slightly oxidized residuals from asphaltic-based crude oils do not contain these flecks and are classed as "homogeneous." The heterogeneous or cracked asphalts have been viewed with disfavor by many engineers on the grounds that they weather badly; their use is prohibited or controlled by a majority of agencies.[86] This is accomplished by ruling out those asphalts that show a positive reaction to the Oliensis spot test or one of its modifications.

For the Oliensis spot test using the standard naphtha solvent, a drop of mix-

[86]There is some doubt whether cracked asphalts are inferior for road-building purposes, particularly when the amount of cracking is limited. Conclusions drawn from the Bureau of Public Road tests were that the positive reaction to the spot test does not definitely indicate that the product will weather badly and that identification tests restricting material to limited sources or processes of manufacture cannot predict weather-resisting properties with accuracy. However, it has been established that cracked asphalts become hard and brittle more rapidly than uncracked ones. See J. T. Pauls and J. Y. Welborn, *Public Roads,* Aug. 1953.

ture of the asphalt and naphtha is placed on a filter paper and the uniformity of the resulting spot is observed. If the entire spot is uniform in color on the first test and on repetition after the mixture stands for 24 hr, the tested asphalt is said to be homogeneous, and the test result is reported as negative. If the center of the spot is dark, surrounded by a lighter-colored ring, the asphalt is classed as heterogeneous, and the test result is reported as positive.

Some uncracked asphalts and high-sulfur crudes react positively to the spot test using the standard naphtha or a heptane solvent. Through research it was found that, if a solvent, in part heptane or naphtha and in part xylene, were used, some measure of the degree of heterogeneity could be obtained, since severely cracked residuals show negative results on the spot test when the percentage of xylene in the solvent approaches 100. A change from a positive result at some intermediate xylene percentage to a negative result at a slightly higher one serves to indicate the extent of cracking. Based on these and other findings, a number of highway agencies have modified the spot test to control the amount of cracked material by specifying that the spot test be negative with a solvent containing a stated percentage of xylene. For example, several western states stipulate a negative result (uniform spot) with a solvent containing 35% xylene and 65% heptane. In this way they permit the inclusion of some cracked products, but limit the amount.

SPECIAL TESTS FOR EMULSIFIED ASPHALTS. *(AASHTO Designation T59).* Tests previously described, some in modified form, serve to measure the asphalt content and consistency of emulsions and the consistency, solubility, and ductility of the suspended asphalt. Others are needed, however, to assure suitable behavior of the emulsion as a carrying agent for the asphalt; for, if it breaks down in an improper manner, construction will not be successful. Also, tests are needed to assure that the asphalt globules in cationic emulsions have positive surface charges.

Test for Demulsibility. The speed with which rapid and medium setting emulsion breaks down into its components is extremely important. To illustrate, if the binder for a seal coat were a rapid-setting emulsion and the emulsion failed to break down, the road could not be opened to traffic. On the other hand, if the binder for a road-mix pavement broke down before mixing and placing were completed, the asphalt–aggregate mixture would become hard and unworkable and be a total loss. To test demulsibility, calcium chloride solution (for anionic) and sodium dioctylsulfosuccinate (for cationic) is added to the emulsion to accelerate its breaking down into asphalt and water. If the emulsion breaks, the asphalt will solidify and adhere to the vessel, to the rod used for stirring, or to a screen through which the water and chemical solution are poured off. Demulsibility, which is stated as a percentage, is the ratio between the weight of asphalt separated from a sample by the demulsibility test and that obtained by distilling a sample of equal weight. A maximum value for demulsibility is set, as here it is essential that the emulsions be resistant to breaking. The tendency for slow-setting emulsions to break is measured by the cement-mixing test (see below).

Test for Settlement. If, after an emulsion has been in storage for some time, the percentage of asphalt at various levels in the container is considerably different, serious difficulty will be encountered in controlling the asphalt content of the finished pavement. To test for settlement, a sample is permitted to stand for 5 days, after which the percentage of asphalt at the top and at the bottom of the container is determined by distillation. Settlement is defined as the numerical difference in percentages of asphalt in the two samples.

Sieve Test. The emulsion is poured through a 20-mesh (0.850-mm) sieve that has been wet with sodium oleate (anionic) or distilled water (cationic).

Aggregate Coating Tests. These tests, for medium-setting emulsions, are intended to measure their ability to produce and maintain an asphalt coating on coarse aggregates. Samples of both dry and previously wetted aggregate are vigorously mixed with the emulsion and then repeatedly sprayed with water from a standard nozzle under constant head. Results may be evaluated in either of two ways:

1. Under *AASHTO Designation T59* coating ability for both anionic and cationic varieties with a standard aggregate is evaluated subjectively. Dry aggregate must show "good" coating, dry aggregate after spraying and wet aggregate before and after spraying must retain a "fair" coating.
2. For other agencies, the ability of a cationic asphalt to coat aggregate from the actual project is appraised by inspecting individual aggregate particles. To be acceptable, 80% of the dry but sprayed and 60% of the premoistened and sprayed aggregates must be coated.[87]

Cement-Mixing Test. For most uses of slow-setting emulsions, it is important that the asphalt remain in suspension in the water after contact with the aggregate; otherwise mixing cannot be accomplished. The emulsion is first mixed with high-early-strength cement, then distilled water is added and the mixing continued. This mixture is then poured through a No. 14 (1.40-mm) sieve. Any material retained is dried at 163°C and weighed. The weight of dried material retained is reported as the percentage emulsion broken. A maximum of 2% is allowed.

Particle Charge and pH Tests. The particle charge test is called for in the AASHTO specifications for all cationic emulsions. To conduct it, a 12-V direct current is passed between two electrodes suspended in the emulsion. Test results are classified as positive and the emulsion as cationic if an appreciable deposit of asphalt appears on the cathode (negative) electrode, while the anode remains clean.

AASHTO recommends that the pH test be substituted in the case of slow-curing cationic emulsions. The difference in potential between two glass electrodes indicates the acidity of the sample. As indicated in Table 14–10 a pH less than 6.7 is specified.

Miscibility and Freezing Tests. For certain applications, it is desirable to dilute

[87]See paper by R. G. Hicks et al., *TRB Record 754* and *MS-19, The Asphalt Institute.*

emulsions with water. Again, under some circumstances, they may be stored at freezing temperatures. The AASHTO specifications also outline standard tests which simulate these conditions and assure that emulsions subjected to them will not break prematurely. Space limitations preclude giving details here.

Problems Related to Bituminous Binders[88]

Specifications and tests for bituminous binders, as for all other materials, are intended to control their quality so that satisfactory pavements can be constructed from them. In certain instances, the requirements may be too strict and exclude more economical materials that would be satisfactory. As already indicated, this objection may apply, at least in some instances, to the requirement that asphalts be homogeneous, as measured by the spot test.

On the other hand, the usual specifications and test may not exclude all unsatisfactory materials. Research on the aging of asphalts, leading to modified specifications for several West Coast states, is a case in point. In this instance it was found that pavements built with asphalts from crude oil from one particular field were unusually susceptible to cracking and breaking up. Investigation showed that, as compared to satisfactory materials, the penetration of the troublesome asphalts decreased markedly both during mixing and in service (see Fig. 14–24). As a result of these findings, specifications have been changed so that asphalts must contain fewer volatiles, lose less ductility on heating, and show smaller changes in consistency under variations in temperature. Before the new asphalts were instituted, the affected agencies had been using softer grades of asphalt, such as 120–150 or even 200–300 penetration, to postpone the onset of excessive brittleness. Currently they feel free to specify the more viscous grades when this appears desirable.

Asphalt film thickness in a pavement is in the order of 0.005–0.010 mm. It has now been demonstrated conclusively that the thicker asphalt films age less rapidly; aging is also less rapid where pavements have lower percentages of voids (in the range of 2%).[89] As indicated on Fig. 14–24 then, too "dry" a mix is more likely, in time, to crack and break up. On the other hand, higher asphalt contents and lower percentages of voids may cause instability in the pavement. This problem is discussed further in Chapter 19.

Improper construction procedures also can "age" asphalts prematurely. For example, as shown in Fig. 14–24, this can result from mixing with overheated aggregate. Storing the combined materials for an extended period at mixing temperatures likewise apparently produces a substantial reduction in the penetration of the asphalt.[90]

[88]*NCHRP Report 195* is an excellent overview on methods to minimize early distress in asphalt pavements. See also paper by J. E. Wilson et al., *Proceedings Association of Asphalt Paving Technologists,* 1979, for a description of short-term and construction problems.

[89]See, e.g., J. Y. Welborn, *Public Roads,* Feb. 1970, and B. A. Vallerga and W. J. Halstead, *HRB Record 361*.

[90]See P. H. Wright and R. J. Paquette, *HRB Record 132*. See also *TRB Record 515, 544* and *595*.

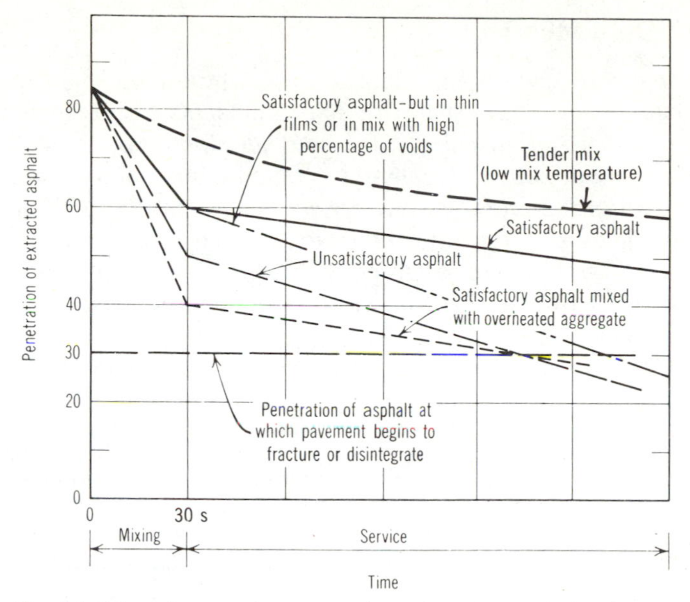

Fig. 14–24. Changes in penetration of paving asphalts during mixing and service periods (all values shown are very approximate.)

Climatic conditions also can seriously affect the behavior of bituminous binders. For example, in Canada, serious shrinkage cracking has occurred during cold weather. To date, prevention has come primarily by selecting the softer grades of asphalts.[91]

Attention is also being focused on the variability in behavior of asphalts from different sources. For example, the viscosity of some asphalts varies with temperature far more than that of others. Asphalts meeting a stated penetration requirement at 77°F, but with high viscosity–temperature ratios, will become excessively hard and brittle at low temperatures; again, they will be overly fluid at mixing temperatures. The test for viscosity at 275°F in the specifications for viscosity-graded asphalt cements (see Table 14–7) is intended to control this problem. Certain specifications also include a "penetration ratio" comparing penetrations at two temperatures as a similar control.[92]

[91]See R. C. G. Haas and T. H. Topper, *HRB Special Report 101,* N. W. McLeod, ibid., R. C. G. Haas and W. A. Phang, *HRB Record 313,* and H. J. Fromm and W. A. Phang, *HRB Record 350.* See also *TRB Record 521, 544,* and *777.*

[92]See J. Y. Welborn and W. J. Halstead, *Public Roads,* Aug. 1959, and Oct. 1960, for detailed comparisons of the properties of some 300 samples of paving asphalts from over 100 refineries. See also *HRB Record 178,* and all papers in *HRB Record 231* and *350.* See also an article by V. P. Puzinauskas, *Proceedings, Association of Asphalt Paving Technologists,* 1979 for a discussion on temperature viscosity properties of asphalts since 1960.

PORTLAND CEMENTS

Portland cement is made from combinations of limestone, marl, or other calcareous material and clay, shale, or like argillaceous substances. These are crushed and pulverized, mixed in carefully determined proportions, and burned to a clinker at about 2800°F. Finally, the cooled clinker and a small amount of gypsum to control the rate of setting are intimately ground until almost all the individual particles are far finer than the No. 200 mesh sieve. The finished product is often packaged in paper sacks holding 94 lb, which represent a cubic foot loose measure.[93] However, cement for large projects is shipped in bulk by rail or ship or in trucks with special bodies.

The AASHTO specifications for portland cement *(AASHTO Designation M85)* list eight types, as follows:

Type I or IA (air entraining)	For general concrete construction when the special properties of the other four types are not required.
Type II or IIA (air entraining)	For general concrete construction exposed to moderate sulfate action, or where moderate heat of hydration is required.
Type III or IIIA (air entraining)	For high early strength.
Type IV	For low heat of hydration.
Type V	For high sulfate resistance.

In addition there are AASHTO specifications covering blended cements *(AASHTO Designation M240).*

The properties expected in the different types of cements are carefully set out in the specifications of the various transportation agencies. Frequently this is done by reference to those prescribed by AASHTO (see *AASHTO Designation M85*). To meet these specifications, samples must pass a number of chemical and physical tests which can only be conducted in a well-equipped laboratory. The chemical tests constitute, in effect, a chemical analysis to determine whether the various strength-giving compounds appear in proper quantity, and if there are excessive amounts of certain undesirable substances. The physical tests include those for fineness, soundness, time of set, air content, and compressive strengths of mortars made using Ottawa sand.[94] Each cement mill operates a complete testing laboratory and maintains close control on its product. Rarely are cements shipped that do not meet specifications.

Chemically, portland cement has four principal constituents. These are tricalcium silicate ($3CaO \cdot SiO_2$ abbreviated C_3S), dicalcium silicate ($2CaO \cdot SiO_2$ or C_2S), tricalcium aluminiate ($3CaO \cdot A1_2O_3$ or C_3A), and tetracalcium alumino ferrite ($4CaO \cdot A1_2O_3 \cdot Fe_2O_3$ or C_4AF). The proportions of these in typical cements of the five types are shown in Table 14–11. The rates at which test specimens

[93]Often cement quantities or prices are stated in terms of barrels of four sacks each or in 100 lb units.
[94]Ottawa sand is a natural silica sand of great uniformity from Ottawa, Ill. It is widely used for testing purposes.

TABLE 14–11. Composition and Strength Characteristics of the Various Types of Portland Cement

Type of Cement	Compound Composition (%)				Compressive Strength, Normal Portland-Cement Concrete (%)			
	C_3S	C_2S	C_3A	C_4AF	1 Day	7 Days	28 Days	3 Months
I—Normal	50	24	11	8	100	100	100	100
II—Modified	42	33	5	13	75	85	90	100
III—High early	60	13	9	8	190	120	110	100
IV—Low heat	26	50	5	12	55	55	75	100
V—Sulfate resistant	40	40	4	9	65	75	85	100

made with the different cements gain strength also appear in the table. The variations result because the speed of chemical reaction between each compound and water is different. Both tricalcium silicate and tricalcium aluminate hydrate quite rapidly while the others react more slowly. As would be expected, Table 14–11 shows that high early strength cement has high percentages of the faster-acting compounds. With this rapid gain in strength goes fast liberation of the heat of hydration. An added factor is that the total heat liberated per unit of weight is greater for the faster-acting compounds. Thus, the rapid gain of strength may under some circumstances produce undesirably high temperatures in recently placed concrete.[95] For this reason high early strength cements are used in pavements primarily to meet special conditions.

GEOTEXTILES (FABRICS)

Geotextiles are relatively new construction materials. As recently as 1970, the only installations were experimental. However, by 1976, some 15 million yd^2 (12.5 million m^2) were installed and estimated used in 1980 was 75 million yd^2 (63 million m^2). In Europe, fabric use is more than double that in the United States. This growth has brought a large number of manufacturers and a variety of products into the market. In the United States there are over two dozen sources, each of which has several products of different characteristics, and new lines are being added constantly. And only recently have publications become available to assist the engineer in selecting the best product for any given use.[96]

[95]In massive structures like large gravity dams, the expansion which would be caused by these high temperatures and the contraction which would come later when the concrete cooled would cause serious concern. In some cases elaborate cooling systems have been installed to remove this liberated heat. In others a lean mix of low-heat cement has been used, and the aggregates and mixing water have been precooled to hold temperature rise to a minimum.

[96]The FHWA report *RD 80-021* by J. R. Bell et al., titled "Evaluation of Test Methods and Use Criteria for Geotechnical Fabrics in Highway Applications" is excellent. See also R. M. Koerner and J. P. Welsh, *Construction and Geotechnical Engineering Uses of Synthetic Fabrics,* Wiley, New York, 1980.

Uses

Geotextiles have been employed in a variety of ways to reinforce soil or pavement and provide drainage and erosion control for both highways and railroads. They serve the following functions.

1. Filter. They retain soil in place while allowing water to escape easily to be carried away by some form of drain.
2. Drainage. Heavy geotextiles themselves provide channels for carrying water away from the soil to be drained. This application is not common.
3. Separation. They prevent dissimilar materials from mixing.
4. Reinforcement. They add mechanical strength to the soil or pavement system.
5. Armor. They protect the soil from surface erosion or tractive forces.

Most applications of geotextiles (see Table 14–12) require a combination of these functions; however, one would normally predominate. For example, a fabric placed between the subgrade and base course to prevent mixing might also serve to drain water away from the subgrade, dissipate excess pore water pressures, and provide reinforcement to reduce tensile stresses in the bottom of the base course.

Types of Textiles

Many geotextiles have been introduced and additional ones are being added. They vary greatly in both basic materials and construction.

MATERIALS. Currently, geotextiles are manufactured from polypropylene, polyester, nylon, polyethylene, and polyvinylidene chloride. Of these, polypropylene and polyester are the most common. In addition to the differences among chemical compositions, the processes by which the fibers are produced will also bring changes in properties.

Important differences in properties include specific gravity, strength, strain at failure, modulus of elasticity, creep resistance, temperature stability, and resistance to ultraviolet light, chemicals, and biological effects.

CONSTRUCTION. There are a variety of methods for forming geotextile materials into fabrics. Each (see Table 14–13) has advantages and disadvantages. These may be classified as woven, knitted, or nonwoven.

For woven geotextiles the filaments or yarns are oriented in two perpendicular directions and overlapped. For transportation applications, the patterns are usually simple, uniform, and rectangular with a relative constant pore-size distribution. Monofilament geotextiles are woven from single strands; multifilament ones are made from yarns having many fine strands; and ribbon filaments are made from strands that have widths many times their thickness.

Knitted fabrics are made up of loops of fibers or yarns connected by straight segments. They may be stretched in either direction without significantly stressing the fibers. They may be knitted in tubes and serve well as filters around drain tile, particularly for agricultural purposes.

TABLE 14–12. Typical Applications and Uses of Fabrics

Application	*Use*
Drain Filter	Trench drains Base course drains Structural drains Pipe wrappings
Subgrade stabilization	
Light	Parking lot pavements Highway pavements Airport pavements
Heavy	Haul roads Storage yards Railroads
Reinforcement	
Light	Low fill foundations Low retaining walls
Heavy	High embankments High retaining walls
Erosion control	
Light	Ditch armor Culvert outlet protection
Medium	Small to medium wave protection
Heavy	Large wave protection
Silt fence	Construction sites
Reflection crack control	Asphalt pavement overlays Asphalt pavement over cement stabilized bases Canal lining repair
Drainage	Foundation consolidation Embankment consolidation

Source: Modified after J. R. Bell et al., *FHWA Report RD 80-021*.

Nonwoven geotextiles are neither woven nor knitted. Rather the arranged fibers or strands are held together in one of the following manners:

1. Needle punching. Needling is accomplished by punching many barbed needles through the fabric. This produces a loose fabric, thick for its weight, which has the appearance of a felt mat and a very complex pore structure.
2. Heat bonding. Heat or melt bonding subjects the fabric to a relatively high temperature which bonds the filaments together at contact points. The fabric is thin with relatively discrete and simple pores.

TABLE 14–13. Fabric Construction

Structure	*Filament*	*Bonding*
Woven	Monofilament Multifilament Ribbon filament	Heat bonding None None
Knitted	Multifilament	None
Nonwoven	Staple or continuous filaments	Needlepunched Heat bonded Resin bonded Combinations
Combination (woven–nonwoven)	All combinations of above	
Special	Other processes	

Source: After J. R. Bell et al., *FHWA Report RD 80-021.*

3. Resin bonding. The fabric is impregnated with a resin which cements the fibers together. Thickness and structure is intermediate between needle punching and heat bonding.
4. Combination bonding. Two or more of the processes described above are combined to produce certain characteristics. Needling a mat onto a woven scrim (mesh) provides both strength and filtering ability.

Nonwoven fabrics, as a group, are relatively inexpensive, have low to medium strengths, and medium to high elongations before failure. They are widely used in filtering, separation, and light reinforcement.

From this brief discussion it should be clear that selecting geotextiles for various applications is difficult.[97]

[97]For added information on the properties of geotextiles and procedures for selecting them for particular applications, see the previous references and *Civil Engineering,* Oct. 1979. *TRB Circular 204* offers an extensive bibliography.

PROBLEMS

14–1*A*. Specimens of two soils were compacted in the laboratory employing the AASHTO standard test method. Unit weights for varying moisture content were as follows:

Soil A		*Soil B*	
Moisture Content, (% by wt of Dry Soil)	*Wet Weight of Soil (lb/ft³)*	*Moisture Content, (% by wt of Dry Soil)*	*Wet Weight of Soil (lb/ft³)*
4.43	134.1	12.6	122.6
6.77	144.0	14.2	125.0
7.37	145.2	16.3	128.8
8.36	145.2	17.0	128.8
8.95	144.0	19.7	127.5

On a single graph, plot the wet weight and dry weight curves for these soils and determine the optimum moisture content for each. If one of these soils is well-graded sandy soil and the other is a clay type, which soil is the clay type? Explain your answer.

14–1*B*. Convert the wet weights of soils *A* and *B* in Problem 14–1*A* to g/cm³ and complete that problem (see Table A inside the front cover for traditional–SI conversion units).

14–2*A*. On the graph prepared for Problem 14–1*A*, plot the zero air-voids curve for the soils, assuming that the soil particles have a specific gravity of 2.68.

14–2*B*. Carry through Problem 14–2*A* in g/cm³.

14–3. Compute the energy in foot-pounds per cubic foot of compacted soil that is applied to laboratory samples in conducting:

a. The AASHTO standard impact compaction test.
b. The modified AASHTO impact compaction test.
c. The California impact test (use 11-in. sample height).

14–4. A soil identical with sample number 3 of Fig. 14–6 has been compacted in an embankment. A test hole of 6 in. diameter and 8 in. depth is dug in the fill. The material, as taken from the hole, weighs 16.7 lb and has a moisture percentage of 10, based on dry weight. What is the dry weight per cubic foot and the relative compaction of the soil in the fill?

14–5. Suppose that very precise measurement of the volume of a hole dug in an embankment to determine density leads to a relative compaction of 95%. If this volume had been measured by means of a small water balloon:

a. What value of relative compaction higher than 95% would be exceeded 1 time in 20?

b. What value of relative compaction lower than 95% would be too high 1 time in 20?

14–6. Work Problem 14–5 if the test is made by the sand, glass jar, and cone methods.

14–7. (This question is for students who have had a first course in statistics.) Nuclear gauges do not measure soil densities exactly. Rather, many tests have shown that the standard deviation of measurements is about 2½%. Suppose that the required density for a given soil is 102 lb/ft^3 and a single reading on the gauge shows the density to be 104.5 lb/ft^3. What percentage chance is there that the true density is less than required?

14–8. Find the textural classification for each soil listed in Table B (inside the back cover). Note that this classification considers only sand and smaller particles.

14–9. Find the AASHTO classification for each soil listed in Table B (inside the back cover).

14–10. Find the Group Index for soils numbered 1–6 inclusive of Table B (inside back cover).

14–11. Determine for your state (or country) the major types of aggregate used. For each, obtain an estimate of the quantities used in base and subbase, bituminous concrete, and portland-cement concrete.

14–12. For your state (or country) obtain copies of the aggregate specifications for base and subbase, bituminous concrete, and portland-cement concrete. Develop a table that compares the grading and other requirements for each end use.

14–13. By contacting a local paving contractor, determine the kinds and grades of bituminous binders readily available in your local area.

14–14. An SC-250 road oil was loaded into a transport tank at a temperature of 250°F(t_2). The amount loaded was 8490 gal. The coefficient of expansion per degree F (K) is 0.00036.

a. Calculate the invoice volume at 60°F(t_1) for this shipment.

b. Compute the cost of this shipment if the purchase price is 80 cents per gallon based upon the volume at 60°F. Hint:

$$V_{60°F} = \frac{V\ 250°F}{1 + K(t_2 - t_1)}$$

14–15. Contact your city or state agency and determine the type and extent of pavement problems related to bituminous binders.

14–16. By contacting a local contractor, determine the types of admixtures used in portland-cement-concrete pavements.

14–17. From the local transportation agency, determine the types of geotextiles used, their applications, and their performance.

15 CONSTRUCTING THE ROADBED

Roadbeds underlie highway pavement structures and the ballast and track on which trains move. Unless there is a bridge, tunnel, or other special structure, this roadbed is constructed of in-situ soils or on earth embankments. Building it is the first step in producing a finished facility.

This chapter deals first with the contractual and management arrangements for building the roadbed and other elements of a completed facility. It then treats construction procedures, including methods and payment arrangements for site preparation, excavation, borrow, haul, and embankment placement. Finally it examines the techniques for constructing roadbeds in marshy areas.

CONTRACTUAL, MANAGEMENT, AND QUALITY ASPECTS OF CONSTRUCTION

Organizing To Carry Out Construction

In the United States, almost all major construction of highway and mass transit facilities is carried out by independent contractors on the basis of competitive bids under which they do projects set out in plans and specifications.[1] Commonly, applicable laws require that such work be awarded to "the lowest responsible bidder who submits a regular bid."[2] Payment to contractors usually is on a *unit price* basis; this is determined by summing the products of measured units of work times the bid price per unit. Examples are excavation at a stated price per cubic yard or cubic meter; borrow at so much per cubic yard, cubic meter, or ton; paving on the basis of tons, cubic yards or cubic meters, or square yards or square meters of completed surface; and pipe, guard rail, or fencing in terms of linear feet or meters, installed. Some items such as demolition may be done for a bid lump sum.[3] This unit price payment plan contrasts

[1]See Chapter 6 and *Guide Specifications for Highway Construction* and *Construction Manual for Highway Construction,* AASHTO.
[2]Some small projects or exceptionally difficult ones and all except specialty maintenance are done by the agency's forces under a procedure referred to as "force account." In many other countries, governmental forces do all or a large portion of construction work.
[3]Details of contract and construction law are treated in depth in books and courses on that subject and are outside the scope of this work. See also *Compendium 16* of Transportation Technology Support for Developing Countries, TRB.

with payment in building work which, for public agencies, is commonly on a lump sum basis.

Construction contracts are directed and carried out by the management staff and workers of contractors or by subcontractors employed by them.[4] Managing this work effectively calls for skills in areas such as planning and scheduling, often by techniques such as the critical path method (CPM), and directing the activities of people and equipment. Such topics are covered by educational programs and literature on construction engineering and management.

To represent the transportation agency on construction sites, supervising engineers, inspectors, and possibly surveyors are employed; their function is to see that the project is carried out in accordance with the plans and specifications and to keep records, including those involving changes, and to see that appropriate payments are made to the contractor.[5]

Owner agencies, in estimating the approximate cost of projects, rely heavily on unit prices submitted on earlier projects where conditions were similar. There are hazards in placing too great confidence on these estimates, since peculiarities of the project, the competitive position of the construction market, the cost and efficiency of labor, and many other factors may influence prices. Furthermore, contractors very commonly unbalance their bids on individual items in varying degrees. The purposes, among others, are to hide their true costs, to secure larger payments early in the project, or to protect their overhead and profit by placing it in nonvariable items. Some, but by no means all, contractors will unbalance their bids to capitalize on errors in the owner's estimate. This is accomplished by placing high unit prices on those items where the owner's quantity estimate appears to be lower than the actual and by bidding low on items when the estimate is too high. At the same time prices on other items are adjusted to bring the desired total bid for the project.

Recently, because of rapidly increasing labor costs and materials and equipment prices, some agencies have introduced price escalation clauses into their contracts. With them, payment to the contractor is adjusted upward in a predefined manner. The alternative is, of course, for contractors to include estimates of these cost increases in their bids. Opinion is divided over which of these alternatives is better.

From the records of contractors' unit prices, many highway agencies have developed *price indexes* to reflect changes in price levels with time. These indexes sometimes are made by repricing a *composite mile* of typical highway. The cost of this composite mile during a given interval is compared with the cost during a base period to give the index. The cost index of the Federal Highway Administration is based on national total quantities and the average unit prices for them. Percentages of total cost assigned to the various items are as follows: excavation; 21, portland-cement-concrete pavement, 12, bituminous concrete,

[4]Highway contracts often limit to (say) 50% the amount of work that can be subcontracted.
[5]For details of preconstruction and construction staffing from the agency point of view see *NCHRP Synthesis 51* or two abstracts from it by H. R. Thomas Jr. et al., *Journal of the Construction Division of ASCE,* June 1980 and *TRB Record 742*.

24; reinforcing steel, 7; structural steel, 15; and structural concrete, 21. The current base is the year 1977 with its index set at 100. On this base the 1968 cost index was 47.8 and by April 1981 it stood at 162.[6]

Quality Assurance in Construction

Construction plans and specifications stipulate many requirements that are to be met by materials or construction procedures. There are two approaches for obtaining them. One, called "manner and method," is to spell out in detail the materials to be supplied or the procedures to be followed. The other, called "result," or "end result" stipulates the desired end product and leaves manner and method to the contractor. Sometimes both may be written into the contract documents, but, in this case, the agency's ability to enforce the "result" clauses may be in doubt, since the contractor cannot be held to both. Each approach offers advantages in certain instances, and most specifications carry some of each.

Another "quality" problem is the inherent variability in materials and the accuracy possible in carrying out construction procedures. In the real world it is physically impossible to precisely meet any stipulated requirement since materials, sampling and testing procedures, and workmanship cannot be controlled exactly. Furthermore, to set requirements that are too exacting is both uneconomical and infeasible. Transportation agencies have become increasingly aware of the "variability" problem and have developed specifications and provisions for acceptance and payment which incorporate them. For example, the 1979 FHWA *Standard Specifications for Construction of Roads and Bridges on Federal Highway Projects* for portland-cement-concrete pavements first spell out in detail the manner and method requirements for materials and construction procedures in order to assure that acceptable practices will be followed. In addition, they stipulate certain results by setting up provisions for acceptance and payment reductions when these are not met. For pavement thickness, these pavement adjustments are as follows:

Deficiency in the Thickness of 5 cores (in.)	*Payment as a % of Contract Price*
0–0.10	100
0.11–0.25	95
0.26–0.50	80
0.51–1.00	50
More than 1.00	Remove

This measure is applied separately to each 3000 ft of pavement lane. Concrete compressive strength in 28 days is targeted at 3000 psi for plain and 3500 psi for reinforced pavements. Manner is stipulated with a maximum water-cement

[6]*The construction cost issues of Engineering News-Record* give details of this and several other highway cost indexes as well as the ENR general indexes and those representative of other segments of construction.

ratio by weight of 0.49. In addition, if the strength "result" is too low payment is reduced. This reduction is based on the average and range of strengths in a specified number of test cylinders by computations too detailed to give here.

A statistically based specification of the California Department of Transportation for the strength of concrete typifies current approaches. It requires that (a) the moving average strength of the last five sets of two test cylinders each must equal the specified value and (b) the minimum strength of individual tests of any pair of cylinders shall be 95% of that value. It further stipulates that if tests of any individual pair of cylinders indicate a strength less than 85% of that specified, the concrete represented by that sample will be rejected and must be removed unless coring shows that it has reached the required strength.

The *control chart* is a statistical technique for quality control widely employed in industry and by some transportation agencies. It provides a means for monitoring continuous processes such as aggregate production. The chart has as its ordinate individual test results; these are plotted in time sequence along the abscissa. Nearly horizontal lines on the chart show the average results of all tests to date and the upper and lower control limits. Commonly, these control limits are so placed that, purely by chance, the plot of only one test result in 100 will fall outside these limits. (In statistical terms, the control limits are set three standard deviations above and below the mean). As long as individual test results plot between the upper and lower limits, the process being monitored by the test is considered to be under control; variations within these limits are to be expected. However, if a point falls outside these limits, this indicates that the process is out of control because some basic change has occurred so that an adjustment may be needed.[7]

Increasingly, transportation agencies are assigning almost all routine quality control to contractors. Under such schemes, the contract stipulates that sampling, testing, and reporting of results is to be done by the contractors. Also, they are required to provide a quality control plan and qualified staff to do sampling, testing, to keep records, and to report and certify the findings. The agency itself will do only a small amount of check sampling and acceptance testing to see that the specified results are being obtained and if not, to provide a basis for appropriate correction or reduction in payment. This approach challenges the traditional concept that the combination of contractors interested in doing the work cheaply with engineers concerned about quality produces satisfactory work at a lower cost. It follows that this approach will require a substantial rethinking and adjustment in the traditional roles of contractor and engineer.

Other Construction Management Problems

The safety of those traveling through construction sites and of those working on them is a primary concern of both contractor and agency. These problems are discussed briefly in Chapter 12. Energy consumption and environmental conse-

[7]For a fundamental presentation on statistical quality control, see E. L. Grant and R. S. Leavenworth, *Statistical Quality Control,* McGraw-Hill, New York, 1980. Other recent references include *NCHRP Syntheses 38* and *65, TRB Record 539, 652, 691, 697, 745,* and *792* and J. H. Willenbrock and S. Shepard, *Journal of the Construction Division of ASCE,* Sept. 1980.

quences including soil erosion, stream pollution, dust, and noise also are of high priority and must be considered (see Chapters 4 and 13, respectively). In addition, contractors and agencies must comply with and report their performance to other governmental agencies on a host of federal and state socially orientated laws and regulations. Among them are employee safety, payment of stipulated minimum wages, providing equal opportunity in the work force for women and minorities, and setting aside contracts for minority firms. All these add complexities to construction, but details of their effects will not be presented here.

Weather Effects on Construction

Precipitation as rain or snow, high winds, and extremely hot or very cold temperatures can slow or halt certain construction operations. The effects differ; for example, heavy precipitation will halt earthmoving and embankment construction; however, it usually can proceed in hot or cold conditions. Hot weather will not stop paving operations but heavy rain or cold weather will. As will be pointed out later in this book, specifications may be written to prevent the contractor from carrying out certain operations when defined adverse conditions prevail. In many situations, however, the decision regarding whether to work falls to the judgement of contractor or engineer and can be one of many sources of disagreement.

Weather and its effects vary widely both within and among countries and generalizations are difficult. It has been reported that, for the United States, 45% of highway construction is weather sensitive.[8]

CLEARING THE SITE

Clearing the site precedes all grading and most other construction operations. It may consist merely of removing and disposing of coarse grass and small bushes and shrubs or involve considerable work, some of it calling for special skills. Problems encountered differ substantially between rural and urban areas and for that reason they are discussed separately below.

Site Clearing in Rural Areas

Site clearing in rural areas may sometimes merely require that grass, shrubs, and other small plants or crops be removed. However, it sometimes can involve removing trees and tree stumps and disposing of the debris. The accepted procedure is to remove practically all vegetable matter from the original ground and from fill material, since, if allowed to remain, it may decay and leave voids that result in settlement. Selective clearing in adjoining areas may at times be required.

In heavily timbered areas, clearing may involve considerable cost. Great care must be exercised to avoid fires, particularly when logs, brush, and slashings

[8]See *NCHRP Synthesis 47* for a brief discussion.

are burned. Contractors may be required to employ special fire guards and to supply fire fighting equipment. If the trees are suitable for lumber, it is common to specify that the logs be cut into merchantable lengths and piled at designated locations. Concern for the environmental effects of waste disposal may bring prohibitions against burning or require special techniques for smoke reduction.

Heavy clearing operations may considerably lengthen the time needed to complete a road. For example, if debris can be burned only in prescribed seasons, progress of the entire project may be held up. Consequently, clearing operations often are set up as special wet-season or winter contracts.

To avoid arguments regarding procedures and payment, specifications must clearly state what is to be done. For example one provision of the 1979 FHWA specifications requires removal of stumps, roots, and nonperishable solid items to 3 ft below subgrade or embankment slopes.

Clearing and grubbing is usually paid for by the acre of land cleared, at the bid unit price for "clearing and grubbing" or as a single lump sum covering all such work on the project. In special circumstances, as where trees are to be removed, payment may be under a heading of "Removal of Trees" at a bid price per tree, or per tree in stated diameter ranges.

Where a project is in settled areas and lies along an existing right of way, overhead lines for power and telephone may be in place along the roadway margins and at crossings. Commonly, arrangements are made for the appropriate utility agency to carry out necessary removals or relocations of such facilities. Payment for this work, if required (see Chapter 7) usually is on a force-account (cost reimbursement) basis.

Site Clearing in Urban Areas

Site clearing in urban areas can bring many difficult problems. As discussed in more detail in Chapter 7, before construction can begin, relocations must be arranged for people and businesses that are displaced and details worked out for moving utilities without disrupting service. Then buildings, pavements, sidewalks, and like obstructions must be removed and surface utilities such as telephone and power lines that parallel or cross the right of way relocated or adjusted. For depressed urban highways, the clearing problems become extremely complex. Rearrangements must be made not only for surface installations, but also of underground utilities. In the central area of cities, installations under a typical city street include pipelines or conduits for water, sewer, gas, storm drainage, telephone, power, and at times steam, oil, or chemicals. Much time and expense is involved in their rearrangement and in providing uninterrupted service.

Close liaison among the transportation agency, contractor, and utility personnel is imperative if utility relocation is to proceed smoothly and not bring project delays. Over the years, excellent organizational relationships have been developed, but details cannot be given here.[9]

[9]See, for example, *HRB Special Report 77* and articles in *TRB Record 483, 571,* and *631.*

GRADING OPERATIONS

Grading is an all-inclusive term to describe construction operations between site clearing and paving. All hauling, spreading, and compacting activities are included. Preceding or accompanying grading is the installation of storm drains, culverts, and bridges (See Chapter 11).

In the earlier days of highway construction, contractors and construction engineers carried out grading and paving operations with little regard for the motorists who, of necessity, passed through the construction area. Likewise, little attention was given to the inconvenience, dirt, dust, noise, and other nuisances that construction activities brought to the occupants of adjoining property or to the general public. Today, however, these problems receive careful attention. Detours are carefully planned, signed, and operated so that traffic flows freely. In fact, one entire section of the *Manual on Uniform Traffic Control Devices* is given to the subject. In congested areas, project specifications may limit the contractor to working hours when traffic volumes are low, or when noise will be least annoying to local business or residents. Seldom is access to adjacent property disrupted except for short periods. The use of water, hygroscopic salts, or other means of laying dust is widespread. Erosion control to prevent contamination of streams and provisions so that fish may have access to upstream spawning beds may at times be a requirement (See Chapter 13).

Environmental control and antipollution legislation enacted by Congress and some state legislatures make the problem of disruption of normal activities a legal as well as public relations one that demands substantial attention from contractors and construction engineers.

Excavation

Excavation is the process of loosening and removing earth or rock from its original position in a cut and transporting it to a fill or to a waste deposit. Selection of equipment depends on the nature of the material, how far it is to be moved (hauled), and the method of disposal.

Materials are usually described as "rock," "loose rock," or "common," with "common" signifying all material not otherwise classified. Rock, sometimes called "solid rock," nearly always must be drilled and blasted, then loaded with a front-end loader or power shovel into trucks or other hauling units (see Fig. 15–1). Blasted rock may be moved or drifted for short distances by means of a bulldozer, which is, in effect, a huge tractor-mounted blade (see Fig. 15–2). Loose rock includes materials such as weathered or rotten rock, or earth mixed with boulders, and often is dug with loaders or shovels without any previous blasting. At times, however, further loosening by blasting may permit faster loading and decrease equipment wear, with an attendant reduction in total cost. Loaders and shovels easily dig common excavation without blasting.

In recent years large rippers mounted on huge crawler tractors (see Fig. 15–2) and pushed by one or more added tractors have been used successfully

Fig. 15–1. Large rubber-tired loader excavating and placing material in a dump truck. (Courtesy, Caterpillar Tractor Co.)

to break up loose or fractured rock. The loosened rock is then handled by tractor-scraper units as is done with "common" excavation (see Fig. 15–3).

Grading procedures in "common" or earth excavation are governed by cost. Where material is moved less than about 200 ft or steeply downhill, drifting with a track or wheel type bulldozer (see Fig. 15–2) is cheapest. For moderate and longer hauls, self-loading scrapers pulled by rubber-tired hauling units and push-loaded by tractors offer lower costs (see Fig. 15–3). Sometimes it is more economical for scrapers to self-load by attaching long, power-driven blades that pull the dirt into the scraper bowl. For hauls of considerable length or over the public highways where axle loads are limited, rear- or bottom-dump trucks loaded by front-end loaders, power shovels, or belt conveyors may be cheapest. At times, weather may be a controlling factor. For example, the rubber-tired tractor units, as shown in Fig. 15–3, have difficulty in operating on wet, slippery roadbeds. Thus, in areas subject to heavy rainstorms, scrapers pulled by track-type tractors may be more satisfactory.

Earthmoving has undergone a revolution since 1925, when the most used tool was a scraper of 1/2 yd^3 or less capacity pulled by two or four horses or mules. For example, combinations of 15-yd^3 loaders and 125-ton trucks are now employed. Scrapers with 32-yd^3 struck capacity are marketed with a second engine mounted on the rear of the scraper to supply added tractive force. A unit similar to that shown in Fig. 15–3, with the added engine on the rear, provides a total

Fig. 15–2. Track-type tractor bulldozing broken rock. Note larger ripper tooth mounted on rear of tractor. (Courtesy, Caterpillar Tractor Co.)

of 950 hp. If two of the 410 hp pusher tractors shown in Fig. 15–2 are added for loading, a total of 1770 hp is brought to bear. Several other manufacturers make comparable equipment. In at least one instance, a 100-yd^3 scraper has been used. Some feel that the "size" revolution in earthmoving equipment is over, except in special situations. They point to the decreased number of large grading contracts, high investment costs, and the difficulty in transporting such large machines from project to project.

There are many variables in earthmoving such as the size and complexity of the project, the nature of the material, climatic conditions, and the skill and knowledge of equipment operators and supervisors. For this reason, there are no easy answers on equipment selection.[10]

Payment for excavation commonly is made at a bid price per cubic yard measured "in place" in the space originally occupied. This unit price includes payment for loosening and loading the material, transporting it any distance less than the "free haul limit" (see below), and spreading and compacting it in the fill. Payment for light clearing and grubbing and for trimming cut and fill slopes is also included in the unit price for excavation. At times, the contractor may be required to separate out and stockpile topsoil for later use or to excavate and dispose of unusable material. Pay for such operations may be based on bid unit prices or on a cost-plus basis.

[10]For information on current earthmoving projects, see the several construction magazines. J. Douglas, *Construction Equipment Policy,* McGraw-Hill, New York, 1975. is authoritative on its subject. Recent discussions on equipment economics appear in *HRB Record 316* and *454*. Equipment manufacturers and their dealers are excellent sources of information on their product lines.

Fig. 15–3. Large twin-engine rubber-tired scraper being push-loaded by a track-type tractor. (Courtesy, Terex. Corporation.)

Excavation usually is listed as "unclassified," which means that the contractor bids and is paid at one unit price, whether the material is solid rock, loose rock, or earth. Sometimes, however, excavation is segregated into "rock" and "common," and a separate unit price is established for each.

There are several reasons for using only one classification for excavation. First are the arguments that arise between the contractor and engineer in classifying borderline material. On the job, it frequently is difficult to determine the level at which rock begins and common excavation ends. Often unexpected material appears that is neither common nor solid rock. For example, even when rock excavation is defined as "ledge rock" that requires blasting, the contractor may demand and deserve payment for "rock" if he encounters a hardpan that, for economy in handling, should be blasted. Conversely, some rotten granites can be excavated with a large loader or by ripping and loading to scrapers without blasting. If a considerably higher price is to be paid when blasting precedes excavation, the contractor and engineer may often disagree on whether blasting is really required. Sometimes, in an attempt to escape such difficulties, an intermediate classification called "loose rock" is introduced, but this may increase rather than decrease the uncertainty.

Some agencies avoid the argument over classified versus unclassified excavation and still indirectly secure largely separate prices for rock and common. This is done by dividing the project into segments that are predominantly "rock" or "common" and obtaining separate unit prices for each.

Almost all materials change volume in movement from cut to fill. Excavated solid rock will expand so that 1 yd^3 of rock in the cut will occupy 1.15–1.50 yd^3 in the fill. If, however, the voids in the rock embankment are filled with earth or other fine material, the volume in the fill will just about equal the combined volumes in the two source locations. Excavated earth will expand beyond its original volume in the transporting vehicle, but it will shrink below the excavated volume when placed in the fill. To illustrate, 1 yd^3 of earth in the cut may use 1.25 yd^3 of space in the transporting vehicle, and finally occupy only 0.85–0.65 yd^3 in the embankment, depending on its original density and the amount of compaction applied. These changes in volume are referred to as "swell" or "shrinkage." Excavation, however classified, is commonly but not always measured in the space originally occupied. But, because of swell or shrinkage, the place and method of measurement must be carefully defined in the specifications.

Before earthmoving for the roadway itself and for ramps, drainage channels, and dikes can begin, points at right angles to the appropriate centerline must be staked out on the ground to mark where cut or embankment slopes intersect the original surface. At the same time, break points in the cross section are noted. By plotting and planimetering these sections, the area of the roadway prism at each section is found. At times, these "end areas" can be computed directly from the field notes by the "criss-cross" method which eliminates plotting and planimetering the sections. Then the quantity of earth between adjacent cross sections is usually computed by the "average-end-area" method which assumes that the true volume between the sections is equal to the average of the end areas multiplied by the linear distance between them measured horizontally along the centerline of the road. Usually the average-end-area method gives an answer greater than the true one. If adjacent cross sections are somewhere near the same size this difference is small; but if one area is considerably larger than the other, the difference becomes appreciable. The discrepancy can be reduced by taking intermediate cross sections or by using more exact methods of computation such as the prismoidal formula. Detailed procedures for making these computations appear in books on route surveying.

Some larger agencies have replaced these traditional methods for measuring and computing earthwork quantities with procedures based on photogrammetry and computers. These are discussed briefly in Chapter 6.

Slides and Slipouts

The side slopes of all cuts are steeper than the original ground surface and those of high fills commonly are relatively steep in order to reduce the amount of embankment. Thus, even though careful soil investigations are made as a part of the location and design procedures (see Chapters 6 and 14), slides and slipouts may occur during construction. And, from the point of view of overall economy, it may be better to accept some such failures than to employ designs that are very conservative.

Payment for slides on cut slopes commonly is made under the excavation item for the project. In rough country the estimated quantity of excavation should allow extra yardage to cover the cost of slides and to permit flattening of unstable slopes. At times, payment is more appropriately made on a "cost-plus" basis, under which the contractor is reimbursed for all costs plus an allowance for overhead and profit.[11] If and how payment is made for correcting embankment slipouts depends on the circumstances in each case.

Overbreak–Presplitting

Overbreak occurs in rock cuts when material outside the staked backslopes becomes loosened and falls or is removed along with the intended excavation. It develops either because the cleavage surfaces of the rock lie at unfortunate angles or because of overshooting. Unavoidable overbreak, all or to a fixed percentage, often is paid for by the owner at the unit price for excavation or at a separately bid unit price. If overbreak results from the contractor's carelessness or error, no payment is made.

Presplitting is a drilling and blasting procedure that may be employed or even specified to control overbreak and give a uniform face to the backslope of rock cuts. Holes are drilled at equal spacings at the desired position and slope. They are loaded with the kind and amount of explosives that, when detonated simultaneously, weaken or fracture the rock continuously along the line of drill holes. After presplitting, subsequent blasting is held within the limits set by the presplitting. Payment for presplitting commonly is included in the price for excavation.

Embankment

The term "embankment" describes the fill added above the low points along the roadway to raise the level to the bottom of the pavement structure. Material for embankment commonly comes from roadway cuts or designated borrow areas. Modern practice requires that embankment construction be carefully executed and controlled. Added detail is given later in this chapter.

Free Haul–Overhaul–the Mass Diagram

As indicated above, the unit price for excavation includes payment for transporting material from cut to fill for a distance up to a "free-haul" limit. When the material is moved a distance greater than this free-haul distance, it is usual to pay for the added transportation under a bid item called "overhaul." Units for overhaul are commonly the station-yard or cubic yard-mile. The 1979 FHWA specifications define overhaul as "the number of cubic yards of overhauled material multiplied by the overhaul distance." This distance is measured along the centerline of the roadway between centers of gravity of cut and fill.

[11]The term "force-account" is used by some agencies to describe payment on a cost-plus basis. Work done by the owner's own forces also is referred to as force-account, which sometimes creates confusion.

Overhaul distance is the total haul distance minus the free-haul distance, which today is set by agencies at values such as 1000, 1500, or 2000 ft. Overhauled volume is measured in the space originally occupied in the cut. Until the advent of motorized earthmoving equipment, the free-haul distance was commonly 500 ft, but, as noted above, it is now longer. It can be noted on Fig. 6–4 that the free-haul distance of the Arizona Department of Transportation is 1500 ft. and that overhaul is computed in cubic yard-miles rather than station-yards.

Where grading for a project involves heavy earthwork and overhaul, it is helpful to plot a "mass diagram," which provides a convenient means for studying haul and overhaul and for computing payment. Figure 15–4 shows the profile for a short section of road with the corresponding mass diagram directly under it. Free-haul distance has been set at 500 ft in order to shorten the example.

A profile has engineer's stations for its abscissa and elevation for its ordinate. The mass diagram has the same abscissa, but its ordinate represents the algebraic sum of excavation aad embankment between a selected point of beginning and any station in question. Since 1 yd^3 of excavation rarely occupies exactly 1 yd^3 of space in the fill, either excavation or embankment quantities must be adjusted before the ordinates of the mass diagram are computed.

The adjustment to put excavation and embankment quantities on a common basis is made by using a "shrinkage or swell factor," defined as the volume occupied in the embankment by material that occupied 1 yd^3 before it was excavated. For uniform material it is convenient to apply the adjustment to the embankment quantities by *dividing* them by the shrinkage or swell factor. Then the ordinates of the mass curve are in terms of excavated volume. Where the shrinkage or swell factors differ considerably between cuts, the adjustments often are applied instead to the excavation quantities by *multiplying* them by the shrinkage or swell factor for each kind of material. Then embankment quantities remain in terms of measured net volumes.[12]

Shrinkage and swell are often expressed as "percent shrinkage" or "percent swell," which represent the percent volume change between cut and fill. The volume occupied in the cut is always considered as unity. Thus if 1 yd^3 from the excavation swells to 1.20 yd^3 in the embankment, the percent swell equals $(1.20-1) \times 100$ or 20%. Again, if 1 yd^3 in the excavation makes only 0.80 yd^3 of embankment, the percent shrinkage equals $(1-0.80) \times 100$ or 20%.

Table 15–1 shows a typical computation for obtaining mass curve ordinates. The embankment yardages have been adjusted; thus the mass curve is in terms of excavated volumes. Figure 15–4 shows the mass curve as plotted from Table 15–1 and added data for the remaining length. Notice that the mass curve does not represent the total yardage but the summation of the differences between cut and fill.

Certain characteristics of the mass curve must be understood before it can be used successfully. They are:

[12]Where this procedure is followed, the ordinates of the mass curve must be adjusted when the pay item of overhaul is computed, since the volume of overhauled material is, by definition, measured in its original position.

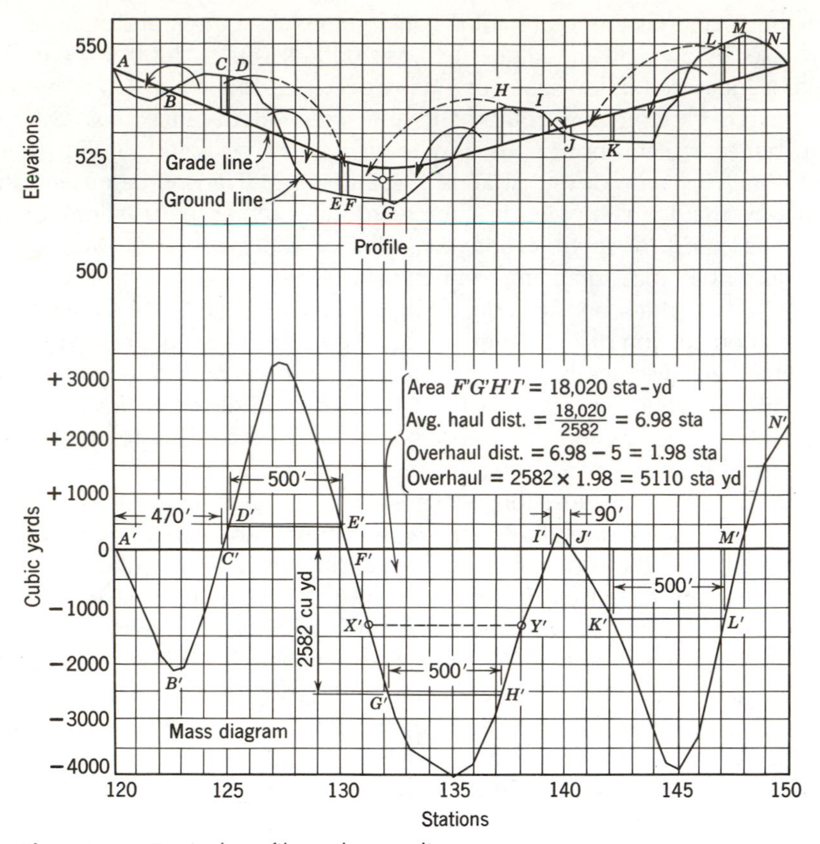

Fig. 15–4. Typical profile and mass diagram.

1. A *rising* mass curve denotes excavation at that point on the roadway; a *falling* curve denotes embankment. Where the roadway lies on a sidehill, the same cross section often shows both excavation and embankment. In such cases, a rising curve indicates an excess of excavation and a falling curve an excess of embankment.
2. Steep slopes of the mass curve reflect heavy cuts or fills; flat slopes indicate small earthwork quantities.
3. Points of zero slope on the mass curve represent points where the roadway goes from cut to fill, or vice versa. These low or high points on the mass curve may not come at the exact station at which the profile goes from cut to fill. There may be a net excess of excavation or embankment at this point if the cross slope is irregular.
4. The difference in ordinate between two points on the curve represents the net excess of excavation over embankment between those points, or conversely, the net excess of embankment over excavation.
5. If a horizontal line intersects the mass curve at two points, the excavation and embankment are in balance (equal in amount) between those points.

Use of the mass curve to compute overhaul, based on a free-haul distance of 500 ft, is illustrated in Fig. 15–4. The procedure is as follows:

TABLE 15–1. Typical Computations for a Mass Diagram.

Swell or Shrinkage Factor = 0.82

Station	Excavation (yd^3)	Embankment (yd^3)	Embankment plus Shrinkage (yd^3)	Excess material in Section (yd^3)		Mass Curve Ordinate (yd^3)
				Excavation	Embankment	
120						0
	0	321	391	—	391	
+50						−391
	0	401	489	—	489	
121						−880
	0	483	589	—	589	
+60						−1,469
	0	318	388	—	388	
122						−1,857
	0	271	330	—	330	
+65						−2,187
	205	73	89	116	—	
123						−2,071
	421	0	0	421	—	
+50						−1,650
	593	0	0	593	—	
124						−1,057
	1,421	0	0	1,421	—	
125						+364
	1,543	0	0	1,543	—	
126						+1,907
	832	0	0	832	—	
+60						+2,739
	514	0	0	514	—	
127						+3,253
	81	12	15	66	—	
+20						+3,319
	125	153	187	—	62	
+60						+3,257
	0	241	294	—	294	
128						+2,963
	0	336	410	—	410	
+40						+2,553
	0	628	766	—	766	
129						+1,787
	0	1,123	1,370	—	1,370	
130						+417
	0	1,162	1,417	—	1,417	
131						−1,000
	0	1,141	1,391	—	1,391	
132						−2,391
	0	516	630	—	630	
+50						−3,021
	0	427	521	—	521	
133						−3,542
Totals	5,735	7,606	9,277	5,506	9,048	

1. Determine all the free-haul sections. (These are sections not exceeding 500 ft in length in which the cuts just balance the fills.) This is done by locating horizontal lines exactly 500 ft in length whose ends lie on the mass curve. In Fig. 15–4 these are *D′E′*, *G′H′*, *K′L′*. Also locate any sections less than 500 ft in length within which the mass curve intersects the zero ordinate twice, as lines *A′C′* and *I′J′* of Fig. 15–4. These free-haul lines must lie either on the zero ordinate of the mass curve (*A′C′* and *I′J′*) or inside the loops formed by the zero line and the mass curve (*D′E′*, *G′H′*, and *K′L′*). If the lines are drawn *outside* the loops, the completed earthmoving plans will include "cross haul," which will mean hauling material in one direction, followed later by moving material over it in the opposite direction. The actual movements of free-haul excavated material are indicated on the profile by arrows with solid lines as stems.
2. Determine the disposition of the remaining excavation. To illustrate, the mass curve shows that the excavation and embankment are in balance between *C* and *F*. Then cut *CD* should be deposited in fill *EF*. Similarly, cut *HI* makes fill *FG*, and cut *LM* makes fill *JK*, all as indicated by dotted arrows on Fig. 15–4. Any other haul plan will result in cross haul.

   For this particular mass diagram, excavation within the section being considered exceeds embankment by about 2200 yd³, as shown by the ordinate of the mass curve at station 150. This material is not needed for fill and might be wasted or used on some other section of the road. In practice, the grade line probably would be raised to secure a closer balance between cut and fill. Inclusion of waste in this illustration merely serves to indicate an added problem that is often encountered.
3. Determine the haul distance. Any *area* such as *F′G′H′I′* on the mass curve, when corrected for scale, is in station yards, which are the units for haul and overhaul. Thus, the area *F′G′H′I′* represents the station yards of haul needed to move cut *HI* into fill *FG*. Similarly, areas *C′D′E′F′* and *J′K′L′M′* give the hauls to place cuts *CD* and *LM* in fills *EF* and *JK*, respectively.

   The average haul distance is found by dividing the appropriate area on the mass curve (station yards) by its ordinate (cubic yards of excavation).[13]
4. Compute the overhaul. The overhaul distance is the haul distance less the free-haul distance. Overhaul, the pay item, is the product of overhaul distance and cubic yards of excavation. Typical calculations for overhaul for area *F′G′H′I′* are shown on Fig. 15–4.

Mass diagrams often are more complex than that shown in this illustration. For example, if material is wasted or borrowed, the base line is shifted up or down, respectively. Again, unusual haul arrangements to provide for the placement of the more desirable materials near the finished roadway surface or to dispose of poor materials in deep fills or under the side slopes sometimes must be introduced.

Borrow and Waste

In grading for a highway, "borrow" material often must be brought into the roadway from outside the grading prism. Thus, in flat country, it often is desir-

[13]The width of the overhaul area at midheight (line *X′Y′* on Fig. 15–4) gives a close approximation of the haul distance. The shorter solution gained by using this approximation is satisfactory for many purposes.

able to raise the roadway surface 4 or 5 ft above the groundwater table, or above the reach of flood waters. On prairies subject to snowfall, the wind usually will keep a slight fill free of snow. Approaches to major bridges or elevated urban freeways often traverse high fills for a considerable distance. All such hauled-in materials are commonly classified as "borrow."

In the past, material for fills in level country has often been obtained as "side borrow" from the roadside within the limits of the right way. Poor side-borrow practices have produced many troublesome conditions. For example, where rights of way were narrow, deep ditches were left along the roadside to become an accident hazard and a trap for rainwater. Correct practice permits side borrow only when a safe and well-drained cross-section can be maintained.

Much borrow today comes from large "borrow pits" located away from the right of way and from widened sections in adjacent cuts. In some instances, urban freeways have been designed in alternate sections of cut and fill, each a mile or more in length. At times, the agency may designate the source of borrow; again the contractor may select the pit, provided the material meets the specifications. Designation of the source of borrow can, of course, assure a superior material. Some specifications stipulate that all borrow pits must be wholly invisible from the road; others that all pits must be neatly trimmed and dressed after the material has been removed.

Particularly in urban or suburban areas, zoning restrictions may preclude an agency from opening borrow pits or disposing of excess excavation. Again, hauling over the streets may be prohibited. These limitations make construction considerably more expensive.

Pay for borrow usually is by the cubic yard, measured in the space originally occupied in the borrow pit. Occasionally, the borrow volume is determined in the completed fill. Some agencies have been paying for borrow on a weight basis, at a bid price per ton. This practice is especially common when the borrow is for "select" material for the roadway or for topsoil.

Excavated material which sometimes remains after all fills within an economical haul distance have been completed is designated as "waste." Waste is often used to widen fills or flatten fill slopes of adjacent sections. No payment other than for excavation is involved when material is wasted, except for occasional overhaul.

CONSTRUCTION OF FILLS AND EMBANKMENTS

Principles of Embankment Construction

At one time, fills, whether of rock or earth, were constructed to full height by "end dumping" from the transporting vehicle. In this procedure the material slid or rolled into place down the face of the progressing fill. No attempt was made to control moisture content or to secure compaction. It was anticipated that such fills would settle for some time and that initial pavement, if any, was temporary. Construction of permanent pavement over high fills often was deferred for a year or more after completion of the fill to allow this settlement to occur.

In the 1930s engineers found that superior embankments could be constructed by spreading the material in relatively thin layers and compacting it at a moisture content close to optimum. The improvement resulted largely because greater density was obtained, which resulted in higher "strength" in the soil mass and in decreased settlement and rutting. Layered construction also produced greater uniformity in the material itself and in its density and moisture content. This was beneficial since any subsequent consolidation or swelling would be relatively uniform. By contrast, in fills constructed by end dumping or by placing in thick layers, material, density, and moisture content could vary greatly from one spot to another. Volume change would be nonuniform and would result in differential settlement or swell between adjacent areas. Today, engineers agree that proper embankment construction requires that the soil be spread in layers, moistened or dried to something near an optimum moisture content, and compacted. As will be shown subsequently, the standards that have been set and the testing and control mechanism for reaching these desirable ends vary widely among agencies, with each adapting its practices to its local conditions.

Recommendations for minimum densities of embankments and subgrades from the 1978 *AASHTO Specifications for Transportation Materials* for the various soil classifications are given in Table 15–2. These are not accepted by all agencies. For example, the 1979 FHWA specifications stipulate 95% of AASHTO standard density (*T-99*) for all subgrades and embankments. A rough comparison with the requirements of the California Department of Transportation, which has its own compaction device, calls for densities in the range of 98 to 108% of the AASHTO standard for different soil types.

The AASHTO minimum density requirements in the early (1955) specifications (*M57–55*) gave users the choice between standard density (*T-99*—5½-lb hammer—3 lifts) and modified density, (*T-180*—10-lb hammer—5 lifts). This represents differences of possibly 6 to 16% for setting standard density. The current standard is based on *T-99* alone. This change might be construed as a suggestion that the density requirements of some agencies were too high. Another interpretation is that, with greater knowledge of the wide variability in the results of tests for density (see Chapter 14), attention should be focused on other approaches for assuring uniform quality and performance of embankments. Even so, it seems clear that control by density testing will, for the time, remain the primary means of assuring quality.

Proposals for changing the AASHO density requirements, made in 1961 by the (then) AASHO-ARBA Joint Subcommittee on Compaction of Earthwork (reported in *American Road Builder*, Sept. 1961), but never implemented, offer sensible alternatives, in light of what is known of the behavior of soils. Briefly paraphrased, these are:

1. Granular soils or those with only a slight degree of plasticity should be compacted to 95–100% of Modified AASHTO density. This approaches their maximum compactibility.
2. Fine-grained soils possessing a low degree of plasticity should be compacted to ap-

TABLE 15–2. Recommended Minimum Requirements for Compaction of Embankments and Subgrades.*

Class of soil (AASHTO M145)	*Minimum Relative Density Requirement+*		
	Embankments		
	Under 50 ft High	*Over 50 ft High*	*Subgrade‡*
A-1, A-3	95	95	100
A-2–4, A-2–5	95	95	100
A-2–6, A-2–7	95	§	95‖
A-4, A-5, A-6, A-7	95	§	95‖

*Based on *AASHTO Designation M57–64* (1974).
+Densities to be determined under *AASHTO Designation T-99* (5½ lb hammer).
‡*AASHTO Designation M146* defines subgrade (basement soil) as the prepared and compacted soil immediately below the pavement system and extending to such depth as will affect the structural design.
§Use of these materials requires special attention to design and construction.
‖Compaction at 95% of *T-99* optimum moisture content.

proximately 100% of Standard AASHTO at a moisture content near laboratory optimum.

3. Densities for soils possessing moderate to high plasticity should be determined by the Standard AASHTO Method. Densities should not be so great as to lead to subsequent swelling. Compactive procedures should not exceed the soil's shear strength, thereby producing parallel particle orientation.
4. Clay soils should not be overcompacted (beyond about 90% Standard AASHTO density) to avoid postconstruction swelling.

Control of Embankment Construction

There are two basically different control procedures for assuring a specified embankment density. The first is to state the expected end result which is the minimum acceptable value for relative compaction, and to make sure, by field density tests, that the specified density is obtained. Techniques and devices for field density control and their usefulness and accuracy were discussed in Chapter 14. The second procedure is to state the manner and method for constructing the embankment. In this case, layer thickness, moisture control, and the number of passes by a roller of specified type and weight are predetermined. Field control is largely a matter of conducting the specified procedure. An in-between combination of "manner and method" and "result" is set out as an alternative in the 1979 FHWA specifications. They stipulate that for each situation a 400 yd^2 (330/m^2) control strip will be constructed and compacted until "no discernible increase in density can be obtained by additional compactive effort." The mean density of this strip becomes the target density for the work, provided it reaches 98% of the density of the stipulated laboratory test. If it does not, a new test strip must be constructed.

Theoretically, control of embankment construction is easy, but, as a practical matter, it offers many problems. Getting and maintaining a suitable and uniform moisture content is difficult, given such problems as variable soil wetness to begin, problems in evenly distributing and mixing in water, and evaporation or rainfall during compaction. Some soils may not easily assimilate moisture unless wetting agents are employed. Often soil types cannot be segregated during construction as they can in the laboratory. For example, efficient equipment operation may demand that loads of unlike materials from several sources be combined in a single embankment. Again, the handling of poor materials is often troublesome. To illustrate, soils that are unsatisfactory as embankment material should be placed deep in the fill or under the side slopes where they will carry little load, and construction operations must be planned accordingly. Aeration with a motor grader or traveling mixer of soils that are too wet is another troublesome and expensive operation. All these as well as other circumstances which confront construction personnel make for many difficulties.

Special care to ensure stability and hold settlement to acceptable limits must be exercised in constructing high embankments.[14]

Current Practices in Embankment Construction

Where embankments provide the supporting structure for pavements or tracks, it has been almost universal practice since World War II to build them up in relatively thin horizontal layers, each of which has been compacted at a moisture content near "optimum." Within this general method, individual agencies have developed procedures which differ in certain details. Reports from the late 1960s and early 1970s[15] are typical of today's practices, which can be briefly summarized as follows:

DENSITY REQUIREMENTS. A large majority of agencies specify the end result, which is that embankments for some or all projects be compacted to a stated minimum relative compaction. Certain others prescribe manner and method by requiring a certain number of passes of a roller of given characteristics.

MOISTURE CONTROL. Most of the agencies stipulate moisture control, but in a qualitative way, which leaves judgment in the hands of the contractor, engineer, or inspector. But to meet the density requirement, overly wet soils must be aerated; moisture must be added to dry ones. There are, however, a few agencies that specifically stipulate that moisture percentages be within, say, 2% of optimum.

Field-moisture determinations, to be useful, must be accomplished quickly and done often. For this reason, nuclear devices (see Chapter 14) have come into widespread use. Drying samples in an oven, as is commonly done in the laboratory, is less satisfactory as several hours are required. On the other hand, with specially built forced-draft or microwave ovens, or by igniting an alcohol-covered sample, answers can be found in 30–45 min. The Proctor penetration

[14]See articles by E. B. Hall, T. Smith, and W. F. Kleinman, *HRB Record 345* for precautions taken in constructing an embankment 383 ft high. See also J. C. Chang and R. A. Forsyth, *HRB Record 457*.

[15]See, for example, *NCHRP Synthesis 8* and *HRB Record 177*.

needle is used by some. With it, calibration curves are prepared showing the relationship between penetration resistance and moisture content of compacted samples. Field samples can be compacted and penetration resistance and moisture content determined in 10 min. Also, a device has been developed which utilizes the reaction of calcium carbide and water to produce acetylene gas. Gas pressure is then related to soil moisture content. Results can be obtained in 5–10 min; however, they are subject to error in heavy clays unless the soil is pulverized (see *AASHTO Designation T-217*).

Experienced engineers, after becoming familiar with soils, can often judge moisture content quite closely by examination. General guides are that friable soils at optimum moisture content (AASHTO Standard) contain sufficient moisture to permit forming a strong cast by compressing the soil in the hand. For some clays, optimum moisture approximates the plastic limit, and can be judged by forming a ribbon, thread, or cube of a sample.[16]

COMPACTION EQUIPMENT AND PROCEDURES. Rollers or other compaction devices increase soil density by expelling air from the voids in the soil and by rearranging or forcing the soil grains into more intimate contact. Water aids compaction up to the optimum moisture content. In porous soils the individual particles are rearranged by pressure and vibration, but in cohesive soils much effort is required. Because of this, heavy cohesive materials must be placed in thin layers if the charges between individual particles are to be overcome.

There is a practical limit to the compaction that can be obtained with a given roller, for the added compaction obtained with repeated loads soon becomes very small. Repetitions of load are particularly effective in increasing the density of fine-grained soils but have less effect on coarse-grained materials. Frequently, then, heavier rollers or a different method rather than many load repetitions may offer a better way to produce increased densities. However, heavy rollers are not a cure-all for compaction problems. The cost and trouble of moving heavy rolling equipment over the highways from one project to another may preclude its use on small projects. More important, many fine-grained soils are remolded and weakened by heavy rollers but can be satisfactorily compacted in thin layers with lighter equipment. With such soils, it may sometimes prove advisable to replace heavy tractor-scraper earthmoving equipment with lighter units.

Compactors include tamping, sheepsfoot, grid, pneumatic-tired, and smooth steel-wheeled rollers; segmented plate compactors; vibratory compactors in the forms of vibrating pads, steel wheels, or pneumatic tires; and finally, the hauling and spreading equipment itself. In some instances two types of compactors, such as smooth steel wheels and pneumatic tires are combined in a single unit. With most compactor types, units are available in several weights or provision is made to change the weight readily. There is small wonder then that compactor selection is a difficult task.[17]

[16]See J. W. Spencer, *HRB Record 284* and *Text 9* of *Compendium 10; Transportation Technology* Support for Developing Countries, TRB.

[17]*HRB Bulletin 272* gives a detailed report and bibliography on many performance tests for different types of compactors. These were conducted in Great Britain and several other countries as well as in the United States. See also *Compendium 10,* op. cit.

Fig. 15–5. A self-propelled tamping-foot soil compactor with attached blade for spreading and leveling loose material. (Courtesy, Galion Manufacturing Division, Dresser Industries Inc.)

The commonly used "result" specifications, which stipulate only density, leave the choice of rollers and rolling procedures to the contractor. Even so, agencies may also specify equipment details; for tamping rollers these have included weight, area of tamping feet, or foot loading in psi. A relatively simple set of requirements, from the FHWA 1979 specifications, calls for minimums as follows: for sheepsfoot, smooth-wheel, and grid rollers, 250 lb per linear inch of drum; for pneumatic-tired rollers, 80 psi; and for vibratory machines, 6000 lb weight.

Tamping rollers (see Fig. 15–5) consist of hollow metal drums to which tamping feet have been attached. A *sheepsfoot roller* differs in that the individual feet cover a smaller area. One or more drums are mounted on an articulated frame which permits uniform load on all the feet. Often rollers are self propelled; again, an assembly may be pulled by a tractor. Pressure on the feet may be increased by filling the drum with water, sand, or a high-density fluid, such as a water-baroid slurry. The tamping feet first penetrate the soil and compact the bottom of the layer; then as consolidation proceeds, they "walk out" until the layer is consolidated and the feet barely penetrate. Tamping rollers vary from light units which weigh 6000–10,000 lb for an 8-ft width, to giants which, when fully loaded, weigh up to 75,000 lb for a 10-ft width. Pressures on the feet vary correspondingly. Tamping rollers are widely used, but their effectiveness in coarse-grained soils is questioned as the feet tend to tear and displace the material rather than compact it.

Grid rollers are basically a hollow metal cylinder, but the facing is a grid of metal bars spaced a few inches on centers. They also penetrate the uncompacted soil, but to a lesser degree than a sheepsfoot. They are particularly effective in breaking down clods and soft rock.

Pneumatic-tired rollers (see Fig. 15–6) consist of rubber tires mounted on an articulated frame which provides uniform load on each tire. The roller is so built that the load can be increased by filling the body with water or wet sand and the tires with calcium chloride solution. Motive power sometimes is provided by a tractor or, as in Fig. 15–6, the unit is self-propelled. Some units are so constructed that the operator can change tire pressure without leaving the driver's seat. Compaction under pneumatic tires comes from shearing and vibration which are particularly effective on loose, sandy soils. Pneumatic rollers for highway work are generally 8 tons or more. For airport work, rollers weighing up to 200 tons have been used. For example, a mammoth, two-wheeled, rubber-tired roller is reputed to have compacted a sandy subgrade on the Baltimore International Airport to a depth of 5 ft.

Smooth-tired rollers, of the two-wheeled (tandem) or three-wheeled type, long have been used to compact bases and bituminous surfaces (see Chapter 19). Such rollers compact the soil from the top downward, whereas the sheepsfoot works from the bottom upward. Many engineers prefer smooth rollers for compacting fills composed of glacial till or like material in which large rocks are combined with finer material.

Vibratory Compactors. Several manufacturers have developed pneumatic or smooth-wheeled compactors equipped with devices for vibrating the wheels. In addition, a variety of plate type units are available. These may be self-propelled or mounted as auxiliary equipment on rollers or other units. Generalizations on

Fig. 15–6. Pneumatic roller compacting fill. The weight of such rollers can be varied by filling the tires with water or calcium chloride solution and the body with sand or scrap iron. (Courtesy, Bros Division of American Hoist and Derrick Co.)

the performance of vibratory compactors are difficult, since frequencies and amplitudes differ widely among makes. They have, however, proved highly effective in compacting stones, as in macadams, and with clean sands and gravels. Performance in clays and other fine-grained soils has been less predictable.

Hauling and Spreading Equipment. Sometimes sufficient compaction can be obtained by properly routing the heavy hauling equipment. If no other compaction is provided, strict control must be maintained to see that the entire fill is covered, otherwise some areas will be missed and a nonuniform result obtained. Because control is so difficult, many engineers are reluctant to rely solely on this method.

THICKNESS OF SOIL LAYERS. The ideal layer thickness would be that which gained the required density at the least total cost. Variables would include the cost of placing and spreading the material, correcting its moisture condition, and compacting it with a roller of the proper type and weight. Each of these costs would be different for each soil type. With so many alternatives and so many factors still unknown, wide variation in current practice should be expected and does exist. Among the state highway departments, maximum permissible layer thickness for embankments before compaction ranges between 6 and 12 in., with 8 in. the most prevalent.

COMPACTION OF ORIGINAL GROUND. Present-day practice requires that the original ground be compacted before the overlying layers are placed. Depths of the compacted layer range from 4 to 12 in., with 6 in. the most popular.

A study of pavement failures reveals that many of them occur in cuts or on low fills. Often these failures result because materials of low density lie close under the pavement surface. It must be concluded that a specification that merely requires compaction of the subgrade is not enough; instead, subgrade compaction must be related to depth below the surface of the pavement. At times it may even be necessary to remove the soil and replace it in compacted layers. At locations where the grade line passes from cut to fill, the further precaution of replacing the topsoil with suitable borrow material may be warranted. These contingencies are covered in the California specifications by requiring a relative compaction of 95% (California impact test) to a depth of at least 2.5 ft below the pavement surface.

COMPACTION FOR TRENCH AND STRUCTURE BACKFILL. Compaction of backfill for trenches and structures is important since later settlement will produce roughness in the finished roadway. Yet the backfill areas often are too confined to permit the use of large-scale compaction equipment. To meet this need manufacturers have produced a wide variety of hand-operated mechanical tampers, rammers, and plate type vibrators. These are propelled by compressed air, explosions of fuel, or electricity. Some of them may be designed to attach to backhoes or other equipment.

The principles of compaction as already stated apply equally to backfill problems, except that the scale of operations is different. Almost all highway agencies require that backfill be compacted. Many stipulate moisture and density control; the others specify layer thickness and the procedures and equipment for tamping or vibrating. Water settling by flooding the trench or by jetting which, until the recent past, was the accepted method is almost never accepted if the location falls under pavement. There is a growing trend toward importing granular backfill materials if those from the excavation for the trench or structure would be difficult to compact.

Payment for Embankment Construction

Methods of payment for embankment construction vary. Some agencies specify that the price per cubic yard of excavation or of borrow shall include all charges for compaction, for water as needed, or for aerating the soil. Others, particularly in the arid parts of the country, pay separately for water. This payment may be for water alone at a price per 1000 gal of water, or in two items: one for installing a water supply and the other for water per 1000 gal. Still others pay for rolling, either at a bid price per hour of rolling with an acceptable roller or at a bid price per cubic yard of material. None pay for aeration or other drying unless by special provision on individual contracts.

It should be noted that if the payment for compaction is covered solely in the price for excavation or borrow there is no financial incentive for contractors to gain high densities. Such incentives do exist if they receive separate payment for rolling.

FINISHING THE ROADBED

Finishing describes a number of operations associated with shaping the roadbed and the rest of the cross section. Surfaces that will underlie the traveled way and shoulders are compacted and smoothed to close tolerances. Roadside and other ditches, backslopes in cuts, sideslopes on embankments, and roadside areas will be dressed to the requirements set by the plans and specifications. Generally there is no specific pay item covering these finishing operations; rather the contractor is instructed to include these costs in other bid items.

SUPPLEMENTARY CONSTRUCTION

Major projects such as freeways, expressways, and arterials usually include not only grading but bases, pavements, and a variety of supplementary construction activities. Among them are drainage facilities such as storm drains, culverts, and bridges (see Chapter 11), pumps and conduits to remove water from depressed roadways, signs, traffic signals, and lighting systems (see Chapter 10), and roadside planting (see Chapter 13). Because coordination is required between all construction activities, they are sometimes lumped into a single contract. The prime contractor who bids the work in turn relies on specialty subcontractors to carry out their portions of it.

Payment for such supplementary work, whether in a single contract or separate ones, is on a unit-price or lump sum basis for the specified work items. Where a single contract includes most of or all the items listed above, the number of bid items often exceeds 100.

EMBANKMENTS THROUGH MARSHY AREAS

Highways and rail facilities sometimes must pass through swampy or marshy areas where the mucks and peats will not provide stable support for the fills. Here conventional construction is unsatisfactory and special procedures are necessary. These include the removal and replacement of unsuitable material and various methods for its displacement; and surcharging, vertical sand drains, fabric reinforcement, and weight reduction.

Removal and Replacement Method

The removal and replacement method is suitable where the unstable material is shallow. It calls for removal to the level of the underlying stable material before fill construction is begun. Sometimes, after excavation, the road site lies below water, but many such installations have been completed satisfactorily. On a project through tidal lands along San Francisco Bay, excavation was performed with floating clamshell dredges. The succeeding hydraulic dredger fill was of sand from the bay bottom, transported up to 5 mi through the discharge pipeline.[18]

Displacement Methods

Many fills have been built over marshes by displacing the unsuitable muck with better material. Displacement may be caused by the weight of the fill and possibly a surcharge or by this weight supplemented by explosives or water jetting. Satisfactory fills constructed in 80 ft of unstable material are on record, although success at this depth was under extremely favorable conditions.

For shallow fills through mucks up to about 10 or 12 ft deep, the imported material is carefully placed so that it slides along the advancing fill slope, flows under the less dense muck, and displaces it sidewise. Again, a trench about the width of the embankment may be blasted out and immediately backfilled with stable embankment material. The blast throws part of the muck out of the trench, and an added amount is liquefied so that it is easily displaced.

Where the muck is deeper, the "underfill" method often is used. A trench is blasted and a large portion of the fill material placed. Explosive charges, set off in the underlying muck, force it from under the embankment, which settles into place. A refinement of the underfill method is the relief method. After the fill material is placed, relief ditches are blasted along the sides of the fill to make

[18]For a summary of this and certain other methods, see L. H. Moore, *HRB Record 133* and *NCHRP Synthesis 29*.

displacement of the underlying muck easier. In this way, the explosive charge can be reduced and disruption to the fill minimized.[19]

Surcharge Method

The instability of peats and mucks results largely from their extremely high-moisture content, which is often 100% or more of dry weight. These materials contain relatively little soil; for example, at 100% moisture content a typical cubic foot contains only 45 lb of solids. If a sufficient portion of this moisture can be dispelled and the soil grains forced into more intimate contact, the supporting power may become adequate.

The surcharge method is begun by laying a working table of fill material over the muck to support construction equipment. Where the top of the muck is soft it may be displaced by careful fill placement. The fill then is built in compacted layers until its height, allowing for settlement, approximates final grade. A surcharge of uncompacted fill material is then placed. This added weight accelerates the flow of water from the muck and speeds up consolidation. Care must be exercised in order that loads do not exceed the shearing strength of the muck; otherwise it will be displaced, and disastrous settlement of the fill will result. After a period of consolidation, the surcharge material is removed for other purposes, and base course and temporary pavement are placed.

The surcharge method is used for low fills over shallow muck up to 12 or 15 ft in depth. It may not provide for complete consolidation during the construction period, but it does obtain most of it. If the muck is of uniform consistency and moisture content, subsequent slow settlement will be fairly uniform and the surface will develop only minor roughness. However, large differential settlement and a rough roadbed will develop where subsurface conditions vary greatly. Even in these instances, releveling after several years have passed and consolidation is largely completed produces highly satisfactory results.

At times, laymen and even engineers complain about the roughness of roads constructed by the displacement and surcharge methods. These individuals have failed to recognize that this temporary roughness represents a small price as compared with the cost of construction by the removal and replacement method.

Vertical Drain Method

Vertical sand drains provide rapid consolidation of deep layers of muck. They were first used in 1934 by the (then) California Division of Highways; since then many others in the United States and abroad have also made successful installations. An outstanding example is the New Jersey Turnpike, where the combined length of sand drains underlying the 113 mi road as originally constructed is 947 mi.

[19]For added information on displacement and surcharge methods, see *A. W. Root, HRB Bulletin 173;* O. Stokstad and K. Allemeier, *HRB Bulletin 254;* M. A. J. Matich and F. C. Brownridge, *HRB Record 57;* and S. J. Johnson, *Journal of the Soil Mechanics Division of ASCE,* Jan. 1970.

Sand drains are vertical columns of sand that penetrate the muck almost to solid material. Across their tops is a horizontal sand blanket extending through the side slopes of the fill (see Fig. 15–7). Under the pressure induced by fill and surcharge, subsurface water flows up the drains and out of the fill. Rapid consolidation of the muck results.

It has been proved that, other things being equal, the time required to force water from a fine-grained soil varies about as the square of the distance the water must travel. Without sand drains, this distance is at least half the thickness of the compressible layer; with them, it is half the distance between drains, often less than 5 ft. Furthermore, most mucks have greater permeability in a horizontal than in a vertical direction, and the drains provide an easy means for the vertical movement of water. It has been demonstrated that sand drains produce the same consolidation in one-tenth to one-hundredth of the time. Stated differently, 10–100 yr settlement is gained in 1 yr. Figure 15–8 gives time-settlement data on an installation in New York.

Sand-drain construction begins with the placing of a working table of fill material (see Fig. 15–7). Special equipment as described subsequently operates from this table to install the vertical columns of sand. Placing of the sand blanket, compacted fill, and surcharge is done by conventional methods.

As fill and surcharge are placed over the sand drains, pore water pressure develops in the muck, and, if this becomes too great, the muck will be displaced from under the fill. It is possible to measure these pressures by gauges connected to well points driven to proper depths. When the gauge readings approach the predetermined limiting value for the muck, filling is halted and not resumed until the pressure has fallen to a safe level. Brief descriptions of several ingenious methods for installing the vertical sand columns follow.

DRIVEN-MANDREL METHODS. A hollow steel tube with a hinged bottom is driven down by a pile driver. The bottom is closed during driving to keep the mandrel empty. After the driven tube has been filled with sand, it is slowly withdrawn, and the sand flows out through the bottom and fills the hole. Often,

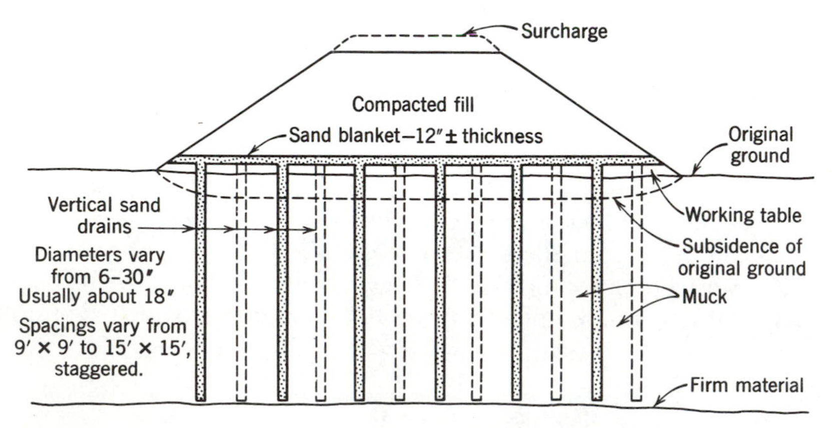

Fig. 15–7. Typical vertical sand drain installation.

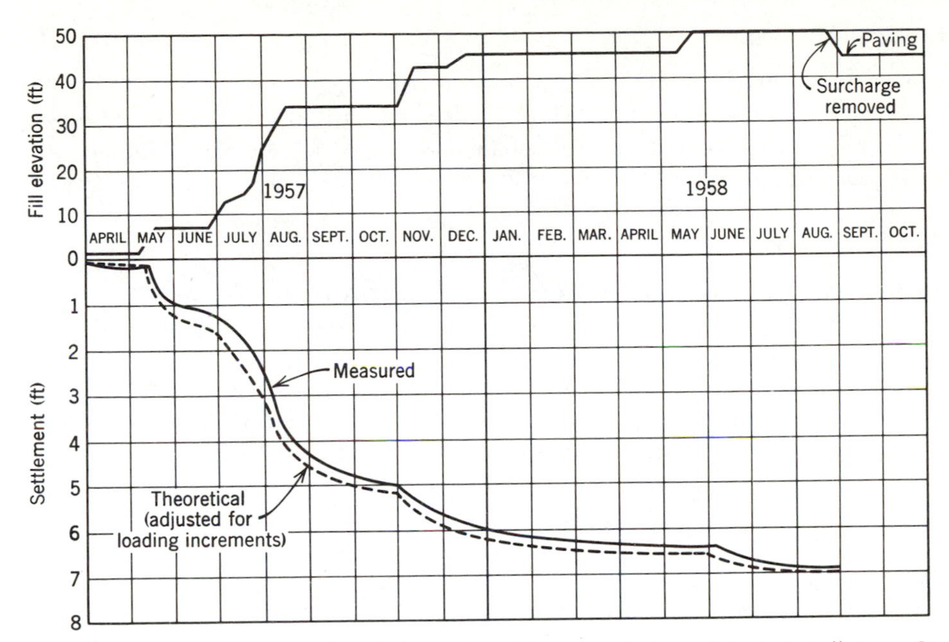

Fig. 15–8. Embankment and settlement record for a vertical sand-drain installation, Oswego Blvd., Syracuse, N.Y. (After L. D. Moore, *HRB Record 133*.).

compressed air is charged into the top of the tube to force the sand to flow more rapidly. Drains over 100 ft deep have been installed in this manner. The driven-mandrel method is more widely used than any other. However, it is unsatisfactory in cohesive soils that will cling to the mandrel, as this may distort and block the channels of moisture flow or prevent withdrawal of the mandrel. Neither is it satisfactory where the jarring action of the pile driver will damage adjacent structures.

JETTED-MANDREL METHOD. Water jets in the base of the mandrel open the ground for it. Sand is placed as the mandrel is withdrawn. Both "closed-end" and "double-walled" types of jetted mandrels have been successful.

AUGER METHODS. Various forms of soil augers also have been employed in sand drain installations in soils that will not collapse or squeeze rapidly into the opening. One of these, the flight auger, has a continuous helical fin mounted on the outside of a hollow shaft. The auger penetrates as it is turned, so that the muck is distorted as little as possible. When the auger reaches the desired depth, a further twist, with vertical movements prevented, shears the muck. As the auger is withdrawn, sand to fill the void is fed through the center shaft.

Rotary drills and rotary jets also have been employed to open a space for the sand. These are not satisfactory where the muck will collapse or squeeze into the hole before the sand can be placed.[20]

[20]Recent articles on sand drains, with references to earlier studies include T. R. Kuesel et al., *HRB Record 457*, H. Su et al., *HRB Record 463*, and articles by R. E. Landau, R. D. Holtz, R. P. Long and P. J. Carey, and A. A. Semour-Jones, *TRB Record 678*, and S. J. Johnson, op cit.

Fabric Reinforcement

Embankments through marshes have been constructed by first covering the muck with a permeable fabric. Through its strength, the fabric adds tensile reinforcement to the bottom of the embankment. Design procedures for such installations are beyond the scope of this book.[21]

Weight Reduction

Installations have been made over marshes employing blocks of styrofoam to reduce the weight of the fill. In other instances, sawdust, flyash, which is a lightweight residue from the burning of coal, or foamed concrete formed by adding a liquid concentrate of hydrolyzed protein to a cement and water slurry have been substituted for earth. To date only small, experimental projects have been constructed and applications probably are limited to special situations.

ROADBED CONSTRUCTION IN DEVELOPING COUNTRIES

This chapter has focused on construction management, methods, and equipment appropriate for developed countries where large projects and a high level of mechanization are appropriate. It does not follow that these approaches are suitable for developing countries. Often, in them, labor is plentiful and its cost is low. Furthermore, equipment acquisition, operation, and maintenance is unattractive because the costs of machines and spare parts are high, foreign exchange to purchase them is scarce, and skilled operators and mechanics may be lacking. These factors argue strongly for labor-intensive methods. For example, short-distance earthmoving by wheelbarrow or animal-drawn wagon may be appropriate, possibly with alignment and grade altered to eliminate long hauls. Often local materials such as stone and timber may be substituted for concrete or steel and this will provide employment for local workers. These examples can be extended into many other areas, but a discussion is outside the scope of this book.[22]

PROBLEMS

15–1. Determine the Federal Highway Administration construction cost index for the current year and 5 and 10 yrs back by referring to the most recent construction cost issue of *Engineering News-Record*. From these, find the percentage change in the index over the last 5 and 10 yr.

15–2. From the standard specifications of the state highway or transportation

[21]See Chapter 14 for a brief description of fabrics and for references.

[22]The World Bank, AID, and the International Labor Organization are among the agencies that have studied this problem in depth. They are the sources of much valuable information. Helpful publications in the United States include *Synthesis 3* in the TRB Series Transportation Technology Support for Developing Countries, *TRB Special Report 160,* and *TRB Record 702*. Other valuable publications have been released by the agencies named above and the Transport and Road Research Laboratory of Great Britain.

department where your college is located, make a list of the items for which acceptance and payment are based on "statistical" rather than single "acceptance or rejection" criteria.

15–3. Obtain a set of plans, specifications, and contract documents for a state or local highway project. From them determine the basis of payment, if they are paid for separately, for the following grading items: clearing and grubbing or site clearance, excavation, overbreak, overhaul, borrow, watering, and compaction. If possible, determine the unit prices for these items for one or more typical projects.

15–4. By a study of the specifications of the state highway or transportation department of your state, determine its practices for controlling embankment construction. Consider each of the topics discussed in the textbook.

15–5. Investigate and report briefly on the procedures followed to construct a highway or street through a marsh or other difficult condition in the area where your college is located.

16 GRAVEL AND CRUSHED ROCK ROADS—STABILIZED ROADS

Roadway fills are constructed of material from adjacent cuts or nearby borrow pits. On these fills are placed the base courses and wearing surfaces on which vehicles travel (see Fig. 14–1). The nature, thickness, and composition of these upper layers will vary widely with the volume and character of traffic, the cost and availability of materials, and the opinions of roadway designers. Common to all is the fact that they are composed largely of soil, sand, and rock. Often a small percentage of bituminous material, cement, or a salt may be added to the surface or the underlying layers, but the bulk always comes from the earth's crust, usually from some location fairly near the project under construction. No other material presently available can compete with these in cost.

Materials in bases and wearing surfaces must be of higher quality than those suitable for the subgrade or embankment as greater demands are placed on them. This higher quality is obtained by careful selection of the source material and (usually) by special processing such as crushing or screening. Sometimes materials from two or more sources are combined to secure a better product.

Given the worldwide variety of materials, climates, and service needs, many methods and processes for producing good road surfaces have been developed. At one extreme is the earth road, constructed by shaping the native soil to the cross section of the finished road. At the other are found high type concrete and bituminous pavements separated from the subgrade by a considerable thickness of relatively expensive and carefully selected base course. It is the engineer's responsibility to decide which of the many possible choices best meets the requirement of lowest overall cost for each job.

In the United States (1979) some 1.1 million mi of graded rural roads have surfaces of untreated soil mixtures or gravels (see Fig. 16–1). In addition, another 650,000 mi are unimproved or graded and drained but unsurfaced, and they present the same problems. The bulk of these mileages are under the jurisdiction of local rural road agencies. However, 84,000 mi (11%) of state-

Fig. 16–1. Motor grader blading a rough road surface. (Courtesy, Galion Manufacturing Division, Dresser Industries, Inc.).

administered highways also fall under these classifications, and so the problem is not entirely a local one. Urban areas have some 90,000 mi of ungraded streets or those surfaced with untreated soil mixtures. The quality and maintenance of these surfaces varies widely, depending on the policies and affluence of the agency involved.

This chapter describes the design and construction of road surfaces from soils or other natural material and also deals briefly with certain methods for stabilizing these natural materials by blending those from two or more sources or by treating them with agents such as calcium or sodium chloride.

ROAD SURFACES OF UNTREATED SOIL MIXTURES

General Requirements

For roads of low traffic volume, surfaces of untreated soil mixtures described as gravel or crushed rock are widely used. They consist largely of stone pebbles or crushed-rock particles combined with clay, lime, iron oxide, or other fine material in sufficient amount to bind the coarse particles together. Many locally occurring materials such as lime rock, shells, caliche, chert, and volcanic cinders have also been satisfactory.

The upper limit of traffic volumes for which untreated road surfaces are economical varies, but is low; possibly in the range of 100 to 250 vehicles per day when vehicle operating costs are considered (see Chapter 4). At higher traffic volumes, surface pitting, the formation of transverse corrugations, the cost of replacing materials that erode or blow away, and the dry-weather dust problem all mitigate against their use.[1]

Requisites of road surfaces of untreated soil mixtures are as follows: First, stability—that is, they must support the superimposed loads without detrimental deformation. Second, they must stand the abrasive action of traffic. Third, they should shed a large portion of the rain which falls on the surface since a large amount of water penetrating the surface might cause loss of stability in the wearing course or softening of the subgrade. Fourth, they should possess capillary properties in amount sufficient to replace the moisture lost by surface evaporation and thus maintain the desirable damp condition in which the particles are bound together by thin moisture films. Fifth, they must be free of large rocks or stones (above 1 in. diameter) so that they can be maintained by blading or dragging. Finally, they must be of low cost, as funds for the improvement of low-traffic roads are limited. This, in turn, limits the sources of materials to the immediate locality, because of high costs involved in transporting them any great distance.

Further criteria are:

1. Control of grading is essential to ensure stability of granular mixtures containing soil binders. Limits on the percentages passing the No. 40 and No. 200 sieves offer a suitable check on the part passing the No. 10 sieve.
2. Control of the plasticity index is essential. Road-surface materials with a low but measurable plasticity index are to be preferred to absolutely nonplastic ones and are decidedly superior to those having appreciably higher plasticity index values.

The quantity and character of the clay or other fines is very important, as it serves as both a binder and a moisture regulator. In dry weather, the moisture film on the clay particles should bind the entire mass together. In wet weather, the first rain that falls on the surface should cause the clay to expand and close the pores, thereby preventing water from entering and softening the material. Traffic over the wet surface merely creates a thin layer of nonslippery mud which rebinds when drying. The effect of excessive amounts of highly expansive clays, when wet, is to swell and unseat the coarser materials, thus weakening the road.[2]

Where no underlying source of capillary moisture is present, long dry periods without rainfall result in deterioration of untreated surfaces. When the moisture film around the clay particles disappears, the binding power of the clay also disappears, and the surface disintegrates under traffic.

[1]See *NCHRP Report 63* for procedures for economic comparisons among surface types.

[2]Excellent early studies by E. R. Willis and others are reported in *HRB Proceedings, 1938, Part 2* and *Public Roads,* June 1940 and Oct. 1942. *Compendium 7* of Transportation Technology Support for Developing Countries, TRB, offers more recent discussions and references applying both in temperate climates and in the tropics where laterites are common road-building materials.

Materials for Untreated Road Surfaces

The aim of specifications for materials for untreated road surfaces is to produce the important properties described above. Thus they control the grading of the material, the plasticity of its fines, and the strength of the aggregate particles.

Gradings recommended by AASHTO for surface and bases are shown in Table 16–1. It is to be noted that, although a wide variety of materials are acceptable, the limits assure a fairly uniform distribution of particle size from coarse to fine.

The AASHTO specifications include an added general requirement with a like purpose. It is that the fraction passing the No. 200 sieve shall not be greater than two-thirds of the fraction passing the No. 40 sieve. A further stipulation for surface courses is that all materials pass a 1-in. sieve. Larger materials are troublesome in that they lodge under the blade of the motor grader used for maintenance and tear up the surface.

As indicated already, some clay binder of controlled quality is desirable in untreated surface courses. The AASHTO specifications call for a maximum liquid limit of 35, a plasticity index range from 4 to 9, and a minimum of 8% passing the No. 200 sieve if the surface course is to be maintained for several years without bituminous-surface treatment or other impervious surfacing. It should be particularly noted that some plasticity is desirable in the fines of surface courses. As will be emphasized in Chapter 17, this is usually undesirable in bases.

Regarding strength and soundness of aggregates, the AASHTO specifications read in part as follows:

> Coarse aggregates retained on the No. 10 sieve (2.00 mm) shall consist of hard, durable particles or fragments of stone, gravel, or slag. Materials that break up when alternately frozen and thawed or wet and dried shall not be used. They shall have a percentage of wear, based on the Los Angeles Rattler Test (see Chapter 14) of no more than 50. Fine aggregate (passing No. 10 sieve) shall consist of natural or crushed sand,

TABLE 16–1. Grading Requirements for Soil-Aggregate Materials (AASHTO Designation M147)

	Percentage by Weight Passing Square Mesh Sieves					
Sieve Designation	*Grading A*	*Grading B*	*Grading C*	*Grading D*	*Grading E*	*Grading F*
2 in. (50 mm)	100	100	—	—	—	—
1 in. (25 mm)	—	75–95	100	100	100	100
3/8 in. (9.5 mm)	30–65	40–75	50–85	60–100	—	—
No. 4 (4.75 mm)	25–55	30–60	35–65	50–85	55–100	70–100
No. 10 (2.00 mm)	15–40	20–45	25–50	40–70	40–100	55–100
No. 40 (0.425 mm)	8–20	15–30	15–30	25–45	20–50	30–70
No. 200 (0.075 mm)	2–8	5–20	5–15	5–20	6–20	8–25
			← Suitable for surface courses →			
	← Suitable for bases and subbases →					

and fine mineral particles passing the No. 200 sieve. The composite soil–aggregate mixture shall be free from vegetable matter and lumps or balls of clay.

Specifications for untreated surface courses are developed to fit local conditions and agency procedures. Many stipulate that all the large or oversized material in the gravel pit be crushed. In this way they include all the tougher, harder stones, and also gain the angular shapes and rough surfaces so important to stability. A cheaper alternative may be to screen out the oversize since it should never be included in the final product. To ensure stability, agencies that use the California bearing-ratio test often require a minimum *CBR* of 80; troublesome fines are excluded by limiting the expansion of the compacted sample on soaking to 1%.

Grading and Draining for Untreated Surfacings

For some modern road construction, untreated or "dry-gravel" surfaces serve temporarily as a "stage-construction"[3] surfacing. This idea applies particularly where traffic must be carried economically for long distances until a more durable type of surface can be financed. In such situations high design standards for grade and alignment of the roadbed are demanded at the outset.

Good practice also dictates that the gravel surface course be full depth across the entire roadbed width. Trench cross sections, with shoulders constructed of embankment material, or "feather-edge" designs, thick at the centerline and thin at the shoulders, have generally been abandoned.

Cross slopes must be sufficient so that rainwater will not pond and soften the surface. Slopes as low as ¼ in./ft may be satisfactory if the material is sufficiently stabı‹ and watertight; but ⅜ to a maximum of ½ in./ft may sometimes be in order. If such high crowns are selected, they must be reduced before any kind of bituminous surface is applied. Also, if the road is on a steep grade, a substantial cross slope may be needed to direct the water to the sides.[4]

Construction of Untreated Surfacing

Untreated surfacing is seldom placed in less than 8 in. loose depth, which, on compaction is about 6 in. thick. Some agencies use loose thicknesses of 10 to 12 in. Commonly, material is distributed from moving dump trucks, spreader boxes, or self-powered spreading machines (see Chapter 19). Distribution in one or two windrows and spreading with a motor grader is sometimes permitted; however this can lead to segregation of coarse from fine sizes. Similarly, dumping the surfacing in large piles on the subgrade before spreading results in segregation.

[3]"Stage construction" is a step-by-step improvement of a roadway structure as the expenditures are justified by the demands of increased traffic. To illustrate, stage 1 might be an earth surface; stage 2, untreated gravel; stage 3, bituminous-surface treatment over the gravel; stage 4, some higher type of pavement with the gravel serving as part or all of the base course. For added discussion see *Synthesis 2* of Transportation Technology Support for Developing Countries, TRB, printed in English, French, and Spanish.

[4]For a detailed study of the relationships between cross slope, longitudinal grade, and pothole formation, see J. W. Spencer, *HRB Record 91*. This paper is reprinted in *Compendium 7*, op. cit.

Rolling the surface after placement is standard practice, but rolling without moisture is scarcely worthwhile except in climates where there is intermittent rain or where the material itself is moist. Rollers can be sheepsfoot, grid, pneumatic, smooth-wheel, or vibratory. Compaction begins at the edges and progresses toward the center with each passage lapping about half a width. Rolling, with watering and blading if needed, should continue until the rolled course or layer is thoroughly consolidated. Although power rolling with alternate watering is preferable, it is not indispensable. When rolling is omitted, it is good practice to extend the successive layers of surfacing material away from the source of supply and to insist that the hauling vehicles spread their tracking entirely across the full roadbed width and not in ruts. Repeated light blading will then in time obliterate ruts and consolidate the layers of gravel.

Some agencies, to save money, use streambed material as surfacing without crushing or screening it. Attempts then may be made to remove the oversize by hand or by breaking individual stones on the roadway by hand or with machines. The unfortunate consequences of incorporating large stones in the surfacing are, first, the impossibility of constructing and maintaining a smooth surface and second, very poor driving conditions with damage to vehicles and their tires.

Payment for Untreated Surfacing

The principal pay item for untreated surfacing is for the material itself. Measurement is commonly by the cubic yard, measured loose in the truck, or by the ton. This single pay item may include: the cost of labor, materials, and equipment required to reshape the existing roadway and to transport, place, compact, and shape the surfacing into a completed roadway. Additional bid items are sometimes included; for example, reshaping the existing roadway may be paid for at a price per mile, water by the thousand gallons, and rolling by the hour.

Other Surfacing Materials

Numerous other materials, including mine and industrial wastes if available close at hand, serve as surfacings for low-traffic roads. In general, the same cross sections, thicknesses, and construction and maintenance procedures as for untreated soil surfaces are used. As a rule, special specifications must be written for them. Among the most common are:

LIMEROCK (MARL). In some of the southeastern, Gulf Coast, and Mississippi Valley states there are deposits of a soft limerock which has served as a base and road surface. As a surface course it has certain objectionable qualities such as a white, glaring surface under sunlight and a tendency to dust under traffic and soften in continued rainy weather.

SHELLS. Deposits of oyster, clam, and similar shells occur along the eastern and Gulf Coasts. These shells, usually obtained by dredging, are used as a surfacing for lightly traveled roads. The mud that is dredged with the shells serves as binder.

CALICHE. In parts of Texas, New Mexico, and Arizona there occurs a calcium formation consisting of sands and gravels cemented together by coats of calcium carbonate. This chemical was carried in solution by groundwater and remained when the water evaporated. In most cases caliche is essentially soft limestone with varying percentages of clay. It is widely used on low-traffic roads as a surface material and often as a substitute for gravel.

CHERT. Chert gravels, found in some quantity in the southeastern United States, consist of mixtures of coarse chert particles and fine material from the dust of fracture and clay. They have given satisfactory service as surface courses; but, when they were used as base courses, numerous failures resulted because of excessive amounts of active binder.

VOLCANIC CINDERS. Volcanic cinders occur widely throughout the western United States and long have been used successfully as a surface for lightly traveled roads. Careful selection is required since some have turned to clay on exposure to air and water.[5]

LATERITES. In tropical areas, laterites are widely employed as road surfacing. Grading and plasticity controls for them parallel those for untreated surfacings cited earlier in this chapter. However, applying rules developed for soils from the temperate zone to laterites is dangerous since their behavior may be quite different.[6]

Maintenance of Soil-Surfaced Roads

Routine care of soil surface roads is concerned primarily with maintaining the smoothness of the road surface. This is done periodically by first cutting off a thin layer of the surfacing with a motor grader (see Fig. 16–1), drag, or other device and then redistributing this layer uniformly over the roadway surface. Depth of cut should be sufficient to completely remove any developing corrugations or potholes. A well-maintained surface probably will show some loose or float material, such as appears in Fig. 16–1. In addition, occasional heavy maintenance will be needed.

The frequency with which surfaces are bladed varies tremendously among agencies. It ranges from once or twice a year to every week or so depending on such factors as finances, traffic, and equipment availability.

The most effective routine maintenance is done immediately after a rain. Then the surface is soft and can be cut with the blade or drag. The loosened material fills holes and corrugations and is fixed in place by the moisture, coupled with the compactive action of traffic. Blading in dry weather is essential also, since the uniform cover of floating materials retards or prevents the formation of washboard and potholes.

An important rule is to maintain the cross slope (see above). Otherwise water will stand and soak and soften the road surface. Furthermore, the standing water is troublesome to motorists. Electronic controls for motor graders offer decided

[5]See L. G. Hendrickson and J. W. Lund, *HRB Record 307* and *311* for added detail.
[6]See *Compendium 7,* op cit., for added detail.

advantages. These devices maintain a constant slope on the blade regardless of the position of the grader carriage. With them, there is greater certainty of maintaining the optimum slope.

A frequent adverse condition is the transverse corrugation or "washboard" effect. This is associated with higher volumes of motor traffic, but may appear when the count is low. The transverse shallow waves tend to average about 30 in. from crest to crest and not to exceed 1½ in. in depth from crest to trough. Sometimes they reach entirely across the roadway. Frequent blading is the most effective treatment, but constant vigilance is necessary. At higher volumes, however, some form of surface treatment is the only complete correction.

Early spring or post rainy-season maintenance demands attention in areas where weather is severe, since deep ruts may have formed. It may be necessary to loosen the surface with the scarifying teeth of the grader. Then heavy blading of the full width and side ditches follows. Early trips are made with the blade set at near right angles to the roadway centerline.

Every few years it will be necessary to supply additional surfacing material to fill ruts and replace material that has worn or blown away. Losses generally range from ¼ to 1 in. of thickness per year, but will vary with a number of factors such as traffic volume and type, rainfall and wind intensity and frequency, and maintenance practices.

One of the most serious defects of untreated road surfaces is the annoying and dangerous dust. Some form of surface protection or dust palliative is required to cure this ill.[7]

STABILIZED ROAD SURFACES

The term "stabilized road" has long denoted a class of surfacing built by combining soil and mineral material or by adding calcium or sodium chloride or certain organic compounds to the material or its surface. Not included in this classification are asphalt- or cement-stabilized layers where the surface is protected by a seal coat.

Sand–Clay Roads

Essentially, sand–clay roads are a favorable mixture of clay, silt, fine and coarse sand, and possibly some fine gravel. In the United States, they first developed in the South Atlantic states and later in the Middle West as a substitute for gravel roads in areas where coarse gravel was not available.

Successful practice with sand–clay construction must be based on the principles previously laid down for untreated-surface courses. Furthermore, such roads are economical only in areas where suitable sources of sand and clay are available locally.

[7]Several publications in the series *Transportation Technology Support for Developing Countries,* TRB, contain details of good maintenance practices. *Synthesis 1* titled "Maintenance of Unpaved Roads," published in English, French, and Spanish, is a general overview with particular emphasis on maintenance management. *Text 4* of *Compendium 5* titled "Roadside Drainage" is a detailed operating manual for motor graders. *Text 1* of *Compendium 12* deals with dust abatement. Equipment manufacturers and agencies such as the World Bank and Transport and Road Research Laboratory also have developed helpful information.

EARLY DEVELOPMENT OF THE SAND–CLAY ROAD. Observations in the South Atlantic states during the light traffic of the period of 1900–1910 showed that road surfaces of "topsoil," "gray soil," "gray grit," "black tobacco soil," "upland soil," and "rotten granite" were usually successful. Roads were built by simply hauling the soil onto the roadbed and depositing a thick layer which consolidated and cured under traffic. It was recognized that the clay binder should just about fill the voids in the sand, but usually no precise measurement of the voids was made. It was further observed that excess clay deformed the roads and that insufficient clay caused ravelling. It soon became apparent that good surfacings could be produced by combining sand and clay from separate sources.

When the subgrade was clay, it was covered with a few inches of sand and plowed and disk-harrowed, preferably when moist. When the subgrade was sand, an even layer of dry clay was spread, beginning next to the clay pit so that hauling broke up the clay lumps. Another layer of sand was added; then usually it was considered necessary to plow and harrow. Advantage was taken of rains, and sand or clay was added as conditions developed.

Such early roads were very cheap. When labor in the South could be had for $1/day and teams for $3, and the average haul did not exceed a mile, a 12-ft-wide sand–clay road, 6 in. deep, consolidated, could be built for about $600/mi. Sometimes roads were built for as low as $300/mi. Figure 16–2 illustrates the construction practices on these early roads.[8]

MODERN SAND–CLAY ROADS. In areas where gravel or crushed stone is expensive, the sand–clay road may be relatively economical and gives good service; also a well-built sand–clay road may serve as a first stage of construction or as a base under superior surfaces. A favorable grain-size distribution for sand–clay mixtures is grading *F* of Table 16–1. The plasticity requirements and construction and maintenance procedures set forth for untreated soil mixtures also apply to materials for sand–clay road surfaces.

Granular Stabilized Roads

Granular stabilized roads are combinations of crushed rock or gravel, sand, and binder soil that come from two or more sources. These are mixed either on the roadway or in a plant, sometimes under quite elaborate control. Compaction and finishing procedures are the same as for other untreated soil mixtures. At times, processing may include adding calcium or sodium chloride (see below).

Stabilization with Calcium or Magnesium Chloride

Calcium chloride ($CaCl_2$) and magnesium chloride ($MgCl_2$) are white, deliquescent salts. Deliquescence is the ability of a material to readily absorb moisture from the air, dissolve, and become liquid.[9] In addition, they reduce the repul-

[8]An excellent summary of early practice appears in L. I. Hewes, *American Highway Practice,* Vol. 1, Wiley, New York, 1942.

[9]At 40°F, absorption begins at a relative humidity of 40%; at 100°F, it begins when the relative humidity reaches 20%.

Fig. 16–2. Hauling topsoil to an early demonstration sand-clay road in Yancyville, N.C. (Courtesy, FHWA.)

sive forces between particles and strengthen the bond of the water film. Also, in solution, they lower the freezing temperature of water. These properties make calcium chloride in particular useful in several ways.

The most common sources of calcium chloride are as by-products from the manufacture of sodium carbonate by the ammonia-soda process as well as other chemical procedures. It comes in flake, pellet, and granular forms.

In dry weather, calcium chloride in or on the road results in a higher moisture content in the materials than if they were untreated. Two factors are responsible. First, because of higher vapor pressure, evaporation of water from a calcium chloride solution occurs at a somewhat slower rate than if the water were untreated. Second, because of its deliquescent properties, the calcium chloride will replace lost moisture at night or under other favorable humidity conditions. It has been reported that if humidity is high enough, calcium chloride will absorb four to ten times its own weight in water and retain one-third to two-thirds of it through the heat of the day. It follows, then, that, in areas of fairly high relative humidity, calcium chloride will act as a dust palliative since it will hold moisture to bind the road surface together, whereas without it the road would be dusty and ravel badly. Surfacing-material losses are reported to be only half as great for treated as for untreated surfaces.

Calcium chloride has served as a dust palliative for more than 50 yr. The usual application is about 1½ to 2½ $lb/yd^2/yr$. After a road has been maintained for a year or so, the total yearly amount may drop to 1½ lb or even as low as 1 lb/yd^2. Best results are obtained if, immediately after a rain, the surface is bladed and patched and then the chemical is distributed in flake form. Mechanical spreaders give more uniform distribution. If application is made before a rain, a great part of the calcium chloride may be washed off the road and lost.

Calcium chloride as a dust palliative is most effective where the soil binder of the surface is clayey rather than silty or sandy. However, slick clay surfaces become even slicker when treated.

Mixing of calcium chloride into surface and base-course materials began about 1932 when it was discovered that it not only acted as a dust palliative but also had a beneficial effect on stability. As previously mentioned, it aids in maintaining a moisture film to bind the particles together. Furthermore, this moisture film has greater cohesion than water alone, because calcium chloride solutions have greater surface tension than water.

Calcium chloride is mixed with the surfacing material on the roadway (road mix) or at a mixing plant (plant mix) The optimum amount is reported to be about 1% of aggregate weight. Plant mixing has many advantages. Not only is greater uniformity obtained, but also delays from unfavorable weather and interruptions to traffic are held to a minimum. The mixture, at proper moisture content, is hauled to the road, placed, and compacted at once.

Compaction to high density is important for calcium chloride stabilization just as in other cases. Conventional rollers are normally used. Numerous tests have indicated that these mixtures are much more easily compacted than untreated ones. The explanation is that calcium chloride increases the surface tension, viscosity, and lubricating properties of water.

Maintenance of calcium chloride-treated surfaces consists of blading and patching. Blading should be done immediately after a rain or even commenced near the end of the rain. Calcium chloride migrates through the soil and during a dry spell tends to concentrate near the surface so that blading during that time is likely to cause loss of the chemical. Under ordinary conditions, much less blading is required than for untreated materials. Moisture is maintained by light applications of calcium chloride whenever the surface begins to show signs of drying or dusting. Patching is normally necessary only during long dry spells. Properly maintained, calcium chloride treated surfaces lose less aggregate and need fewer bladings than untreated surfacings.

For some time, there have been complaints that calcium and sodium chloride cause serious and costly corrosion of automobile bodies. To this has been added the fear that they will contaminate surface and groundwater. For these reasons, there have been serious proposals that their uses as soil-stabilizing, dust-control, and de-icing agents be curtailed.[10] (See Chapter 21 for further discussion.)

Treatment with Sodium Chloride

Sodium chloride (NaCl or common salt) is hygroscopic; it absorbs and retains moisture. It also is used for stabilization of road surfaces and bases. Like calcium chloride, sodium chloride controls the moisture content of graded mixtures and thereby effects a decrease in volume change and an increase in den-

[10]For an excellent discussion of the properties and uses of both calcium and sodium chloride as stabilizing and dust-control mechanisms and an extensive bibliography, see T. H. Thornburn and R. Mura, and F. O. Wood, *HRB Record 294*.

sity and stability. It has electrolytic and crystal-forming properties but is not strongly deliquescent; it absorbs moisture only when the relative humidity is above 75%. It also reduces changes in moisture content by forming a barrier to the movement of water in the liquid phase. Generally, surfaces treated with sodium chloride are harder, with a drier appearance and slightly more dust, for in dry weather the sodium chloride forms fine crystals which give a hard crust of salt and aggregate. For this reason, they are more difficult to blade.

Construction and maintenance methods for sodium chloride stabilization closely parallel those for the calcium chloride treatment. Sodium chloride may be applied as either rock salt or brine; the quantities used are comparable.

Other Dust-Preventative Treatments

A variety of liquid asphalts and petroleum resins have been used as dust palliatives either on lightly traveled roads or to prevent wind erosion of soils in areas not subject to traffic. These are usually applied by a distributor truck (see Chapter 19) at a rate of 0.1 to 0.25 gal/yd.[2] Also, lignin sulfonate, a water-soluble by-product of the sulfite wood-pulping process, has been used successfully as a dust palliative. It normally is mixed with materials from the loosened roadway, after which the combination is spread and compacted. About 0.2 to 0.5 gal/yd^2 of lignin sulfonate is applied in a 10 to 30% solution in water.[11]

Problems

16–1. Which of the soils listed in Table B (inside the back cover) meet the AASHTO grading requirements for an untreated surface course? In which of the grading classifications do they fall?

16–1. Of the soils which meet the grading requirements as determined in Problem 16–1, which also meet the plasticity requirements of AASHTO for surface courses?

[11]See, for example, *Text 1* of *Compendium 12,* op. cit.

17 BASE COURSES

A base course is the layer immediately under the wearing surface (see Fig. 14–1). This definition applies whether the wearing surface is bituminous or portland-cement concrete 8 in. or more thick, or but a thin bituminous surface treatment. Because the base course lies close under the pavement surface, it is subject to severe loading. It follows that the materials in a base course must be of extremely high quality and construction must be carefully done.

Until about 1940 base courses almost always consisted solely of mineral aggregates. The most common type, called "granular base course," was a mixture of soil particles ranging in size from coarse to fine. Processing involves crushing oversized particles and screening where necessary to secure the desired grading. "Macadam" type bases involving successive layers of crushed rock bound with rock dust were also used. Since about 1940 "treated bases" composed of mineral aggregates and additives to make them stronger or more resistant to moisture have become increasingly common. The most-used treating agents are bitumen, portland cement, and lime. In some instances, flyash, calcium and sodium chloride, and organic chemicals have been employed.

In areas of deep frost penetration, subbases of clean noncapillary materials often underlie the regular base course.[1] For clarity of presentation this chapter subdivides base courses into granular and treated types. Macadam bases are discussed in Chapter 18.

GRANULAR BASE COURSES

General Requirements

The requirements of a satisfactory soil–aggregate surface have previously been given as stability, resistance to abrasion, resistance to penetration of water, and capillary properties to replace moisture lost by surface evaporation (see Chapter 16). Upon the addition of a wearing surface, some of these no longer apply. The first, stability, which is the ability to transfer loads to the underlying layers

[1]The term subbase rather than base is commonly employed to describe the layer immediately underlying portland-cement-concrete pavements.

without permanent deformation, is still absolutely necessary. The second, resistance to abrasion, usually disappears, as the wearing surface now performs this function. The importance of the third and fourth requirements, concerned with moisture penetration and capillary action, are dependent on the type of wearing surface to be used.

Suppose, for example, that a soil–aggregate base course is to be protected by an open-mix pavement through which water may penetrate freely. The soil–aggregate combination then must contain enough clay to seal its surface against the penetration of this water in order that the base remain stable in wet weather. On the other hand, capillary moisture offers no great problem, as in dry weather it can evaporate freely through the open pavement. Thus, a base course under an open pavement should react to moisture much as a good soil–aggregate surface course does. If, on the other hand, the wearing surface is impervious to water or water vapor, a lower clay content is advisable. In this case, surface moisture cannot penetrate to the base course and presents no problem. However, the evaporation of capillary moisture is prevented, so that in time both base and underlying subgrade become saturated. If either material, when saturated, becomes too weak as the result of high clay content, inadequate compaction, or other causes, failure will occur.

The changes in service characteristics between surfaces and bases are of particular importance when stage construction of a road over a period of years is planned, for then a material that meets the service requirements of both surface and base must be selected. For example, a bituminous surface treatment on an existing soil–aggregate road may be considered to control dust in dry weather. But if the present surface course is high in clay and the surface treatment prevents evaporation, then the softened clay will not support the superimposed wheel loads and potholes will develop. In such situations, it is either necessary to remove the surface treatment or to provide a suitable base course before the surface treatment is applied.

Specifications for Granular Base Courses

Base courses must be of higher quality than the underlying "basement" soils. Typical of the many different specifications for them is *AASHTO Designation M147*. Grading requirements for six different types are given in Table 16–1. Four of these are also suitable for soil–aggregate surface courses. Strength and soundness standards are also the same. However, the provisions covering plasticity characteristics are different. For base courses, the maximum liquid limit is set at 25, and the maximum plasticity index at 6. As pointed out earlier, for surface courses the maximum liquid limit is 35 and the plasticity-index range is 4 and 9. As lower values of liquid limit and plasticity index indicate lower clay content, this change in the specification reflects the difference in service requirements discussed in previous paragraphs. Specifications for subbases correspond to those just cited for bases, with the added proviso that, on the basis of local experience, the percentage passing the No. 200 may be lowered to prevent frost damage.

Agencies other than AASHTO have somewhat different specifications for base courses. In general, grading requirements are about the same, but some lower the plasticity index (PI) to 4 or less to place a more stringent limit on troublesome fines. Again, physical tests other than for plasticity index may be used to control the properties of the soil mortar. For example, one specification requires that the *CBR* for base courses be not less than 80 and that expansion when a compacted sample is soaked not exceed 1%. As another illustration, for its Class 2 (highest type) granular base course the California Department of Transportation sets minimum resistance values (Hveem stabilometer) at 78 for a single test; respective minimum and running average "sand equivalent" ratings are 28 and 31.

No attempt will be made here to set out requirements for other base course materials such as shell, caliche, chert, cinders, or laterite. They vary greatly from the normal materials already described in specific gravity, absorption, or grading, so that appropriate limits based on local experience must be set for each.

Under some conditions two layers of base course, one of higher type than the other, may be used to good advantage and will result in a lower overall cost. The lower course often is designated "select material" or "subbase."

The problems to be met in producing base courses are the same as those for producing materials for untreated surface courses. Where demands are less exacting, they may be loaded at the pit and placed on the road without any processing whatsoever. At the other extreme, if close control is demanded, crushing, separating into various sizes, proportioning by weight, and recombining, along with a controlled amount of water, may be specified. As indicated in *NCHRP Report 98,* a variety of tests may be employed to evaluate the material's resistance to degradation.

The thickness of base-course layers is controlled by the character of the underlying subgrade over which it distributes the wheel loads delivered from above. Several design methods will be presented in Chapter 19. Minimum thickness, which is commonly between 4 and 6 in., is controlled by the limitations of common construction procedures. Sometimes special considerations such as frost action may control the base-course depth. For example, northern states use sand and gravel subbase thicknesses ranging from one-half to the full depth of frost penetration.

Equipment and procedures for compacting granular base courses parallel those for embankments. Vibratory rollers are becoming increasingly popular. Requirements usually include observation of moisture-density relationships and compaction to some predetermined relative compaction. Controls are set quite closely as the demands on the base course are higher than on underlying layers. However, procedures vary considerably between agencies, just as they do for embankment compaction.[2]

[2]Where density is the control, the most commonly stipulated value is 100% *AASHTO standard (T–99). AASHTO Specification M147* calls for moisture equal to or slightly below optimum but does not stipulate the method by which it is to be determined.

Some agencies "slurry" the base as the final step in compaction, although this practice is being discontinued. Slurrying is accomplished by heavily wetting the surface followed by rolling to flush some of the fines to the surface. This produces a hard, tight surface on which to lay the pavement; on the other hand it concentrates any troublesome clays in a most vulnerable location. Others apply a bituminous prime coat which protects the surface and provides a good working table for subsequent paving operations.

Payment for Granular Base Course

Payment for base courses is usually made on a unit basis at a bid price per cubic yard or per ton, with payment by weight gaining increasing favor. This price may include water and/or rolling, or these may be paid for under separate bid items. Some organizations pay for bases at a bid price per completed square yard of a specified thickness or include payment for the base in the price per square yard of completed pavement. Where payment is by weight, there is a stipulation that the tonnage of water in the material exceeding some fixed percentage of the dry weight or alternatively, the optimum moisture content, will be deducted in computing the pay quantity. The 1979 FHWA specifications also stipulate pay reduction for failure to meet certain statistically determined variations in grading.

TREATED BASE COURSES

As indicated above, soil–aggregate base course materials are often combined with bitumen, cement, lime, or other products to make "treated or "stabilized" base courses. Again, materials such as clean sand or even the better clays, which alone would be unsatisfactory as base, may furnish the soil portions of these mixtures. These treatments are adopted when, in the opinion of the designer, a satisfactory result can be obtained at a lower overall cost.

Figure 17–1 is a useful guide for choosing among the most common type of additive for treated bases. It shows well-proven combinations with acceptability of the soil defined by the plasticity index and percent passing the No. 200 sieve. Methods for designing and constructing each of these treated bases are described below.[3]

Bitumen-Treated Granular Soils

A variety of bituminous materials has been employed to waterproof and bind natural materials ranging from untreated granular bases and granular soils through sands and even clays. This discussion considers only a few of the many workable approaches; they are classed as bases because they usually require a

[3]Among the many excellent references dealing with the various treated bases are *FHWA Implementation Package 80–2, Compendium 8* of Transportation Technology Support for Developing Countries, TRB, articles in *TRB Special Report 160* and *TRB Record 702,* and the publications of the Asphalt Institute, Portland Cement Association, and National Lime Association.

protective surface treatment or overlay. The many other combinations of aggregate and bitumen are classed as pavements and are treated in Chapter 19.

The highest type of bituminous-treated soil is untreated granular base further upgraded by mixing it with some form of bitumen. By so doing, the base is waterproofed so that its water content stays uniform and low. Some design methods (see Chapter 19) recognize that the structural strength of the base is increased by bituminous treatment and permit decreases in the overall depth of the pavement section.

On the AASHO Test Road an asphalt base performed extremely well. Aggregate for it was the same as in the sand-gravel subbase which corresponded closely but not exactly to an *A*-1-*a* soil and gradings *D* and *E* of Table 16–1. This material was plant mixed at 290°F with 5.2% of 85–100 penetration asphalt. Compaction followed procedures for bituminous pavements. In many regards, this base can be considered to be a relatively low type of asphalt concrete.

Open-graded aggregates also have been bound with bitumens, including emulsions, to give excellent bases or, without sealing, as surface courses. These are discussed under bituminous macadams in Chapter 19.

Two aggregate properties distinguish bituminous-treated granular soils from bituminous pavement (see Fig. 17–1). The first is the fairly high percentage (up to 25) passing the No. 200 sieve. This is compared with a maximum of less than 10% for bituminous pavements. The second is that some plasticity is acceptable in this relatively large percentage of fines, as shown by the PI maximum of 6 or PI times percentage passing the No. 200 of 72. It follows that a primary function of the bitumen in treated soils is waterproofing. In other words, it prevents water

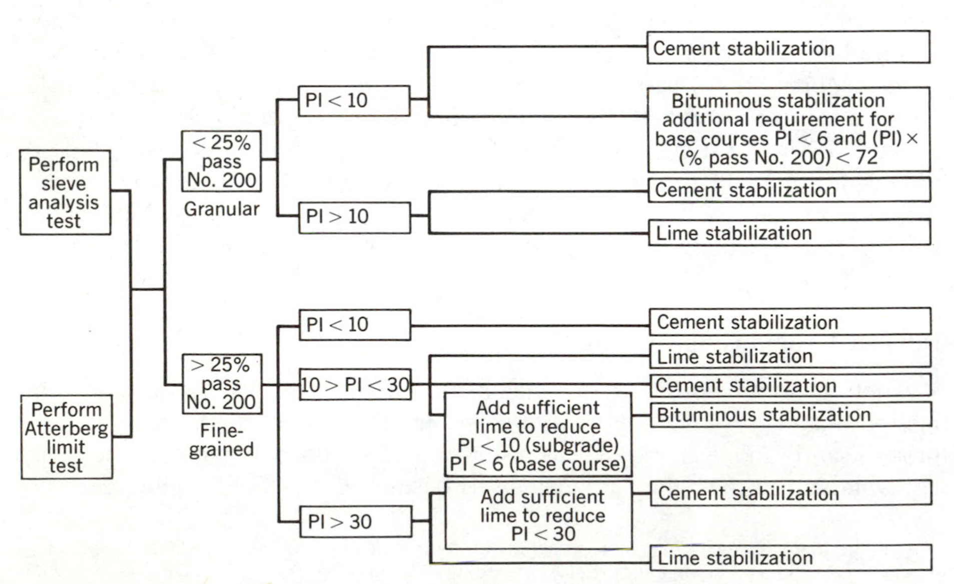

Fig. 17–1. Suggested procedure for selecting a stabilizing agent for base courses. (Source: FHWA Implementation Package 80–2.)

penetration with the resulting weakening from lubrication and a possible disruptive expansion of the clay fines. There will also be some binding of the soil mass, giving greater stability.

Selection among bituminous binders for bitumen-stabilized granular soils depends on a variety of factors. Among them is the mixing procedure; is it on the road or in a plant and, if in a plant, is the aggregate heated? Common selections include asphalt cements, cutbacks, or emulsions. Percentage of binder may be determined by either stability or waterproofing criteria. If stability is the control, it is measured in the laboratory by, for example, the Marshall or Hveem stabilometer procedures. In this instance, the percentage by weight will probably be in the order of 5 to 7%. For waterproofing, 2 to 3% of binder is added. With emulsions, enough water is included in the mix to permit compaction at near optimum moisture content.

Figure 17–1 shows the distinguishing properties of the aggregates of two forms of less-expensive bituminous-treated soils. They can be separated into "granular" with less than 25% passing the No. 200 sieve, and "fine-grained" having a higher percentage. Aggregates for sand-bitumen mixtures fall in the first group. They, and bitumen-stabilized fine-grained soil bases are discussed separately below.

Sand–Bitumen Base Courses

Sand–bitumen base courses consist of loose sand from beach, dune, pit, or river cemented with bituminous materials. As already mentioned, sand–clay mixtures have long been used as road surfaces and for base courses. In more recent years, bitumens, consisting of cutback asphalts, emulsified asphalts, or tars have often been substituted for the clay binder to produce excellent base courses for highways and airports. Sand–bitumen bases have particular application in the southern, Gulf, and certain midwestern states where aggregates from streams or quarries are costly, but deposits of clean sand are available within reasonable distances.

Sands, to be suitable for sand–bitumen bases, must be relatively clean. Grading is not critical, but the sand must be stable; that is, the surface properties and grain shape must be such that they will resist displacement under load. If the sand is not satisfactory, it may be blended with sharp angular particles such as crushed aggregates, stone or slag screenings, stone dust, loess, cement, or other substantially noncohesive mineral matter to produce a stable mixture. Testing has been by the Hubbard–Field, Marshall, and Hveem stabilometer procedures.

Bituminous binders range from the paving grades of asphalt cements for hot plant mixes through the medium viscosity, rapid- or medium-curing asphalts, slow-setting emulsified asphalts, or tars, grades *RT*-6 to *RT*-10. The percentage of binder, by weight, ranges from 4 to 10.

Combination of the sand with the binder has been achieved by "mix-in-place" methods using cultivators, harrows, and blades, with traveling mixers, and at a central mixing plant. Which method is selected depends on the peculiarities of the job. Compaction is done with pneumatic-tired or smooth-wheeled rollers. Where the surface is tight and nonfriable, sealing may be omitted.

Fine-Grained Soil–Bitumen Bases

As indicated in Fig. 17–1, bitumen-stabilized bases and subgrades have been formed of fine-grained soils having PIs up to 6 and 10, respectively. Also, soils with PIs up to 30 have been processed if pretreated with lime. Others have reported that soils with up to 50% passing the No. 200 sieve and PIs up to 18 can be stabilized without pretreatment. However, the high percentages of binder called for, difficulties in pulverizing the soil and getting an intimate mixture, and the need to protect the friable surface all mitigate against widespread use of this approach.[4]

Cement Stabilization of Soils and Base Courses[5]

Stabilizing soils and aggregate base courses by adding portland cement is a widespread practice, first carried out in 1915. Such mixtures, protected by a bituminous surface treatment, serve as pavements for lightly traveled roads. For major thoroughfares, they may replace the untreated bases under bituminous or portland-cement-concrete pavements. As with bituminous bases, some of the design methods allow a reduction in the total depth of the flexible pavement section when cement treatment is used.

Cement stabilization calls for an intimate mixture of natural materials and portland cement, compacted at optimum moisture content and cured to hydrate the cement. It forms a strong, stable base that has limited susceptibility to changes in moisture and temperature. It is considerably less rigid than portland-cement concrete. Its modulus of elasticity can range from 100,000 for clay soils with little cement up to 1 million for the strongest mixtures; compressive strength is in the range of 300–600 psi and flexural strengths about 20% of compressive values. In contrast, the modulus for portland-cement concrete ranges between 3 and 6 million and its compressive strength from 3000 to 5000 psi.

Cement reacts with fine-grained soil particles in two ways. First, by surface chemical action it quickly produces flocculation and reduces the moisture affinity of clays. Second, and more slowly, it promotes cementation, thereby producing a semirigid soil framework. Observations of cement-clay mixtures through the electron microscope indicate that, at first, the fabric is one of separate cement grains distributed throughout the clay. Then, as hydration of the cement proceeds, a gel forms along the edges of the clay particles. Eventually, the soil and cement can no longer be distinguished, indicating that clay and cement have reacted chemically.

The surface of cement-treated natural materials is friable. It must be protected

[4]See *FHWA Implementation Package 80–2* and H. F. Winterkorn, *HRB Proceedings,* 1957 for added detail.

[5]The literature on cement stabilization is extensive. For example, the Portland Cement Association has issued a number of publications covering design, laboratory procedures, and construction methods. *FHWA Implementation Package 80–2, Compendium 8,* op. cit., and *TRB Record 702* are among the other recent reports with extensive bibliographies.

from the weather and from direct contact with the tires of motor vehicles or it will scuff and pit. A bituminous surface treatment offers sufficient protection for low volumes of light traffic. Otherwise, the function of soil-cement is that of a base course, to be covered by a suitable pavement.

Cement-treated materials, particularly those of higher cement content, may develop shrinkage or fatigue cracks. These can be particularly troublesome when they reflect upward through an overlying bituminous surface.[6] Again, erosion of the soil cement under pumping action at the joints of portland-cement-concrete pavements can lead to joint faulting (see Chapter 20).

DESIGN OF CEMENT-STABILIZED MIXTURES. As shown in Fig. 17–1, satisfactory cement-stabilized mixtures, commonly called soil-cement, have been produced with a wide range of native materials. They fall into three general subdivisions: (a) sandy and gravelly soils containing less than 25% silt and clay (roughly AASHTO classifications *A*–1 and *A*–2); (b) sands deficient in fines, such as beach sands, glacial sand, and windblown sand (AASHTO classification *A*–3); and (c) silty and clayey soils (AASHTO classifications *A*–4, *A*–5, *A*–6, and *A*-7). Seldom are materials with a PI greater than 30 used, unless lime also is added. Calcium clays are easy to stabilize; sodium clays are difficult. Materials with a high pH or sulfate content are usually unsatisfactory.

Cement contents vary from 5 to 14% by volume or 3 to 16% by weight of dry aggregates. Sandy and gravelly soils require lower amounts, whereas silty and clayey soils call for the higher percentages. The quantity of cement needed to stabilize heavy clays is even higher.[7] As indicated, they are seldom treated because of the high cost of cement and the difficulty in pulverizing and processing them.

Dry densities for cement-stabilized mixtures compacted by the AASHTO standard method range from 135 lb/ft^3 for well graded gravel down to 85 lb/ft^3 for silty or clayey soils. They, and optimum moisture contents, may be either higher or lower than for the untreated material. The recommended field density is usually about 95% of AASHTO standard.

The minimum cement content for a soil–cement mixture is set by laboratory tests for maximum density and optimum moisture content, wetting and drying, and freezing and thawing. The procedure for determining maximum density and optimum moisture content parallels the AASHTO standard method for soils as already outlined. For the other tests, samples at optimum moisture and maximum density are moist cured for seven days and then subjected to either 12 cycles of wetting and drying or 12 cycles of freezing and thawing. After each cycle the samples are brushed on all surfaces with two firm strokes of a wire brush. The quality of the soil–cement mixture is measured by its ability to resist the abrasive and disintegrating action to which it is subjected. This is expressed as "percent loss" by weight.[8]

[6]See, for example, L. Raad et al., *TRB Record 690*.

[7]Numerical data cited here are rough poolings of data from the many references.

[8]See *AASHTO Designations T134, T135,* and *T136* for details of these tests. *AASHTO Designation T144* is the procedure for determining cement content of hardened soil–cement mixtures by chemical analysis.

To determine the cement content to be used with a particular soil, samples containing various percentages of cement are tested as outlined. Percentages of loss permitted in either test, as recommended by *FHWA Implementation Package 80–2* are:

AASHTO Soil Classification M–145	*Maximum Loss in Test (%)*
A-1, A-2, A-3	14
A-4, A-5	10
A-6, A-7	7

Other methods, such as unconfined and triaxial compression and flexure tests also are employed for establishing cement content. There are also less elaborate procedures.[9]

CONSTRUCTION OF CEMENT-STABLIIZED BASES.[10] Construction of cement-stabilized bases involves spreading, conpacting, and curing an intimate mixture of natural material, cement, and water on a prepared subgrade or subbase. In some instances, the existing road surface is plowed and pulverized to provide the soil portion; again, it may be brought in from a nearby borrow site. Where the soil is to be imported, "plant mixing" in a continuous or batch type mixer has certain advantages. In particular, closer control can be held over proportioning and mixing, and the possibility of damage to the mix or interruption to construction by bad weather is minimized. On the other hand, when the soil is already in place on the road or when a mixing plant is not readily available, processing on the road, called "road mixing," may be more satisfactory.

Before soil–cement processing is begun, the roadway must be shaped to proper grade and cross slope. Otherwise, the more expensive soil–cement layer must serve as a leveling course of variable thickness if the roadway surface is to be smooth.

Cement may be distributed in a variety of ways. On early projects, it was delivered in sacks and spread over the surface by hand. Today, most deliveries are in bulk in dump or hopper trucks. Distribution on the roadway is by spreader box or some other means for obtaining uniform amounts. If the soil to be processed is in a windrow, cement may be placed in a trough that has been formed along its top.

Early soil cements were processed by the "train-processing" method, where mixing was accomplished with a number of pieces of farm type equipment operating in sequence. This procedure still is economical under some circumstances. Figure 17–2 shows one of several modern machines that picks up soil and cement from the roadway, adds water to them, and discharges the mixture ready for compacting. It contains rotating blades that pulverize and mix in a single pass. Figure 19–23a shows another machine often used to mix soil cement.

[9]See *Soil-Cement Laboratory Manual* issued by the Portland Cement Association.
[10]See *Soil-Cement Construction Handbook* of the Portland Cement Association for greater detail.

Fig. 17–2. Single-pass soil stabilizer processing soil-cement base. Cement has been spread on roadway ahead of mixer. Tank truck supplies water (Courtesy, Seaman Corporation).

Soil cements often are first compacted with sheepsfoot rollers, beginning at the edges and working toward the center. Very sandy soils that cannot be compacted with a sheepsfoot are processed by pneumatic-tired rollers instead. Vibratory compactors are also used. Compaction of the top 1 or 2 in. is commonly by pneumatic-tired roller, preceded by shaping with a blade grader or subgrading maching. Final compaction is usually with a smooth-wheel roller. If a tight surface is desired, the final operation consists of sprinkling lightly, followed by compaction with a pneumatic-tired roller.

Hydration of the cement begins when water and cement come together. To prevent setting before the mix is compacted, specifications may limit the length of the mixing, compacting, and surface-finishing interval. The 1979 FHWA specifications set a 2-hr limit after water is added for road-mix projects and 1 hr between the beginning of mixing and the start of compaction on the roadway when materials are processed in a plant.

CURING. Evaporation of moisture from the completed soil–cement must be prevented until hydration of the cement is complete. Bituminous seals of light *RC* or *MC* grades, emulsions, and tars have all been used satisfactorily. The past practice of curing with moist straw or earth has largely disappeared because of high labor costs.

CEMENT-MODIFIED SOILS. Cement modification implies adding sufficient cement to a fine-grained soil to reduce its plasticity to meet a particular requirement; however, the amount added is less than is needed to produce soil cement. For example, as with lime stabilization (see below), it may be less costly to control the expansion of in-place soils by adding cement than to remove them or to cover them with nonexpansive material. There are no standard procedures for cement modification; rather each situation is considered separately.

Small amounts of cement also may be added to granular soils which are substandard because the soil mortar fraction has higher plasticity or moisture susceptibility than is desired. For example, in this manner a base course suitable for use under an asphalt or portland-cement-concrete pavement may be developed from granular materials that would otherwise be unsatisfactory.

CEMENT-TREATED GRANULAR BASES. Cement added to granular base courses increases their tensile and flexural strength, binds the particles more tightly, and provides excellent waterproofing. At the same time, the material has a relatively low strength and modulus of elasticity as compared to regular concrete. It therefore is better able to adjust to settlement of the underlying embankment or subgrade because of its greater flexibility and the closer spacing of cracks. Under portland-cement-concrete pavements, a cement-treated base is less likely than a granular base to erode under the pumping of water by the overlying slab. The fact that cement treatment adds strength is recognized in most pavement-design methods.

The California Department of Transportation has made extensive use of both road-mix and plant-mix cement-treated bases under bituminous and portland-cement-concrete pavements. One of two classes may be chosen, depending on the design and construction situation. Aggregate gradings are about the same as for granular bases. A somewhat poorer quality fines is acceptable, with the sand-equivalent minimum set at a moving average of 21. Cement content of class *A* base is not to exceed 5% by weight of dry aggregate; however, samples must develop a seven-day compressive strength of 750 psi minimum. For the California class *B* base, the aggregate must have a minimum *R* (stabilometer) value of 60; this must exceed 80 after mixing with 2½% or less cement. Since the cement content of these bases is set by the engineer, payment to contractors for cement is adjusted so that they are paid fairly.

Construction procedures for cement-treated granular bases closely parallel those for soil–cements. It should be emphasized again, however, that where a highly uniform quality is desired, plant-mix methods are superior, since proportioning and mixing can be more carefully controlled. Also, particular care must be exercised in finishing and curing the surface of bases underlying portland-cement-concrete pavements in order to secure a nonerodible surface. California practice adds yet another assurance against surface erosion by sealing the surface with an *MC*-250 liquid asphalt. This penetrates and tightly seals the surface of the treated base. Also overall pavement performance was improved by extending the treated base at least 1 ft outside the edge of the overlying pavement.

A cement treated base performed well under a special section of flexible pavement on the AASHO Test Road. This base consisted of the granular subbase material with 4% cement by weight added to develop a seven-day compressive strength of 650 psi.

Lime-Stabilized Bases and Subbases

Lime as a stabilizing agent was used in construction of the Appian Way and many other Roman roads. It was also employed for this purpose in ancient Greece, India, and China. Particularly since World War II its beneficial effects on soil behavior have been increasingly recognized. It is now widely employed both to make clay-bearing soils suitable as subbases and to enhance the strength and other properties of potentially useful base course materials which contain clay. As indicated, lime is only effective when the natural material contains suitable amounts and types of clay. For example, Fig. 17–1 calls for a minimum PI of 10. The stabilization process results because calcium hydroxide ($CaOH_2$), commonly called lime,[11] reacts favorably with some but not all clays. The reaction causes the clay's properties to change substantially. Plasticity and the tendency to expand with added moisture are immediately reduced as the troublesome cations of sodium and potassium are replaced by those of calcium or magnesium. Also, lime causes flocculation, a lumping together of clay particles which increases their effective grain size. Over a longer period of time, there can be a substantial strength gain through a pozzolanic reaction with any available silica or alumina.

Because of the chemical complexity of clays, lime stabilization has not always been effective. Also, lime produced from different raw materials will react differently. It follows that careful laboratory study or the construction of test sections should precede any large-scale program of lime stabilization. Furthermore, there is evidence that lime-stabilized bases may break up or at least lose strength under freezing and thawing, which might limit their usefulness in areas where the ground freezes. But even in these areas, the residual strength may be higher than that of untreated materials.[12]

Lime is commonly delivered in slaked form, $Ca(OH)_2$, either as a powder or a slurry. At times, bulk unslaked "quicklime," CaO, is used, but because it is caustic and also generates heat when "slaked" with water, care must be taken to protect workers from burns.

As indicated, lime stabilization of natural materials for bases and subbases is intended both to reduce plasticity and the accompanying volume changes with moisture content and to increase strength. The common test for reduction in plasticity and volume change is the PI. Results with one group of clays showed values reduced by one-third to two-thirds with 3% lime and to a nonplastic condition with 5%. Strength increases are often measured by standard tests such as unconfined or triaxial compression, the Hveem stabilometer, *CBR,* or flexure. Higher strengths result almost immediately, with the increases almost in proportion to lime content up to 6%. Also, strengths usually increase over time as pozzolanic action takes place.

[11]Dolomite (magnesium) lime is also effective. To be acceptable, not over 36% by weight may be magnesium, calculated as magnesium oxide. See *AASHTO Designation M-216* for specifications for both products.

[12]See, for example, E. K. Sauer and N. F. Weimer, *Transportation Engineering Journal of ASCE,* Mar. 1978.

Figure 17–1 indicates that treating soils with lime to make them less plastic may precede bitumen or cement stabilization. Also, on occasion, lime may be employed on muddy work sites to make them accessible to construction equipment.

Lime-stabilization construction procedures and equipment generally follow those described earlier for bitumen or portland-cement stabilization. Density, which is usually less than for the untreated natural material, is often set at 95% of AASHTO standard. Where very troublesome clays are being processed by road-mix methods, a first application of 2 to 3% lime may be made and worked in, the layer sealed, and permitted to cure for a week. This will make pulverization and processing easier when the remaining lime is added.

Moisture is required for curing lime-stabilized materials; so the surface, after shaping, is commonly sealed with an appropriate bitumen.

Delays during the processing and compacting of lime-stabilized mixtures must be avoided. Otherwise, carbon dioxide from the air will react with the lime forming a weak calcium carbonate.[13]

Lime–Flyash and Volcanic-ash Stabilized Bases and Subbases

Flyash is the particulate matter in the stack gas that results with the burning of coal, lignite, or like fuels. It is removed from the gas by cyclonic or baghouse collectors or electronic precipitators in a finely divided powdery form as a waste product. Chemically it is complex, but in the dry state is inactive and can be stored. In soil stabilization, the glassy calcium silicate and similar compounds react with lime or lime and cement to create strength through pozzolanic reactions. As the chemical composition of flyash, lime, and natural volcanic materials all vary greatly, only careful testing similar to that described for cement or lime stabilization will assure satisfactory results.

The natural material in a typical lime–flyash–stabilized crushed stone or gravel base contains less than 15% passing the No. 200 sieve and a PI of 6 or less. Added would be 10 to 18% flyash and 2.5 to 5% lime. Construction procedures are as described for lime stabilization.[14]

Research has been conducted on the effects of ground expanded shale, volcanic ash, pumice, and diatomaceous earth as substitutes for lime and flyash and on the influence of various salts and chemicals on the properties of lime and lime–flyash–stabilized materials. These results cannot be reported here.[15]

Calcium and Sodium Chloride Treated Base Courses

Calcium chloride is sometimes used as a stabilizing agent for base courses. It assists in the compactive process, making it possible to obtain greater densities

[13]Authoritative publications on lime stabilization include *FHWA Implementation Package 80–2, TRB Circular 180* (reprinted as *Text 3* of *Compendium 8, op. cit.),* and M. M. Aly Sabry and J. V. Parcher, *Transportation Engineering Journal of ASCE.* Jan. 1979. Fundamental studies of clay–lime reactions using the electron microscope and x-ray defraction are reported by W. C. Ormsby and E. B. Kinter in *Public Roads,* Mar. 1973.

[14]For added detail on lime–flyash stabilization see *FHWA Implementation Package 80–2, NCHRP Synthesis 37,* and articles in *TRB Record 501* and *725.*

[15]*See, for example,* T. H. Thornburn and R. Mura, *HRB Record 294.*

and greater strengths with normal compactive effort or to get usual densities with greatly decreased rolling. Construction methods are like those outlined earlier for calcium chloride stabilized surface courses.

Sodium chloride also has been used satisfactorily as a base stabilizer. It is reported to reduce the shrinkage, increase the strength, and reduce the moisture loss of certain montmorillonitic clays. In certain instances, it further improves lime-modified soils. Construction methods are as outlined for other stabilized bases.[16]

Other Soil-Stabilization Methods

Research, mainly before 1955, sponsored in part by the Armed Forces of the United States, investigated several inorganic materials such as sodium silicate and acidic phosphorus compounds, several resinous waterproofers, resinous bonding agents, such as aniline furfural and calcium acrylate, and other organic cations that have hydrophobic (water hating) properties. None of these have had extensive use. One of the problems with the organic stabilizing agents is that they may be consumed by soil bacteria. For example, some of the airfields in Great Britain during World War II were successfully stabilized with resins, only to return to their original condition in a short time through bacterial action.[17]

Payment for Treated Bases, Subbases, and Subgrades

Payment for the various forms of treatment for upgrading natural or prepared materials generally follow the same pattern. Hauled-in materials are paid for by the cubic yard, ton, or other appropriate measure. Additives and sealing materials usually are priced in tons or, for bitumens, possibly in gallons. Often, shaping the roadbed and processing, compacting, and shaping the final surface will be paid for separately at a price per square yard or linear foot or mile.

PROBLEMS

17–1. Which of the soils listed in Table B (inside the back cover) meet the AASHTO grading requirements for granular base courses as given in Table 16–1? In which classification do they fall?

17–2. Of the soils which meet the requirements for granular base courses as determined in Problem 17–1, which also meet the plasticity requirements of AASHTO for base courses?

17–3. According to Fig. 17–1, what stabilizing agent or agents are suitable for treating soils 3, 4, and 8 of Table B (inside the back cover)?

[16]See Thornburn, op. cit.

[17]See *HRB Bulletins 108, 129, 241,* and *318* for research reports and detailed bibliographies.

18 MACADAM SURFACES—MACADAM BASES

The term "macadam" originally designated a road surface or base in which clean broken or crushed ledge stone was mechanically locked by rolling and bonded by stone screenings which were worked into the voids and "set" with water. With the beginning of the use of bituminous material, the term "plain macadam," or "ordinary macadam," and more frequently "waterbound macadam" was used to distinguish the original type from "bituminous macadam" in which the binder was a bituminous material. The original macadam was for 100 yr the highest type of road surface known. In Massachusetts, in 1914, for example, it was the surface on 95% of the state highways. Many miles of macadam, protected by bituminous blankets or surface treatments, are still in service. Their principal disadvantages are not related to structural soundness, but to their narrowness and excessive crown.

Waterbound macadam surfaces are almost never built in the United States today. In the first place, the combined effects of the vacuum under a moving motor vehicle and the thrust of its wheels will quickly remove the binder and reduce the surface to a pile of loose rocks. Secondly, macadam construction of either surfaces or bases calls for a comparatively large amount of skilled and expensive labor. However, macadams may still be economically and technically sound in areas of the world where labor is less costly and labor-intensive methods of construction are desirable. But the details of macadam construction are historically interesting. Furthermore, practicing engineers may encounter macadams when they cut into or undertake repairs to existing roads.

Macadam roads developed originally in France and England and are named after John Louden MacAdam.[1] MacAdam deserves great credit for the work he did; but, when MacAdam was only 21 yr of age, Tresaguet[2] was presenting in France to the Assembly of Bridges and Highways a report that was in fact a trea-

[1]John Louden MacAdam, 1756–1836, was a famous Scottish road builder and engineer. Probably his greatest contribution was the development of the road surface type that bears his name.
[2]Pierre-Marie Jerome Tresaguet, 1716–1796, a great French engineer, improved the methods for construction and maintenance of stone roads. He made it possible for Napoleon to build the great system of French highways and may be called the father of modern road building.

Fig. 18–1. Laying telford foundations in Mercer County, PA., 1911.

tise on road construction. Tresaguet improved the drainage, crowned the grade and the stone foundation, and reduced the depth of broken stone to 10 in. Some 30 yr later, Telford[3] in Scotland, constructed roads like those developed by Tresaguet. His foundation course was of stones of 3 in. minimum thickness, 5 in. breadth, and 7 in. height (see Fig. 18–1). Smaller stones were driven into the top voids with mauls and the surface was trued by breaking the projecting points. This base was covered with a broken-stone wearing surface 4–7 in. thick. He used a flat subgrade and obtained a slight crown with stones of varying heights. The old Cumberland Road or National Turnpike from Baltimore through Cumberland, Maryland to Wheeling, West Virginia on the Ohio River, completed in 1818, followed the Tresaguet–Telford method.

MacAdam omitted the telford foundation of large stones and introduced small, broken stones not exceeding 1 in. in diameter, on the contention that they should lock together because of their angularity. He developed the function of the wearing course of broken stone and demonstrated the adequacy of well-drained earth subgrades when covered with proper surfaces. His road surfaces were never more than 10 in. thick and had crowns of 3 in. for roads 30 ft wide. Consolidation was by traffic traveling the road, but continued raking kept the surface smooth until it had set.

The invention of the stone crusher in 1858 by Eli Blake greatly advanced macadam construction. Almost simultaneously came the invention of the steam roller in France, first used on the Bois de Boulogne in 1860. The first roller in England was used in Hyde Park 6 yr later. In 1867 Aveling and Porter made their first steam roller, a machine of 30 tons, and one was imported into the United States the next year. However, horse-drawn rollers were used for many more years. By the early 1900s, macadams were firmly established as the finest of road surfaces.

[3]Thomas Telford, 1757–1834, a Scot, born in Dumfriesshire and buried in Westminster Abbey, founded the Institution of Civil Engineers and was president of it until he died.

WATERBOUND MACADAM SURFACES[4]

Subgrade and Foundation

Waterbound macadams were built on earth subgrades, which were carefully excavated by hand after the road had been rough graded. This subgrade developed in the form of a shallow vertical-edged trench between earth shoulders. It was thoroughly rolled, and all soft spots were removed and replaced by sound material. This trenched subgrade frequently required drains through the shoulders. When rolled to an unyielding condition by a 10-ton or heavier roller, the subgrade was ready for the broken stone of the macadam course.

Material for Macadam Roads

Hard and tough broken stone was essential to good macadam construction. The first scientific testing for good broken stone was in France. Beginning in 1878, standard stone tests were applied in the laboratory of Ponts et Chausées in France. French practices were adopted and somewhat elaborated in America.[5] The tests were to determine hardness, toughness, and binding power of various rocks. A test for the cementing value of broken stone dust was introduced by Page, who also devised a supplemental test for toughness by the Page impact machine. Since hardness and toughness of rock generally were associated, it gradually became standard practice to rate stone for macadam and other surfaces according to the *percent of wear,* determined by the Deval test (see *AASHTO Designation T3*) and later by the Los Angeles rattler test (*AASHTO Designation T–96*) with a percentage of wear less than 40. Soundness under the sodium sulfate test (*AASHTO Designation T–104*) was 12% for 5 cycles.

In general, the trap-rock group, of igneous origin, proved to be best; they were denser and more fine-grained than granite and had an interlocking crystalline structure yielding high toughness. Next in order of suitability were certain types of granite, limestone, and dolomite, and a few harder sandstones. Coarse aggregate gradings were in three size groupings: 3½–1½ in., 2½–1½ in., and 2–1 in. Grading for screenings was: passing a 3/8 in. sieve, 100%; passing No. 4, 85–100%; passing No. 100, 10–30%.

Constructing Waterbound Macadam

Waterbound macadams were constructed in three courses, called base, surface, and binder. It was laid on a carefully leveled and rolled subgrade. At times, this was covered with about 1 in. of rock dust before the first rock was laid. The clean stone from one of the coarser sizes, as indicated above, was dropped onto steel or wooden dumping boards, which necessitated the complete rehandling

[4]See L. I. Hewes, *American Highway Practice,* Vols. I and II, Wiley, New York, 1942, for added detail.

[5]The first American testing laboratory was established in 1893 by the Lawrence Scientific School of Harvard University, in the charge of Logan Waller Page, afterward first director of the U.S. Bureau of Public Roads.

by shovelers. To avoid segregation, dumping in piles on the subgrade was prohibited. Before any rolling was undertaken, the base course was inspected thoroughly for low and high spots and for segregated or dirty stone, and the spread stone was shifted considerably by shovels and rakes in order to true the surface. To produce a 4-in. compacted layer, it usually was necessary to place nearly 6 in. of loose stone, as the broken stone shrank about one-third during rolling.

Rolling was done by a power roller weighing not less than 10 tons. Early rollers had the rear wheels coned to fit the road crown, a practice that has now disappeared. Rolling began at the outer edge with the rear wheel overlapping the shoulder. When the broken stone became firm, the roller was shifted to the opposite side of the road and the operation repeated. After both edges were rolled moderately firm, the roller was gradually moved toward the center until the entire base or lower course was thoroughly compacted. Sometimes during rolling depressions formed, and these were brought to true section and grade by the addition of more base stone.

After the base was thoroughly compacted, the earth shoulders were rebuilt to line. Later, wooden forms, gauged to the compacted thickness of the top course, were used. It was good practice to have the base course a foot wider than the top course in order to ensure solid support for the edges of the finished road. The finer stone for the top course was deposited about 3 in. deep, loose, and rolled to produce a solidly locked, and true surface. Smoothness was judged by eye until the 10 ft straightedge was introduced. Depressions or high spots were eliminated by moving the rolled stone or by adding additional stone.

The binder was spread either directly from carts by hand or from piles deposited along the shoulders (see Fig. 18–2). Binder was cast over the surface in successive layers by a lateral motion of the shovels. It disappeared upon rolling.

Fig. 18–2. Spreading stone screenings in three coats on a macadam road.

The use of brooms to distribute the dry screenings was very effective in producing uniform results. Only when the voids began to fill with screenings was it time to sprinkle. Then alternate applications of binder and rolling continued until a batter of stone dust and water began to flood ahead of the roller wheels. With good stone it was possible to produce a final surface that was completely waterproof and that would ring under horses' hoofs when the road was dry. As indicated above, many miles of waterbound macadam, protected by bituminous blankets or seal coats, are serving well.

WATERBOUND MACADAM BASE COURSES

The combination of tightly keyed coarse aggregate with the bond produced by stone chips and dust creates a base course equally as good as other untreated bases. Thus, in areas where cheap aggregates were not available in stream beds or other gravel deposits so that quarrying and crushing of ledge rock was required, macadam bases competed favorably. Today, where crushing of ledge rock is required to produce aggregate, macadam bases may be less costly than "graded" granular ones, since crushing to produce the intermediate sizes is expensive.

After World War II, macadam bases fell into disfavor, and today they have been dropped from the specifications of many highway agencies. In certain instances, and particularly on low-traffic roads, some agencies have gone back to them since they are equally as good and less costly than bitumen- or cement-treated bases.

Modern macadam base construction generally follows the steps outlined above for macadam surfaces, except that a blanket of stone screenings 1 to 3 in. thick is almost always placed over the subgrade. Also, the degree of mechanization is high in the United States. Materials are spread with devices fitted with vibrating or tamping screeds (see Chapter 19). Vibratory compactors generally replace rollers for consolidation and for working in the screenings. On the other hand, where manpower is plentiful and cheap, the hand-labor methods described above may be appropriate. As with other treated and untreated bases, protection of the surface with some form of seal or pavement is essential.

19 BITUMINOUS PAVEMENTS

Bituminous pavements consist of combinations of mineral aggregates with bituminous binders. Under this broad heading are included a multitude of pavement types, ranging from inexpensive surface treatments 1/4 in. or less thick to asphalt concretes, which are comparable in cost to portland-cement-concrete pavements.

Bitumen as a binder for road surfaces appeared first in recorded history in Babylon about 625 BC, when the Procession Street was paved with asphalt and burned brick. Mention of bituminous surfacings also appears in Greek and Roman documents. Later, although bitumen apparently was employed for a variety of purposes and the natural deposit of bituminous sandstone at the Val de Travers in Switzerland was reported about 1720, bitumen in road surfaces did not again appear until 1834. In 1852 an asphalt road from Paris to Perpigan on the Mediterranean Sea was constructed; in 1869 Threadneedle Street in London was paved. In the United States, although earlier pavements composed of coal tar, pitch, crushed rock and sand had been built, the first true asphalt pavement was placed in Newark, N.J. in 1870. Somewhat later, in 1876, a sheet asphalt with a Trinidad Lake asphalt binder was placed on Pennsylvania Avenue in Washington, D.C. By the turn of the century, bituminous pavement was in wide use.[1] As of 1979, something more than 1.8 million mi or 94% of the paved roads in the United States have bituminous surfaces.

The uninformed often refer to all bituminous surfaces as *blacktop,* because of their appearance. Much unwarranted criticism has resulted from this connotation, as failures of the cheapest temporary pavements have been construed by them to mean that all bituminous pavements are unsatisfactory. On the other hand, such a diversity of names has grown up among engineers that the person who tries to distinguish among them often becomes bewildered. Hveem, materials and research engineer, California Department of Transportation, retired, has stated the problem well, as follows:

> Mixtures of these two simple ingredients, rock particles and asphalt, have masqueraded under a number of names, such as Asphalt Macadam, Asphaltic Concrete, Sheet Asphalt, Topeka Mix, Mastic; proprietary, process, or trade names such as Tarvia, Warrenite, Bitulithic, Willite, Ameisite, Durite, Permatite, National Paving; and type names, such as plant mix, road mix, armor coat, retreads, penetration, inverted pen-

[1]For authorative presentations on the history of asphalt pavements and a list of references see F. N. Hveem, *HRB Highway Research News,* Winter 1971, and *Proceedings, Association of Asphalt Paving Technologists,* Vol. 43A, 1974.

etration, multiple lift, oil mat, and so on. This is by no means a complete list, but represents some of the most commonly used. These various names and terms do serve a purpose in identifying certain types of mixtures or methods of construction, but tend to make the engineer forget that virtually all bituminous pavements consist of nothing but mineral aggregate and asphalt.

The subject of bituminous pavements is further complicated for the student because there are many right answers; that is, there are many combinations of aggregates and binders that make good pavements, even under restrictive local conditions. This idea is alien to much of the engineer's earlier training which has been devoted to solving problems having but one correct solution.

If good service is to be received from bituminous pavement it must, for its full life, retain the following qualities: freedom from surface raveling or cracking from shrinkage or fatigue failure; resistance to weather, including the effects of surface water, heat, cold, and oxidation; resistance to internal moisture, particularly to water vapor; tight, impermeable surface, or porous surface (if either is needed for continued stability of underlying base or subgrade); and a smooth-riding and nonskid surface. The design of a pavement that will meet all these demands for a considerable number of years is an exacting task. It requires careful selection and control of materials and close supervision of each step of construction. Proper design and construction of subgrade and base course are a "must"; otherwise pavement failure is a foregone conclusion.

Pavements meeting all the requirements outlined here have been produced by four distinctly different construction processes as follows:[2]

1. Heat a viscous bituminous binder to make it fluid; then, in a plant, mix it with heated aggregate. Place and compact the mixture while it is still hot.
2. Use a fluid bituminous binder such as a liquid asphalt or emulsion. Mix it with aggregates at normal temperatures. Mixing may be done in a plant (plant mix) or on the prepared roadway base (road mix). Spread and compact the mixture at normal temperatures before the solvent evaporates or the emulsion breaks.
3. Spread and compact clean crushed aggregate as for waterbound macadam. Over it spray heated, dissolved, or emulsified bituminous binder which penetrates open areas of the rock and binds the aggregate together. This is commonly called the *penetration* method.
4. Spread a bituminous binder over the roadway surface; then cover it with properly selected aggregate. This is commonly called the *inverted-penetration* method.

In recent years, old asphalt concrete has been recycled by either adding a rejuvenating agent (see Chapter 14) and then mixing hot as in (1) above or cold as in (2) above. Recycling is discussed later in this chapter and in Chapter 21.

Not all the listed methods serve on roads carrying large volumes of heavy vehicles; nor can those that are most costly be economical on roads of low traffic volume. Availability of suitable aggregate is often a controlling factor. Personal preference and experience of the engineer in charge plays a large part in the final choice of method and of the details of that method.

[2]This breakdown deliberately oversimplifies current American practice, which introduces many variations within the processes here outlined. A combination of some heat with partial liquefaction of the binder is particularly common.

In this book bituminous pavements are classed as dense-graded mixes, open mixes, penetration types, and inverted-penetration types. Such an approach makes for an orderly presentation of design and laboratory procedures.

The remainder of this chapter presents, in a limited way, details of current bituminous-pavement practice in the United States. It is suggested that students, when reading it, keep firmly in mind the previously mentioned qualities of successful pavements and the common procedures for constructing them. If they do so, the need for each of the many materials and the purposes of the many tests (discussed in Chapter 14) will become apparent; the subject will become an integrated whole rather than a mass of unrelated details.

PRINCIPLES OF DESIGN FOR THE PAVEMENT STRUCTURE

Causes of Pavement Failure

The problems of pavement design parallel those of structural design of a bridge. A bridge must support a vehicle by transferring its load through successive members to the foundation beneath. Similarly, a pavement structure must support the vehicle load on its surface and transfer this load through successive layers of surface, base course, and subgrade to the undisturbed soil on which it rests. Bridge structures are usually built of steel, concrete, or timber, the properties of which are reasonably predictable. Pavements, however, are built of materials whose properties vary widely, and about which much is still unknown.

To the pavement structure are applied the wheel loads of motor vehicles which may number several million over a period of years. Each time a load passes, some deflection of the surface and the underlying layers occurs. If the load is excessive or the supporting layers are lacking in strength, repeated applications cause roughening and cracking that ultimately lead to complete failure. This deflection of the pavement may result from elastic deformation, from consolidation of the base and subsoils, or from a combination of elastic and plastic deformation.

Elastic deformation occurs as the wheel load temporarily deforms the foundation materials and compresses the air that fills the voids of the base and subgrade. In truly elastic deflection, the surface returns to its original position after the load passes, so that permanent unevenness does not occur even under repeated applications of load. Elastic deformation in the subgrade structure has been observed to depths of more than 20 ft, although most occurs within a few feet of the surface. If these elastic deformations are small, they do not result in damage to the roadway. With highly resilient soils, however, deflections under repeated heavy wheel loads can cause fatigue failures in the bituminous surface. This is evidenced by alligator or map cracking of the pavement surface, but without rutting.

Permanent deformation occurs when the load produces large enough stresses in the soil, base, or pavement to either densify the material or cause shear deformation (plastic flow with no volume change). Although the densification that results from one application of a moving wheel load is small, it is permanent and progresses with additional load repetitions. Deformation resulting from plastic flow is also progressive under load repetitions. Figure 19–1 illustrates the usual method of occurrence. Each is the result of a shear failure accompanied by movement in the affected layers. Pavement, base, and subgrade are all susceptible; however, the perimeter along which failure occurs has the least length if the shortcoming is in the surface layer and becomes longer as the source of trouble moves to greater depths (see Fig. 19–1). Since the total applied load in a given instance is constant, it follows that the unit shearing stresses that may be developed become smaller as the distance below the surface increases. This in turn indicates that the materials in successive layers downward from the surface may be progressively weaker without increased likelihood of plastic deformation and failure.[3]

As indicated earlier, soils are weaker at higher moisture contents, and to date, efforts to keep moisture out of the subgrade and base course have generally not been fruitful. This situation is recognized by most design methods which rely on strength measurements (see Chapter 14). Testing is carried out after samples have been brought to the moisture content anticipated in service or to a saturated condition if reliable long term moisture content is unknown.

The Flexible Pavement Concept

Pavements with bituminous surfaces are often referred to as *flexible* in contrast to *rigid* pavements of portland-cement concrete. As indicated by Fig. 19–1, a bituminous surface may deform and not completely recover under sustained or repeated loads. Within limits, it adjusts to densification of the underlying layers. In contrast, slabs of portland-cement concrete are rigid, in that they are elastic and return to their original conformation when loads are removed. In this instance, if the underlying layers do not recover completely, the slab rises away from them and spans the depressed area. In time, if a large area is unsupported and loads are large and frequent enough, the slab fails in fatigue.

The concept that bituminous pavements are flexible, although helpful in distinguishing between pavement types, must not be taken too literally. Bituminous materials have beam strength which increases with thickness. Without question this beam action assists in supporting the imposed loads. Again, certain other pavements might be classed as rigid, for example, when the base course has been cement-treated. Such situations are recognized in some design methods, as will be outlined subsequently.

Principles of Pavement Design

Basically, pavement design involves measuring the strength and other important properties of the pavement surfacing and the individual underlying layers and setting the respective thicknesses of pavement surface, base course and subbase (if any), and other imported materials which must overlie the native soil. Often, any one of several combinations of materials and layer thicknesses will meet the

[3]For an excellent and detailed exposition on causes of pavement failures, including many photographs, see F. N. Hveem, *HRB Bulletin 187*. See also *HRB Special Report 113, FHWA Report RD–79–66,* and *TRB Record 616* and *715*.

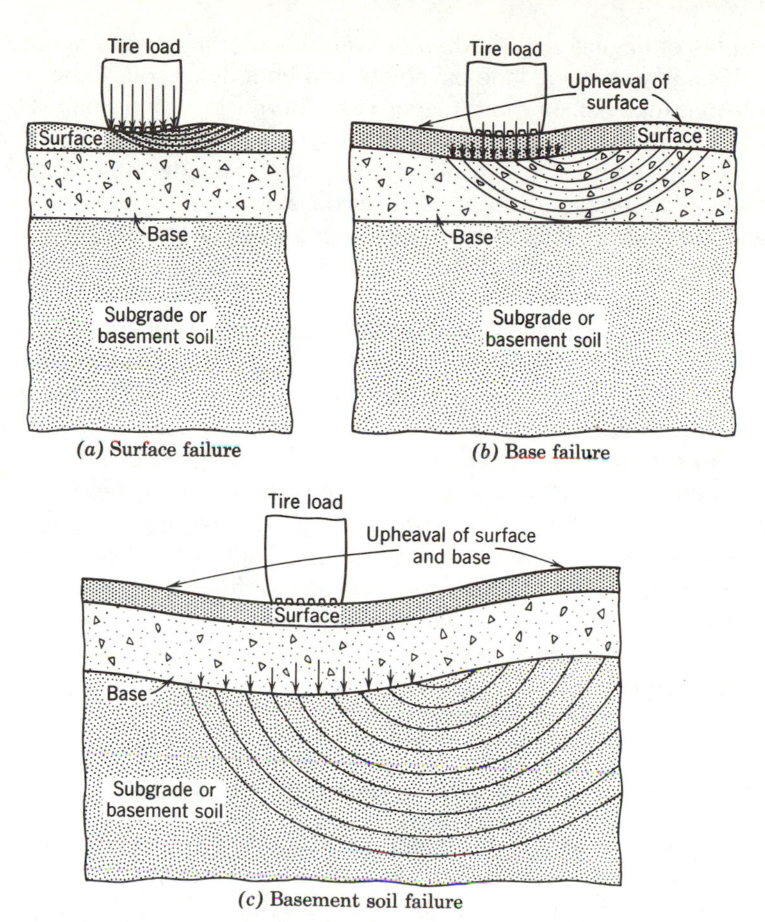

Fig. 19–1. Results of plastic deformation in basement soil, base, or surface. (After F. N. Hveem.)

requirements of a particular design method. At times, variables such as weather and soil-moisture conditions dictate more conservative treatment than usual. Without question, pavement design involves much more than substituting data into a formula or taking values from a design chart.

The aim of pavement design is to select the combination of materials and layer thicknesses that give the desired service at the least cost in the long run. Such an analysis usually considers the costs of construction, maintenance, and resurfacing. Where alternatives are being compared, any of the procedures outlined in Chapter 4 may be followed. Several imaginative approaches to these problems have been proposed,[4] most of which rely on the use of computers. As with all computer-based models, the results of such analyses are only as good as the design and cost parameters on which they are based. The discussions of design methods and construction practices which follow should indicate that large areas of uncertainty still exist.

Theoretical analyses to determine stresses in layered pavement systems from

[4]See *TRB Record 512* and *572* and *NCHRP Reports 139* and *160*.

the principles of engineering mechanics were begun many years ago by Bouissinesq. His analysis assumed the pavement and underlying soils to be single layered, isotropic material. In more recent years Burmister, Monismith, Finn, and others have modified and extended this analysis to include multilayer systems and to reconcile theoretical and observed results.[5] Among the complexities of the theoretical approach are the nonelastic behavior of soils and pavement materials and the fact that differences in materials and construction procedures mean sizable variations in mechanical properties and bond both in and between the layers.

Without question theoretical analysis and computer approaches are playing an increasingly important role in pavement design. Even so, the methods in common use are either entirely or partially empirical and are founded primarily on careful observation of past successes and failures, supplemented at times by findings from test tracks or experimental roads. Some methods rely on physical tests or soil classifications to relate past performances to the problem at hand. Layer thicknesses are found by substituting these test results in formulas or charts. Descriptions of a few are presented here; information on numerous others will be found in the various publications of the Transportation Research Board and in the *Proceedings* of the International Conference on the Structural Design of Asphalt Pavement, held every five years since 1962 under the auspices of the Department of Civil Engineering, University of Michigan, Ann Arbor.

DESIGN METHODS FOR BITUMINOUS PAVEMENTS

All pavement design methods begin with an estimate of expected traffic volume and character over the design life of the pavement. One approach is to classify traffic by such descriptive terms as heavy, medium, or light. For pavement design, as contrasted to capacity studies, this classification may include only commercial vehicles, since passenger cars do not contribute significantly to pavement failures. Increasingly, traffic projections are being combined with the results of loadometer studies to provide estimates for "equivalent" wheel or axle loads. For example, the effect of a "five-axle" truck is translated into a stated number of 5000-lb wheel loads or 18,000-lb axle loads.

Some design methods are based on estimates of traffic traveling in both directions; others use "single-direction" totals. Either approach is, of course, satisfactory if the user recognizes which is intended. Another problem arises with multilane highways. On them most of the truck traffic and pavement damage is concentrated in the outside lanes. For pavement design, some agencies assign all traffic to a single or "design" lane; others somewhat reduce the traffic estimate in the "design" lane. For example, the procedure of the Illinois Depart-

[5]See, as a beginning, *HRB Proceedings 1943, 1950,* and *1953; HRB Bulletin 177* and *342;* and *HRB Record 13, 71, 77, 228, 239, 337, 345, 362, 407* and 466. More recent studies appear in *HRB Special Report 126* and *140; NCHRP Reports 139, 140, 195* and *213; TRB Record 572* and *602*; and *Transportation Journal of ASCE,* Nov. 1970 and Jan. 1977. See also the U. of Michigan *Proceedings of the International Conference on Structural Design of Asphalt Pavements,* 1962, 1967, 1972 and 1977.

ment of Transportation, for rural highways, reduces the estimated number of 18,000-lb axle loads in the "design" lane by 10% for four-lane highways and by 20% where total lanes in both directions are six or more.

Several of the many pavement design methods now in use in the United States are presented in abbreviated form in the pages that follow.

Design by Precedent

Many agencies, particularly those of small cities and counties that do not have laboratory equipment or personnel, rely almost entirely on precedent in making pavement designs. The rule for residential subdivisions of a western city of moderate size furnishes an illustration. It calls for 6 in. of compacted base course from a local quarry topped by 2 in. of asphalt concrete surfacing. This standard design is followed for all residential streets, regardless of the subgrade conditions. On completion, the city assumes all responsibility for maintenance. Performance has for many years been generally satisfactory, although failures over particularly bad subgrades caused concern and led to some revision in the design. For business streets and other traffic arteries, layers are thicker, but their dimensions are likewise determined by precedent.

Many satisfactory pavements have been designed by such "rule-of-thumb" methods. However, dimensions appropriate for one set of soil, moisture, climatic, and traffic conditions are not necessarily appropriate under different circumstances. Thus rule-of-thumb design methods should be used only over small areas where like conditions exist, or where previous experience indicates that good results can be expected. Design by precedent often may be uneconomical. If a particular design proves satisfactory on all the streets or roads in an area, it is so because the design is appropriate for the worst condition of that area. Other sections probably are overdesigned and therefore more costly than necessary. On the other hand, the "price" of even an occasional failure in terms of public confidence is so great that engineers understandably are reluctant to depart too far from practices that have been satisfactory.

California (Hveem) Method[6]

This method, used widely in the western part of the United States, was developed from limited test road information, experience with in-service pavements, and some theory. Originally developed in the 1940s, it has been modified several times to adjust for changes in traffic and other conditions. The original intent was to preclude plastic deformation and to prevent expansive forces in the materials from distorting the pavement surface. The method was later modified slightly to minimize early fatigue cracking under heavy wheel loads.

Three factors that effect permanent deformation are considered in this method. They are: (1) the effect of traffic, normally expressed as number of equivalent 18,000-lb axle loads; (2) the strength characteristics (*R*-value) of the

[6]*See California Department of Transportation Design Manual, Section 7–651*. For development, see *HRB Record 13* and its bibliography. California converted a 5000 lb wheel load to an 18,000 lb equivalent axle load in 1975. Minnesota's use of this method is described in *HRB Record 329*.

soil and base (or subbase) materials as measured in the stabilometer test (see Chapter 14); and (3) the tensile strength characteristics of the materials above the subgrade as measured in the Hveem cohesivemeter, stated as a gravel equivalency factor (G_f) (see below). The general design equation is as follows:

$$GE = 0.0032(TI)(100 - R) \quad (19\text{–}1)$$

where

R = measure of the soil resistance of the material (see Chapter 14 for test method) normally determined at an exudation pressure of 300 psi.

GE = thickness of material required above a given layer in terms of gravel equivalent, in feet.

TI = traffic index, or a measure of the number of 18000-lb equivalent axle loads (EAL) expected in the design lane over the pavement life. TI is related to EAL as follows.

$$TI = 9.0 \left[\frac{EAL}{10^6}\right]^{0.119} \quad (19\text{–}2)$$

Once the required gravel equivalent above each layer is determined, the individual layer thickness (D) is set as shown in Fig. 19–2, employing either Equation 19–1 or Fig. 19–3. The example that follows is based on data from the *California Design Manual*.

First of all, the total number of 18,000-lb equivalent axle loads in one direction for the expected service life (typically 10 or 20 yr) must be estimated using highway planning data on traffic volumes, character, and wheel loadings. Passenger cars and pickup trucks are not considered in the analysis. For a typical California situation, yearly axle-load equivalents for the daily one-way traffic, by axles per vehicle, are as follows:

Number of axles	2	3	4	5 or more
Yearly 18,000 = lb equivalent axle loads per daily vehicle	1020	3120	4480	13,780

For two- and four-lane highways this number of wheel loads is assumed to travel solely in the right-hand lane for which the design is made. For six- and

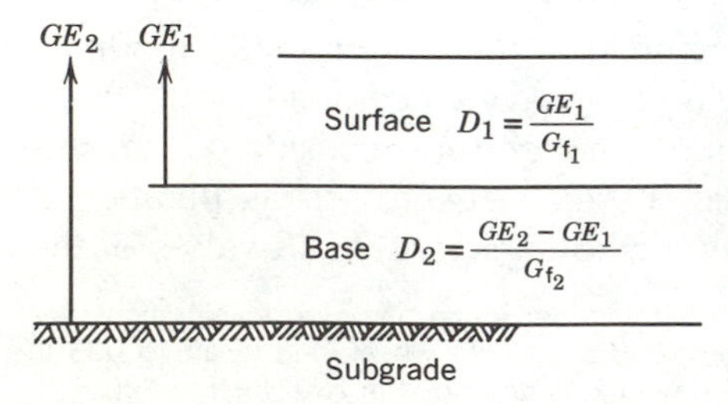

Fig. 19–2. Procedure for determining layer thickness, Hveem method. GE = gravel equivalent (ft); D = layer thickness (ft); G_f = gravel equivalency factor for a given layer.

eight-lane facilities, this number is normally reduced to 80% assuming some heavy vehicles will travel in other lanes. From the equivalents it can be shown, for example, that five-axle trucks are most destructive, with each truck passage contributing the effect of 4.4 three-axle or 13.5 two-axle vehicles.

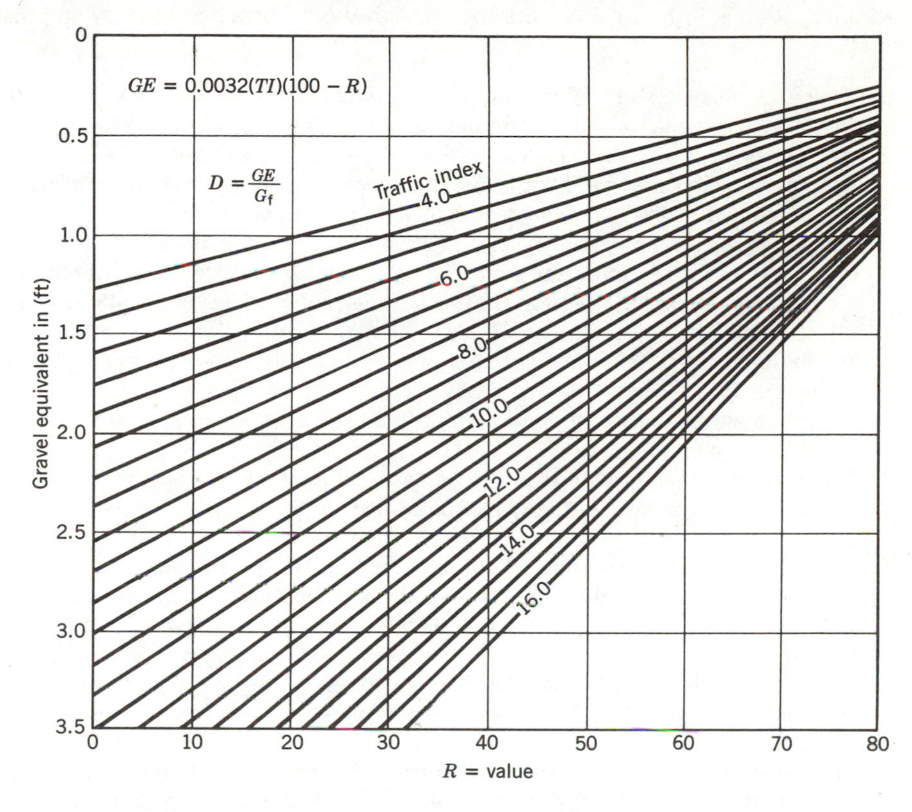

Material	G_f
Cement treated base	
Class A	1.7
B	1.2
Lime-treated base	1.2
Untreated aggregate base	1.1
Aggregate subbase	1.0
Asphalt concrete for T.I. of	
≤5.0	2.50
5.5–6.0	2.32
6.5–7.0	2.14
7.5–8.0	2.01
8.5–9.0	1.89
9.5–10.0	1.79
10.5–11.0	1.71
13.5–14.0	1.52

Fig. 19–3. Design chart for California method for setting thickness of pavement layers.

To illustrate the stabilometer design method, assume that a pavement cross section is to consist of (a) asphalt concrete of 4 in. (0.33 ft) thick and (b) an untreated granular base with an *R* value of 78 at 300 psi exudation pressure. These overlie a basement soil having the characteristics shown on Fig. 19–4. The traffic index is 8.0. For this situation, design would proceed step by step as follows:

1. Determine the thickness of the asphalt concrete. From Fig. 19–3 with *R* value of 78 for the base course and *TI* of 8.0, determine the gravel-equivalent thickness required for the pavement. It is 0.56 ft. Note that the gravel-equivalent factor (see Fig. 19–3) is 2.01. The required pavement thickness for strength is 0.28 ft. Pavement thickness, then, is controlled by the minimum requirement of 0.33 ft, given above.
2. Determine the thickness of granular base, if strength controls the design.
   a. Take *R* values for basement soil from Fig. 19–4. Determine gravel equivalents for each *R* value from Fig. 19–3 for *TI* of 8.0. These, for *R* values of 53, 43, and 14 are, respectively, 1.20, 1.45, and 2.20 ft.
   b. Plot these thickness values against the appropriate exudation pressures (see curve *A*, Fig. 19–4).
   c. Determine gravel-equivalent cover for subgrade soil at exudation pressure of 300 psi. This is 1.83 ft (point 1 on Fig. 19–4).
   d. Determine gravel-equivalent depth to be supplied by the base course. This is the total gravel equivalent less the gravel equivalent of the asphalt concrete. This is $1.83 - 0.33 \times 2.01 = 1.17$ ft.
   e. Determine the actual thickness of the base course. This is the gravel-equivalent thickness (1.17 ft) divided by the gravel-equivalent factor of 1.1. Base thickness is then 1.06 ft. Note that the total actual thickness of pavement and base is $0.33 + 1.06$ ft $= 1.39$ ft.
3. Check to see that the design for strength meets expansion pressure requirements.
   a. Plot curve *B* for gravel-equivalent thickness of pavement and base versus percent moisture. (Fig. 19–4)
   b. Determine from curve *B* the moisture content that matches the gravel-equivalent design thickness. In this case the percent moisture is 23.2 for a thickness of 1.83 ft (point 2). This will be the moisture content on which the "design for strength" is based.
   c. Plot curve *C* of expansion pressure thickness versus moisture content from the soil data.
   d. Compare total thickness of pavement and base for strength with curve *C*. It can be seen that the weight of pavement and base is more than enough to prevent expansion. Therefore, the strength requirement controls the design.

Figure 19–4 also shows two other situations where expansion pressure may enter the design. First, consider a soil whose characteristics are portrayed by curves *A, B,* and *D*. Here it can be seen that the design for strength balances expansion pressure at about 22.8% moisture (point 3). But at least theoretically it will be restrained from expansion if its moisture content when compacted in the embankment is greater than 22.8%. However, expansion could be expected if it were compacted on the dry side. The second condition can be illustrated by a soil having the characteristics given by curves *A, B,* and *E* of Fig. 19–4. In this instance the thickness that will satisfy the strength (stabilometer) require-

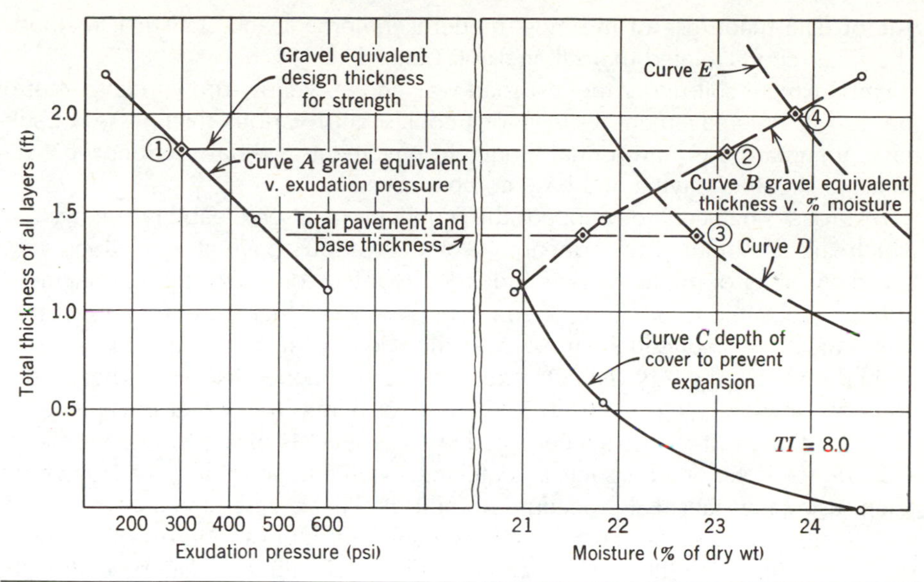

Text Data for Soil

Moisture (%)	*R Value*	*Exud. Press. (psi)*	*R Value Thickness (ft)*	*Exp. Press. (psi)*	*Exp. Pr. Thickness (ft)*
20.9	53	600	1.20	1.25	1.38
21.8	43	450	1.45	0.50	0.55
24.5	14	170	2.25	0.00	0

Fig. 19–4. Example of pavement design based on the Hveem stabilometer method.

ment will not prevent expansion. Rather the total thickness of pavement and base must be increased to 2.0 ft and the percent moisture to 23.6% to produce a balance between the strength requirement (curve *B*) and that for expansion (curve *E*).

AASHTO Method

The AASHO Test Road was a $27 million cooperative project sponsored by the (then) American Association of State Highway Officials. It involved both bituminous and portland-cement-concrete pavements and a group of 50 ft single-span bridges of steel with composite and noncomposite design, prestressed concrete, both pretensioned and posttensioned, and reinforced-concrete T beams. Planning the project began about 1950; the site near Ottawa, Ill., was selected in 1954; construction was carried on in the years 1956–1958; testing began in October 1958 and ended in late 1960; data analysis and final reporting were completed in 1962. In all, the test road contained six loops, each with two lanes. Single-axle loadings ranged from 2000 to 30,000 lb; tandems from 24,000 to 48,000 lb. Field testing and measurement, laboratory work, and analysis of data made use of the most modern equipment and statistical methods. The final reports totaled more than 1600 pages.[7]

Embankment soil under the test road was an *A*-6 yellow-brown clay. Most of the 450 flexible-pavement test sections consisted of various thicknesses of subbase, granular base, and asphalt concrete pavement. Special sections were devoted to shoulder paving and base comparisons.

Test-road engineers developed the concept of "serviceability ratings" to which the smoothness and rideability of the various pavement sections were keyed. A panel of highway users individually rated some 74 highway segments of bituminous and 64 sections of concrete pavements that were in various states of repair. These individuals drove over the sections at high speed and, if they wished, examined the pavements critically. They independently rated each section on a scale ranging from 0 to 5, with lower ratings for poorer pavements. At the same time, test-road engineers made measurements of roadway deformation and surface deterioration such as cracking, spalling, potholing, and patching. Then, using the statistical procedure of multiple linear regression analysis, equations were developed to express the subjective user ratings in terms of the measured properties. For flexible pavements, the final rating equation, called the Present Serviceability Index (*PSI* or *p*) was:

$$p = 5.03 - 1.91 \log_{10}(1 + \overline{SV}) - 1.38\, \overline{RD}^2 - 0.01\sqrt{C + P} \qquad (19\text{–}3)$$

where

p = present serviceability index
$\overline{SV}$ = slope variance along the wheel path (as integrated on a 9 in. base by the test road or CHLOE profilometer)[8]
$\overline{RD}$ = depth of wheel path rut in inches as measured from a 4 ft straightedge
C = cracked area in $ft^2/1000\ ft^2$ of pavement area
P = patched area in $ft^2/1000\ ft^2$ of pavement area

The value of *p* on typical new sections of flexible pavement on the test road was 4.2; when the value fell to 1.5, testing was stopped. Recommended *p* values at which highway pavements should be resurfaced are normally 2.5 for primary and 2.0 for secondary roads.

As would be expected the present serviceability index *p* decreased progressively during testing (see Fig. 19–5). The left-hand curves are smoothed plots of individual *p* values appraised biweekly. The curves in the graph in the right-hand diagram have been weighted to reflect the greater deterioration per wheel load during the spring thaw. Finally, exponential equations were developed which expressed the shape of the curve in terms of wheel loads and beginning and ending *p* values.

[7]See *HRB Special Reports 61-A* through *61-G, 66,* and *73*. For an earlier road test sponsored by WASHO (Western Association of State Highway Officials), see *HRB Special Report 18* and *22*. This test proved conclusively that asphalt concrete thickness and shoulders contributed greatly to pavement strength.

[8]Several other roughness or slope-measuring devices have been developed. Most now operate at the speed of prevailing traffic. See e.g., *HRB Special Report 133* and R. Haas and W.R. Hudson, *Pavement Management Systems,* McGraw-Hill, New York, 1978.

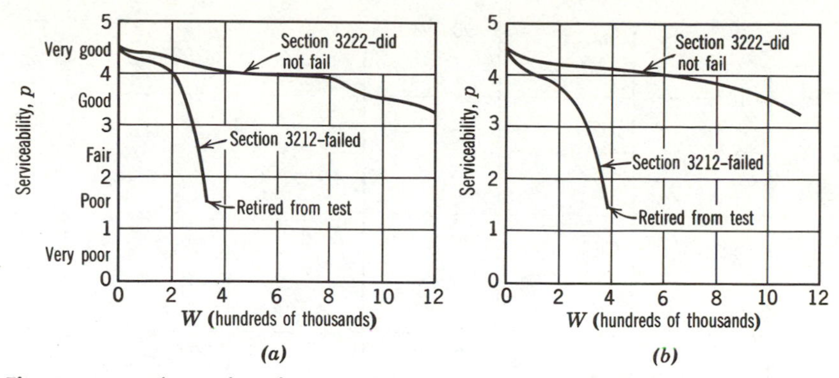

Fig. 19–5. Relationships between repetitions of axle load and serviceability for two typical sections of the AASHO Test Road: *(a)* actual performance; *(b)* performance weighted to reflect weather effects.

The embankment and overlying layers of the test road were made of local materials. Material quality and construction procedures were carefully controlled; weather conditions were peculiar to the area. Loading for each loop was for a selected axle load rather than mixed, as on ordinary highways. It follows that test-road results cannot be used directly; rather, each highway agency must adapt the test-road findings to its own local situation.

An interim guide for the design of flexible pavements has been published by AASHTO.[9] This procedure relies heavily on test-road findings and on correlations that were developed with several soil-rating procedures used in the United States. Test-road findings have already led numerous agencies to change design methods or parameters.[10]

Very briefly the interim design procedure is as follows:

1. The soil support value *S* for the subgrade, embankment material, or other layer in question is determined from stabilometer *R* value, *CBR,* or group index. Conversion from them to *S* can be done with a horizontal line on Fig. 19–6.
2. The number of "equivalent" daily 18,000-lb single-axle load applications is computed. An equation relating other axle loads to the 18,000-lb single-axle load was developed from test-road findings. A limited number of factors, computed from this formula, are given in Table 19–1. Estimated equivalent daily 18,000-lb axle-loadings are found by multiplying the estimated daily one-direction axle-loadings by the appropriate factors from Table 19–1. "Cut and try" solutions for several values of *SN,* the weighted structural number, may be necessary since it is required in the solution and is also the answer being sought. Incidentally, the destructive effects of various axle loads, as represented in Table 19–1, help in settling such troublesome matters as permissible maximum axle loads and the relative magnitude of taxes for vehicles of different sizes and weights under the increment-cost or consumption-rate theories (see Chapter 5).
3. Given *S,* the soil support value, and the equivalent daily 18,000-lb single-axle load

[9]See *Interim Guide for Design of Pavement Structures,* AASHTO, 1972. A detailed evaluation of it has been issued as *NCHRP Report 128.*
[10]*See e.g., HBR Record 71, 90, 189* and *291. NCHRP Reports 2* and *2A* offer guidelines for satellite studies.

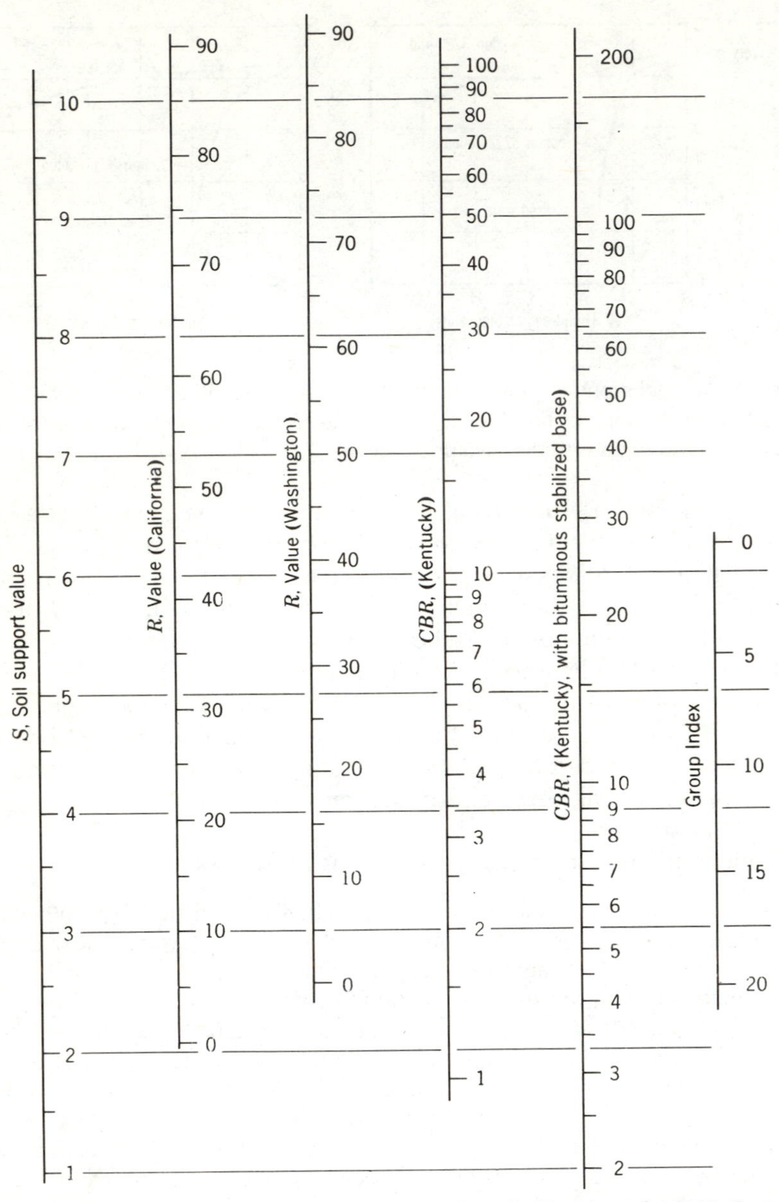

Fig. 19–6. Flexible pavement-design procedure based on AASHO Road Test results. Chart for finding the soil support value S from stabilometer R value, *CBR,* or group index (group index correlation based on procedure outlined in *HRB Proceedings, 1945,* pp. 375–392 and not on data from cooperative testing program).

TABLE 19–1 AASHO Road Test Table for Converting Single- and Tandem-Axle Loads to Equivalent 18,000-lb Single-Axle Loads

	Equivalence Factors, $p=2.5$					
	Single Axles Structural Number SN			*Tandem Axle Sets Structural Number SN*		
Loads (kips)	*1*	*4*	*6*	*1*	*4*	*6*
2	0.0004	0.0002	0.0002			
6	0.01	0.01	0.01			
10	0.08	0.10	0.08	0.01	0.01	0.01
14	0.33	0.39	0.34	0.03	0.03	0.02
18	1.00	1.00	1.00	0.07	0.09	0.07
22	2.48	2.09	2.30	0.16	0.21	0.17
26	5.33	3.91	4.48	0.33	0.40	0.34
30	10.31	6.83	7.79	0.61	0.70	0.63
34	18.41	11.34	12.51	1.06	1.11	1.08
38	30.90	18.06	18.98	1.75	1.68	1.73
40	39.26	22.50	23.04	2.21	2.03	2.14
44				3.41	2.88	3.16
48				5.08	3.98	4.49

Note: In abstracting values for this table from more complete data, some accuracy has been lost. Straight-line interpolation may therefore introduce relatively small errors.

applications, the "unweighted" structural number $\overline{SN}$ is determined. Figure 19–7 offers a graphical method for doing so with a straightedge.

4. The "regional factor" R is determined. This factor reflects the fact that during a spring thaw or at other times when the ground is saturated, a load does more damage than when the ground is frozen or dry. The regional factor can also be used to recognize high water table or other destructive factors.

   The recommended procedure is to appraise the regional factor for appropriate time periods throughout the year. Then a weighted factor is computed, based on the periods of time that the various conditions exist. Suggested ranges for the regional factors are given on Fig. 19–7.
5. Given $\overline{SN}$ and the regional factor R, determine SN, the weighted structural number. This is done with a straightedge on the right-hand side of Fig. 19–7.
6. Determine the combination or combinations of pavement, base, and subbase thicknesses that will develop the required weighted structural number SN. This is done with the equation

$$SN = a_1D_1 + a_2D_2 + a_3D_3 \tag{19–4}$$

where

a_1, a_2, a_3 = coefficients of relative strength
D_1, D_2, D_3 = thicknesses in inches, respectively of bituminous surface, base course, and subbase

Suggested values for the coefficients of relative strength are:[11]

[11]Individual agencies have developed their own values for these coefficients; e.g., *HRB Special Report 117* indicates that the thickness ratings of bituminous bases, as compared to granular bases, range from 1 to 1 to 4½ to 1. See also *NCHRP Report 128*.

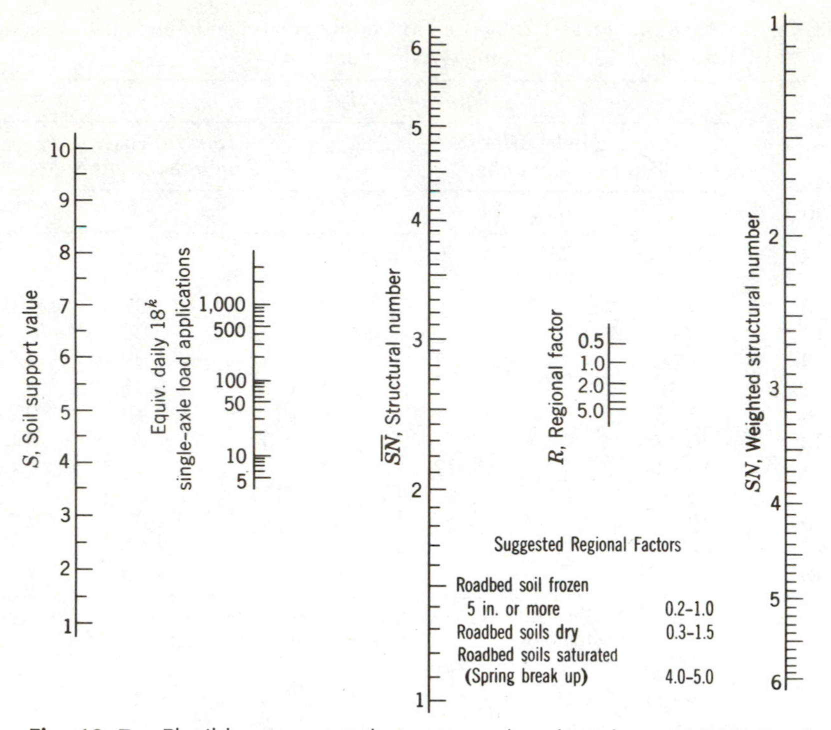

Fig. 19–7. Flexible pavement-design procedure based on AASHO Road Test results. Chart for determining the weighted structural number *SN*. Terminal serviceability index = 2.5 (for main highways). Design for 20 yr period.

a_1 (pavements)—road mix (low stability), 0.20; plant mix (high stability), 0.44; sand asphalt, 0.40

a_2 (base course)—sandy gravel, 0.07; crushed stone, 0.14; cement-treated (650 psi or more), 0.23; cement-treated (400–650 psi), 0.20; cement-treated (400 psi or less), 0.15; coarse-graded bituminous-treated, 0.34; sand asphalt, 0.30; lime-treated, 0.15–0.30

a_3 (subbase, if used): sandy gravel, 0.11; sand or sandy clay, 0.05–0.10

An important feature of the AASHO Test Road was a cooperative testing program. Identical samples of test-road materials were sent to various highway agencies for testing in their individual laboratories. A comparison of results indicated significant differences between laboratories, which introduces yet another variable in applying test-road results. A few of these differences are presented in Table 19–2.

Asphalt Institute Design Method

A proposed Thickness Design Method (*MS*-1) of the Asphalt Institute applies elastic layered theory to pavement design. This method is quite different from the AASHTO and California methods in that it relies on the laws of mechanics

TABLE 19–2. Characteristics of Embankment Soil from AASHO Road Test; Partial Results of Cooperative Testing Program*

	Atterberg Tests			*CBR at 0.1-in. Penetration*			
					Dynamic Compaction		
	LL	*PL*	*PI*	*Static Compaction*	*(a)*	*(b)*	*Hveem Stabilometer†*
Average	27.7	15.1	12.6	4.0	5.8	8.4	28
Highest	34.0	18.0	18.0	6.7	7.2	13	32
Lowest	22.5	13.0	9.5	0.8	5.0	5	26
Standard deviation	1.91	1.40	1.77	—	—	—	—

(a) 5½-lb hammer, 12 in. fall, 55 blows on each of three lifts.
(b) 10-lb hammer, 18 in. fall, 55 blows on each of five lifts.
*Data from J. F. Shook and H. Y. Fang, *HRB Special Report* 66.
†*R* value at 400-psi exudation pressure. One reported value of 50 has been excluded from the tabulation.
Note: 57 agencies reported the Atterberg test results. Others are based on six of fewer reports.

to predict critical stresses and strains rather than on empirical relationships relating soil strength and traffic conditions to pavement thickness.

In this method, the materials of each layer are characterized by their modulus of elasticity and Poisson's ratio. Traffic is expressed in terms of an 18,000-lb single-axle load applied to a pavement on two sets of dual tires (Fig. 19–8).

The method can be used to design asphalt pavements composed of asphalt concrete surface and base, emulsified asphalt surface (with surface treatment) and base, and asphalt concrete surface with untreated base and/or subbase. For full depth asphalt, the pavement is regarded as a three-layer system; for systems with untreated aggregate, a four-layer system is assumed. The subgrade is assumed to be infinite in the vertical direction. All layers are infinite in the horizontal direction.

LIMITING STRAINS. Two strains are considered critical for design (Fig. 19–8):

1. The horizontal tensile strain (ε_t) on the underside of the asphalt stabilized layer. If this strain is excessive, fatigue cracking of the surface layer will result.
2. The vertical compressive strain ε_c at the surface of the subgrade. If this strain is excessive, permanent deformation (rutting) will result.

Limiting values for each critical strain are shown in Figs. 19–9 and 19–10.

MATERIALS. All materials in the pavement system are characterized by their modulus of elasticity (see Chapter 14) and Poisson's ratio. Values for asphalt concrete are selected based on the mean annual air temperature; the higher the temperature, the lower the modulus. For emulsion mixes, modulus values are based on temperature and curing times. Only one untreated granular base is considered. A variety of subgrade modular values are included, ranging from 1000 psi (poor subgrade) to 100,000 psi (very good subgrade).

ENVIRONMENTAL CONSIDERATIONS. Different environments are repre-

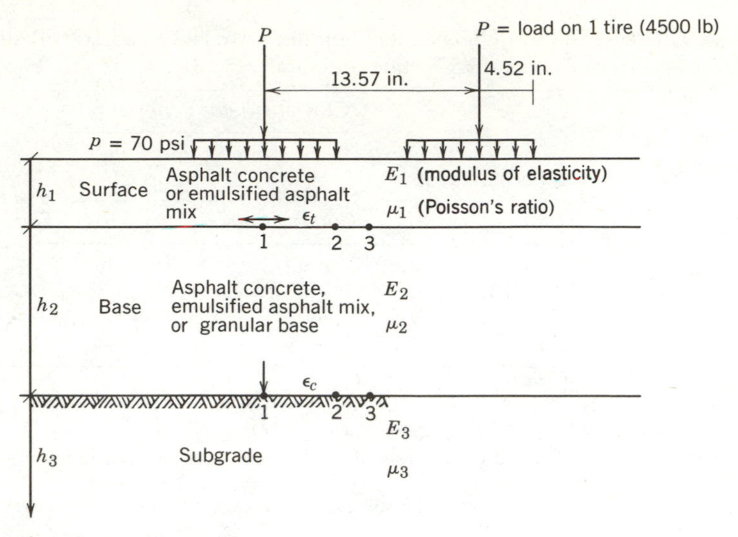

Fig. 19–8. Loading on a layered pavement system—Asphalt Institute Design Method. Applied loads are distributed uniformly over a circular area. Critical strains (ε_t or ε_c) may occur at points 1, 2 or 3.

sented by temperature, termed mean annual air temperature *(MAAT)*.[12] Frost effects are included for two temperature conditions by increasing subgrade modulus to represent the freezing period and reducing it to represent the thaw period. Similar techniques are used to represent environmental effects on untreated granular bases.

TRAFFIC. All traffic is converted to equivalent 18,000-lb single axle load applications *(EAL)*. The *EAL* is calculated in a manner similar to that described in

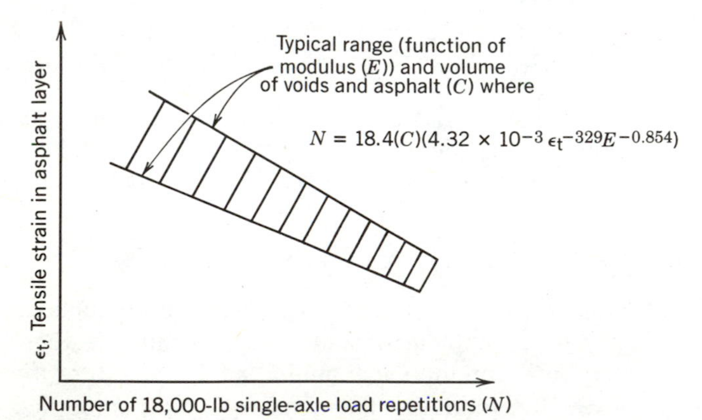

Fig. 19–9. General relationship between ε and number of loads to cause fatigue cracking.

[12] $MAAT = \Sigma \dfrac{\text{(mean monthly air temperature)}}{12}$

the AASHTO procedure by multiplying the number of vehicles in each weight class by an appropriate truck factor and obtaining the sum or

$$EAL = \Sigma \text{ (number of vehicles } \times \text{ truck factor)}$$

where

$$\text{truck factor} = \frac{\Sigma \text{ (number of axles} \times \text{load equivalency factor)}^{13}}{\text{number of vehicles}} \qquad (19\text{–}5)$$

Truck factors for typical situations in the United States are given in Table 19–3; however, changes in legal weights will result in changes in these factors. In summary, the following steps are used to determine *EAL:*

1. Determine the average number of each class of vehicle expected during the first year of traffic.
2. Determine, from axle data, or select from Table 19–3, a truck factor.
3. Select from Table 19–4, a single growth factor for all vehicles or separate factors for each vehicle type.
4. Multiply the number of vehicles of each type times the truck and growth factors.
5. Sum the values in (4) to obtain design *EAL*.

TABLE 19–3. Typical Truck Factors for Different Classes of Highway, All Trucks Combined (After *Asphalt Institute MS-1*)

	Truck Factors	
Type of Facility	*Average*	*Range*
Interstate rural	0.49	0.34–0.77
Other rural	0.31	0.20–0.52
All rural	0.42	0.29–0.67
All urban	0.30	0.15 0.59
All systems	0.40	0.27–0.63

DESIGN CHARTS: Design charts, based on the criteria given in Figures 19–9 and 19–10, are available for several conditions as follows:

1. Full depth asphalt concrete.
2. Emulsified asphalt mixes (three types).
3. Asphalt concrete surface over untreated aggregate base and subbase.

Three temperature ranges *(MAAT)* are considered for each type of section: 45°F (frost exists), 60°F (possible frost), and 75°F (no frost).

Figure 19–11 shows design charts for two material types for the possible frost condition. Knowing the *EAL, MAAT,* and subgrade modulus, the thickness of asphalt concrete over the subgrade or over 6 in. of aggregate base can be determined. Similar charts for a variety of other situations are given in the manual.

MINIMUM THICKNESSES. Minimum thicknesses of asphalt concrete over subgrade, emulsion mixes, and untreated aggregate bases vary with traffic level as follows:

[13]Load equivalency factors are from *AASHTO Interim Guide for Design of Pavement Structures,* Table 19–1.

EAL	Thickness (in.) Over Subgrade and Aggregate Base	Over Emulsion Mixes
10^4	4	2
10^5	4	2
10^6	5	3
10^7	6	4

EXAMPLE. Given a situation where the expected number of 18,000-lb load applications is 1.5 million, the subgrade modulus is poor ($M_R = 5500$ psi), and $MAAT = 60°F$ with frost possible, the design thickness from Fig. 19–11*a* is 10½-in. asphalt concrete over the subgrade or from Fig. 19–11*b*, 9½ in. of asphalt concrete over 6 in. of aggregate base.

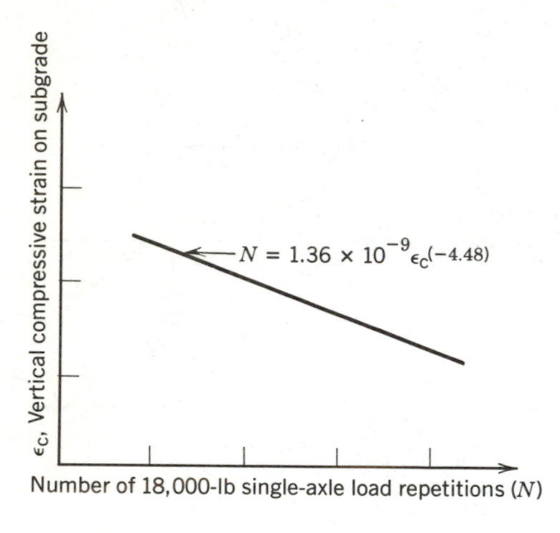

Fig. 19–10. General relationship between ε_c and number of load repetitions to failure.

TABLE 19–4. Growth Factors*

Design Period, n (yr)	Annual Growth Rate r (%) 0	2	4	6	8	10
1	1.0	1.0	1.0	1.0	1.0	1.0
5	5.0	5.20	5.42	5.64	5.87	6.11
10	10.0	10.95	12.01	13.18	14.49	15.94
15	15.0	17.29	20.02	23.28	27.15	31.77
20	20.0	24.30	29.78	36.79	45.76	57.28

* Factor $= \dfrac{(1+r)^n - 1}{r}$, where $r = \dfrac{\text{rate}}{100}$ and is not zero.
If annual growth is zero, growth factor = design period.

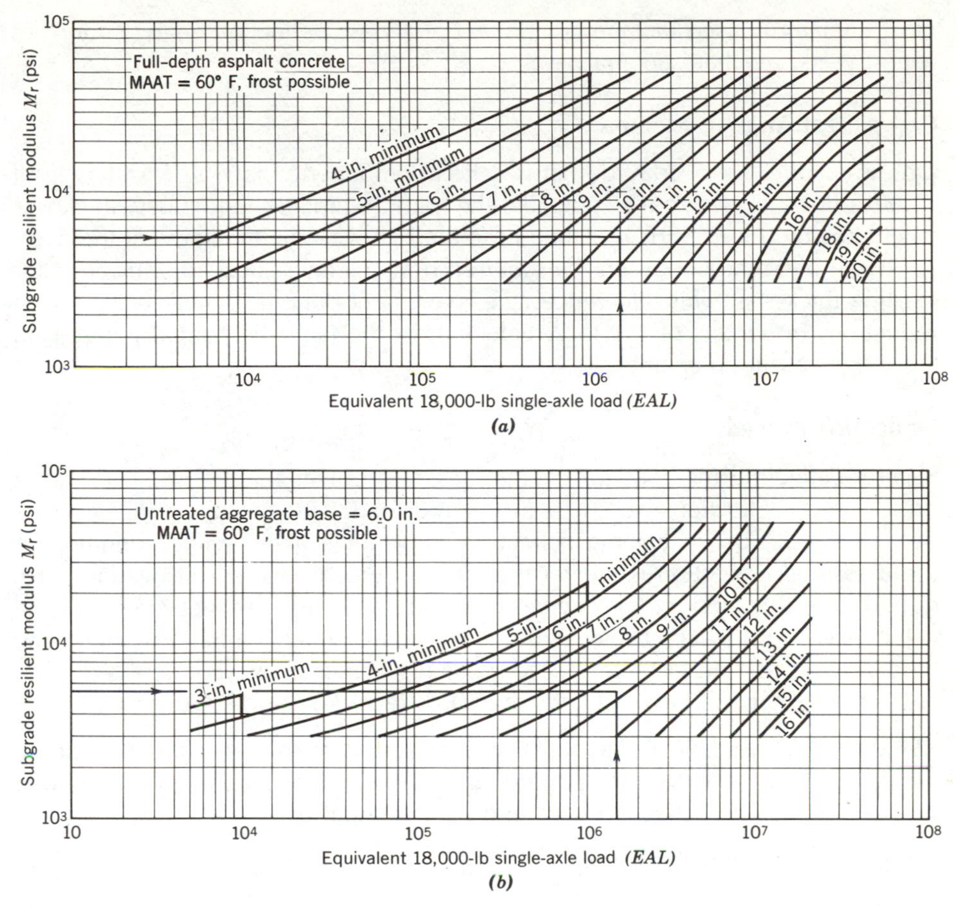

Fig. 19–11. *(a)* Full-depth asphalt concrete design chart, slightly modified from *Asphalt Institute MS-1*. *(b)* Asphalt concrete over aggregate base design chart, slightly modified from *Asphalt Institute MS-1*.

OVERLAY DESIGN PROCEDURES FOR FLEXIBLE PAVEMENTS[14]

As the present serviceability index of an asphaltic concrete pavement reaches an unacceptable level (normally 2.5), the pavement can be reconstructed, recycled, or overlaid. For asphaltic concrete pavements the overlay is normally asphaltic concrete, but can be portland-cement concrete. Design procedures

[14]For details, see *AASHTO Interim Guide*. See also R. W. Bushey et al., *TRB Record 572* for a description of the California Method and *The Asphalt Institute MS-17* for details of its procedure. Other approaches are discussed in *TRB Record 632* and *700*.

most commonly used include component analyses, deflection analysis, and, in recent years, mechanistic analyses.

Component Analysis Method

One procedure for overlay design is called component analysis. A soil support value is first assigned to the subgrade, the traffic estimated, and a new thickness calculated by the AASHTO method described above. The thickness of the overlay is the difference between the required new thickness and an allowance for existing thickness. This allowance takes pavement condition into account with reductions by factors of 1.1 to 2. Selection of the factor is normally based on the judgment of the engineer.[15]

Deflection Procedures

Several deflection methods for overlay design are available. All rely on measured surface deflection caused by a standard load. This measured deflection is compared with an acceptable deflection value for a given traffic condition. If the measured deflection is greater than the acceptable value, an overlay is required. The thickness of the overlay is a function of the amount the measured value exceeds an acceptable value. Because of space limitations, only one method, The Asphalt Institute Procedure, is discussed below. For details on others see the *AASHTO Interim Guide* and *TRB Record 572* and *700*.

REPRESENTATIVE DEFLECTION. For overlay design of a section of road, a number of surface deflection tests are taken by one of the techniques described in Chapter 14. The deflection used in design is equal to

$$\text{representative deflection} = (\bar{X} + 2s)fc \qquad (19\text{–}6)$$

where

$\bar{X}$ = the arithmetic mean of the individual values.
s = standard deviation.
f = pavement temperature adjustment factor (0.8 at 95°F, 1.0 at 70°F, and 1.6 at 40°F). Actual deflections are adjusted to a 70°F temperature.
c = critical period adjustment factor ($c = 1$ if tests are made in spring, greater than 1 if tests are made in summer).

The $2s$ factor in the equation gives a design deflection which is greater than or equal to 95% of all measurements adjusted to a pavement temperature of 70°F.

ALLOWABLE DEFLECTION. Figure 19–12 shows the allowable (or design) deflection as a function of traffic (*DTN*—daily 18,000-lb single-axle loads).[16] This relationship, developed from test roads, recognizes the relationship between load repetition and pavement strength.

[15]For typical values, see E.J. Yoder and M.W. Witczak, *Principles of Pavement Design*, Wiley, New York, 1975.
[16]The *DTN* can be determined by dividing total expected 18,000-lb single-axle loads by 365 × design period (in years).

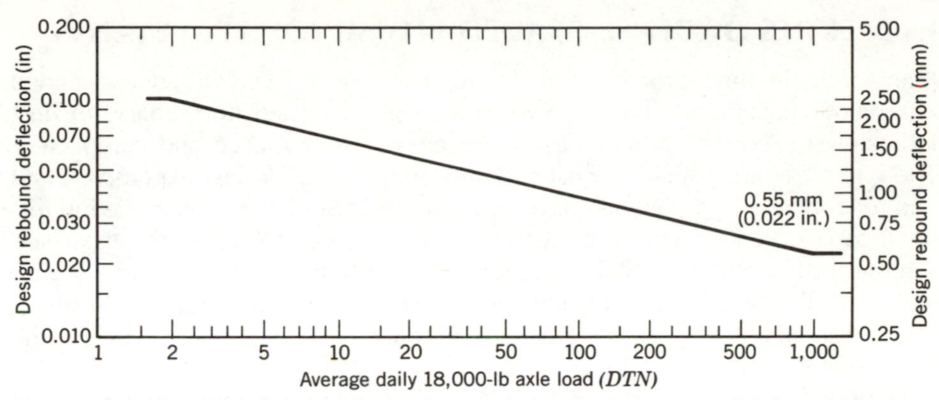

Fig. 19–12. Design deflection chart (after *Asphalt Institute MS-17*) slightly modified.

DESIGN CHART. Figure 19–13 presents the design chart used to determine overlay thickness. Developed from layered theory, it sets overlay thickness, given design deflection and expected traffic.

Mechanistic Approaches

Several methods have recently been developed which rely on deflection tests, layered theory, and failure criteria to preclude fatigue cracking and permanent deformation. For a discussion of these, refer to *TRB Record 700*.

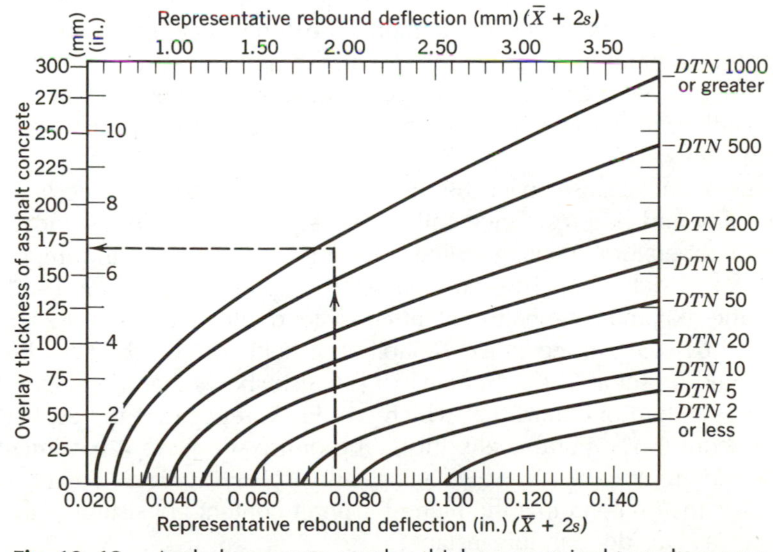

Fig. 19–13. Asphalt concrete overlay thickness required to reduce pavement deflection value (Example: *DTN* of 985; representative deflection of 0.076 in.; overlay required to reduce deflection to an acceptable level (0.022 in.) is 7 in.)

PROPERTIES OF DENSE-GRADED BITUMINOUS PAVEMENTS

Many of the important roads in the United States are surfaced with dense-graded bituminous pavements. They consist of aggregates graded from coarse to dust intimately mixed with bituminous binder before being placed and compacted. The highest type is called asphalt concrete; others include less expensive plant and road mixes. Dense-graded pavements of all classes have been treated in this book as one general group because the same considerations apply; these are summarized in Table 19–5. This table attempts to pull together, in a very general way, the most important findings of intensive research that began about 1940. For added detail the reader is referred to the references given on the pages that follow.

The two left-hand columns of Table 19–5 list certain desirable properties and the primary modes of failure of dense-graded bituminous pavements. The next two columns give specific characteristics of the separate constituents, bitumen and mineral aggregate, that will best produce the particular desirable property. These characteristics and specifications and tests to assure that they are obtained have already been discussed in Chapter 14. But it can be seen in Table 19–5 that, for the binder, high viscosity of the bitumen is desirable for stability and resistance to fatigue failure. On the other hand, it has been shown that a low-viscosity asphalt is less likely to become so hard and brittle that the pavement will fracture or disintegrate or break up under cold weather conditions. It follows that the pavement designer must compromise among these concerns in choosing the binder to use in a given situation.

Strong, tough, nonstripping, and nonpolishing aggregate particles with rough surfaces and angular shapes produced by crushing have all the desirable properties needed. With aggregates, the compromise comes between quality and the cost of producing and transporting them. For example, for a major traffic artery the high cost of an excellent crushed aggregate, transported a long distance, may be justified; for a low-volume road it may be better to use local aggregates of lower quality.

A serious dilemma arises over the percentage of binder to include in a pavement. Stability and skid resistance call for a "lean" mix, while resistance to fatigue, low-temperature cracking, aging, raveling and stripping, and imperviousness call for a rich one. This problem is illustrated by Fig. 19–14. In this example, the maximum asphalt content consistent with stability is 6.2%. If this is decreased by 0.3% to recognize variability in field control, the actual asphalt content would be set at 5.9%. But Fig. 19–14 also shows that the fatigue life of this mix is less than one-third that which would be reached if the asphalt content were about 6.6%. Fortunately, this compromise still gives a pavement having a nonskid surface, resistance to aging, and imperviousness. Today, paving technologists recommend that the highest asphalt content consistent with stability be used, as was done in this instance.

TABLE 19–5. Factors Important in Obtaining Satisfactory Performance from Dense-Graded Bituminous Pavements

		Desired Characteristic in Particular Element		
Desirable Properties	*Mode of Failure*	*Bitumen*	*Mineral Aggregate*	*Mix Proportions*
Stability of mix	Distortion: Rutting or shoving	High viscosity	Rough particle surfaces—angular or cubical shape	Low asphalt and void contents; High percentage of fines (passing No. 200 sieve)
Resistance to fatigue failure	Fracture in flexure	High viscosity for pavements thicker than 6 in.; low viscosity for pavements 2 in. and thinner	Not subject to stripping	High asphalt, content, dense, gradation
Resistance to low-temperature cracking	Fracture or tension	Low viscosity		High asphalt content; low void content
Resistance to aging	Fracture or disintegration	Low viscosity, resistant to aging	Not subject to stripping	High asphalt content; low void content
Resistance to raveling or stripping	Disintegration—asphalt—aggregate bond broken	Suitable surface charge and chemical properties	Not subject to stripping	High asphalt, content
Imperviousness to water and air	Weakening of underlying layers—or disintegration or instability of pavement		Not subject to stripping	High asphalt content, high percentage of fines (passing No. 200 sieve), low voids
Skid resistance	Surface becomes slick		Rough surfaces, resistant to polishing, not subject to stripping	Low asphalt content, coarse gradation

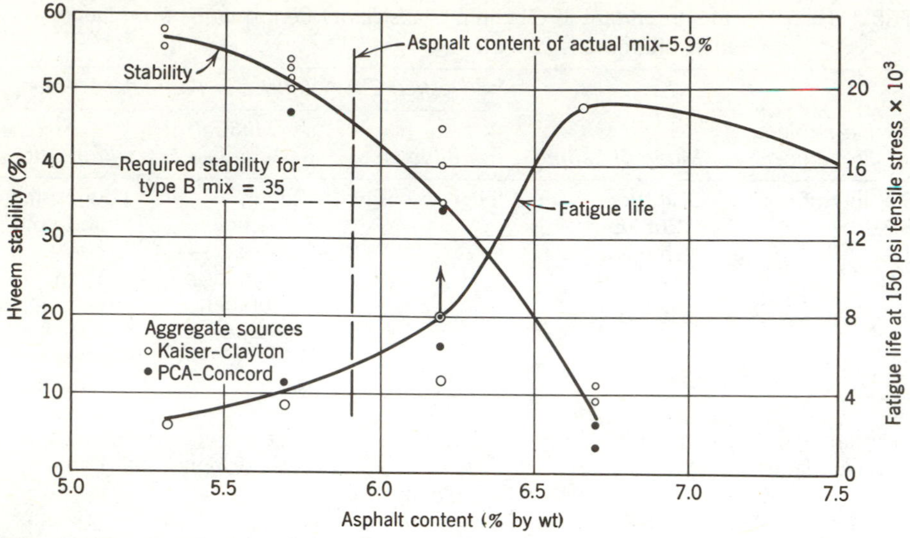

Fig. 19–14. Results of stabilometer and fatigue tests on asphalt concrete—Ygnacio Valley Road Project. (After C. L. Monismith.)

Stability

Stability of bituminous mixes has been the subject of intensive theoretical analysis and laboratory research which began before World War II and was largely completed before the mid-1950s. Today, careful attention to stability is considered a "must" by all pavement designers.[17]

Figure 19–1 indicates how failure of a bituminous mix occurs over adequate support: a shearing action along one or more surfaces permits material to flow from under the load. The forces that cause this action are repeated each time a wheel passes; on heavily travelled roads, there will be several million such loadings. At points such as stop signs or traffic signals where vehicles are braking, the tire load also has a substantial horizontal component. But before these shearing forces can cause failure, three resisting elements must be overcome: namely the friction between aggregate particles which might be called the sliding resistance, the cohesion introduced by the bituminous binder, and the inertia of the pavement mass.

Research has indicated that frictional resistance provides the most important element of stability and that this frictional resistance is largely dependent on aggregate qualities. Granular materials, without binder, can distort under load only if the particles slide over each other or become displaced in some other way. Resistance to sliding, like static friction in mechanics, depends largely on the total load normal to the direction of motion and on the angle of sliding friction between particle surfaces. It is largely independent of the rate at which load

[17]Researchers are attempting to predict resistance to permanent deformation using analytical techniques. For descriptions of these procedures see *TRB Record 616*, K. Majidzadeh, *TRB Record 715*, and J. B. Rauhut, *TRB Record 777*.

is applied and the size of the loaded area. From this it can be concluded that aggregates with rough surfaces offer greater resistance to displacement than those with smooth surfaces. Upon the introduction of a lubricant, either bituminous material or water, the basic characteristics of static friction remain, but frictional resistance is reduced by the separation and lubrication of the contact surfaces, with added lubricant causing added reduction. Too much lubricant will lower frictional resistance to such a level that instability of the mix will result (see Fig. 19–14). In these lubricated mixtures, friction is further affected by the degree and method of compaction because of their influence on the proximity and arrangement of particles.

Addition of bituminous binders to granular materials provides cohesion, which also increases stability. This cohesive strength is relatively independent of normal pressure, but varies almost directly with the surface area of the mineral particles and the viscosity of the binder, as controlled by selection or temperature. The asphalt binder offers great resistance to loads of short duration but practically none to static loads. Cohesive strength increases with bituminous content until the particles are well covered with a film of binder, after which it changes little. It increases also with greater roughness of particle surfaces. From these ideas it follows that cohesive resistance of a mixture may be increased by an addition of fine material such as mineral filler and by the use of more viscous binders; on the other hand, high temperatures and slow rates of loading cause large decreases.

Stability is adversely affected unless a small percentage of air voids remains after compaction. For heavy loadings, the suggested lower limit is 3% of the total volume.[18]

TESTING FOR STABILITY. The factors most commonly considered in stability testing are type and percentage of binder, amount and type of mineral filler, aggregate frictional and shape characteristics, and aggregate grain-size distribution. A recent survey of state practices for plant-mixed bituminous bases indicated that 18 agencies used the Marshall, 11 the Hveem stabilometer, and 3 an unconfined compression test to control stability. The remainder relied on controls on aggregate types and gradation and asphalt content.

Four stability-testing procedures used either for research or control purposes are described briefly. It is beyond the scope of this work to discuss their relative merits.

Hubbard–Field Stability Test (old AASHTO Designation T169). This early method, now replaced by the Marshall and Hveem tests (see below), can be classed as an extrusion test. A cylindrical briquette of compacted paving mixture or a cored specimen from an existing pavement is heated to 140°F and placed in a close-fitting cylindrical mold. The bottom of this mold is fitted with a standard circular orifice of smaller diameter than the mold proper (see Fig. 19–15.) Load is applied rapidly at the rate of 1 in. every 25 s, and the maxi-

[18]See e.g., N. W. McLeod and discussion by C. R. Foster, *HRB Record 158*. See also *The Asphalt Institute Manual Series 2,* "Mix Design Methods for Asphalt Concrete."

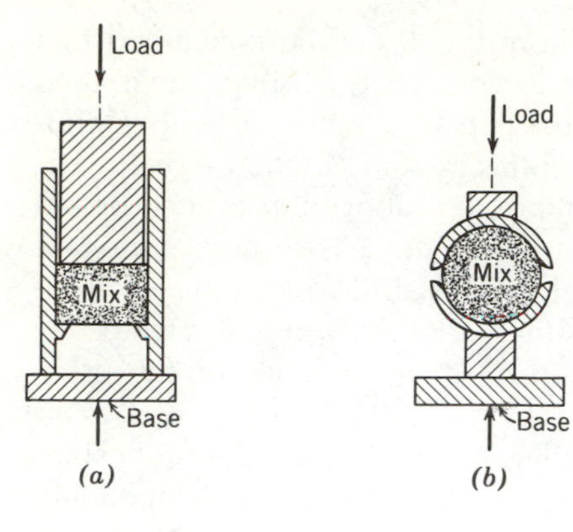

Fig. 19–15. Schematic drawings of two devices for measuring stability of dense-graded bituminous mixes: *(a)* Hubbard–Field, *(b)* Marshall.

mum load developed in forcing the mixture through the orifice is recorded as the stability. For sheet asphalts or sand asphalts, the specimen is 2 in. in diameter and 1 in. high; for asphalt concrete or other mixes containing large aggregate, the briquette is 6 in. in diameter and 2 in. high.

Stability as determined by the Hubbard–Field test actually measures some combination of friction and cohesion but does not express them separately. High stability values may be obtained by using too little binder. Also, mixes showing high stabilities may be made by combining hard asphalts with aggregates of poor frictional characteristics. Under static loads, where cohesion is not effective, pavements so constructed will shove and distort.

Marshall Test (AASHTO T-245). For this test a cylindrical specimen, 4 in. in diameter and 2½ in. long and of controlled proportions, is compacted by impact methods. The specimen at 140°F is placed in the collar-like testing device which offers no confinement at the ends and is loaded rapidly at a speed of 2 in./min (see Fig. 19–15b). Stability is measured as the maximum load applied in deforming the specimen; flow value is the deformation of the sample at the instant of maximum load.

Criteria set by the Asphalt Institute for acceptable mixes for heavy loads are as follows: stability, minimum of 750 lb; flow, range of 0.08–0.16 in.; percentage of air voids in mix 3–5. Like the Hubbard–Field test, the Marshall test measures frictional and cohesive resistance in combination.[19]

Hveem Stabilometer (AASHTO T-246). Development of this device (see Fig. 14–14) by the then California Division of Highways began in 1930. Specimens are short cylinders of compacted paving mixtures of 4 in. diameter and 2½ in. in height. This size specimen was selected so that undisturbed samples, cut from pavements in place, could be tested. It is common practice first to make

[19]For a summary of an elaborate testing program by the Corps of Engineers, see *HRB Research Report, 7-B*. A detailed description of the equipment is given in *HRB Bulletin 105*. Numerous articles giving the results of testing with the Marshall device appear in *HRB Record 158, 178, 256, 273,* and *307*. Interpretation of test data is given in *The Asphalt Institute Institute Manual Series No. 2,* op. cit.

preliminary determinations of asphalt content for laboratory specimens using the centrifuge kerosene equivalent test (see below).

In the stabilometer, the load is applied vertically to the ends of the cylinder, and the horizontal pressure developed in the fluid confining the sides of the specimen is recorded. For bituminous-paving mixtures, results are stated in terms of "Hveem stability" rather than as "resistance value" as is done with soil–aggregate mixtures. Hveem stability is expressed by the empirical formula

$$\text{Hveem stability} = \frac{22.2}{\dfrac{P_h D_2}{P_v - P_h} + 0.222} \tag{19–7}$$

where

P_v = vertical pressure, typically 400 psi
P_h = horizontal pressure in psi, taken at the instant P_v is recorded
D_2 = turns of the displacement pump to change the horizontal pressure from 5 to 100 psi; this factor corrects for the different degrees of smoothness or roughness in the exterior surface of the test specimens.

As with the equation for R value (see Chapter 14) liquids will show a relative stability of 0; for perfect solids it is 100. Values of Hveem stability greater than 35 are considered entirely satisfactory. Mixes usually give trouble at values lower than 30.

Compaction for stabilometer specimens is accomplished by a special tamping shoe or kneading compactor which applies many repetitions of load (see *AASHTO Designation T-247*). This method was adopted because stabilometer tests on laboratory samples so compacted and those of the same mix cored from completed pavements gave comparable results. Like correlation was not found between field samples and those compacted in the laboratory by static or impact methods.

The stabilometer test is so conducted that it primarily measures the frictional resistance of the mix. The sample is tested at 140°F; load is applied slowly at 0.05 in./min. Under these conditions the binder and any water present serve as lubricants but develop little cohesive resistance.

Along with the stabilometer was developed the *cohesiometer*. The same briquette is transferred from the stabilometer to the cohesiometer, which, by bending it around a diameter of the base, subjects the specimen to tension. Although this test is still used, researchers have now turned increasingly to fatigue tests (see below).[20]

Triaxial Tests for Bituminous Mixtures. A description of the open triaxial test as applied to soils appears in Chapter 14. There are similar tests to measure the stability of dense-graded bituminous mixtures. These are primarily research rather than testing tools. One of them, a "closed-system" device, uses a tall

[20]For details on the stabilometer and a bibliography see F. N. Hveem et al., *HRB Proceedings, 1948,* 101–136 and B. A. Vallerga, *HRB Proceedings, 1955,* pp. 173–183. Details of the cohesiometer are given in *HRB Bulletin 105*. See also *The Asphalt Institute Manual Series No. 2, op. cit.*

sample (4 in. in diameter by 8 in. in height) to eliminate the effects of arching of the aggregates at the ends of the specimen and thereby permit failure in shear. The fluid is confined so that lateral pressure increases as vertical load increases. Samples are compacted by a kneading device like that for preparing stabilometer samples. Testing is at 140°F. Load is applied by increments; after each application, load and lateral pressure readings are delayed until they become stabilized. This procedure gives the equivalent of static loading.[21]

Fatigue

As discussed earlier, pavement-design methods based on the Hveem stabilometer and AASHTO procedures recognize that dense-graded bituminous mixtures have beam strength. This assists in carrying the superimposed loads and permits a reduction in the total thickness of the roadway structure overlying the subgrade. On the other hand, in the discussion of elasticity and resilience in Chapter 14, it was indicated that excessive deflections of the pavement resulting from compression and relaxation of the underlying layers would cause the pavement to fail in fatigue. Thin pavements were shown to tolerate greater deflections; thicker pavements smaller ones. In theory, a balanced design would consider, for various pavement thicknesses, the tradeoffs between pavement beam strength in fatigue as related to deflections on the one hand and the elasticity of the underlying layers on the other. In recent years, this subject has received intensive study which may, in time, result in rational pavement design procedures to replace the present empirical ones. The Asphalt Institute's procedure described earlier is one example of such a procedure.[22]

There are inherent properties of bituminous mixtures that make application of theoretical approaches extremely difficult. These include the following:

1. Bituminous mixtures are viscoelastic rather than elastic. This means that, although they will respond elastically to quickly applied and withdrawn loads, they will flow under static loads or under those that are applied slowly. F.N. Finn recommends that because of their complexity, the present theories of viscoelasticity should defer to those of elasticity in the design of asphaltic surfacings.[23]
2. Tensile and beam strength in fatigue, tolerable strain, and modulus of elasticity all vary greatly. Among the external factors are temperature and frequency, rate, and intensity of loading. Internal influences include source, viscosity, and amount of binder, aggregate grading and shape, void content, and the amount and nature of

[21]For a description of work with triaxial tests see V. R. Smith, *Proceedings, Association of Asphalt Paving Technologists,* 1949, and N. W. McLeod, *ibid.,* 1950. Other studies involving the triaxial approach include Y. H. Huang, *HRB Record 178* and A. E. Z. Wissa et al., *HRB Record 256.*

[22]See *Proceedings, Fourth International Conference on Structural Design of Asphalt Pavements,* 1977 for a description of several other methods. See also *TRB Record 512, 602,* and *700* for description of new and overlay design methods. For recent work on fatigue characteristics of conventional and sulfur asphalt mixtures see *TRB Record 659* and *712* and *NCHRP Report 195.* See also the *Proceedings of the Association of Asphalt Paving Technologists* and *American Society for Testing and Materials STP 508* and *561.*

[23]See F. N. Finn, *NCHRP Report 39,* for greater detail on this and the factors listed below. At present, only the FHWA design program VESYS is considered a viscoelastic one; most of the others are based on the materials behaving in a linear elastic manner. See F. N. Finn et al., *TRB Record 602.*

compaction and the temperature at which it is carried out. To illustrate two of these variations: tensile strength may range from 500 to 1400 psi and modulus of elasticity between 300,000 and 1,800,000 psi between the temperature extremes reached in a moderate climate.

In spite of the difficulties just enumerated, researchers have made great strides toward evaluating the effects of many of these variables and in developing analytical methods to handle them. However, a discussion of these efforts is far beyond the scope of this book.[24]

Density and Percentage of Voids

The theoretical specific gravity (density) G_M for voidless bituminous paving mixtures is determined by

$$G_M = \frac{100}{\frac{W_1}{G_1} + \frac{W_2}{G_2} + \frac{W_3}{G_3} \cdots \frac{W_n}{G_n}} \tag{19–8}$$

where

W_1 = the percentage by weight of bitumen
G_1 = the specific gravity of the bitumen (see *AASHTO Designation T228)*
$W_2, W_3, \ldots, W_n$ = percentages by weight of the different aggregate fractions
$G_2, G_3, \ldots, G_n$ = specific gravities of the respective aggregate fractions (see *AASHTO Designations T84 and T85)*[25]

The procedure for determining d, the bulk specific gravity of a compacted specimen, is given in *AASHTO T-166* as

$$d = \frac{A}{B-C} \tag{19–9}$$

where

A = weight of the dry specimen in air, in grams
B = weight of saturated surface-dry specimen in air, in grams
C = weight of saturated specimen in water, in grams

The formula for determining d, the bulk specific gravity of a compacted asphaltic mixture when the specimen is coated with paraffin, is as follows:

[24]*NCHRP Report 139, 140, 160,* and *213* and *HRB Special Report 126* and *140* give excellent introductions to this topic and contain detailed bibliographies. See also *Proceedings, Fourth International Conference on Structural Design of Asphalt Pavements,* 1977 and *TRB Record 512, 549, 572, 602, 616* and *725.*

[25]Details of the laboratory procedures for determining specific gravities *(AASHTO Designation T84* and *T85)* are normally presented in courses in concrete or engineering materials and will not be repeated here.

$$d=\frac{A}{D-E-\frac{(D-A)}{F}} \tag{19–10}$$

where
A = weight of the dry specimen in air, in grams
D = weight of the specimen plus paraffin coating in air, in grams
E = weight of the specimen plus paraffin coating in water, in grams
F = bulk specific gravity of the paraffin

The percentage of voids, V, in an actual paving mixture is determined by

$$V=\frac{(G_M-d)\times 100}{G_M} \tag{19–11}$$

It is also important to check the density and void content of compacted bituminous mixtures in the completed pavement. Samples are cut from it; testing and computations are as outlined above (see *AASHTO Designation T230*). Higher densities improve the stability and beam strength of the mixture and lessen the tendency for ruts to form in the wheel paths under traffic. Also a lower void content reduces the tendency of the binder to age. On the other hand, overrich mixes compacted so that the void content is below roughly 3% by volume often are unstable; if it is as low as 1%, the bitumen may flush to the surface under the traffic, making it slippery.[26]

Correct determination of the actual percentage of voids in a bituminous paving mixture is considerably more difficult than the formulas given here would indicate. First of all, a 1% difference could come about because of permissible variability in the tests for aggregate specific gravity alone. In addition, almost all aggregate is absorptive; that is, water or asphalt will enter its surface pores. This absorption varies. To illustrate, for a granite it was 0.27% by weight for either water or 85–100 penetration asphalt cement, whereas for a limestone the percentages were 2.58 and 2.46 and for a sandstone 3.56 and 2.54, for water and asphalt cement, respectively. Thus, to correctly find the void space in the mix proper an estimate must be made for asphalt absorption into the voids in the aggregate.[27]

Other Properties

As indicated in Table 19–5, several other mix properties are important to ensure good performance. These include resistance to low temperature cracking, aging and raveling or stripping, imperviousness, and skid resistance. It is beyond the scope of this book to discuss these important properties. For extensive discus-

[26]See N. W. McLeod and discussion by C. R. Foster, *HRB Record 158* for an excellent presentation of this subject. See also R. D. Barksdale, *Proceedings Association of Asphalt Paving Technologists,* 1978 for a discussion of the effect of density on fatigue life and permanent deformation.
[27]For added discussion and references on this subject, see *HRB Bulletin 105* and N. W. McLeod, *HRB Proceedings, 1957*. See also *Asphalt Institute Manual Series 2,* op. cit., for a procedure to correct for absorption.

sions of these topics the reader is referred to various publications of TRB, ASTM, and the *Proceedings,* Association of Asphalt Paving Technologists.[28]

Other Approaches for Setting the Asphalt Content of Dense-Graded Bituminous Mixes

It was not until the 1950s that more than a handful of agencies began using laboratory tests for stability to determine bitumen content of high type dense-graded mixes such as bituminous concrete and sheet asphalt. Rather, the decision was largely by precedent; often the percentage of binder was prescribed in the specifications. Fortunately these procedures brought few apparent failures. First of all, aggregate proportions and quality were closely controlled. Then there was generally a wide spread between the percentage of asphalt required to coat the particles and that which would produce instability. Furthermore, the effect of asphalt film thickness on durability was largely unknown. Serious problems of dry mixes and unstable or bleeding pavements did develop in connection with the cheap road and plant mixes that became popular, particularly in the West, beginning in the 1920s, and procedures were developed to handle them (see below). Today, however, the problems related to having stable and durable mixes to carry increasing numbers of heavy wheel loads while retaining resistance to fatigue cracking and aging have focused attention on the importance of carefully determining the bitumen content. As shown earlier, stability testing is the common approach, although some agencies still rely primarily on experience.

BITUMEN CONTENT BY SURFACE AREA AND CENTRIFUGE KEROSENE EQUIVALENT. The common stability-testing methods are complex and time-consuming. It follows that a short, easily performed test is helpful to field engineers. Also, it is useful in the laboratory as a means for decreasing the number of stability tests that must be conducted. One such test developed by the California Department of Transportation is described here.[29]

This procedure considers the surface area, surface roughness, and absorption of the aggregate; it also recognizes the effects of asphalt viscosity. The final result is stated as the "bitumen ratio," which is the percentage of asphalt or road oil, by weight, in the mix.

Aggregate surface area is determined from sieve analysis based on the total percentage passing a stipulated group of sieves. Finer materials have much greater surface areas than coarse ones. For example, material passing the No.

[28]For examples on low-temperature cracking see *TRB Record 632, 659,* and *777, NCHRP Report 195,* and *ASTM STP 628;* on raveling and stripping see G. W. Maupin Jr., *TRB Record 712* and *NCHRP Report 192* and Chapter 14; on aging (or asphalt durability) see E. L. Green et al., *TRB Record 595* and *NCHRP Synthesis 59;* on permeability see A. Kumar et al., *TRB Record 659;* and on skid resistance, see Chapt. 8, J. J. Henry et al. and J. Ryell et al. *TRB Record 712, TRB Record 788,* and *ASTM STP 530* and *583*..

[29]See *Materials Manual,* California Department of Transportation, Test 303, or the *Asphalt Institute MS-2* for added details of the *CKE* method. Several other surface-area methods are discussed by J. A. Epps et al., in *HRB Record 351.* See also *NCHRP Synthesis 30* for use with emulsion mixes.

200 sieve has a surface area constant of 160; that passing the No. 4 and retained on the No. 8 sieve has a constant of 2. Results of surface area computations give for an asphalt concrete of ¾ in. maximum size aggregate and 12% passing the No. 200 sieve, a value of about 35.

Aggregates with rough surfaces demand more asphalt to coat them; those that are porous absorb binder into the pores. These tendencies are measured jointly. For coarse aggregates, a sample in the ⅜ in. to No. 4 size range is immersed in SAE 10 lubricating oil for 5 min; then it is drained for 15 min at a temperature of 140°F and the percentage of oil by weight of aggregate determined. Materials passing the No. 4 sieve are saturated in kerosene, subjected to a force of 400 times gravity for 2 min in a centrifuge, and the percentage of kerosene retained by weight of aggregate determined. Findings for coarse and fine aggregates are weighted to give a final centrifuge kerosene equivalent *(CKE)* value. By means of a special chart, surface area, aggregate specific gravity, and *CKE* are combined to give the recommended oil ratio for an *SC, MC,* or *RC* 250 binder. An added chart relates the oil ratio for 250 grade oil to bitumen content for other grades of road oil and for asphalt cement. This conversion recognizes the higher densities of more viscous asphalts and also that mixes low in fines can tolerate thicker films of viscous asphalts before they become unstable.

BITUMINOUS CONTENT BASED ON SIEVE ANALYSIS. As already indicated, a given volume of small particles of a given shape and roughness has a greater surface area than the same volume of larger particles of like characteristics. It follows that, as the number of fine particles increases, the amount of bituminous binder needed to coat these surfaces also increases. Based on this concept, formulas expressing percentage of bitumen in terms of grain-size distribution were developed. Such formulas were little used with the carefully controlled gradings of early-day high type of pavements. However, the advent of road mix, which used the old road or other material of widely varying grading as mineral aggregate, made the problem of the percentage of binder a serious one.

An early formula developed in California to determine oil content using liquid asphalts, stated in terms of sieve analysis was

$$P = 0.02a + 0.045b + 0.18c \tag{18–6}$$

where by weight, P is the percentage of oil required, and a, b, and c are, respectively, the percentages of material retained on the No. 10 sieve, passing the No. 10 sieve and retained on the No. 200 sieve, and passing the No. 200 sieve. A number of like formulas were developed in other states. Similar equations are available for asphalt emulsions.[30]

The use of these formulas has been confined mainly to low-cost road- and plant-mix projects. At times they have proved unsatisfactory, as when the mineral aggregate contained large amounts of blowsand or other hard, slick, fine material. Under these conditions, the percentage of oil called for by the formula gave a mix so rich that, under traffic, free oil came to the surface to streak ve-

[30]For details on both see *Asphalt Institute Manual Series 14,* "Asphalt Cold-Mix Design."

hicles traveling the roads. Such mixes also were more often unstable. Experienced engineers checked formula results by rubbing unmixed samples between the hands and observing the relative ease of coating the particles and by carefully observing completed sections. These visual methods would not permit the determination of the oil content at which stability was lost since, once the particles were coated, added binder produced no change in appearance. Yet many miles of serviceable pavements were and are still being laid with visual methods as the only guide.

ASPHALT OR BITUMINOUS CONCRETE PAVEMENTS

The term "asphalt concrete" denotes a dense-graded road surface made of hot mineral aggregates plant-mixed with hot asphalt. "Bituminous concrete" is a more general term which includes both asphalt concrete and similar mixtures made with refined tar. The coarse aggregate is generally crushed stone, crushed slag, or crushed gravel, to which is added sand or sand and filler. By contrast, sheet asphalt involves the use of sand and filler only.

Asphalt concrete is the highest type of dense-graded bituminous pavement and is suitable for the most heavily traveled roads. It is mixed and normally laid at high temperatures, around 275–300°F, and requires an asphalt cement binder (see Table 14–6).[31] Mineral aggregates are of high quality and are proportioned within tight limits. Specifications for mixing, placing, final density, and accuracy of surface finish provide close control over all phases of construction. Bituminous concrete has the important advantage that traffic may use it immediately after construction.

History Of Asphalt Concrete[32]

Sheet asphalt was quite commonly used before asphalt concrete was developed. Some of the early pavement mixtures in Washington, D.C., however, were essentially asphalt concrete. Hamilton, Ontario, used a refined tar and stone mix as early as 1880.

In 1903 a patent (No. 727, 505) was issued to Warren of Massachusetts for a paving mixture of graded aggregate and bituminous materials. The Warren patents had a marked influence on the design of asphalt–concrete mixtures because certain favorable gradings of coarse aggregate could be employed only in these proprietary mixes. For a number of years the use of aggregates larger than ½ in. for other dense mixtures practically ceased, although open mixes consisting of coarse aggregates without smaller material were developed.

In 1910, by court decision, it was held that certain asphalt–concrete paving mixtures laid in Topeka and Emporia, Kansas, did not infringe on the Warren patent. This "Topeka mix" was essentially a sheet asphalt with about 20% of mineral aggregate passing the ½ in. sieve but retained on the No. 10 sieve,

[31]In recent years lower temperatures (240–275°F) have been used in an attempt to save energy.

[32]For an excellent history on asphalt pavements see *Proceedings, Association of Asphalt Paving Technologists,* Vol. 43A, 1974.

whereas the Warren patent used stone 2 in. or more in size. Following this court ruling, a variety of gradings came into use. Although the Warren patent expired in 1920, its effect is still reflected in present-day practice and many of the mixes developed to circumvent it are still in use and performing well.

Materials for Bituminous Concrete

As already indicated, asphalt cements are usually specified as the binder for bituminous concrete. In earlier practice, harder grades, such as 30–40 or 40–50 penetrations, were commonly used. Today, softer binders than these are generally employed since hardening from weathering and oxidizing is slowed. The 85–100, *AC*–10 (or *AR*–4000) grade is most popular. But, as indicated in Table 19–5, there may be advantages to using an even softer binder such as 120–150, *AC*-5 (or *AR*-2000) penetration for thin pavements or where the weather is extremely cold. Under some circumstances tars of the heaviest grades are also used.

Ranges in percentage of binder by weight for typical mixes are shown in Table 19–6. As indicated earlier, the recommended practice is to incorporate as much asphalt as possible consistent with stability. Furthermore, in thick pavement, it is often advisable to richen the mix even more in the bottom layers where stability demands are less stringent and fatigue cracking is initiated.

Aggregates for asphalt concrete are carefully selected and have the desirable characteristics outlined in Chapter 14. Coarse aggregate[33] is generally broken stone or crushed gravel, although uncrushed gravel has sometimes been used. Some specifications exclude uncrushed material entirely by stating that coarse aggregate be made from large stones (about 2 in. minimum size) recrushed to meet grading requirements. Considerable strength and toughness are desirable and usually specified, although softer aggregates which are normally unsuited for open mixes or macadams have been used successfully. A large majority of agencies have placed specific limitations on the amounts of deleterious materials, such as organic substances, clay, soft or elongated particles, coal, or shale. Generally aggregates that will strip or polish are excluded; however, aggregates that strip may be used if antistrip agents are added.[34]

Fine aggregates are the particles ranging in size from the No. 10 or No. 4 sieve downward and consist of sand or stone screenings or a combination of the two. Most of the state agencies specify that they be free from large amounts of clay, organic matter, or other deleterious materials. Individual particles are to be clean, uncoated, and strong. Few agencies attempt to control particle shape by specification.

In some areas of the country, the supply of suitable aggregates is growing short and processes have been developed to utilize deposits where good materials are mixed with unacceptable ones. One procedure is called "heavy media separation"; it operates by floating out objectionable materials in a fluid having

[33]Coarse aggregates are defined by many agencies as those retained on the No. 10 sieve. Others set the division at the No. 4 or No. 8 sizes. Actually this break point is of little practical significance.
[34]See G. W. Maupin Jr., *TRB Record 712* for evaluation of various anti-strip agents.

TABLE 19–6. Typical Examples of Proportioning for Asphalt Concrete Surfacings and Bases by Percentage Passing

	Weight Passing Sieve (%)									
	AASHO Test Road		*California Department of Transportation*				*Federal Highway Administration*		*Black Bases**	
			Coarse		Medium					
Sieve Openings for Aggregates	*Binder Course*	*Surface Course*	*Individual*	*Moving Average*	*Individual*	*Moving Average*	*Grading A*	*Grading D*	*Iowa*	*Virginia*
1½ in.							100	—		100
1 in.	100	—	100	100	100	100	95–100	—	100	—
¾ in.	88–100	100	87–100	90–100	90–100	95–100	78–95	100	90–100	72–87
½ in.	55–86	86–100	—	—	—	—	—	95–100	—	—
⅜ in.	45–72	70–90	55–80	60–75	60–85	65–80	54–75	74–92	60–94	—
No. 4	31–50	45–70	35–60	40–55	40–65	45–60	36–58	48–70	40–80	35–50
No. 8	—	—	22–45	27–40	25–50	30–45	25–45	33–53	30–65	28–38
No. 10	19–35	30–52	—	—	—	—	—	—	—	—
No. 20	12–26	22–40	—	—	—	—	—	—	—	—
No. 30	—	—	8–26	12–22	11–29	15–25	11–28	15–30	15–40	—
No. 40	7–20	16–30	—	—	—	—	—	—	—	—
No. 80	4–12	9–19	—	—	—	—	—	—	—	—
No. 200†	0–6	3–7	0–10	3–7	0–11	3–8	0–8	4–9	3–10	2–9
Asphalt % by wt.	4–6.5	4–7.5			determined for each project				4.0	4.5

*Source: *HRB Special Report 117*. Some of the black bases now in use are, in reality, expensive, high-quality binder courses of asphalt concrete; some are aggregate base courses upgraded by the addition of bitumen.

†Deficiencies in this size are made up by adding mineral filler.

a specific gravity higher than the poor materials but lower than the acceptable ones. Another process is the "bounce method"; objectionable materials that have a low coefficient of restitution (rebound) do not bounce as far from an inclined polished steel plate as do the acceptable ones. Good and poor materials land in separate bins.[35]

Some agencies use mineral filler as a separate ingredient of bituminous concrete; others specify that it be added to make up deficiencies in the finer sizes; others do not specify it at all. Some require it in surface courses but not in the underlying base, binder, or leveling layers.

Aggregate Grading for Bituminous Concrete

Many different aggregate gradings have given excellent bituminous–concrete pavements. However, the shape of the curve of percentage passing against sieve opening conforms in a general way to that plotted in Fig. 19–16. Chief variables between mixes are maximum size of aggregate, percentages of coarse and fine aggregate, and percentage passing the No. 200 sieve. Maximum aggregate size is limited by the thickness of the pavement layer and rarely exceeds 75% of that value, but is often much less. Surface courses commonly have smaller maximum sizes than underlying layers. Maximum sizes in all mixes have decreased in recent years as engineers have learned how to produce stable mixes without the larger sizes that make finishing difficult. The percentage passing the No. 200 (0.075 mm) sieve is usually higher for surface courses than for the lower layers, as the added fines make for easier manipulating and finishing and give a more impervious surface, but at greater cost.

On Fig. 19–16 are also indicated the problems that may arise if grading departs too far from that of the usual asphalt concrete. Gradings that plot outside the dotted lines of the figure present the problems indicated. Of particular importance is the danger of instability in mixes having high percentages of fines. In such cases, an undetected decrease will result in an overasphalted and unstable pavement.

Two methods for stating the desired aggregate gradings for bituminous concrete are in use. One (see Table 19-6) specifies limits to the total percentages by weight smaller than given sieve sizes; the other, not included here, specifies limiting percentages of material within given size ranges. With either method. consistent gradings can be obtained.[36] By studying this table, which presents only a part of the gradings used by a few agencies, the student can begin to realize the diversity of practice in the United States. *HRB Special Report 117* gives much more complete data to illustrate this point.

Statistical approaches for setting standards for and control of aggregate grading are in common use. One example is shown in Table 19–6 by the dual requirements of the California Department of Transportation. One set gives grad-

[35]For other methods of upgrading low quality aggregates, see *NCHRP Report 207*.
[36]See J. B. Dalhouse, *Public Roads,* Apr. 1953, for a suggested method of converting gradings based on percentages within given size ranges to percentages passing.

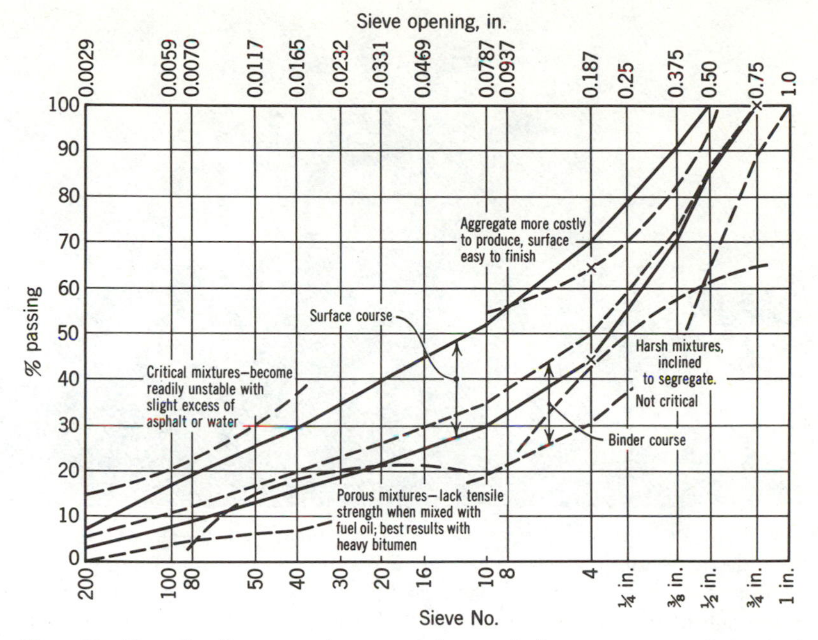

Fig. 19–16. Grading requirements for asphalt–concrete surface and binder courses as used on the AASHO Test Road.

ing limits for a single sample, the other for the running average. More sophisticated approaches are outlined in *NCHRP Reports 17, 34,* and *69.*[37]

Where two or more aggregate sizes must be combined to meet given specifications, it is necessary to determine the percentage of each to use. This may be done by trial and error; however, graphical or analytical procedures have been developed that are considerably more satisfactory.[38]

Preparation of Bituminous–Concrete Mixtures[39]

Elaborate, expensive "hot plants" are required to prepare bituminous–concrete mixtures. An overall view of such a plant is shown as Fig. 19–17 and schematic drawings of batch, continuous, and dryer drum types as Fig. 19–18. In the vicinity of all large cities where the demand for paving materials is continuous, permanent installations which serve the respective metropolitan areas are found. For individual projects in rural areas a portable plant, moved to the site over railroad or highway, will probably be set up. Plants are continually growing larger in size; a capacity of 600 tons/hr or more is now common. A typical specification for such plants is given under *AASHTO Designation M-156.*[40]

Bituminous materials are usually transported from refinery to "hot plant" in

[37]For added references, see the section on Quality Assurance in Chapter 15 and *American Society for Testing and Materials, STP 709.*
[38]See the *Asphalt Institute MS-2, op. cit.*
[39]See the *Asphalt Plant Manual, MS-3,* of the Asphalt Institute for a far more detailed presentation.
[40]Though not included in *AASHTO M-156,* dryer drum plants have been used extensively in recent years, mainly because of their simplicity and economy.

Fig. 19–17. A modern batch-type hot plant. Conveyor in right foreground elevates cold aggregate to dryer. Heated aggregate is conveyed vertically to top of mixing plant. Dust-laden exhaust gases from dryer pass through "cyclone" dust collector (to left of dryer) and then through filter cloth in "bag house" (to right of dryer) before being discharged. (Courtesy, Stansteel Corporation.)

steam-coil-heated tank cars, but are sometimes moved by tank truck. Small quantities are shipped in barrels or light steel drums, and in some foreign countries all binders are transported in this manner. Binders are loaded at elevated temperatures; however, temperature losses in transit are about 15–25°F/day, depending on atmospheric conditions. At times, therefore, it may be necessary to reheat the car or truck before the material is unloaded. At the plant, storage normally is in large tanks. These tanks often have capacity equal to that of several transport vehicles in order that plant operation not be halted by delivery failures. Heating is accomplished by coils carrying heated oil or steam, or by electric heating elements. Heated binder is forced from the storage tanks to the mixing platform by positive action pumps or by air pressure. Provision is made for recirculation through return pipes to prevent cooling or solidification overnight or during shutdowns.

Almost all agencies require that supplies of coarse and fine aggregates be stored separately (see Fig. 19–18). Materials, in the approximate proportions needed, are drawn from storage onto a belt which leads to the cold elevator. For batch and continuous type plants, the combined aggregate enters the drier, in which the aggregate falls repeatedly through hot gases until any moisture is driven off and the aggregate reaches mixing temperature (usually 300–325°F.)[41] The hot, combined aggregates then go up the hot elevator to the top of the mixing plant, where they are separated into several (usually three or more) sizes on

[41]As indicated previously, the kinematic viscosity of the asphalt for mixing should fall between 150 and 300 centistokes. Aggregate temperature should not exceed that needed to meet this requirement.

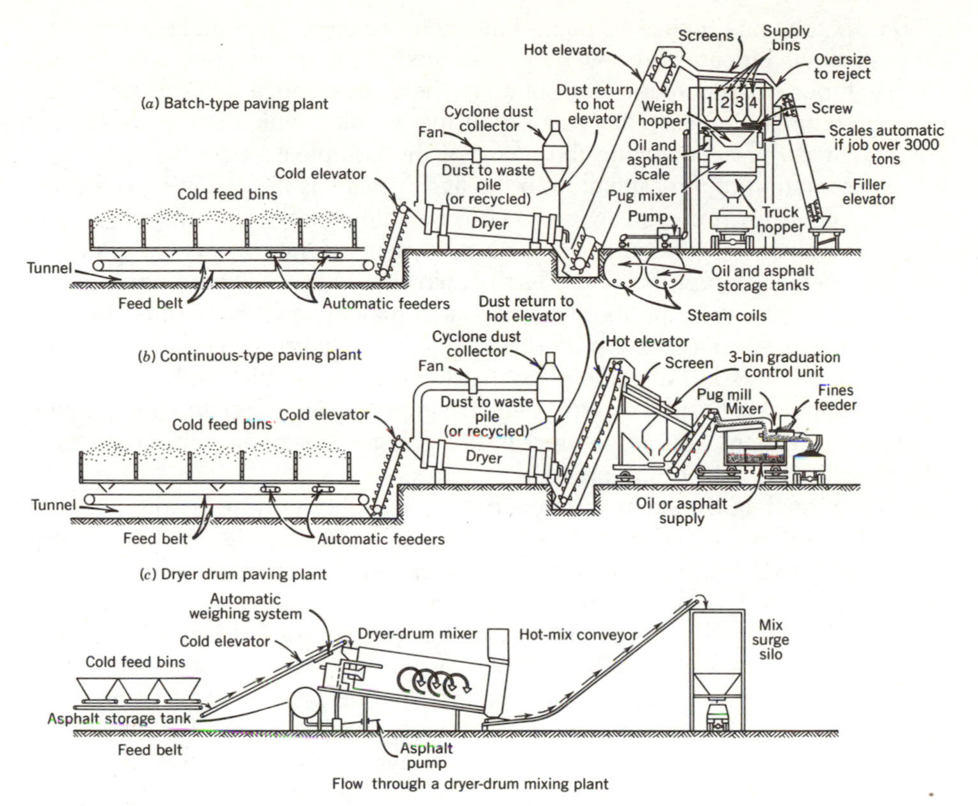

Fig. 19–18. Schematic diagrams of "hot plants" for preparing bituminous concrete and plant mix (diagram does not show devices for cleaning exhaust gases). *(a)* Batch-type paving plant. *(b)* Continuous-type paving plant. *(c)* Dryer drum paving plant.

shaking or rotary screens. Temporary "hot storage" is provided in bins located directly below the screens. For the dryer drum plant, there are no screens or hot-aggregate bins. The combined aggregates enter the dryer drum and are mixed with the asphalt prior to complete drying. The mix processes for each are discussed in more detail below.

For batch mixing (see Fig. 19–18 *a*), the prescribed amount of each of the hot aggregates is successively drawn from the bins into a "weigh box" located just below the "hot-storage" bins. Proportioned aggregates fall from the weigh box into the mixer, which is known as a "pug mill." Pairs of blades revolving in opposite directions throw the material upward between them and also knead it against the walls of the mixer. Commonly the dry aggregates and filler, which are introduced separately to the pug mill, are mixed dry for a short period, after which a weighed amount of binder is added uniformly across the width of the mixer. Mixing then continues until the binder is distributed throughout the mass and all aggregates are coated. Mixed materials are released through a gate in the bottom of the mixer into a waiting truck or into a truck hopper which holds one or more batches and permits continued mixer operation even though a truck is not immediately on hand to receive each batch.

Some batch and continuous plants have been equipped with large hoppers (or surge silos) which can store several hours' need of the placing operation or supply numerous small demands for hot mix. There is evidence that this procedure may substantially increase the viscosity of the bitumen which can have adverse consequences by accelerating the aging of the completed pavement. In open mixes, separation of the asphalt from the aggregates has also been reported.

It has been shown previously that a considerable variety of gradings have produced satisfactory bituminous–concrete pavements. Once a mix for a project has been decided upon, however, each batch should be of like proportions and temperature; otherwise the resulting pavement may be defective. To ensure uniformity, many specifications prescribe not only the permissible variation from the designated mix but also the plant that will be acceptable, and, in addition, rigidly control each step of the heating, proportioning, and mixing process. Other specifications are not so precise, but give the engineer authority to regulate closely all phases of the process.

Uniform feeding of materials through the plant is of utmost importance, and controlled proportioning from the stockpiles, as shown in Fig. 19–18, is commonly specified to produce this result. Feeders employing moving belts of rubber or segmented steel, reciprocating plates, or vibrators are common. If the proper relative amounts of coarse and fine aggregate are fed through together, the temperature of the heated aggregate can be held nearly constant. Furthermore, the hot screens will be uniformly effective in separating the material to the various sizes. If, however, the feed consists largely of sand for a time, and then is largely coarse aggregate, some of the aggregate will be too cold and that which follows it too hot, since more heat per unit of weight is usually required to dry sand and bring it to the desired temperature. Then if the materials are too cold, improper mixing, placing, and compaction may result; if too hot, the binder may be excessively hardened. Also, heavily loaded screens do not operate as effectively as lightly loaded ones, so that considerable variation in the aggregate grading of the final mix results.

The drier consists of a long hollow steel cylinder lined with firebrick and containing projecting radial fins. For batch and continuous plants, the inlet end is set slightly higher than the outlet. Heat is supplied by a jet of flame, fed by steam- or air-vaporized oil, coal dust, or gas, which is directed into the lower end of the drier. As the cylinder rotates, the aggregate is carried to the top by the fins, from where it falls downward and a little toward the outlet. By the time the aggregates have traveled the length of the drier, they are free of moisture and heated to the desired temperature. Drier capacity is greatly reduced and heating costs are increased if aggregates are wet, so that both greater capacity and operating economy result from protecting aggregate supplies from rain.

The problem of aging of bituminous binders has already been discussed. Many tests have shown that asphalts mixed with overheated aggregates harden and become brittle in the short mixing period (see Fig. 14–24).[42] One authority

[42]This also depends to a great extent on asphalt source. See J. E. Wilson et al., *Proceedings, Association of Asphalt Paving Technologists,* 1979.

estimates that the penetration is decreased by one-half during a 30-s mixing period. For this reason, most engineers favor holding aggregate temperatures at the lowest level consistent with proper mixing and placing. Overheating the binder in the storage tanks may not be so serious because the liquid is in bulk rather than in thin films; no air is present, and oxidation cannot take place.[43] Almost all agencies require that contractors provide pyrometers or thermometers that continuously record the temperature of the heated aggregates. Mixes that are too hot or cold are rejected.

In older hot plants, smokestacks were used to create a draft through the drier. This draft carried with it much of the fine dust introduced with the aggregate, which not only wasted the dust but also created a public nuisance. Often these lost fines were replaced by expensive mineral filler. Modern plants are equipped with a blower to produce the draft and a dust collector or "cyclone" which traps the fine particles and returns them to the heated aggregates below the drier. Exhaust gas from the cyclone is washed or passed through fabric filters to remove the last traces of dust. The dust collected by washing cannot be reused; that collected with the filters is normally introduced back into the mix.

Revolving circular screens once were used to separate the hot aggregates into different sizes. They have largely been supplanted by flat shaking screens. Screen sizes are sometimes left to the option of the contractor; again they may be specified. To illustrate, the Arizona and California Departments of Transportation require that the plant be equipped with screens to separate the material into at least three sizes. These are, with three-bin plants, larger than 3/8 in., 3/8 in. to No. 8 or No. 10, and passing the No. 8 or No. 10. Carryover of smaller sizes into the bins for coarser material is limited to values in the range of 10–15%. Many agencies also require that a scalping screen be used to reject oversize aggregate.

Batching aggregates into the weigh box of batch mixers is done carefully; the AASHTO specification requires that weights be correct to 1/2%. Actual proportioning is sometimes done by a weighman employed by the contractor. Many engineers think that this procedure does not provide proper control over the contractor's operations, as proportions can be altered without permission. Some agencies now require automatic devices which remove control of proportioning from the weighman. With them, once the desired weights have been set, merely pressing a button actuates, and possibly records the weighing process.

Some agencies no longer require that the heated aggregates be separated by size into several bins. Arguments favoring omission of the "hot screens" are that the capacity of the entire plant may be limited to that of the screens. Alternatively, if they are overloaded, some of the aggregate will "carry over" into the bin for the next larger size and alter the intended grading. Furthermore, it has been demonstrated that with accurate feeders at the stockpiles, satisfactory grading control can be obtained without this rescreening.[44]

[43]See *HRB Special Report 54* for a detailed discussion of many of the problems of temperature in bituminous mixtures. See also *NCHRP Synthesis 59* for an excellent discussion plus bibliography.
[44]See, e.g., J. L. Farrell et al., *HRB Record 316*.

Pug-mill mixers range in capacity up to 16 tons or more. Mixing procedures are not uniform among agencies. Many agencies require that aggregate and filler be mixed dry for (most usually) 15 s followed by addition of the binder and a wet mixing period, commonly 30 or 45 s. Others specify about the same total mixing time but do not require dry mixing. In a few instances, the specifications merely require complete coating of the aggregate and leave mixing time to the discretion of the engineer.

Pug-mill designs have been continuously improved over the years. For example, installation of a spray bar to replace an asphalt bucket has brought a more uniform distribution of binder through the mix. A Swiss mixer throws aggregate into vaporized asphalt that is introduced at high pressure.

Almost all the state highway departments permit continuous mixers (see Fig. 19–18 *b*) as well as the batch mixers already described. Materials are proportioned by volume through specially calibrated devices which are correlated with the meter that measures and supplies the binder. Materials flow continuously through the mixer and are discharged in a steady stream. Properly controlled continuous plants can produce mixtures of very uniform grading and binder content.

With continuous feed mixing, problems may develop in maintaining a constant flow of fines or binder. Safeguards against such happenings may include vibrators or other special devices on the feeder for the fines and common electrical circuits for aggregate and asphalt feeds to ensure that they operate in concert.

Many agencies now permit hot plants in which the bituminous binder is introduced directly into the dryer rather than in a separate operation. This is referred to as the *dryer drum process* (Fig. 19–18*c*). Output is continuous; it can be as high as 600 tons/hr. The first advantage claimed for it is a far simpler and less costly hot plant. All the equipment downstream from the dryer, except for a storage hopper, is eliminated. It has been reported that well-controlled mixes can be prepared and placed at lower temperatures than those common in traditional plants. Neither does exposure of the binder in the dryer seem to greatly "age" it by substantially reducing its penetration. Furthermore, the problem of dust control is at least partially eliminated, since a substantial fraction of the fines are absorbed into the mix rather than being blown out of the dryer; however, to meet air pollution requirements, most dryer drum plants require washing or filtering of the dust.[45]

Periodic sampling to ensure that bitumen and aggregate meet the specifications, that aggregate gradings remain consistent, and that the proportions of the final bituminous mix are proper is a must. Aggregates are commonly sampled at the stockpile and at the discharge gates of the hot bins. Samples of the mix are taken from the hauling vehicles and cut from the completed pavement. With

[45]See E. C. Granley et al., *Public Roads,* Sept. 1973, and T. W. Kennedy et al., *TRB Record 712* for test results. See also R. L. Terrel, *Proceedings, Association of Asphalt Paving Technologists,* 1976 and C. R. Foster, *Transportation Engineering Journal of ASCE,* Nov. 1977. For a discussion of some problems with dryer drums see paper by J. E. Wilson et. al., *Proceedings, Association of Asphalt Paving Technologists,* 1979.

these, the asphalt is dissolved and recovered; asphalt content and aggregate grading can then be determined. It is essential that samples be truly representative, and great care must be taken in this regard. Particularly troublesome is the tendency of aggregates to segregate, with the coarser particles predominating on the surface or at the base of piles or at the top surface of material on conveyor belts. Detailed procedures for accurate sampling appear as *AASHTO Designation T2* for aggregates, *T40* for bituminous materials, and *T168* for bituminous-paving mixtures. Recommended practices for mixing-plant inspection, including sampling, are given by *AASHTO Designation T172*. Results of research on sampling procedures at hot plants appear in *NCHRP Reports 34* and *69*.

Hot-plant operation has also been simulated on the computer as an approach to quality assurance. Quality was predicted using inputs generated by a Monte Carlo (random number) technique.[46] Modern quality assurance systems, including statistically based specifications, are being adopted by many states. As of 1978, 21 states used quality-assurance type specifications for asphalt concrete construction and several others are developing statistical specifications.[47]

It is common to specify that mixing time and mixer loading shall be such that the sizes of aggregates are uniformly distributed and all particles are thoroughly and uniformly coated with binder. Visual examination for the coating of coarse particles is a common control (see *AASHTO Designation T195*). The Asphalt Institute suggests that, for surface courses, 95% of the particles be fully coated without even a speck of uncoated aggregate showing.

With the increased attention to pollution which has developed and been put into legislation by the Federal Clear Air Act of 1967 (amended in 1970) and parallel legislation at the state level, stringent controls are being placed on emissions from hot plants. Not only is it necessary to remove almost all the dust from the gases exhausted from the drier by washing them or passing them through a "bag house" filter, but also the plant and its appurtenances must be dust tight. Other controls are being imposed to cover water, noise, odor, and noxious gas pollution. Ways of accomplishing these ends and a model local ordinance have been proposed by National Asphalt Paving Association.[48] Achieving these desirable results require that costly devices be installed on existing or new plants. Also, on occasion, conflicts are bound to arise as balances between cost and acceptable pollution levels are hammered out.

Production of Recycled and Sulfur Asphalt Mixtures[49]

High quality asphalt mixtures can also be made with recycled materials and with sulfur. The equipment to produce mixtures using these materials is similar to that for conventional asphalt concrete.

[46]See N. F. Bolyea and W. H. Clark III, *HRB Record 316*.
[47]See S. P. Lahue, *American Society for Testing and Materials, STP 709*. See also *NCHRP Synthesis 38* and *65*.
[48]See, e.g., its publication titled *Environmental Pollution Control at Hot Mix Asphalt Plants*.
[49]For details see *NCHRP Synthesis 54* and Report 224 on recycled asphalt, and *TRB Record 741* and *FHWA RD-78-95* for sulfur asphalt mixes, including extensive bibliographies.

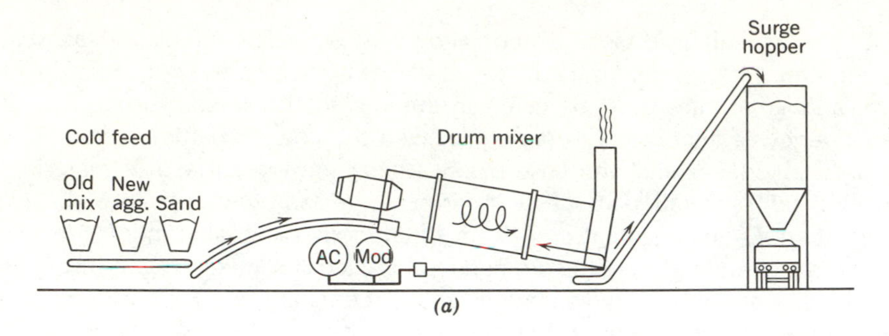

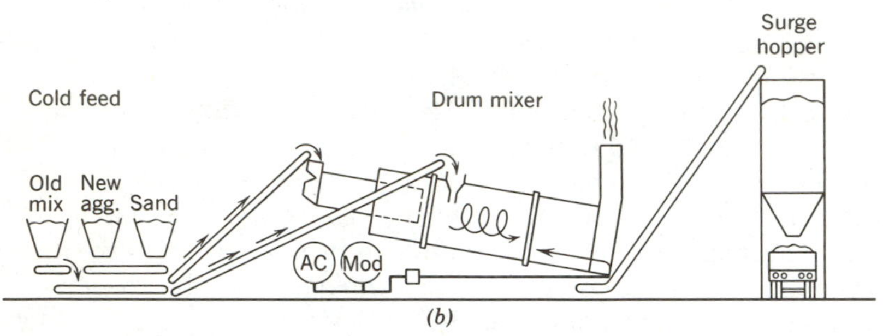

Fig. 19–19. Schematic diagrams of "hot plants" for preparing recycled asphalt mixtures. (After *NCHRP Synthesis 54*.) *(a)* Drum mixer with heat dispersion shield. *(b)* Drum-within-a-drum plant.

RECYCLED MIXES. Figure 19–19 shows two methods that have been used to recycle asphalt concrete. Figure 19–19*a* portrays the direct flame heating process where all materials are mixed simultaneously. This method initially resulted in air-quality problems, which led to the use of a heat shield to reduce the temperature of the hot gases in the drum and thus reduce the amount of smoke. A drum-within-a-drum (Fig. 19–19*b*) has also been employed. New or "virgin" aggregate is introduced into the inner drum and heated to 300 to 500°F. Reclaimed materials come in through the outer drum. The reclaimed and heated virgin aggregate meet at the discharge point, where heat transfer occurs. Also a variety of other techniques have been used; these are described in *NCHRP Syntheses 54*. The major concern in such mixing operations is control of air pollution which is normally achieved by altering production rate, amount of virgin rock aggregate, mix moisture content, or mix temperature.[50]

SULFUR ASPHALT MIXES. A number of mixtures incorporating elemental sulfur have been placed in recent years. Though the first use dates back to 1935, it was not until the 1970s that large quantities of sulfur were made available to the construction industry. Sulfur has been used in three ways as follows:

1. Addition of sulfur to hot sand-asphalt mixtures.

[50]For descriptions of typical project results see *TRB Record 695, 712,* and *780*.

2. Preblending of sulfur and asphalt to produce sulfur extended asphalt (SEA).
3. Pug-mill blending of sulfur and asphalt to produce SEA.

The basic approach in the first method is to add hot elemental sulfur, essentially as a filler, to a hot sand-asphalt mix after the asphalt has been mixed with the aggregate. In the second method, the hot sulfur is dispersed in asphalt to create a SEA binder which is then mixed with aggregate. The third method creates a SEA binder by the separate addition of the components in the pug mill rather than by preblending.

A major concern with the use of sulfur is the creation of noxious gases such as hydrogen disulfide (H_2S) or sulfur dioxide (SO_2). These gases are a potential danger to workers in the plant or at the paving operations.

Placing Bituminous–Concrete Mixtures[51]

It is common practice to apply a "prime coat" (see below) over untreated and some treated bases before asphalt concrete or other plant-mix bituminous surfaces are laid. This serves not only to bind any loose particles of base, but also to act as a bond between base and pavement and to deter rising moisture from penetrating the pavement. The total thickness of compacted asphalt concrete may range between 2 in. for lightly traveled roads up to 16 in. or more where traffic is very heavy and the pavement is laid directly on a very poor subgrade. Designs vary greatly among agencies, depending on the design method employed, and whether untreated or treated bases separate paving from subgrade. Successive layers of asphalt concrete of controlled thickness and smoothness are placed and compacted to give the total desired depth.

Hot bituminous–concrete mixtures are usually transported from plant to roadway in end or bottom dump trucks. If the weather is cool or the haul distance long, the load is covered with canvas to prevent loss of heat. Sometimes trucks with insulated bodies are used. Most agencies prohibit placing bituminous concrete in wet or cold weather. It is common to prescribe minimum air temperatures (ranging from 32 to 60°F, with 40°F most common) at which work may be carried out. During World War II, much satisfactory pavement was placed under wet and cold conditions, but engineers generally are not in favor of extending this practice into normal operations. It has been proposed that a more logical control would be the laydown temperature and cooling rate of the paving mixture, since this would determine whether or not adequate compaction could be achieved. Variables in cooling rate would include layer thickness, air temperature, and wind velocity. By delivering the mix to the laydown site at somewhat higher temperatures it would then be possible to pave in colder weather without damage to the completed pavements.[52]

Methods for placing asphalt concrete have changed as new equipment has

[51]See the *Asphalt Paving Manual, MS-8,* published by the Asphalt Institute, for a more detailed presentation on placing and compaction.

[52]See C. R. Foster, *HRB Record 316.* See D. C. Colony and R. K. Wolfe, *Transportation Engineering Journal of ASCE,* May 1978 for cooling rates of asphalt concrete mats. A paper by N.D. Shah et al., *TRB Record 549,* discusses preheating the base for cold weather paving.

Fig. 19–20. Finishing machine spreading and partially compacting an asphalt–concrete surface course. Note dump truck discharging materials into hopper at front of machine. Electronic devices for controlling the height and cross slope of the strike-off screed are often employed. (Courtesy, Barber-Greene Company.)

been developed. Originally the hot mix was dumped on wood or sheet-metal platforms from which it was moved and placed by hand shoveling. Leveling of the uncompacted mixture was done by skilled "rakers," many of whom displayed almost uncanny ability to produce a smooth, true surface. Later it was common to deposit the mix in a movable hopper called a "spreader box" which was drawn along on the base by the truck as it unloaded. In cases where no curb or gutter had been placed, pavement edges and finished grades were defined by side forms of wood or steel similar to those used for concrete pavement (see Chapter 20). By 1928, mechanical-spreading machines operating on the side forms had been developed. These machines were similar in many ways to those for portland-cement-concrete pavements. With these developments, the need for shovelers and rakers disappeared, except for placing occasional small, irregular, or warped areas.

Today, self-propelled finishing machines which operate directly on the base or other underlying layers have been almost universally adopted (see Fig. 19–20). Since they make side forms unnecessary, substantial savings result. These machines consist essentially of a hopper into which the truck dumps its load and a heated strike-off bar for spreading the mixture to uniform thickness. The machines also have a tamping or vibrating bar which partially consolidates the pavement. The machines are mounted on tracks or pnuematic-crawlers for traction. Although they bridge some base irregularities, leveling is normally accomplished with automatic screed controls that maintain depth and cross slope.[53]

[53]See, e.g., L. C. Bower and B. B. Gerhardt, *HRB Record 316*. European developments are reported by M. Blumer in *HRB Record 132. See also, The Asphalt Institute, MS-8,* op. cit.

Manual adjusting devices permit gradual changes in the depth of material placed, so that layers of varying depth, often needed to level old surfaces, may be laid. Small areas or extremely warped surfaces still must be placed and raked by hand.

Some agencies permit dumping of the hot mix directly on the base or underlying layer rather than into the hopper of the laydown machine. It is then picked up by a loader which discharges into the hopper. This procedure permits hauling in bottom-dump trucks which can unload immediately on arrival at the paving site, rather than employing dump trucks which must wait their turns to discharge into the laydown machine. Hazards that may offset the potential cost reduction are that the hot mix may segregate during the extra handling and the chance that the mix will get cold if breakdowns or other delays of the pickup or laydown machines occur.

Output of paving materials has increased almost continuously. In 1924 the 8-hr output averaged about 250 tons. With the advent of mechanical spreading machines the figure rose to about 600 tons. Each advance in plant capacity or improvement in finishing machines has led to matching improvements in the other. Today a production of 2400 tons is common.

Compacting Bituminous Concrete

As indicated earlier, stability of bituminous mixtures increases as the density of the mix increases, up to a limit in the range of 3% voids. Furthermore, by obtaining high densities during construction, rutting in the wheel tracks under traffic is minimized.

Compaction is obtained by means of smooth-wheeled, vibratory, and pneumatic rollers. The roller pattern once consisted of a large three-wheeled machine going first with its large diameter wheels ahead for initial or breakdown rolling. The tandem did the final smoothing, with intermediate rolling using pneumatic-tired machines (see Fig. 19–21). Not only can the weight of these machines be varied, but also tire pressure in some instances can be altered by the operator while the machine is running. Also available, and sometimes required by specification, are three-axle tandem rollers. On these, the center wheel can be locked into the plane of the other two wheels or moved vertically to place a larger fraction of the roller's weight on it. Now it is not uncommon to see only a vibratory compactor on construction projects. This compactor is operated initially in the vibratory mode to consolidate, followed by the static mode to finish.[54]

Field control to assure suitable density is obtained by either stipulating the method (the number and weight of rollers) or by end result (specifying a standard such as unit weight or percentage of voids). Nuclear devices are used to measure densities and asphalt content in place,[55] while cores are taken to de-

[54]See R. J. Nittinger, *TRB Record 659* for New York experience with vibratory compactors. See also *Proceedings, Association of Asphalt Paving Technologists,* 1977 for an excellent discussion and bibliography on the use of vibratory compaction for asphalt mixtures.

[55]See, e.g., M. M. Varma et al., *HRB Record 66;* W. R. Brown, *HRB Record 107;* H. W. Walters, *HRB Record 117;* and R. L. Grey, *HRB Record 361*. See also P. S. Kandhal et al., *TRB Record 695* for use of nuclear gauges to measure moisture content in asphalt mixtures.

Fig. 19–21. Rollers compacting a bituminous pavement. Three-wheel unit, left, serves as a breakdown roller. Pneumatic-tired device, center, provides further compaction by its kneading action. Tandem, right, does final smoothing. (Courtesy, Galion Manufacturing Division, Dresser Industries, Inc.)

termine asphalt content and gradation. Statistical specifications and pay adjustment factors can be found in the literature.[56]

Among the states, allowable thicknesses of each compacted layer range from 2 to 6 in. The number of rollers sometimes is expressed as a function of both roller types and tonnage of mix placed per roller hour; again the number and types of rollers to be furnished may be the sole control. Roller weights often are stipulated; usual ranges are 8–12 tons for tandems, 10–12 tons for three-wheeled units and 6–17 tons for vibratory rollers. Roller weight may also be expressed in terms of pounds per inch of wheel width, with a range of 200–350.

Rolling is commenced as soon as the spread mixture will sustain the roller without excessive displacement or checking and is continued until roller marks are no longer perceptible on the surface. It is usual to require that the edges of the newly placed material be rolled first, after which rolling progresses toward the center. Some agencies require rolling only along the length of the roadway; others also require "cross rolling" or "diagonal rolling." Roller wheels are kept moist to prevent the hot mix from adhering to them. Water is carried in a tank on the roller. With rubber tires, a release agent often is added to the water.

Compaction of bituminous mixes has been the subject of intensive study since, as indicated earlier, it affects stability, strength, durability, and the tendency to rut under traffic. One finding is that increased density can be obtained only when the mix is relatively hot. It follows that specifying thin layers may not give high densities, since they will cool rapidly while thicker layers will retain their heat. It also appears that intermediate rolling with pneumatic tires does not greatly increase density. On the other hand, it seems to reorient the individual

[56]See *NCHRP Synthesis 38 and 65; TRB Record 652, 691,* and *697.*

aggregate particles in a manner which decreases permeability and reduces later rutting under traffic.[57]

The surface of a properly designed and compacted bituminous concrete is highly resistant to permeation by water and has and maintains a high coefficient of friction. Some propose developing further skid resistance with open graded plant mix seals.[58] Where seal coats are applied, the intention usually is to provide a lighter color or a contrasting surface texture. On occasion, where it is desired to discourage the use of a given paved area, a coarse aggregate seal, which creates a rumble unpleasant to vehicle occupants, is applied.

Surfaces of bituminous concrete can be finished to a high degree of smoothness. This is reflected in the requirements of the various agencies. The most rigid specification permits only ⅛ in. variation from a 16-ft straightedge. Variations up to ⅛ in. from a 10-ft straightedge represent a more usual control.

OPEN-GRADED MIXES[59]

Open-graded asphalt concrete and plant mix differ from dense-graded ones in that they contain little fine aggregate or dust. Typical aggregate grading specifications are given in Table 19–7. It is common to stipulate that aggregates for open mixes be tougher than for dense-graded mixes, since the coarse particles are not protected by a matrix of fine particles.

For stability, open-graded mixes depend on friction and interlocking of the aggregates and the cohesiveness of a heavy binder, as do penetration macadam pavements. They differ from macadam in that aggregates and binder are combined by mixing rather than by penetration methods and are often laid as one course, whereas macadams are constructed by a series of operations performed on the road.

Asphalt content of open-graded mixes generally is set as high as possible, short of having it drain off during handling and placing. This does not affect stability, since point-to-point contact between aggregate surfaces is established during compaction.

Open-graded asphalt concrete is often employed in a thin (±1 in.) overlay over old bituminous or concrete pavements. It leaves a nonskid and relatively noiseless surface. Being porous, it temporarily absorbs water from light rains which might cause slickness on dense-graded surfaces. Experience indicates that, compared with dense mixes, this open mix better resists the reflection of cracks from the old pavement upward to the surface.

[57]For a few of the many research reports on compaction of bituminous mixtures, see R. C. Swanson, et al., *HRB Record 117;* R. J. Schmidt et al., *HRB Record 132;* N. W. McLeod, *HRB Record 158;* P. Arena et al., *HRB Record 178;* J. A. Epps et al., *HRB Record 313;* and *HRB Special Report 131.* See also the *Proceedings, Association of Asphalt Paving Technologists.* See Chapter 14 for a discussion and references on nuclear devices for measuring density.

[58]See V. Adam et al., *TRB Record 523* and *NCHRP Synthesis 49.*

[59]For detailed treatment, see *NCHRP Synthesis 49.* For experiences in Virginia and Pennsylvania see *TRB Record 595* and *659. TRB Record 695* describes uses of open graded mixes with both asphalt cements and emulsions. See *TRB Record 712* and *754* for properties of open graded emulsion mixes.

TABLE 19–7. Grading Specifications for Open-Graded Asphalt Concrete

	Percentage Passing California Department of Transportation				NCHRP Synthesis 49
	3/8 in. Maximum		*1/4 in. Maximum*		*3/8 in. Maximum*
Sieve Sizes	*Individual Test Result*	*Moving Average*	*Individual Test Result*	*Moving Average*	*Individual Test Result*
1/2 in. (12.5 mm)	100	100	—	—	100
3/8 in. (9.5 mm)	88–100	90–100	100	100	95–100
1/4 in. (6.25 mm)	—	—	84–100	85–100	—
No. 4 (4.75 mm)	23–42	25–40	—	—	30–50
No. 8 (2.36 mm)	4–22	5–20	4–22	5–20	5–15
No. 16 (1.19 mm)	0–12	0–10	0–12	0–10	—
No. 200 (.075 mm)	0–4	0–3	0–4	0–3	2–5

SHEET ASPHALT

Sheet asphalt, a mixture of sand, filler, and asphalt cement, makes a comparatively noiseless and easily cleaned surfacing. But, with an excess of binder, it can become unstable and slick when wet. It was used extensively as a surface course on city streets, even before the beginning of this century. However, the high binder content (9–12% by weight) makes it too expensive to compete with other serviceable types unless quality aggregates are not available.

Sheet asphalts commonly were about 1½ in. thick. Underlying them were asphalt layers similar to base or binder courses for asphalt concrete or, alternatively, a slab of portland-cement concrete. Preparation was in a hot plant similar to that for asphalt concrete. Mixing methods and problems were the same. Spreading and compacting were done by methods already described for asphalt concrete.

Through the years much study has been devoted to such problems as the grading of sands for sheet asphalts, the percentage and kind of filler to use, and the determination of proper type and amount of asphalt cement.[60]

OTHER BITUMINOUS PAVEMENTS

The bulk of the bituminous concrete pavement construction in the United States is hot laid. However, many agencies use or permit the substitution of cold-laid mixes.

Cold-laid asphalt mixtures consist of coarse and fine aggregate, liquid asphalts, or emulsions. The mixtures can be prepared in a central plant or road mixed. Normally 5 to 10% liquid asphalt is required to permit coating. These

[60]See L. I. Hewes, *American Highway Practice,* Wiley, New York, 1942, Vol. II, Chapt. II, for an extensive discussion of early practices. See R. D. Barksdale, *TRB Record 741,* for construction and performance information on recent sand–asphalt mixes.

mixes are particularly advantageous for low-volume roads; however, they may take time to set up, particularly in cooler climates.[61]

Plant-Mix Pavements

The term "plant mix" could well designate any bituminous surface for which the materials had been mixed in a plant rather than on the road. However, as the term is commonly used, asphalt concrete, sheet asphalt, and open-graded mixes are excluded, and "plant mix" generally denotes the cheaper and less rigidly controlled products.

Plant mixing of local aggregates with binder began in the 1920s soon after road mixing. By mixing the materials at a plant and placing them as soon as they were delivered to the road, many of the delays to road mix caused by inclement weather were avoided. Often these early plants consisted only of a semiportable pug mill and the appurtenant scales, elevators, storage bins, and tanks. Aggregates were used as they came from gravel pit or quarry; the oversize was scalped off, crushed, and reintroduced. Provision was also made to dry the aggregates. Some of the later plants had dryers, hot screens, several storage bins, and weigh boxes so that proportioning could be accurately controlled. Also the range of binders was extended to include more viscous grades which could be mixed only with hot aggregates. At present, requirements for some plant mixes demand much of the equipment found in asphalt concrete plants, and controls over aggregate quality and grading are also equally stringent. Others, however, specify only that aggregates be placed in a single stockpile, proper proportioning, and mixing at the plant. As of today, then, a pavement called plant mix by one agency may be equal in almost every regard to the asphalt concrete of another.

The cost of producing plant-mix materials increases each time some additional refinement is introduced and as controls over grading and other characteristics are tightened. However, by introducing these refinements and controls the engineer gains added assurance that the completed surface will give satisfactory service.

Heated plant mixes using the heavier grades of binders generally are placed with finishing machines, as described for asphalt concrete. Compaction procedures are similar.

For "cold-laid" and similar plant mixes having binders of cutbacks or emulsions, aeration before laying may be needed to permit evaporation of some of the solvent or water. Otherwise the pavement may be overlubricated and unstable. In these cases the mix is spread along the roadway in a uniform windrow and then bladed back and forth across the road by a power grader as is done for road mix. After aeration, the mix is spread with the grader and compacted with rollers. An earlier practice of trusting to traffic for compaction has largely disappeared as the importance of high densities has become more generally appreciated.

Some agencies place seal coats over plant mixes; others do not. Where they

[61]See *The Asphalt Institute, MS-14* and *MS-19, NCHRP Synthesis 30,* and *TRB Record 702, 712* and *754* for additional information on dense- and open-graded cold mixes.

are used over mixes made with cutbacks or emulsions, application must be delayed until most of the fluidizing agent has evaporated.

Road-Mix (Mixed-In-Place) Pavements

In 1915, J. S. Bright, the county engineer of San Bernardino County, California, mixed light oil and desert sand, using plows and disk harrows, and, with several gallons of oil per square yard of area, produced a surface several inches thick. This roadbed gave excellent service. Generally, however, sheet asphalt and bituminous concrete were the only types of dense-graded bituminous mixtures in common use until about 1926. Because of their high first cost, application was restricted to city streets or heavily traveled rural roads as was also true of pavements of portland-cement concrete. Surface treatments, as discussed later, were also used. None of these, however, filled the need for a cheap, dustless, and relatively permanent surface for the many miles of road carrying low or moderate volumes of traffic. Road mix was the first step toward filling this need. Once introduced, its success, particularly in the far West, was phenomenal. Road-mix methods are still widely used, especially in the construction of less heavily traveled roads, for reworking rough pavements that must be torn up and relaid, and for processing patching material for maintenance operations.

Aggregates for the first road-mix projects were the dense-graded gravel or fine-crushed material already in place as the road surface. The top 2 or 3 in. of this surface was loosened and pulverized using scarifiers and tooth harrows, after which it was bladed back and forth across the road to produce uniformity. A total of about 1½–2½ gal of asphaltic oil per square yard (at 100°F or hotter) was then applied in successive increments by pressure distributors and each application turned under by spring tooth or disk harrows. After all of the oil had been added, the mixture was "processed" by motorized blade graders (see Fig. 19–22). This operation involved blading the mixture back and forth across the road until it became homogenous and of uniform color. The motor grader traveled fast enough (usually about 4 mph) so that the material rolled rather than slid in front of the blade. After processing, the mix was spread to uniform thickness and cross section by a blade grader operated by a particularly skillful bladesman called a "laydown man." Sometimes the completed surface was rolled, at times only the edges were rolled, but in many cases compaction was left to the vehicles using the road. It was common practice to apply a seal coat to the completed surface to prevent the intrusion of surface moisture. Often this operation was postponed until a curing period of several weeks after laying had transpired.

Many variations to the original road-mix process have been developed. One of the most important has been to import mineral aggregate from local pits or quarries instead of using the loosened material from the existing road surface. In this manner it is possible to control more closely its quality and grading, particularly the percentage of dust. Furthermore, a constant amount of mineral aggregate per unit of roadway length can be assured by sizing the aggregate windrow with a spreader box. This results in a more uniform oil content and fewer "fat" and "lean" spots in the completed roadway.

Fig. 19–22. Motor grader processing road mix. (Courtesy, J. D. Adams Mfg. Co.)

Road-mix processing is now often performed by a single machine which picks up the aggregate from the roadway or receives it from a truck and adds and mixes in the oil in specified amount (see Fig. 19–23). Spreading of the processed mix is commonly done with blade graders, or with a finishing machine.

Dense-graded and open-graded aggregates, sands, and sandy soils have all been successfully road-mixed. Materials high in fines and without fines and with or without coarse aggregate have been used. Sometimes roadside pits, even without screening, offer suitable materials. At times, no processing other than the rejection of oversize or of excess fines is needed. Angular, broken particles which produce added stability are to be preferred, but crushing to produce them adds to the cost and is avoided unless it is necessary to produce the desired grading.

A typical specification for mineral aggregate for road mix is that of the Federal Highway Administration. It requires that 100% pass the ¾-in. sieve, 45–65% pass the No. 4, 33–53% pass the No. 8, 10–20% pass the No. 50, and 3–8% pass the No. 200.

As indicated by Table 14–6, a wide variety of asphalt products have been successfully employed in road mixes. The principal control is that the binder's viscosity at the expected air temperatures must be such that mixing and placing can be carried out. For combination with aggregates containing high percentages of fines, slow-curing oils (*SC*-250 or 800) or emulsions (*CSS*-1 or *SS*-1) are a common choice. Cohesive strength is provided by the oil film surrounding the many fine particles. For open-graded aggregates that contain small amounts of

Fig. 19–23. (a) Single-pass mixer processing road mix asphalt pavement. (Courtesy, Pettibone Road Machinery Division of Pettibone Corporation.)

dust, medium-curing cutbacks (*MC*-250 or 800) or emulsions (*MS*-2 or *CMS*-2) are generally preferred, with the heavy residual asphalt supplying most of the cohesive strength. For clean sand or coarse open-graded aggregates, medium-curing emulsions and rapid-curing oils, grades *RC*-250 and 800, provide a heavy binder after curing. Liquid tars, grades *RT*-5 and *RT*-7, have also been used successfully for road mix. Binder type (for example *SC* versus *MC*) normally is dictated by the character of aggregates found in local deposits, for, if aggregates are transported any great distance, the low-cost advantage of road mix is lost. Choice of grade (as *SC*-250 versus *SC*-800) depends largely on the air temperature at which mixing is accomplished, the lighter grades being used in cooler climates. In general, engineers prefer the heaviest grade that can be successfully processed. Between 5–10% of oil, by weight, is commonly required; this percentage varies with aggregate characteristics. As indicated for plant mixes, if the binder contains volatiles or water, aeration before final spreading may be required.

Road mixes, like other pavements, must be placed on a sound roadbed of width somewhat greater than the pavement itself. Satisfactory results cannot be obtained over yielding subgrades or those that become unstable when the evaporation of capillary moisture is prevented by the moistureproof bituminous blanket. Furthermore, the roadbed must be carefully leveled and consolidated before road-mix operations are begun; otherwise variations in the thickness of the bituminous blanket or undulations in the finished surface must result. Maintaining constant surface thickness is particularly troublesome on superelevated curves. Unquestionably, the electronic devices for maintaining constant cross slope have proved helpful in these operations. As would be expected, smooth-

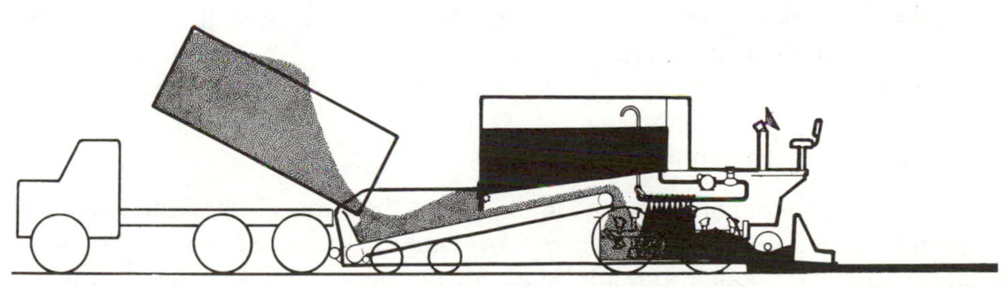

Fig. 19–23. (b) Paving train for on-site mixing of cold laid emulsion mixtures. (Courtesy, Midland Machinery Company, Inc.)

ness requirements for road mix are less stringent than for asphalt concrete. For example, the California Department of Transportation permits a variation of ¼ in. under a 12-ft straightedge, as contrasted with ⅛ in. for asphalt concrete.

During the rainy season, in the late fall, or in areas subject to thundershowers, processing by road-mix methods is done under a serious handicap. Mixing in the rain soon soaks the spread-out materials; therefore, if it begins to rain, the mix must quickly be pushed into a windrow. All operations are then suspended until the materials are again dried. This is sometimes hastened by spreading and blading the mixture. If wet mixtures are laid down, they will probably be unstable. This is commonly controlled by a specification which states that no "laydown" may be commenced if the moisture content of the mixture exceeds some maximum such as 1 or 1.5% by weight. Winter construction often is prohibited in areas where the weather is severe.[62]

PENETRATION MACADAM-BITUMINOUS MATS[63]

Penetration or bituminous macadam bases or surfaces consist of two or three layers of progressively smaller, clean, sharp angular stones bound with bitumen. Each layer is consolidated and keyed by rolling, after which it is sprayed with the binder. Commonly, surface courses are sealed by sprayed-on binder blotted with fine crushed stone.

Bituminous macadam represents the adaptation of waterbound macadam (see Chapter 18) to the motor vehicle. It originated about 1907 and was soon used in Massachusetts, New York, and Ohio. Later it was adopted in other New England states and in California. Pavement for the New Jersey Turnpike, completed in 1952, was 7½ in. of bituminous-macadam base under 4½ in. of asphalt concrete. Today, bituminous macadam has been largely supplanted by other pavement types.

Bituminous macadam contains a large percentage of voids, particularly in the lower part of the layer. Its strength comes from aggregate interlock which holds the individual stones together. For these reasons, an unyielding and well-drained base that will not displace or squeeze upward into the pavement voids is absolutely necessary.

Expensive equipment is required to produce the mineral part of macadam pavements, since all aggregates must be crushed and then screened to size. Furthermore, good riding qualities can be produced only with skilled and experienced workers. For these reasons, macadams are costly and difficult to construct.

Materials for Penetration Macadam

The binder for bituminous macadam must, at the time of spraying, be fluid enough to penetrate into and coat the aggregate particles. In place, it must in itself be cohesive enough to hold the aggregate mass tightly together and to resist abrasion by traffic. Asphalt cements are commonly used where openings between the stones are large. Medium grade *RC* oils, and tars in grades *RT*-10 to *RT*-12 or *RS* emulsions serve where voids are small and the binder must be more fluid. Binders are sprayed over the aggregate by pressure-distributor trucks although in the past hand-pouring methods were common.

The type and amount of binder are greatly influenced by the source. To illustrate, California asphalts, used in the western states, are more sensitive to heat and are relatively sticky compared with those from other fields. Because of these properties, smaller amounts and softer grades are appropriate.

[62]See R. R. Biege, Jr. et al, *HRB Record 51,* for a discussion of roadmix problems. See also *Asphalt Institute MS-14* and *MS-19*.
[63]See L. I. Hewes, op. cit. Vol. II, for a detailed exposition.

Individual lifts of modern bituminous macadams tend to be one-stone thick; in other words, the maximum size of coarse aggregate approximates the thickness of the completed pavement. This condition provides stability, because the larger stones engage the base course or lower layer and the surface has no tendency to "roll" or "wave." If the stone is soft, a somewhat larger maximum size may be specified to offset the effect of breakage under rolling. Individual particles should tend to be cubical in shape. "Pencil" or "slab" shapes are undesirable. The variation in size of stone for each course is small. For example, for a 3 in. thick, three-course macadam, the first stone laid would be in the 2½–1¼ in. size range, the second from 1¾ to ¾ in., and the third from 1¼ to ½ in. For a seal coat, a range of ¾–¼ in. would be appropriate.

Aggregates for macadams should be strong and tough, as excessive crushing under the roller will spoil the interlock between individual stones. Also, the finer crushed particles tend to close the surface so that penetration by the binder is retarded and "fat" spots in the completed pavement result. Requirements for hardness and toughness commonly set the maximum permissible percentage of wear (Los Angeles Abrasion Test) at about 30 after 500 revolutions.

Aggregate surfaces must be clean so that excellent coating by the binder can be obtained. Some agencies specify that dirty aggregates be washed before use. At times aggregates are treated either by plant mixing or spraying on the road with some liquid designed to improve coating properties. These products, which are patented, are commonly called "facilitating agents" or "mobile oil."

Constructing Penetration Macadam

The sequence of operations for constructing bituminous macadam is as follows:

1. *Spread the coarsest aggregate.* Some specifications require that the base course be first covered with about 1 in. of rock dust. Then mechanical stone spreaders or asphalt-paving machines (see Fig. 19–20) spread the stone to the designated depth. Since surface smoothness depends largely on the uniform size distribution of this course, great care must be taken in handling and placing the stone to prevent segregation. If pockets of finer stones occur, they must be removed and replaced.
2. *Compact the coarsest aggregate.* Either smooth-wheeled rollers or vibratory devices serve well. The aim is to key the stones into a tight interlocking mass. Compaction is begun at the edges and progresses toward the center. Breakage must be avoided, otherwise the binder will not penetrate evenly and fat spots will develop. Smoothness is checked after compaction and any low spots are corrected by removing, replacing, and recompacting the stone.
3. *Application of bitumen.* Spreading the right amount of bitumen at the appropriate temperature is essential. If the binder is too viscous, penetration will be insufficient and the lower stones will remain uncoated; if it is too fluid, it will run through onto the underlying layer. In some instances, it may be necessary to tow the distributor truck, if its drive wheels tend to spin and displace the stone.
4. *Constructing the keystone course or courses.* The sequence of operations just outlined is followed for each succeeding course. Only enough of the smaller stone is spread to fill the upper voids in and interlock with the larger stone. Compaction and binding with bitumen follows.

5. *Sealing the surface.* Sealing the surface against moisture involves spreading the selected binder and applying and rolling in the cover aggregate. The binder for sealing is not necessarily that for the earlier courses. If a light-reflecting surface is desired, care must be taken to drop the cover aggregate in such a way that it does not roll over and become coated with binder.

From this brief description it can be seen that dry, warm, and constant weather conditions are essential for successful macadam construction. If the weather conditions are less than favorable, the chances of success are better if emulsified asphalt is selected as the binder. Its fluidity is less affected by contact with the cool stone; also it is better suited for coating the aggregate. With emulsions, it is common to "choke" the coarse aggregate with smaller stone before applying the binder.

SURFACE TREATMENTS[64]

The term "surface treatment" encompasses a variety of procedures intended to dustproof or otherwise upgrade untreated surfaces or to rejuvenate or improve existing pavements. In most instances, surface treatments involve the *"inverted penetration"* method of construction under which the binder is first sprayed over the prepared surface and then covered or blotted with aggregate. This contrasts with the penetration method where binder spreading follows aggregate placement.

Surface treatments are commonly subdivided on the basis of the purpose to be accomplished. The main subdivisions are: dust palliatives for dust control; prime coats or tack coats for treating surfaces on which a new wearing course is to be constructed; surface treatments and armor coats for providing protection for untreated mineral surfaces; and seal coats, including slurry seals, for protecting, skid-proofing, or otherwise improving existing pavements.

The key to successful surface treatment construction is the pressure distributor (see Fig. 19–24). This machine usually consists of a truck-mounted tank fitted with a spray bar which, through special nozzles, spreads the binder over a prescribed width of roadway. The liquid is subjected to pressure by a power pump, usually of the gear or gravity type. The pump sometimes is powered by the truck motor, although drive by an auxiliary engine is preferable as it offers better pressure control. There is a return pipe to the tank from the spray bar; flow through it is controlled by an adjustable relief valve which maintains constant pressure on the nozzles. Recirculation also keeps the hot binder in the spray bar and nozzles from cooling and solidifying between spreads. For a given binder at a given temperature, the rate of application will depend on the nozzle opening, pressure, and truck speed. With good equipment and skilled operators, quantities very close to those specified can be spread.[65]

[64]The literature on surface treatments is extensive. Three excellent references, each with extensive bibliographies, are M. Herrin et al., *HRB Special Report 96, MS-13, Asphalt Surface Treatments,* published by the Asphalt Institute, and *Compendium 12,* Transportation Technology Support for Developing Countries, TRB.

[65]The actual amount of binder delivered to an area has been checked by spreading strips of paper or cloth, or placing special pans over a surface to be sprayed by a distributor truck, and then determin-

Fig. 19–24. Bituminous distributor truck spraying binder for a surface treatment. Spray bar can be raised and its ends folded up for highway travel. On and off controls can be operated either from truck cab or from back platform. Note that individual spray patterns overlap to provide multiple coverage of all areas. (Courtesy, Rosco Manufacturing Company.)

Distributors with capacities ranging from 800 to 5500 gal are available on truck or semitrailer mountings. Spray bars covering a width of 20 ft are often provided. On some units the tanks and appurtenances are insulated; on others they are not. Some have provisions for heating the load. Controls on the spray bar often permit application through part or all of the nozzles, as needed for the particular spread. Thus, in special cases, shoulders, or half or partial widths can be covered. Often an auxiliary spray is provided for touching up lean or white streaks caused by clogged or slow-flowing nozzles. Some specifications require a trough arrangement that can be swung under the spray bar to check the application suddenly or to prevent dripping.

Nozzles may be either the cone or slotted type and are available in various aperture sizes. They must be of proper design and kept scrupulously clean; otherwise coverage by individual sprays will be uneven, or the adjacent sprays will overlap and leave wet streaks. To assure uniform coverage, the distributor truck should be traveling at prescribed speed before the nozzles are opened; otherwise at the junctions of succeeding spreads a double application or none at all may result. This is often prevented by covering the last few feet of the com-

---

ing the amount of binder deposited on them. In many cases this test has indicated that, although the proper distribution was obtained for an entire load, considerable variation occurred between individual sections along the length and width of the area. Careful checking on the performance of distributor trucks before using them seems to be indicated; see, e.g., D. C. Mahone and S. N. Runkle, *HRB Record 236*.

pleted section with a temporary cover of building paper or sand on which spraying is started. Also, the height of the spray bar above the surface must be held to close tolerances.

Dust Palliatives

Wind and air disturbances caused by motor vehicles often stir up dust and fine sand from untreated surfaces or shoulders. This creates an accident hazard by limiting sight distance and causes a generally annoying condition for motorists and roadside activities. Wind also often erodes cut or fill slopes of sandy materials containing little binding soil. Treatment to control these conditions consists of a small application (about 0.2 gal/yd^2) of a light, slow-curing oil, usually *SC*–70 but occasionally *MC*–30 or 70. This oil penetrates the surface for a depth of about ½ in. and provides a film which surrounds the individual particles and binds them together. Slow-curing oils are selected because they remain soft for long periods of time. Cutbacks or emulsions are often unsatisfactory, as they turn into harder asphalts. These, on setting, produce a brittle surface that would soon crack up and disintegrate. Successful results have been had with slow-setting emulsions diluted in four–nine parts of water. Also used crankcase oil from motor vehicles has been employed as a dust palliative. Sometimes a light application of sand is spread after the oil is placed; again it may be used only to blot up excess oil in sections where penetration is poor.

The oil, heated to around 175°F, is spread with a distributor truck or with a hand sprayer in less accessible areas. Treatment should be made in warm, calm weather. If the soil is slightly moist, penetration is improved.

Prime Coats

Often, before the placing of a bituminous pavement over a base of earth, gravel, or waterbound macadam, the surface is "primed" by spraying on an initial application of bituminous material. The intent is to plug capillary voids to halt the upward movement of water and to coat and bind dust and loose mineral particles, thus hardening and toughening the surface. Adhesion between the base and the surface course is also improved. Where traffic must use half the roadway during construction, the prime coat also serves to protect the base before the placing of pavement. If paving is by road-mix methods, the prime coat provides a table on which mixing may be done.

The lighter, medium-curing cutback oils have generally been chosen for prime coats. They are fluid enough to penetrate into the base but leave a viscous asphalt in the pores of the treated surface. The more fluid oils are recommended for tight surfaces; heavier grades may be used for looser surfaces. Slow setting emulsions are increasing in use because of EPA regulations limiting the use of cutbacks (see Chapter 14). Light tars, grades *RT*–1 to *RT*–3, also have been successful. The quantity of binder to be applied depends on the tightness of the surface being primed. As little as 0.3 or as much as 0.8 gal/yd^2 may be needed.

Before the prime coat is placed, the surface must be shaped, moistened, and rolled to make it solid and uniform. Variations in surface texture will result in nonuniform penetration and will leave wet and dry spots. If, when the binder is applied, the base is slightly moist, better penetration will result. The oil is spread by a pressure-distributor truck at a prescribed temperature (see Table 14–8). After application, traffic should be detoured until the surface is no longer sticky and will not be picked up by traffic. If this cannot be done, a blotter course of sand must be applied.

Tack Coats (Bituminous Paint Binders)

Often, old bituminous or concrete pavements are resurfaced with a bituminous blanket of plant mix, asphalt concrete, or sheet asphalt. Under these circumstances it is common to precede actual resurfacing with a tack coat to bond or "tack" the old and new layers together thoroughly.

For tack coats, the binder should penetrate and soften the surface of an old bituminous mix or firmly attach itself to a concrete surface. It must also be cohesive enough to bind old and new layers tightly together. Rapid-curing cutbacks in lighter grades and water-diluted slow-setting emulsions are commonly specified. Some specifications include fairly heavy tars, grades *RT*–8 and *RT*–9. The usual rate of application ranges between 0.05–0.15 gal/yd^2 of surface. Application temperatures are fitted to the product (see Table 14–8). Tack coats are usually applied soon before the resurfacing is laid. Traffic must be kept from them.

Surface Treatment or Armor Coats

The term "surface treatment" ordinarily designates a thin bituminous surface of binder covered by mineral aggregates, applied to an earth, gravel, or waterbound macadam surface, or to a stabilized base. On lightly traveled roads, bases protected by surface treatment provide a relatively permanent, cheap pavement which solves the problems of dust control, formation of corrugations, and loss of surfacing materials created by the abrasive action of vehicle tires. For highways carrying heavy traffic, surface treatments can serve only temporarily because they lack the strength of more expensive pavements. They are used often by agencies whose funds are limited as a temporary pavement and protection for the base material. Their use for a period before permanent pavement is placed allows easy correction of weaknesses in base or drainage.

Surface treatments were first used about 1907 to prevent ravel and wear and to control dust on waterbound macadam roads. By the 1920s bituminous surface treatments over waterbound macadams, sand–clay gravel, and fine-crush surfaces became general.

Because surface treatments are thin, a firm bond between them and the aggregate courses which they protect is essential. It is common, therefore, after shaping the surface, to apply prime coats, just as in constructing many other pavements.

The surface treatment proper, which follows the prime coat, may be as thin as ¼ in., or as thick as 1 in. Usually the thinner types (sometimes called "one-shot'' types) consist of about 0.20–0.40 gal/yd^2 of fairly heavy liquid bituminous material covered with 10–25 lb of clean stone screenings or fine screened gravel or slag that is free of dust. Grading generally ranges between the ⅜ in. and No. 10 sizes, although sand may be employed. Larger cover material with maximum sizes up to ½ or ⅝ in. will usually require 35–40 lb/yd.^{2} For binding coarser aggregates, a more cohesive binder is needed, and the choice is confined to the heavier-grade cutbacks, a high-penetration asphalt cement, a rapid-setting emulsion, or a heavier-grade tar (see Table 14–6).

Bituminous material is spread at the proper application temperature (see Table 14–8) from a pressure-distributor truck. Cover materials are ordinarily spread from dump trucks, usually through some form of spreading device which assures the prescribed distribution (see Fig. 19–25). In order to avoid picking up the uncovered bituminous material, it is common to spread from a backing truck, so that the tires roll on the freshly applied screenings. Light brooming with drag brooms follows to assure uniform distribution of screenings. Light rolling with pneumatic-tired or smooth-wheeled rollers to set the screenings firmly usually follows the brooming.

Surface treatments should be applied during good sunny weather when there is no wind and when the surface is dry and clean. The chances of failure are high if construction is attempted during cold or wet weather. In very hot weather, however, the roadway surface temperature may rise as high as 150°F. Then the aggregates will tend to absorb even heavy liquid asphalts and may not leave enough on the surface to tie down the intended cover coat. Under such conditions the liquid application should be increased.

Surface treatments applied in two or more lifts are designated by such terms as ''armor coats'' and ''double (or triple) surface treatments.'' For two-course treatments cover material for the first lift often has a maximum size of ¾–1 in.; for three-course treatments the maximum size commonly is 1–1¼ in. Maximum size of the aggregate for each subsequent lift is reduced. Binder is spread before each application of screenings, and each course is broomed and rolled. Such multiple-lift surface treatments resemble macadams in that strength is gained through interlocking the coarser aggregate by rolling and by keying it with subsequent applications of smaller stone.

Seal Coats and Retreads for Existing Pavements

Applications of binder to pavement surfaces, covered with aggregates, are designated as ''seal coats.'' In many specifications and other presentations no distinction is made between seal coats and surface treatments, as the materials and construction processes and problems are almost identical. Although seal coats are thought of as temporary, their average life is reported to be about 10 yrs.

Seal coats will serve one or more distinct purposes, including the following:

1. To provide an abrasion- and water-resistant surface. This applies particularly where the treatment is to protect soil or friable materials such as stabilized aggregate bases or waterbound macadam.

Fig. 19–25. Self-propelled spreader distributing cover aggregate over freshly sprayed bituminous binder for surface treatment. Note that spreader and aggregate truck travel over newly spread aggregate. (Courtesy, Road Division, Koehring Co.)

2. To improve the skid resistance of bleeding bituminous surfaces or of bituminous or concrete pavements where the surface has polished under traffic.[66]
3. To improve light reflection and no-glare characteristics of the pavement surface. A light-colored blotter aggregate, placed carefully to prevent coating with binder, gives good light reflectance. Again, because the surface of the seal coat is irregular, it will not reflect light and produce glare when wet as do slick or polished surfaces.
4. To provide lane, shoulder, or other demarcations. As an example, traffic lanes can be distinguished from shoulders by applying a seal blotted with light-colored aggregate to one surface or the other. Again, a properly designed seal can discourage continuous occupation of certain areas, such as the center lanes of three-lane highways. In this instance, a seal coat having a very large (say 1 in. maximum) cover aggregate produces an unpleasant rumble in the vehicle, so that drivers will only use the lane for passing.
5. To alert drivers that unusual or hazardous conditions lie ahead. These include stop signs or traffic signals in isolated or unusual locations or unexpected left-lane exit or entry ramps on freeways. In these instances, several "rumble strips" consisting of a very coarse seal are placed at intervals across the affected roadway. Often such installations substantially reduce accidents.[67]

Since seal coats generally are thin, they add little shear or beam strength to the pavement structure. It follows that failures of the base or waves caused by instability in the pavement itself cannot be corrected by applying them.

[66]See B. M. Gallaway, *TRB Record 523*.
[67]See, e g., articles by D. W. Hoyt, *Traffic Engineering*, Nov. 1968, and W. R. Bellis, ibid., Apr. 1969.

Binders similar to the ones used for surface treatments have been successfully applied as seal coats (see Table 14–6); *RC*–800 is a favorite.[68] Use of *RS* emulsions has increased in recent years for energy savings and environmental reasons.

For single-treatment seal coats, the most common aggregate consists of clean stone chips or gravel distributed in size from ½ in. to the No. 8 sieve. Aggregates in this size range are small enough so that the resulting seal is watertight and relatively noiseless, but the particles are large enough to provide a nonskid surface. For situations where general use of the paved area is to be discouraged, or if it is desired to get the driver's attention, larger aggregates are often specified.

Strong, tough crushed aggregates with a Los Angeles rattler test percent of wear less than 35 are much preferred. The angular shape gives better point to point contact between individual particles, and the fresh surfaces tend to be rough, which gives good skid resistance. Light-colored surfaces can be produced with nearly white aggregates, placed carefully to prevent coating with binder. Sand or other fine, graded aggregates are not widely used for sealing highway surfaces, as the nonskid properties are not as good. However, they are preferred for sealing airport runways since coarser aggregates would cause excessive tire wear. Cover materials containing fine particles must be placed very carefully to obtain uniform depth and to prevent the formation of ripples.

Seal coats should be constructed during favorable weather. A survey in the western states indicated that, for the projects included, between 85 and 90% of the satisfactory seal coats were placed during June and July and that 60% of those placed after October 1 were failures.

For the ½-in. maximum-size cover aggregate, the amount of binder to be applied may vary between 0.15 and 0.30 gal/yd^2, depending on the surface treated and on the shape, grading, roughness, and porosity of the cover material. If a prime coat is omitted, the quantity of binder must be raised to allow for penetration into the old surface. Aggregates should be embedded for about 0.5–0.7 of the thickness of the seal coat, which approximates the average least dimension of the aggregate. About 20–30 lb/yd^2 of cover aggregate is needed, including an allowance for that lost by whipping off the road. Quantities are, of course, reduced if smaller aggregate is specified.[69]

Multiple-lift seal coats, sometimes referred to as retreads, are at times specified. Design and construction closely follow the procedures already set out for armor coats. Seal coats of hot plant-mixed binder and chips, or binder and sand, are also used at times. These are spread over a hot plant mix or asphalt-concrete surface course after the first rolling and are incorporated into that surface by subsequent compaction.

[68]See R. A. Crawford, *HRB Record 236,* for an evaluation of seal coats made with *RC* cutbacks, emulsions, and rubberized asphalts. See also, *American Society for Testing and Materials STP 724* for a discussion of seal coats using rubberized asphalts.

[69]See *HRB Special Report 96,* for several procedures for setting the quantities of materials needed in seal coats. See also *Compendium No. 12,* op. cit.

Slurry Seals and Rejuvenators

A "slurry seal," composed of sand, crushed stone screenings, emulsified asphalt, and water, has proved very effective as a means for filling cracks and rejuvenating the surface of badly deteriorated pavements. A typical mix contains 3000-lb fine sand, 3000-lb rock dust, and 140-gal *SS*–1 or *SS*–1*h* emulsified asphalt diluted with around 80 gal of water. Materials are mixed in a transit-mix concrete truck or a portable mixer. Application to the dampened pavement is by means of a drag-like frame with a rubber squeegee mounted along the trailing edge. The aim is to fill all cracks and leave a coating about ⅛ in. thick over the entire pavement surface. If the seal is too thick or overasphalted, it may become slick under traffic.[70]

At times, a pavement surface is sprayed with a light coating of *SS*–1*h* emulsified asphalt or an emulsified petroleum resin to rejuvenate it or improve its appearance. This treatment goes under such names as black seal, fog, and color coats. Also, special binders containing rubber or epoxy resins have been used in seal coats, sometimes in combination with silica sand or other skid-resistant aggregates.

Bituminous Overlays.

Overlays are relatively thick layers of bituminous-bound aggregate placed over existing pavements. A primary purpose always has been to level out a distorted surface or to cover joints and cracks that have become rough and annoying to motorists. More recently, overlays have been considered as a means of increasing the load-carrying capacity of existing bituminous pavements.

Open-graded mixes, discussed earlier, often are employed for leveling and joint concealment. If the aim is to strengthen the existing pavement, the choice is commonly asphalt concrete, tightly bonded to the old pavement with a tack coat.[71] Asphalt-rubber layers or fabrics are also often used prior to the overlay to reduce the chances of reflection cracking.[72]

At times, existing narrow pavements are widened, with or without an overlay covering the existing pavement. This requires careful planning to avoid later settlement of the widened portion and formation of a crack at the junction. Special machinery is often employed.[73]

Many agencies have undertaken surveys of the riding qualities of their roads using rating schemes similar to the present serviceability index and of their strength based on deflection measurements. Based on these findings, rational decisions can be made as to the type of rehabilitation needed (see Chapter 21 for more detail).

[70]For more detail on slurry seals, see the various publications of the Asphalt Institute; also W. J. Harper et al., *HRB Record 104*.
[71]See, e.g., C. W. Beagle, *HRB Record 173*, and several articles in *HRB Record 300* and *327* and *HRB Special Report 116*.
[72]See *TRB Record 572, 595, 632* and *700*. See also *NCHRP Report 195* and *ASTM STP 724, op. cit.*
[73]See, for example, *NCHRP Synthesis 28*.

On some bridge decks and in other situations where weight cannot be added to the pavements, thin overlays may be employed. For example, the San Francisco-Oakland Bay Bridge was resurfaced with ¼ in. of open-graded iron slag and granite, bound with epoxy asphalt.[74]

PAYMENT FOR BITUMINOUS PAVEMENTS

Payment for materials, equipment, labor, overhead, and profit for contract construction of bituminous pavement is either at a bid price per unit of completed pavement surface or per unit of material supplied. A number of highway agencies pay for all types by the square yard of completed surface; some incorporate charges for subgrade preparation, base, and pavement in a single item. Payment per unit of material varies from agency to agency and also with pavement type.

On a weight basis, plant mixes are usually paid for by the ton of paving mixture, including charges for hauling, placing, and compacting. Some agencies, however, pay under two items: mixed material at a big price per ton *plus* binder at a unit price per ton or per gallon.[75] Payment under the two items makes it possible for the engineer to change the percentage of binder to fit variations in the aggregate without penalizing or overpaying the contractor. Some agencies pay for rolling separately, by the hour. Prices for road-mix, penetration-macadam, and inverted-penetration construction (if paid on a weight basis) are normally set in two items: aggregates at a price per ton and binder at a price per ton or per gallon. These prices include all charges for hauling, processing, and compacting. Separate items for each class of aggregate are provided at times. Sometimes separate payment for preparation of the underlying surface or for processing, in the case of road mix, is made.

PROBLEMS

19–1. Observe and analyze a bituminous-pavement failure on a highway or street near your campus. Then with a sketch and explanation, prepare a brief report. In this report give:

a. Exact location.

b. Description of the existing pavement and a diagnosis of the form or forms of distress. Explain the failure in terms of basic design, materials, or construction deficiencies.

c. Recommended corrective action for the failed area.

d. Explain how you would prevent a similar failure if you were to do a complete redesign.

[74]See R. A. Brewer, *HRB Special Report 116;* see also *Engineering News Record,* Oct. 14, 1976.

[75]Weights per unit volume vary with grade. For example, a ton of *SC*–70 at 60°F contains about 251 gal, whereas a ton of *SC*–3000 at the same temperature contains only 241 gal. Volume–weight relationships also change with temperature; a gallon of liquid asphalt or asphalt cement at 300°F weighs 8% less than at 60°F. If payment is by the gallon, the specifications will then include specific-gravity and temperature-conversion tables from which measured quantities are converted into pay quantities.

19–2. A new two-lane rural highway is to traverse an area for which the native soil corresponds to No. 2 of Table B. This soil will be the subgrade or basement soil in cuts and will be used for embankments. A crushed rock base course comparable to soil No. 11, Table B, can be produced from a stream bed within an economical haul distance of the project. Pavement is to be asphalt concrete. Traffic is light and totals 300 vehicles per day in each direction; 15% of the vehicles are trucks; and, of these, 50% are two-axle, 20% are single-unit and 20% are combination three-axle, 5% are four-axle, and 5% are five axle. Pavement life is 20 yr.

For the situation outlined in the preceding paragraph, develop typical roadway cross sections like Fig. 14–1. Show the depth of each individual layer to the nearest ½ in. or 0.04 ft.

a. Base your design on the Hveem stabilometer method. Minimum pavement thickness is 4 in. (0.33 ft.). Assume that the soils do not develop expansion pressures.

b. Extend the design of part *a* (Hveem stabilometer method) to recognize expansive forces in the basement soil. Base and pavement weigh 130/ft^3. Expansion test results on soil No. 2 are as follows:

Moisture (% of dry wt)	12.3	14.0	16.3
Resistance value, Hveem stabilometer	17	14	11
Expansion pressure (psi)	1.5	1.1	0.8

19–3. Work Problem 19–2, assuming that the total daily traffic volume is 1200 vehicles per day in each direction. Distribution of the 15% of trucks is as in Problem 19–2.

19–4. Work Problem 19–2, assuming that the surface consists of 3 in. of asphalt concrete underlain by a 6 in. class *B* cement-treated base. Pavement life is 20 yr.

19–5. Work Problem 19–2, changed as follows:

The native soil, which is the basement soil in cuts and is used for embankments, is like soil No. 1 of Table B. The base course is like soil No. 10. Expansion tests for soil No. 1 give the following results:

Moisture (% of dry wt)	18.0	20.4	22.7
Resistance value, Hveem stabilometer	5	4	3
Expansion pressure (psi)	2	1	0.7

19–6. Solve Problem 19–2, part *a*, under the assumption that the native soil corresponds to soil No. 5.

19–7. Work Problem 19–2, parts *a* and *b*, assuming that the asphalt concrete is laid directly on native soil No. 2.

19–8. A new six-lane divided highway traverses an area for which the native soil corresponds to soil No. 2 of Table B. This soil will be the subgrade or basement soil in cuts and will be used for embankments. Granular base course comparable to soil No. 11 of Table B can be purchased from a nearby commercial plant. Pavement is to be asphalt concrete. Traffic totals 25,000 vehicles per day in each direction, and is 20% trucks. Of these, 40% are two-axle six tires, 10% are three-axle single unit and 20% three-axle combinations, and 15% are four-axle, and 15% are five-axle.

For the situation outlined in the preceding paragraph, develop typical roadway cross sections for each roadway like Fig. 14–1. Show the depth of each individual layer to the nearest ½ in. Thicknesses will be the same for all lanes; but, since most of the trucks will travel in the right-hand lane, dimension the design lane to accommodate 80% of all traffic, including passenger cars. Pavement life is 25 yr.

a. Base your design on the Hveem stabilometer method. Pavement thickness is 6 in. (0.5 ft).

b. Extend the design of part *b* (Hveem stabilometer method) to recognize expansive forces in the basement soil. Base and pavement weigh 130 lb/ft^3. Expansion test results for soil No. 2 appear in part *b* of Problem 19–2.

19–9. Work parts *a* and *b* of Problem 19–8, assuming that the 6-in. asphalt concrete pavement is underlain by 6 in. of class *A* cement-treated base.

19–10. Work Problem 19–8, changed as follows: the native soil, which is the basement soil in cuts and is used for embankments, is like soil No. 1 of Table B. Expansion test results for soil No. 1 appear in Problem 19–5.

19–11. Work Problem 19–8, assuming that the asphalt concrete is laid directly on native soil No. 2.

19–12. A new two-lane highway is to traverse an area where the native soil corresponds to No. 2 of Table B. This soil will be covered by a crushed rock base and a 4-in. asphalt concrete surface. Estimated traffic is the same as in Problem 19–2. Assume that, from loadometer studies, typical vehicles have shown the following weight characteristics:

	Axle Loads (kip)			
Vehicle Type	*Front Axle*	*Other Single Axles*	*Tandem Axles*	*Total Weight*
Passenger cars	2	1 @ 2		4
Trucks				
Two-axle—6 tires	6	1 @ 10		16
Three-axle—single unit	8		1 @ 22	30
Three-axle—tandem	8	2 @ 10		28
Four-axle—tandem	8	1 @ 12	1 @ 20	40
Five-axle—tandem	8		2 @ 26	60

Using the *CBR* value (Kentucky) to determine the soil support value and a regional factor of 0.8, determine how thick the crushed rock base must be, using the method developed from AASHO Test Road. In computing the value of *SN*, the weighted structural number, use straight-line interpolation in Table 19–1 when intermediate values are needed.

19–13. Work Problem 19–12, assuming that the crushed rock base is overlain by 6 in. of cement-treated base (400 psi of less) and 4 in. of asphalt concrete.

19–14. Work Problems 19–12 and 19–13 with the native soil conforming to No. 1 of Table B.

19–15. Determine the required pavement thickness for a 20-yr design life for the following subgrade soil conditions and pavement material characteristics using the AASHTO method (assume $P_t = 2.5$, $R = 1.8$). Determine alternate sections using:

a. Full depth asphalt concrete.
b. Asphalt over base rock.
c. Asphalt over base rock over subbase.

The native soil conforms to No. 1, Table B; the subbase is soil No. 6; and the base is soil No. 8. The design traffic is as follows:

Axle of Gross Load (kips)		*Total Expected Repetitions During 20-Yr Period*	
Single Axle	*Tandem Axle*	*Single Axle*	*Tandem Axle*
34	48	2,400	10,000
30	44	2,100	2,200
26	40	6,500	9,000
22	38	10,000	20,000
20	34	38,000	34,000
18	32	1,500,000	180,000
14	30	4,600,000	780,000
10	26	5,000,000	1,500,000
6	22	12,000,000	2,600,000
	18		1,500,000

19–16. For the traffic condition given in 19–15, determine the pavement thickness using the Asphalt Institute procedure if subgrade modulus is 3000 psi, *MAAT* = 60°F using:

a. Full depth asphalt concrete.
b. Asphalt concrete over 6 in. aggregate base.

19–17. Repeat Problem 19–16 for subgrade modulus of 50,000 psi and *MAAT* = 60°F.

19–18. Repeat Problem 19–16 for subgrade modulus of 10,000 psi and *MAAT* = 60°F. Discuss effect of subgrade modulus on design thickness.

19–19. Data on a particular asphalt concrete paving mixture are as follows:

Material	*Specific Gravity*	*Percent by Weight*
Asphalt cement	1.02	6.3
Limestone dust	2.82	9.0
Sand	2.65	40.0
Crushed gravel	2.65	44.7

For this mixture determine:

a. The theoretical specific gravity of a voidless mixture.

b. The percent voids, if the measured specific gravity is 2.37.

19–20. The proportions by weight and specific gravities of each of the constituents of a particular sheet asphalt paving mixture are as follows:

Material	*Specific Gravity*	*Percent by Weight*
Asphalt cement	1.04	10.0
Limestone dust	2.82	16.5
Sand	2.66	73.5

A cylindrical specimen of the mixture was molded in the laboratory and weighed in air and in water with the following results:

Weight of dry specimen in air, grams	111.95
Weight of saturated, surface-dry specimen in air, grams	112.09
Weight of saturated specimen in water, grams	61.20

a. Calculate the bulk specific gravity of the compacted specimen.

b. Compute the maximum theoretical specific gravity of the sheet asphalt paving mixture.

c. Determine the percentage of voids in the laboratory molded specimen.

d. When this mixture was placed and rolled on the street, a core of compacted pavement was removed and its specific gravity was found to be 2.13. The specifications require a minimum density in completed pavements equaling or exceeding 95% of that obtained in a standard laboratory specimen. Does the core meet this requirement? Show calculations.

e. Calculate the weight of a square yard of 1½ in. wearing surface composed of this sheet asphalt mixture.

19–21. A core of compacted asphalt concrete pavement was tested for specific gravity. The following weights were obtained:

Weight of the dry specimen in air, grams	2007.5
Weight of the specimen plus paraffin coating in air, grams	2036.5
Weight of the specimen plus paraffin coating in water, grams	1135.0
Bulk specific gravity of the paraffin	0.903

Calculate the bulk specific gravity of the core.

19–22. During a working day of 8 hr, a particular hot plant produces enough asphalt concrete to lay 15,000 yd^2 of wearing course 3 in. thick, compacted. Specific gravity of the compacted pavement is 2.37. Mix proportions, by weight, the specific gravities of each of the materials, and the unit weights of sand and stone are as follows:

	Proportions (% by wt)	*Specific Gravity*	*Weight per ft^3*
Asphalt cement (100% bitumen)	6	1.02	—
Limestone dust	8	2.75	—
Sand	41	2.66	106
Crushed stone	45	2.77	106

The capacity of the mixer, counting all ingredients, is 10 tons per batch. The sand and crushed stone are run through the drier, but the limestone dust is not. For this situation:

a. What is the weight, in pounds, of a square yard of this 3-in. asphalt concrete wearing course?

b. Calculate the required daily capacity of the drier in cubic yards of loose material. The dust does not go through the drier.

c. Calculate the required daily capacity of the melting tanks for asphalt cement, in gallons. (Note: There are 7.48 gal in a cubic foot.)

d. Determine the number of batches that must be run to produce the 15,000 yd^2 of surfacing.

e. What would be the percentage of voids in the mixture as compacted on the street?

19–23. A crushed gravel (specific gravity 2.65) having the same grading as soil No. 10 of Table B (inside the back cover) is employed as aggregate for a road-mix pavement. How many units (pounds or kilograms) of *SC*-250 oil are required per unit of aggregate, based on the early California formula?

19–24. Work Problem 19–23 using soil No. 9 of Table B as the aggregate. Specific gravity is 2.61.

19–25. Secure a sheet of four cycle semilog × 10 to the inch graph paper. Mark the sheet horizontally on the logarithmic scale with sieve openings and sieve numbers. Mark it vertically on the natural scale with percent passing. (See Fig. 19–16 as an example and for the required data.) On the ruled sheet plot, in different colors:

a. The grading limits of the Federal Highway Administration specification for asphalt concrete coarse surface course (see Table 19–6).

b. The grading limits of the Virginia black base (see Table 19–6).

c. The grading of soil No. 11, Table B (inside the back cover of this book). Determine which, if either, of the specifications soil No. 11 passes.

d. Determine which of the three grading specifications is most likely to lead to stability problems. Explain.

19–26. A bituminous road-mix wearing course to be laid over an existing gravel surface is 3 in. thick when compacted, and 24 ft wide. The aggregate is graded gravel and sand. The binder is *SC*-250 road oil; its weight is 4% of the weight of the aggregate. The compacted mixture weighs 140 lb/ft^3. The road oil has a specific gravity of 0.94 and the aggregate weighs 105 lb/ft^3, loose.

a. How many cubic yards of loose aggregate are required (1) per square yard of pavement and (2) per mile of road?

b. How many gallons of road oil are required (1) per square yard of pavement and (2) per mile of road? (There are 7.48 gal in a cubic foot.)

c. The loose aggregate costs \$6/yd^3, delivered; the road oil costs 65 cents/gal, delivered; and the charge for shaping the existing surface and mixing, laying, and compacting the oil mix is \$4000/mi. What, then, is the cost of a mile of complete surface?

19–27. A tack coat is to be applied on top of an old concrete pavement 24 ft wide that is to be resurfaced with a bituminous plant-mix pavement.

a. Select a cutback asphalt for the tack coat and determine the average quantity needed for a project that is 1.3 mi long.

b. What application temperature would be suitable for the selected bituminous material?

19–28. A thin bituminous-surface treatment is to be constructed on top of a gravel road 24 ft wide which has a tight surface. A prime coat is to be applied using *MC*-70 liquid asphaltic material. *RC*-800 liquid asphaltic material has been selected for the surface treatment to be covered with clean stone screenings of grading ranging between the ⅜ in. and No. 10 sieve sizes. Estimate the approximate quantities of materials needed for a mile of road.

20 PORTLAND-CEMENT-CONCRETE PAVEMENTS

Portland-cement-concrete or "concrete" pavement consists of a relatively rich mixture of portland cement, sand, and coarse aggregate laid as a single course. When properly designed and constructed, it has long life and relatively low maintenance cost. In 1978, some 120,000 mi of roads and streets in the United States had surfaces of portland-cement concrete.

There are records of a short section of concrete road base outside London in 1828 and of a concrete foundation course for sheet asphalt in Paris in 1858. Concrete was used as a base in New York in 1888. In Bellefontaine, Ohio, in 1892 a concrete street was constructed in two courses and marked in squares like a sidewalk. In 1909 concrete pavements began to emerge from the trial period, when Wayne County, Michigan, paved Woodward Avenue, which leads from Detroit. By 1912 there were sections in Milwaukee County, Wisconsin and the then State Highway Department of California had adopted it as a standard paving.

At the beginning of World War I (1914), concrete pavement was still in a transitional stage. The then California Highway Department was building 4-in. concrete "bases" 15 ft wide. In 1917 Pennsylvania standardized on the alternatives of 5-in. slabs or slabs 5 in. thick at the edge and 7 in. at the center on flat subgrades. Maricopa County, Arizona, in 1918, under an $8 million bond issue, built an extensive mileage of thickened-edge pavements, some of which are still rendering service. Precast slabs were tried at Casper, Wyoming, in 1920 and in California at Suisun in 1922.

Test tracks at Bates, Illinois, and Pittsburg, California in 1920–1922 tested thickened-edge and other concrete pavement designs. Since that time, portland-cement concrete has been accepted as one of the high type pavements. Thus, from the crude, relatively weak, rough-finished roads, there have now developed first-class pavements built under rigid inspection to carefully drawn specifications.

A wide variety of satisfactory designs for concrete pavements have been or are now in use. For heavily traveled roads, these range from 8 to 13 in. in thickness and can be plain jointed, simply reinforced, continuously reinforced, or prestressed. In earlier years many lightly traveled roads and streets were of concrete in relatively thin slabs. In the past 5 yr, there has been a resurgence of interest in them because of such factors as cost and energy conservation.[1]

It is beyond the scope of this book to explore fully all phases of concrete pavement design, construction, and maintenance. The following pages represent an overview of past and current practice and of research on the subject.

STRESSES IN CONCRETE PAVEMENTS

The behavior of concrete pavement slabs subject to the environment and loading depends on the properties of the concrete and of the underlying subgrades and base courses. Concrete will withstand relatively high compressive stresses but has little tensile strength. Because of its low tensile strength, flexural or beam strength of the slabs is also relatively low. Concrete, like other materials, expands or contracts as its temperature increases or decreases. Like wood, it expands on wetting and contracts on drying. It also shrinks soon after placing as the mortar hardens and the cement hydrates. When made with certain aggregates, its volume increases with age. Because of these properties, and because concrete pavements are exposed to the elements, they change length with time of day, with the seasons, and with variations in the weather. In addition, daily and seasonal temperature and moisture differences between tops and bottoms of slabs introduce a tendency to curl. Further complications arise because the base or subgrade supporting the slabs yields (and sometimes erodes) when loads are imposed and recovers, at least partially, when they are removed. For these reasons theoretical determination of the stresses in concrete pavement slabs is extremely involved and less than precise.

Because of the complexities just mentioned, almost all analysts make simplifying assumptions before attempting to compute the stresses in and determine slab depths for portland-cement-concrete pavements. Among the most important of these assumptions are:

1. Pavement slabs can be treated as beams of plain concrete. If reinforcing steel is employed, its contribution to flexural strength is ignored.
2. Transverse cracking of concrete paving slabs under combined flexural and direct tensile stresses is inevitable. However, with suitable crack control, structural integrity and beam action of large segments of the pavement is preserved. Three distinct jointing practices have been effective; these are:
   a. With unreinforced slabs, cracks are confined to weakened-plane joints spaced at 15–20 ft centers. Faulting (vertical offsetting) across the narrow cracks is prevented by aggregate interlock or with dowel bars.

[1]*NCHRP Synthesis 27* offers an excellent summary of practices in the design, construction, and maintenance of concrete pavements for low-volume situations. See also *FHWA Report TS-78-202* "Portland Cement Concrete Pavements Performance Related to Design-Construction-Maintenance" for an excellent summary of current design, construction, and maintenance practices in the United States. See also article by G. Van Heystraeten, *Public Roads,* Dec. 1976, entitled "Cement Concrete in Belgium" for an example of European practice and article by G. K. Ray, *Transportation Engineering Journal, ASCE,* Sept. 1977, for a historical look at concrete construction in transportation.

b. With simply reinforced slabs, major cracks occur only at weakened-plane joints spaced at 40–70 ft intervals. Hair cracks, held tightly together by the steel, develop between joints, but beam strength of large sections of the slabs is maintained. Although relatively wide cracks develop at the joints, faulting is prevented by dowel bars.
c. With continuous reinforcing, transverse joints are omitted. Hair cracks, held tightly together by the steel, develop at close intervals. Faulting is prevented by aggregate interlock and the reinforcing steel.

3. Longitudinal cracking of slabs more than one lane wide also is inevitable. Undesirable consequences are avoided by jointing arrangements which cause these cracks to occur between traffic lanes. Differential slab settlement is prevented by the use of tie bars.
4. Pavement slabs are supported on foundations which deflect under load and recover when the load is removed. A variety of foundation materials ranging from poor to excellent can be employed. For analytical purposes, their behavior may be assumed to be elastic, or like a dense liquid; their depths are either finite or infinite.

The sections that follow support the statements just made by demonstrating how and to what degree flexural, moisture, or temperature stresses develop in concrete pavement slabs.

Flexural Stresses in Concrete Pavements

Over the years many theoretical studies and field observations of stresses in concrete pavements have been made. Even so the work of Westergaard, beginning in 1925, is still accepted as fundamental.[2] He presented formulas giving the flexural stresses in slabs of uniform thickness resulting from loads and the effects of slab curling under temperature differences. Modifications of his formulas based on experimental findings have been suggested by various investigators, but the basic considerations remain unchanged.[3]

To develop his equations for load stresses, Westergaard placed loads at three critical locations as follows (see Fig. 20–1):

Case A. Load applied close to the rectangular corner of a large slab. Such a condition exists at the intersection of the pavement edge with a transverse joint if no provision is made to transfer a portion of the load across the joint to the adjoining slab. With present-day wide lanes, this condition is no longer critical and is not discussed further.

Case B. Load applied to the interior of a large slab at a considerable distance from its edges.

Case C. Load applied at the edge of the slab at a considerable distance from any corner.

Another set of formulas gives theoretical stresses created at the same three positions by temperature differences between tops and bottoms of slabs. Since,

[2]See *HRB Proceedings 1925,* Part 1; and *Public Roads,* Apr. 1926.

[3]See A. S. Vesic and S. K. Saxena, *NCHRP Report 97,* for an excellent summary of earlier studies and extensive bibliography. This report also presents the procedures for applying the finite element method of analysis to slabs resting on soils. See also *Principles of Pavement Design,* by E. J. Yoder and M. W. Witczak, Wiley, New York, 1975.

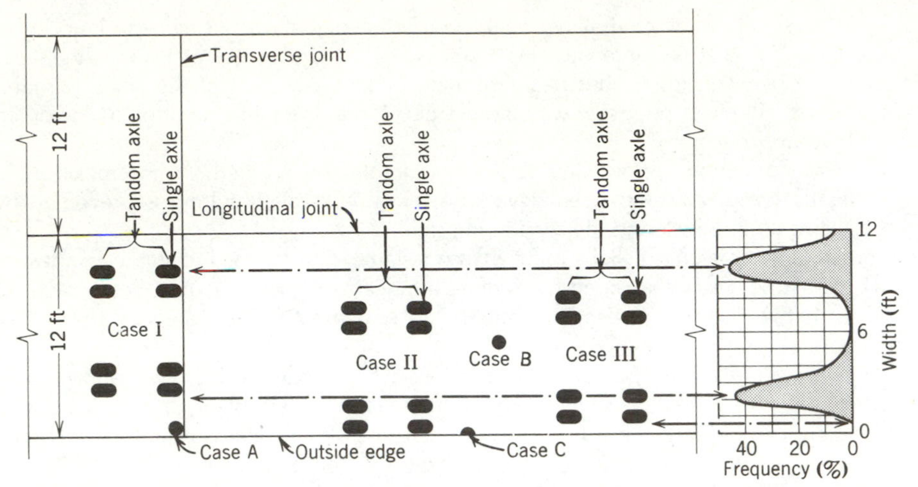

Fig. 20–1. Critical locations for loads on concrete pavement slabs. Cases A, B, and C are for Westergaard's analysis; cases I, II, and III are for modern wide lanes.

during the day, temperature at the surface is higher than in the elements below it, the slab edges curl downward in relation to the central section. This tendency, which is resisted by the weight of the slab, creates bending stresses: tension at the bottom and compression at the top. At night, the temperature gradient is reversed, so that the direction of temperature curling and the sense of temperature curling stresses are reversed.

Differences in moisture content between tops and bottoms of slabs also cause curling and flexural stresses but these are usually omitted from theoretical analysis. Kelly[4] has stated that

> During hot summer days, when moisture and temperature differentials are both a maximum, the curvature caused by one is in the opposite direction to that caused by the other and such stress as may be produced by moisture serves to reduce rather than to increase the stress due to temperature warping. . . . To ignore moisture warping appears to add some factor of safety of unknown magnitude and importance.

These conclusions are supported by the findings of others.

Theoretical flexural stresses produced by loads and by restrained temperature curling, computed for slabs of uniform thickness by Formulas 20-1 through 20-5, given later, are shown in Fig. 20–2. Slab width is 10 ft, or about that of a single early-day traffic lane. Variables are thickness and length of pavement slab. Average values for load and impact, concrete properties, and temperature gradient have been assumed and are stated on the figure. The supporting subgrade would be rated as relatively weak.

The combined stresses shown in Fig. 20–2 are approximately equal to the ultimate static flexural strength of paving concrete, which ordinarily is in the range of 550–750 lb/in.^{2} for static loading conditions.[5] Under repeated or fa-

[4]See *Public Roads,* July and Aug. 1939.

[5]Static flexural strengths cited here are based on the modulus of rupture for third-point loading.

tigue loadings, however, failure can occur at much lower stresses. For an infinite number of load repetitions, this fatigue strength is about 55% of the static strength. But some finite number of loadings which produce tensile stresses greater than the 55% of static strength will cause failure.

Although Fig. 20–2 is based on studies made over 40 yr ago, it still serves to illustrate certain fundamental principles that are applicable today. Among the most important are:

1. Cracking caused by flexural stresses alone is almost inevitable in slabs of any great length. Figures 20–2*b* and *20–2c* show conclusively that combined stresses at slab edges and interiors, for lengths of 30 ft, far exceed the fatigue strength and approach the static strength of plain concrete. This applies for all reasonable slab thicknesses, since reduced load stresses gained by increasing slab thickness are largely offset by increased temperature stresses.
2. For constant loads on slabs of reasonable lengths (say 15 ft), load stresses and total stresses decrease as slab thickness increases. Slabs of this length, if made thick enough, will retain their integrity as beams of plain concrete.

Formulas for Flexural Stresses in Concrete Pavement Slabs of Uniform Thickness

The formulas for flexural stresses in concrete pavements on which Fig. 20–2 is based are those of Westergaard, as modified by full-scale tests by the Bureau of Public Roads. These formulas are as follows:

Edge loading when the edges of the slab are warped upward at night (empirical equation)

$$\sigma_e = \frac{0.572P}{h^2}\left[4\log_{10}\left(\frac{\ell}{b}\right) + \log_{10}b\right] \tag{20–1}$$

Edge loading when the slab is unwarped or when the edge of the slab is curled downward in daytime (Westergaard equation, based on $\mu = 0.15$)

$$\sigma_e = \frac{0.572P}{h^2}\left[4\log_{10}\left(\frac{\ell}{b}\right) + 0.359\right] \tag{20–2}$$

Interior loading (Westergaard equation, based on $\mu = 0.15$)

$$\sigma_i = \frac{0.316P}{h^2}\left[4\log_{10}\left(\frac{\ell}{b}\right) + 1.069\right] \tag{20–3}$$

Curling stress along the edge of a slab

$$\sigma_{xe} = \frac{C_x Eet}{2} \tag{20–4}$$

Curling stresses in the interior of a slab

$$\sigma_x = \frac{Eet}{2}\left(\frac{C_x + \mu C_y}{1 - \mu^2}\right) \tag{20–5}$$

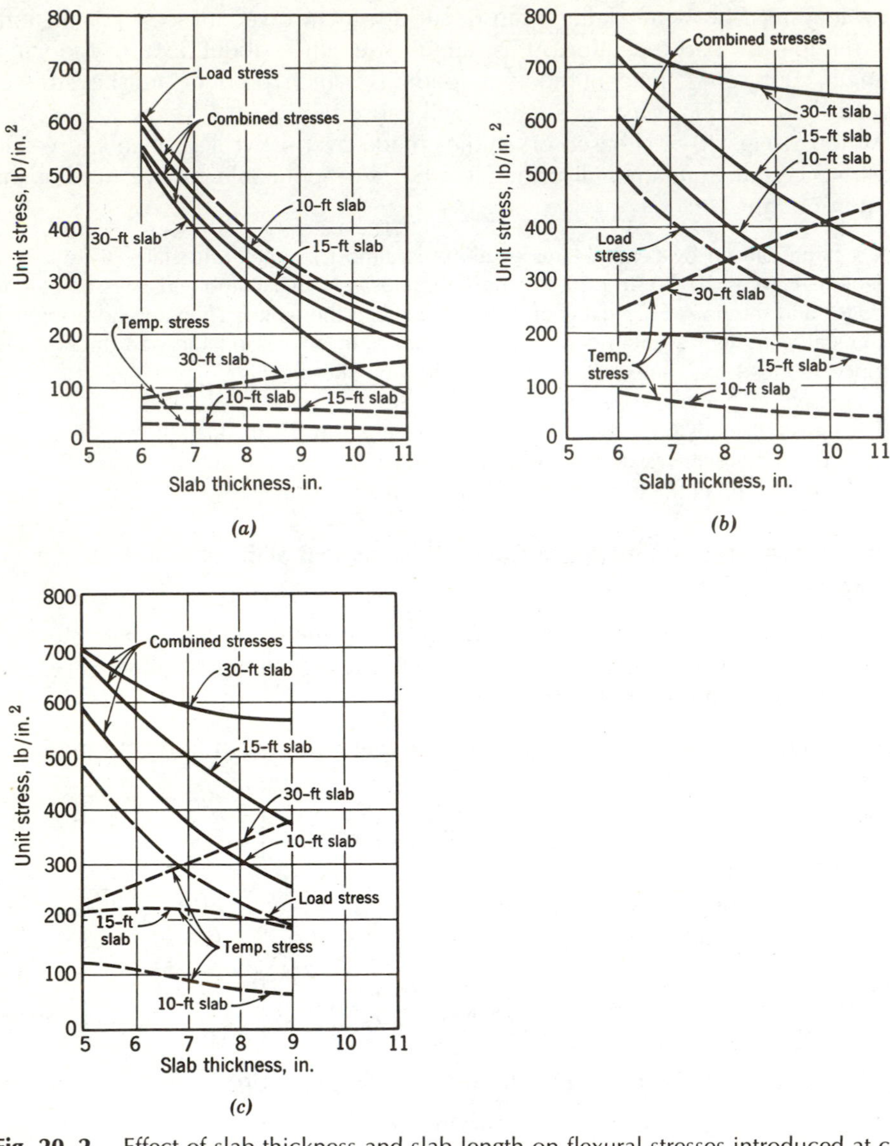

Fig. 20–2. Effect of slab thickness and slab length on flexural stresses introduced at certain critical points by loads and temperature curling. Slab width = 10 ft. The beams for which stresses are computed run the length of the slab. (*a*) Edge loading, edges curled upward (night). (*b*) Edge loading, edges curled downward (day). (*c*) Interior loading, edges curled downward (day). *Assumed conditions:* Static wheel load in dual, high-pressure tires = 8000 lb; including impact load = 11,800 lb. Radius of area of load contact $(a) = 7.8$ in. Modulus of elasticity of concrete $(E) = 5{,}000{,}000$ lb/in.² Poisson's ratio for concrete $(\mu) = 0.15$. Coefficient of expansion for concrete $(e) = 0.000005$/°F. Temperature difference in degrees Farenheit between top and bottom of slab—day, −3°F/in. thickness—night, +1°F/in. thickness. Modulus of subgrade reaction $(k) = 100$ lb/in³. (Source: *Public Roads,* Jul. 1939.)

Symbols in these formulas are defined as follows:

σ_e = maximum load induced tensile stress in the bottom of the slab directly under the load P at the edge, and in a direction parallel to the edge, stated in pounds per square inch

σ_i = maximum load-produced tensile stress in the interior of the slab stated in pounds per square inch; this occurs in the bottom of the slab directly under the load

σ_{xe} = maximum temperature-curling stress at the edge of the slab in the direction of slab length, stated in pounds per square inch

σ_x = maximum temperature-curling stress in the interior of the slab in the direction of slab length

P = load in pounds, including an allowance for impact

h = thickness of slab in inches

ℓ = radius of relative stiffness, which measures the stiffness of the slab in relation to that of the subgrade; it is expressed by the equation

$$\ell = \sqrt[4]{\frac{Eh^3}{12\,(1-\mu^3)\,k}}$$

E = modulus of elasticity of concrete in pounds per square inch

μ = Poisson's ratio for concrete

k = the subgrade modulus, or resistance of subgrade to deformation, expressed in pounds per cubic inch

b = radius of equivalent distribution of pressure, expressed in inches; this factor recognizes the influence of slab thickness on tensile stress directly under an applied load; it is expressed by the equations

$$b = \sqrt{1.6a^2 + h^2} - 0.675h \qquad \text{(when } a < 1.724h\text{)} \qquad (20\text{–}6a)$$

$$b = a \qquad \text{(when } a > 1.724h\text{)} \qquad (20\text{–}6b)$$

a = radius of area of load contact, in inches; this area is assumed as circular in the case of corner and interior loads and semicircular for edge loads

C_x and C_y = coefficients that relate slab length and the relative stiffness of slab and subgrade to temperature-curling stresses; values of C_x and C_y are given in Fig. 20–3.

e = thermal coefficient of expansion and contraction of concrete per degree Fahrenheit

t = difference in temperature between top and bottom of slab in degrees Fahrenheit

It is impossible to assign single values to such terms as load, concrete and subgrade properties, and temperature variation since they change with place and situation. The following paragraphs briefly describe the design variables and their complexities.

LOAD AND IMPACT, *P*. Permissible static wheel loads differ widely. Many states currently limit single-axle loads to 18,000 lb, which in turn limits wheel loads to 9000 lb. Some states permit greater loads. For example, several eastern

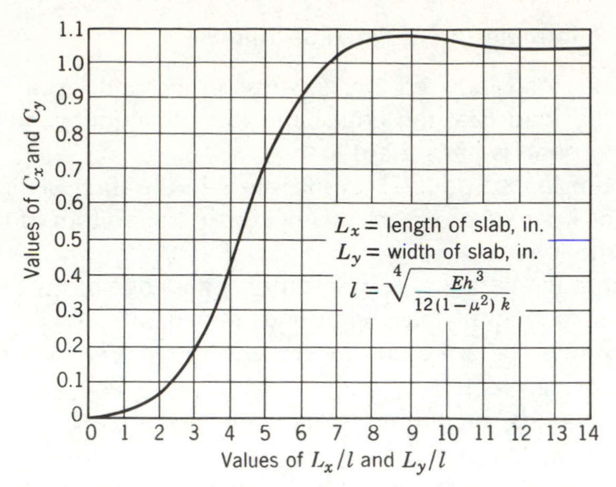

Fig. 20–3. Values of C_x and C_y for use in formulas for flexural stresses in concrete pavements.

and midwestern states permit axle loads of 22,400 lb, or wheel loads of 11,200 lb.[6] Interested parties are continuously applying pressure on Congress and state legislatures to authorize higher load limits. In addition to differences in static loads, the influence of impact varies with speed, condition of pavement surface, weight and springing of vehicle, and tire pressure and stiffness. In one group of tests, dynamic reactions of vehicles to surface irregularities of the pavement produced loads more than double the static weight.[7]

MODULUS OF ELASTICITY OF CONCRETE, E. A summary of many tests indicates that the modulus of elasticity of concrete of the character generally used in pavements is roughly 1000 times its compressive strength and ranges from 2 to 6 million lb/in.^{2}. The modulus varies not only with strength, but also with age, moisture state, stress conditions, and other factors. A higher modulus results in higher curling stress, since it increases directly with modulus of elasticity.

POISSON'S RATIO FOR CONCRETE, μ. For concrete, no definite relationship exists between Poisson's ratio and strength. Based on the results of many tests, it has been determined that the range to be expected lies between 0.10 and 0.20. The average figure of 0.15 is usually adopted. The maximum error introduced by variation between this average and the stated limits is 4.3% for interior stresses and 2.5% for edge stresses.

MODULUS OF SUBGRADE REACTION, k. This modulus is a measure of the stiffness of the subgrade and is stated in terms of load in pounds per square inch per inch of deflection measured under a rigid plate 30 in. in diameter. The use

[6]For a discussion of state practices see *TRB Record 656*. See also *NCHRP Report 198* for an excellent summary of state laws and regulations on truck size and weight.

[7]For an in-depth study of the effects of dynamic loads on stresses in pavements on the AASHO Test Road, see *HRB Special Report 61F*. A more recent detailed study is reported in *NCHRP Report 105*.

of a single value of *k* in analysis assumes that the subgrade or subbase is elastic—that is, that the support it supplies is directly proportional to the deflection. As shown in Fig. 20–4, values of *k* range from about 50 lb/in.[3] for very poor subgrades, such as the worst *A-7-5* and *A-7-6* soils, up to about 225 for the best materials in these classes. They may be 700 or more for extremely good soils. This figure also shows approximate relationships between *k*, bearing values measured under a 30 in. diameter plate, AASHTO soil classification, California bearing ratio, and Hveem stabilometer *R* values.

For the plate-bearing test to determine *k*, load is applied to the surface of the prepared subgrade or native soil through a rigid steel plate. Diameters ranging from 6 to 30 in. are employed; however, plates should have an area not less than that of the tire print. One test procedure *(AASHTO Designation T221)* involves applying and completely removing successively increasing increments of load. Another *(AASHTO Designation T222)* calls for application of successive increments of load, with pauses after each increment until penetration ceases. Load deformation data are taken; from them the modulus of subgrade reaction is computed.

To conduct the bearing test with a 30-in. plate, forces as large as 30 tons, acting downward, are required. Commonly the upward thrust of the test device is transferred to the frames of two heavily loaded trucks placed some distance on either side of the bearing plate. Special care also must be taken to ensure that the plate bears uniformly over the loaded area. It follows, then, that such tests are expensive and time-consuming. For this reason, other means for predicting the modulus of subgrade reaction, such as those shown in Fig. 20–4, are widely employed.[8]

If untreated or treated subbases are placed between the subgrade and pavement slab, values of *k* are increased substantially. Examples of these effects are shown in Table 20–1. For cement-treated subbases with a subsoil having a *k*

TABLE 20–1. Effects of Thickness of Untreated or Treated Subbases on the Combined Modulus of Subgrade Reaction *k*.*

*Untreated Subbases**					*Treated Subbases*				
Subgrade k Value	*Subbase Thickness*				*Subgrade k Value*	*Subbase Thickness*			
	4 in.	*6 in.*	*9 in.*	*12 in.*		*4 in.*	*5 in.*	*6 in.*	*7 in.*
50	65	75	85	110	100*	300	450	550	600
100	130	140	160	190	100†	220	260	330	350
200	220	230	270	320	300†	480	530	600	700
300	320	330	370	430	600†	800	900	1050	1150

*Data from Portland Cement Association.
†Approximate values from Fig. D.4–1 of *AASHTO Interim Guide for Design of Pavement Structures*.

[8]Results of a recent survey indicated that only one state used in-situ plate-bearing tests. Fifteen used correlations with the *CBR*, seven with *R* values, and four with soil classifications. For further detail and references see *Thickness Design for Concrete Pavements*, Portland Cement Association, pp. 2–4. A. C. Estep and P. I. Wagner, *HRB Record 239*, report on the relationship between *k* and *R* values and give data on the increases in *k* with subbase thickness increases.

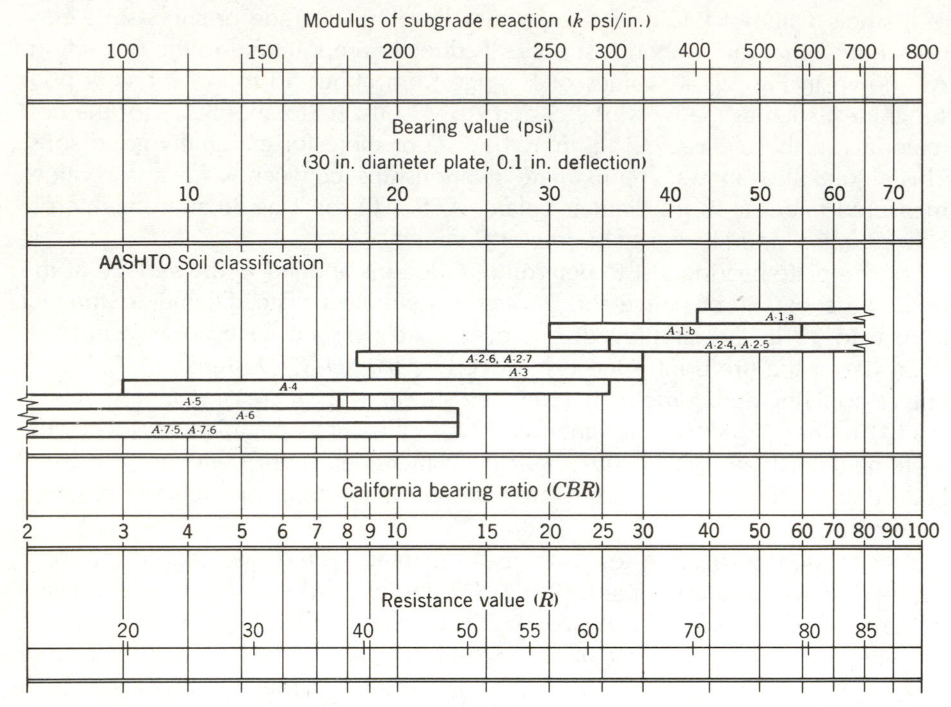

Fig. 20–4. Approximate relationships between modulus of subgrade reaction k and plate bearing value, AASHTO soil classification, *CBR*, and Hveem stabilometer R value. (Data from Portland Cement Association.)

value of 100, the Portland Cement Association recommends that k range from 300 lb/in.^{3} for a subbase 4 in. thick to 600 lb/in.^{3} for a 7-in. thickness.

THERMAL COEFFICIENT OF EXPANSION, e. For concretes like those for pavements, values for the thermal coefficient of expansion range from 0.000004 to 0.000007 per degree Fahrenheit. The highest coefficients have been found with siliceous aggregates and the lower values with those of granite, limestone, or diabase. It is important to note that curling stresses vary directly with this coefficient. Thus they will be higher in slabs made of concretes with high coefficients of expansion.

TEMPERATURE GRADIENT FROM TOP TO BOTTOM OF SLAB. Observations of temperature differences between tops and bottoms of slabs made at Arlington showed variation nearly proportional to depth of slab. Temperature differences in daytime, particularly in summer after the sun had heated the upper surface, ranged up to 4°F per inch of depth. At night, after cooling, the temperature of the surface was below that of the bottom by as much as 1.5°F per inch of depth. As shown in Fig. 20–2, commonly used values are 3°F per inch of depth in the daytime and 1°F per inch of depth at night.

Direct Tensile and Compressive Stresses in Concrete Pavement Slabs

Changes in temperature and moisture content not only create slab curling and flexural stresses, but also cause overall lengthening and shortening of slabs. If slabs were perfectly free to move, these volume changes would take place without creating stresses. However, the subgrades on which pavements rest offer considerable resistance to horizontal movement. Thus, the tendency for slabs to shorten because of temperature drop or drying creates tensile stresses, whereas the tendency to lengthen from temperature rise or increased moisture creates compressive stresses.

Tensile stresses in unreinforced slabs vary more or less directly with the daily temperature drop, the distance between joints, and the coefficient of subgrade resistance, which is the average coefficient of friction between the slab and its supporting subgrade, Kelley reports[9] that stresses slightly larger than 100 lb/in.^{2} may be expected for the average temperature drop of 40°F, a slab length of 100 ft, and an average coefficient of subgrade resistance of slightly over 2. Stresses are proportionally lower for shorter slabs. He concludes that, for unreinforced pavements provided with transverse joints at reasonable intervals, crack formation (unless at early ages) generally cannot be attributed to direct tensile stresses. Rather it must be assigned to the warping stresses mentioned earlier. Tensile stresses resulting from decreased moisture content are ignored in design as they usually are opposite in sense and partly compensatory for those caused by temperature change.

The length of an unconfined concrete slab will increase if its temperature rises or its moisture content increases. If this expansion is in part or wholly restrained, compressive stresses are created. Such stresses, considered alone, are much below the compressive strength of the concrete. In combination with other factors, however, they may lead to "blowups," in which short sections of slab abruptly buckle upward.

With continuously reinforced pavements, in particular, considerable movement occurs near the ends; if they are restrained, substantial forces develop. Structures that abut these pavements may be damaged unless provisions are made for their protection. Two approaches have been used. One is to accom-

[9]Kelley *(Public Roads,* July and Aug. 1939) states that tensile stresses are dependent not on seasonal temperature change, but on the temperature drop that occurs during one single period of continuously falling temperature or, at most, during a relatively few cycles of temperature change in which the minimum temperatures are decreasing. Based on a 3-yr observation period, the maximum 24-hr difference in air temperatures recorded in 22 cities in the United States was 60°F, at Denver, Colorado. Temperature variation was smallest in Miami, Florida, where the maximum change was 27°F. Changes in slab temperatures would be less than these values. The coefficient of subgrade resistance varies with subgrade type, amount of slab displacement, and slab thickness. Typical values are 0.34 for loam subgrade and a movement of 0.001 in.; 2.18 for a 3-in. crushed stone subgrade and a displacement of 0.05 in., and 3.5 for a slab only 2 in. thick supported on silt-loam soil and displaced 0.10 in. For reports on the coefficient of subgrade resistance and means of reducing it developed in conjunction with studies of prestressed concrete pavements, see papers by P. Melville and T. Cholnoky, *HRB Bulletin 179.*

modate movement by providing expansion joints near the structures. The other is to restrain the ends with a terminal treatment such as "anchor lugs."[10]

Effects of Edge Support on Design of Concrete Pavements

When load is applied near the edge or corner of a concrete slab, the slab deflects downward. If the slab is isolated or free from connection with its neighbors, the load is carried by the beam action of the slab plus the support of the underlying subgrade or base course. However, when adjacent slabs are interconnected so that the edges of both deflect together, the load is distributed between them. Some agencies now use concrete shoulders which are tied to the pavement to reduce distress occurring at the longitudinal shoulder joint.[11] There are several methods for providing for this load transfer (see the discussion of joints). Some design methods recognize the stress reduction from load transfer; others do not.

Effects of Thickened Edge on Design of Concrete Pavement

When a load is applied near an unsupported pavement edge, the flexural stress in a slab of uniform thickness may be higher than that resulting when the load is applied at other points (see Fig. 20–2). If the depth is set to keep the stress near the unsupported edge at a safe level, the remainder of the slab may be overdesigned. To make flexural stresses equal over the entire slab, "thickened-edge" pavements were developed in the 1920s to provide extra slab thickness close to all unsupported edges.

Today, thickened-edge designs are primarily of historical interest. Lanes on present-day highways are wide, usually 12 ft, and most of the vehicles are so positioned (see Fig. 20–1) that the gain from a thickened-edge design largely disappears. Furthermore, preparing the subgrade for a thickened-edge pavement is more expensive, since it calls for more complicated machinery or hand labor, and good compaction is difficult to achieve on the sloped section of subgrade.

DISTRESS TYPES

Because of the stresses created by load, moisture, and temperature, concrete pavements deteriorate. Numerous types of distress can be experienced, but can generally be grouped into three categories: distortion, cracking, or disintegration.[12]

Distortion, which consists primarily of faulting (vertical displacement of concrete slabs at joints or cracks) is often considered a main weakness of jointed concrete pavements. Two factors have been reported as causes of faulting; one is loss of slab support and the other is erosion of the subbase. For faulting to occur, there must be free water on top of the base and pavement deflection

[10]For practices in 1977, see *TRB Special Report* 173.

[11]For an excellent discussion of the shoulder joint problem refer to *NCHRP Report 202*. For design of concrete shoulders see *TRB Records 594*, *666* and *725* and *NCHRP Synthesis 63*.

[12]For details on the types of distress refer to *FHWA Report RD-79-66* "Highway Distress Identification Manual" and *TRB Record 602* and *715*. See also *NCHRP Syntheses 56* and *60*.

across the joint from heavy axle loads. This results in "pumping" fines out of the joints or moving the fines to the "leave slab" (the slab before the joint). It occurs even with cement treated bases; this emphasizes the tremendous forces that develop under the pavement.

Cracking can take many forms in concrete pavements and can result either from the applied loads or from temperature or moisture changes. The most common types are: corner cracks, associated with excessive corner deflection; transverse cracks, associated generally with moisture or temperature stresses; and longitudinal cracks, associated with either temperature, load stresses, or poor construction practices.

Disintegration, usually in the form of *D*- (durability) cracking, scaling, or spalling, is generally the result of mix design or construction related problems. *D*-cracking usually results from freeze–thaw action. This can generally be prevented by air entrainment (see below) and proper testing of the aggregates (see Chapter 14). Scaling refers to a network of shallow, fine hairline cracks which extend through the upper surface of the concrete. It may be caused by deicing salts, improper construction, freeze–thaw cycles, or steel reinforcement too close to the surface. Spalling of cracks or joints is the breaking or chipping of the joint edges. It usually results from excessive stresses at the joint, weak concrete, or poorly designed or constructed joints.

Details of the many other distress types are beyond the scope of this book. For additional discussion the reader is referred to FHWA report RD-79-66. Methods of coping with these problems are discussed in subsequent sections of this chapter.

SETTING SLAB THICKNESSES FOR CONCRETE PAVEMENTS

The preceding paragraphs listed the many variables that should be considered in designing concrete pavement slabs. They also indicated that many of these variables are not subject to precise evaluation so that design calculations give only approximate answers. Partially, at least, this uncertainty explains why agencies have adopted one, two, or possibly three standard designs. However, some agencies analyze each project before making this decision. Brief summaries of two design methods are presented in the sections which follow.[13]

Portland Cement Association Design Method[14]

The fundamental assumptions underlying the Portland Cement Association design method are:

[13]In the past, the expense of providing side forms in a variety of heights has deterred engineers from changing pavement thicknesses. This is not as serious a problem with slip form pavers, which can be set to produce pavements of any reasonable thickness.

[14]For greater detail than is presented here and for references to underlying source material, see *Thickness Design for Concrete Pavements* published by the Portland Cement Association or P. Fordyce and W. A. Yrjanson, *Transportation Engineering Journal*, ASCE, Aug. 1969. For a flow diagram listing all important considerations, see R. K. Kher et al., *HRB Record 362*. The Portland Cement Association has also published manuals detailing design procedures for city streets, industrial driveways, and airports. A computer program based on the PCA method is detailed by C. E. Warnes in *Transportation Engineering Journal, of ASCE,* Feb. 1972.

1. Slabs will be of uniform thickness. Because few load applications occur at longitudinal edges, their effect is not critical. Thickened edges are both costly and impractical with present-day equipment and are not considered.
2. Critical stresses occur when tires are positioned at the edge of the transverse joint and directly under the point where load repetitions are most frequent (case I in Fig. 20–1). Stresses produced by other load positions (cases II and III) are smaller.
3. Maximum tensile stress occurs in the bottom of the slab directly under the load; the moments producing it act in a vertical plane parallel to the joint edge. Figures 20–5 and 20–6 show, for single and tandem axles, the relationships between load, *k*, slab thickness, and tensile stress.
4. Although provision for effective load transfer across transverse joints is essential to prevent faulting, no credit is taken for the resulting reductions in stress.
5. The design aims to prevent fatigue failure from flexure under repeated loads. Thus, the effects of wheel loads which produce stresses less than 50% of the static modulus of rupture can be ignored. However, the fatigue effects of loads producing stresses greater than this 50% value are cumulative.

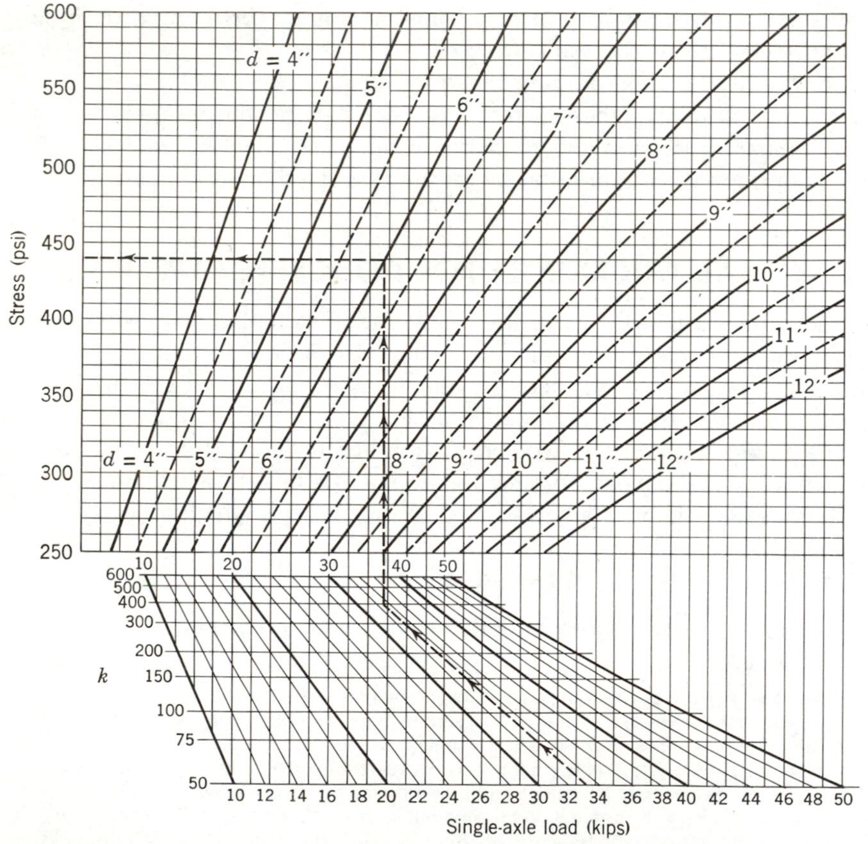

Fig. 20–5. Design chart for single-axle loads for case I load placement. (Courtesy, Portland Cement Association.)

Figure 20–7 is a typical design calculation for the Portland Cement Association method using the relationships shown in Figs. 20–5 and 20–6. For each situation, several parallel computations are made until the most desirable combination of concrete strength, slab depth, and subbase and subgrade properties are determined. The analysis begins with assumptions regarding highway type, the predicted characteristics of subgrade, subbase, and concrete, and trial depths for subbase and pavements. Also, projections are made of the total number of heavy single- and tandem-axle loads in several weight groups that will occur during the design life of the pavement (columns 1 and 6 of Figure 20–7). Normally these would be developed by the planning division of the transportation agency from loadometer studies, as described in Chapter 3. On multilane roads, most of the heavy vehicles travel in the outside lane, but some do not. It is usual, then, to reduce the number of axle loads by 5–10% where traffic volume is low and by as much as 25% when it is heavy. It should be noted also

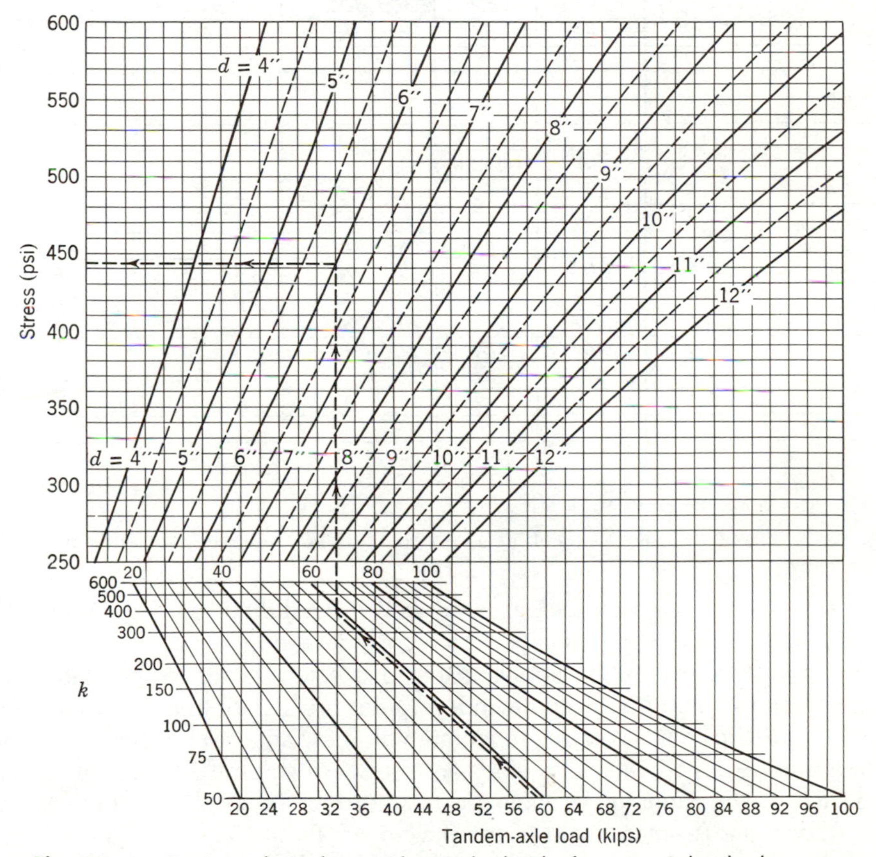

Fig. 20–6. Design chart for tandem-axle loads for case I load placement. (Courtesy, Portland Cement Association.)

CALCULATION OF CONCRETE PAVEMENT THICKNESS
(Use with Case I Single & Tandem Axle Design Charts)

Project DESIGN FOUR-A
Type Rural Primary-Rolling Terrain No. of Lanes 2
Subgrade k 125 pci., Subbase 5-in Granular Cement-Treated
Combined k 400 pci., Load Safety Factor 1.1 (L.S.F.)

1	2	3	4	5	6	7
Axles Loads kips	Axle Loads X/./L.S.F. kips	Stress psi	Stress ratios	Allowable Repetitions No.	Expected Repetitions No.	Fatigue Resistance Used percent

Trial depth 6.0 in. M.R.* 700 psi k 400 pci

SINGLE AXLES

1	2	3	4	5	6	7
30	33.0	440	.63	14,000	680	5
28	30.8	417	.60	32,000	680	2
26	28.6	390	.56	100,000	1350	1
24	26.4	366	.52	300,000	43,160	14
22	24.2	342	<.50	Unlimited	168,530	0
20	22.0		"	"	509,330	0

TANDEM AXLES

1	2	3	4	5	6	7
54	59.4	443	.63	14,000	680	5
52	57.2	431	.62	18,000	900	5
50	55.0	418	.60	32,000	7690	24
48	52.8	403	.58	57,000	7690	14
46	50.6	392	.56	100,000	12,650	13
44	48.4	376	.54	180,000	38,180	21
42	46.2	363	.52	300,000	44,890	15
40	44.0	344	<.50	Unlimited	66,180	0
38	41.8		"		70,170	0

Total = 119

*M.R. Modulus of Rupture for 3rd pt. loading

Fig. 20–7. Typical design calculation based on Portland Cement Association method.

that all the axle loads listed in Fig. 20–7 well exceed the usual legal limits, indicating that overloads are a major design consideration.

The calculation begins by applying the load safety factor to the static axle loads shown in column 1. Recommended values are 1.2 for freeways and major highways, 1.1 for roads carrying moderate volumes, and 1.0 for those with low truck traffic. Adjusted values for loads are shown in column 2. Given these adjusted loads, the modulus of subgrade reaction, *k* (Table 20–1 shows typical values), and an assumed depth of the pavement, Figs. 20–5 and 20–6 are entered to determine the tensile stresses. The dashed lines on the figures demonstrate solutions for a 33-kip single-axle and for 59.4-kip tandem axles, which give stresses of 440 and 443 psi, respectively.

In column 4 is recorded the ratio of the calculated stresses and the assumed modulus of rupture. For example, in the first line, 440 divided by 700 equals 0.63. In column 5 have been entered the total allowable load repetitions before fatigue failure will occur. These are taken from Table 20–2, which is based on the results of extensive tests. For example, as shown in the top line of Fig. 20–7, it is predicted that the slab in question will fail in flexure under 14,000 stress repetitions of 440 psi each.

Column 7 states the percentage of total fatigue resistance consumed by the estimated number of axle load repetitions. For example, the 5% consumed by the 30-kip single-axle load equals the ratio, expressed as a percentage, between 680 and 14,000. Finally, Fig. 20–7 indicates that the total fatigue resistance consumed by all single and tandem axles is 119% which, on first glance, says that the design is inadequate, and that another analysis is called for. However, the Portland Cement Association procedure proposes that, in some situations, designs may be satisfactory until the fatigue resistance reaches 125%. An example would be where the modulus of rupture for design is based on 28-day strength, but the pavement is not to be opened to traffic for 90 days.

TABLE 20–2. Relationship Between Stress Ratios* and Allowable Number of Load Repetitions (Portland Cement Association Data)

Stress Ratio	*Repetitions*	*Stress Ratio*	*Repetitions*	*Stress Ratio*	*Repetitions*
0.50	Infinite	0.62	18,000	0.76	360
0.51	400,000	0.64	11,000	0.78	210
0.52	300,000	0.66	6,000	0.80	120
0.54	180,000	0.68	3,500	0.82	70
0.56	100,000	0.70	2,000	0.84	40
0.58	57,000	0.72	1,100	0.85	30
0.60	32,000	0.74	650		

*Stress ratio is that between actual stress and the modulus of rupture for third-point loading.

AASHTO Interim Design Method

The AASHTO Committee on Design has recommended a method[15] for setting slab thickness of concrete pavements based on AASHO Test Road findings. A number of agencies have adopted similar procedures, although they may alter certain parameters or the values assigned to certain design variables. The paragraphs that follow are intended primarily to summarize the approach; details will be found in the references.

The design begins by estimating the number of equivalent 18-kip single-axle loads which the design lane will carry in its projected life. These combine traffic

[15]Details of this method are given in the *AASHTO interim Guide for the Design of Pavement Structures, 1972,* with revisions in 1980. Application of the AASHTO design method for evaluating surface condition and remaining service life are offered by I. E. Corvi and B. G. Bullard in *Public Roads,* Dec. 1970. *NCHRP Report 128,* which was issued concurrently with the *AASHTO Interim Guide,* gives much basic data and suggests future changes.

estimates with loadometer studies. Table 20–3 gives suggested multipliers to convert single- and multiple-axle loads to 18-kip equivalents. Some agencies have developed their own equivalency factors for direct application to traffic estimates. For example, the Illinois Department of Transportation for main highways has developed 18,000-lb equivalents as follows: passenger cars, 0.0004; single-unit trucks, 0.123; multiple-unit trucks, 1.155. On four-lane highways it assigns 45% of the total estimated single or multiple truck units in both directions to the design lane; on six-lane facilities the assignments are 40% for rural and 37% for urban situations.

Given traffic estimates; the desired present serviceability index at the end of the pavement design life (see Chapter 19); expected values for the working stress (flexural strength/factor of safety, with factor of safety normally 1.33 but as high as 2.0); modulus of elasticity of the concrete; and the modulus of subgrade reaction; substitutions are made in a series of equations to determine the design pavement thickness. These are based on the "Spangler" equation for corner loading, which assumes some edge support from adjacent slabs. Figure 20–8 is a nomograph solving the equation for one set of conditions; a typical solution is indicated on the figure.[16]

An alternative design procedure that directly recognizes the serviceability index and the effectiveness of load transfer also is presented in the *Interim Guide* and *NCHRP Report 128*.

The AASHTO method and that of the Portland Cement association outlined earlier both recognize dynamic rather than static loading. However, the AASHTO method defines "failure" in terms of overall pavement performance as measured by the present serviceability index, whereas the PCA method relates "failure" to crack formation resulting from fatigue.

OVERLAY DESIGN PROCEDURES FOR RIGID PAVEMENTS[17]

As the serviceability index of a concrete pavement reaches an unacceptable level (normally about 2.5), the pavement can be reconstructed, recycled, or overlaid. For concrete pavements, the overlay can either be portland cement or asphaltic concrete.

Asphaltic Concrete over Portland-Cement-Concrete Pavements

The *AASHTO Interim Guide* can also be used to design the asphalt overlay. For this procedure, a soil support value is first assigned to the subgrade, the traffic is estimated, and a new thickness calculated. The thickness of the overlay is determined by subtracting the existing structure thickness from the calculated thickness. The layer coefficient of the concrete is usually assumed equal to 0.3 to 0.4; subbase coefficients are those given in Chapter 19 for flexible pavement design.

[16]For verification of the AASHTO method for continuously reinforced pavements, see *TRB Record 632*.

[17]For details refer to the *AASHTO Interim Guide*. For descriptions of recently developed design procedures see *TRB Record 700*, particularly the paper by C. L. Monismith.

TABLE 20–3. Multipliers for Converting Single- or Tandem-Axle Loads to Equivalent 18-kip Single-Axle Loads—AASHTO Interim Design Method, Present Serviceability Index, p=2.5.

Load on Axles (kips)	*Slab Thickness (in.)* Single Axle 6	Single Axle 8	Single Axle 10	Single Axle 11	Tandem Axles 6	Tandem Axles 8	Tandem Axles 10	Tandem Axles 11
2	0.0002	0.0002	0.0002	0.0002				
4	0.003	0.002	0.002	0.002				
6	0.01	0.01	0.01	0.01				
8	0.04	0.03	0.03	0.03				
10	0.10	0.08	0.08	0.08	0.01	0.01	0.01	0.01
12	0.20	0.18	0.18	0.17	0.03	0.03	0.03	0.03
14	0.38	0.35	0.34	0.34	0.06	0.05	0.05	0.05
16	0.63	0.61	0.60	0.60	0.10	0.08	0.08	0.08
18	1.00	1.00	1.00	1.00	0.16	0.14	0.13	0.13
20	1.51	1.55	1.58	1.58	0.23	0.21	0.20	0.20
22	2.21	2.28	2.38	2.40	0.34	0.31	0.30	0.30
24	3.16	3.23	3.45	3.50	0.48	0.45	0.44	0.44
26	4.41	4.42	4.85	4.95	0.64	0.63	0.62	0.62
28	6.05	5.92	6.61	6.81	0.85	0.85	0.85	0.85
30	8.16	7.79	8.79	9.14	1.11	1.13	1.14	1.14
32	10.81	10.10	11.43	11.99	1.43	1.47	1.50	1.51
34	14.12	12.94	14.59	15.43	1.82	1.87	1.95	1.96
36	18.20	16.41	18.33	19.52	2.29	2.35	2.48	2.51
38	23.15	20.61	22.74	24.31	2.85	2.91	3.12	3.16
40	29.11	25.65	27.91	29.90	3.52	3.55	3.87	3.94
42					4.32	4.30	4.74	4.86
44					5.26	5.16	5.75	5.92
46					6.36	6.14	6.90	7.14
48					7.64	7.27	8.21	8.55

The overlay thickness determined by the Corps of Engineers method, also presented in the *AASHTO Guide,* is as follows.

$$t = 2.5(Fh_d - h) \tag{20–7}$$

where

t = thickness of the overlay, in inches
F = a factor related to the strength of the existing subgrade ranging from 0.6 for excellent subgrade to 1.0 for poor subgrade.
h_d = the design thickness determined from the PCA or AASHTO methods, in inches
h = the existing pavement thickness, in inches

Several other methods based on layered elastic theory or finite element analysis have been proposed.[18] Discussion of these methods is beyond the scope of this book.

[18]For a discussion of some of this work, see *TRB Record 700.*

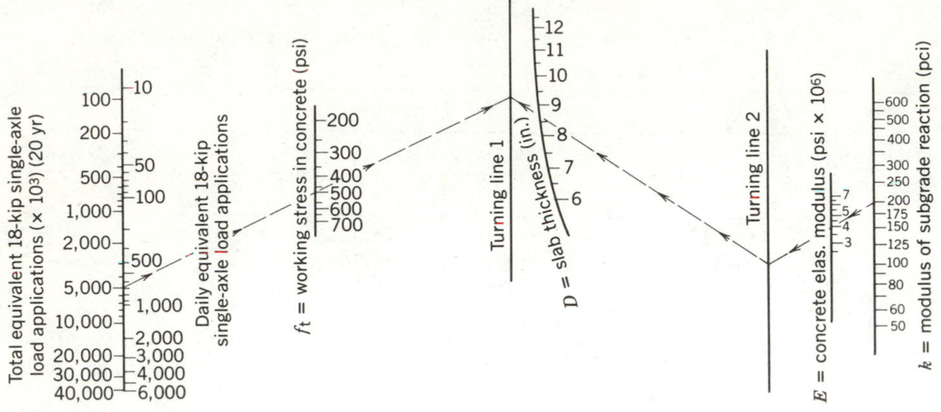

Fig. 20–8. AASHTO design chart for rigid pavements. Condition: Working stress = modulus of rupture ÷ factor of safety; terminal serviceability index = 2.5. Thicknesses are for unreinforced or lightly reinforced pavements with protected corners; with continuous reinforcement, thickness may be decreased, based on local experience. (Source: *Revision to AASHTO Guide for Design of Pavement Structures,* 1980).

Example: Equivalent 18-kip wheel loads, 5,000,000; working stress in concrete 670/1.33 = 500 psi; modulus of subgrade reaction, 200 lb/in.³; concrete modulus of elasticity, 4,000,000 psi. From this, D = 8.7 in.

Portland Cement Concrete Overlays for Portland-Cement-Concrete Pavements

The required thickness of concrete overlays over concrete pavements can be determined using the following empirical formulas:

Partially bonded (continuous case):

$$h_o = \sqrt[1.4]{h_d^{1.4} - Ch^{1.4}} \tag{20–8}$$

Unbonded (noncontinuous case):

$$h_o = \sqrt[2]{h_d^{2} - Ch^{2}} \tag{20–9}$$

where

h_o = thickness of overlay, in inches
h_d = thickness of new pavement from PCA or AASHTO design methods, in inches
h = thickness of existing pavement, in inches
C = coefficient ranging from 0.35 for badly cracked pavement to 1.0 for slabs in excellent condition.

To ensure complete bond, it is necessary to repair the old concrete. All loose material must be removed, the joints cleaned, and a bonding grout placed before constructing the overlay. Unbonded conditions exist where an intentional bond breaker (e.g., granular base or soft asphalt layer) is placed before construction of the overlay. The slab thickness for the unbonded case is normally greater than that for the bonded situation.

Improved techniques for overlay design include deflection tests, testing the component layers for modulus (see Chapter 14), and layered elastic theory. For a description of recent developments, the reader is referred to *TRB Record 700*.

CRACK CONTROL FOR CONCRETE PAVEMENTS[19]

Transverse Contraction Joints

As mentioned earlier, stresses in concrete slabs of any great length will exceed the tensile strength of the material. Therefore, cracking of long slabs is bound to occur either in a random, uncontrolled, and unsightly manner, or at pre-established locations in the pavement where it can be dealt with effectively. It was also indicated that in American highway practice, cracking that occurs across the pavement at right angles to the direction of vehicle travel (transverse cracking) is commonly accommodated in one of three ways.

1. With unreinforced slabs, provide weakened sections (joints) at close intervals (around 15–20 ft). Cracking and minor length changes from temperature and moisture variation and other causes then will occur only at these joints.
2. Provide relatively light reinforcing in the slab and space the weakened sections more widely, generally in the range of 40–100 ft. With this design it is anticipated that fine, hairline cracks will develop at intervals, but that the reinforcing steel will prevent them from opening. Adjustments in slab length brought on by temperature and moisture change and other factors will occur at the joints.
3. Provide relatively heavy continuous reinforcing and omit joints entirely. This procedure causes fine cracks to develop at such close intervals that there is almost no slab movement from temperature and moisture variation, except near the ends of sections.

For the "weakened plane" or "dummy" type contraction joint (see Fig. 20–9a), a plane of weakness is made across the slab by creating a notch into the top of the concrete. When tension develops, a crack occurs at this point rather than at some other location. Most agencies now require or permit the use of special saws that cut the joints into the concrete after it has stiffened. Under proper conditions, joints made in this manner are very neat and trim. However, careful control of the time and circumstances of sawing must be observed. For example, sawing too soon, particularly through hard coarse aggregate, will tear the pavement surface and require patching that may spall off later. If sawing is delayed too long, however, random cracks will occur in the pavement and, once formed, will not heal. Again, where a new slab is being poured against an existing one, movement in the joints of the existing pavement may bring a random crack across the new slab before its joints can be cut. Because of the many

[19]*NCHRP Synthesis 19* is an authoritative statement on the design, construction, and maintenance of joints in concrete pavements. Its Appendix B summarizes the design and jointing practices of the state and provincial highway agencies in the United States and Canada as of 1973.

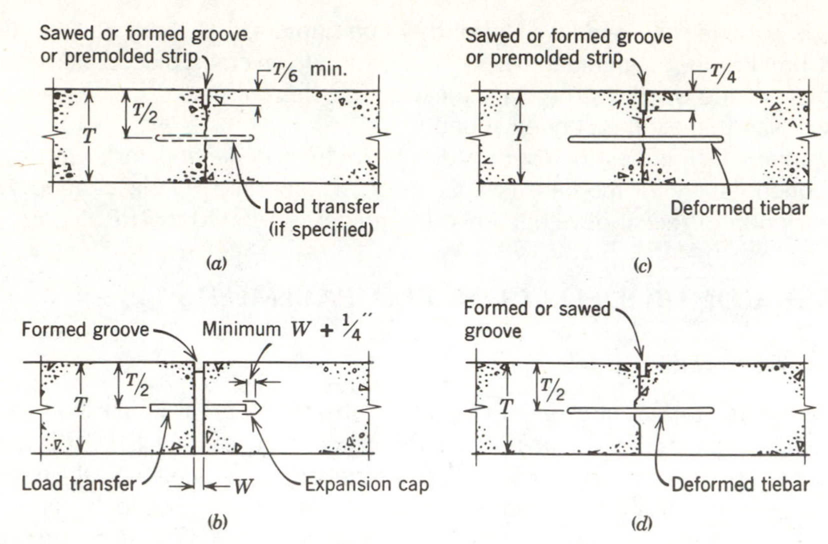

Fig. 20–9. Basic types of concrete pavement joints: (*a*) contraction joint; (*b*) expansion joint; (c) longitudinal joint; (*d*) construction joint. (After W. G. Westall, *HRB Record 80*.)

variables, few rules for sawing times and procedures can be laid down; rather they must be developed for each set of circumstances.[20] Vibrating beams are sometimes forced into the fresh concrete along the line where sawing will be done later. The aim is to force the coarse aggregate away from the joint. With this procedure, the joint is less likely to tear on sawing; on the other hand, the concrete abrades easily and is subject to spalling.

Some agencies insert plastic, fiber, or metal strips in the fresh concrete to create the weakened plane. The trend has been away from such practices, in part because of high labor costs.

Load transfer and joint alignment is sometimes obtained solely by aggregate interlock. This results because crack formation at a weakened-plane joint leaves an irregular and rough break in the concrete. As long as the joint opening is small, in the range of 0.03–0.04 in., the projecting aggregate faces stay tightly keyed, and, when the edge of one slab deflects under load, the adjoining edge is forced down also. Thus interlock can furnish edge or corner support between adjacent slabs. If the coarse aggregates are weak or the joint opens too widely, aggregate interlock is ineffective as a means of load transfer. It follows that reliance on it commonly occurs with plain concrete pavements having close joint spacing and in areas where strong aggregates are plentiful. In all, some 18 states have employed designs based on aggregate interlock.[21] Some have experienced faulting and have begun to use load transfer devices (dowels) at the joints.

Smooth steel bars called dowels are very widely used as load-transfer devices (see Fig. 20–9a). Dowels can be greased, painted, or coated with asphalt for

[20]For more on sawed joints, see G. K. Ray, *HRB Bulletin 229*.

[21]For a report of an investigation of aggregate interlock, see B. E. Colley and H. A. Humphrey, *HRB Record 189*.

one-half their length to break the bond with the concrete and permit the dowel to slip within one of the abutting slab ends.[22] Extreme care must be taken during construction to keep dowels in careful alignment; otherwise they will prevent proper joint action. Almost all agencies have special holding devices for this purpose.

In the past, a wide variety of dowel lengths, diameters, and spacings has been used by transportation agencies, and some variation still exists. Dimensions recommended by the *AASHTO Interim Guide* are given in Table 20–4.

Skewed transverse contraction joints in plain concrete pavements are now specified by many agencies. Typically, joints are slanted 2 ft in each 12-ft lane from the normal position. This offset places the tire pattern of the inside dual wheels of a truck just forward of the joint at the same instant that the outside wheels are just behind the joint. Some agencies also employ a nonuniform pattern of contraction joint spacing in plain concrete pavements. For example, the Arizona Department of Transportation repeats the sequence of 15, 13, and 17 ft. The aim is to avoid any tendency for resonance to develop in the springing systems of vehicles.

Transverse Expansion Joints for Concrete Pavements

Expansion joints provide space into which the ends of pavement slabs can protrude when the slabs lengthen or shift position. In early days, vertical openings filled with mastic or premolded filler were placed at intervals ranging from 30 to 150 ft along the roadway. When these joints failed to stop cracking at random spacings along the pavement, some engineers decided they were unnecessary and omitted all joints except those placed at the end of each day's run. Many thousands of miles were built without joints, and most of them gave good service; 5 yr or more after construction, however, with some of these pavements short sections abruptly buckled upwards. These "blowups" were taken as conclusive evidence that expansion joints were needed. By 1934, the installation of expansion joints at spacings of 100 ft or less was almost universal practice.

TABLE 20–4. Recommended Dimensions of Dowels for Joints in Portland-Cement-Concrete Pavements*

Pavement Thickness (in.)	*Dowel Diameter (in.)*	*Dowel Length (in.)*	*Dowel Spacing (in.)*
6	3/4	18	12
7	1	18	12
8	1	18	12
9	1 1/4	18	12
10	1 1/4	18	12
11	1 1/4	18	12
12	1 1/4	18	12

*Source: *AASHTO Interim Guide* and proposed revisions in 1980.

[22]See *TRB Record 535*, for a discussion on plastic coated dowels. *AASHTO Specification M-254* provides details for coated dowel bars.

Since then the trend has reversed, and all but a few state transportation departments now construct pavements without expansion joints, except where pavements join structures or intersect other concrete pavements.

The principal argument advanced for installing expansion joints was to prevent blowups by relieving direct compressive stresses before they reach dangerous levels. One group of engineers[23] attributed blowups to the infiltration of dirt and other foreign materials into joints and cracks. Others assigned primary responsibility for blowups to growth of the concrete itself rather than to infiltration of dirt into joints and cracks. On several projects in California, alkali–aggregate reaction (see Chapter 14) was blamed. In Indiana, growth was ascribed to swelling of the coarse aggregate under repeated freezing and thawing. In both these instances the agencies involved adopted tests to eliminate unsuitable aggregates and eliminated expansion joints from their pavement designs. Based on a study of 517 blowups in Indiana[24], it was concluded that blowups occurred predominantly in midafternoon at a temperature above 90°F—during a period which was usually preceded by varying amounts of precipitation. The infiltration of dirt into cracks, of course, would accentuate the difficulty by eliminating some of the otherwise available room for expansion.

By careful attention to aggregate and cement selection to eliminate concretes that increase in length over time, the blowup problem seems to have largely disappeared, although it has occurred recently where joints have failed.[25]

A detail of the most commonly used form of expansion joint is given in Fig. 20–9*b*. The filler is generally in plank form so that it can be handled easily. In earlier times, asphalt-impregnated felt was widely used. However, it extruded when the joint closed as the temperature increased; this caused a bump in the pavement that maintenance forces had to cut away. On cooling, the joint opened and the void filled with dirt. These extrusive joint fillers have been supplanted to a large extent by nonextrusive fillers of cane fiber, cork, or sponge rubber. Redwood or cypress boards and a variety of telescoping metal devices have also been employed.[26]

One-half of the dowels for expansion joints must be coated with grease or asphalt to break the bond with the concrete so that the dowel may slide in and out. The same end must be covered with a grout-proof cap to provide free space in the concrete. Also, expansion joints are sealed at the top as described below.

Longitudinal Joints for Concrete Pavements

All transportation agencies place longitudinal joints between adjacent traffic lanes. They are designed as hinges, which means that they provide edge support but permit rotation between the slabs. In this way, flexural stresses are relieved which otherwise might cause irregular and unsightly cracks along the length of

[23]See, e.g., A. A. Anderson, *Transactions, ASCE,* 1949.
[24]See K. B. Woods, *Transactions ASCE,* 1949.
[25]The Virginia Department of Transportation has recently reported on the use of 4-in.-wide expansion joint to reduce blowups. See *TRB Record 632*.
[26]See *AASHTO Designations M33, M46, M153, M213,* and *M220* for specifications for the various forms of joint fillers. *AASHTO Designation T42* offers detailed test procedures for all preformed joint fillers.

TABLE 20–5. Recommendation for Tiebars for Longitudinal Joints in Concrete Pavements with 12-ft Lane Widths*

			½ in. Bar Diameter		⅝ in. Bar Diameter	
Type and Grade of Steel	*Working Stress (psi)*	*Pavement Thickness (in.)*	*Length (in.)*	*Maximum Spacing (in.)*	*Length (in.)*	*Maximum Spacing (in.)*
Grade 40 billet or axle steel	30,000	6	25	48	30	48
		8	25	40	30	48
		10	25	32	30	48

*(Condensed from *AASHTO Interim Guide*).

the pavement. Generally, longitudinal joints have not been troublesome. Two explanations are (a) that there are few demands for heavy load transfer across them and (b) that the expansive and contractive movements and forces developed across the width of a pavement are relatively small.

When the concrete for all lanes is placed simultaneously, as is usually done with slip-form paving methods, longitudinal joints are much like those for transverse contraction joints (see Fig. 20–9c). When the lanes are constructed at different times using side forms, the construction joint between them commonly provides a keyway in the first slab to assure load transfer (see Fig. 20–9*d*).

In longitudinal joints, deformed tiebars replace smooth steel dowels, since the aim is to hold the slabs tightly together, rather than to permit the joints to open and close. Diameters and spacings of tiebars, then, are determined by the force necessary to pull the narrower pavement slab over the subgrade to the joint (see the discussion on reinforcement for further detail on design). Length of tiebar is determined from the embedment in the concrete needed to develop the strength of the bar. Tiebar diameters, spacings, and lengths recommended by the *AASHTO Interim Guide* are given in Table 20–5.

Construction Joints for Concrete Pavement

When, for any reason, concrete placement is interrupted long enough that a "cold" joint will occur, it is common practice to install a transverse construction joint (see Fig. 20–9*d*). This can be accomplished by creating a vertical face by ending the concrete placement against a header board of wood or steel. Typically, a keyway on the header protrudes into the concrete to provide for load transfer. Deformed tiebars may be used to hold the joint tightly closed. On the other hand, if the construction joint replaces a contraction joint, dowels may be substituted.

Joint Sealing for Concrete Pavements[27]

Many agencies carefully clean all concrete pavement joints or cracks and then seal them to prevent the infiltration of water to the subgrade and to keep mud and dirt out of the joints. Others seal only expansion joints. Materials for poured

[27]Joint sealing is a major problem in concrete-pavement maintenance as well as in design. Much laboratory research and field study has been devoted both to joint sealers and to sealing practices. See *NCHRP Report 38 and 202*. Also, *HRB Record 80, 320 and 389* have several reports and exten-

joints include the harder paving and air-blown asphalts, sometimes mixed with mineral filler, rubber asphalts, and a variety of rubber compounds. Some of these are poured hot and become stiff in cooling; others are placed cold. *AASHTO Designation M173* commonly is the specification for poured joint materials. It stipulates the performance to be expected rather than the materials themselves.

Other states employ preformed seals consisting of strips of extruded neoprene. These are compressed for insertion into the joint groove. After insertion, they expand to completely fill the space. An adhesive causes the strips to adhere to the opposing joint faces.

Reinforcement for Jointed Concrete Pavements

Reinforcing steel in jointed concrete pavement prevents the widening of cracks produced by shrinkage or thermal contraction and holds the fractured faces in intimate contact. In this way, aggregate interlock is preserved, and the intrusion of dirt or water is prevented. Seldom if ever is reinforcing counted on to resist flexural stresses produced by loads or curling.

Among the agencies that use reinforcing steel in pavements, there are wide differences in the kinds and amounts of reinforcing specified. Welded-wire fabric and mats of reinforcing bars both are employed. Weights required generally ranged between 40 and 80 lb/100 ft^2 of pavement surface.

Welded-wire fabric is made from cold-drawn steel wires.[28] Minimum-permitted tensile strength is 80,000 psi; the yield strength is 70,000 psi. Reinforcing bars of billet, rail, or axle steel are available.[29] Yield strengths among them range from 40,000 to 75,000 psi. The reinforcing steel is placed in one layer, at or slightly above middepth of the slab. Where deicing salts are used, the reinforcing should be deeper.

In designing reinforcement for jointed pavements, the assumption is made that the reinforcing steel must be strong enough to drag both ends of each individual slab over the subgrade toward its center. This idea, expressed as a formula, becomes

$$A_s = \frac{LfW}{2S} \qquad (20\text{–}10)$$

where

A_s = square inches of steel cross section per foot of slab width
L = length of the slab between joints, in feet
f = coefficient of friction between the slab and the subgrade—also called

sive bibliographies. *HRB Special Report 112* offers a glossary of terms related to joint sealing. See also *TRB Record 535, 604* and *752*. The American Concrete Institute's *Manual of Concrete Practice Part 5*, "Guide to Joint Sealants for Concrete Structures," 1980, is an excellent reference.

[28]Specifications for the wire itself appear as *AASHTO Designation M32*. Those for the completed fabric are *AASHTO Designation M55*.

[29]See *AASHTO Designation M31, M42,* and *M53*.

the coefficient of subgrade resistance; assumptions for it usually range from 1 to 2, with 1.5 recommended by the *Interim Guide*

W = weight of the slab per square foot of pavement surface, in pounds

S = working stress in the reinforcing steel, in pounds per square inch; the *AASHTO Interim Guide* suggests working stresses ranging from 30,000 to 45,000 psi, depending on the type and grade of steel

Experimental pavements reinforced with short steel fibers mixed into the concrete have also been constructed. The individual fibers are 1 in. long and 0.01–0.06 in. in diameter; the steel occupies 1½% by volume (5% by weight).[30] These are not yet in common use.

CONCRETE PAVEMENTS WITHOUT JOINTS

Careful measurement of slab movements on experimental pavements laid in Indiana in 1938 showed that 500–1000 ft at the ends of long, reinforced pavement slabs moved on the subgrade, but that central portions of the slabs did not. Rather, minute cracks spaced 6–8 ft developed in the concrete. In most instances these center sections gave excellent performance. It therefore appeared that a heavily reinforced continuous pavement, without joints, should perform like these center sections. Soon after World War II, experimental sections to test this theory were constructed in numerous states, and their performance carefully observed. By 1979, some 13,000 two-lane mi of road were in service and many agencies had experimental installations or used them on some or all heavy-duty pavements.[31]

The behavior of continuously reinforced pavements is substantially different than that of slabs of unreinforced or lightly reinforced concrete. This difference is not recognized explicitly in the usual design procedures. However, many agencies do so implicitly. For example, in the early design procedures of the Illinois Department of Transportation a 7-in thickness of continuously reinforced pavement was considered to be the equivalent of a 10-in. lightly reinforced slab. Because this reduced thickness resulted in early distress under heavy loads, most agencies now use the same thickness irrespective of whether the pavement is plain, simply reinforced, or continuously reinforced (between 8 and 11 in.). Part of the cost difference associated with the steel is recovered through not sawing joints and reduced joint maintenance. Furthermore, continuous pavements are considered to provide a smoother ride because of reduction in joint roughness and faulting.

The *AASHTO Interim Guide* recommends that the cross-sectional area of longitudinal reinforcing steel must equal or exceed 0.4–0.8% (most use 0.6–0.7%)

[30]For a status report on fiber reinforced concrete, see *American Concrete Institute SP-51, Roadway and Airfield Pavements,* 1975.

[31]For reports on continuously reinforced pavements and bibliographies listing earlier studies, see *NCHRP Syntheses 16* and *60* and articles in *HRB Record 112, 131, 239, 291, 407,* and *466* and *TRB Record 485, 572, 602, 632* and *756*. Investigations of terminal anchorages for continuously reinforced slabs are reported by B. F. McCullough in *HRB Record 362* and *TRB Special Report 173.*

of the concrete cross section to develop continuity. At lesser amounts of steel, wider cracks develop at less frequent intervals and the advantage is lost. For an 8-in. slab, this requirement calls for 130–260 lb of steel per 100 ft^2 of pavement surface, as contrasted to the 40–80 lb used in conventionally reinforced slabs. Commonly, sufficient transverse steel is tied onto the heavy longitudinal reinforcing to hold the main steel at the intended spacing. In some instances, however, installation of the longitudinal bars is done in such a manner that the proper spacing is maintained without the transverse bars. This reduces the amount of steel required and greatly simplifies its handling and placing.

In Europe, Belgium has, since 1970, used continuous reinforcement in their concrete roads. The slab thickness is normally 8 in.[32] with 0.85% longitudinal steel placed 3.5 in. below the surface. Transverse steel, spaced 28 in. on centers is placed at a 60° angle to the longitudinal steel to reduce transverse cracking.

In jointed but reinforced pavements, the continuity of the longitudinal reinforcing usually is broken at each joint. Construction procedures in Switzerland and experimental projects elsewhere in Europe have employed a so-called "elastic" joint. These designs have contraction joints about 100 ft apart. Regular deformed reinforcing bars are placed in four segments, each slightly less than 25 ft long. At each break in this steel, plain bars ranging in length from 2.5 to 6 ft, coated with a bond breaker of asphalt or epoxy, are installed. Also, narrow strips are placed at the top and bottom of the slab at each break to create weakened planes where cracks can form. It is claimed that "elastic" joint designs have the advantages of continuous reinforcing, but use roughly half as much steel.[33]

PRESTRESSED CONCRETE PAVEMENTS

Concrete pavements placed in compression by means of highly tensioned steel wires or strands have been used on a number of airports and experimental roads in Europe. A few airport sections have also been constructed by both the Corps of Engineers, U. S. Army and the Naval Facilities Engineering Command, U. S. Navy. Highway test sections have been built in Mississippi, Pennsylvania, on the access road to Dulles airport, Washington, D.C. for the "Transpo 72" show, and in Arizona.

Prestressing involves placing entire pavement sections in compression. Wide joints where the prestressing devices operate separate these sections. These thin prestressed sections show performance equal to considerably thicker ones of plain or conventionally reinforced concrete. In effect, the question is primarily one of cost: "Will the savings by reduction in the quantity of concrete pay for the cost of prestressing materials and labor?" In the case of airports for large planes, the heavy wheel loads demand thick pavements, and substantial concrete savings are possible from prestressing. Pavements for highways are much

[32]See article by G. Van Heystraeten, *Public Roads,* Dec. 1976.
[33]See B. O. E. Persson and B. F. Friberg, *HRB Record 291,* for added detail.

thinner, so that the saving in concrete is less substantial. For example, the reduction may be from about 10 to 5 in. Furthermore, airport runways and aprons are straight and on relatively constant grades; highways often must have curved horizontal and vertical alignment. It follows that prestressed highway pavements, other than as experiments, still remain to be proved.[34]

UNDERCOURSES (SUBGRADES AND SUBBASES) FOR CONCRETE PAVEMENTS[35]

Before World War II, little attention was paid to the soils on which concrete pavements were laid. They were classed as *rigid* pavements that could spread loads over a considerable area of subgrade by beam action. Trouble developed, however, in pavements laid directly over expansive soils. Sometimes water from the fresh concrete would be drawn into the subgrade, and it would swell and disrupt the newly placed pavement. Again, water percolating through joints and cracks of completed pavements would lead to differential expansion of the subgrade, and the pavement would curl upward at the joints and become exceedingly rough. Uneven surfaces also resulted as moisture adjustments over several seasons brought differential subgrade expansion. Pumping (see below) was a common ailment of heavily traveled roads.

Today construction standards for subgrades and embankments are exacting and are the same for all pavement types (see Chapter 15). In addition, an undercourse of carefully selected material, often referred to as "subbase" in concrete pavement literature, is almost always imposed between the subgrade or embankment and the concrete. The *AASHTO Interim Guide* recommends that this extend 1–2 ft outside the pavement to provide edge support.

As of 1978, practices were far from uniform in the United States and Canada. For the heaviest traveled facilities, most agencies use a treated (cement or asphalt) subbase; only ten specify untreated materials (see Chapter 17 for details of these subbase types). With concrete pavements, these undercourses can be relatively thin, since they contribute little to pavement strength. However, the presence of a good subbase increases fatigue life and reduces joint faulting.

The widespread inclusion of undercourses under portland-cement-concrete

[34]A design procedure is offered by M. Sargious and S. K. Wang in *Journal,* American Concrete Institute, July–Aug. 1971. A report and bibliography of the Pennsylvania section are given by J. R. Smith and R. K. Lightholder in *HRB Record 60,* and by J. R. Smith and J. L. Evanko in *HRB Record 131.* The "Transpo 72" pavement is described in *HRB News,* Winter 1972, *Public Roads,* June 1972, pp. 16–21, and by B. F. Friberg and T. J. Pasko, Jr. *HRB Record 466.* Another experimental section in Pennsylvania is detailed by R. J. Brunner in *TRB Record 535.* G. R. Morris, *Civil Engineering,* March 1978, describes and shows cost comparisons between conventional concrete and prestressed concrete construction. Experiments with self-stressing reinforced concrete pavements using expansive cement are reported by C. E. Dougan in *HRB Record 112,* and *HRB Record 291.* Performance of a pavement consisting of prestressed concrete panels covered by an asphalt concrete overlay is examined by E. R. Hargett in *HRB Record 239* and L. J. Larson and W. R. Haug in *HRB Record 389.* For a recent review, see *FHWA Report RD 77-8.*

[35]The literature on this subject is extensive. For an excellent summary and bibliography see the Portland Cement Association publication, *Subgrades and Subbases for Concrete Pavements.* For current practices see *FHWA Report TS–78–202,* op. cit.

pavements has resulted mainly from extensive investigation of the "pumping" of concrete paving slabs. Pumping is the ejection of water and subgrade soil through joints and cracks and along the edges of concrete pavements. It has been found that repeated depression of pavement joints by heavy axle loads is the primary activating element in pumping. Then, where free water is present and the subgrade is fine-grained, churning of the water and soil occurs, forming a slurry which is expelled to the surface. As pumping continues, the supporting soil is flushed from under the pavement at the affected locations. Faulting of the joints and finally transverse cracking or breaking of the corners results. The formation of one crack offers new opportunity for pumping action, so that joint faulting and cracking is progressive.[36]

Pumping is reduced where the paving slab is underlaid by a granular undercourse of proper grading or where the undercourse has been made resistant to erosion by cement or bituminous treatment. On the other hand, no sure and effective means of preventing joint movement or for excluding free water from beneath pavement joints, cracks, and edges has been developed. Thus the use of undercourses with provisions for drainage seems the most reliable method for preventing pumping.

The merits of various gradings and thicknesses of granular undercourses have been the subject of extensive research. Some investigators have reported that open-graded bases do not perform well because a softened subgrade intrudes into the base from below; they also state that very dense gradings (3% or more passing the No. 200 sieve) are troublesome because the fines pump out. The findings of others do not entirely agree.[37] There is good evidence from the AASHO Test Road (see below) that relatively thin layers perform well.

TEST-ROAD FINDINGS APPLIED TO RIGID PAVEMENT DESIGN

As mentioned, the Pittsburg, California, and Bates, Illinois, test roads of the early 1920s led to the widespread use of thickened-edge concrete pavements. Two other tests, one carried out in 1950 and the other completed in 1962, also have had far-reaching consequences. Their findings are described briefly below.

Road Test One-MD[38]

Much knowledge useful in concrete-pavement design came from Road Test One-MD conducted in 1950 in Maryland. This cooperative project was supported by 11 transportation agencies, the transportation department of the District of Columbia, the U. S. Bureau of Public Roads, various truck manufactur-

[36]The fundamental studies that led to an understanding of pumping and its causes are summarized, along with an extensive bibliography, in *HRB Proceedings 1948,* pp. 281–310. Pumping also occurs along the longitudinal shoulder joint. For design details to minimize this, see *NCHRP Report 202*.
[37]See *AASHTO Interim Guide,* Chapter III for recommendations on subbases for concrete pavements.
[38]See *HRB Special Report 4* for a detailed record. *Special Report 14* covers a supplemental investigation of the structural effects of a heavy-duty trailer on the same pavement.

ers, the petroleum industry, the Department of Defense, and the Highway Research Board, and was supervised by the Highway Research Board. The test section was a 1.1 mi length of U. S. 301, approximately 9 mi south of La Plata, Maryland. The pavement, constructed in 1941, was in excellent condition at the start of the test. It was thickened edge, with a 9–7–9 in. parabolic cross section. It had been laid directly on the subgrade, 15% of which was granular and the remainder fine-grained and plastic. Accelerated loading on separate, representative sections were carried out using single-axle loads of 18 and 22.4 kip and tandems of 32 and 44.8 kip.

Among the findings were the following:

1. There was a definite correlation between soil type and pavement behavior. The higher the granular content and the lower the plasticity of the soil, the better the pavement performed.
2. For these particular test conditions, the destructive effect of increased axle loads is many times greater than the increase in the axle loads themselves. Two of several evidences of this were:
   a. The 44,800-lb tandem-axle loads caused 11 times as much cracking (lineal feet) as the 32,000-lb tandem-axle load. Also, the 22,400-lb single-axle load caused approximately six times as much cracking as the 18,000-lb single-axle load.
   b. After 92,000 tandem-truck passes, 96% of the slabs under the heavier truck contained cracks that were analyzed as structural failures. For the lighter tandem truck the percentage was 27. For 238,000 passes of the heavier and lighter single-axle trucks these percentages were 64 and 28, respectively.
3. Variations in joint deflection under load which served as a clue to the stress in the pavement were:
   a. Normally much greater at night when the pavement edges were curled upward than in the daytime when the edges were curled downward.
   b. Two to three times greater at pumping than nonpumping joints.
   c. Larger at pumping expansion joints than at pumping contraction joints.

AASHO Test Road[39]

Conception, organization, and the testing program for this project have been described in Chapter 19. Also, the equation for rating flexible pavements by the serviceability index was outlined. A similar equation for portland-cement-concrete pavements also was derived from test data. It is

$$p = 5.41 - 1.80 \log_{10}(1 + \overline{SV}) - 0.09\sqrt{C + P} \qquad (20\text{–}11)$$

where p is the serviceability index, and $\overline{SV}$, C, and P are measured slope variance in the wheel track, and measured amounts of cracking and patching, respectively.

Comparison of the equations for rigid and flexible pavements points up two of their differences. First, the rating equation for rigid pavements carries no term for rutting in the wheel tracks since rutting does not develop in a concrete slab; second, cracking and patching are much more damaging to the rideability of

[39]For added detail, see *HRB Special Reports 61B, 61E, 61G, 66,* and *73.*

rigid than flexible pavements, as indicated by the multipliers of the last term in the equations.

The principal variables in the portland-cement-concrete experiment were (a) slab thickness, (b) subbase thickness, (c) nonreinforced with doweled contraction joints at 15 ft centers and reinforced with doweled contraction joints at 40 ft centers, and (d) paved versus unpaved shoulders. A part of the testing program involved elaborate, carefully instrumented studies of pavement stresses, strains, and deflections.

A few of the conclusions drawn by engineers on the Test Road staff, and others who have studied the results carefully are as follows:

1. For the specific variables evaluated by the Test Road, its results generally confirm the design procedures for slab thickness in common use.
2. The value of a granular undercourse (subbase) was firmly established. Without it, failures occurred far more quickly. However, undercourses 3 in. thick performed equally as well as those 6 and 9 in. thick.
3. Nonreinforced slabs with doweled contraction joints at 15 ft centers and reinforced slabs with doweled contraction joints at 40 ft centers performed equally well.
4. No increased pavement life resulted from paving the shoulders.
5. The granular undercourse (subbase) will pump from beneath thin slabs. Almost all this pumping occurred at the slab edges; there was very little through the doweled joints.

The Test Road reports indicated that added research, including satellite test roads and comparison with local successful and failed pavement, was needed to adapt Test Road findings to local conditions. Even this elaborate project could not encompass all the variables that should be evaluated. Among those on which data were not developed are:

1. The effect of time and climatic factors. Any accelerated test may not fully evaluate the time element. Similarly, the climate is that of a single area.
2. Added information on jointing and reinforcing. As indicated, the Test Road examined but a few of the possible alternatives.
3. Evaluation of other subbases, including those treated with cement or bitumen. Only one material was used. It complied with *A–1–a* grading requirements, with the percentage passing the No. 200 sieve in the 5–9 range and a plasticity index *less* than 6, but with a *CBR less* than 60.

Among the several research studies based on AASHTO Test Road data was an analysis by Vesic and Saxena which is reported in *NCHRP Report 97*. Among other findings were the following useful relationships:

1. Slab thickness should be increased as the fifth root of the number of load applications.
2. A 10% increase in the strength of the concrete can mean a 50% increase in pavement life.
3. A 20% overload (increase in axle load) may reduce pavement life by 50%.

PROPORTIONING CONCRETE FOR PAVEMENTS

The fundamental rules governing the selection of the basic materials—cement, aggregate, and water—must be observed to obtain good concrete for pavements. In addition, the unique properties demanded for pavements as contrasted with those for structures or other engineering works must be recognized. The following paragraphs are devoted to these subjects.[40]

Materials

A variety of material combinations are available for concrete mixtures. The relative proportion of portland cement, water, coarse and fine aggregate, and the type and amount of admixture greatly affect how the pavement will behave.

PORTLAND CEMENTS. As discussed in Chapter 14, several types of cement are available. Normally type I or II cements are specified for concrete pavements. Type III is used only when high early strength at particular locations is desired. Some agencies gain the same objective by enriching their usual mixes or by adding an accelerator (e.g., calcium chloride).[41] Air entraining admixtures are often used to improve resistance to freeze–thaw action. This is discussed further under the topic of admixtures.

MIXING WATER. Mixing water must be free of acids, alkalies, and oil. Waters containing decayed vegetable matter are to be particularly avoided as they may interfere with the setting of the cement. Generally water that is suitable for drinking purposes is satisfactory for concrete, with the possible exception of drinking water containing large amounts of sulfates. *AASHTO Designation T26* gives methods for testing mixing water for excessive acidity, alkalinity, and for solids or inorganic matter.

AGGREGATES. As discussed in Chapter 14, mineral aggregates form about 75% of the volume or roughly 80% of the weight of normal paving concrete. It follows that, if concrete is to be strong, sound, and durable, the aggregates must have like properties. This is of particular importance in areas where winter brings freezing temperatures, snow, and sleet. Under these conditions the concrete is subject to freezing and thawing, the pounding and wear from tire chains and studded tires (if permitted), as well as the action of calcium chloride, salt, or other deicing or antiskid agents. To make this high quality certain, highway departments have almost without exception prescribed that aggregates pass appropriate tests for strength, soundness, wear, or combinations of these three.

[40]The fundamentals of concrete-mix design are presented in detail in *Design and Control of Concrete Mixtures,* published by the Portland Cement Association, and in textbooks treating concrete as a material. Also, the publications of the American Concrete Institute and the American Society for Testing and Materials contain many authoritative research reports. It has been assumed that students have some acquaintance with this subject, and attention here has been directed mainly to the problems peculiar to concrete for pavements.

[41]Calcium chloride should not be used if reinforcement or dowels are provided as it will cause them to corrode.

The maximum permitted size of coarse aggregate for paving concrete is generally 2 or 2½ in. These maximums greatly exceed those for structural concrete, where closely spaced reinforcing steel limits aggregate size. This use of large aggregates in concrete for paving has rested on the precept that with them, stronger and more durable concrete resulted with the same cement content per unit volume. A corollary was that less cement would be required to reach a prescribed strength. This precept rests on the demonstrated fact that, with larger aggregate, less water is needed to lubricate the concrete; thus the ratio of cement to water is increased. Under the water–cement ratio rule, then, the concrete will be stronger. However, it has been found that this strength gain with increasing aggregate size did not actually come about when the maximum size of aggregate exceeded ¾ in. Suggested explanations are that the larger aggregates may have planes of weakness in them, that with larger aggregate the surface area available for bonding between aggregate and paste is smaller, and that control of grading is more difficult with the larger sizes. Not all researchers agree with these findings. In any event, highway agencies are considering or have permitted a reduction in maximum size aggregate for pavements. In California, for example, the contractor is permitted to choose between 2½ and 1½ in. maximum sizes with the idea that the selection will be based on relative cost.

Grading of coarse and fine aggregate is subject to careful control (Chapter 14) and the methods for combining them extend this size control to the concrete itself.

As aggregates are being processed, or concrete is batched and mixed, the materials pass across or through a series of screens, travel along the conveyor belts, and are dropped into hoppers, bins, or stock piles. Much of the energy necessary to carry out these operations is dissipated in surface wear and breakage of the individual aggregate particles. If the resulting fines are plastic, they can be detrimental to concrete. Some agencies subject samples of both coarse and fine aggregates to wear by agitating them in water; then the plasticity of the resulting fines is determined using procedures similar to the sand-equivalent test. One such test procedure is *AASHTO Designation T210*.

ADMIXTURES[42]. Many admixtures, substances that are added to change the characteristics of the concrete mix, are available. The most common is an air entraining admixture, but others, such as water reducers, retarders, accelerators, pozzolans, and superplasticizers are also used. Only air entraining admixtures are discussed here.

Air entrainment is the entrapment of air in concrete in the form of well-distributed, minute bubbles. It was first used to increase the resistance of con-

[42]*HRB Special Report 119* titled *Admixtures for Concrete* gives a brief but authoriative presentation not only on the gains and losses accompanying the use of air entrainers, but also of accelerators, water reducers, retarders, and pozzolans. It also contains an extensive bibliography. Experiments with silicones as an additive to increase resistance to freezing and thawing and to deicing chemicals have also been reported (see B. C. Carlson et al. *HRB Record 62)*. See also *TRB Record 564* for four papers on admixtures and *TRB Record 720* for seven papers on superplasticizers.

crete pavements to alternate freezing and thawing and to the surface scaling caused by deicing with calcium or sodium chloride. Other advantages such as improved workability and reduced bleeding in fresh concrete have led to widespread use of air entrainment in both pavements and structures in nonfrost areas. It also permits surface finishing soon after the concrete is placed, an important advantage from an efficiency and cost standpoint. The effectiveness of air entrainment in increasing durability is influenced by a number of factors. Among these are the percentage of air, the grading of aggregates, and the size and distribution of air bubbles.

Air may be entrained in concrete with an air-entraining admixture added to the batch at the time of mixing, or by using an air-entraining portland cement. Numerous commercial air-entraining admixtures are available. They may be tested for suitability under *AASHTO Designation T157*. Three types of air-entraining cements are manufactured. They must be capable of forming an adequate air-void system as measured for a standard mortar of cement and Ottawa sand. (See *AASHTO Designation T137*.) Concrete made with the cement must also show adequate durability under freezing and thawing as measured by changes in its dynamic modulus (see *AASHTO Designation T188*).

Employing an air-entraining portland cement avoids the need for an automatic dispensing device or the services of another worker with the ever-present chance of error. On the other hand, the practice of adding the air-entraining admixture at the mixer permits closer control over the air content of the mix. With this procedure the amount of agent can be varied to recognize the factors that influence air content, which include type and gradation of aggregates, type of mixer, mixing time, consistency of the concrete, and temperature.[43] Opinion is divided over which method is better.

To be effective against freezing and thawing, the mortar portion of the concrete should contain about 10% air in very small bubbles spaced not greater than 0.008 mm from any point in the concrete to the nearest void. To obtain this air content in the mortar, the recommended air content in concretes having 2 in. maximum sized aggregate is 4%; with ⅜ in. maximum size it is 8%. Mix designs for air-entrained concretes are discussed below.

Properties of Cement–Water Paste (Water–Cement Ratio)

For a given combination of materials, the strength and other desirable properties of concrete mixtures vary almost directly with the ratio of cement to mixing water. This idea, developed about 1920 by Abrams, is basic to proper concrete design. It says that, for mixes of reasonable proportions and workability, the quality depends on the richness of the cement–water paste. To illustrate, a non-air entrained concrete with a water–cement ratio by weight of 0.44 (5 gal of water per sack of cement) may have a compressive strength in 28 days of about

[43]Temperature has a large effect so that the amount of air entraining admixture should be varied during the day to maintain a constant amount of air in the mix. These effects are difficult to account for with an air entrained cement.

5300 psi. Another, similar to the first in every regard except for a water–cement ratio of 0.62 (7 gal per sack) will develop a 28-day strength of only 3700 psi. Durability also is dependent on the water–cement ratio and places a second control on the richness of the paste. In consideration of these factors, the *AASHTO Guide Specifications for Highway Construction* set maximum water–cement ratios at 0.53 (6 gal per sack of cement) for normal conditions and 0.49 (5½ gal per sack) for severe freezing and thawing. Many transportation agencies specify similar maximum water–cement ratios, and the remainder impose like controls by carefully specifying the amounts of cement, aggregate, and water.

The structure and properties of cement–water paste and of other factors that affect concrete strength have been the subject of intensive research for half a century. Space limitations preclude further discussion here.[44]

Concrete Mix Design

PRINCIPLES OF MIX DESIGN. It is helpful in mix design to picture concrete as a space filled with closely packed coarse aggregate; in turn the voids in the coarse aggregate are filled with sand, and the voids in the sand filled with a cement–water paste rich enough to meet the specified strength and durability requirements. Since cement cost far exceeds that of aggregate, the aim is to use as much aggregate and as little cement–water paste as possible while still maintaining the workability necessary for successful placing and consolidation. For structural concrete, which must be placed in relatively inaccessible areas and around reinforcing steel, it is necessary to overfill the voids in the aggregate and to use a relatively wet and free-flowing cement–water paste to permit successful placing. However, for pavements, where the slabs are thin and the concrete can be manipulated from the surface, harsher, drier mixes can be placed successfully. Thus a higher percentage of aggregate and a less fluid cement–water paste can be employed.

Four means for reducing the amount of cement–water paste and the cost of the mix are:

1. Permit the largest size of aggregate that can be accommodated.
2. Make sure that aggregate is uniformly graded from coarse to fine.
3. Use the largest reasonable percentage of coarse aggregate consistent with proper workability.
4. Demand the minimum lubrication (lowest slump) consistent with proper placing and finishing. For paving concrete, commonly specified values of slump are 1–2 or 2–3 in.[45]

[44]Beginning points for further study on his topic are *HRB Special Reports 90* and *127*. These publications contain extensive bibliographies. See also T. C. Powers, *Properties of Fresh Concrete,* Wiley, New York, 1968, and *HRB Records 210* and *370*. See also *TRB Records 423* and *504*.

[45]*The slump test (AASHTO Designation T119)* is the traditional and most widely used for determining the consistency of concrete. A truncated cone of sheet metal, 12 in. high with base and top diameters 8 and 4 in., respectively, is filled in three layers with fresh concrete. Each layer is rodded 25 times. Then the cone is lifted off vertically, permitting the concrete to subside. The slump is the distance in inches that the top of the specimen falls.

The Kelly ball *(ASTM Designation C-360)* is also used. The apparatus for it is a metal cylinder of

Concrete mixes commonly are designed by the trial-batch method. A first batch weighing on the order of 20–25 lb min is proportioned using past experience or suggested values (see below) for similar applications and ingredients. Proportions of subsequent batches are varied until a trial batch that seems best to fit the particular application is found. Often, further adjustments in mix proportions are made on the job.[46]

DESIGN FOR A FIXED-CEMENT FACTOR. The cement factor is the number of sacks (or barrels) of cement in a cubic yard of concrete. A mix of fixed-cement factor, then, must produce a prescribed volume of concrete per sack of cement. Furthermore, the concrete must be workable, and the slump must be within specified limits. Mix design involves finding that combination of cement, aggregates, and water which meets these requirements with the lowest reasonable water–cement ratio. Actual design is usually by the trial-batch method. Suggested proportions for the first trial mix for concrete pavements are shown in Table 20–6. Specimens for proving the strength are prepared from the trial batch.[47] It is common practice to permit minor alterations in the proportions during construction if they are necessary to secure proper workability.

Most agencies design pavement concrete for a fixed cement factor, with cement contents ranging from 5 to 7 sacks (470–658 lb) per cubic yard. It is worth noticing that, legally, if the owner's representative prescribes the exact mix design, the contractor cannot then be held responsible if the concrete does not reach a prescribed minimum strength.

Some engineers argue that the fixed-cement-factor method is wasteful, as generally the strengths obtained are considerably above the usual permitted minimums. Those who favor it declare that it provides a highly desirable factor of safety at small added cost. They further contend that it makes proportioning easier and simplifies payment to contractors.

Calculation of the cement factor for concrete mixes is by the method of "absolute volumes." The space occupied by each ingredient is found by dividing its weight by the product of specific gravity times 62.4, the weight of a cubic foot of water. For the first mix shown in Table 20–7, the calculations are given in Table 20–7. Air entrapped in the concrete usually is from 1 to 3%, if no air entraining admixture is used.

---

6 in. diameter and 4⅝ in. height with the bottom shaped as a hemisphere; the weight is 30 lb. A graduated handle rising from the top of the ball passes through a metal frame that has feet 12 in. apart. The ball is set on the surface of the fresh concrete and its penetration is measured by comparing its position with that of the frame. The Kelly ball has the advantage that readings can be taken quickly on the concrete as placed in the roadway. On the other hand, the slump test requires more time and can be conducted only on selected samples.

W. E. Grieb and R. A. Marr, Jr. (see *Public Roads*, Feb. 1956, or *HRB Bulletin 132*) report quite consistent ratios of slump to Kelly ball penetration ranging between 1.5 and 1.6 to 1.

[46]For details of several mix design methods see the references at the beginning of this section. Possible slight modifications in trial-batch proportions recommended by American Concrete Institute Committee 221, including metric-system weights, appear in *ACI Manual of Concrete Practice, Part 1*, 1980.

[47]*AASHTO T-22* is commonly followed to make specimens for compressive strength and *T-23* for flexural strength.

TABLE 20–6. Suggested Trial Proportions for Concrete of Specified Cement Content and Slump*

Type of Concrete†	Type of Coarse Aggregate	*Mixture Proportions per 94 lb Bag of Cement‡*				
		Water		*Aggregates (lb)*		
					Coarse‖	
		(lb)	*(gal)*	*Fine§*	*Small*	*Large*
Plain	Round gravel	41	5.0	185	145	200
Plain	Crushed gravel or stone	46	5.5	200	135	200
Plain	Crushed slag	50	6.0	220	105	155
Air-entrained	Round gravel	37	4.5	160	145	220
Air-entrained	Crushed gravel or stone	41	5.0	180	135	200
Air-entrained	Crushed slag	46	5.5	195	105	155

*Proportions intended to produce concrete containing 6.0 bags of cement per cubic yard with slump of 1½–3 in., suitable for normal machine placement. When vibration is used, slump may be reduced to about ½–1½ in. and batch quantities adjusted accordingly to maintain same yield and cement factor.

†Air content assumed to be 1% for plain mixes and 5.5% for air-entrained mixes.

‡Aggregate weights based on assumed bulk specific gravity, saturated-surface-dry, of 2.65 for sand, gravel, and stone and 2.25 for slag. For other specific gravities, aggregate weights should be adjusted in direct proportion to the specific gravity. Aggregate weights and quantity of added mixing water must be adjusted to allow for free moisture on aggregates.

§Fine aggregate assumed to be well-graded natural sand of average fineness (fineness modulus about 2.6–2.9).

‖Coarse aggregates assumed to be well-graded from 1½ in. to No. 4 or 2 in. to No. 4, furnished in two sizes separated on the ¾ or 1 in. sieve, respectively, and used in the proportion of approximately 40% of the small and 60% of the large size.

Concrete-mix proportions are normally given in terms of saturated surface-dry aggregates, the condition that exists when all the interstices of the individual particles are filled with water but their surfaces are dry. Aggregates, as used, are wetter or drier than this theoretical state, and job adjustments from the saturated, surface–dry proportions are always required. The free water in sands as delivered to the job site normally ranges from 2 to 6% by weight, but may reach 8% or more if the sand is extremely wet. Coarse aggregates seldom contain over 2% of free water by weight. If, on the other hand, the aggregates are air-dry, they will absorb up to 1% of their weight of water before reaching a saturated, surface-dry condition. Dry aggregates that are extremely porous will, of course, absorb several times this amount of water.

DESIGN TO A WATER–CEMENT RATIO. Mix design by the water–cement ratio method starts with requirements for strength and slump. Trial batches and test specimens from them are made with the cement and aggregates that will be used in construction. The objective is to find the combination that uses a minimum of cement and still provides the prescribed strength and consistency. Previous experience with the same materials offers the best guide for the proportions of the first trial mix. Otherwise a beginning can be made from the

TABLE 20–7. Calculation of Cement Factor for Concrete Mixes

Material	*Amount Used*	*Specific Gravity*	*Absolute Volume (ft³)*
Cement	1 sack – 94 lb	3.15*	$\frac{94}{3.15 \times 62.4} = 0.48$
Water	5 gal† – 42 lb	1	$\frac{42}{1 \times 62.4} = 0.67$
Fine aggregate	185 lb	2.65‡	$\frac{185}{2.65 \times 62.4} = 1.12$
Coarse aggregate			
Small (¾ × No. 4)	145 lb	2.65‡	$\frac{145}{2.65 \times 62.4} = 0.88$
Large (1½ × ¾)	200 lb	2.65‡	$\frac{200}{2.65 \times 62.4} = 1.21$
			4.36
Entrapped air	1%		$0.01 \times 4.36 = 0.04$
Total weight	666 lb		Absolute volume 4.40
Factors	Air-free value		Value considering air
Yield (ft³ concrete/sack of cement)	4.36		4.40
Weight of concrete (lb/ft³)	$\frac{666}{4.36} = 153$		$\frac{666}{4.40} = 151$
Cement factor (sacks/yd³)	$\frac{27}{4.36} = 6.2$		$\frac{27}{4.40} = 6.1$§

*See *AASHTO Designation T133* for laboratory test to determine the specific gravity of cement.
†A gallon of water weighs 8.33 lb.
‡See *AASHTO Designation T84*, and *AASHTO Designation T85* for tests for specific gravity and absorption of fine and coarse aggregates.
§The cement factor of 6.1 does not agree exactly with the 6.0 shown in Table 20–6. If differences exist, they are normally associated with errors in air content measurements.

suggestions given in Table 20–6. Secondary controls are the minimum number of sacks of cement per cubic yard and the maximum water–cement ratio for durability requirements.

Tests for flexural strength (modulus of rupture) of simple beams rather than for compressive strength are widely used as measures of the quality of paving concrete. They are not standardized among agencies; some use third-point loading *(AASHTO Designation T97)*; others employ center loading *(AASHTO Designation T177)*. Specimens are usually 6 in. square in cross-section. Actual span length is set at three times the depth with a minimum overhang of 1 in. Flexural strength is computed by the formula for stress in the extreme fibers of a rectangular beam. For third-point loading and failure within the middle third, this becomes

$$R = \frac{P\ell}{bd^2} \tag{20–12}$$

where R is the modulus of rupture in pounds per square inch, P is the total applied load, and ℓ, b, and d are, respectively, span length, beam width, and beam depth, all in inches. The flexural test is particularly well suited for field use, since the testing device is relatively small and light. Accelerated methods for measuring the strength of concrete abound, but will not be discussed here.[48]

Data gathered by the Portland Cement Association indicate that the apparent flexural strength as determined by center loading is some 12% greater than by third-point loading. It recommends the third-point method because it reflects the strength of the center one-third of the beam rather than for a single point.

DESIGN OF AIR-ENTRAINED CONCRETE.[49] For properly designed concrete, air entrainment produces no serious loss in strength, if recommended air contents are not exceeded. Air-entrained mixes of the same water–cement ratio are weaker than plain mixes by about 5% for each percent of air; however, this strength loss can be largely offset by reductions in water and sand made possible by the improved workability produced by the air bubbles.

Suggested trial mixes for air-entrained concrete of fixed cement factor appear in Table 20–6. Changes from plain concrete are decreases of 0.5 gal of water and 20–25 lb of sand per sack of cement. The entrained air, in minute bubbles, takes the space formerly occupied by the water and sand and provides about the same lubrication.

Several factors other than the air-entraining agent influence the percentage of entrained air and must be recognized in mix design and in subsequent field control. Each of the factors and its influence appears in Table 20–8.

Three methods for determining the percentage of air in concrete are given in the *AASHTO Specifications for Transportation Materials*. These are the "pressure method" *(AASHTO Designation T152),* the "volumetric method" *(AASHTO Designation T-196),* and the "gravimetric method" *(AASHTO Designation T121).* The pressure method is used by all but a few agencies. The apparatus for it is a special pressure vessel with a large bowl of known volume at the bottom and a slender, graduated standpipe rising vertically from the top. The bowl is filled with concrete; then water is carefully added to cover the surface of the concrete and fill the standpipe. Air pressure is applied to the top of the water column and this compresses the air entrained in the concrete, which results in a lowering of the surface of the concrete and of the top of the water column in the standpipe. The change in volume of the concrete under a stipulated pressure as shown by the change in level of the water column is then converted to percent air.

The volumetric method is similar in concept to the pressure method. After the bowl is filled with concrete, water is added to a calibration level. The entire application is then agitated or rolled to allow the water to displace the air in the mix. This results in a drop in water level which is read directly as percent air.

[48]See, for example, *TRB Record 558* and *American Concrete Institute SP-56, Accelerated Strength Testing,* 1978.

[49]See *HRB Special Report 119* and earlier references in this chapter for further discussion and extensive bibliographies.

TABLE 20–8. Factors, Other Than Air-Entraining Agent, That Influence Air Entrainment

Factor	*Usual Effect on Air Entrainment*
Amount of sand	Increasing the amount of sand will increase the amount of air.
Richness of mix	Rich mixes entrain less air than lean mixes.
Consistency of mix	Wet mixes entrain more air than dry, stiff mixes. Air content increases with slump up to about 7 in., but with further increases in slump it decreases rapidly.
Type of mixing	Machine mixing will entrain more air than hand-mixing. Different mixers may entrain different amounts of air.
Length of mixing time	During initial mixing period, entrained air increases. Extending mixing beyond a point producing the maximum air content causes a reduction in entrained air.
Temperature of concrete	Amount of air entrained decreases as temperature of concrete increases.
Vibration	Internal vibration usually will reduce air content, but has little effect on smaller bubbles and the spacing factor.

For the gravimetric method, the actual weight per cubic foot of the mixed concrete is determined by filling and weighing a measure of ½ or 1 ft^3 capacity. This actual weight is compared to the voidless weight computed as shown in Table 20–7.

A pocket-sized device called the AE-55 or Chace air meter has also been developed. In operation, it displaces the air in a small sample of mortar from the concrete with alcohol and indicates the resulting volume change. This method *(AASHTO T-199)* is approximate and should not be used to replace the methods described above.

Polymer Concrete

In recent years polymers have been used alone or in combination with portland cement to bind aggregates together in concrete. Depending on the materials used and method of mixing the polymer, the mixtures are referred to as follows:

1. *Polymer-impregnated concrete (PIC)*—This is a portland-cement concrete which, after curing, is impregnated with a monomer, that is polymerized *in situ*.
2. *Polymer-concrete (PC)*—This is an aggregate bound with a polymer binder.
3. *Polymer-cement-concrete (PCC)*—This is a premixture of cement paste and aggregate to which a monomer is added prior to curing.

Generally the polymer imparts higher strengths, but because of its relatively high cost, its use has been limited to bridge decks, overlays, or repair of spalled joints. Polymer-impregnated concrete also has greater resistance to frost action and deicing salts.

Preparation and curing of polymer-concrete mixtures call for revisions to the processes described here for normal concrete.[50]

[50]For added detail see *TRB Record 542, NCHRP Report 190, American Concrete Institute Special Publications, Polymers in Concrete, SP-40,* 1973, and *SP-58, 1978, and FHWA Report RD-75-507, Introduction to Concrete-Polymer Materials.* See also several articles in the *Transportation Engineering Journal of ASCE,* Feb. 1975 and May 1977.

CONCRETE PAVEMENT CONSTRUCTION

Concrete pavement on major projects is constructed with highly specialized mechanical equipment operated by skilled workers. The entire operation is carefully organized, with each worker assigned definite tasks and responsibilities. Inspection is exacting. The result of these efforts is a finished pavement of uniformly high quality and excellent smoothness.

Until the early 1970s, equipment and procedures similar to those illustrated by Fig. 20–10 were almost universally followed on large concrete pavement projects. Ingredients, except water for the concrete, were batched into compartmentalized dump trucks at a central plant. At the paving site, the trucks discharged the batches, one at a time, into a mixer that traveled along the roadway. Concrete was placed between steel side forms that were fastened to the subgrade; machines that spread the concrete and finished its surface rolled on these side forms.

Currently, two modifications to procedures involving side forms and job-mixing are being widely used. The first is the substitution of mixing at the plant for mixing at the job site. "Central mixing" is conducted at the batch plant and the concrete is hauled to the paving site in dump trucks. The second innovation is the slip-form paver. This machine carries the side forms with it along the roadway. It also mounts elements that spread and vibrate the concrete and strike off the surface to grade as the paver goes by.

Fig. 20–10. Typical concrete paving operation. Tribatch paver in right foreground with skip raised. Pavement in rear has been sprayed with opaque curing compound. (Courtesy, Koehring Co.)

To construct concrete pavements with central-mix slip-form equipment requires a very substantial investment. For this reason it is economical primarily on projects where paving runs are long and continuous. On smaller projects, on ramps at interchanges, or on some city streets it may be cheaper to job-mix the concrete or to supply it in ready-mix trucks and to place it between side forms.[51] Where pavements are of variable width or have warped surfaces, hand strike off and finishing also may be appropriate.

Each of the steps in concrete-pavement construction is discussed briefly in the pages that follow. Methods and standards for the construction of embankments and undercourses have been presented in Chapters 15 through 17.[52]

Side Form Job-Mix Procedures

SIDE FORMS. Side forms outline the sides and define the top surface of the concrete slab. Heavy wooden planks were used first. With the advent of heavy finishing machines that traveled on the side forms there was need for greater strength and durability, and steel forms were developed. Steel forms have a "channel" cross section with a broad flange at the base. The outer edge of the top flange is turned down to provide added rigidity. In addition, the section is stiffened at intervals. Provision is made for steel stakes to pass through the form into the subgrade, and stake and side form are fastened tightly together by steel wedges that can be tightened or loosened with a few blows of a hammer. Individual sections are commonly 10 ft long and are keyed together by a sliding steel tongue.

Side forms are set carefully to line and grade on the prepared undercourse, the steel stakes are driven, and forms and stakes are locked together.

FINAL SUBGRADE PREPARATION. After the side forms are set, the subgrade between them is dressed to close tolerances. Originally this was done by hand-labor methods. Today a machine called a "subgrade planer," which travels on the side forms, shapes the subgrade and by a conveyor system deposits the excess material outside the forms. The subgrader is followed by a roller that recompacts the surface. Finally a template shaped to the exact cross section is dragged along to knock off any remaining high spots or to mark them for removal.

Where the design calls for assembled joints and dowels or other load-transfer devices, a crew placing them follows the subgrading operation. Extreme care in placing and aligning them is a must. Before concrete is placed, the subgrade should be carefully moistened to prevent absorption of water from the concrete.

PROPORTIONING JOB-MIXED PAVING CONCRETE. Materials for paving concrete are carefully proportioned at a "batching plant" set up at the aggregate

[51]Rapidly increasing asphalt prices have resulted in an increase in the use of concrete for city streets. *NCHRP Synthesis 27* gives an excellent discussion of construction practices for low-volume roads and city streets.

[52]Requirements governing the construction of concrete pavements will be found in specifications of individual state agencies. See also the *AASHTO Guide Specifications for Highway Construction*.

source or some other carefully selected location. Batching plants usually consist of a group of elevated aggregate bins equipped with a weigh box and multiple-beam scales for weighing aggregates. They include a separate silo and scales for storing and weighing bulk cement. Dump trucks with beds divided into units holding a single batch haul the proportioned materials from plant to mixer.

Care must be exercised not to lose bulk cement from the batch trucks. The usual operation places the cement in the truck between layers of aggregate. If the cement is placed on top of the load, it must be covered with a tarpaulin. Batches combining cement with moist aggregate must be used within a few hours; otherwise the cement will harden in the truck.

MIXING AND PLACING JOB-MIXED CONCRETE. Paving concrete commonly is mixed at the paving site in large crawler-mounted machines that produce a cubic yard or more of concrete per batch (see Fig. 20–10). The early machines were of the "single-drum" type, with all mixing accomplished in a single horizontal cylinder. Then mixers with two and three drums end to end were developed. Partial mixing in one drum is followed by transfer to and added mixing in the next.

A mixer is loaded by backing a batch truck into the large "skip" at its forward end. After the truck has discharged a batch and pulled away, the skip is raised, and the material slides into the mixer. At the same time, a measured quantity of water flows into the drum. After the concrete is mixed for the prescribed time, it is discharged into a bottom dump bucket that travels out a horizontal boom and spreads its load uniformly over the subgrade. Observation of minimum mixing periods is enforced by automatic timing devices that lock the discharge lever during the mixing period and ring a bell to indicate its end. Most specifications ban mixers with defective timing mechanisms. The effects of mixing time on concrete quality and of this and other factors on paving costs have been carefully investigated.[53]

Water for the mixer is supplied either by a flexible hose connected to a pipeline that extends along the project or by a truck or trailer moving along with the mixer. All specifications require accurate devices for metering mixing water. Generally measurement is by volume in a calibrated tank that can be set to supply any desired amount, although weight proportioning devices also are employed.

Transit mixing of paving concrete is often done for small projects which do not warrant job mixing or setting up a special central mix plant. Sometimes the weighed but unmixed aggregates, cement, and water are charged directly to the drum of the ready-mix truck and mixing is accomplished as the vehicle travels from batch plant to job site. Again, partial mixing called "shrink mixing" may be done at the batch plant in order to charge a larger load to the transit mix truck.

The usual transit-mix truck has a capacity of approximately 7 yd^3, since larger loads plus the weight of truck and drum will exceed the maximum axle loads

[53]See D. O. Woolf, *Public Roads,* Apr. 1960, or *HRB Bulletin 265, Public Roads,* Feb. 1962 and M. J. Kilpatrick, *Public Roads,* Apr. 1960, or *HRB Bulletin 265.*

permitted on public highways and streets. There are, however, specially designed trucks with more axles that haul about 14 yd^3. Mixing is controlled by stipulating the number of revolutions of the mixer drum, with minimums ranging from 50 to 70. Maximum retention times in the mixer range from 45 to 90 min.

SPREADING AND FINISHING CONCRETE WITHIN SIDE FORMS. As shown on Fig. 20–10, several machines in sequence, each traveling on the side forms, place and finish the concrete pavement. Originally, these spreading and finishing operations were performed entirely by hand-labor methods. However, because of increasing labor costs and demands for uniform quality, hand-labor methods are now used only on transitions or other sections with variable widths or warped surfaces.

The three machines in most widespread use with side forms are the mechanical spreader, the transverse finishing machine, and the longitudinal finisher. All are self-powered.

The mechanical spreader distributes the concrete deposited by the mixer uniformly and without segregation. Where mats of bar or mesh reinforcing steel must be placed in the slab, one approach is to place one lift of concrete to the level of the reinforcing. After the mats are laid in place, the remainder of the concrete is placed and leveled by a second trip of the spreader. This method can result in delaminations between lifts. Other agencies use single lifts where the mat is vibrated in from the surface or secured with chairs to hold the bars at the correct height. Spreaders often mount vibrators to ensure that the concrete is densified and that no voids occur against the side forms. Next in the sequence comes a transverse finishing machine. At the forward edge of the machine proper is a strike-off screed that moves back and forth transversely and levels the concrete surface flush with the tops of the side forms. Some concrete must remain piled up ahead of the screed to ensure that no depressions remain in the surface. At the rear is a tamping bar that works the coarse aggregate down from the surface in order to make subsequent finishing operations easier. Among the features of most transverse finishers are adjustments for placing a slight arch in the screeds if a parabolic or other curved surface is desired. Following the transverse finishing machine is a longitudinal finisher, often called a longitudinal float. The strike-off bar is set parallel to the roadway centerline and moves across the roadway in a sawing motion. This action cuts off high spots and fills in low ones along the direction of vehicular travel. After a transverse pass is completed the machine moves ahead approximately half a float length to its next position. The overlap in succeeding passes eliminates abrupt changes in the surface in the direction of traffic. After the longitudinal float has passed, surface texturing is completed and the curing membrane applied.[54] These procedures are identical or very similar to those employed with the slip-form paver, as described below.

[54]Care must be taken not to overwork the concrete. Excess finishing will cause bleeding which makes the concrete more susceptible to scaling and other surface deterioration.

Central-Mix Slip-Form Procedures[55]

As indicated above, long, continuous runs of concrete pavement (either new construction or overlays) can be placed more quickly and economically by central-mix, slip-form procedures than by the side-form job-mix or transit-mix methods. Concrete mixers with capacities up to 15 yd^3 and pavers that can place a 48-ft-wide strip in a single pass are in common use. Slipform techniques are also common for curbs and gutters.

SUBGRADE AND SUBBASE PREPARATION. Base-course materials, whether treated or plain, are more expensive volume for volume than those in the underlying layers; portland-cement concrete is even more costly. It follows that carefully trimming the surfaces of subgrade and subbase to close tolerances will reduce the quantities of the more expensive materials and reduce pavement cost. Today, machines are readily available to do this surface trimming. One such machine is pictured in Fig. 20–11.[56]

As shown by Fig. 20–11, subgrade or subbase trimmers have a width greater than the finished pavement. Commonly they are mounted on segmented tracks which travel on the compacted layer being trimmed. Rotating teeth or blades cut the surface to the desired level; then the loosened material is collected in a windrow by an auger or similar device for picking up by a loader or scraper. Alternatively the surplus is placed on a belt and deposited outside the working area or loaded directly to hauling units. A smooth wheel roller may follow the trimmer to recompact the surface.

Line and grade are maintained by electronic sensors controlled by a tight guide wire or polyethylene cord stretched on one side (or on both sides) of the work area. Signals from them steer the machine and set the elevation of the trimmer blades. With these devices, undulations of the machine do not affect the elevation of the finished surface. It probably can be anticipated that in the future laser beams or other means may replace wire or cord controls on at least straight stretches of pavement.

As discussed earlier, it is now common practice to provide a granular or treated subbase over the subgrade. A variety of procedures for placing and trimming these subbases have been found acceptable. The aim is to place slightly more than the required amount of material in a manner that prevents segregation. Spreader boxes, blade graders, or special spreading machines all can produce acceptable results. Granular bases or plant-mixed cement- or bituminous-treated bases are compacted and then trimmed to precise grade as noted above.

Econocrete (lean concrete) bases are gaining in popularity.[57] They can be placed using slip-form pavers (without compaction) and provide an accurate

[55]For detailed descriptions of early slip-form paving methods, see *HRB Record 98* and *ARBA Technical Bulletin 263*. More recent developments are reported in occasional articles in construction periodicals and in brochures prepared by equipment manufacturers. See also G. K. Ray et al., *Transportation Engineering Journal of ASCE,* Nov. 1975.

[56]Several manufacturers make a complete line of concrete paving equipment; Fig. 20–11 and those that follow each illustrate one among several machines for carrying out the particular operation.

[57]For current practices, plus bibliography, see *W. A. Yrjanson* et al., *TRB Record 741*.

Fig. 20–11. Machine for trimming surface of subgrade or subbase. Conveyor deposits excess outside the work area. Notice extension arm which supports electronic guide. (Courtesy, R. A. Hansen Co., Inc.)

grade which does not require planing as do other bases. An advantage of econocrete over cement-treated base is a more resistant surface that is less likely to erode from "pumping" at the pavement joints which will result in faulting.

At times, the subbase may be produced by cement treating the in-place subgrade material or by adding cement to an already-placed granular base. This process also has been highly mechanized and a single machine may pick up the native material and cement, add water, mix the ingredients, and leave them spread ready for compaction and trimming. On smaller projects, this sequence of operations may be performed by several less expensive machines.

CENTRAL PROPORTIONING AND MIXING THE CONCRETE. On large projects, aggregates are proportioned and the concrete is mixed at a central plant and hauled to the paving site in dump trucks. A typical batching and mixing arrangement is shown in Fig. 20–12. It consists of separate stockpiles of aggregates by sizes from which material is transferred to a series of bins above the weigh hopper. Cement commonly is delivered in bulk in rail- or truck-mounted hoppers and stored in silos. Predetermined weights of each size of aggregate and water[58] are charged into the mixer, which is usually of the tilting type. The

[58]Volume measurement of water is generally permissible. Nuclear methods for measuring water content also are available. See F. A. Iddings et al., *TRB Record 539*.

Fig. 20–12. General view of two central mix plants for paving concrete. Aggregates in three sizes are delivered by separate conveyors to the aggregate storage bins (upper left). They are batched by scales under the hopper and delivered by conveyor belt to the horizontal drum mixer. Cement from horizontal, ground-level storage bins (foreground, left and right) is moved by compressed air through multiple pipes to vertical cement storage bin (top of plant). Cement scales below bin weigh cement for delivery to horizontal-drum mixer. After preliminary mixing in horizontal drum, batch is transferred to tilting mixer. It is shown dropping a completed batch of concrete into a dump truck for delivery to the paving site. A second plant appears in the right background of the photograph. (Courtesy, Construction Machinery, Rexnord.)

entire process must be carefully controlled to produce concrete of uniform quality and consistency.

Control begins with attention to aggregate handling to prevent segregation. It has been found that casting or dumping the material into horizontal layers not over 3 ft in thickness is the most suitable procedure. On the other hand, forming single large cones of aggregate at the end of conveyors or by dropping from clamshell buckets in the same spot gives poor results. Where materials are transferred from stockpile to batch plant with front-end loaders or are being shifted with bulldozers, both segregation and aggregate degradation can result if precautions are not taken.[59]

Accuracy in proportioning concrete materials is demanded. For example, *AASHTO Guide Specifications for Highway Construction* stipulate tolerances of 1% for cement, 2% for aggregate, and 3% for admixtures. It is common to require that the weighing devices be set and locked, be fully automatic, and that

[59]See *NCHRP Reports* 5 and 46 for added detail and extensive bibliographies.

the individual quantities charged to each batch be recorded. Mixing time also is carefully and automatically controlled, although minimums differ substantially among agencies, ranging from 50 to 90s. Some permit lowering the stipulated times when tests indicate suitable concrete strengths can be obtained. Excessively long mixing times are also to be avoided with air-entrained concrete, since they lead to reduced air contents.

Particularly with slip-form pavers, careful moisture control is essential, since even a slight increase in fluidity will cause the unsupported edges of the pavement to slump. It is common to require that the aggregates be retained in the stockpiles for a long enough period that their moisture content stabilizes.

HAULING CONCRETE. Almost all agencies permit delivery of central mix concrete to the job site in dump (nonagitating) trucks. For relatively stiff and cohesive air-entrained paving concrete, segregation is not a problem. However, controls ranging from 30 to 60 min are imposed on the length of time the concrete remains in the truck. Some agencies require that the concrete be discharged directly into the spreading or paving machine; others permit dumping on the prepared subgrade or subbase ahead of the spreader or paver.

PLACING REINFORCING STEEL AND LOAD-TRANSFER DEVICES. Reinforcing steel must be placed at the stipulated height in the paving slab. For simply reinforced concrete pavements, one method is to support the tied mats or mesh securely on special chairs that are driven into the subgrade and to spread and consolidate the concrete under and over the steel. The second involves spreading the concrete to slightly over full depth and slightly under full width; then the preassembled mats of steel are placed on the surface of the fresh concrete and pushed down to correct position by a mesh-depressing device that travels ahead of the slip-form paver. For continuous reinforcement the steel is supported on chairs or the bars are placed in the fresh concrete through tubes in the paving machine. With tubes the transverse steel is eliminated. Selection among the procedures rests with each agency's specification or the contractor.

Where steel dowels are specified to provide load transfer, it is common to group them in a single rigid assembly made up of the dowels and reinforcing steel. This, then, is firmly supported on chairs driven into the subbase. Other agencies permit dowel installation by a device that positions them in the fresh concrete. Extreme care is necessary to keep dowels in exact alignment.

PLACING AND CONSOLIDATING THE CONCRETE.[60] Figure 20–13 shows one among several available slip-form paving spreads, including spreader, reinforcing steel placer and depressor, slip-form paver, and diagonal finisher. Figure 20–14 is a rear view of a slip-form paver, showing the appearance of the slab before final smoothing. The distinguishing feature of all such machines is that the side forms progress with the paver. These side forms support the vertical faces of fresh concrete during the spreading, placing, vibrating, and striking off of the surface. In earlier machines, the slip forms were very long and projected

[60]*NCHRP Synthesis 44* is an excellent reference on consolidation of concrete.

Fig. 20–13. Typical slip-form paving operation. Concrete from dump truck (foreground) is distributed by track-mounted spreader. Strands of reinforcing steel cross left side of spreader and are spaced and depressed into concrete by second machine, which is mounted on rubber tires. Track-mounted slip-form paver (third machine) strikes off and consolidates concrete. Diagonal float (tube finisher) provides final surface leveling. (Courtesy, Construction Machinery, Inc.)

far to the rear of the machine. But this feature was abandoned when it was found that concrete of the proper consistency did not need support after the machine had passed.

In place of a separate spreading machine, some slip-form pavers carry a hopper into which the fresh concrete is dumped from the transporting vehicles. Conveyors, augers, or other devices for uniformly spreading the concrete are provided. All pavers incorporate vibrators. These may be a series of individual spud vibrators that project down into the concrete; some machines also have horizontal tube or pan vibrators. Following the vibrators are oscillating screeds which move at right angles to the direction of travel. These strike off the surface of the concrete to its final elevation. Finally, a transverse float smooths the surface to remove marks left by the strike-off mechanism. When the machine has passed, the surface is ready for final finishing.[61]

Figure 20–9c details a typical longitudinal joint. These are installed between

[61]L. C. Bower and B. B. Gerhardt, *HRB Record 357,* report that better vibration increases concrete strength by as much as 10% as well as improving resistance to surface polishing.

Fig. 20–14. Rear view of slip-form paver. The side forms have already passed, but the edges of the slab stand without support. The photograph clearly shows the track mounting and electronic grade and direction controls. (Courtesy, Construction Machinery, Inc.)

the individual pavement lanes. With multilane slip-form machines, a notched wheel device can be employed to force deformed tiebars into the concrete to the proper depth. Also, a continuous premolded plastic strip can be fed from a reel and inserted vertically just below the surface of the concrete to provide a weakened plane longitudinal joint.

SURFACE FINISHES FOR CONCRETE PAVEMENTS. After the paver has passed, any remaining surface undulations are removed with a diagonal float or tube finisher (see Fig. 20–15). This step is required to meet the exacting smoothness requirements (see below). This machine makes several passes over the surface. It forces the coarse aggregate slightly below the surface and also removes high spots and fills in low ones. Often the machine can supply water to provide a very fine mist in case the surface has become too dry to finish.

Particular care should be exercised in the surface-finish stage to avoid weakening the mortar by increasing its water–cement ratio. Excess working of the surface by the finishing machine or with the hand-operated bull float (a wide board attached to a long handle) can bring water up from the concrete below and add it to the mortar. Excess sprinkling will have a like effect. In time the weak mortar will wear away in the wheel tracks and the pavement will become slick.

Originally, concrete pavements were given a "broom" finish. This was obtained by lightly brushing the surface transversely with a heavy, stiff broom. Later, a "belt" finish was used. For it, workers drew a narrow strip of canvas or

Fig. 20–15. Self-propelled diagonal tube finisher. Machine is rubber mounted; horizontal control for travel to either front or rear by guide wire. Spray bars provide added water, if needed. The tube itself floats on the surface of the concrete. (Courtesy, R. A. Hansen Co., Inc.)

burlap back and forth across the surface in short strokes. A few inches of forward progress were made with each stroke. Both brooming and belting required hand labor, and have largely been abandoned on major work. A common practice is to attach a burlap drag to the rear of the finishing machine. This drag is lowered onto the pavement surface as the finishing machine makes its last forward pass. At times, the surface may be further textured by brooming, but the skid resistance it produces may not be lasting.[62] Transverse grooving with metal tines is now often required in finishing to provide a more durable skid-resistant texture.[63] Another control on surface finish is to stipulate the minimum coefficient of friction that the hardened surface must show. For example, the specifications might stipulate that the surface must develop some minimum coefficient of friction or the surface be ground or scored to produce it.

Checking Concrete Pavement Smoothness and Density

Almost all agencies check surface smoothness by means of long straightedges placed in successive positions parallel to the roadway centerline. Advance along the roadway is in increments measuring half a straightedge length. Most

[62]See D. L. Spellman and L. S. Spickelmire, *HRB Special Report 101*, J. G. Rose and W. B. Ledbetter, *HRB Record 357*, W. P. Chamberlin and D. E. Amsler, *HRB Record 389*, and D. C. Malone et al., *TRB Record 652* for further information and bibliographies on surface finishes for concrete pavements and means for developing and maintaining skid resistance. For British practice see W. E. Murphy and D. P. Maynard, *Transportation Engineering Journal of ASCE*, Feb. 1975.
[63]See D. C. Malone et al., *TRB Record 652* for texturing experiments in Virginia.

agencies use a straightedge 10 ft long and limit variations under it to ⅛ in. Tolerances more and less severe than these are specified by some. Straightedges are swung from handles which permit easy inspection of soft concrete. The first check is made while the concrete is still plastic. Low spots are filled with fresh concrete, and high ones are cut down, after which the surface is refinished. A second check is conducted after the concrete has set, usually on the succeeding day. Only minor corrections are permitted at that time. More serious irregularities are cause for grinding or removal and replacement of the affected sections. A few agencies also check transverse smoothness. For example, the California Department of Transportation limits transverse variations under a 12-ft straightedge to 0.02 ft.

Some agencies check the roughness of concrete and other pavements by means of road-roughness indicators. These devices, when towed or pushed along the pavement, summarize the irregularities in the road surface. Results commonly are stated in inches of roughness per mile.[64]

Obtaining proper consolidation of portland-cement concrete during construction is a prerequisite for good quality pavements. Nuclear devices are now available to continuously and automatically monitor the degree of consolidation of newly placed concrete by back-scatter techniques (see Chapter 14) which measure density at surface speeds of abour 2 ft/min.[65]

Curing Concrete Pavements[66]

Concrete gains strength as the chemical action between cement and water (hydration) proceeds. If the concrete dries out, hydration and strength gain stop. When water again becomes available, strength gain is resumed. Thus, even with sporadic wetting, concrete will eventually gain considerable strength. With pavements, however, rapid drying of the fresh concrete results in crazing or cracking of the surface. Where these conditions are extremely unfavorable because of hot weather or dry winds, severe cracking occurs. Although hydration will resume with rewetting, the fine hair cracks caused by drying will not heal. From these facts it is clear that any curing method that prevents drying of the newly finished surface and that keeps the concrete continuously moist is satisfactory.

Curing may be accomplished by a variety of methods, with most of the agencies permitting contractors to choose among two or more alternatives. Basically, all the methods fall into two categories: (a) those that keep the surface wet or cover it with a water absorbent material that is rewet from time to time and (b)

[64]For an account of the development of these "profilometers" and a detailed description of the one used in California, see F. N. Hveem, *HRB Bulletin 264*. Details of its use are given by L. R. Gillis and L. S. Spickelmire in *HRB Record 98*. See also J. E. Haviland and R. W. Rider, *HRB Record 316*. For New York's experience, see J. E. Bryden, *TRB Record 535*. See also R. M. Weed, *TRB Record 539*.

[65]For a discussion of this device, see T.M. Mitchell et al., *Public Roads,* Mar. 1979.

[66]See *TRB Circular 208* for an excellent discussion of various curing practices plus a bibliography. See *TRB Record 539* for California practices and products.

those that prevent evaporation of the water already in the concrete which, if retained, is sufficient for hydration.

The early pavements were often cured by outlining the surface with low earthen dikes and filling these ponds with water. Alternately, the surface was covered with earth, hay, straw, or sawdust, which was kept wet. However, high labor costs have generally outmoded these procedures. Later, covers of felt or cotton mats, or two layers of burlap, kept wet by sprinkling, were widely used. The difficulties with all these methods are, first, that the cover must be both placed and removed and second, the numerous repetitions of sprinkling to keep the cover wet.

Covers of waterproof paper or polyethylene plastic are also used. These are very effective in preventing evaporation. At times, there is difficulty in holding them in place against the wind; also they must be removed after the curing period. With care, several reuses are possible, which reduces the cost of the covering material.

A sprayed-on, impervious membrane is now most widely used. It can be applied immediately after finishing, and, once spread, requires no further attention. (Wheel-mounted spray carts straddling the completed pavement appear at the top of Figs. 20–10 and 20–15.) At first the "clear" curing compounds were most popular. These contain a fugitive dye that assures uniform coverage but disappears after spreading. Currently, pigmented compounds (including chlorinated rubber) that leave a white or light gray surface are standard. Their main advantage is that they reflect the radiant heat of the sun. This reduces the temperature rise in the slab which, in turn, minimizes crack formation when the slab cools.[67]

Harsh brooming or tined grooving creates more surface to be coated with curing compound and the application rate should be increased to compensate for it. Also, the amount should be increased on hot, dry days, at least for pavements placed in the morning.

Both climate and time of placement affect early cracking in reinforced pavements. Observations in Oregon found that reinforced pavements with contraction joints at 61.5-ft centers cracked between joints if the concrete was placed during the morning on hot days. Concrete laid later so that the critical stage of curing went into the cool evening did not crack. Few early cracks developed in the fall or on cloudy days. Cracking also was less frequent if the curing compound amount was increased by 50%.

Most agencies set the minimum permissible curing period between 3 and 7 days, but a few specify intervals greater than a week. Also, construction or regular traffic must be kept from concrete pavements until they can carry the superimposed loads without suffering damage. This is assured by specifications that defer use for a stated number of days after placing or until test specimens develop a stated flexural or compressive strength. The shortest period is, for normal concrete, 5 days, and the lowest acceptable flexural stress is 450 lb/in.^{2} If

[67]Specifications for liquid membranes, waterproof paper, polyethylene sheeting, and burlap are given as *AASHTO Designations M148, M171,* and *M182*. Tests for impermeability appear as *AASHTO Designation T155*.

the design requires sealing of joints, they must be cleaned and the sealer placed before the pavement is opened to traffic.

In cold climates, even with air-entrained concrete, freezing and thawing and deicing salts may cause deterioration of the surfaces of concrete paving and bridge decks. Research indicates that added protection can be gained by lightly coating the completed and cured surface with linseed oil or other substances.[68]

Cold-Weather Concreting

Concrete hydrates and gains strength very slowly at temperatures near the freezing point. Furthermore, severe damage results if fresh concrete freezes. For these reasons, many agencies prohibit paving when the air temperature is below some minimum such as 40°F. Sometimes two minimum temperatures are set, for example, 35°F when the air temperature is rising and 40°F when it is falling. Where concrete placement is permitted during cold weather, heating of the aggregates, mixing water, or both is usually required. The *AASHTO Guide Specifications* state that "the temperature of the mixed concrete shall not be less than 50°F nor more than 90°F at the time of placing in the forms." After placing, heat generated by hydration helps to keep the slab warm so that strength will develop. Some agencies take the added precaution of insulating the finished slab to prevent heat loss.

QUALITY CONTROL AND PAYMENT FOR CONCRETE PAVEMENTS

Payment for concrete pavements is usually on a unit price basis, most commonly stated as square yards of completed pavement surface. Compensation for subgrade preparation and watering, furnishing and placing side forms, curing, and other work are included in this price. In most instances, pay for joint materials, dowels, and like items is also included in the unit price of the pavement, but some agencies treat them as separate items. Compensation for reinforcing bars or welded-wire fabric is sometimes included in the price of the pavement and sometimes made separately at a price per pound or per square yard of pavement. Many agencies check the thickness of completed pavements by taking cores or measuring thickness by nuclear methods at intervals along the roadway.[69] No added payment is given if the slab is thicker than specified; however, payment may be reduced or denied when these tests show the slab to be thinner than specified.[70]

MAINTENANCE OF CONCRETE PAVEMENTS

Routine maintenance of good concrete pavements consists largely of sealing cracks and transverse and longitudinal contraction and expansion joints. Wide

[68]See *TRB Records 504, 539,* and *652.*
[69]See *NCHRP Report 168.*
[70]See Chapter 15.

cracks or spalled joints must first be blown out with compressed air or cleaned in some other manner and then be sealed to prevent the intrusion of extraneous material and to block the downward penetration of surface water. There is serious question, however, whether sealing narrow cracks is effective, and many maintenance engineers recommend against it.[71]

Where concrete pavement is in good condition except for small, scattered areas that have broken, patching with concrete is economical. At affected locations, the old concrete should be removed by vertical straight-line cuts parallel and perpendicular to the roadway centerline. For corner repairs, the angle from the roadway centerline should be greater than 30° and less than 60°. Sections removed should have minumum areas of 20 ft^2. The new slab should always be as thick as the original pavement, and, if the subgrade is questionable, it should be replaced with suitable material, properly compacted. It is excellent practice to provide extra thickness adjoining the old concrete and to extend the patch under the old pavement to a width and depth of at least 4 in. High early-strength concrete produced with extra cement, high early-strength cement, or an accelerator such as calcium chloride, is commonly used for patching in order to gain early use of the pavement, often by the evening of the day the repair is made.[72]

Areas that have spalled from freezing and thawing or salt action often are leveled with bituminous mixes or surface treatments; another approach calls for a patch of concrete bonded to the old concrete with a cement–water mixture or epoxy. Spalled joints also may be built up with cement grout containing epoxy.[73]

A maintenance technique known as "mud-jacking" is widely used to restore subgrade support under pumping pavements and to level uneven slabs. A liquid filler is forced under the slab through previously drilled holes. This "mud" or "slurry" fills all vacant spaces and, if injection continues, applies hydraulic pressure which forces the slab upward. Subsided pavements have been raised several inches without damage to the concrete. Slurrys for mud-jacking usually consist of mixtures of fine-grained soil, portland cement, and water; or of these three ingredients plus a small amount of asphalt. Slurrys having one part of cement and three-four parts of ground limestone also have served. The mixture must pass through the pump without gumming and flow readily under the pavement and into small voids. It should harden quite rapidly without excessive shrinkage and without becoming hard and brittle. Spacing of holes through the concrete should be carefully planned. Where slabs are to be lifted by pumping slurry into several holes, the nozzle should be moved frequently so that lifting is in small increments.

An alternative to mud-jacking is "undersealing." Here a heated asphalt, often an air-blown product, is substituted for the slurry. An advantage of undersealing

[71]See *TRB Record 598* for a discussion on resealing joints.
[72]*TRB Special Report 113* illustrates the more common failures. See also *NCHRP Syntheses 9, 25, 56* and *60* for methods to repair concrete pavements.
[73]Specifications for epoxies for various purposes appear as *M234* through *M237*.

is that the asphalt forms an effective seal against the penetration of surface water into the subgrade. Many highway agencies have developed special equipment combinations solely for mud-jacking or undersealing purposes. These are truck mounted and proceed down the highway or airport runway as a train.

Where concrete surfaces have become uneven, the joints have faulted, or slight cracking occurs, it is common to resurface them with a bituminous blanket (Chapter 19). In time, however, the joints and cracks in the old concrete break through the new surface forming "reflection cracks." Among the solutions explored to prevent them from forming are to reinforce the bituminous layer with mesh[74] and the use of asphalt–rubber (Chapter 14) interlayers to break the bond. Fabrics also have been widely employed but are not yet proven for portland-cement-concrete pavements.[75]

At times reflection cracking has been reduced by breaking the old concrete with a heavy roller or by placing a layer of granular base before bituminous resurfacing.[76] Reflection cracking can also be slowed by using asphalt–concrete overlays thicker than 4 in. Recycling (discussed in Chapter 21) has been and will continue to be a method for repair of worn concrete pavements.

As indicated earlier, the surface of some concrete pavements may polish under traffic, reducing the coefficient of friction below acceptable levels. One procedure for correcting this condition is to cut longitudinal grooves in the surface. Favorable dimensions are ⅛ in. deep and ⅛ in. wide spaced at ¾ in. on centers.[77] The relationship between grooving and coefficient of friction is uncertain. It is clear, however, that grooving helps to eliminate "hydroplaning" and skidding.

Another technique, using the British Klarcrete machine, hammers the surface and removes ⅛ to ¼ in., improving the coefficient of friction substantially.[78]

TWO-COURSE PAVEMENTS WITH CONCRETE BASES

Many miles of roads and streets with brick wearing surfaces supported on concrete bases have been built in the United States. Although brick pavements are no longer commonly constructed, some of them remain in service. A large additional mileage has been covered with bituminous material and so is actually still in use. New brick pavements are now limited to decorative applications like sidewalks, driveways, and transit malls.

Brick pavements are in three courses; the concrete base, a mastic or bedding

[74]See E. Tons et al., *HRB Bulletin 290,* for a 5-yr report on the performance of two test projects and a detailed bibliography on earlier work. See also W. E. Chastain, Sr. et al., *HRB Record 61,* and F. Copple et al., *HRB Record 239.*

[75]See articles by G.R. Morris et al., *TRB Record 595,* and N.F. Coetzee and C.L. Monismith, *TRB Record 700,* for discussion on rubber–asphalt interlayers. See K.H. McGhee, *TRB Record 700,* for the use of fabrics.

[76]See *HRB Record 239* and *327.*

[77]See J.L. Beaton et al., *HRB Special Report 101,* and E.E. Farnsworth, *HRB Special Report 116.* See also *TRB Record 633.*

[78]See *TRB Record 484.*

course, and the brick surfacing held in place with a filler. In the earliest pavements, the mastic or bedding course was of sand, and later of cement grout. More recent practice used a mastic composed of sand and 5–8% of fluid bituminous binder, usually *RC*- or *MC*-250 cutback asphalt or a cutback tar. Paving brick is hard burned and extremely tough. It also differs from conventional brick in that special lugs are formed on one side and one end to hold the placed brick slightly apart to leave room for the filler.

Clean sand was originally employed as filler between the bricks. Later, cement grout was used, but it produced such rigidity that the pavement was subject to severe cracking. Finally, bituminous filler became standard. It generally was a hard grade petroleum asphalt, often of 23–32 or 30–45 penetration. Before the heated filler was poured over the laid brick the exposed surface was treated with a separating agent in order that the excess asphalt could be stripped away easily.[79]

Stone blocks are a principal paving material in many European cities. However, only a limited mileage of them were constructed in the United States. Some are still in service, although commonly covered with an asphalt-bound surface layer.

PROBLEMS

20–1. For a concrete paving slab 10 in. thick, and with conditions and properties as given in Fig. 20–2, find the maximum combined flexural stress in the slab:

a. When the slab is 30 ft long.
b. When the slab is 15 ft long.
c. When the slab is 10 ft long.

Compare these stresses with the safe flexural stress in plain paving concrete.

20–2. In an attempt to reduce flexural stresses in a concrete pavement, the thickness was increased from 8 to 11 in. What would be the reduction in combined stresses in the daytime for edge loading:

a. If the slab were 30 ft long?
b. If the slab were 15 ft long?

20–3. The current practice of a certain highway agency is to space transverse joints in concrete pavements at 40 ft centers. The suggestion has been made that curling stresses along the edge of the slab can be substantially reduced by reducing the joint spacing. If $E=5\times10^6$, $\mu=0.15$, $h=8$ in., $k=100$, $e=0.000005$, and $t=3$, find the curling stress along the edge of the slab for:

a. The present 40-ft joint spacing.
b. A 30 ft-joint spacing.

[79]For a detailed description of materials and construction practices for brick pavements, see L. I. Hewes, op. cit., Vol. II, pp. 337–374.

c. A 15-ft joint spacing.

Where possible check your answers against the plotted values on Fig. 20–2.

20–4. Work Problem 20–3 for a 10-in. slab thickness.

20–5. A highway agency has a substantial mileage of 8-in. portland-cement-concrete pavement for which the flexural static strength is 700 psi. It was placed on a 6-in. granular subbase. At the time of construction, permissible axle loads were 20,000 lb single and 32,000 lb tandem, but they have now been raised to 24,000 lb single and 40,000 lb tandem.

A given stretch of six-lane divided highway, built on a relatively poor subgrade (modulus of subgrade reaction k of 100) now carries 25,000 vehicles daily. It is estimated that one effect of the change in load limits will daily add 20 single-axle loads of 30,000 lb and 10 tandem-axle loads of 50,000 lb (both illegal) to the outer traffic lane.

For this situation, calculate the increase in the percentage of fatigue resistance that will be consumed in the next 20 yr. Follow the Portland Cement Association design procedure. Set the load safety factor at 1.2. (*Note:* The student should make a critical evaluation of the accuracy of the assumptions underlying this problem and the sensitivity of the results to these assumptions.)

20–6. Do Problem 20–5, assuming that the 6-in. subbase is cement-treated and that Portland Cement Association values for the combined modulus of subgrade reaction k apply.

20–7. Using the *AASHTO Interim Design Method,* select the thickness of portland-cement-concrete pavement for a six-lane urban freeway. As in Problem 19–8, estimated daily traffic is 25,000 vehicles and is 20% trucks. Of these, 40% are two-axle six tire, 10% are three-axle single unit and 20% three-axle combinations, 15% are four-axle, and 15% are five-axle. Load distribution among axles is as shown in Problem 19–12. Assume that 80% of all axle loads will travel in the right-hand (design) lane. Static flexural strength of the concrete is 650 psi; the native soil has a k value of 100; it is overlain by a 6-in. untreated subbase. The modulus of elasticity for the concrete is 4×10^6 psi.

20–8. Work Problem 20–7 assuming that the subgrade with k value of 100 is overlain by a 6-in.-thick cement-treated subbase. Use the *AASHTO Interim Guide* value for the combined modulus of subgrade reaction.

20–9. What total cross-sectional area of longitudinal reinforcing steel is required for a slab 12 ft wide, 40 ft long, and 10 in. thick? The concrete weighs 150 lb/ft^3; working stress in the steel is 35,000 psi. Employ the recommendations of the *Interim Guide* to obtain other needed data.

20–10. On the basis of data offered in Table 20–6 determine the weight of cement, sand, small and large coarse aggregates, and water to fill an 8 yd^3 tilting mixer to capacity, under the following conditions:

a. Plain concrete, crushed-stone aggregate, 6 bags cement/yd^3 concrete, all aggregates saturated surface dry.

b. Air-entrained concrete, crushed slag aggregate, 6 bags cement/yd^3 concrete, all aggregates saturated surface dry.

c. Air-entrained concrete, crushed stone aggregate, 6 bags cement/yd^3, free water in sand 6%, based on saturated-surface-dry weight; coarse aggregates, 1% free water based on saturated, surface-dry weight.

20–11. Materials and their conditions as used in a batch of paving concrete are as follows:

Material	Batch Quantities			Specific Gravity (Aggregates on Saturated, Surface-Dry Basis)	Moisture Content, Based on Oven-Dry Weights	
					For Saturated, Surface-Dry Condition	As Used
Cement	75	bags	(7050 lb)	3.15	—	—
Sand	14,200	lb		2.70	1.2	5.0
Crushed stone	24,700	lb		2.66	0.8	1.4
Water	300	gal	(2500 lb)	1.00	—	—

For the mix shown above:

a. Calculate the cement factor on an air-free basis.

b. Calculate the water–cement ratio in gallons per sack, on the basis of saturated, surface-dry aggregates.

c. If the actual weight of a cubic foot of wet concrete is 147.0 lb, what is the percentage of entrapped air?

d. What is the cement factor including the entrapped air?

21 HIGHWAY MAINTENANCE AND REHABILITATION

Once construction, reconstruction, or rehabilitation of a highway (or transit) facility is completed, responsibility for it is assigned to the "maintenance" organization. Its function, as defined by AASHTO, is "the preservation and keeping of each type of roadway, roadside structure, and facility as nearly as possible in its original condition as constructed or as subsequently improved, and the operation of highway facilities and services to provide satisfactory and safe transportation."

As shown in Chapter 5, highway activities at the state level, as measured by expenditures, have been strongly focused on the construction of new facilities and the reconstruction of existing ones in an attempt to keep up with traffic demands. Maintenance has had a secondary role. In contrast, maintenance has had high priority with local rural and urban agencies. On the horizon at all levels is a greater attention to maintenance, and particularly to pavement rehabilitation, as new construction declines.

Expenditures for maintenance, excluding projects for major rehabilitation,[1] account for 30 cents of the highway dollar. For the various systems, expenditures per mile were state rural, $4700; state urban, $6500; local rural, $1200; and local urban, $3800. Expenditures per vehicle-mile were, respectively, about 0.6 cent, 0.1 cent, 3.0 cents, and 1.4 cents.

Routine maintenance, except for specialized services, is usually done by the agency's own forces. This is in contrast to construction and reconstruction, which, in the United States, is commonly done by contract (see Chapter 15). Probably most of the pavement rehabilitation also will be done by contract. Planning and managing this rehabilitation function are discussed later in this chapter as separate topics.

Some maintenance organizations also do "betterment" projects. Usually these are small and include grading and paving for short alignment changes to

[1]State expenditures are for 1978 and others for 1977. Source: *Highway Statistics,* 1978, FHWA.

eliminate steep grades or sharp curves, resurfacing, and mulching, planting, and erosion control. In some jurisdictions, the dollar amounts of betterment projects may be limited by law.

As discussed earlier in this book, there is a close relationship between design and construction practices and maintenance problems. For example, insufficient pavement or base thickness or improper construction of these elements soon results in expensive patching or surface repair. Pavement-edge and shoulder care becomes a serious problem where narrow lanes force heavy vehicles to travel with one set of wheels near the edge or off the pavement. Improperly designed drainage facilities mean erosion or deposition of materials and costly cleanup operations or other corrective measures. Sharp ditches and steep slopes require hand maintenance whereas flatter ditches and slopes permit machines to do the work more cheaply. In snow country, unsuitable locations, low fills, and narrow cuts that leave no room for snow storage can create extremely difficult snow-removal problems. In many instances, high maintenance costs resulting from poor design or construction practices offer the most compelling reason for reconstruction.

Routine maintenance activities and the efforts devoted to each vary widely among agencies. Nationally, state highway agencies divided their $4 billion total maintenance expenditures as follows: road and roadside, 63%; bridges, 6%; snow removal and sanding, 12%; traffic control, 12%; and operation of toll facilities, 7%. However, these averages are deceiving. Some agencies care for a large mileage of primarily low-volume rural roads; others have the responsibility for many miles of urban freeways and other high-volume urban arteries. There are many other variables; for example, Vermont, a rural state with a cold climate, spends 49% of its budget for snow removal and sanding while southern and southwestern states may not have this problem at all. At the other extreme, densely populated New Jersey allocated 18% of its budget to traffic control and 23% to toll operations, levels far above the national average.

Another measure of maintenance activity is the personnel devoted to it. Extremes among the state agencies are 39 lane-miles per employee in largely rural North Dakota and 6 in urbanized Connecticut.[2]

Highway maintenance forces have the responsibility for keeping roads open and traffic moving under all conditions. Often they are called on in time of flood, heavy snow, or other disasters to rescue stranded motorists or residents of afflicted areas. In carrying out these duties, maintenance workers have performed many difficult and sometimes heroic feats.

Traffic control while maintenance or rehabilitation operations are carried out creates serious difficulties including delays and accident hazards for motorists and maintenance personnel alike. Often, to minimize these problems, major operations must be carried out at night or over weekends when traffic volumes

[2]These and much of the other data cited here are from the Nov. 1980 *Progress Report on Maintenance and Operating Personnel* issued annually as a circular by TRB.

are lighter. Careful advance planning as well as signing and traffic control at the work site are a must for all maintenance operations.[3]

The courts have in many instances held highway agencies and their employees legally responsible for injury or property damage resulting from improper highway maintenance. Rough, slick, or dirty pavements, inadequately marked obstructions and equipment, and negligent acts of maintenance employees or supervisors are among the charges in such claims. This topic is treated in more detail in Chapter 7.

All highway agencies must carry out certain common maintenance operations. Managing and performing them is the topic of the next two sections of this chapter. It is followed by a separate discussion of pavement rehabilitation and management.

MAINTENANCE MANAGEMENT

In the past, maintenance was often considered as secondary to other agency functions and maintenance personnel often were, or felt they were, second-class citizens. In many agencies the spoils system prevailed, and maintenance positions were used to pay political debts or buy support. Unfortunately, this situation still prevails in a few agencies. However, beginning in the 1950s with the realization of the importance of and level of expenditures for maintenance, it became a subject for serious concern and study. Today, maintenance and its management is one of the most rapidly changing highway technologies.

Many highway officials point to a cooperative study carried out in 1959 and 1960 by the (then) Iowa Highway Commission and (then) U.S. Bureau of Public Roads as the first breakthrough.[4] It was concentrated on state highways in three representative counties and focused the best management and industrial engineering tools on time utilization, productivity, methods, and management on maintenance.

The Iowa study clearly indicated the need for improvement in the following areas: education in management techniques for foremen and maintenance engineers, more detailed and careful planning and scheduling of work, established standards for the level of maintenance, major improvements in maintenance equipment, better communications through a two-way radio system, and clearly established lines of authority and responsibility. Studies of roadside mowing, for example, revealed these among other conditions: the best equipment could mow three to six times as much area as the poorest equipment; marking obstacles could greatly lessen mowing time; two mowers working in conjunction ac-

[3]A discussion of the safety aspects of maintenance operations, with references, is presented in Chapter 12. Part VI of the *Manual on Uniform Traffic Control Devices*, FHWA, *Work-Zone Traffic Control-Standards and Guidelines*, FHWA, *Traffic Control for Construction and Maintenance Operations*, ITE, *NCHRP Synthesis 25*, and articles in *TRB Record 484*, and *703* and *TRB Special Report 153* offer a few of the many discussions of practices for traffic handling during maintenance. M. Karan and R. C. G. Haas, *TRB Record 554*, offer a user-delay cost model.

[4]See *HRB Special Report 65* and its *Supplement 1* for details.

complished less than when they worked separately; 1.3 hr of nonproductive time was spent in preparatory operations, travel, and at the job site for each hour of productive work; there was no set number of times per year to mow shoulders, it varied from one to four. As a result of these findings, the Iowa Highway Commission drastically modified its procedures and practices.

As of November 1980, 42 states and numerous other highway agencies had installed maintenance management and work improvement programs. However, instituting them is only the beginning; for them to succeed requires education and commitment to change at all levels in the organization and in the controlling commission, board, or administrator.

Sometimes it may be necessary to make haste slowly. For example, one of the many adjustments is to reduce the number of maintenance stations substantially. With motorized equipment it probably is possible to extend the miles covered by general purpose crews consisting of a foreman and a few workers and to develop special groups trained and equipped to carry out specific tasks over longer road mileages. To avoid political repercussions, these reorganizations and accompanying reductions or reassignments of personnel sometimes are carried out concurrently with retirements or resignations.[5]

Staffing with permanent, trained, and dedicated personnel is required if a maintenance management program is to succeed. In all but a few states and many local agencies, great strides have been made in this direction. For example, by 1980, maintenance employees in 45 state agencies were under civil service or some other merit system and all state agencies had retirement, vacation, and sick-leave and full or partial hospitalization plans. In 42 agencies, there was full or partial agency-paid life insurance and in 29 protection against lawsuits. In all but one state there were job-classification plans; in 35, salary increases were automatic or automatic with recommendation. In 36 states, all permanent employees were salaried. In 27 states, nonsupervisory employees were unionized; in 29, supervisory employees also had collective bargaining rights. These and other employment conditions indicate that a framework for and, in many cases, implementation of, good personnel policies are in effect.

Efficient, accurate, and fair budgeting and reporting procedures also are requirements of an effective management system, and a number of them are in use.[6] Implementing them is difficult, as many administrative and behavioral problems must be overcome.

At the task level, maintenance management demands attention to at least five areas, as follows:

1. Setting the standard or level of maintenance. For example, if mowing the roadside is required, decisions must be made as to how often it should be done and whether the entire roadside or only the areas close to the shoulder are to be mowed.
2. Setting performance standards for workers, crews, and maintenance units. Here, using time studies or more advanced techniques such as Methods Time Measurement

[5]See G. L. Russell et al., *TRB Record 727*.
[6]*NCHRP Report 131* offers details on performance budgeting and *NCHRP Synthesis 46*, on recording and reporting methods.

(MTM), expected output for respective operations are determined. For example, the actual or theoretical time required to mow an acre of roadside would be projected. Then after making allowances for travel and personal time, expected daily output is set.

3. Developing an accurate and fair reporting system so that workers, crews, foremen, and management can compare actual and expected production.
4. Developing a management system that provides for the efficient scheduling of work assignments.
5. Developing better procedures, methods, and machinery for carrying out individual maintenance operations. Opportunities here generally are plentiful. One of the most promising usually involves adjustments in crew size, work assignments, or both. An essential ingredient to any such approach is to gain the enthusiastic support of workers, crews, and supervision by teaching them methods improvement techniques, involving them in the search for new approaches, and acknowledging their efforts by personal recognition and, when feasible, by financial rewards.[7]

Approaches to and developments in maintenance management are still evolving. As indicated, some agencies are far advanced, others have yet to begin.[8]

COMMON MAINTENANCE OPERATIONS

Maintenance, as defined earlier, involves keeping facilities as near the "constructed or reconstructed" condition as possible. Depending on circumstances, it calls for periodic or almost immediate correction of unfavorable or unacceptable situations. For the brief discussion that follows, maintenance has been categorized into surface, shoulder and approach, roadside and drainage, snow and ice control, bridge, and traffic service.

In the United States and other developed countries where labor is costly, many maintenance operations have been mechanized. For example, power tools have augmented or replaced the traditional pick, shovel, hoe, scythe, axe, or machete. As is pointed out below, there are many special-purpose machines or attachments that fit general purpose ones and new approaches are being developed continually.[9]

Surface Maintenance

About one-half of the state highway maintenance dollar goes for routine care of the roadway surface. With other road agencies also, surface maintenance is a

[7]For details about developing such methods see, for example, *NCHRP Report 161,* H. W. Parker and C. H. Oglesby, *Methods Improvement for Construction Managers,* McGraw-Hill, New York, 1972, M. E. Mundel, *Motion and Time Study,* 5th ed., 1978, Prentice-Hall, Englewood Cliffs, N.J. and B. F. Niebel, *Motion and Time Study,* 6th ed. 1976, R. D. Irwin, Homewood, Ill.

[8]Among the many references are *NCHRP Report 223,* titled "Maintenance Levels of Service Guidelines", *Maintenance Management,* Vol. VIII of the Action Guide Series of the National Association of County Engineers, *HRB Special Report 100, HRB Record 347,* and papers in *HRB Record 391* and *451,* and *TRB Record 506, 554, 598, 674, 727, 774, 776,* and *781.* For publications directed particularly at maintenance management in developing countries see *Synthesis 1* and *Compendium 11* of Transportation Technology Support for Developing Countries, TRB.

[9]For up-to-date information on maintenance equipment and its use see articles in such periodicals as *Civil Engineering, Highway and Heavy Construction,* and *Public Works.*

major expense. For gravel roads, blading and occasional reshaping or material replenishment are necessary (see Chapter 16). For surface treatments and low-type bituminous surfaces, patching, seal coating, or possibly loosening, remixing, and relaying are involved (see Chapter 19). For high-type surfacings such as bituminous or portland-cement concrete, removal and replacement or filling in failed areas may be called for (see Chapter 19 and 20). Joint sealing, and slab- or mud-jacking or undersealing of portland-cement-concrete pavements is a relatively common operation (see Chapter 20). At times, slick bituminous surfaces require roughening, burning, or a nonskid seal (see Chapter 19). Almost all these operations may be carried out by either general- or special-purpose crews of the highway agency.

In general, the same materials are used for surface maintenance as for surface construction and reconstruction. However, machinery and methods for construction are designed for high-volume output, whereas maintenance operations usually involve small quantities at widely separated locations. Furthermore, maintenance operations must be planned for rapid performance and to cause the least possible disruption and hazard to traffic. As an illustration, small patches in bituminous surfaces are often made with premixed aggregate and binder that has been stockpiled along the road or at the maintenance station ready for immediate use. The mix is hauled from the stockpile by truck and placed by hand-labor methods. Compaction is with hand tampers, power-driven tampers or vibrators, or truck tires. Another method utilizes a small tank trailer for bituminous binder towed by a dump truck loaded with aggregate. Pressure for operating the hand spray for the binder is provided by a small gasoline-driven pump.[10]

Particularly on heavily traveled roads, pavement failures may become too extensive for correction by maintenance forces. Surfaces may become slick from aggregate polishing or asphalt concentrations on the surface. In other instances, depressions may develop in the wheel tracks from compaction by traffic or wear from "studded" tires. Correcting such major deficiencies is usually beyond the capacity of the normal maintenance forces.

Shoulder and Approach Maintenance

Shoulder and approach maintenance procedures depend on the surface character of these areas. Sod shoulders must be mowed and occasionally bladed down to the level of the roadway so that water is not trapped in the traveled way. The grass must be fertilized, reseeded, and otherwise treated to keep it in good condition. Care of shoulders protected by bituminous blankets or surface treatments is the same as for roadways with like surfaces. Gravel and earth shoulders are maintained by blading under proper weather conditions.[11]

Rutting or settling of the shoulders that leaves a gap or dropoff at the pave-

[10]For added details on materials and procedures see *NCHRP Synthesis* 64 and Chapter 2 of the *AASHTO Maintenance Manual*.
[11]*See AASHTO Maintenance Manual*, Chapter 3, for more details.

ment edge creates a serious accident hazard: If this condition develops, it should be corrected as soon as possible by reconstruction, resurfacing, or other appropriate means.[12]

Roadside and Drainage Maintenance

The character of the roadside determines what maintenance is required. Where the roadside is grass, this must be mowed (at least close to the shoulder), fertilized, and sometimes treated with lime. Reseeding or resodding may be necessary in some instances. If weeds are troublesome, cutting, plowing, or spraying with weed killer may be required. If dry grass along the roadside and on adjacent lands constitutes a fire hazard, burning or plowing may be in order. Where backslopes are covered with brush, trimming to maintain clearance and sight distance is needed occasionally. Control of slope erosion by mulching, seeding, or other means often becomes a maintenance operation. Where there are shrubs and trees, spraying and mulching and occasional tree surgery will be required. Picking up litter thrown or blown along the roadside or into wayside areas is another annoying but necessary chore.

Many ingenious machines have been developed to reduce the cost of roadside maintenance. Included are mechanical sod cutters, combined seed and fertilizer spreaders, sprayers for distributing seed and fertilizer in suspension in water, power mowers, motor-driven but portable grass cutters for use in close quarters, brush mowers for cutting heavy bushes along the roadside, and brush choppers that reduce the stalks to chips suitable for mulch. For roadside cleanup, some agencies have machines that pick up papers, cans, bottles and other roadside trash and sort them for disposal.[13]

Drainage maintenance involves keeping ditches, culverts, structures, and appurtenances such as drop inlets and catch basins clean and ready to carry the next flow of water. Sediments deposited during periods of heavy flow must be removed. Brush, branches, and other debris that collect in trash racks or at culvert and structure entrances must be disposed of. Badly eroded channels and dikes must be repaired, and paving, seeding, sodding, riprap, bank protection, or other means must be adopted to prevent recurrence. After extreme storm damage, maintenance forces may be called upon to reconstruct much of the drainage system.[14]

Snow and Ice Control[15]

Snow removal is the major winter maintenance problem in affected areas. To solve it properly requires careful organization and advance training. Often "dry runs" are held in the fall to discover and correct deficiencies in equipment or plans that would cause serious trouble during a storm. Quick communication

[12]See *NCHRP Report 202* for a detailed study of this problem.

[13]For added detail on roadside mowing see *AASHTO Maintenance Manual,* Chapter 5, and articles in *TRB Record 647, 674,* and *776.*

[14]See *AASHTO Maintenance Manual,* Chapter 4, for more detail.

[15]See Chapter 8, *AASHTO Maintenance Manual,* and *TRB Record 776* for more detail.

by means of two-way radio equipment has been a major forward step in coordinating snow-removal operations.

Measures to minimize the formation of snowdrifts across the roadway should be taken before the coming of winter. *Snow fences* offer an effective and economical means for causing snowdrifts to form adjacent to rather than on the traveled way. Snow fences are placed before the ground freezes and are removed in the spring. Proper location is on the windward side at right angles to the prevailing wind and about 15 fence heights from the road. The most common fence consists of wooden slats about 4 ft long woven together with galvanized wire. Sections can be rolled up for handling and storage. Support is furnished by posts or angle irons driven into the ground. Others have used fences up to 10 ft high.[16] Trees, particularly of the coniferous variety, or other vegetation at a sufficient distance from the roadway are also effective in controlling drift formation. However, guard rail, fences, or planting close to the road aggravate the drift problem and should be avoided whenever possible.

In rural areas, snow that falls on the roadway is bladed or thrown to the roadside. Removal operations should start soon after snow begins to fall. Where annual snowfall is less than about 30 in., regular maintenance trucks equipped with blade plows are effective. Larger trucks, usually four-wheel drive and equipped with wing plows or hydraulically operated blade or V plows, are widely used in areas of heavier snowfall. In mountain passes and other locations where heavy snowfall and drift formation create extreme snowdrifts, rotary plows that throw the snow a considerable distance are most effective.

In the downtown areas of cities, snow from heavy falls is generally loaded into trucks and hauled away. There are a few instances where ramps or short stretches of pavement or sidewalk have been heated to melt the snow as it falls, but this approach is too costly for general use.[17]

Ice forming on the roadways after rains or snow may reduce the coefficient of friction between tires and surface to 0.05 or less and makes proper vehicle control almost impossible. To minimize the danger thus created, most highway agencies apply abrasives to heavily traveled roads and streets. Suitable materials are clean, sharp sand, cinders, and washed stone screenings. Normally the abrasives are treated with calcium or sodium chloride in solid or brine form before storage. As salt solutions have a lower freezing point than water, these additives prevent freezing in the stockpile or bin and ensure penetration of the particles into the icy surface. Experience has shown that untreated aggregates, even if heated before application, whip or blow off the roadway. Used alone, salts can prevent ice formation, melt that which has already formed, or prevent buildup.

For pretreating aggregates, salts are mixed with the aggregate in proportions suited to the conditions; commonly 50 to 100 lb/yd^3 of abrasives. These mixtures generally are stored in large elevated bins ready for gravity-loading into trucks when needed. Distribution is with a mechanical spreader or by hand.[18]

[16]See, for example, *TRB Record 506* and *776*.
[17]See G. P. Williams, *TRB Record 576*, for a discussion and bibliography.
[18]See *NCHRP Synthesis 24* for added detail on proportioning and distribution.

Demands of the public that roads be kept open for travel at high speed have brought dramatic increases in salt use and with them serious problems. In addition to deterioration of concrete pavements, walks, and bridges (see below), salts corrode automobile bodies, although the situation is not as serious as in earlier years.[19] In addition, great concern has been expressed over the effects of salts on the quality of surface and groundwater. Research to date indicates that contamination of rivers is not a problem but means for controlling contamination of roadside areas are needed. Also, the salts kill certain roadside plantings; this problem is being solved in part by developing salt-resistant varieties.[20]

Bridge Maintenance

Most bridge maintenance is of a specialized nature. On structures having exposed steelwork, cleaning by sandblasting, flame, or other means followed by repainting usually represents the biggest maintenance item.[21] At times cleaning, freeing, and painting joints that are designed to move may be required. Deck joints may extrude or become filled with dirt so that cleaning and resealing is necessary. On occasion, vehicles out of control strike handrails or other appurtenances and these must be repaired. If bridge decks become rough or slick, resurfacing is in order; in this instance surfacing weight may be a primary control on what is used. Remedial measures are sometimes required to correct serious scour around and under piers and abutments. Because bridge maintenance requires special skills, it is common practice among highway agencies to have traveling crews exclusively for bridge work. Often painting and other specialty work is carried out under contract.[22]

Deterioration of concrete bridge decks under the combination of freezing and thawing and deicing salts has become a major maintenance item. Commonly, the problem begins when the salt penetrates to and corrodes the reinforcing, resulting in spalling of the overlying concrete. Correction may require removal of the concrete, cleaning the steel, and applying new material, possibly polymer concrete. Sometimes sealants or overlays of asphaltic materials are employed to provide further protection. Such major repairs not only are costly but may cause serious delays and inconvenience to motorists.

Means developed to prevent or check bridge deck deterioration include applying waterproof membranes, placing overlays of dense or latex-modified concrete, impregnating the deck with polymers, or passing an electric current

[19]See, for example, papers by D. M. Murray, M. C. Belangie, and F. O. Wood in *TRB Record 647, 651* and *762*.

[20]Sources of information on the environmental impacts of salt use include *NCHRP Report 170* and articles in *HRB Record 425* and *TRB Record 506* and *647*. A discussion of the economics of snow and ice control by J. C. McBride appears in *TRB Record 674*. Snow removal and ice control research is the topic of *TRB Special Report 185*.

[21]In the fiscal year 1978–79, expenditures for painting the 350 acres of exposed steel on the 3.9-mi steel portions of the San Francisco–Oakland Bay Bridge totaled $1.3 million. Approximately 55 workers were employed, full time.

[22]See *Bridge Maintenance Manual* and *Manual for Maintenance Inspection of Bridges,* AASHTO.

through the reinforcing steel.[23] In some instances, solar cells have provided the small amounts of power required.[24] If decks are replaced, epoxy-coated reinforcing steel may be employed.

Traffic Service[25]

Traffic service includes such continuing functions as striping, sign repair, and maintenance of street and highway lights and signals. Generally these are performed by special crews of the highway agency although street-light and even traffic-signal maintenance may be turned over to the local utility company.

At times, signal systems malfunction because of power or other failures. And as traffic-control devices and systems become more sophisticated and complex the probability of breakdown increases. At the same time, the consequences of breakdowns can be very severe; for example, if a computer-controlled, area-wide system of signals fails, traffic can be snarled for hours. To keep such systems operating and to correct breakdowns quickly calls for highly skilled personnel. In fact, one argument against the more elaborate traffic-control systems has been the fear that they may malfunction and create chaotic conditions. It is to be expected that, as experience with such systems accumulates, they will become more reliable.[26]

Traffic service also includes coping with emergencies. For example, during heavy storms maintenance personnel will be on patrol to try to keep the roads open, sign or barricade washouts, and rescue stranded motorists.

PAVEMENT REHABILITATION

Pavements are wearing out faster than they can be repaired. This situation was recognized in the 1976 Federal-Aid Highway Act which broadened the definition of "construction" to include resurfacing, restoration, and rehabilitation commonly referred to as the 3 R's. The intent of this legislation was to permit use of federal aid to rehabilitate pavements to extend their useful lives. Geometric standards lower than current ones for new construction are permitted.[27] This has led to controversy with proponents of high standards to promote safety.

Primary objectives of pavement rehabilitation as indicated in the AASHTO policy are to: (1) improve surface smoothness, (2) extend pavement life, (3) improve skid resistance, (4) reconstruct sections having poor foundations, and (5) improve drainage. Deficiencies in the existing highway system are to be identi-

[23]See *NCHRP Report 180* and *TRB Record 604, 651, 692,* and *762.*

[24]See, for example, *Transportation Research News,* TRB, Jan–Feb. 1977.

[25]See *AASHTO Maintenance Manual,* Chapter 9, for more detail.

[26]The *ITE Journal* is a fruitful source of information on traffic signal maintenance and operation. See also *NCHRP Synthesis 22* and D. A. Randolph and T. K. Datta, *TRB Record 727.*

[27]See *Geometric Design Guide for Resurfacing, Restoration, and Rehabilitation (R-R-R) of Highways and Streets,* AASHTO, 1977.

fied by sufficiency ratings, accident data, skid tests, maintenance reports, and in some cases, suggestions by the public.

Pavement rehabilitation is commonly distinguished from pavement maintenance as shown in Fig. 21–1. Maintenance, as discussed earlier, refers to actions that are corrective or preventative and are commonly performed by the agency. Rehabilitation is performed primarily to provide better ride and skid resistance or to improve the structural adequacy of the pavement. FHWA defines these terms as follows:[28]

1. *Resurfacing, Restoration and Rehabilitation Work:* Improvements and associated engineering on existing roadway pavements and bridge decks to restore them to their original safe usable condition without expansion of the original capacity.
2. *Resurfacing:* The addition of a pavement layer over the existing roadway or bridge deck surface to provide additional structural capacity. Pavement layer thickness shall be ¾-in. or greater.
3. *Restoration and Rehabilitation Work:* This includes "replacement of malfunctioning joints, repair of spalled joints, substantial pavement undersealing. . . , reworking or strengthening of bases or subbases, recycling or reworking existing materials to improve their structural integrity, adding underdrains, improving shoulders, removing and replacing contaminated or deteriorated materials. . . "

As of the present time, the techniques commonly used to rehabilitate pavements include those shown in Fig. 21–1; namely reconstruction, overlays, or recycling.

Reconstruction and Overlays

Reconstruction consists of removing the existing pavement system and replacing it with a completely new asphaltic or portland-cement-concrete pavement and undercourses. Overlays can be either portland-cement or asphalt concrete placed on the existing pavement system. Chapters 19 and 20 give methods for determining the thickness of overlays and for placing the materials.

Recycling[29]

This solution is to recycle existing pavement materials using either a bituminous material, portland cement, or lime. Recycling is expected to help stabilize costs, conserve scarce material resources, and reduce energy requirements.

Pavement recycling can be categorized by the procedure used, type of materials, and the structural benefit to be gained. The discussion that follows breaks down recycling by procedure as shown on Fig. 21–2.

Surface recycling, which consists of reworking about the top inch of asphalt

[28]*FHWA Notice N 5040.19,* "Resurfacing, Restoring, and Rehabilitation (R-R-R) Work," June 28, 1976.

[29]*NCHRP Synthesis 54* is an excellent reference on recycling. See also *NCHRP Report 224, TRB Records 695, 712,* and *780,* and Gene K. Fong, *Public Roads,* June 1981. *ASTM STP 662* and the *Proceedings of the Association of Asphalt Paving Technologists* are also excellent sources.

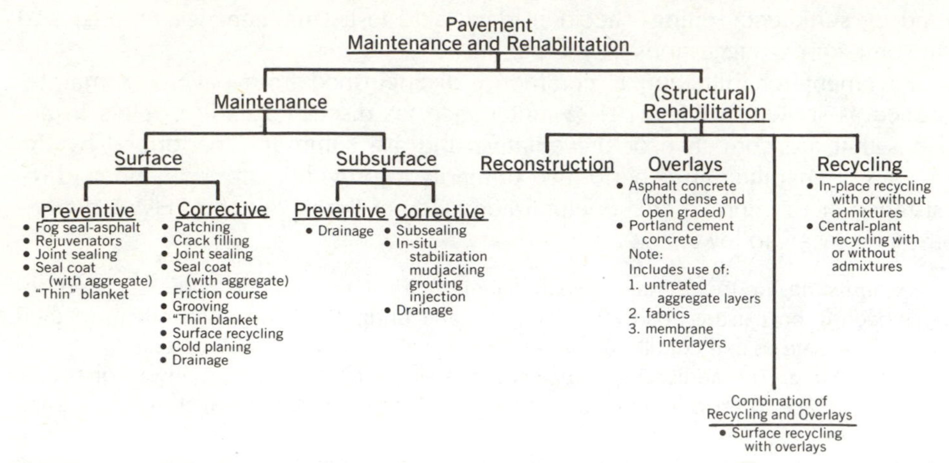

Fig. 21–1. Maintenance and pavement rehabilitation alternatives. (After Monismith, *TRB Record 700.)*

pavements, is currently the most common. It treats raveling, wheel-track rutting, flushing, and corrugations. Techniques include heater-planer, heater-scarifier, cold planers, and cold milling. Some of these techniques were developed in the 1930s and have been employed by cities so that existing curb and gutter grades could be met. Their principal disadvantages are noise, heat, and air pollution. Table 21–1 identifies other advantages and disadvantages.

In-place surface and base recycling involves pulverizing all of the existing pavement followed by reshaping and compaction. It has, in the past, been performed with bulldozers or rollers, and more recently with specialized mechanized equipment. Often, before relaying, the materials are upgraded with lime, portland cement, asphalt, or chemicals. Both asphalt and portland-cement-concrete pavements have been recycled using this technique. The major advantage of in-place recycling is the ability to increase the load-carrying capacity of the pavement without a major change in the grade. Other advantages and disadvantages are given in Table 21–1.

Central-plant recycling consists of removal of the material from the roadway, crushing it, mixing it in a plant, then placing and compacting it with conventional equipment. Both portland-cement concrete and bituminous material have been reprocessed to make aggregates for stabilized or unstabilized base layers. Bituminous materials have also been processed into surfacing materials.

When used as base, the processed material is blended, without heat, with additives such as portland cement, lime, or cutback or emulsified asphalt. For surfacing, it is heated and then blended with new asphalt cement or recycling agents and often with new aggregate.[30]

[30]It is estimated that by 1985, 10% of the asphalt concrete supplied will have been recycled.

TABLE 21–1. Major Advantages and Disadvantages of Recycling Categories*

Recycling Categories	*Advantages*	*Disadvantages*
Surface	• Reduces reflection cracking • Promotes bond between old pavement and thin overlay • Provides a transition between new overlay and existing gutter, bridge, pavement, etc. that is resistant to raveling (eliminates feathering) • Reduces localized roughness • Treats a variety of types of pavement distress (raveling, flushing, corrugations, rutting, oxidized pavement, faulting) at a reasonable cost • Improved skid resistance • Minimum disruption to traffic	• Limited structural improvement • Heater-scarification and heater planing have limited effectiveness on rough pavement without multiple passes of equipment • Limited repair of severely flushed or unstable pavements • Some air quality problems • Vegetation close to roadway may be damaged • Mixtures with maximum size aggregates greater than 1-in. cannot be treated with some equipment
In-place	• Significant structural improvements • Treats all types and degrees of pavement distress • Reflection cracking can be eliminated • Frost susceptibility may be improved • Improve ride quality	• Quality control not as good as central plant • Traffic disruption • Pulverization equipment repair requirement • Cost • Cannot be easily performed on PCC pavements
Central-plant	• Significant structural improvements • Good quality control • Treats all types and degrees of pavement distress • Reflection cracking can be eliminated • Improved skid resistance • Frost susceptibility may be improved • Geometrics can be more easily altered • Better control if additional binder and/or aggregates must be used • Improve ride quality	• Increased traffic disruption • May have air quality problems at plant site

*From *NCHRP Synthesis 54*

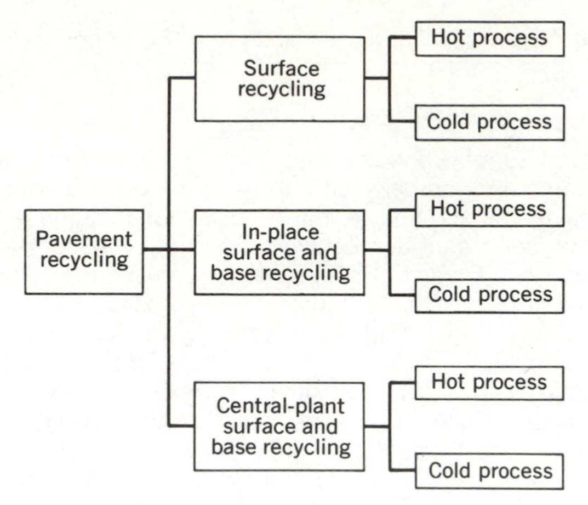

Fig. 21–2. Categorization of recycling approaches based on recycled procedures (from *NCHRP Synthesis 54*).

Equipment to centrally process recycled material is presently available. *NCHRP Synthesis 54* describes in detail current construction techniques. Chapter 19 briefly treats equipment for recycling asphalt concrete.

PAVEMENT EVALUATION AND MANAGEMENT

Highway agencies have for years collected pavement condition data to make maintenance and rehabilitation decisions. This generally was done on a project by project basis. The data were used to determine which projects to maintain or rehabilitate, and what action was required to correct the observed pavement deficiencies. Decisions were made on a year to year basis, generally in an environment where resources (both manpower and money) were more plentiful then they are today.

In the 1940s and 1950s, reliance was on visual inspections to establish type, extent, and severity of distress and to establish maintenance and rehabilitation programs. In the late 1950s and early 1960s, roughness meters to measure ride quality, deflection test equipment to measure structural adequacy, and skid test devices to measure surface friction came into play so that objective data could be collected and used together (or separately) with visual surveys to make maintenance and rehabilitation decisions. In the 1970s many highway agencies began to realize that they could no longer manage their roadways on the basis of field observations alone. As a result they have now developed objective methods of evaluation to establish:

1. What projects are presently in need of maintenance or rehabilitation.
2. What type of maintenance or rehabilitation is required now.
3. What maintenance or rehabilitation strategy should be undertaken now and in the future to minimize life cycle costs (construction, maintenance, and user costs) or maximize net benefits.

Types of Pavement Data Collected

Several types of data are collected to develop maintenance and rehabilitation programs. In general, they include ride (surface roughness), surface distress, surface deflection, and surface friction (skid resistance).

Surface roughness is generally defined as irregularities in the pavement surface which adversely affect ride quality. Equipment can be towed or mounted in a vehicle.[31]

Surface distress data are collected by many agencies. Distress is defined as[32]

> Any indication of unfavorable pavement performance or signs of impending failure; any unsatisfactory performance of a pavement short of failure.

Types of distress normally observed include fracture (load and nonload associated cracking), distortion (permanent deformation or faulting), and disintegration (spalling, raveling, etc.) For each distress type, the corresponding amount, severity, and location are noted. All data are collected subjectively following some form of survey guide.[33]

Surface deflection measurements of the change in pavement surface level between loaded and unloaded states provide a means for designing specific rehabilitation strategies for pavement segments. They measure structural adequacy, or the ability of a pavement to support traffic without developing appreciable distress.[34] Observations may be of static deflection (Benkelman Beam), steady-state deflection (Road Rater or Dynaflect), or impact load response (Falling Weight Deflectometer).

Surface friction data identify pavement segments with low coefficients of friction. Several types of equipment have been used to obtain a single value known as the skid number (*SN*). As pointed out in Chapter 8, this number approximates the surface coefficient of friction multiplied by 100. It is normally measured using a locked wheel trailer (ASTM Method E 274) or sometimes with a "yaw" mode trailer where the wheels are skewed with respect to the direction of travel.[35]

[31]For details on surface roughness measuring devices, see R. C. G. Haas and W. R. Hudson, *Pavement Management Systems,* McGraw-Hill, New York, 1978; *FHWA Report RD-74-60,* "Surface Evaluation of Pavements: State of the Art"; *HRB Special Report 133; TRB Record 633* and *715; FHWA Report RD-79-68,* "Pavement Condition Measurement Needs and Methods" and *NCHRP Report 228.*

[32]*HRB Special Report 113.*

[33]For details of rating procedures, see Haas and Hudson, op. cit., *TRB Records 700* and *715, Pavement Management Guide,* Road and Transport Association of Canada, 1977, and *FHWA Report RD-79-66,* "Highway Pavement Distress Identification Manual," Mar. 1979.

[34]For techniques for measuring deflection, see *TRB Circular 189* and Haas and Hudson, op. cit. For the use of pavement deflection to design overlays, see Chapter 19: *TRB Record 572* and *700; MS-17, Asphalt Overlays and Pavement Rehabilitation,* The Asphalt Institute; and *Asphalt Concrete Overlay Design Manual,* California Department of Transportation, Jan. 1979.

[35]For details on test equipment see *NCHRP Synthesis 14.* See also *AASHTO Guidelines for Skid Resistant Pavement Design,* 1976 and *TRB Record 788.*

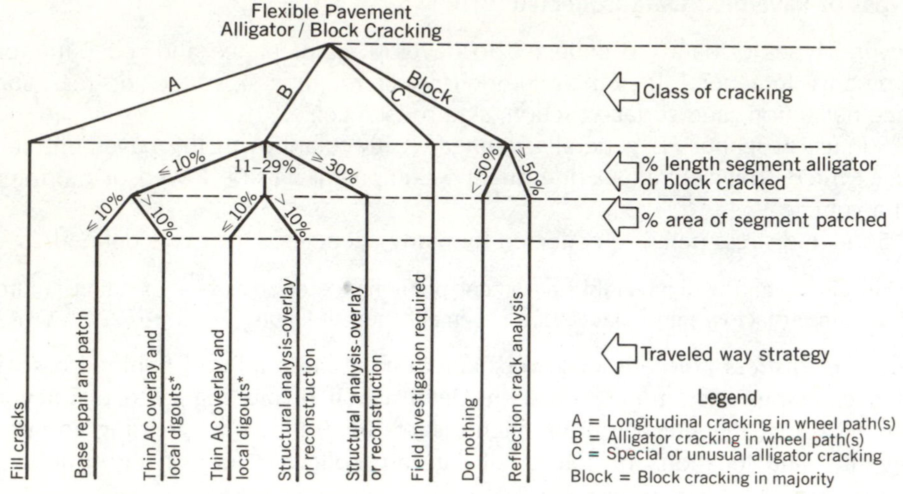

*Thin AC overlay = < 0.10 ft dense graded or open graded mix

Fig. 21–3. Example of California method for selecting action plan for pavement rehabilitation. (Source: *FHWA Report TS-79-206,* slightly modified.)

Use of Pavement Condition Data[36]

At present, there are three specific uses of pavement data. All of these have been loosely termed *pavement management* but they are quite different. They include:

1. *Establish Priorities.* Data such as ride, distress, surface friction, and deflection are used to establish which projects are most in need of maintenance or rehabilitation. Often ride and distress are combined into a single rating which may or may not consider the present volume of traffic. Once identified, the projects with the low ratings are more closely evaluated to establish repair strategies.
2. *Establish Rehabilitation Strategies.* Data such as type, extent, and severity of distress are often used to establish an "action plan," normally on a year to year basis. This allows for a systematic determination of the maintenance or rehabilitation strategy (repairs such as surface treatments, overlays, type of recycling methods) most appropriate for a given condition. An example for selecting the repair strategy for flexible pavements is given in Fig. 21–3; one for choosing the recycling method is given in Table 21–2.
3. *Project Future Pavement Performance.* A few agencies project performance (or some measure of it) in an effort to determine the future condition of the pavement system or network. From this the budget needed to maintain the pavement system at a given

[36]For details see Haas and Hudson, op. cit.; *FHWA Report-TS-79-206;* NCHRP Reports 160 and 215; *Transportation Engineering Journal of ASCE,* Jul. 1979 and Jan. 1980; *TRB Record 512, 572, 598, 633, 679, 700* and *702; NCHRP Synthesis 48, 58* and *76;* and *Compendium No. 11,* Transportation Technology Support for Developing Countries, TRB.

TABLE 21–2. Selection of Recycling Techniques Based on Roadway Conditions*

	Recycling Methods / Condition of Existing Pavement →	TYPE OF DISTRESS																																							
		RUTTING										RAVELING										ALLIGATOR CRACKING										TRANSVERSE CRACKING									
			% AREA										% AREA										%AREA										NO PER STA								
			1–15			16–30			30				1–15			16–30			30				1–5			6–25			25				1–4			5–9			10		
		NONE	SLIGHT	MODERATE	SEVERE	SLIGHT	MODERATE	SEVERE	SLIGHT	MODERATE	SEVERE	NONE	SLIGHT	MODERATE	SEVERE	SLIGHT	MODERATE	SEVERE	SLIGHT	MODERATE	SEVERE	NONE	SLIGHT	MODERATE	SEVERE	SLIGHT	MODERATE	SEVERE	SLIGHT	MODERATE	SEVERE	NONE	SLIGHT	MODERATE	SEVERE	SLIGHT	MODERATE	SEVERE	SLIGHT	MODERATE	SEVERE
Surface	HEATER PLANER WITHOUT ADDITIONAL AGGREGATE				X			X			X		X	X	X	X	X	X	X	X	X		X	X	X	X	X	X	X	X	X		X	X	X	X	X	X	X	X	X
Surface	HEATER PLANER WITH ADDITIONAL AGGREGATE				X			X			X		X	X	X	X	X	X	X	X	X		X	X	X	X	X	X	X	X	X		X	X	X	X	X	X	X	X	X
Surface	HEATER SCARIFY				X			X			X							X			X			X	X	X	X	X	X	X	X		X	X	X	X	X	X	X	X	X
Surface	HEATER SCARIFY + THIN OVERLAY																								X	X	X	X	X	X	X				X		X	X	X	X	X
Surface	HEATER SCARIFY + THICK OVERLAY																																								
Surface	SURFACE MILLING																						X	X	X	X	X	X	X	X	X		X	X	X	X	X	X	X	X	X
Surface	SURFACE MILLING + THIN OVERLAY																							X	X	X	X	X	X	X	X				X		X	X	X	X	X
Surface	SURFACE MILLING + THICK OVERLAY																																								
In Place	THIN ASPHALT CONCRETE—MINOR STRUCTURAL IMPROVEMENT WITHOUT NEW BINDER																								X		X	X	X	X	X										
In Place	THIN ASPHALT CONCRETE—MINOR STRUCTURAL IMPROVEMENT WITH NEW BINDER																								X		X	X	X	X	X										
In Place	THIN ASPHALT CONCRETE—MAJOR STRUCTURAL IMPROVEMENT WITHOUT NEW BINDER																																								
In Place	THIN ASPHALT CONCRETE—MAJOR STRUCTURAL IMPROVEMENT WITH NEW BINDER																																								
In Place	THICK ASPHALT CONCRETE—MINOR STRUCTURAL IMPROVEMENT WITHOUT NEW BINDER																								X		X	X	X	X	X										
In Place	THICK ASPHALT CONCRETE—MINOR STRUCTURAL IMPROVEMENT WITH NEW BINDER																								X		X	X	X	X	X										
In Place	THICK ASPHALT CONCRETE—MAJOR STRUCTURAL IMPROVEMENT WITHOUT NEW BINDER																																								
In Place	THICK ASPHALT CONCRETE—MAJOR STRUCTURAL IMPROVEMENT WITH NEW BINDER																																								
Central Plant	COLD PROCESS—MINOR STRUCTURAL IMPROVEMENT WITHOUT NEW BINDER																								X		X	X	X	X	X										
Central Plant	COLD PROCESS—MINOR STRUCTURAL IMPROVEMENT WITH NEW BINDER																								X		X	X	X	X	X										
Central Plant	COLD PROCESS—MAJOR STRUCTURAL IMPROVEMENT WITHOUT NEW BINDER																																								
Central Plant	COLD PROCESS—MAJOR STRUCTURAL IMPROVEMENT WITH NEW BINDER																																								
Central Plant	HOT PROCESS—MINOR STRUCTURAL IMPROVEMENT WITHOUT NEW BINDER																								X		X	X	X	X	X										
Central Plant	HOT PROCESS—MINOR STRUCTURAL IMPROVEMENT WITH NEW BINDER																								X		X	X	X	X	X										
Central Plant	HOT PROCESS—MAJOR STRUCTURAL IMPROVEMENT WITHOUT NEW BINDER																																								
Central Plant	HOT PROCESS—MAJOR STRUCTURAL IMPROVEMENT WITH NEW BINDER																																								

*After *NCHRP Report 224*

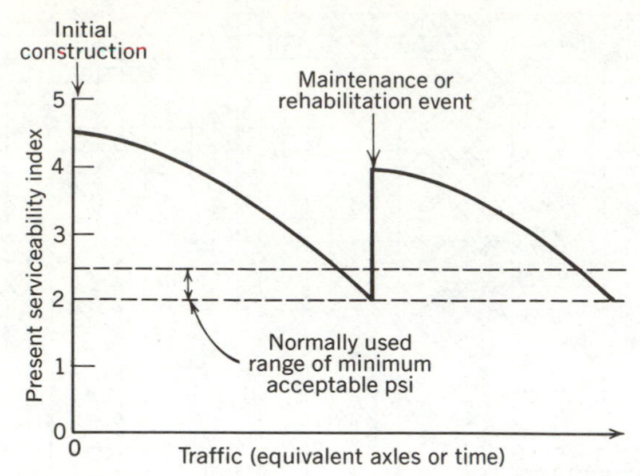

Fig. 21–4. Concept of projecting pavement performance using the present serviceability index.

serviceability level (see Fig. 21–4) or the serviceability level with a given budget can be determined.

In all of the above applications, it should be recognized that whatever pavement management system is used, it serves only as a tool for decision makers; it does not replace them.

PROBLEMS

21–1. Determine for your local highway agency, the annual maintenance budget. Of this, what is spent on surface maintenance, shoulder maintenance, roadside and drainage maintenance, snow and ice control, bridge maintenance, and traffic service?

21–2. Determine the methods used to manage maintenance activities. If maintenance management procedures are employed, determine their effectiveness.

21–3. How does the highway agency in your area distinguish between surface maintenance and rehabilitation?

21–4. Visit a local contractor to determine the types of pavement recycling techniques used in the vicinity and their advantages and disadvantages. If none, determine why recycling is not used.

21–5. Determine for your local state highway agency (or county), the types of pavement data collected. What percentage of the highway system is monitored each year? What equipment is used to collect the data? How often are the data collected? What problems, if any, have been encountered in the data collection process?

21–6. How are the data collected in Problem 21–5 used to make decisions regarding surface maintenance and rehabilitation? Estimate the annual costs and benefits resulting from the pavement management system.

INDEX

NAME OF HIGHWAY
NAME OF SECTION
NAME OF COUNTY

① TOTAL CURVE
Δ = 41° 22'00" Lt
R + "0" = 2870.02'
② T = 1383.44'
L = 2668.30'
Ext = 202.96'
MAIN CURVE
Δ = 29° 22'00"
D = 2° 00'00"
R = 2864.79'
L = 1468.30'
Super = 0.074'/Ft

N

T.12S ⊕ R.7E ⑤

29 | 28
32 | 33

Sct Cor ③
Fnd 2" Iron Pin

1/4 Cor Fnd 1" Iron Pipe
15.10' North Of Survey ℄
New Survey Monument
℄ Cover Std C-21.01

New 30'×8'×150' Channel Rt
New 30'×8'×150' Channel Lt
See Detail "A" For Section
℄ Profile

P.O.C. 2980+33.0
Section Tie

New Type 2 Gate
Std C-12.01

Sta 2983
New 3-
Std C

150'×150'
Drn Esm

C.S.
2983+24.00

Sta
New
③

② S.C. 2968+56.60

⑤ 2970

100'
100'
35'
45'

New R/W

Toe Of Slope

Survey ℄ Cst ℄

2975

2980

2985

+50

100'
100'

New R/W

Sta 2970+50
New 24"×78' Pipe Clr ①

① Sta 2973+10
New 54"×90' Pipe Clr ②

Lt Sta 2970+00 To 2979+85
Rt Sta 2970+00 To 2980+70
2055 Lin Ft New 4-Wire Game Fence
Std C-12.02

M.P. 97

Exst R/W
Exst 20' Bit Rdwy
Exst R/W

Exst R/W Markers
Remove

Obliterate And Remove
Exst Bit Pvmt Within New
R/W And Outside New Cst

150'×150'
Drain Easc

LA FONG WASH ①

+50

New 10'×6'×
Type "A" Std

Rt
New
Type A

⑤
STA. 2969+50
BEGIN PROJECT (Project Number)
END PROJECT (Project Number Or Existing Project)

BM #298, 96' Lt Sta 2980+00
1/2" Bolt, Elev 4003.95

RURAL 2-WAY ROADWAY

DA Rt 0.01 Sq Mi

② DA Rt 0.06 Sq Mi

DA Rt 1.00 Sq Mi

⑤ 40

② PI Elev 4000.00
Corr = -1.30

① New 450' Crown Ditch Lt
Std C-3.01

① Channel Lt
390 CY Drn

①
Q_{25} = 16 cfs ①
Design HW 3994.0
F.L. Elev 3991.5
Slope = 1.0%

②
Q_{25} = 94 cfs
Design HW 3995.5
F.L. Elev 3991.0
Slope = 1.25%

Exst Ground line

+0.4500%

Fn ℄ Grade

25' Cut Ditch Rt

+50

① 800' VC
SD_s = 925'

Sta 2983+01.67
New 3-10'×8' Box Clr
Q_{25} = 1280 cfs
Design HW 3993.50
℄ Floor Elev 3985.80
Slope = 1.0%

New 600' Grader Ditch Rt
Std C-3.01

+75
2411 CY to Sta 2998+05

Exc	6570
15% Shr	986
Emb	3143
Haul out	2411
Grd Comp	30

+74

Channel
340 CY

4010
④ 4000
3990
3980

2970 71 72 73 74 2975 76 77 78 79 2980 ④ 81 82 83 84

SURVEY NOS.	FINISHED PLANS	REVISIONS	LOCATION	DATE

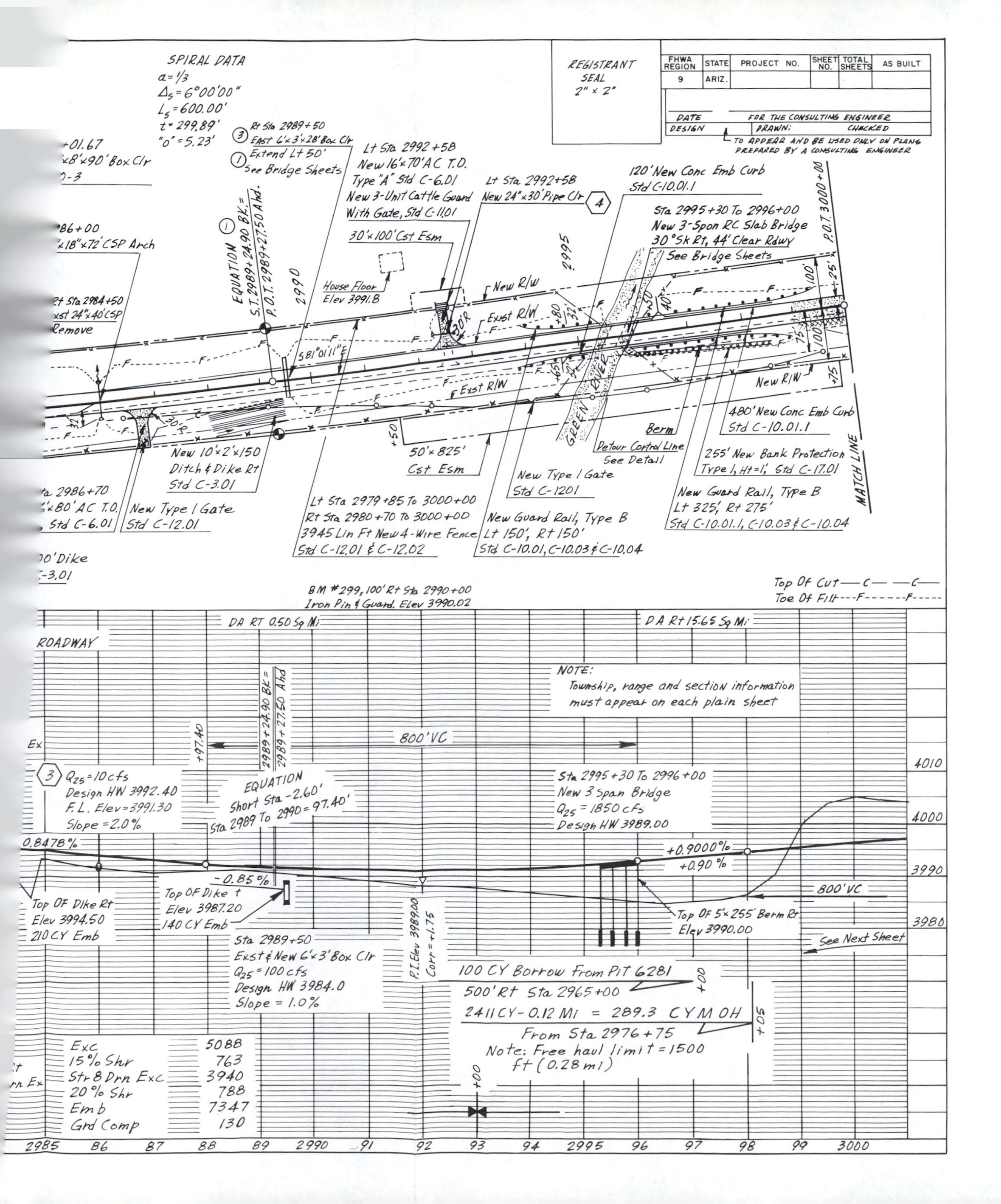

SPIRAL DATA
a = 1/3
Δs = 6°00'00"
Ls = 600.00'
t = 299.89'
"o" = 5.23'
REGISTRANT SEAL 2" × 2"
FHWA REGION | STATE | PROJECT NO. | SHEET NO. | TOTAL SHEETS | AS BUILT
9 | ARIZ.
DATE | FOR THE CONSULTING ENGINEER
DESIGN | DRAWN: | CHECKED
TO APPEAR AND BE USED ONLY ON PLANS PREPARED BY A CONSULTING ENGINEER
Rt Sta 2989+50 East 6'×3'×28' Box Clr
Extend Lt 50'
See Bridge Sheets
Lt Sta 2992+58
New 16'×70' AC T.O.
Type "A" Std C-6.01
New 3-Unit Cattle Guard
With Gate, Std C-11.01
Lt Sta 2992+58
New 24"×30' Pipe Clr
120' New Conc Emb Curb
Std C-10.01.1
Sta 2995+30 To 2996+00
New 3-Span RC Slab Bridge
30° Sk Rt, 44' Clear Rdwy
See Bridge Sheets
P.O.T. 3000+00
EQUATION
S.T. 2989+24.90 BK. =
P.O.T. 2989+27.50 Ahd.
2990
2995
30'×100' Cst Esm
House Floor Elev 3991.8
New R/W
Exst R/W
S81°01'11"E
New 10'×2'×150
Ditch & Dike Rt
Std C-3.01
50'×825'
Cst Esm
New Type I Gate
Std C-12.01
Lt Sta 2979+85 To 3000+00
Rt Sta 2980+70 To 3000+00
3945 Lin Ft New 4-Wire Fence
Std C-12.01 & C-12.02
New Type I Gate
Std C-1201
New Guard Rail, Type B
Lt 150', Rt 150'
Std C-10.01, C-10.03 & C-10.04
Berm
Detour Control Line
See Detail
GREEN RIVER
480' New Conc Emb Curb
Std C-10.01.1
255' New Bank Protection
Type I, Ht=1', Std C-17.01
New Guard Rail, Type B
Lt 325', Rt 275'
Std C-10.01.1, C-10.03 & C-10.04
MATCH LINE
BM #299, 100' Rt Sta 2990+00
Iron Pin & Guard, Elev 3990.02
Top Of Cut — C — C —
Toe Of Fill - - - F - - - F - - -
DA RT 0.50 Sq Mi
DA Rt 15.65 Sq Mi
ROADWAY
NOTE:
Township, range and section information
must appear on each plain sheet
800' VC
+97.40
2989+24.90 BK =
2989+27.50 Ahd
Q25 = 10 cfs
Design HW 3992.40
F.L. Elev = 3991.30
Slope = 2.0%
EQUATION
Short Sta -2.60'
Sta 2989 To 2990 = 97.40'
Sta 2995+30 To 2996+00
New 3 Span Bridge
Q25 = 1850 cfs
Design HW 3989.00
0.8478%
-0.85%
+0.9000%
+0.90%
800' VC
Top OF Dike Rt
Elev 3994.50
210 CY Emb
Top OF Dike Rt
Elev 3987.20
140 CY Emb
Top Of 5'×255' Berm Rt
Elev 3990.00
See Next Sheet
Sta 2989+50
Exst & New 6'×3' Box Clr
Q25 = 100 cfs
Design HW 3984.0
Slope = 1.0%
P.I. Elev 3989.00
Corr = +1.75
100 CY Borrow From Pit 6281
500' Rt Sta 2965+00
2411 CY - 0.12 Mi = 289.3 CY M OH
From Sta 2976+75
Note: Free haul limit = 1500
ft (0.28 mi)
Exc 5088
15% Shr 763
Str 8 Drn Exc 3940
20% Shr 788
Emb 7347
Grd Comp 130
4010
4000
3990
3980
2985 86 87 88 89 2990 91 92 93 94 2995 96 97 98 99 3000